What's Inside?

The goals of this text are to show that mathematics is a lively, interesting, useful, and surprisingly rich human activity; to instill an overall appreciation for mathematics as a discipline; and to expose readers to the subtlety and variety of its many facets. In particular, a few topics that might pique your interest are:

How are elections decided? (Chapter 1: The Mathematics of Voting)

How can power be measured? (Chapter 2: The Mathematics of Power)

How do we schedule a project so that it is completed as early as possible? (Chapter 8: The Mathematics of Scheduling)

How do you get rich? (Chapter 10: The Mathematics of Money)

How should data be collected? (Chapter 13: Collecting Statistical Data)

The connection between the mathematics presented and down-to-earth, real-life problems is immediate and obvious. Read the book and go beyond the classroom for your own mathematical adventure.

Excursions in
Modern Mathematics

Instructor's Edition

7th edition

Excursions in Modern Mathematics

Peter Tannenbaum

California State University—Fresno

Prentice Hall

New York Boston San Francisco
London Toronto Sydney Tokyo Singapore Madrid
Mexico City Munich Paris Cape Town Hong Kong Montreal

Executive Editor: *Anne Kelly*
Acquisitions Editor: *Marnie Greenhut*
Project Editor: *Elizabeth Bernardi*
Assistant Editor: *Leah Goldberg*
Editorial Assistant: *Leah Driska*
Senior Managing Editor: *Karen Wernholm*
Senior Production Supervisor: *Sheila Spinney*
Photo Researcher: *Beth Anderson*
Supplements Supervisor: *Kayla Smith-Tarbox*
Associate Media Producer: *Nathaniel Koven*
Software Development: *Eileen Moore and Marty Wright*
Executive Marketing Manager: *Becky Anderson*
Marketing Coordinator: *Bonnie Gill*
Senior Author Support/Technology Specialist: *Joe Vetere*
Senior Prepress Supervisor: *Caroline Fell*
Senior Manufacturing Buyer: *Evelyn Beaton*
Senior Media Buyer: *Ginny Michaud*
Text Design, Production Coordination, Composition: *Nesbitt Graphics, Inc.*
Illustrations: *Scientific Illustrators*
Cover Design: *Beth Paquin*
About the Cover: *Cowboy—Lincoln Rogers/Shutterstock; woman biking—Geir Olav
    Lyngfjell/Shutterstock; woman climbing—Danny Warren/Shutterstock; flippers—
    Gina Smith/Fotolia; surfer—NorthShoreSurfPhotos/Fotolia; marche dans l'herbe—
    Richard Villalon/Fotolia; male snowboarding—ant236/Fotolia*

For permission to use copyrighted material, grateful acknowledgment is made to the copyright holders listed on page 709, which is hereby made part of this copyright page.

Many of the designations used by manufacturers and sellers to distinguish their products are claimed as trademarks. Where those designations appear in this book, and Prentice Hall was aware of a trademark claim, the designations have been printed in initial caps or all caps.

Library of Congress Cataloging-in-Publication Data
Tannenbaum, Peter, 1946–
    Excursions in modern mathematics/Peter Tannenbaum.—7th ed.
        p.   cm.
    Includes bibliographical references and index.
    ISBN 13: 978-0-321-56803-8 (student edition)—ISBN 13: 978-0-321-56508-2 (instructor edition)
    ISBN 10: 0-321-56803-6 (student edition)—ISBN 10: 0-321-56508-8 (instructor edition)
1.   Mathematics.   I. Title.
    QA36.T35 2009
    510—dc22
                                                                                    2008039968

1 2 3 4 5 6 7 8 9 10—QWT—10 09 08

Prentice Hall
is an imprint of

www.pearsonhighered.com

ISBN 10: 0-321-56803-6
ISBN 13: 978-0-321-56803-8

To the members of the board of Last Tango

Contents

3 The Mathematics of Sharing 76
Fair-Division Games

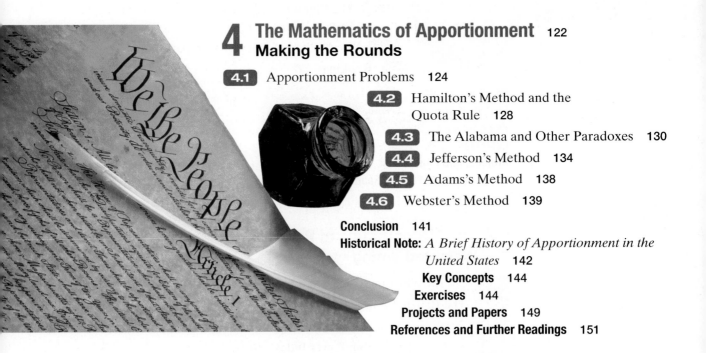

4 The Mathematics of Apportionment 122
Making the Rounds

Mini-Excursion 1
Apportionment Today 152
The Huntington-Hill Method

part 2 Management Science

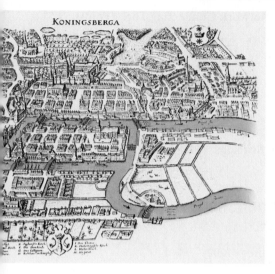

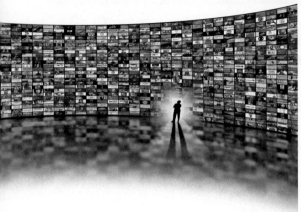

7 The Mathematics of Networks 240
The Cost of Being Connected

8 The Mathematics of Scheduling 280
Chasing the Critical Path

Mini-Excursion 2
A Touch of Color 318
Graph Coloring

part 3 Shape, Growth, and Form

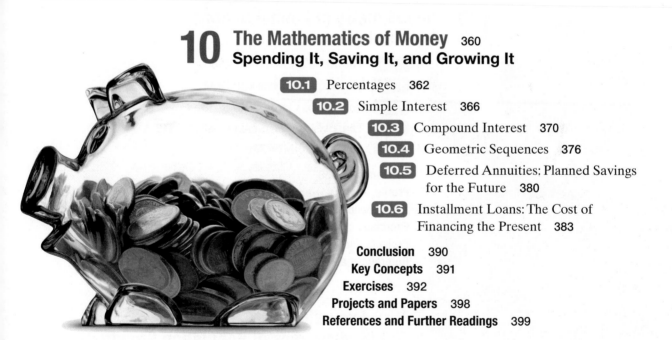

11 The Mathematics of Symmetry 400
Beyond Reflection

12 The Geometry of Fractal Shapes 436
Naturally Irregular

Mini-Excursion 3
The Mathematics of Population Growth 474
There Is Strength in Numbers

part 4 Statistics

16 The Mathematics of Normal Distributions   586
The Call of the Bell

Mini-Excursion 4
The Mathematics of Managing Risk   612
Expected Values

Preface

> " To most outsiders, modern mathematics is unknown territory. Its borders are protected by dense thickets of technical terms; its landscapes are a mass of indecipherable equations and incomprehensible concepts. Few realize that the world of modern mathematics is rich with vivid images and provocative ideas. "
>
> —Ivars Peterson,
> *The Mathematical Tourist*

Excursions in Modern Mathematics is, as I hope the title might suggest, a collection of "trips" into that vast and alien frontier that many people perceive mathematics to be. While the formal purpose of this book is quite conventional—it is intended to serve as a textbook for a college-level liberal arts mathematics course—its philosophy and contents are not. My goal is to show the open-minded reader that mathematics is a lively, interesting, useful, and surprisingly rich human activity.

The "excursions" in this book represent a collection of topics chosen to meet the following simple criteria.

- **Applicability.** There is no need to worry here about that great existential question of college mathematics: What is this stuff good for? The connection between the mathematics presented in these excursions and down-to-earth, concrete real-life problems is transparent and immediate.

- **Accessibility.** As a general rule, the excursions in this book do not presume a background beyond standard high school mathematics—by and large, intermediate algebra and a little Euclidean geometry are appropriate and sufficient prerequisites. (In the few instances in which more advanced concepts are unavoidable, an effort has been made to provide enough background to make the material self-contained.) A word of caution—this does not mean that the excursions in this book are easy! In mathematics, as in many other walks of life, simple and basic is not synonymous with easy and superficial.

- **Modernity.** Unlike much of traditional mathematics, which is often hundreds of years old, most of the mathematics in this book has been discovered within the last 100 years, and in some cases within the last couple of decades. Modern mathematical discoveries do not have to be the exclusive province of professional mathematicians.

- **Aesthetics.** The notion that there is such a thing as beauty in mathematics is surprising to most casual observers. There is an important aesthetic component in mathematics, and just as in art and music (which mathematics very much resembles), it often surfaces in the simplest ideas. A fundamental objective of this book is to develop an appreciation of the aesthetic elements of mathematics.

Outline

The material in the book is divided into four independent parts. Each part consists of four individual excursions plus a shorter "mini-excursion." If one thinks of a chapter in this book as a snorkeling excursion along a colorful coral reef, then a mini-excursion is a free dive into somewhat deeper waters along a particularly interesting section of the reef.

- **Part 1 Social Choice.**   This part deals with mathematical applications in social science. How are elections decided? (Chapter 1); How can the power of voting blocs be measured? (Chapter 2); How can competing claims on property be resolved in a fair and an equitable way? (Chapter 3); How are seats apportioned in a legislative body? (Chapter 4); and, What is the current method for apportioning seats in the United States House of Representatives? (Mini-Excursion 1).

- **Part 2 Management Science.**   This part deals with methods for solving problems involving the organization and management of complex activities—that is, activities involving either a large number of steps or a large number of variables. Typically, these problems revolve around questions of efficiency: some limited or precious resource (time, money, raw materials) must be managed in such a way that waste is minimized. Among the many applications discussed in this part are: How do we cover a network of roads with the least amount of backtracking? (Chapter 5); How do we find the least expensive route that passes through a set of locations? (Chapter 6); How can we connect a set of points to create an optimal network? (Chapter 7); How do we schedule a project so that it is completed as early as possible? (Chapter 8); and, How do we assign roles (colors) to the members of a network using the fewest number of colors? (Mini-Excursion 2).

- **Part 3 Shape, Growth, and Form.**   This part deals with nontraditional notions of growth and shape. What are the mechanisms that govern spiral growth in natural shapes? (Chapter 9); How does time affect the value of money? (Chapter 10); How can we analyze the symmetries of an object or a pattern? (Chapter 11); What type of geometric shape is a head of cauliflower, our circulatory system, or a mountain range? (Chapter 12); and, How can mathematics be useful in predicting the rise and fall of populations? (Mini-Excursion 3).

- **Part 4 Statistics.**   In one way or another, statistics affects all of our lives. Government policy, insurance rates, our health, our diet, and public opinion are all governed by statistical data. This part deals with some of the most basic aspects of statistics. How should data be collected? (Chapter 13); How can large amounts of data be summarized so that the data make some sense? (Chapter 14); How can we measure the uncertainty produced by random phenomena? (Chapter 15); How can we use statistical laws to predict long-term patterns in random events? (Chapter 16); and, How do we separate foolish risks from risks that are worth taking? (Mini-Excursion 4).

Exercises and Additional Resources

An important goal for this book is that it be flexible enough to appeal to a wide range of readers in a variety of settings. The exercises, in particular, have been designed to convey the depth of the subject matter by addressing a broad spectrum of levels of difficulty—from the routine drill to the ultimate challenge. For convenience (but with some trepidation) the exercises are classified into three levels of difficulty:

- **Walking.** These exercises are meant to test a basic understanding of the main concepts, and they are intended to be within the capabilities of students at all levels.

- **Jogging.** These are exercises that can no longer be considered as routine—either because they use basic concepts at a higher level of complexity or they require slightly higher-order critical thinking skills, or both.

- **Running.** This is an umbrella category for problems that range from slightly unusual or slightly above average in difficulty to problems that can be a real challenge to even the most talented of students.

Most chapters conclude with a biographical **Profile** that highlights the human side of mathematics. After each chapter's exercise set, a **Projects and Papers** section offers some potential topics and ideas for further exploration and enrichment. In most cases, the projects are well suited for group work, be it a handful of students or an entire small class. Each chapter concludes with an extensive list of **References and Further Readings**.

New in the Seventh Edition

The most significant change to this seventh edition is the inclusion of a new Chapter 10, "The Mathematics of Money." It would be hard to find a subject that is more universal, applicable, and relevant to people's lives than money. My goal for this chapter is to introduce some of the basic concepts of financial mathematics—simple and compound interest, annuities, and installment loans—in a style and at a level consistent with this book's overall approach and philosophy.

Other changes to this edition include the following:

- Chapter 9, "The Mathematics of Spiral Growth," now includes an expanded discussion of Fibonacci sequences and their properties.

- An all-new Mini-Excursion 3, "The Mathematics of Population Growth," includes a discussion of linear, exponential, and logistic growth.

- Approximately 25 percent of the exercises in the text are new or revised from the sixth edition.

- Many more examples have been provided, and approximately 20 percent of all examples are new or revised.

- An all-new media package includes a *MyMathLab*® course for this text and a new video program.

- A new *Instructor's Edition* provides complete answers in the back of the text for all walking and jogging exercises.

- A new *Insider's Guide to Teaching with Excursions in Modern Mathematics* provides helpful teaching tips, advice, and classroom resources from current users of the text to offer new ideas and help faculty with course preparation.

- An online *Instructor's Testing Manual* provides four test forms per chapter—two in multiple-choice format and two for free-response answers.

Supplementary Materials

Student Supplements

Student Resource Guide

- By Dale R. Buske, St. Cloud State University
- In addition to the worked-out solutions to odd-numbered exercises from the text, this guide contains "selected hints" that point the reader in one of many directions leading to a solution and keys to student success, including lists of skills that will help prepare for the chapter exams.

ISBN 13: 978-0-321-57519-7; ISBN 10: 0-321-57519-9

New! Video Lectures on DVD with Optional Subtitles

- The video lectures for this text are available on DVD, making it easy and convenient for students to watch the videos from a computer at home or on campus. The videos feature engaging chapter summaries and example solutions. The format provides distance-learning students with critical video instruction for each chapter, but also allows students needing only small amounts of review to watch instruction on a specific problem type. The videos have optional subtitles; they can be easily turned off or on for individual student needs.

ISBN 13: 978-0-321-57522-7; ISBN 10: 0-321-57522-9

Instructor Supplements

New! Instructor's Edition

- This special edition of the text includes answers to all walking and jogging exercises in a separate section in the back of the book.

ISBN 13: 978-0-321-56508-2; ISBN 10: 0-321-56508-8

Instructor's Solutions Manual

- By Dale R. Buske, St. Cloud State University
- Contains solutions to all the exercises in the text.

ISBN 13: 978-0-321-57516-6; ISBN 10: 0-321-57516-4

Instructor's Testing Manual (download only)

- By Joseph P. Kudrle, University of Vermont
- This manual includes four alternate tests per chapter. Two have multiple-choice exercises, and two have free-response exercises.

Available for download from *www.pearsonhighered.com/irc*.

New! Insider's Guide

- Includes resources designed to help faculty with course preparation and classroom management. Includes learning outcomes, skill objectives, ideas for the classroom, and worksheets from current users of the text.
- Offers helpful teachings tips.

ISBN 13: 978-0-321-57620-0; ISBN 10: 0-321-57620-9

PowerPoint® Lecture Slides (download only)

■ Available through *www.pearsonhighered.com/irc* or inside your *MyMathLab* online course, these classroom presentations cover all important topics from the text.

TestGen® (download only)

■ TestGen enables instructors to build, edit, print, and administer tests using a computerized bank of questions developed to cover all the objectives of the text. TestGen is algorithmically based, allowing instructors to create multiple but equivalent versions of the same question or test with the click of a button. Instructors can also modify test bank questions or add new questions. Tests can be printed or administered online.

Available for download through *www.pearsonhighered.com/irc* or inside your *MyMathLab* course.

Pearson Math Adjunct Support Center

■ The Pearson Math Adjunct Support Center (*www.pearsontutorservices.com/ math-adjunct.html*) is staffed by qualified instructors with more than 50 years of combined experience at both the community college and university levels. Assistance is provided for faculty in the following areas:

  ■ Suggested syllabus consultation

  ■ Tips on using materials bundled with your book

  ■ Book-specific content assistance

  ■ Teaching suggestions, including advice on classroom strategies

Online Resources

MyMathLab

New! *MyMathLab*

■ *MyMathLab* is a series of text-specific, easily customizable online courses for Pearson Education's textbooks in mathematics and statistics. Powered by CourseCompass™ (our online teaching and learning environment) and MathXL® (our online homework, tutorial, and assessment system), *MyMathLab* gives you the tools you need to deliver all or a portion of your course online, whether your students are in a lab setting or working from home. *MyMathLab* provides a rich and flexible set of course materials, featuring free-response exercises that are algorithmically generated for unlimited practice and mastery. Students can also use online tools such as video lectures, animations, and a multimedia textbook to independently improve their understanding and performance. Instructors can use *MyMathLab*'s homework and test managers to select and assign online exercises correlated directly to the textbook, and they can also create and assign their own online exercises and import TestGen tests for added flexibility. *MyMathLab*'s online gradebook—designed specifically for mathematics and statistics—automatically tracks students' homework and test results and gives the instructor control over how to calculate final grades. Instructors can also add offline (paper-and-pencil) grades to the gradebook. *MyMathLab* is available to qualified adopters. For more information, visit our Web site at *www.mymathlab.com*, or contact your sales representative.

New! *MathXL*

- MathXL is a powerful online homework, tutorial, and assessment system that accompanies Pearson Education's textbooks in mathematics and statistics. With MathXL, instructors can create, edit, and assign online homework and tests using algorithmically generated exercises correlated at the objective level to the textbook. They can also create and assign their own online exercises and import TestGen tests for added flexibility. All student work is tracked in MathXL's online gradebook. Students can take chapter tests in MathXL and receive personalized study plans based on their test results. The study plan diagnoses weaknesses and links students directly to tutorial exercises for the objectives they need to study and retest. MathXL is available to qualified adopters. For more information, visit our Web site at *www.mathxl .com* or contact your sales representative.

Acknowledgments

This edition benefited from the contributions and opinions of a large number of people. I owe a special debt of gratitude to "consigliere" Dale Buske of St. Cloud State University, who read early drafts of each chapter and provided useful feedback and ideas. Dale also contributed extensively to the new and updated exercise sets. Special thanks also to Deirdre Smith of the University of Arizona, who allowed us to use a modified version of her house-buying project in Chapter 10.

Adding the new financial mathematics chapter to the book was a significant undertaking, and I am greatly indebted to the reviewers of this edition for their feedback, suggestions, and encouragement. The following is a list of reviewers for the last two editions (asterisks indicate reviewers of this seventh edition).

*Lowell Abrams, *George Washington University*
* Erol Barbut, *Washington State University*
  Guanghwa Andy Chang, *Youngstown State University*
  Kimberly A. Conti, *SUNY College at Fredonia*
*Lauren Fern, *University of Montana*
  Abdi Hajikandi, *Buffalo State College, State University of New York*
  Cynthia Harris, *Triton Community College*
  Peter D. Johnson, *Auburn University*
*Karla Karstens, *University of Vermont*
  Stephen Kenton, *Eastern Connecticut State University*
  Katalin Kolossa, *Arizona State University*
*Randa Lee Kress, *Idaho State University*
  Thomas Lada, *North Carolina State University*
*Margaret A. Michener, *University of Nebraska at Kearney*
*Mika Munakata, *Montclair State University*
*Kenneth Pothoven, *University of South Florida*
  Shelley Russell, *Santa Fe Community College*
*Salvatore Sciandra Jr., *Niagara County Community College*
*Deirdre Smith, *University of Arizona*
  Hilary Spriggs, *Colorado State University*
*W. D. Wallis, *Southern Illinois University*

Many people contributed to previous editions of this book. Special thanks go to Robert Arnold; to Dale Buske (again); and to the following reviewers: Teri

Anderson, Carmen Artino, Donald Beaton, Terry L. Cleveland, Leslie Cobar, Crista Lynn Coles, Ronald Czochor, Nancy Eaton, Lily Eidswick, Kathryn E. Fink, Stephen I. Gendler, Marc Goldstein, Josephine Guglielmi, William S. Hamilton, Harold Jacobs, Tom Kiley, Jean Krichbaum, Kim L. Luna, Mike Martin, Thomas O'Bryan, Daniel E. Otero, Philip J. Owens, Matthew Pickard, Lana Rhoads, David E. Rush, Kathleen C. Salter, Theresa M. Sandifer, Paul Schembari, Marguerite V. Smith, William W. Smith, David Stacy, Zoran Sunik, John Watson, and Sarah N. Ziesler.

For this edition I inherited a new production team at Pearson, and I'd especially like to thank the real movers and shakers that made it all possible: Acquisitions Editor Marnie "the Marine" Greenhut, Executive Marketing Manager Becky Anderson, Senior Production Supervisor Sheila Spinney, Project Editor Elizabeth Bernardi, Assistant Editor Leah Goldberg, Editorial Assistant Leah Driska, Marketing Assistant Bonnie Gill, Associate Media Producer Nathaniel Koven, and Intern Extraordinaire Chelsea Pingree.

A Final Word

This book grew out of the conviction that a liberal arts mathematics course should teach students more than just a collection of facts and procedures. The ultimate purpose of this book is to instill in the reader an overall appreciation of mathematics as a discipline and an exposure to the subtlety and variety of its many facets: problems, ideas, methods, and solutions. Last, but not least, I have tried to show that mathematics can be interesting and fun.

Peter Tannenbaum

Social Choice

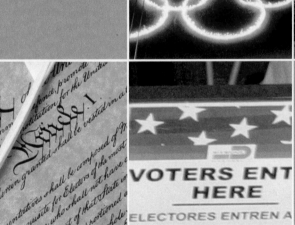

VOTERS ENT
HERE
ELECTORES ENTREN A

1 The Mathematics of Voting

The Paradoxes of Democracy

Vote! In a democracy, the rights and duties of citizenship are captured in that simple one-word mantra. And, by and large, as citizens of a democracy, we should and do vote. We vote in presidential elections, gubernatorial elections, local elections, school bonds, stadium bonds, *American Idol* selections, and initiatives large and small. The paradox is that the more opportunities we have to vote, the less we seem to appreciate and understand the meaning of voting. Why should we vote? Does our vote really count? How does it count?

> **❝** It's not the voting that's
> democracy; it's the counting. **❞**
> —Tom Stoppard

We can best answer these questions once we understand the full story behind an election. We have elections because we don't all think alike. Since we cannot all have things our way, we vote. But *voting* is only the first half of the story, the one we are most familiar with. As playwright Tom Stoppard suggests, it's the second half of the story—the *counting*—that is at the very heart of the democratic process. How does it work, this process of sifting through the many voices of individual voters to find the collective voice of the group? More important, how well does it work? Is the process always fair? Answering these questions and explaining a few of the many intricacies and subtleties of *voting theory* are the purpose of this chapter.

But wait just a second! Voting theory? Why do we need a fancy theory to figure out how to count the votes? It all sounds pretty simple: We have an election; we count the ballots. Based on that count, we decide the outcome of the election in a consistent and fair manner. Surely, there must be a reasonable way to accomplish this! Surprisingly, there isn't.

In the late 1940s mathematical economist Kenneth Arrow discovered a remarkable fact: For elections involving three or more candidates, there is no consistently fair democratic method for choosing a winner. In fact, Arrow demonstrated that *a method for determining election results that is democratic and always fair is a mathematical impossibility*. This fact, the most famous in voting theory, is known as *Arrow's impossibility theorem*.

This chapter is organized as follows. We will start with a general discussion of elections and ballots in Section 1.1. This discussion provides the backdrop for the remaining sections, which are the heart of the chapter. In Sections 1.2 through 1.5 we will explore four of the most commonly used *voting methods* and look at how they work, their implications, and how they measure up in terms of a few basic tests of fairness called *fairness criteria*. In Section 1.6 we generalize these methods to elections in which we want to produce not just a winner but a complete or partial ranking of the candidates. The entire excursion will be sprinkled with plenty of examples depicting real-life situations or metaphors thereof. At the end of the chapter, with some gained insight, we will come back to Arrow's impossibility theorem.

■ In 1972 Kenneth Arrow was awarded the Nobel Prize in Economics for his pioneering work in what is now known as social-choice theory.

For more details, see the biographical profile of Arrow at the end of this chapter.

Preference Ballots and Preference Schedules

We kick off our discussion of voting theory with a simple but important example that we will revisit several times throughout the chapter. (You may want to think of this example as a mathematical parable, its importance being not in the story itself but in what lies hidden behind it. As you will soon see, there is a lot more than first meets the eye behind this simple example.)

EXAMPLE 1.1   **The Math Club Election**

The Math Appreciation Society (MAS) is a student organization dedicated to an unsung but worthy cause, that of fostering the enjoyment and appreciation of mathematics among college students. The Tasmania State University chapter of MAS is holding its annual election for president. There are four candidates running for president: Alisha, Boris, Carmen, and Dave (*A*, *B*, *C*, and *D* for short). Each of the 37 members of the club votes by means of a ballot indicating his or her first, second, third, and fourth choice. The 37 ballots submitted are shown in Fig. 1-1. Once the ballots are in, it's decision time. Who should be the *winner* of the election? Why?

Ballot	Ballot	Ballot	Ballot	Ballot	Ballot	Ballot	Ballot	Ballot	Ballot	Ballot	Ballot	Ballot	Ballot	Ballot	Ballot	Ballot	Ballot
1st *A*	1st *B*	1st *A*	1st *C*	1st *B*	1st *C*	1st *A*	1st *B*	1st *C*	1st *A*	1st *C*	1st *D*	1st *A*	1st *A*	1st *C*	1st *A*	1st *C*	1st *D*
2nd *B*	2nd *D*	2nd *B*	2nd *B*	2nd *D*	2nd *B*	2nd *B*	2nd *D*	2nd *B*	2nd *B*	2nd *D*	2nd *C*	2nd *B*	2nd *B*	2nd *B*	2nd *B*	2nd *B*	2nd *C*
3rd *C*	3rd *C*	3rd *C*	3rd *D*	3rd *C*	3rd *D*	3rd *C*	3rd *C*	3rd *D*	3rd *C*	3rd *B*	3rd *B*	3rd *C*	3rd *C*	3rd *D*	3rd *C*	3rd *D*	3rd *B*
4th *D*	4th *A*	4th *D*	4th *A*	4th *A*	4th *A*	4th *D*	4th *A*	4th *A*	4th *D*	4th *A*	4th *A*	4th *D*	4th *D*	4th *A*	4th *D*	4th *A*	4th *A*

Ballot	Ballot	Ballot	Ballot	Ballot	Ballot	Ballot	Ballot	Ballot	Ballot	Ballot	Ballot	Ballot	Ballot	Ballot	Ballot	Ballot	Ballot	Ballot
1st *C*	1st *A*	1st *D*	1st *D*	1st *C*	1st *C*	1st *D*	1st *A*	1st *D*	1st *C*	1st *A*	1st *D*	1st *B*	1st *A*	1st *C*	1st *A*	1st *A*	1st *D*	1st *A*
2nd *B*	2nd *B*	2nd *C*	2nd *C*	2nd *B*	2nd *B*	2nd *C*	2nd *B*	2nd *C*	2nd *B*	2nd *B*	2nd *C*	2nd *D*	2nd *B*	2nd *D*	2nd *B*	2nd *B*	2nd *C*	2nd *B*
3rd *D*	3rd *C*	3rd *B*	3rd *B*	3rd *D*	3rd *D*	3rd *B*	3rd *C*	3rd *B*	3rd *D*	3rd *C*	3rd *B*	3rd *C*	3rd *C*	3rd *B*	3rd *C*	3rd *C*	3rd *B*	3rd *C*
4th *A*	4th *D*	4th *A*	4th *A*	4th *A*	4th *A*	4th *A*	4th *D*	4th *A*	4th *A*	4th *D*	4th *A*	4th *A*	4th *D*	4th *A*	4th *D*	4th *D*	4th *A*	4th *D*

FIGURE 1-1 The 37 anonymous ballots for the Math Club election.

Before we try to answer these two deceptively simple questions, let's step back and take a brief look at elections in general. The essential ingredients of every election are a set of *voters* (in our example the members of the Tasmania State chapter of MAS) and a set of *candidates* or *choices* (in our example *A*, *B*, *C*, *D*). Typically, the word *candidate* is used when electing people and the word *choice* is associated with nonhuman alternatives (cities, colleges, pizza toppings, etc.), but in this chapter we will use the two words interchangeably.

Individual voters are asked to express their opinions through *ballots*, which can come in many forms. In the Math Club election, the ballots asked the voters to rank each of the candidates in order of preference, and ties were not allowed. A ballot in which the voters are asked to rank the candidates in order of preference is called a **preference ballot**. A ballot in which ties are not allowed is called a **linear ballot**.

In this chapter we will illustrate all our examples using linear preference ballots as the preferred (no pun intended) format for voting. While it is true that the preference ballot is not the most typical way we cast our vote (in most elections for public office, for example, we are asked to choose the top candidate), it is also true that it is one of the best ways to vote since it allows us to express our opinion on the relative merit of *all* the candidates.

A quick look at Fig. 1-1 is all it takes to see that there are many repeats among the ballots submitted; there are, after all, only a few different ways that the four candidates can be ranked. Thus, a logical way to organize the ballots is to group together identical ballots (Fig. 1-2), which leads in a rather obvious way to Table 1-1, called the **preference schedule** for the election.

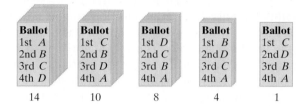

FIGURE 1-2 The 37 Math Club election ballots organized into piles.

TABLE 1-1	Preference Schedule for the Math Club Election				
Number of voters	14	10	8	4	1
1st choice	A	C	D	B	C
2nd choice	B	B	C	D	D
3rd choice	C	D	B	C	B
4th choice	D	A	A	A	A

If we think of an election as a process that takes an *input* (the ballots) and produces an *output* (the winner or winners of the election), then the preference schedule is the input, neatly packaged and organized and ready for us to do the counting. From now on, all our election examples will be given in the form of a preference schedule.

Transitivity and Elimination of Candidates

There are two important facts to keep in mind when we work with preference ballots. The first is that a voter's preferences are **transitive**, a fancy way of saying that a voter who prefers candidate A over candidate B and prefers candidate B over candidate C automatically prefers candidate A over candidate C. A useful consequence of this observation is this: *If we need to know which candidate a voter would vote for if it came down to a choice between just X and Y, all we have to do is look at where X and Y are placed on that voter's ballot—whichever is higher would be the one getting the vote.* We will use this observation throughout the chapter.

Ballot
1st C
2nd B
3rd D
4th A

FIGURE 1-3

The other important fact is that the relative preferences of a voter are not affected by the elimination of one or more of the candidates. Take, for example, the ballot shown in Fig. 1-3 and pretend that candidate B drops out of the race right before the ballots are submitted. How would this voter now rank the remaining three candidates? As Fig. 1-4 shows, the relative positions of the remaining candidates are unaffected: C remains the first choice, D moves up to the second choice, and A moves up to the third choice.

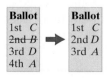

FIGURE 1-4

Let's now return to the business of deciding the outcome of elections in general and the Math Club election in particular.

1.2   The Plurality Method

Perhaps the best known and most commonly used method for finding a winner in an election is the **plurality method**. With the plurality method, all we care about is first-place votes—the candidate with the *most* first-place votes (called the **plurality candidate**) wins. Thus, in an election conducted under the plurality method we don't really need each voter to rank the candidates. The only information we need is the voter's first choice. The vast majority of elections for political office in the United States are decided using the plurality method, so most citizens are aware of the procedure but rarely aware of its many drawbacks. As we will soon see, other than its utter simplicity, the plurality method has little else going in its favor.

(**EXAMPLE 1.2**)   **The Math Club Election (Plurality)**

If the Math Club election (Example 1.1) is decided under the plurality method, then

> *A* gets 14 first-place votes,
>
> *B* gets 4 first-place votes,
>
> *C* gets 11 first-place votes,
>
> *D* gets 8 first-place votes,

and the results of the election are clear—the winner is *A* (Alisha).

With political elections the allure of the plurality method lies in its simplicity (voters have little patience for complicated procedures) and in the fact that plurality is a natural extension of the principle of majority rule: *In a democratic election between two candidates, the candidate with a majority (more than half) of the votes should be the winner.* (Not surprisingly, the candidate with a majority of first-place votes is called the **majority candidate**.)

With two candidates a plurality candidate is also a majority candidate and everything works out well. The problem with the majority rule is that when there are *three* or more candidates there is no guarantee that there is going to be a majority candidate. In the Math Club election, for example, to get a majority would require at least 19 first-place votes (out of 37). Alisha, with 14 first-place votes, had a *plurality* (more than any other candidate) but was far from being a majority candidate. In closely contested elections with many candidates, the percentage of the vote needed to win under plurality can be ridiculously low.

■ See, for example, Exercises 15 and 16.

The Majority Criterion

One of the most basic expectations in a democratic election is the notion that if there is a majority candidate, then that candidate should be the winner of the election. This notion is the basis for our first basic fairness criterion known as the **majority criterion**.

┌─ **THE MAJORITY CRITERION** ─────────────────────

If candidate *X* has a majority of the first-place votes, then candidate *X* should be the winner of the election.

└──

Under the plurality method a majority candidate is guaranteed to be the winner of the election by the simple fact that a majority always implies a plurality. (If you have more than half of the votes, you automatically have more than anyone else.)

As we will soon see, there are widely used voting methods that can produce *violations* of the majority criterion. Specifically, a violation of the majority criterion occurs in an election in which there is a majority candidate but that candidate does not win the election. If this can happen under some voting method, then we say that the voting method itself *violates* the majority criterion. Be careful how you interpret this—it only means that violations *can* happen, not that they *must* happen! Thus, there is a subtle but important difference between an election violating a criterion and a voting method violating a criterion. The former is an individual instance, the latter is a general property of the voting method itself, and an undesirable one at that! (Hold this thought because it will come up again!)

The plurality method satisfies the majority criterion—that's good! Other than that, the method has very little going for it and is generally considered a very poor method for choosing the winner of an election when there are more than two candidates.

The Condorcet Criterion

The principal weakness of the plurality method is that it fails to take into consideration a voter's other preferences beyond first choice and in so doing can lead to some very bad election results. To underscore the point, consider the following example.

(**EXAMPLE 1.3**)   **The Marching Band Election**

Tasmania State University has a superb marching band. They are so good that this coming bowl season they have invitations to perform at five different bowl games: the Rose Bowl (R), the Hula Bowl (H), the Fiesta Bowl (F), the Orange Bowl (O), and the Sugar Bowl (S). An election is held among the 100 members of the band to decide in which of the five bowl games they will perform. A preference schedule giving the results of the election is shown in Table 1-2.

Under the plurality method the winner of the election is the Rose Bowl (R), with 49 first-place votes. It's hard not to notice that this is a rather bad outcome, as there are 51 voters that have the Rose Bowl as their last choice. By contrast, the Hula Bowl (H) has 48 first-place votes and 52 second-place votes. Simple common sense tells us that the Hula Bowl is a far better choice to represent the wishes of the entire band. In fact, we can make

TABLE 1-2	Preference Schedule for the Band Election		
Number of voters	49	48	3
1st choice	R	H	F
2nd choice	H	S	H
3rd choice	F	O	S
4th choice	O	F	O
5th choice	S	R	R

the following persuasive argument in favor of the Hula Bowl: If we compare the Hula Bowl with any other bowl on a *head-to-head* basis, the Hula Bowl is always the preferred choice. Take, for example, a comparison between the Hula Bowl and the Rose Bowl. There are 51 votes for the Hula Bowl (48 from the second column plus the 3 votes in the last column) versus 49 votes for the Rose Bowl. Likewise, a comparison between the Hula Bowl and the Fiesta Bowl would result in 97 votes for the Hula Bowl (first and second columns) and 3 votes for the Fiesta Bowl. And when the Hula Bowl is compared with either the Orange Bowl or the Sugar Bowl, it gets all 100 votes. Thus, no matter which bowl

Marie Jean Antoine Nicolas Caritat, Marquis de Condorcet (1743–1794). Condorcet was a French aristocrat, mathematician, philosopher, economist, and social scientist, and one of the great rationalist thinkers of the eighteenth century.

we compare the Hula Bowl with, there is always a majority of the band that prefers the Hula Bowl.

A candidate preferred by a majority of the voters over every other candidate when the candidates are compared in head-to-head comparisons is called a **Condorcet candidate** (named after Le Marquis de Condorcet, an eighteenth-century French mathematician and philosopher). Not every election has a Condorcet candidate, but if there is one, it is a good sign that this candidate represents the voice of the voters better than any other candidate. In Example 1.3 the Hula Bowl is a Condorcet candidate—it is not unreasonable to expect that it should be the winner of the election.

In 1785 Condorcet introduced a basic criterion for fairness in an election: If there is a Condorcet candidate, then that candidate should be the winner of the election. This important fairness criterion is now known as the **Condorcet criterion**.

THE CONDORCET CRITERION

If candidate X is preferred by the voters over each of the other candidates in a head-to-head comparison, then candidate X should be the winner of the election.

We can now see the problem with Example 1.3 in a new light—the example illustrates a violation of the Condorcet criterion. The fact that this can happen using the plurality method means that *the plurality method violates the Condorcet criterion*.

Once again, an important point to remember: Just because violations of the Condorcet criterion *are possible* under the plurality method does not imply that they *must happen*. The marching band example was extreme in the sense that there was one very polarizing choice (the Rose Bowl) that was either loved or hated by the voters. One of the major flaws of the plurality method is that it sometimes leads us to pick such a choice even when there is a much more reasonable choice—one preferred by a majority of the voters over each of the others.

Insincere Voting

We will return to the concept of head-to-head comparisons between the candidates shortly. In the meantime we conclude our discussion of the plurality method by pointing out another one of its major flaws: the ease with which election results can be manipulated by a voter or a block of voters through insincere voting.

The idea behind **insincere voting** (also known as **strategic voting**) is simple: If we know that the candidate we really want doesn't have a chance of winning, then rather than "waste our vote" on our favorite candidate we can cast it for a lesser choice who has a better chance of winning the election. In closely contested elections a few insincere voters can completely change the outcome of an election.

To illustrate how easy it is to manipulate the results of an election decided under the plurality method, let's take a second look at Example 1.3, the marching band example.

> **EXAMPLE 1.4**   **The Marching Band Election Gets Manipulated**

We know that given the preference schedule shown in Table 1-2 (shown again as Table 1-3A) under the plurality method the winner of the marching band election is the Rose Bowl, a result sure to disappoint most of the band members. It so happens that three of the band members (last column of Table 1-3A) realize that there is no chance that their first choice, the Fiesta Bowl, can win this election, so rather than waste their votes they decide to make a strategic move—they cast their votes for the Hula Bowl by switching the first and second choices in their ballots (Table 1-3B).

TABLE 1-3A	The Real Preferences		
Number of voters	49	48	3
1st choice	*R*	*H*	*F*
2nd choice	*H*	*S*	*H*
3rd choice	*F*	*O*	*S*
4th choice	*O*	*F*	*O*
5th choice	*S*	*R*	*R*

TABLE 1-3B	The Actual Votes		
Number of voters	49	48	3
1st choice	*R*	*H*	*H*
2nd choice	*H*	*S*	*F*
3rd choice	*F*	*O*	*S*
4th choice	*O*	*F*	*O*
5th choice	*S*	*R*	*R*

This simple switch by just three voters changes the outcome of the election—the new preference schedule gives 51 votes, and thus the win, to the Hula Bowl.

While all voting methods are manipulable and can be affected by insincere voting (see Project D in Projects and Papers), the plurality method is the one that can be most easily manipulated, and insincere voting is quite common in real-world elections. For Americans, the most significant cases of insincere voting occur in close presidential or gubernatorial races between the two major party candidates and a third candidate ("the spoiler") who has little or no chance of winning. These are exactly the circumstances describing the 2000 and 2004 presidential elections—in both cases the race was very tight and in both cases the third-party candidate, Ralph Nader, lost many votes from insincere voters who liked him best but did not want to "waste their vote."

Insincere voting not only hurts small parties and fringe candidates, it has unintended and often negative consequences on the political system itself. The history of American political elections is littered with examples of independent candidates and small parties that never get a fair voice or a fair level of funding (it takes 5% of the vote to qualify for federal funds for the next election) because of the "let's not waste our vote" philosophy of insincere voters. The ultimate consequence of the plurality method is an entrenched two-party system that often gives the voters little real choice.

■ The idea that the plurality method invariably leads to a two-party system is known as *Duverger's law* after the French sociologist Maurice Duverger.

1.3   The Borda Count Method

A commonly used method for determining the winner of an election is the **Borda count method**, named after the Frenchman Jean-Charles de Borda. In this method each place on a ballot is assigned points. In an election with N candidates we give 1 point for *last* place, 2 points for *second from last* place, and so on. At the top of the ballot, a *first-place* vote is worth N points. The points are tallied for each candidate separately, and the candidate with the highest total is the winner. We will call such a candidate the *Borda winner*.

> ❝ My scheme is only intended for honest men. ❞
>
> – Jean-Charles de Borda

■ Jean-Charles de Borda (1733–1799). Borda was a brilliant scientist (he made contributions on such diverse subjects as mathematics, astronomy, and voting theory) as well as a captain in the French navy who fought against the British in the American War of Independence.

While not common, two or more candidates could tie with the highest point count, and in these cases we must either break the tie using some predetermined tie-breaking procedure or allow the tie to stand. To simplify the presentation, unless otherwise noted we will assume in this chapter that ties are allowed to stand. A more detailed discussion of ties and tie-breaking procedures can be found in MyMathLab.*

(**EXAMPLE 1.5**)   **The Math Club Election (Borda Count)**

Let's use the Borda count method to choose the winner of the Math Appreciation Society election first introduced in Example 1.1. Table 1-4 shows the point values under each column based on first place worth 4 points, second place worth 3 points, third place worth 2 points, and fourth place worth 1 point.

TABLE 1-4	Borda Points for the Math Club Election				
Number of voters	14	10	8	4	1
1st choice: 4 points	A: 56 pts	C: 40 pts	D: 32 pts	B: 16 pts	C: 4 pts
2nd choice: 3 points	B: 42 pts	B: 30 pts	C: 24 pts	D: 12 pts	D: 3 pts
3rd choice: 2 points	C: 28 pts	D: 20 pts	B: 16 pts	C: 8 pts	B: 2 pts
4th choice: 1 point	D: 14 pts	A: 10 pts	A: 8 pts	A: 4 pts	A: 1 pt

When we tally the points,

A gets $56 + 10 + 8 + 4 + 1 = 79$ points,
B gets $42 + 30 + 16 + 16 + 2 = 106$ points,
C gets $28 + 40 + 24 + 8 + 4 = 104$ points,
D gets $14 + 20 + 32 + 12 + 3 = 81$ points.

The Borda winner of this election is Boris! (Wasn't Alisha the winner of this election under the plurality method?) ⬭

*To use MyMathLab, you need a course ID and a student access code. Contact your instructor for more information.

What's Wrong with the Borda Count Method?

■ See Exercise 68 for details.

In contrast to the plurality method, the Borda count method takes into account all the information provided in the voters' preference ballots, and the Borda winner is the candidate with the best average ranking—the best compromise candidate if you will. On its face, the Borda count method seems like an excellent way to take full consideration of the voter's preferences, so indeed, what's wrong with it?

The next example illustrates some of the problems with the Borda count method.

> **EXAMPLE 1.6**   **The School Principal Election**

The last principal at Washington Elementary School has just retired and the School Board must hire a new principal. The four finalists for the job are Mrs. Amaro, Mr. Burr, Mr. Castro, and Mrs. Dunbar (A, B, C, and D, respectively). After interviewing the four finalists, each of the 11 school board members gets to rank the candidates by means of a preference ballot, and the Borda winner gets the job. Table 1-5 shows the preference schedule for this election.

TABLE 1-5	Preference Schedule for Example 1.6		
Number of voters	6	2	3
1st choice	A	B	C
2nd choice	B	C	D
3rd choice	C	D	B
4th choice	D	A	A

A simple count tells us that Mr. B is the Borda winner with 32 points. (You may want to check this out on your own—it will just take a minute or two.) According to the school bylaws, he gets the principal's job. But what about Mrs. A? She is the majority candidate, with 6 out of the 11 first-place votes, as well as the Condorcet candidate, since with 6 first-place votes she beats each of the other three candidates in a head-to-head comparison. Needless to say, Mrs. A feels very strongly that the only reasonable outcome of this election is that she should be getting the principal's job. After consulting an attorney, who encourages her to sue, she consults a voting theory expert, who advises her that she has no case—Mr. B won the election fair and square.

The sad tale of the unlucky Mrs. A in Example 1.6 is really a story about the Borda count method and how the method violates two basic criteria of fairness—the *majority criterion* and the *Condorcet criterion*. It is tempting to conclude from this that the Borda count method is a terrible voting method. It is not! In fact, despite its flaws, experts in voting theory consider the Borda count method one of the best, if not the very best, method for deciding elections with many candidates. Two things can be said in defense of the Borda count method: (1) Although violations of the majority criterion can happen, they do not happen very often, and when there are many candidates such violations are rare; and (2) violations of the Condorcet criterion automatically follow violations of the majority criterion, since a majority candidate is automatically a Condorcet candidate. (If there is no violation of the majority criterion, then a violation of the Condorcet criterion is still possible, but it is also rare.)

■ See, for example, Exercises 69, 70, and 71.

In real life, the Borda count method (or some variation of it) is widely used in a variety of settings, from individual sports awards (Heisman Trophy winner, NBA Rookie of the Year, NFL MVP, etc.) to college football polls, to music industry awards, to the hiring of school principals, university presidents, and corporate executives.

1.4   The Plurality-with-Elimination Method (Instant Runoff Voting)

In most municipal and local elections a candidate needs a majority of the first-place votes to get elected. When there are three or more candidates running, it is often the case that no candidate gets a majority. Typically, the candidate or candidates with the fewest first-place votes are eliminated, and a runoff election is held. Since runoff elections are expensive to both the candidates and the municipality, this is an inefficient and a cumbersome method for choosing a mayor or a county supervisor.

A much more efficient way to implement the same process without needing separate runoff elections is to use preference ballots, since a preference ballot tells us not only which candidate the voter wants to win but also which candidate the voter would choose in a runoff between any pair of candidates. The idea is simple but powerful: From the original preference schedule for the election we can eliminate the candidates with the fewest first-place votes one at a time until one of them gets a majority. This method has become increasingly popular and is nowadays fashionably known as **instant runoff voting** (IRV). Other names had been used in the past and in other countries for the same method, including *plurality-with-elimination* and the *Hare method*. For the sake of clarity, we will call it the **plurality-with-elimination method**—it is the most descriptive of the three names.

A formal description of the plurality-with-elimination method goes like this:

PLURALITY-WITH-ELIMINATION METHOD

- **Round 1.** Count the first-place votes for each candidate, just as you would in the plurality method. If a candidate has a majority of first-place votes, then that candidate is the winner. Otherwise, eliminate the candidate (or candidates if there is a tie) with the *fewest* first-place votes.

- **Round 2.** Cross out the name(s) of the candidates eliminated from the preference schedule and recount the first-place votes. (Remember that when a candidate is eliminated from the preference schedule, in each column the candidates below it move up a spot.) If a candidate has a majority of first-place votes, then declare that candidate the winner. Otherwise, eliminate the candidate with the fewest first-place votes.

- **Rounds 3, 4, ...** Repeat the process, each time eliminating one or more candidates until there is a candidate with a majority of first-place votes. That candidate is the winner of the election.

> **EXAMPLE 1.7**   **The Math Club Election (Plurality-with-Elimination)**

TABLE 1-6	Preference Schedule for the Math Club Election				
Number of voters	14	10	8	4	1
1st choice	A	C	D	B	C
2nd choice	B	B	C	D	D
3rd choice	C	D	B	C	B
4th choice	D	A	A	A	A

Let's see how the plurality-with-elimination method works when applied to the Math Club election. For the reader's convenience Table 1-6 shows the preference schedule again. It is the original preference schedule for the election first shown in Table 1-1.

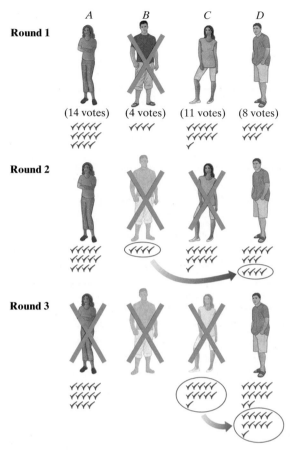

Round 1

(14 votes) (4 votes) (11 votes) (8 votes)

Round 2

Round 3

FIGURE 1-5 Boris is eliminated first, then Carmen, and then Alisha. The last one standing is Dave.

Round 1.

Candidate	A	B	C	D
Number of first-place votes	14	4	11	8

Since B has the fewest first-place votes, he is eliminated first (Fig. 1-5).

Round 2. Once B is eliminated, the four votes that originally went to B in round 1 will now go to D, the next-best candidate in the opinion of these four voters. The new tally is shown below:

Candidate	A	B	C	D
Number of first-place votes	14		11	12

In this round C has the fewest first-place votes and is eliminated.

Round 3. The 11 votes that went to C in round 2 now go to D (just check the relative positions of D and A in the second and fifth columns of Table 1-5). This gives the following:

Candidate	A	B	C	D
Number of first-place votes	14			23

We now have a winner, and lo and behold, it's neither Alisha nor Boris. The winner of the election, with 23 first-place votes, is Dave!

Applying the Plurality-with-Elimination Method

The next two examples are intended primarily to illustrate a few subtleties that can come up when applying the plurality-with-elimination method.

(**EXAMPLE 1.8**)   **Electing the Mayor of Kingsburg**

TABLE 1-7	Preference Schedule for the Kingsburg Mayoral Election

Number of voters	93	44	10	30	42	81
1st choice	A	B	C	C	D	E
2nd choice	B	D	A	E	C	D
3rd choice	C	E	E	B	E	C
4th choice	D	C	B	A	A	B
5th choice	E	A	D	D	B	A

Five candidates (A, B, C, D, and E) are running for mayor in the small town of Kingsburg. Being at the cutting edge of electoral reform, Kingsburg chooses its mayor using instant runoff voting, in other words, plurality-with-elimination. Table 1-7 shows the preference schedule for the election. Notice that since there are 300 voters voting in this election, a candidate needs 151 or more votes to win. We now eliminate candidates one by one until someone gets 151 or more votes.

Round 1.

Candidate	A	B	C	D	E
Number of first-place votes	93	44	40	42	81

Here C has the fewest number of first-place votes and is eliminated first. Of the 40 votes originally cast for C, 10 go to A (look at the third column of Table 1-7) and 30 go to E (from the fourth column of Table 1-7).

Round 2.

Candidate	A	B	C	D	E
Number of first-place votes	103	44		42	111

In this round D has the fewest first-place votes and is eliminated. The 42 votes originally cast for D would next go to C (look at the fifth column of Table 1-7), but C is out of the picture at this point, so we dip further down into the column. Thus, the 42 votes go to E, the next top candidate in that column.

Round 3.

Candidate	A	B	C	D	E
Number of first-place votes	103	44			153

At this point we are finished, since E has a majority of the votes and is therefore the winner.

⊂⊃

(**EXAMPLE 1.9**)   **The School Principal Election Revisited**

We are going to take another look at the controversial election for the principal's job at Washington Elementary School (Example 1.6). Due to a change in the school's bylaws, the principal is now going to be chosen using the plurality-with-elimination method. The preference schedule is shown again in Table 1-8.

This one is easy: Mrs. A is the majority candidate and thus automatically wins round 1 under the plurality-with-elimination method.

⊂⊃

The simple but important moral of Example 1.9 is that *the plurality-with-elimination method satisfies the majority criterion.*

TABLE 1-8	Preference Schedule for the Principal Election		
Number of voters	6	2	3
1st choice	A	B	C
2nd choice	B	C	D
3rd choice	C	D	B
4th choice	D	A	A

What's Wrong with the Plurality-with-Elimination Method?

The main problem with the plurality-with-elimination method is quite subtle and is illustrated by the next example.

EXAMPLE 1.10 **There Go the Olympics**

Three cities, Athens (*A*), Barcelona (*B*), and Calgary (*C*), are competing to host the Summer Olympic Games. The final decision is made by a secret vote of the 29 members of the Executive Council of the International Olympic Committee, and the winner is to be chosen using the plurality-with-elimination method.

Two days before the actual election is to be held, a straw poll is conducted by the Executive Council just to see how things stand. The results of the straw poll are shown in Table 1-9.

> "Straw poll: An unofficial vote or poll indicating the trend of opinion on a candidate or issue."
>
> —*American Heritage Dictionary*

Based on the results of the straw poll, Calgary is going to win the election. (In the first round Athens has 11 votes, Barcelona has 8, and Calgary has 10. Barcelona is eliminated, and in the second round Barcelona's 8 votes go to Calgary. With 18 votes in the second round, Calgary wins the election.)

Although the results of the straw poll are supposed to be secret, the word gets out that it looks like Calgary is going to host the next Summer Olympics. Since everybody loves a winner, the four delegates represented by the last column of Table 1-9 decide as a block to switch their votes and vote for Calgary first and Athens second. Calgary is going to win, so there is no harm in that, is there? Well, let's see.

The results of the official vote are shown in Table 1-10. The only changes between the straw poll in Table 1-9 and the official vote are the 4 votes that were switched in favor of Calgary. (To get Table 1-10, switch *A* and *C* in the last column of Table 1-9 and then combine columns 3 and 4—they are now the same—into a single column.)

When we apply the plurality-with-elimination method to Table 1-10, Athens gets eliminated in the first round, and the 7 votes originally going to Athens go to Barcelona in the second round. Barcelona, with 15 votes in the second round gets to host the next Summer Olympics!

How could this happen? How could Calgary lose an election it was winning in the straw poll just

TABLE 1-9	Results of the Straw Poll			
Number of voters	7	8	10	4
1st choice	*A*	*B*	*C*	*A*
2nd choice	*B*	*C*	*A*	*C*
3rd choice	*C*	*A*	*B*	*B*

TABLE 1-10	Official Election Results		
Number of voters	7	8	14
1st choice	*A*	*B*	*C*
2nd choice	*B*	*C*	*A*
3rd choice	*C*	*A*	*B*

because it got additional first-place votes in the official election? While you will never convince the conspiracy theorists in Calgary that the election was not rigged, double-checking the figures makes it clear that everything is on the up and up—Calgary is simply the victim of a quirk in the plurality-with-elimination method: the possibility that you can actually do worse by doing better! Example 1.10 illustrates what in voting theory is known as a violation of the **monotonicity criterion**.

THE MONOTONICITY CRITERION

If candidate X is a winner of an election and, in a reelection, the only changes in the ballots are changes that favor X (and only X), then X should remain a winner of the election.

■ See Exercises 33 and 60.

Based on Example 1.10, we know that the plurality-with-elimination method *violates the monotonicity criterion*. It is not hard to come up with examples showing that the plurality-with-elimination also *violates the Condorcet criterion*.

Despite its flaws, the plurality-with-elimination method is used in many real-world elections. While Example 1.10 was just a simple dramatization, it was realistic to the extent that the International Olympic Committee does use the plurality-with-elimination method to choose the Olympic host cities. These high-stakes elections take place every four years and are typically surrounded by much drama and intrigue. In addition, plurality-with-elimination (under the name *instant runoff voting*) is becoming an increasingly popular method in municipal elections for mayor, city council, district attorney, and so forth. In 2002 voters in San Francisco approved the use of instant runoff voting in all San Francisco municipal elections, and in 2005 a similar proposition was passed by the voters of Burlington, Vermont. Many other municipalities across the United States (for example Berkeley, California, and Ferndale, Michigan) are in the process of implementing instant runoff voting in their local elections. Instant runoff voting is also used in Australia to elect members of the House of Representatives.

■ The most current information on the use of instant runoff voting can be found at www.fairvote.org, the Web site of the Center for Voting and Democracy.

1.5 The Method of Pairwise Comparisons

So far, all three voting methods we have discussed violate the Condorcet criterion, but this is not an insurmountable problem. It is reasonably easy to come up with a voting method that satisfies the Condorcet criterion. Our next method does. The method is known as the **method of pairwise comparisons** (and sometimes as *Copeland's method*).

Much like a round robin tournament in which every player plays once against every other player, in this method every candidate is matched head-to-head against every other candidate. Each of these head-to-head matches is called a **pairwise comparison**. In a pairwise comparison between X and Y every vote is assigned to either X or Y, *the vote going to whichever of the two candidates is listed higher on the ballot*. The winner is the one with the most votes; if the two candidates split the votes equally, the pairwise comparison ends in a tie. The winner of the pairwise comparison gets 1 point and the loser gets none; in case of a tie each candidate gets $\frac{1}{2}$ point. The winner of the election is the candidate with the most points after all the pairwise comparisons are tabulated. (Overall point ties are common under this method, and, as with other methods, the tie is broken using a predetermined tie-breaking procedure or the tie can stand if multiple winners are allowed.)

Once again, we will illustrate the method of pairwise comparisons using the Math Club election.

> **EXAMPLE 1.11** **The Math Club Election (Pairwise Comparisons)**

Let's start with a pairwise comparison between A and B. For convenience we will think of A as the *red* candidate and B as the *blue* candidate—the other candidates do not come into the picture at all. (Don't read too much into the colors—they have no political significance and their only purpose is to help us visualize the comparison.) The first column of Table 1-11 represents 14 red voters (they rank A above B); the last four columns of Table 1-11 represent 23 blue voters (they rank B above A). Consequently, the winner is the blue candidate B. We summarize this result as follows:

TABLE 1-11 Pairwise Comparison Between A and B

Number of voters	14	10	8	4	1
1st choice	A	C	D	B	C
2nd choice	B	B	C	D	D
3rd choice	C	D	B	C	B
4th choice	D	A	A	A	A

> A versus B: 14 votes to 23 votes (B wins). B gets 1 point.

Let's next look at the pairwise comparison between C and D. We'll call C the *red* candidate and D the *blue* candidate. The first, second, and last columns of Table 1-12 represent red votes (they rank C above D); the third and fourth columns represent blue voters (they rank D above C). Here red beats blue 25 to 12.

TABLE 1-12 Pairwise Comparison Between C and D

Number of voters	14	10	8	4	1
1st choice	A	C	D	B	C
2nd choice	B	B	C	D	D
3rd choice	C	D	B	C	B
4th choice	D	A	A	A	A

> C versus D: 25 votes to 12 votes (C wins). C gets 1 point.

If we continue in this manner, comparing in all possible ways two candidates at a time, we end up with the following scoreboard.

> A versus B: 14 votes to 23 votes (B wins). B gets 1 point.
> A versus C: 14 votes to 23 votes (C wins). C gets 1 point.
> A versus D: 14 votes to 23 votes (D wins). D gets 1 point.
> B versus C: 18 votes to 19 votes (C wins). C gets 1 point.
> B versus D: 28 votes to 9 votes (B wins). B gets 1 point.
> C versus D: 25 votes to 12 votes (C wins). C gets 1 point.

■ The reader is encouraged to independently confirm these results. It's always a good idea!

The final tally produces 0 points for A, 2 points for B, 3 points for C, and 1 point for D. Can it really be true? Yes! The winner of the election under the method of pairwise comparisons is Carmen! ⬭

The method of pairwise comparisons satisfies all three of the fairness criteria discussed so far in the chapter. A majority candidate automatically wins every pairwise comparison and thus will win the election under the method of pairwise comparisons, so the method *satisfies the majority criterion*. A Condorcet candidate wins every pairwise comparison by definition, so the method clearly *satisfies the Condorcet criterion*. It can also be shown that the method of pairwise comparisons *satisfies the monotonicity criterion*.

■ See Exercise 65.

So, What's Wrong with the Method of Pairwise Comparisons?

So far, the method of pairwise comparisons looks like the best method we have, at least in the sense that it satisfies the three fairness criteria we have studied. Unfortunately, the method does violate a fourth fairness criterion. Our next example illustrates the problem.

EXAMPLE 1.12 **The NFL Draft**

As the newest expansion team in the NFL, the Los Angeles LAXers will be getting the number-one pick in the upcoming draft. After narrowing the list of candidates to five players (Allen, Byers, Castillo, Dixon, and Evans), the coaches and team executives meet to discuss the candidates and eventually choose the team's first pick on the draft. By team rules, the choice must be made using the method of pairwise comparisons.

Table 1-13 shows the preference schedule after the 22 voters (coaches, scouts, and team executives) turn in their preference ballots.

TABLE 1-13 **LAXer's Draft Choice Election**

Number of voters	2	6	4	1	1	4	4
1st choice	A	B	B	C	C	D	E
2nd choice	D	A	A	B	D	A	C
3rd choice	C	C	D	A	A	E	D
4th choice	B	D	E	D	B	C	B
5th choice	E	E	C	E	E	B	A

The results of the 10 pairwise comparisons between the five candidates are:

A versus *B*: 7 votes to 15 votes. *B* gets 1 point.

A versus *C*: 16 votes to 6 votes. *A* gets 1 point.

A versus *D*: 13 votes to 9 votes. *A* gets 1 point.

A versus *E*: 18 votes to 4 votes. *A* gets 1 point.

B versus *C*: 10 votes to 12 votes. *C* gets 1 point.

B versus *D*: 11 votes to 11 votes. *B* gets $\frac{1}{2}$ point, *D* gets $\frac{1}{2}$ point.

B versus *E*: 14 votes to 8 votes. *B* gets 1 point.

C versus *D*: 12 votes to 10 votes. *C* gets 1 point.

C versus *E*: 10 votes to 12 votes. *E* gets 1 point.

D versus *E*: 18 votes to 4 votes. *D* gets 1 point.

■ This is a good place to take a break from your reading and verify these calculations on your own!

The final tally produces 3 points for *A*, $2\frac{1}{2}$ points for *B*, 2 points for *C*, $1\frac{1}{2}$ points for *D*, and 1 point for *E*. It looks as if Allen (*A*) is the lucky young man who will make millions of dollars playing for the Los Angeles LAXers.

The interesting twist to the story surfaces when it is discovered right before the actual draft that Castillo had accepted a scholarship to go to medical school and will not be playing professional football. Since Castillo was not the top choice, the fact that he is choosing med school over pro football should not affect the fact that Allen is the number-one draft choice. Does it?

Suppose we eliminate Castillo from the original preference schedule (we can do this by eliminating *C* from Table 1-13 and moving the players below *C* up one slot). Table 1-14 shows the results of the revised election when Castillo is not a candidate.

TABLE 1-14	LAXer's Revised Draft Choice Election						
Number of voters	2	6	4	1	1	4	4
1st choice	*A*	*B*	*B*	*B*	*D*	*D*	*E*
2nd choice	*D*	*A*	*A*	*A*	*A*	*A*	*D*
3rd choice	*B*	*D*	*D*	*D*	*B*	*E*	*B*
4th choice	*E*	*E*	*E*	*E*	*E*	*B*	*A*

Now we have only four players and six pairwise comparisons to consider. The results are as follows:

A versus *B*: 7 votes to 15 votes. *B* gets 1 point.

A versus *D*: 13 votes to 9 votes. *A* gets 1 point.

A versus *E*: 18 votes to 4 votes. *A* gets 1 point.

B versus *D*: 11 votes to 11 votes. *B* gets $\frac{1}{2}$ point, *D* gets $\frac{1}{2}$ point.

B versus *E*: 14 votes to 8 votes. *B* gets 1 point.

D versus *E*: 18 votes to 4 votes. *D* gets 1 point.

In this new scenario *A* has 2 points, *B* has $2\frac{1}{2}$ points, *D* has $1\frac{1}{2}$ points, and *E* has 0 points, and the winner is Byers. In other words, when Castillo is not in the running, then the number-one pick is Byers, not Allen. Needless to say, the coaching staff is quite perplexed. Do they draft Byers or Allen? How can the presence or absence of Castillo in the candidate pool be relevant to this decision?   ⬭

The strange happenings in Example 1.12 help illustrate an important fact: The method of pairwise comparisons violates a fairness criterion known as the **independence-of-irrelevant-alternatives criterion**.

> **THE INDEPENDENCE-OF-IRRELEVANT-ALTERNATIVES CRITERION (IIA)**
>
> If candidate *X* is a winner of an election and in a recount one of the *nonwinning* candidates withdraws or is disqualified, then *X* should still be a winner of the election.

An alternative way to interpret the independence-of-irrelevant-alternatives criterion is to switch the order of the two elections (think of the recount as the original election and the original election as the recount). Under this interpretation, the IIA says that *if candidate X is a winner of an election and in a reelection another candidate that has no chance of winning (an "irrelevant alternative") enters the race, then X should still be the winner of the election.*

The method of pairwise comparisons can sometimes be quite indecisive—it is not unusual to have multiple ties for first place. Our next example shows an extreme case of this.

EXAMPLE 1.13   **The Hockey Team Is Out to Lunch**

The Icelandia State University varsity hockey team is on a road trip. An important decision needs to be made: Where to go out to lunch? In the past this has led to some heated arguments, so this time they decide to hold an election. The choices boil down to three restaurants: Andreychuk's (*A*), Bure's (*B*), and Chelios' (*C*). The decision is to be made using the method of pairwise comparisons. Table 1-15 shows the results of the voting by the 15 players on the squad.

Here is the surprising part: *A* beats *B* 10 votes to 5, *B* beats *C* 9 votes to 6, and *C* beats *A* 8 votes to 7. This results in a three-way tie for first place. What now? Since the coaches insist that the team must eat together, it becomes necessary to break the tie. If plurality is used to break the tie, then the winner is *C*; if the Borda count is used to break the tie, then the winner is *A*; and if plurality-with-elimination is used to break the tie, the winner is *B*. Clearly, we are no better off now than when we started!

TABLE 1-15	Restaurant Election Results			
Number of voters	4	2	6	3
1st choice	*A*	*B*	*C*	*B*
2nd choice	*B*	*C*	*A*	*A*
3rd choice	*C*	*A*	*B*	*C*

How Many Pairwise Comparisons?

One practical difficulty with the method of pairwise comparisons is that as the number of candidates grows, the number of comparisons grows even faster: With 5 candidates we have a total of 10 pairwise comparisons, with 10 candidates we have a total of 45 pairwise comparisons, and (heaven forbid) with 100 candidates we have a total of 4950 pairwise comparisons. Computing all these comparisons by hand would not be fun.

In this section we will establish the exact connection between the number of candidates and the number of pairwise comparisons. In so doing we will develop a simple but elegant mathematical formula that has many other applications. We will start with a seemingly unrelated example.

EXAMPLE 1.14   **Long Sums Made Short**

We would like to find the sum of the first 99 positive integers, namely, $1 + 2 + 3 + \cdots + 98 + 99$. We could certainly do it by adding all these numbers, but that would be tedious and wouldn't teach us anything. Instead, we will call the sum S, and write S twice, once forward and once backward, as follows:

$$1 + 2 + 3 + \cdots + 98 + 99 = S$$
$$99 + 98 + 97 + \cdots + 2 + 1 = S$$

You may have noticed that when we add the numbers column by column, each column totals 100. Thus, adding the 99 columns together totals $99 \times 100 = 9900$. At the same time, this total equals $2S$, so we can conclude that $2S = 9900$ and therefore $S = 4950$.

The argument used in Example 1.14 can be easily generalized to any sum of the form $S = 1 + 2 + 3 + \cdots + L$, where *L* is a positive integer (think of *L* as short for "last" integer). When we write the sum twice, once forward and once backward, we get *L* columns, each of which totals $(L + 1)$. This implies that

$2S = L(L + 1)$ and $S = \dfrac{L(L + 1)}{2}$. This gives us an extremely nice and useful formula for the sum of all integers from 1 to L:

SUM OF CONSECUTIVE INTEGERS FORMULA

$$1 + 2 + 3 + \cdots + L = \frac{L(L + 1)}{2}$$

■ See Exercises 53 and 54.

With a little care the above formula can be also used to find the sum of any set of consecutive integers, even when the starting term is bigger than 1.

(**EXAMPLE 1.15**)   **Counting Pairwise Comparisons**

Consider an election with 10 candidates: A, B, C, D, E, F, G, H, I, and J. Let's count the pairwise comparisons:

- Compare A with each of the other 9 candidates (B through J). This gives 9 pairwise comparisons. We are now done with A.
- Compare B with each of the remaining 8 candidates (C through J). This gives 8 pairwise comparisons. We are now done with B.
- Compare C with each of the remaining 7 candidates (D through J). This gives 7 pairwise comparisons. We are now done with C.
- We keep going this way. At the end, when we get to I, we have one comparison left, between I and J.

This gives a total of $1 + 2 + 3 + 4 + 5 + 6 + 7 + 8 + 9 = (9 \times 10)/2 = 45$ pairwise comparisons.

Generalizing the argument used in Example 1.15 is straightforward: the number of pairwise comparisons with N candidates is given by the sum $1 + 2 + 3 + 4 + \cdots + (N - 1)$. Notice that the "last" number in the sum is $(N - 1)$.

Using the formula for the sum of the first L integers and substituting $(N - 1)$ for L gives us the following:

THE NUMBER OF PAIRWISE COMPARISONS

In an election with N candidates the total number of pairwise comparisons is $\dfrac{(N - 1)N}{2}$.

(**EXAMPLE 1.16**)   **Scheduling a Round-Robin Tournament**

In a round-robin tournament, every player (team) plays every other player (team) once. Imagine you are in charge of scheduling a round-robin Ping-Pong tournament with 32 players. The tournament organizers agree to pay you $1 per match for running the tournament—you want to know how much you are going to make.

Since each match is like a pairwise comparison, we can think of the players as candidates and use the formula for pairwise comparisons. We can conclude that there will be a total of $(31 \times 32)/2 = 496$ matches played. Not a bad gig for a weekend job!

1.6 Rankings

Quite often it is important not only to know who wins the election but also to know who comes in second, third, and so on. Let's consider once again the Math Club election. Suppose now that instead of electing just the president we need to elect a board of directors consisting of a president, a vice president, and a treasurer. The club's bylaws state that rather than having separate elections for each office, the winner of the general election gets to be the president, the second-place candidate gets to be the vice president, and the third-place candidate gets to be the treasurer. In a situation like this, we need a voting method that gives us not just a winner but also a second place, a third place, and so on—in other words, a **ranking** of the candidates.

Extended Ranking Methods

Each of the four voting methods we discussed earlier in this chapter has a natural extension that can be used to produce a ranking of the candidates.

EXAMPLE 1.17   **The Math Club Election (Extended Plurality)**

For the reader's convenience, the original preference schedule for the Math Club election is shown again in Table 1-16.

TABLE 1-16   Preference Schedule for the Math Club Election

Number of voters	14	10	8	4	1
1st choice	A	C	D	B	C
2nd choice	B	B	C	D	D
3rd choice	C	D	B	C	B
4th choice	D	A	A	A	A

The number of first-place votes is

A: 14 first-place votes,

B: 4 first-place votes,

C: 11 first-place votes,

D: 8 first-place votes.

We know that under the plurality method Alisha, with the most first-place votes, is the winner. If we extend the same logic to rank the remaining candidates, then Carmen is second (11 votes), Dave is third (8 votes), and Boris (4 votes) is last. The ranking of all the candidates under the *extended* plurality method is shown in Table 1-17.

TABLE 1-17   Ranking Under Extended Plurality

Place	1st	2nd	3rd	4th
Candidate	A	C	D	B
First-place votes	14	11	8	4

> **EXAMPLE 1.18**   The Math Club Election (Extended Borda Count)

Under the extended Borda count method candidates are ranked according to their Borda point totals. Recall that the Borda point totals for the Math Club election (Example 1.5) were

A: 79 Borda points,

B: 106 Borda points,

C: 104 Borda points,

D: 81 Borda points.

The resulting ranking is given in Table 1-18.

TABLE 1-18	Ranking Under Extended Borda Count			
Place	1st	2nd	3rd	4th
Candidate	B	C	D	A
Borda points	106	104	81	79

> **EXAMPLE 1.19**   The Math Club Election (Extended Plurality-with-Elimination)

Under the *extended* plurality-with-elimination method we rank the candidates in reverse order of elimination: Since Boris was eliminated in the first round (see Example 1.7), we rank Boris last. Carmen was eliminated in the second round, so she gets third place; Alisha was eliminated in the third round, so she gets second place, and Dave takes first place. Table 1-19 summarizes the outcome of the Math Club election under this method. (In elections in which a candidate gets a majority of votes before all the rounds of elimination are completed, we continue the process of elimination to rank the remaining candidates.)

TABLE 1-19	Ranking Under Extended Plurality-with-Elimination			
Place	1st	2nd	3rd	4th
Candidate	D	A	C	B
Eliminated in Round		3	2	1

> **EXAMPLE 1.20**   The Math Club Election (Extended Pairwise Comparisons)

Under the *extended* method of pairwise comparisons we rank the candidates according to the total number of points in their comparisons with the other candidates. (Recall that a win counts as 1 point, a loss as 0 points, and a tie as $\frac{1}{2}$ point.) Table 1-20 shows the results for the Math Club election (see Example 11.1 for the details of the calculations).

TABLE 1-20	Ranking Under Extended Pairwise Comparisons			
Place	1st	2nd	3rd	4th
Candidate	C	B	D	A
Points	3	2	1	0

A summary of the results of the Math Club election using the different extended ranking methods is shown in Table 1-21. The most striking thing about Table 1-21 is the wide discrepancy of results. While it is somewhat frustrating to see this much equivocation, it is important to keep things in context: This is the exception rather than the rule. One purpose of the Math Club example is to illustrate how crazy things can get in some elections, but in most real-life elections there tends to be much more consistency among the various methods.

TABLE 1-21	The Math Club Election: A Tale of Four Methods			
		Ranking		
Method	1st	2nd	3rd	4th
Extended plurality	*A*	*C*	*D*	*B*
Extended Borda count	*B*	*C*	*D*	*A*
Extended plurality with elimination	*D*	*A*	*C*	*B*
Extended pairwise comparisons	*C*	*B*	*D*	*A*

Recursive Ranking Methods

We will now discuss a different, somewhat more involved strategy for ranking the candidates, which we will call the *recursive* approach. (A word to the wise—recursive processes will show up again several times later in this book and are extremely useful and important.) The idea behind a recursive process is similar to that of a feedback loop: At each step of the process the output of the process determines the input to the next step of the process. In the case of an election the process is to find the winner and remove the winner's name from the preference schedule, thus creating a new preference schedule. We then start the process all over again.

The **recursive ranking** approach follows a consistent strategy, but the details depend on the voting method we want to use. Let's say we are going to use the recursive version of some generic voting method *X* to rank the candidates in an election. We first use method *X* to find the winner of the election. So far, so good. We then remove the name of the winner on the preference schedule and obtain a new preference schedule with one less candidate on it. We apply method *X* once again to find the "winner" based on this new preference schedule, and this candidate is ranked second. We can continue this process to rank as many of the candidates as we need to.

We will illustrate recursive ranking with a couple of examples, both based on the Math Club election.

(**EXAMPLE 1.21**)   **The Math Club Election (Recursive Plurality)**

Here is how we rank the four candidates in the Math Club election using the *recursive* plurality method.

Step 1. (Choose the winner and remove.) Table 1-22 shows the original input (the original preference schedule). We already know the winner is *A* with 14 first-place votes. Table 1-23 shows the output of this step—the preference

TABLE 1-22 Step 1 Input

Number of voters	14	10	8	4	1
1st choice	A	C	D	B	C
2nd choice	B	B	C	D	D
3rd choice	C	D	B	C	B
4th choice	D	A	A	A	A

TABLE 1-23 Step 2 Input

Number of voters	14	10	8	4	1
1st choice	B	C	D	B	C
2nd choice	C	B	C	D	D
3rd choice	D	D	B	C	B

TABLE 1-24 Step 3 Input

Number of voters	14	10	8	4	1
1st choice	C	C	D	D	C
2nd choice	D	D	C	C	D

See Exercises 45 and 46.

schedule when A is removed. This will be the input to the next step.

Step 2. (Choose second place and remove.) Table 1-23 is the input for this step—the preference schedule after A has been removed. The "winner" of this election is B with 18 votes. Thus, *second place* goes to B. Table 1-24 shows the result of removing B from Table 1-23. This will be the input to the next step.

Step 3. (Choose third and fourth places.) The last step is to find the results of the two-candidate election shown in Table 1-24. Clearly, C wins with 25 votes. Thus, *third place* goes to C and *last place* goes to D.

The final ranking of the candidates under the *recursive* plurality method is shown in Table 1-25. It is worth noting how different this ranking is from the ranking shown in Table 1-17 based on the *extended* plurality method. In fact, except for first place (which will always be the same), all the other positions turned out to be different.

TABLE 1-25 Ranking Under Recursive Plurality

Place	1st	2nd	3rd	4th
Candidate	A	B	C	D

For our final example we will illustrate how to rank the candidates in the Math Club election using the *recursive* plurality-with-elimination method. Ranking the Math Club election using recursive Borda count and recursive pairwise comparisons are left as end-of-chapter exercises.

EXAMPLE 1.22 **The Math Club Election (Recursive Plurality-with-Elimination)**

We have to be careful to distinguish the two different types of "elimination" that take place in this method. Under the plurality-with-elimination method, candidates are *eliminated* in rounds until there is a winner left. Under the recursive approach, the winner at each step of the recursion is *removed* from the preference schedule so that we can move on to the next step. Thus, each step consists of several rounds of elimination and at the end, the removal of the winner. Here is how it works:

TABLE 1-26   Step 1 Input

Number of voters	14	10	8	4	1
1st choice	A	C	D	B	C
2nd choice	B	B	C	D	D
3rd choice	C	D	B	C	B
4th choice	D	A	A	A	A

TABLE 1-27   Step 2 Input

Number of voters	14	10	8	4	1
1st choice	A	C	C	B	C
2nd choice	B	B	B	C	B
3rd choice	C	A	A	A	A

TABLE 1-28   Step 3 Input

Number of voters	14	23
1st choice	A	B
2nd choice	B	A

Step 1. (Choose the winner and remove.) Table 1-26 shows the original input (the original preference schedule). After several rounds of elimination, the winner is D. (We worked out the details in Example 1.7.) Table 1-27 shows the output of this step—the preference schedule when D is removed. This will be the input to Step 2.

Step 2. (Choose second place and remove.) Table 1-27 shows the input for this step. In this election no rounds of elimination are needed, since C has 19 votes (a majority). Thus, *second place* goes to C. We now remove C from the preference schedule and get the preference schedule shown in Table 1-28 after like columns are combined. This will be the input to the last step.

Step 3. (Choose third and fourth places.) The last step is to find the results of the two-candidate election shown in Table 1-28. Clearly, B wins with 23 votes. Thus, *third place* goes to B and *last place* goes to A.

The final ranking of the candidates under the *recursive* plurality-with-elimination method is shown in Table 1-29. Once again, if we compare this ranking with the ranking under the *extended* plurality-with-elimination method (see Table 1-19), the differences in how second, third, and fourth places are ranked are striking.

TABLE 1-29   Ranking Under Recursive Plurality with Elimination

Place	1st	2nd	3rd	4th
Candidate	D	C	B	A

It is clear from the last two examples that other than first place, recursive ranking methods and extended ranking methods can produce very different results. Which one produces better rankings? As with everything else in election theory, there is no simple answer to this question. It is true, however, that in real-life elections, extended ranking methods are almost always used. Recursive ranking methods, while mathematically interesting, are of little practical use. But don't fret—what we learned about recursion will come in handy in Chapters 9 and 12.

CONCLUSION   Elections, Fairness, and Arrow's Impossibility Theorem

Several broad themes run through this chapter. One of them is that elections are more than just for choosing our president or our governor—"small" elections of various kinds play a pervasive and important role in all our lives, from getting a job to deciding where to go to dinner. The second is that there are many different formal methods that can be used to decide the outcome of an election. In this chapter we focused on four basic methods—mostly because they are simple and commonly used—but there are many others, some quite elaborate and exotic. Third, we saw that the results of an election can change if we change the voting method. The Math Club example dramatically illustrated this point—each of the four voting methods produced a different winner. Since there were four candidates, we can say that each of them won the election (just pick the "right" voting method).

Given an abundance of voting methods, how do we determine which are "good" voting methods and which are not so good? In a democratic society the most important quality we seek in a voting method is that of *fairness*. This leads us to the last, but probably most important, theme of this chapter—the notion of *fairness criteria*. Fairness criteria set the basic standards that a fair election should satisfy. We discussed four of them in this chapter—let's review what they were:

- **The Majority Criterion:** *A majority candidate should always win the election.* (After all, it seems clearly unfair when a candidate with a majority of the first-place votes does not win.)

- **The Condorcet Criterion:** *A Condorcet candidate should always win the election.* (When the candidates are compared two at a time, the Condorcet candidate beats each of the other candidates. How could it be fair to declare a different candidate as the winner?)

- **The Monotonicity Criterion:** *Suppose candidate X is a winner of the election, but for one reason or another there is a new election. If the only changes in the ballots are changes in favor of candidate X (and only X), then X should win the new election.* (Think of the election as a card game. If at some point in the game X has the winning hand and then X's hand improves while all the other players' hands stay the same or get worse, then X should still have the winning hand. By any reasonable standard, if X were to lose under these circumstances, it would not be a fair game.)

- **The Independence-of-Irrelevant-Alternatives Criterion (IIA):** *Suppose candidate X is a winner of the election, but for one reason or another there is a new election. If the only changes are that one of the other candidates withdraws or is disqualified, then X should win the new election.* (Clearly, it would be unfair to penalize the winner of an election because of the mere fact that one of the losers scratches out of the race). The flip side of this criterion is that a winner of the election should not be penalized by the introduction of irrelevant new candidates who have no chance of winning.

■ For additional fairness criteria see Exercises 74–78.

Other fairness criteria beyond the ones listed above have been proposed, but these four are sufficient to establish a minimum standard of fairness—*a fair voting method should clearly satisfy all four of the above*. Surprisingly, none of the four voting methods we discussed in this chapter meets this minimum standard—far from it! The plurality method violates the Condorcet criterion and the IIA criterion; the Borda count method violates the Majority criterion, the Condorcet criterion, and the IIA criterion; the plurality-with-elimination method violates the

Condorcet criterion, the monotonicity criterion, and the IIA criterion; and the pairwise comparisons method violates the IIA criterion.

An obviously relevant question then is, What voting method satisfies the minimum standards of fairness set forth by the above four criteria? For democratic elections involving three or more candidates, the answer is that *there is no such voting method*. At first glance, this fact seems a little surprising. Finding a voting method that satisfies a few simple fairness criteria does not seem like an impossible task, and given the obvious importance of having fair elections in a democracy, how is it possible that no one has come up with such a voting method? Until the late 1940s this was one of the most challenging questions in social-choice theory. Finally, in 1949 mathematical economist Kenneth Arrow demonstrated that this is indeed an impossible task. **Arrow's impossibility theorem** essentially says (his exact formulation of the theorem is slightly different from the one given here) that *it is mathematically impossible for a democratic voting method to satisfy all of the fairness criteria*. Arrow's impossibility theorem is surprising and paradoxical. It tells us that making decisions in a consistently fair way is inherently impossible in a democratic society.

> " The search of the great minds of recorded history for the perfect democracy, it turns out, is the search for a chimera, a logical self-contradiction. "
>
> —Paul Samuelson

PROFILE: Kenneth J. Arrow (1921–   )

Kenneth Arrow is one of the best known and most versatile mathematical economists today. Born and raised in New York City, Arrow completed his undergraduate studies at City College of New York, where at age 19 he received bachelor degrees in both social science and mathematics. He continued his graduate studies at Columbia University, where he received a master of arts in mathematics in 1941. At this time he became interested in mathematical economics, and by 1942 he completed all the necessary course work for a Ph.D. in economics at Columbia. His graduate studies at Columbia were interrupted by World War II, and between 1942 and 1946 he was assigned to serve as a weather officer in the U.S. Army Air Corps. As a result of his weather work for the Army, Arrow was able to publish his first mathematical research paper, *On the Optimal Use of Winds for Flight Planning*. In this paper, Arrow's ability to see significant mathematical ideas in seemingly nonmathematical subjects already comes through. This talent would become one of the hallmarks of his later work.

In 1946 Arrow returned to Columbia to write his doctoral thesis, which he completed in 1949. (By then he was already working as a research associate and Assistant Professor of Economics at the University of Chicago.)

Arrow's doctoral thesis, titled *Social Choice and Individual Values*, became a landmark work in mathematical economics and would eventually lead to his being awarded the 1972 Nobel Prize in Economics. Using the mathematical principles of game theory, Arrow proved his famous *impossibility theorem*, which in essence says that it is impossible for a democratic society to take the individual opinions of its voters (the "individual values") and always make a collective decision (the "social choice") that is fair. As one might imagine for a doctoral thesis and Nobel Prize–quality work, the details are quite technical, but the mathematics itself was not difficult. Arrow's genius was to find a way to redefine the idea of "fairness" in a way that lends itself to mathematical treatment.

In 1949 Arrow was appointed Assistant Professor of Economics and Statistics at Stanford University, eventually becoming Professor of Economics, Statistics, and Operations Research. Between 1968 and 1979 he was a Professor of Economics at Harvard University, but in 1979 he returned to Stanford, where he is currently the Joan Kenney Professor of Economics Emeritus, and Professor of Operations Research. Even in his late 80s Arrow remains active and pursuing new areas of research.

KEY CONCEPTS

Arrow's impossibility
theorem, **28**
Borda count method, **10**
Condorcet candidate, **8**
Condorcet criterion, **8**
extended ranking methods, **22**
independence-of-irrelevant-
alternatives criterion, **19**

insincere voting, **8**
linear ballot, **4**
majority candidate, **6**
majority criterion, **6**
method of pairwise
comparisons, **16**
monotonicity criterion, **16**
plurality candidate, **6**

plurality method, **6**
plurality-with-elimination
method, **12**
preference ballot, **4**
preference schedule, **5**
rankings, **22**
recursive ranking
methods, **24**

EXERCISES

WALKING

A Ballots and Preference Schedules

1. Figure 1-6 shows the preference ballots for an election
   with 21 voters and 5 candidates.

   (a) Write out the preference schedule for this election.

   (b) Which candidate has a plurality of the first-place
   votes?

   (c) Which candidate has the fewest last-place votes?

2. Figure 1-7 shows the preference ballots for an election
   with 17 voters and 4 candidates.

   (a) Write out the preference schedule for this election.

   (b) Which candidate has a plurality of the first-place
   votes?

   (c) Which candidate has the fewest last-place votes?

Ballot	**Ballot**	**Ballot**	**Ballot**	**Ballot**	**Ballot**	**Ballot**
1st *C*	1st *A*	1st *B*	1st *A*	1st *C*	1st *D*	1st *A*
2nd *E*	2nd *E*	2nd *E*	2nd *E*	2nd *E*	2nd *C*	2nd *E*
3rd *D*	3rd *B*	3rd *A*	3rd *C*	3rd *D*	3rd *B*	3rd *C*
4th *A*	4th *C*	4th *C*	4th *D*	4th *A*	4th *E*	4th *D*
5th *B*	5th *E*	5th *D*	5th *E*	5th *B*	5th *A*	5th *E*

Ballot	**Ballot**	**Ballot**	**Ballot**	**Ballot**	**Ballot**	**Ballot**
1st *B*	1st *A*	1st *D*	1st *D*	1st *A*	1st *C*	1st *A*
2nd *E*	2nd *B*	2nd *C*	2nd *C*	2nd *B*	2nd *E*	2nd *D*
3rd *A*	3rd *C*	3rd *B*	3rd *B*	3rd *C*	3rd *D*	3rd *B*
4th *C*	4th *D*	4th *A*	4th *E*	4th *D*	4th *A*	4th *C*
5th *D*	5th *E*	5th *E*	5th *A*	5th *E*	5th *B*	5th *E*

Ballot	**Ballot**	**Ballot**	**Ballot**	**Ballot**	**Ballot**	**Ballot**
1st *B*	1st *C*	1st *A*	1st *C*	1st *A*	1st *D*	1st *D*
2nd *E*	2nd *E*	2nd *B*	2nd *E*	2nd *D*	2nd *C*	2nd *C*
3rd *A*	3rd *D*	3rd *C*	3rd *D*	3rd *B*	3rd *B*	3rd *B*
4th *C*	4th *A*	4th *D*	4th *A*	4th *C*	4th *A*	4th *E*
5th *D*	5th *B*	5th *E*	5th *B*	5th *E*	5th *E*	5th *A*

FIGURE 1-6

Ballot	**Ballot**	**Ballot**	**Ballot**	**Ballot**
1st *C*	1st *B*	1st *A*	1st *C*	1st *B*
2nd *A*	2nd *C*	2nd *D*	2nd *A*	2nd *C*
3rd *D*	3rd *D*	3rd *B*	3rd *D*	3rd *D*
4th *B*	4th *A*	4th *C*	4th *B*	4th *A*

Ballot	**Ballot**	**Ballot**	**Ballot**	**Ballot**	**Ballot**
1st *A*	1st *A*	1st *B*	1st *B*	1st *C*	1st *C*
2nd *D*	2nd *C*	2nd *C*	2nd *C*	2nd *A*	2nd *A*
3rd *B*	3rd *D*	3rd *D*	3rd *D*	3rd *D*	3rd *D*
4th *C*	4th *B*	4th *A*	4th *A*	4th *B*	4th *B*

Ballot	**Ballot**	**Ballot**	**Ballot**	**Ballot**	**Ballot**
1st *A*	1st *A*	1st *C*	1st *B*	1st *A*	1st *C*
2nd *C*	2nd *D*	2nd *A*	2nd *C*	2nd *D*	2nd *A*
3rd *D*	3rd *B*	3rd *D*	3rd *D*	3rd *B*	3rd *D*
4th *B*	4th *C*	4th *B*	4th *A*	4th *C*	4th *B*

FIGURE 1-7

3. An election is held to choose the Chair of the Mathematics Department at Tasmania State University. The candidates are Professors Argand, Brandt, Chavez, Dietz, and Epstein (A, B, C, D, and E for short). The following table gives the preference schedule for the election:

Number of voters	5	3	5	3	2	3
1st choice	A	A	C	D	D	B
2nd choice	B	D	E	C	C	E
3rd choice	C	B	D	B	B	A
4th choice	D	C	A	E	A	C
5th choice	E	E	B	A	E	D

(a) How many people voted in this election?

(b) How many first-place votes are needed for a majority?

(c) If it came down to a choice between Argand and Dietz, which one would get more votes?

4. The student body at Eureka High School is having an election for Homecoming Queen. The candidates are Alicia, Brandy, Cleo, and Dionne (A, B, C, and D for short). The following table gives the preference schedule for the election:

Number of voters	153	102	55	202	108	20	110	160	175	155
1st choice	A	A	A	B	B	B	C	C	D	D
2nd choice	C	B	D	D	C	C	A	B	A	B
3rd choice	B	D	C	A	D	A	D	A	C	C
4th choice	D	C	B	C	A	D	B	D	B	A

(a) How many students voted in this election?

(b) How many first-place votes are needed for a majority?

(c) If it came down to a choice between Brandy and Dionne, which one would get more votes?

5. This exercise refers to the election for Mathematics Department Chair discussed in Exercise 3. Suppose that the election rules are that when there is a candidate with a majority of the first-place votes, he or she is the winner.

Otherwise, all candidates with 20% or less of the first-place votes are eliminated and the ballots are recounted.

(a) Which candidates are eliminated in this election?

(b) Find the preference schedule for the recount.

(c) Which candidate is the majority winner after the recount?

6. This exercise refers to the Eureka High School Homecoming Queen election discussed in Exercise 4. Suppose that the election rules are that when there is a candidate with a majority of the first-place votes, she is the winner. Otherwise, all candidates with 25% or less of the first-place votes are eliminated and the ballots are recounted.

(a) Which candidates are eliminated in this election?

(b) Find the preference schedule for the recount.

(c) Which candidate is the majority winner after the recount?

7. The Demublican Party is holding its annual convention. The 1500 voting delegates are choosing among three possible party platforms: L (a liberal platform), C (a conservative platform), and M (a moderate platform). Seventeen percent of the delegates prefer L to M and M to C. Thirty-two percent of the delegates like C the most and L the least. The rest of the delegates like M the most and C the least. Write out the preference schedule for this election.

8. The Epicurean Society is holding its annual election for president. The three candidates are A, B, and C. Twenty percent of the voters like A the most and B the least. Forty percent of the voters like B the most and A the least. Of the remaining voters 225 prefer C to B and B to A, and 675 prefer C to A and A to B. Write out the preference schedule for this election.

In this chapter we used preference ballots that list ranks (1st choice, 2nd choice, etc.) and asked the voter to put the name of a candidate or choice next to each rank. Exercises 9 and 10 refer to an alternative format for preference ballots in which the names of candidates appear in some order and the voter is asked to put a rank (1, 2, 3, etc.) next to each name.

9. Rewrite the following preference schedule in the conventional format used in the book.

Number of voters	47	36	24	13	5
A	3	1	2	4	3
B	1	2	1	2	5
C	4	4	5	3	1
D	5	3	3	5	4
E	2	5	4	1	2

10. Rewrite the following conventional preference schedule in the alternative format. Assume that the candidates are listed in alphabetical order on the ballots.

Number of voters	47	36	24	13	5
1st choice	A	B	D	C	B
2nd choice	C	A	B	A	D
3rd choice	B	D	C	E	E
4th choice	E	C	E	B	A
5th choice	D	E	A	D	C

B | Plurality Method

11. This exercise refers to the Eureka High School Homecoming Queen election discussed in Exercises 4 and 6. For your convenience, here is the preference schedule again:

Number of voters	153	102	55	202	108	20	110	160	175	155
1st choice	A	A	A	B	B	B	C	C	D	D
2nd choice	C	B	D	D	C	C	A	B	A	B
3rd choice	B	D	C	A	D	A	D	A	C	C
4th choice	D	C	B	C	A	D	B	D	B	A

(a) Use the plurality method to find the winner(s) of the election.

(b) **Tie-breaking rule:** *If there is more than one alternative with a plurality of the first-place votes, then the tie is broken by choosing the alternative with the fewest last-place votes.* Who would be the Homecoming Queen under this tie-breaking rule?

(c) **A different tie-breaking rule:** *If there are two candidates tied with a plurality of the first-place votes, the tie is broken by a head-to-head comparison between the two candidates.* Who would be the Homecoming Queen under this tie-breaking rule?

12. A math class is asked by the instructor to vote among four possible times for the final exam—*A* (December 15, 8:00 A.M.), *B* (December 20, 9:00 P.M.), *C* (December 21, 7:00 A.M.), and *D* (December 23, 11:00 A.M.).

The following table gives the preference schedule for the election:

Number of voters	3	4	9	9	2	5	8	3	12
1st choice	A	A	A	B	B	B	C	C	D
2nd choice	B	B	C	C	A	C	D	A	C
3rd choice	C	D	B	D	C	A	B	D	A
4th choice	D	C	D	A	D	D	A	B	B

(a) Use the plurality method to find the winner(s) of the election.

(b) **Tie-breaking rule:** *If there is more than one alternative with a plurality of the first-place votes, then the tie is broken by choosing the alternative with the fewest last-place votes.* Which would be the winning alternative under this tie-breaking rule?

(c) **A different tie-breaking rule:** *If there are two alternatives tied with a plurality of the first-place votes, then the tie is broken by a head-to-head comparison between the two alternatives.* Which would be the winning alternative under this tie-breaking rule?

13. An election with 4 candidates (*A*, *B*, *C*, *D*) and 150 voters is to be decided using the plurality method. After 120 ballots have been recorded, *A* has 26 votes, *B* has 18 votes, *C* has 42 votes, and *D* has 34 votes.

(a) For *A* to win the election outright, at least how many of the remaining 30 votes need to have *A* in first place? Explain.

(b) For *C* to win the election outright, at least how many of the remaining 30 votes need to have *C* in first place? Explain.

14. An election with 4 candidates (*A*, *B*, *C*, *D*) and 150 voters is to be decided using the plurality method. After 120 ballots have been recorded, *A* has 26 votes, *B* has 18 votes, *C* has 42 votes, and *D* has 34 votes.

(a) For *B* to win the election outright, at least how many of the remaining 30 votes need to have *B* in first place? Explain.

(b) For *D* to win the election outright, at least how many of the remaining 30 votes need to have *D* in first place? Explain.

15. Consider an election with 721 voters.

   (a) If there are 5 candidates, at least x votes are needed to have a plurality of the votes. Find x.

   (b) Suppose that at least 73 votes are needed to have a plurality of the votes. What is the number of candidates in the election?

   (c) Suppose that at least 104 votes are needed to have a plurality of the votes. What is the number of candidates in the election?

16. Consider an election with 1025 voters.

   (a) If there are 4 candidates, at least x votes are needed to have a plurality of the votes. Find x.

   (b) Suppose that at least 129 votes are needed to have a plurality of the votes. What is the number of candidates in the election?

   (c) Suppose that at least 206 votes are needed to have a plurality of the votes. What is the number of candidates in the election?

C Borda Count Method

17. This exercise refers to the Mathematics Department Chair election discussed in Exercises 3 and 5. For your convenience, here is the preference schedule again:

Number of voters	5	3	5	3	2	3
1st choice	A	A	C	D	D	B
2nd choice	B	D	E	C	C	E
3rd choice	C	B	D	B	B	A
4th choice	D	C	A	E	A	C
5th choice	E	E	B	A	E	D

   (a) Use the Borda count method to find the winner of the election.

   (b) Suppose that before the votes are counted candidate E is declared ineligible. Find the new preference schedule for an election without candidate E.

   (c) Use the Borda count method to find the winner of the new election without candidate E.

   (d) The results of (a) and (c) imply a violation of which fairness criterion?

18. This exercise refers to the Eureka High School Homecoming Queen election discussed in Exercises 4, 6,

and 11. For your convenience, here is the preference schedule again:

Number of voters	153	102	55	202	108	20	110	160	175	155
1st choice	A	A	A	B	B	B	C	C	D	D
2nd choice	C	B	D	D	C	C	A	B	A	B
3rd choice	B	D	C	A	D	A	D	A	C	C
4th choice	D	C	B	C	A	D	B	D	B	A

   (a) Use the Borda count method to find the winner of the election.

   (b) Suppose that before the votes are counted D is found to be ineligible because of her grades. Find the preference schedule for the election after D is removed from the original preference schedule.

   (c) Use the Borda count method to find the winner of the new election without candidate D.

   (d) The results of (a) and (c) imply a violation of which fairness criterion?

19. The editorial board of *Gourmet* magazine is having an election to choose the "Restaurant of the Year." The choices are Andre's, Borrelli, Casablanca, Dante, and Escargot ($A, B, C, D,$ and E for short). The winner of the election is to be determined using the Borda count method.

   The following table gives the preference schedule for the election:

Number of voters	8	7	6	2	1
1st choice	A	D	D	C	E
2nd choice	B	B	B	A	A
3rd choice	C	A	E	B	D
4th choice	D	C	C	D	B
5th choice	E	E	A	E	C

   (a) Find the winner of the election.

   (b) This election illustrates a violation of the *majority* criterion. Explain how this happens.

   (c) This election illustrates a violation of the *Condorcet* criterion. Explain how this happens.

20. The members of the Tasmania State University soccer team are having an election to choose the captain of the team from among the four seniors—Anderson, Bergman, Chou, and Delgado. The winner of the election is to be determined using the Borda count method.

The following table gives the preference schedule for the election:

Number of voters	4	1	9	8	5
1st choice	A	B	C	A	C
2nd choice	B	A	D	D	D
3rd choice	D	D	A	B	B
4th choice	C	C	B	C	A

(a) Find the winner of the election.

(b) This election illustrates a violation of the *majority* criterion. Explain how this happens.

(c) This election illustrates a violation of the *Condorcet* criterion. Explain how this happens.

21. An election is held among four candidates (A, B, C, D). Each column in the following preference schedule shows the percentage of voters voting that way. Find the winner of the election under the Borda count method. (*Hint:* The winner is independent of the number of voters.)

Percentage of voters	40%	25%	20%	15%
1st choice	A	C	B	B
2nd choice	D	B	D	A
3rd choice	B	D	A	D
4th choice	C	A	C	C

22. An election is held among four candidates (A, B, C, D). Each column in the following preference schedule shows the percentage of voters voting that way. Find the winner of the election under the Borda count method. (*Hint:* The winner is independent of the number of voters.)

Percentage of voters	48%	24%	16%	12%
1st choice	A	C	B	B
2nd choice	D	B	D	A
3rd choice	B	D	A	D
4th choice	C	A	C	C

23. An election is held with four candidates ($A, B, C,$ and D) and 110 voters. The winner of the election is determined using the Borda count method.

(a) What is the maximum number of points a candidate can receive?

(b) What is the minimum number of points a candidate can receive?

(c) How many points are given out by one ballot?

(d) What is the total number of points given out to all four candidates?

(e) If A gets 320 points, B gets 290 points, and C gets 180 points, how many points did D get?

24. An election is held with five candidates ($A, B, C, D,$ and E) and 20 voters. The winner of the election is determined using the Borda count method.

(a) What is the maximum number of points a candidate can receive?

(b) What is the minimum number of points a candidate can receive?

(c) How many points are given out by one ballot?

(d) What is the total number of points given out to all five candidates?

(e) If A gets 69 points, B gets 70 points, C gets 64 points, and D gets 48 points, how many points did E get?

25. An election is held with four candidates ($A, B, C,$ and D) and 50 voters. The winner of the election is determined using the Borda count method. Suppose you are given the following point totals: 115 points for A, 120 points for B, and 125 points for D. The point totals for C are missing. Find the winner of the election anyway and explain your answer. (*Hint:* If you have trouble with this exercise, try Exercise 23 first.)

26. An election is held with five candidates ($A, B, C, D,$ and E) and 40 voters. The winner of the election is determined using the Borda count method. Suppose you are given the following point totals: 139 points for A, 121 points for B, 80 points for C, and 113 points for D. The point totals for E are missing. Find the winner of the election anyway and explain your answer. (*Hint:* If you have trouble with this exercise, try Exercises 23 or 24 first.)

D Plurality-with-Elimination Method

27. This exercise refers to the Mathematics Department Chair election discussed in Exercises 3, 5, and 17. For your convenience, here is the preference schedule again:

Number of voters	5	3	5	3	2	3
1st choice	A	A	C	D	D	B
2nd choice	B	D	E	C	C	E
3rd choice	C	B	D	B	B	A
4th choice	D	C	A	E	A	C
5th choice	E	E	B	A	E	D

(a) Use the plurality-with-elimination method to determine the winner of the election.

(b) Suppose that before the votes are counted candidate C withdraws from the election. Find the new preference schedule for an election without candidate C.

(c) Use the plurality-with-elimination method to find the winner of the new election without candidate C.

(d) The results of (a) and (c) imply a violation of which fairness criterion?

28. This exercise refers to the Eureka High School Homecoming Queen election discussed in Examples 4, 6, and 18. Find the winner of the election under the plurality-with-elimination method.

Number of voters	153	102	55	202	108	20	110	160	175	155
1st choice	A	A	A	B	B	B	C	C	D	D
2nd choice	C	B	D	D	C	C	A	B	A	B
3rd choice	B	D	C	A	D	A	D	A	C	C
4th choice	D	C	B	C	A	D	B	D	B	A

29. This exercise refers to the "Restaurant of the Year" election discussed in Exercise 19. Here is the preference schedule again:

Number of voters	8	7	6	2	1
1st choice	A	D	D	C	E
2nd choice	B	B	B	A	A
3rd choice	C	A	E	B	D
4th choice	D	C	C	D	B
5th choice	E	E	A	E	C

(a) Find the winner of the election under the plurality-with-elimination method.

(b) Explain why the winner in (a) can be determined in the first round.

(c) Explain why the plurality-with-elimination satisfies the *majority* criterion.

30. This exercise refers to the Tasmania State University soccer team captain election discussed in Exercise 20. Here is the preference schedule again:

Number of voters	4	1	9	8	5
1st choice	A	B	C	A	C
2nd choice	B	A	D	D	D
3rd choice	D	D	A	B	B
4th choice	C	C	B	C	A

(a) Find the winner of the election under the plurality-with-elimination method.

(b) Explain why the winner in (a) can be determined in the first round.

(c) Explain why the plurality-with-elimination satisfies the *majority* criterion. (No need to do this again if you have already done Exercise 29.)

31. An election is held among four candidates ($A, B, C,$ and D). Each column in the following preference schedule shows the percentage of voters voting that way. Find the winner of the election under the plurality-with-elimination method.

Percentage of voters	40%	25%	20%	15%
1st choice	A	C	B	B
2nd choice	D	B	D	A
3rd choice	B	D	A	D
4th choice	C	A	C	C

32. An election is held among four candidates ($A, B, C,$ and D). Each column in the following preference schedule shows the percentage of voters voting that way. Find the winner of the election under the plurality-with-elimination method.

Percentage of voters	48%	24%	16%	12%
1st choice	A	C	B	B
2nd choice	D	B	D	A
3rd choice	B	D	A	D
4th choice	C	A	C	C

33. The Professional Surfing Association executive committee is having an election to choose the site of the next Quicksilver Extreme Surfing Tournament. The candidates are Año Nuevo (California), Banzai Pipeline (Hawaii), Cloudbreak (Fiji), and Dungeons (South Africa). The preference schedule for the election is given in the following table.

Number of voters	10	7	5	5	4
1st choice	A	D	B	C	B
2nd choice	C	B	C	D	C
3rd choice	B	A	A	A	D
4th choice	D	C	D	B	A

(a) Use the plurality-with-elimination method to find the winner of the election.

(b) Find the Condorcet candidate in this election.

(c) The results of (a) and (b) imply a violation of which fairness criterion?

34. The 17 members of a small Iowa precinct are preparing for the Iowa caucuses held in January. The candidates running for president this year are Anderson, Buford, Clinton, and Dorfman. The results of a straw poll taken prior to the official vote are given in the following preference schedule.

Number of voters	4	5	6	2
1st choice	A	B	C	A
2nd choice	D	C	A	C
3rd choice	B	A	D	D
4th choice	C	D	B	B

(a) Use the plurality-with-elimination method to find the winner of the straw poll.

(b) In the official vote everyone votes the same as in the straw poll except for the two voters in the last column of the table—they switch their votes and move Clinton ahead of Anderson in their ballots. Use the plurality-with-elimination method to find the winner of the official vote.

(c) The results of (a) and (b) imply a violation of which fairness criterion?

E Pairwise-Comparisons Method

35. The 26 members of the Tasmania State University chess team are holding an election to choose a senior captain. The five candidates are $A, B, C, D,$ and E. The results of the election are shown in the following preference schedule.

Number of voters	8	6	5	5	2
1st choice	C	A	E	D	D
2nd choice	B	E	C	C	A
3rd choice	A	D	D	A	E
4th choice	D	B	B	E	B
5th choice	E	C	A	B	C

(a) Use the method of pairwise comparisons to find the winner of the election.

(b) Suppose that before the votes are counted candidate A is declared ineligible. Find the new preference schedule for an election without candidate A.

(c) Use the method of pairwise comparisons to find the winner of the new election without candidate A.

(d) The results of (a) and (c) imply a violation of which fairness criterion?

36. Use the method of pairwise comparisons to find the winner of the election given by the following preference schedule. (This is the final exam election discussed in Exercise 12.)

Number of voters	3	4	9	9	2	5	8	3	12
1st choice	A	A	A	B	B	B	C	C	D
2nd choice	B	B	C	C	A	C	D	A	C
3rd choice	C	D	B	D	C	A	B	D	A
4th choice	D	C	D	A	D	D	A	B	B

37. Use the method of pairwise comparisons to find the winner of the election given by the following preference schedule. (This is the Mathematics Department Chair election discussed in Exercises 3, 5, 17, and 27.)

Number of voters	5	3	5	3	2	3
1st choice	A	A	C	D	D	B
2nd choice	B	D	E	C	C	E
3rd choice	C	B	D	B	B	A
4th choice	D	C	A	E	A	C
5th choice	E	E	B	A	E	D

38. Use the method of pairwise comparisons to find the winner of the election given by the following preference schedule. (This is the "Restaurant of the Year" election discussed in Exercises 19 and 29.)

Number of voters	8	7	6	2	1
1st choice	A	D	D	C	E
2nd choice	B	B	B	A	A
3rd choice	C	A	E	B	D
4th choice	D	C	C	D	B
5th choice	E	E	A	E	C

39. An election with five candidates (A, B, C, D, and E) is decided using the method of pairwise comparisons. Suppose that B loses two pairwise comparisons, C loses one, D loses one and ties one, and E loses two and ties one.

(a) How many points did each candidate get? (Remember, one point for a win, half a point for a tie).

(b) Who was the winner of the election?

40. An election with six candidates (A, B, C, D, E, and F) is decided using the method of pairwise comparisons. Suppose that A loses three pairwise comparisons, B and C both lose two each, D and E both lose two and tie one.

(a) How many points did each candidate get? (Remember, one point for a win, half a point for a tie).

(b) Who was the winner of the election?

F Ranking Methods

41. For the election given by the following preference schedule (this is the Professional Surfing Association election discussed in Exercise 33),

Number of voters	10	7	5	5	4
1st choice	A	D	B	C	B
2nd choice	C	B	C	D	C
3rd choice	B	A	A	A	D
4th choice	D	C	D	B	A

(a) rank the candidates using the extended plurality method.

(b) rank the candidates using the extended Borda count method.

(c) rank the candidates using the extended plurality-with-elimination method.

(d) rank the candidates using the extended pairwise-comparisons method.

42. For the election given by the following preference schedule,

Number of voters	10	8	8	6	4	1
1st choice	B	D	C	A	D	A
2nd choice	A	A	B	C	B	B
3rd choice	C	B	A	D	C	D
4th choice	D	C	D	B	A	C

(a) rank the candidates using the extended plurality method.

(b) rank the candidates using the extended Borda count method.

(c) rank the candidates using the extended plurality-with-elimination method.

(d) rank the candidates using the extended pairwise-comparisons method.

43. For the election given by the following preference schedule (this is the election discussed in Exercise 21),

Percentage of voters	40%	25%	20%	15%
1st choice	A	C	B	B
2nd choice	D	B	D	A
3rd choice	B	D	A	D
4th choice	C	A	C	C

(a) rank the candidates using the extended plurality method.

(b) rank the candidates using the extended Borda count method.

(c) rank the candidates using the extended plurality-with-elimination method.

(d) rank the candidates using the extended pairwise-comparisons method.

44. For the election given by the following preference schedule (this is the election discussed in Exercise 22),

Percentage of voters	48%	24%	16%	12%
1st choice	A	C	B	B
2nd choice	D	B	D	A
3rd choice	B	D	A	D
4th choice	C	A	C	C

(a) rank the candidates using the extended plurality method.

(b) rank the candidates using the extended Borda count method.

(c) rank the candidates using the extended plurality-with-elimination method.

(d) rank the candidates using the extended pairwise comparisons method.

Exercises 45 and 46 refer to the MAS election introduced in Example 1.1 and given by the following preference schedule:

Number of voters	14	10	8	4	1
1st choice	A	C	D	B	C
2nd choice	B	B	C	D	D
3rd choice	C	D	B	C	B
4th choice	D	A	A	A	A

45. Rank the candidates in the election using the recursive Borda count method.

46. Rank the candidates in the election using the recursive pairwise-comparisons method.

Exercises 47 through 50 refer to the election given by the following preference schedule:

Number of voters	10	7	5	5	4
1st choice	A	D	B	C	B
2nd choice	C	B	C	D	C
3rd choice	B	A	A	A	D
4th choice	D	C	D	B	A

47. Rank the candidates in the election using the recursive plurality method.

48. Rank the candidates in the election using the recursive Borda count method.

49. Rank the candidates in the election using the recursive pairwise-comparisons method.

50. Rank the candidates in the election using the recursive plurality-with-elimination method.

G Miscellaneous

Find the sums in Exercises 51–54.

51. $1 + 2 + 3 + \cdots + 498 + 499 + 500$

52. $1 + 2 + 3 + \cdots + 3218 + 3219 + 3220$

53. $501 + 502 + 503 + \cdots + 3218 + 3219 + 3220$ (*Hint:* Do Exercises 51 and 52 first.)

54. $1801 + 1802 + 1803 + \cdots + 8843 + 8844 + 8845$ (*Hint:* Do Exercise 53 first.)

55. In an election with 15 candidates,

(a) how many pairwise comparisons are there?

(b) if it takes one minute to calculate a pairwise comparison, approximately how long would it take to calculate the results of the election using the method of pairwise comparisons?

56. Suppose that 21 players sign up for a round-robin Ping-Pong tournament.

(a) How many matches must be scheduled for the entire tournament?

(b) If six matches can be scheduled per hour and the tournament hall is available for 12 hours each day, for how many days would the tournament hall have to be reserved?

57. In an election with 3 candidates, what is the maximum number of columns possible in the preference schedule?

58. In an election with 4 candidates,

(a) what is the maximum number of columns possible in the preference schedule?

(b) how many different ways are there to choose the first and second choices?

59. Consider the election given by the following preference schedule:

Number of voters	7	4	2
1st choice	A	B	D
2nd choice	B	D	A
3rd choice	C	C	C
4th choice	D	A	B

(a) Find the Condorcet candidate in this election.

(b) Use the Borda count method to find the winner of the election.

(c) Suppose C drops out of the race. Use the Borda count method to find the winner of the election after C is removed from the preference schedule.

(d) The results of (a), (b), and (c) show that the Borda count method violates several of the fairness criteria discussed in the chapter. Which ones? Explain.

60. Consider the election given by the following preference schedule:

Number of voters	10	6	5	4	2
1st choice	A	B	C	D	D
2nd choice	C	D	B	C	A
3rd choice	D	C	D	B	B
4th choice	B	A	A	A	C

(a) Find the Condorcet candidate in this election.

(b) Use the plurality-with-elimination method to find the winner of the election.

(c) Suppose D drops out of the race. Use the plurality-with-elimination method to find the winner of the election after D is removed from the preference schedule.

(d) The results of (a), (b), and (c) show that the plurality-with-elimination method violates several of the fairness criteria discussed in the chapter. Which ones? Explain.

JOGGING

61. Two-candidate elections. Explain why when there are only two candidates, the four voting methods we discussed in this chapter give the same winner and the winner is determined by straight majority. (Assume that there are no ties.)

62. Give an example of an election with four candidates $(A, B, C, \text{and } D)$ satisfying the following: (i) No candidate has a majority of the first-place votes, (ii) C is a Condorcet candidate but has no first-place votes, (iii) B is the winner under the Borda count method, (iv) A is the winner under the plurality method.

63. Explain why the plurality method satisfies the monotonicity criterion.

64. Explain why the Borda count method satisfies the monotonicity criterion.

65. An election is held using the method of pairwise comparisons. Suppose that X is the winner of the election under the method of pairwise comparisons, but due to an election irregularity, there is a reelection. In the reelection, the only changes are changes that favor X and only X (specifically, you can interpret this to mean that the only changes involve voters who move X up in their ballot without altering the relative order of any of the other candidates). Explain why candidate X must be the winner of the reelection (i.e., the method of pairwise comparisons satisfies the monotonicity criterion).

66. Equivalent Borda count (Variation 1). The following simple variation of the Borda count method described in the chapter is sometimes used: A first place is worth $N - 1$ points, second place is worth $N - 2$ points, . . . , last place is worth 0 points (where N is the number of candidates). The candidate with the most points is the winner.

(a) Suppose that a candidate gets p points using the Borda count as originally described in the chapter and q points under this variation. Explain why if k is the number of voters, then $p = q + k$.

(b) Explain why this variation is equivalent to the original Borda count described in the chapter (i.e., it produces exactly the same election results).

67. Equivalent Borda count (Variation 2). Another commonly used variation of the Borda count method described in the chapter is the following: A first place is worth 1 point, second place is worth 2 points, . . . , last place is worth N points (where N is the number of candidates). The candidate with the fewest points is the winner, second fewest points is second, and so on.

(a) Suppose that a candidate gets p points using the Borda count as originally described in the chapter and r points under this variation. Explain why if k is the number of voters, then $p + r = k(N + 1)$.

(b) Explain why this variation is equivalent to the original Borda count described in the chapter (i.e., it produces exactly the same election results).

68. The average ranking. The average ranking of a candidate is obtained by taking the place of the candidate on each of the ballots, adding these numbers, and dividing by the number of ballots. Explain why the candidate with the best (lowest) average ranking is the Borda winner.

69. The 2006 Associated Press college football poll. The AP college football poll is a ranking of the top-25 college football teams in the country and is one of the key polls used for the BCS (Bowl Championship Series). The voters in the AP poll are a group of sportswriters and broadcasters chosen from across the country. The top 25 teams are ranked using a Borda count: each first-place vote is worth 25 points, each second-place vote is worth 24 points, each third-place vote is worth 23 points, and so on. The following table shows the ranking and total points for each of the top three teams at the end of the 2006 regular season. (The remaining 22 teams are not shown here because they are irrelevant to this exercise.)

Team	Points
1. Ohio State	1625
2. Florida	1529
3. Michigan	1526

(a) Given that Ohio State was the unanimous first-place choice of all the voters, find the number of voters that participated in the poll.

(b) Find the number of second- and third-place votes for Florida.

(c) Find the number of second- and third-place votes for Michigan.

70. The 2005 National League MVP vote. Each year the Most Valuable Player of the National League is chosen by a group of 32 sportswriters using a variation of the Borda count method. The following table shows the results of the 2005 voting:

Player	1st place	2nd place	3rd place	Total points
Albert Pujols	18	14	0	378
Andruw Jones	13	17	2	351
Derrek Lee	1	1	30	263

Determine how many points are given for each first-, second-, and third-place vote in this election.

71. The 2003–2004 NBA Rookie of the Year vote. Each year, a panel of broadcasters and sportswriters selects an NBA rookie of the year using a variation of the Borda count method. The following table shows the results of the balloting for the 2003–2004 season.

Player	1st place	2nd place	3rd place	Total points
LeBron James	78	39	1	508
Carmelo Anthony	40	76	2	430
Dwayne Wade	0	3	108	117

Determine how many points are given for each first-, second-, and third-place vote in this election.

72. Plurality with a runoff. This is a simple variation of the plurality-with-elimination method. Here, if a candidate has a majority of the first-place votes, then that candidate wins the election; otherwise, we eliminate *all* candidates except the two with the most first-place votes. The winner is chosen between these two by recounting the votes in the usual way.

(a) Use the Math Club election to show that plurality with a runoff can produce a different outcome than plurality with elimination.

(b) Give an example that shows that plurality with a runoff violates the monotonicity criterion.

(c) Give an example that shows that plurality with a runoff violates the Condorcet criterion.

73. The Coombs method. This method is just like the plurality-with-elimination method except that in each round we eliminate the candidate with the *largest number of last-place votes* (instead of the one with the fewest first-place votes).

(a) Find the winner of the Math Club election using the Coombs method.

(b) Give an example showing that the Coombs method violates the Condorcet criterion.

(c) Give an example showing that the Coombs method violates the monotonicity criterion.

74. Bucklin voting. (This method was used in the early part of the 20th century to determine winners of many elections for political office in the United States.) The method proceeds in rounds. **Round 1**: Count first-place votes only. If a candidate has a majority of the first-place votes, that candidate wins. Otherwise, go to the next round. **Round 2**: Count *first- and second-place* votes only. If there are any candidates with a majority of votes, the candidate with the most votes wins. Otherwise, go to the next round. **Round 3**: Count *first-*, *second-, and third-place* votes only. If there are any candidates with a majority of votes, the candidate with the most votes wins. Otherwise, go to the next round. Repeat for as many rounds as necessary.

(a) Find the winner of the Math Club election using the Bucklin method.

(b) Show that the Bucklin method violates the Condorcet criterion.

(c) Show that the Bucklin method satisfies the monotonicity criterion.

RUNNING

75. Consider the following fairness criterion: *If a majority of the voters have candidate X ranked last, then candidate X should not be a winner of the election.*

(a) Give an example to show that the plurality method violates this criterion.

(b) Give an example to show that the plurality-with-elimination method violates this criterion.

(c) Explain why the method of pairwise comparisons satisfies this criterion.

(d) Explain why the Borda count method satisfies this criterion.

76. Suppose that the following was proposed as a fairness criterion: *If a majority of the voters prefer candidate X to candidate Y, then the results of the election should have X ranked above Y.* Give an example to show that all four extended voting methods discussed in the chapter can violate this criterion. (*Hint*: Consider an example with no Condorcet candidate.)

77. The Pareto criterion. The following fairness criterion was proposed by Italian economist Vilfredo Pareto (1848–1923): *If every voter prefers candidate X to candidate Y, then X should be ranked above Y.*

(a) Explain why the extended Borda count method satisfies the Pareto criterion.

(b) Explain why the extended pairwise-comparisons method satisfies the Pareto criterion.

78. Explain why all four recursive ranking methods (recursive plurality, recursive plurality-with-elimination, recursive Borda count, and recursive pairwise comparisons) satisfy the Pareto criterion as defined in Exercise 77.

79. The Condorcet loser criterion. *If there is a candidate who loses in a one-to-one comparison to each of the other candidates, then that candidate should not be the winner of the election.* (This fairness criterion is a sort of mirror image of the regular Condorcet criterion.)

(a) Give an example that shows that the plurality method violates the Condorcet loser criterion.

(b) Give an example that shows that the plurality-with-elimination method violates the Condorcet loser criterion.

(c) Explain why the Borda count method satisfies the Condorcet loser criterion.

80. Consider a variation of the Borda count method in which a first-place vote in an election with N candidates is worth F points (where $F > N$) and all other places in the ballot are the same as in the ordinary Borda count: $N - 1$ points for second place, $N - 2$ points for third place, . . . , 1 point for last place. By choosing F large enough, we can make this variation of the Borda count method satisfy the majority criterion. Find the smallest value of F (expressed in terms of N) for which this happens.

81. Suppose that the candidates in an election are ranked using the extended pairwise-comparisons method and that there is no Condorcet candidate. Explain why if the number of voters is odd, then there must be a tie somewhere in the rankings. Explain why the result does not have to be true with an even number of voters.

PROJECTS AND PAPERS

A Ballots, Ballots, Ballots!

In this chapter we discussed elections in which the voters cast their votes by means of linear preference ballots. There are many other types of ballots used in real-life elections, ranging from the simple (winner only) to the exotic (each voter has a fixed number of points to divide among the candidates any way he or she sees fit). In this project you are to research other types of ballots; how, where, and when they are used; and what are the arguments for and against their use.

B Sequential Voting

Sequential voting is a voting method used by legislative bodies and committees to choose one among a list of alternatives. In sequential voting the alternatives are presented in some order A, B, C, D, and so on. The voters then choose between A and B, the winner is then matched against C, the winner of that vote against D, and so on. In sequential voting, the *agenda* (the order in which the options are presented) can have a critical impact on the outcome of the vote. Write a research paper on sequential voting, its history, and its use, paying particular attention to the issue of the agenda and its impact on the outcome. Illustrate your points with examples that *you* have made up.

C Instant Runoff Voting

Imagine that you are a political activist in your community. The city council is having hearings to decide if the *method of instant runoff voting* (plurality with elimination) should be adopted in your city. Stake out a position for or against instant runoff voting and prepare a brief to present to the city council that justifies that position. To make an effective case, your argument should include mathematical, economic, political, and social considerations. (Remember that your city council members are not as well versed as you are on the mathematical aspects of elections. Part of your job is to educate them.)

D The 2000 Presidential Election and the Florida Vote

The unusual circumstances surrounding the 2000 presidential election and the Florida vote are a low point in American electoral history. Write an analysis paper on the 2000 Florida vote, paying particular attention to what went wrong and how a similar situation can be prevented in the future. You should touch on technology issues (outdated and inaccurate vote-tallying methods and equipment, poorly designed ballots, etc.), political issues (the two-party system, the Electoral College, etc.), and, as much as possible, on issues related to concepts from this chapter (can presidential elections be improved by changing to preference ballots, using a different voting method, etc.).

E Short Story

Write a fictional short story using an election as the backdrop. Weave into the dramatic structure of the story elements and themes from this chapter (fairness, manipulation, monotonicity, and independence of irrelevant alternatives all lend themselves to good drama). Be creative and have fun.

REFERENCES AND FURTHER READINGS

1. Arrow, Kenneth J., *Social Choice and Individual Values*. New York: John Wiley & Sons, Inc., 1963.
2. Baker, Keith M., *Condorcet: From Natural Philosophy to Social Mathematics*. Chicago: University of Chicago Press, 1975.
3. Brams, Steven J., and Peter C. Fishburn, *Approval Voting*. Boston: Birkhäuser, 1982.
4. Dummett, Michael A., *Voting Procedures*. New York: Oxford University Press, 1984.
5. Farquharson, Robin, *Theory of Voting*. New Haven, CT: Yale University Press, 1969.
6. Fishburn, Peter C., and Steven J. Brams, "Paradoxes of Preferential Voting," *Mathematics Magazine*, 56 (1983), 207–214.
7. Gardner, Martin, "Mathematical Games (From Counting Votes to Making Votes Count: The Mathematics of Elections)," *Scientific American*, 243 (October 1980), 16–26.
8. Hill, Steven, *Fixing Elections: The Failure of America's Winner Take All Politics*. New York: Routledge, 2003.
9. Kelly, Jerry S., *Arrow Impossibility Theorems*. New York: Academic Press, 1978.
10. Merrill, Samuel, *Making Multicandidate Elections More Democratic*. Princeton, NJ: Princeton University Press, 1988.
11. Niemi, Richard G., and William H. Riker, "The Choice of Voting Systems," *Scientific American*, 234 (June 1976), 21–27.
12. Nurmi, Hannu, *Comparing Voting Systems*. Dordretch, Holland: D. Reidel, 1987.
13. Saari, Donald G., *Basic Geometry of Voting*. New York: Springer-Verlag, 1995.
14. Saari, Donald G., *Chaotic Elections: A Mathematician Looks at Voting*. Providence, RI: American Mathematical Society, 2001.
15. Saari, Donald G., *Decisions and Elections: Explaining the Unexpected*. Cambridge, U.K.: Cambridge University Press, 2001.
16. Saari, Donald G., "Suppose You Want to Vote Strategically," *Math Horizons*, (Nov. 2000), 5–10.
17. Saari, Donald G., and F. Valognes, "Geometry, Voting, and Paradoxes," *Mathematics Magazine*, 71 (Oct. 1998), 243–259.
18. Straffin, Philip D., Jr., *Topics in the Theory of Voting*, UMAP Expository Monograph. Boston: Birkhäuser, 1980.
19. Taylor, Alan, *Mathematics and Politics: Strategy, Voting, Power and Proof*. New York: Springer-Verlag, 1995.

2 The Mathematics of Power

Weighted Voting

In a democracy we take many things for granted, not the least of which is the idea that we are all equal. When it comes to voting rights, the democratic ideal of equality translates into the principle of *one person–one vote*. But is the principle of *one person–one vote* always fair? Should *one person–one vote* apply when the *voters* are institutions or governments, rather than individuals?

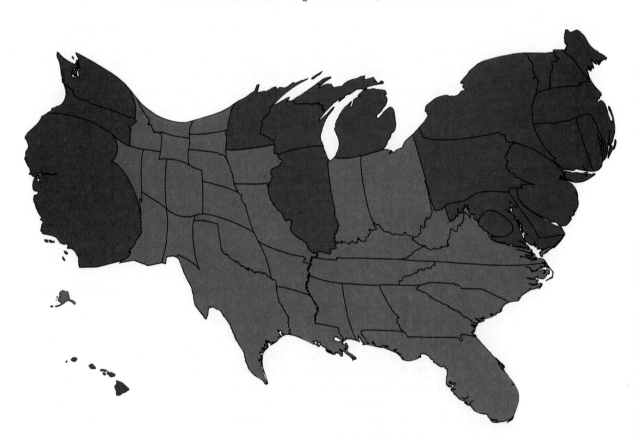

Cartogram of 2004 presidential election with states sized in proportion to their number of electoral votes (red states went to George W. Bush, blue states to John F. Kerry). © 2004 M. T. Gastner, C. R. Shalizi, and M. E. J. Newman.

I n a diverse society it is in the very nature of things that voters—be they individuals or institutions—are not equal, and sometimes it is actually desirable to recognize their differences by giving them different amounts of say over the outcome of the voting. What we are talking about here is the exact opposite of the principle of *one voter–one vote*, a principle best described as *one voter–x votes* and formally called *weighted voting* (the word "weight" here refers to the number of votes controlled by a voter, and not to pounds or kilograms).

Weighted voting in the United Nations Security Council, where permanent members have more weight than nonpermanent members.

Weighted voting is not uncommon—we see examples of weighted voting in shareholder votes in corporations, in business partnerships, in legislatures, in the United Nations, and, most infamously, in the way we elect the President of the United States.

The Electoral College, that uniquely American institution used to elect the President of the United States, offers a classic illustration of weighted voting. The Electoral College consists of 51 "voters" (each of the 50 states plus the District of Columbia), each with a weight determined by the size of its Congressional delegation (number of Representatives and Senators). At one end of the spectrum is heavyweight California (with 55 electoral votes); at the other end of the spectrum are lightweights like Wyoming, Montana, North Dakota, and the District of Columbia (with a paltry 3 electoral votes). The other states fall somewhere in between. (See the appendix at the end of this book for full details.)

The 2000 and 2004 presidential elections brought to the surface, in a very dramatic way, the vagaries and complexities of the Electoral College system, and in particular the pivotal role that a single state (Florida in 2000, Ohio in 2004) can have in the final outcome of a presidential election.

In this chapter we will look at the mathematics behind weighted voting, with a particular focus on the question of *power*. Since the point of weighted voting is to give different voters different amounts of influence in the outcome of the voting (i.e., different amounts of power), how does one go about *measuring* power? We will answer this question not once but twice, and in so doing cover some new and interesting mathematics.

We will start by introducing and illustrating the concept of a *weighted voting system*, a simple mathematical formalism that will allow us to describe most weighted voting situations (Section 2.1). In the remaining sections we will discuss two different approaches for measuring the power of each voter in a weighted voting system. Sections 2.2 and 2.3 deal with what is known as the *Banzhaf* measure of power, and Sections 2.4 and 2.5 deal with what is known as the *Shapley-Shubik* measure of power. Along the way we will introduce several basic but important mathematical ideas—*coalitions, sequential coalitions,* and *factorials.* Enjoy the ride!

2.1   An Introduction to Weighted Voting

We will use the term **weighted voting system** to describe any formal voting arrangement in which voters are not necessarily equal in terms of the number of votes they control. We will only consider voting on *yes–no* votes, known as **motions.** Note that any vote between two choices (say *A* or *B*) can be rephrased as a *yes-no* vote (a *Yes* is a vote for *A*, a *No* is a vote for *B*).

Every weighted voting system is characterized by three elements:

- **The players.** We will refer to the voters in a weighted voting system as players. Note that in a weighted voting system the players may be individuals, but they may also be corporations, governmental agencies, states, municipalities, or countries. We will use *N* to denote the number of players in a weighted voting system, and typically (unless there is a good reason not to) we will call the players $P_1, P_2, \ldots, P_N$. (Think of P_1 as short for "player 1," P_2 as short for "player 2," etc.—it is a little less personal but a lot more convenient than using Archie, Betty, etc.)

- **The weights.** The hallmark of a weighted voting system is that each player controls a certain number of votes, called the *weight* of the player. We will assume that the weights are all positive integers, and we will use $w_1, w_2, \ldots, w_N$ to denote the weights of $P_1, P_2, \ldots, P_N$, respectively. We will use $V = w_1 + w_2 + \cdots + w_N$ to denote the total number of votes in the system.

  A weighted voting system is only interesting when the weights are not all the same, but in principle we do not preclude the possibility that the weights are all equal.

- **The quota.** In addition to the player's weights, every weighted voting system has a **quota**. The quota is the *minimum number of votes required to pass a motion*, and is denoted by the letter *q*. While the most common standard for the quota is a *simple majority* of the votes, the quota may very well be something else. In the U.S. Senate, for example, it takes a simple majority to pass an ordinary law, but it takes a minimum of 60 votes to stop a filibuster, and it takes a minimum of two-thirds of the votes to override a presidential veto. In other weighted voting systems the rules may stipulate a quota of three-fourths of the votes, or four-fifths, or even *unanimity* (100% of the votes).

Notation and Examples

The standard notation used to describe a weighted voting system is to use square brackets and inside the square brackets to write the quota *q* first (followed by a colon) and then the respective weights of the individual players separated by commas. It is convenient and customary to list the weights in numerical order, starting with the highest, and we will adhere to this convention throughout the chapter.

Thus, a generic weighted voting system with N players can be written as:

GENERIC WEIGHTED VOTING SYSTEM WITH N PLAYERS

$$[q: w_1, w_2, \ldots, w_N] \quad (\text{with } w_1 \geq w_2 \geq \cdots \geq w_N)$$

We will now look at a few examples to illustrate some basic concepts in weighted voting.

EXAMPLE 2.1   **Venture Capitalism**

Four partners (P_1, P_2, P_3, and P_4) decide to start a new business venture. In order to raise the \$200,000 venture capital needed for startup money, they issue 20 shares worth \$10,000 each. Suppose that P_1 buys 8 shares, P_2 buys 7 shares, P_3 buys 3 shares, and P_4 buys 2 shares, with the usual agreement that one share equals one vote in the partnership. Suppose that the quota is set to be two-thirds of the total number of votes.

Since the total number of votes in the partnership is $V = 20$, and two-thirds of 20 is $13\frac{1}{3}$, we set the quota to $q = 14$. Using the weighted voting system notation we just introduced, the partnership can be described mathematically as $[14: 8, 7, 3, 2]$.

EXAMPLE 2.2   **Anarchy**

Imagine the same partnership discussed in Example 2.1, with the only difference being that the quota is changed to 10 votes. We might be tempted to think of this partnership as the weighted voting system $[10: 8, 7, 3, 2]$, but there is a problem here: the quota is too small, making it possible for both Yes's and No's to have enough votes to carry a particular motion. (Imagine, for example, that an important decision needs to be made and P_1 and P_4 vote yes and P_2 and P_3 vote no. Now we have a stalemate, since both the Yes's and the No's have enough votes to meet the quota.)

In general when the quota requirement is less than simple majority ($q \leq V/2$) we have the potential for both sides of an issue to win—a mathematical version of anarchy.

EXAMPLE 2.3   **Gridlock**

Once again, let's look at the partnership introduced in Example 2.1, but suppose now that the quota is set to $q = 21$, more than the total number of votes in the system. This would not make much sense. Under these conditions no motion would ever pass and nothing could ever get done—a mathematical version of gridlock.

Given that we expect our weighted voting systems to operate without anarchy or gridlock, from here on we will assume that *the quota will always fall somewhere between simple majority and unanimity of votes*. Symbolically, we can express these requirements by two inequalities: $q > V/2$ and $q \leq V$. These two restrictions define the range of possible values of the quota:

RANGE OF VALUES OF THE QUOTA

$$V/2 < q \leq V \quad (\text{where } V = w_1 + w_2 + \cdots + w_N)$$

(**EXAMPLE 2.4**)   **One Partner–One Vote?**

Let's consider the partnership introduced in Example 2.1 one final time. This time the quota is set to be $q = 19$. Here we can describe the partnership as the weighted voting system $[19: 8, 7, 3, 2]$. What's interesting about this weighted voting system is that the *only way a motion can pass is by the unanimous support of all the players.* (Note that P_1, P_2, and P_3 together have 18 votes—they still need P_4's votes to pass a motion.) In a practical sense this weighted voting system is no different from a weighted voting system in which each partner has 1 vote and it takes the unanimous agreement of the four partners to pass a motion (i.e., $[4: 1, 1, 1, 1]$).

The surprising conclusion of Example 2.4 is that the weighted voting system $[19: 8, 7, 3, 2]$ describes a one person–one vote situation in disguise. This seems like a contradiction only if we think of *one person–one vote* as implying that all players have an *equal number of votes* rather than an *equal say in the outcome of the election.* Apparently, these two things are not the same! As Example 2.4 makes abundantly clear, just looking at the number of votes a player owns can be very deceptive.

(**EXAMPLE 2.5**)   **Dictators**

Consider the weighted voting system $[11: 12, 5, 4]$. Here one of the players (P_1) owns enough votes to carry a motion singlehandedly. In this situation P_1 is in complete control—if P_1 is for the motion, then the motion will pass; if P_1 is against it, then the motion will fail. Clearly, in terms of the power to influence decisions, P_1 has *all* of it. Not surprisingly, we will say that P_1 is a *dictator.*

In general a player is a **dictator** if *the player's weight is bigger than or equal to the quota.* Since there can only be one dictator (if you had two players with weights larger than the quota then the quota would be too small), and since the player with the highest weight is P_1, we can conclude that *if there is a dictator, then it must be P_1.* Or, to put it on a more positive note, if P_1 is not a dictator, then there are no dictators.

When P_1 is a dictator, all the other players, regardless of their weights, have absolutely no say on the outcome of the voting—there is never a time when their votes really count. A player who never has a say in the outcome of the voting is a player who has no power, and we will call such players **dummies**.

When there is a dictator, all the other players are dummies, but there can be dummies even when there is no dictator. In fact, sometimes a player can have a large number of votes and still be a dummy. This is illustrated in our next example.

(**EXAMPLE 2.6**)   **Unsuspecting Dummies**

Four college friends (P_1, P_2, P_3, and P_4) decide to go into business together. Three of the four (P_1, P_2, and P_3) invest \$10,000 each, and each gets 10 shares in the partnership. The fourth partner (P_4) is a little short on cash, so he invests only \$9000 and gets 9 shares. As usual, one share equals one vote. The quota is set at 75%, which here means $q = 30$ out of a total of $V = 39$ votes. Mathematically (i.e., stripped of all the irrelevant details of the story), this partnership is just the weighted voting system $[30: 10, 10, 10, 9]$.

Everything seems fine with the partnership until one day P_4 wakes up to the realization that with the quota set at $q = 30$ he is completely out of the decision-making loop: For a motion to pass P_1, P_2, and P_3 all must vote Yes, and at that point it makes no difference how P_4 votes. Thus, there is never going to be a time when P_4's votes are going to make a difference in the final outcome of a vote. Surprisingly, P_4—with almost as many votes as the other partners—is just a dummy! ⊂⊃

EXAMPLE 2.7   Veto Power

Consider the weighted voting system [12: 9, 5, 4, 2]. Here P_1 plays the role of a "spoiler"—while not having enough votes to be a dictator, the player has enough votes to *prevent a motion from passing*. This happens because if we remove P_1's 9 votes the sum of the remaining votes $(5 + 4 + 2 = 11)$ is less than the quota $q = 12$. Thus, even if all the other players voted Yes, without P_1 the motion would not pass. In a situation like this we say that P_1 has *veto power*. ⊂⊃

In general we will say that a player who is not a dictator has **veto power** if a *motion cannot pass unless the player votes in favor of the motion*. In other words, a player with veto power cannot force a motion to *pass* (the player is not a dictator), but can force a motion to *fail*. If we let w denote the weight of a player with veto power, then the two conditions can be expressed mathematically by the inequalities $w < q$ (the player is not a dictator), and $V - w < q$ (the remaining votes in the system are not enough to pass a motion).

> **VETO POWER**
>
> A player with weight w has veto power if and only if $w < q$ and $V - w < q$ (where $V = w_1 + w_2 + \cdots + w_N$).

2.2 The Banzhaf Power Index

John F. Banzhaf III (1940–) is currently Professor of Public Interest Law at George Washington University as well as founder and executive director of the national antismoking group *Action on Smoking and Health*. Banzhaf was just 25 years old when he first came up with the ideas we will discuss in this section.

From the preceding set of examples we can already draw an important lesson: In weighted voting the player's weights can be deceiving. Sometimes a player with a few votes can have as much power as a player with many more votes (see Example 2.4); sometimes two players have almost an equal number of votes, and yet one player has a lot of power and the other one has none (see Example 2.6).

To pursue these ideas further we will need a formal definition of what "power" means and how it can be measured. In this section we will introduce a mathematical method for measuring the power of the players in a weighted voting system. This method was first proposed in 1965 by, of all people, a law professor named John Banzhaf III. We will start our discussion of Banzhaf's method with a simple example.

EXAMPLE 2.8   The Weirdness of Parliamentary Politics

The Parliament of Icelandia has 200 members, divided among three political parties—the Red Party (R), the Blue Party (B), and the Green Party (G). The Red Party has 99 seats in Parliament, the Blue Party has 98, and the Green Party has only 3. Decisions are made by majority vote, which in this case requires 101 out of the total 200 votes. Let's assume, furthermore, that in Icelandia members of Parliament always vote along party lines (not voting with your party is very unusual in parliamentary governments). Under these assumptions we can think of the Parliament of Icelandia as the weighted voting system [101: 99, 98, 3].

By just looking at the numbers of votes, one would guess that most of the power is shared between the Red and Blue parties, with the Green Party having very little power, if any. But a very different story emerges when we look at the different ways that a motion might pass.

Let's consider the different party combinations needed to pass a motion in Icelandia's Parliament (remember that we are assuming that voting is strictly along party lines). There are only four:

- Reds and Blues vote Yes, Greens vote No. The motion passes 197 to 3.
- Reds and Greens vote Yes, Blues vote No. The motion passes 102 to 98.
- Blues and Greens vote Yes, Reds vote No. The motion passes (just barely) 101 to 99.
- Reds, Blues and Greens all vote Yes. The motion passes unanimously, 200 to 0.

The surprising fact that emerges from the above list is that for a motion to pass it must have the support of at least two of the three parties, and it makes no difference which two.

Before we continue with this example, we will take a break and introduce some important new concepts.

- **Coalitions.** We will use the term **coalition** to describe any set of players who might join forces and vote the same way. In principle, we can have a coalition with as few as *one* player and as many as *all* players. The coalition consisting of all the players is called the **grand coalition**. Since coalitions are just sets of players, the most convenient way to describe coalitions mathematically is to use *set* notation. For example, the coalition consisting of players P_1, P_2, and P_3 can be written as the set $\{P_1, P_2, P_3\}$ [or $\{P_3, P_1, P_2\}$, or $\{P_2, P_1, P_3\}$, etc.—the order in which the members of a coalition are listed is irrelevant].

- **Winning coalitions.** Some coalitions have enough votes to win and some don't. Quite naturally, we call the former **winning coalitions** and the latter **losing coalitions**. A single-player coalition can be a winning coalition only when that player is a *dictator*. (Since our focus is going to be on weighted voting systems with no dictators, our winning coalitions will have a minimum of two players.) At the other end of the spectrum, the grand coalition is always a winning coalition, since it controls all the votes. In some weighted voting systems (see Example 2.4) the grand coalition is the only winning coalition.

- **Critical players.** In a winning coalition a player is said to be a **critical player** for the coalition if the coalition must have that player's votes to win. In other words, if we subtract a critical player's weight from the total weight of the coalition, the number of remaining votes drops below the quota.

CRITICAL PLAYER

A player P in a winning coalition is a *critical player* for the coalition if and only if $W - w < q$ (where W denotes the weight of the coalition and w denotes the weight of P).

Sometimes a winning coalition has no critical players (the coalition has enough votes that no single player's desertion can keep it from winning), sometimes a winning coalition has several critical players, and when the coalition has just enough votes to make the quota, then every player is critical.

(**EXAMPLE 2.8**)   **(continued)**

We will now revisit the voting in the Parliament of Icelandia using the language of coalitions and critical players. The Parliament is formally the weighted voting system $[101: 99, 98, 3]$, and for convenience we will now refer to the three parties as P_1, P_2, and P_3, respectively. In this weighted voting system we have four winning coalitions, listed in the first column of Table 2-1.

In the two-party coalitions, both parties are *critical* players (without both players the coalition wouldn't win); in the grand coalition no party is critical—any two parties together have enough votes to win. Each of the three parties is critical the same number of times, and consequently, one could rightfully argue that each of the three parties has the same amount of power (never mind the fact that P_3 has only 3 votes!).

TABLE 2-1	Winning Coalitions and Critical Players for [101: 99, 98, 3]	
Coalition	Weight	Critical players
$\{P_1, P_2\}$	197	P_1 and P_2
$\{P_1, P_3\}$	102	P_1 and P_3
$\{P_2, P_3\}$	101	P_2 and P_3
$\{P_1, P_2, P_3\}$	200	None

When John Banzhaf introduced his mathematical interpretation of power in 1965, he had a key insight: *A player's power should be measured by how often the player is a critical player.* Thus, the key to getting a measure of a player's power is to count the number of winning coalitions in which that player is critical. From Table 2-1 we can clearly see that in $[101: 99, 98, 3]$ each player is critical twice. Since there are three players, each critical twice, we can say that each player holds two out of six, or one-third of the power.

The preceding ideas lead us to the final definitions of this section.

- **The Banzhaf power index.** (For the sake of simplicity we will describe the calculations for P_1—exactly the same idea applies to each of the other players.) We start by counting how many times P_1 is a critical player in a winning coalition. Let's call this number the *critical count* for P_1, and denote it by B_1. We repeat the process for each of the other players and find their respective *critical counts* $B_2, B_3, \ldots, B_N$. We then let T denote the sum of the critical counts of all the players ($T = B_1 + B_2 + \cdots + B_N$) and compute the ratio B_1/T (critical count for P_1 over the total of all critical counts). The ratio B_1/T is the **Banzhaf power index** of P_1. This ratio measures the size of P_1's "share" of the "power pie," and can be expressed either as a fraction or decimal between 0 and 1 or equivalently, as a percent between 0 and 100%. For convenience, we will use the symbol β_1 (read "beta-one") to denote the Banzhaf power index of P_1.

- **The Banzhaf power distribution.** Just like P_1, each of the other players in the weighted voting system has a Banzhaf power index, which we can find in a similar way. The complete list of power indexes $\beta_1, \beta_2, \ldots, \beta_N$ is called the **Banzhaf power distribution** of the weighted voting system. The sum of all the β's is 1 (or 100% if they are written as percentages).

The following is a summary of the steps needed to compute the Banzhaf power distribution of a weighted voting system with N players.

COMPUTING THE BANZHAF POWER DISTRIBUTION OF A WEIGHTED VOTING SYSTEM

- **Step 1.** Make a list of all possible *winning* coalitions.
- **Step 2.** Within each winning coalition determine which are the *critical* players. (For record-keeping purposes, it is a good idea to underline each critical player.)
- **Step 3.** Count the number of times that P_1 is critical. This gives B_1, the critical count for P_1. Repeat for each of the other players to find B_2, B_3, ..., B_N.
- **Step 4.** Add all the B's in Step 3. Let's call this number T. ($T = B_1 + B_2 + \cdots + B_N$ represents the total of all critical counts.)
- **Step 5.** Find the ratio $\beta_1 = B_1/T$. This gives the *Banzhaf power index* of P_1. Repeat for each of the other players to find β_2, β_3, ..., β_N. The complete list of β's is the *Banzhaf power distribution* of the weighted voting system.

In the next set of examples we will illustrate how to carry out the above sequence of steps.

EXAMPLE 2.9   **Banzhaf Power in [4: 3, 2, 1]**

Let's find the Banzhaf power distribution of the weighted voting system $[4:3,2,1]$ using the procedure described above.

Step 1. There are three winning coalitions in this weighted voting system. They are $\{P_1, P_2\}$ with 5 votes, $\{P_1, P_3\}$ with 4 votes, and the grand coalition $\{P_1, P_2, P_3\}$ with 6 votes.

Step 2. In $\{P_1, P_2\}$ both P_1 and P_2 are critical; in $\{P_1, P_3\}$ both P_1 and P_3 are critical; in $\{P_1, P_2, P_3\}$ only P_1 is critical.

Step 3. $B_1 = 3$ (P_1 is critical in three coalitions); $B_2 = 1$ and $B_3 = 1$ (P_2 and P_3 are critical once each).

Step 4. $T = 3 + 1 + 1 = 5$.

Step 5. $\beta_1 = B_1/T = 3/5$; $\beta_2 = B_2/T = 1/5$; $\beta_3 = B_3/T = 1/5$. (If we want to express the β's in terms of percentages, then $\beta_1 = 60\%$, $\beta_2 = 20\%$, and $\beta_3 = 20\%$.)

Of all the steps we must carry out in the process of computing Banzhaf power, by far the most demanding is Step 1. When we have only three players, as in Example 2.9, we can list the winning coalitions on the fly—there simply aren't that many—but as the number of players increases, the number of possible winning coalitions grows rapidly, and it becomes necessary to adopt some form of strategy to come up with a list of *all* the winning coalitions. This is important because if we miss a single one, we are in all likelihood going to get the wrong Banzhaf power distribution. One conservative strategy is to make a list of *all* possible coalitions and then cross out the losing ones. The next example illustrates how one would use this approach.

⬭ **EXAMPLE 2.10** **Banzhaf Power and the NBA Draft**

When NBA teams prepare for the annual draft of college players, the decision on which college basketball players to draft may involve many people, including the management, the coaches, and the scouting staff. Typically, not all these people have an equal voice in the process—the head coach's opinion is worth more than that of an assistant coach, and the general manager's opinion is worth more than that of a scout. In some cases this arrangement is formalized in the form of a weighted voting system. Let's use a fictitious team—the Flyers—for the purposes of illustration.

In the Flyers draft system the head coach (P_1) has 4 votes, the general manager (P_2) has 3 votes, the director of scouting operations (P_3) has 2 votes, and the team psychiatrist (P_4) has 1 vote. Of the 10 votes cast, a simple majority of 6 votes is required for a yes vote on a player to be drafted. In essence, the Flyers operate as the weighted voting system [6: 4, 3, 2, 1].

We will now find the Banzhaf power distribution of this weighted voting system using Steps 1 through 5.

Step 1. Table 2-2 starts with the complete list of *all* possible coalitions (first column) and their weights (second column). From this information we can immediately determine which are the winning coalitions. (Note that the list of coalitions is organized systematically—two-player coalitions first, three-player coalitions next, etc. Also note that within each coalition the players are listed in numerical order from left to right. Both of these are good bookkeeping strategies, and you are encouraged to use them when you do your own work.)

TABLE 2-2 Coalitions for [6: 4, 3, 2, 1]

Coalition (winning = *)	Weight	Coalition (winning = *)	Weight
{$\underline{P_1}, \underline{P_2}$}*	7	{P_1, P_2, P_3}*	9
{$\underline{P_1}, \underline{P_3}$}*	6	{$\underline{P_1}, \underline{P_2}, P_4$}*	8
{P_1, P_4}	5	{$\underline{P_1}, \underline{P_3}, P_4$}*	7
{P_2, P_3}	5	{$\underline{P_2}, \underline{P_3}, \underline{P_4}$}*	6
{P_2, P_4}	4	{P_1, P_2, P_3, P_4}*	10
{P_3, P_4}	3		

Step 2. Now we disregard the losing coalitions and go to work on the winning coalitions only. For each winning coalition we determine which players are critical. This is indicated in Table 2-2 by underlining the critical players. (Don't take someone else's word for it—please check that these are all the critical players!)

Step 3. We now carefully tally how many times each player is underlined in Table 2-2. These are the critical counts: $B_1 = 5$, $B_2 = 3$, $B_3 = 3$, and $B_4 = 1$.

Step 4. $T = 5 + 3 + 3 + 1 = 12$.

Step 5. $\beta_1 = \frac{5}{12} = 41\frac{2}{3}\%$; $\beta_2 = \frac{3}{12} = 25\%$; $\beta_3 = \frac{3}{12} = 25\%$, and $\beta_4 = \frac{1}{12} = 8\frac{1}{3}\%$.

An interesting and unexpected result of these calculations is that the team's general manager (P_2) and the director of scouting operations (P_3) have the same Banzhaf power index—not exactly the arrangement originally intended. ⬭

A Brief Mathematical Detour

Before we go on to the next example, let's consider the following question: For a given number of players, how many different coalitions are possible? Here, our identification of coalitions with sets will come in particularly handy. Except for the empty subset { }, we know that every other subset of the set of players can be identified with a different coalition. This means that we can count the total number of coalitions by counting the number of subsets and subtracting one. So, how many subsets does a set have?

A careful look at Table 2-3 shows us that each time we add a new element we are doubling the number of subsets—the same subsets we had before we added the element plus an equal number consisting of each of these subsets but with the new element thrown in.

TABLE 2-3	The Subsets of a Set			
Set	$\{P_1, P_2\}$	$\{P_1, P_2, P_3\}$	$\{P_1, P_2, P_3, P_4\}$	$\{P_1, P_2, P_3, P_4, P_5\}$
Number of subsets	4	8	16	32
Subsets	{ } $\{P_1\}$ $\{P_2\}$ $\{P_1, P_2\}$	{ } $\{P_3\}$ $\{P_1\}$ $\{P_1, P_3\}$ $\{P_2\}$ $\{P_2, P_3\}$ $\{P_1, P_2\}$ $\{P_1, P_2, P_3\}$	{ } $\{P_4\}$ $\{P_1\}$ $\{P_1, P_4\}$ $\{P_2\}$ $\{P_2, P_4\}$ $\{P_1, P_2\}$ $\{P_1, P_2, P_4\}$ $\{P_3\}$ $\{P_3, P_4\}$ $\{P_1, P_3\}$ $\{P_1, P_3, P_4\}$ $\{P_2, P_3\}$ $\{P_2, P_3, P_4\}$ $\{P_1, P_2, P_3\}$ $\{P_1, P_2, P_3, P_4\}$	The 16 subsets of $\{P_1, P_2, P_3, P_4\}$ along with each of these with P_5 thrown in.

Since each time we add a new player we are doubling the number of subsets, we will find it convenient to think in terms of powers of 2. Table 2-4 summarizes what we have learned.

TABLE 2-4	The Number of Coalitions	
Players	**Number of subsets**	**Number of coalitions**
P_1, P_2	$4 = 2^2$	$2^2 - 1 = 3$
P_1, P_2, P_3	$8 = 2^3$	$2^3 - 1 = 7$
P_1, P_2, P_3, P_4	$16 = 2^4$	$2^4 - 1 = 15$
P_1, P_2, P_3, P_4, P_5	$32 = 2^5$	$2^5 - 1 = 31$
$\vdots$	$\vdots$	$\vdots$
$P_1, P_2, \ldots, P_N$	2^N	$2^N - 1$

Shortcuts for Computing Banzhaf Power Distributions

We will now return to the problem of computing Banzhaf power distributions. Given what we now know about the rapid growth of the number of coalitions, the strategy used in Example 2.10 (list all possible coalitions and then eliminate the losing ones) can become a little tedious (to say the least) when we have more than a handful of players. Sometimes we can save ourselves a lot of work by figuring out directly which are the winning coalitions.

EXAMPLE 2.11   **Winning Coalitions Rule**

The disciplinary committee at George Washington High School has five members: the principal (P_1), the vice principal (P_2), and three teachers (P_3, P_4, and P_5). When voting on a specific disciplinary action the principal has three votes, the vice principal has two votes, and each of the teachers has one vote. A total of five votes is needed for any disciplinary action. Formally speaking, the disciplinary committee is the weighted voting system [5: 3, 2, 1, 1, 1].

There are $2^5 - 1 = 31$ possible coalitions. If we subtract the five one-player coalitions, we are still left with 26 coalitions to contend with. We can save a fair amount of effort by skipping the losing coalitions and listing only the winning coalitions. A little organization will help: We will go through the winning coalitions systematically according to the number of players in the coalition. There is only one two-player winning coalition, namely $\{P_1, P_2\}$. The only three-player winning coalitions are those that include the principal P_1. All four-player coalitions are winning coalitions, and so is the grand coalition.

Steps 1 and 2. Table 2-5 shows *all* the winning coalitions, with the critical players underlined. (You should double check to make sure that these are the right critical players in each coalition—it's good practice!)

TABLE 2-5   Winning Coalitions in [5: 3, 2, 1, 1, 1] and Critical Players

Winning coalitions	Comments
$\{\underline{P_1}, \underline{P_2}\}$	Only possible winning two-player coalition.
$\{\underline{P_1}, \underline{P_2}, P_3\}$ $\{\underline{P_1}, \underline{P_2}, P_4\}$ $\{\underline{P_1}, \underline{P_2}, P_5\}$ $\{\underline{P_1}, \underline{P_3}, \underline{P_4}\}$ $\{\underline{P_1}, \underline{P_3}, \underline{P_5}\}$ $\{\underline{P_1}, \underline{P_4}, \underline{P_5}\}$	Winning three-player coalitions must contain P_1 plus any two other players.
$\{\underline{P_1}, P_2, P_3, P_4\}$ $\{\underline{P_1}, P_2, P_3, P_5\}$ $\{\underline{P_1}, P_2, P_4, P_5\}$ $\{\underline{P_1}, P_3, P_4, P_5\}$ $\{\underline{P_2}, \underline{P_3}, \underline{P_4}, \underline{P_5}\}$	All four-player coalitions are winning coalitions.
$\{P_1, P_2, P_3, P_4, P_5\}$	The grand coalition always wins.

Step 3. The critical counts are $B_1 = 11$, $B_2 = 5$, $B_3 = 3$, $B_4 = 3$, and $B_5 = 3$.

Step 4. $T = 25$.

Step 5. $\beta_1 = 11/25 = 44\%$; $\beta_2 = 5/25 = 20\%$; $\beta_3 = \beta_4 = \beta_5 = 3/25 = 12\%$.

EXAMPLE 2.12   **Tie-Breaking Power**

The Tasmania State University Promotion and Tenure committee consists of five members: the dean (D) and four other faculty members of equal standing (F_1, F_2, F_3, and F_4). (For convenience we are using a slightly different notation for the players.) In this committee faculty members vote first, and motions are carried by simple majority. The dean votes *only* to break a 2-2 tie. Is this a weighted voting system? If so, what is the Banzhaf power distribution?

The answer to the first question is yes, and the answer to the second question is that we can apply the same steps we used before even though we don't have specific weights for the players.

Step 1. The possible winning coalitions fall into two groups: (1) A majority (three or more) faculty members vote yes. In this case the dean does not vote. These winning coalitions are listed in the first column of Table 2-6. (2) Only two faculty members vote yes, and the dean breaks the tie with a yes vote. These winning coalitions are listed in the second column of Table 2-6.

Step 2. In the winning coalitions consisting of just three faculty members, all faculty members are critical. In the coalition consisting of all four faculty members, no faculty member is critical. In the coalitions consisting of two faculty members plus the dean, all players, including the dean, are critical. Table 2-6 shows the winning coalitions with the critical players underlined in each winning coalition.

TABLE 2-6 Winning Coalitions and Critical Players	
Three or four faculty members	Two faculty members plus the dean
$\{\underline{F_1}, \underline{F_2}, \underline{F_3}\}$	$\{\underline{D}, \underline{F_1}, \underline{F_2}\}$
$\{\underline{F_1}, \underline{F_2}, \underline{F_4}\}$	$\{\underline{D}, \underline{F_1}, \underline{F_3}\}$
$\{\underline{F_1}, \underline{F_3}, \underline{F_4}\}$	$\{\underline{D}, \underline{F_1}, \underline{F_4}\}$
$\{\underline{F_2}, \underline{F_3}, \underline{F_4}\}$	$\{\underline{D}, \underline{F_2}, \underline{F_3}\}$
$\{F_1, F_2, F_3, F_4\}$	$\{\underline{D}, \underline{F_2}, \underline{F_4}\}$
	$\{\underline{D}, \underline{F_3}, \underline{F_4}\}$

Step 3. The dean is critical six times (in each of the six coalitions in the second column). Each of the faculty members is critical three times in the first column and three times in the second column. Thus, all players are critical an equal number of times.

Steps 4 and 5. No counts are really necessary here. Since all five players are critical an equal number of times, they all have the same Banzhaf power index: $\beta_1 = \beta_2 = \beta_3 = \beta_4 = \beta_5 = 20\%$.

---

66 **The Vice President of the United States shall be the President of the Senate, but shall have no Vote, unless they be equally divided.** 99

—Article I, Section 3, U.S. Constitution

The surprising conclusion of Example 2.12 is that although the rules appear to set up a special role for the dean (the role of tie-breaker), in practice the dean is no different than any of the faculty members. A larger-scale version of Example 2.12 occurs in the U.S. Senate, where the vice president of the United States can only vote to break a tie. An analysis similar to the one used in Example 2.12 shows that as a member of the Senate the vice president has the same Banzhaf power index as an ordinary senator.

2.3 Applications of the Banzhaf Power Index

The Nassau County (N.Y.) Board of Supervisors Weighted voting in the state of New York has had a long history going all the way back to the 1800s, but it took a law professor to mathematically demonstrate that the weighted voting systems used in some New York counties were seriously flawed (and unconstitutional). In 1965, John Banzhaf introduced the Banzhaf Power Index in an article entitled *Weighted Voting Doesn't Work* (reference 2). One of Banzhaf's key points was that in weighted voting votes do not necessarily imply power, and he illustrated this point using the Board of Supervisors of Nassau County, New York.

In the 1960s, Nassau County was divided into 6 uneven districts, 4 with high populations and 2 rural districts with low populations. Table 2-7 shows the names of the 6 districts and their weights based on their respective populations. The quota was a simple majority of 58 (out of 115) votes. The idea behind this (and similar weighted voting legislatures) is that a district's power should be in proportion to its population, thus ensuring to all citizens the "equal protection" guaranteed by the Constitution.

Banzhaf's case went something like this: Instead of looking at the weights of the districts on the Nassau County Board, one should focus on which districts are critical players in the many winning coalitions that can be formed on the Board. Banzhaf argued that only the 3 largest districts—Hempstead #1, Hempstead #2, and Oyster Bay—could ever be critical players, and consequently, the other 3 districts had no power whatsoever. Notice that among the three districts that had no power was North Hempstead, never mind its 21 votes on the Board! (For the details of Banzhaf's analysis, see Exercise 21.)

Banzhaf's mathematical analysis of the unfair power distribution in the Nassau County Board set the stage for a series of lawsuits against Nassau and other New York state counties based on the argument that weighted voting violated the "equal protection" guarantees of the Fourteenth Amendment (which was the very reason that weighted voting was instituted to begin with). The final result was a federal court decision in 1993 abolishing weighted voting in New York state (see Project D). In 1996, after a protracted fight, the Nassau County Board of Supervisors became a 19-member legislature, each member having just one vote and representing districts of roughly equal population.

TABLE 2-7	Nassau County Board (1964)
District	Weight
Hempstead #1	31
Hempstead #2	31
Oyster Bay	28
North Hempstead	21
Long Beach	2
Glen Cove	2

The United Nations Security Council The United Nations Security Council is an important international body responsible for maintaining world peace and

security. As currently constituted, the Security Council consists of 15 voting nations—5 of them are the *permanent* members (Britain, China, France, Russia, and the United States); the other ten nations are *nonpermanent* members appointed for a two-year period on a rotating basis. (The current 15-nation arrangement might very well change in the near future, as Germany, Japan, Brazil, India, and possibly one African nation are being considered for permanent membership.)

To pass a motion in the Security Council requires a yes vote from each of the five permanent members (in effect giving each permanent member *veto power*) plus at least 4 additional yes votes from the

10 nonpermanent members. In other words, a winning coalition in the Security Council must include all 5 of the permanent members plus 4 or more nonpermanent members.

All in all, there are 848 possible winning coalitions in the Security Council—too many to list one by one. However, it's not too difficult to figure out the critical player story: In the winning coalitions with 9 players (5 permanent members plus 4 nonpermanent members) *every* member of the coalition is a critical player; in all the other winning coalitions (10 or more players) only the permanent members are critical players. Using a few simple (but not elementary) calculations, one can find that of the 848 winning coalitions, there are 210 with 9 members and the rest have 10 or more members. Carefully piecing together this information leads to the following surprising conclusion: the Banzhaf power index of each permanent member is 848/5080 (roughly 16.7%), while the Banzhaf power index of each nonpermanent member is 84/5080 (roughly 1.65%). (See Exercise 67 for all the details.) Note the discrepancy in power between the permanent and nonpermanent members—a permanent member has more than 10 times as much power as a nonpermanent member. Was this really the intent of the United Nations charter or the result of a lack of understanding of the mathematics behind weighted voting?

The European Union   The European Union (EU) is a political and an economic confederation of European nations, a sort of United States of Europe. As of the writing of this edition, the EU consists of 27 member nations (Table 2-8), with three more countries (Turkey, Croatia, and Macedonia) expected to join the EU in the near future.

The legislative body for the EU (called the EU Council of Ministers) operates as a weighted voting system where the different member nations have weights that are roughly proportional to their respective populations (but with some tweaks that favor the small countries). The second column of Table 2-8

TABLE 2-8	The European Union (2008): Banzhaf Power Distribution		
Member nation	Weight	Percentage of votes	Banzhaf power
France, Germany, Italy, United Kingdom	29	8.41%	7.78%
Spain, Poland	27	7.83%	7.42%
Romania	14	4.06%	4.26%
Netherlands	13	3.77%	3.97%
Belgium, Greece, Czech Republic, Hungary, Portugal	12	3.48%	3.68%
Austria, Bulgaria, Sweden	10	2.90%	3.09%
Denmark, Ireland, Lithuania, Slovakia, Finland	7	2.03%	2.18%
Cyprus, Estonia, Latvia, Luxembourg, Slovenia	4	1.16%	1.25%
Malta	3	0.87%	0.94%

shows each nation's weight in the Council of Ministers. The total number of votes is $V = 345$, with the quota set at $q = 255$ votes. The third column of Table 2-8 gives the relative weights (weight/345) expressed as a percent. The last column in Table 2-8 shows the Banzhaf power index of each member nation, also expressed as a percent.

We can see from the last two columns of Table 2-8 that there is a very close match between Banzhaf power and weights (when we express the weights as a percentage of the total number of votes). This is an indication that, unlike in the Nassau County Board of Supervisors, in the EU votes and power go hand in hand and that this weighted voting system works pretty much the way it was intended to work.

■ An applet that will allow you to compute Banzhaf power in moderately large weighted voting systems is available in MyMathLab.* (MML)

| 2.4 | # The Shapley-Shubik Power Index |

■ Lloyd Shapley is Professor Emeritus of Mathematics and Economics at UCLA (see the biographical portrait of Shapley at the end of this chapter). Martin Shubik is Professor of Mathematical Economics at Yale University.

In this section we will discuss a different approach to measuring power, first proposed by American mathematician Lloyd Shapley and economist Martin Shubik in 1954. The key difference between the Shapley-Shubik measure of power and the Banzhaf measure of power centers on the concept of a sequential coalition. In the Shapley-Shubik method the assumption is that coalitions are formed sequentially: Players join the coalition and cast their votes in an orderly sequence (there is a first player, then comes a second player, then a third, etc.). Thus, to an already complicated situation we are adding one more wrinkle—the question of the order in which the players join the coalition.

Let's illustrate the difference with a simple example.

(**EXAMPLE 2.13**)   **Three-Player Sequential Coalitions**

When we think of the *coalition* involving three players P_1, P_2, and P_3, we think in terms of a set. For convenience we write the set as $\{P_1, P_2, P_3\}$, but we can also write in other ways such as $\{P_3, P_2, P_1\}$, $\{P_2, P_3, P_1\}$, and so on. In a coalition the only thing that matters is who are the members—the order in which we list them is irrelevant.

In contrast to ordinary coalitions, a **sequential coalition** is one in which the order of the players does matter—there is a "first" player, a "second" player, and so on. For three players P_1, P_2, and P_3 we can form six different sequential coalitions: $\langle P_1, P_2, P_3 \rangle$ (this means that P_1 is the first player, then P_2 joined in, and last came P_3); $\langle P_1, P_3, P_2 \rangle$; $\langle P_2, P_1, P_3 \rangle$; $\langle P_2, P_3, P_1 \rangle$; $\langle P_3, P_1, P_2 \rangle$; $\langle P_3, P_2, P_1 \rangle$. The six sequential coalitions of three players are graphically illustrated in Fig. 2-1. (Note the change in notation. From now on the notation $\langle \quad \rangle$ will indicate that we are dealing with a *sequential coalition*, and the order in which the players are listed does matter!) ⊂⊃

The Shapley-Shubik approach to measuring power is parallel to the Banzhaf approach, but it is based on the concept of the *pivotal player* in a sequential coalition.

■ **Pivotal player.** In every sequential coalition there is a player who contributes the votes that turn what was a losing coalition into a winning coalition—we

*To use MyMathLab, you need a course ID and a student access code. Contact your instructor for more information.

Ordinary Coalition	Sequential Coalitions		

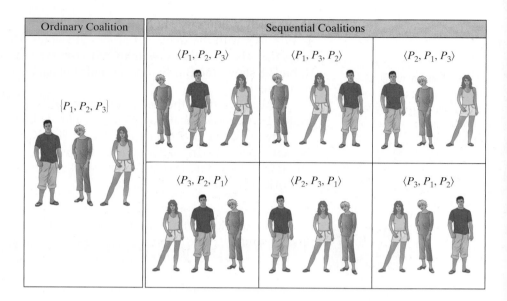

FIGURE 2-1 Three players form six different sequential coalitions.

call such a player the **pivotal player** of the sequential coalition. Every sequential coalition has *one and only one* pivotal player. To find the pivotal player we just add the players' weights from left to right, one at a time, until the tally is bigger or equal to the quota q. The player whose votes tip the scales is the pivotal player. This idea is illustrated in Fig. 2-2.

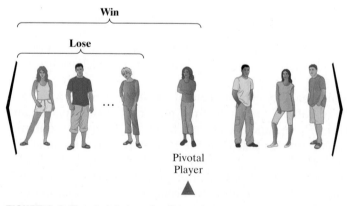

FIGURE 2-2 The pivotal player tips the scales.

■ **The Shapley-Shubik power index.** For a given player P, the **Shapley-Shubik power index** of P is obtained by counting the number of times P is a *pivotal* player and dividing this number by the total number of times *all* players are pivotal (just like what we did with the Banzhaf power index but now using pivotal players instead of critical players and sequential coalitions instead of ordinary coalitions). As in the case of the Banzhaf power index, the Shapley-Shubik power index of a player can be thought of as a measure of the size of that player's "slice" of the "power pie" and can be expressed either as a fraction between 0 and 1 or as a percent between 0 and 100%. We will use σ_1("sigma-one") to denote the Shapley-Shubik power index of P_1, σ_2 to denote the Shapley-Shubik power index of P_2, and so on.

■ **The Shapley-Shubik power distribution.** The complete listing of the Shapley-Shubik power indexes of all the players is called the **Shapley-Shubik power distribution** of the weighted voting system. It tells the complete story of how the (Shapley-Shubik) power pie is divided among the players.

The following is a summary of the steps needed to compute the Shapley-Shubik power distribution of a weighted voting system with N players.

COMPUTING A SHAPLEY-SHUBIK POWER DISTRIBUTION

- **Step 1.** Make a list of all possible sequential coalitions of the N players. Let T be the number of such coalitions. (We will have a lot more to say about T soon!)
- **Step 2.** In each sequential coalition determine *the* pivotal player. (For bookkeeping purposes underline the pivotal players.)
- **Step 3.** Count the total number of times that P_1 is pivotal. This gives SS_1, the *pivotal count* for P_1. Repeat for each of the other players to find SS_2, $SS_3, \ldots, SS_N$.
- **Step 4.** Find the ratio $\sigma_1 = SS_1/T$. This gives the *Shapley-Shubik power index* of P_1. Repeat for each of the other players to find $\sigma_2, \sigma_3, \ldots, \sigma_N$. The complete list of σ's gives the *Shapley-Shubik power distribution* of the weighted voting system.

■ This is the beginning of a brief but extremely important mathematical detour. At the end of this detour we will be able to easily answer the question: How many sequential coalitions are possible?

The Multiplication Rule and Factorials

We will start by introducing one of the most useful rules of basic mathematics, stunning in its simplicity:

THE MULTIPLICATION RULE

If there are m different ways to do X and n different ways to do Y, then X and Y together can be done in $m \times n$ different ways.

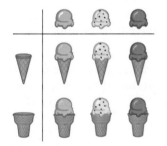

FIGURE 2-3 The scoop on the multiplication rule.

(**EXAMPLE 2.14**)   **Cones and Flavors**

An ice cream shop offers 2 different choices of cones and 3 different flavors of ice cream (this is a boutique ice cream shop!). Using the multiplication rule we can conclude that there are $2 \times 3 = 6$ different choices for a *single* (cone and one scoop of ice cream) order. The choices are illustrated in Figure 2-3.

Figure 2-3 also clarifies the logic behind the multiplication rule: If we had m different types of cones and n different flavors of ice cream, then the corresponding arrangement would show the cone/flavor combinations laid out in m rows and n columns, illustrating the $m \times n$ possibilities. ⬭

(**EXAMPLE 2.15**)   **Cones and Flavors Plus Toppings**

A bigger ice cream shop offers 5 different choices of cones, 31 different flavors of ice cream (this is *not* a boutique ice cream shop!), and 8 different choices of topping. The question is, If you are going to choose a cone and a single scoop of ice cream but then add a topping for good measure, how many orders are possible? The number of choices is too large to list individually, but we can find it by using the multiplication rule twice: First, there are $5 \times 31 = 155$ different cone/flavor combinations, and each of these can be combined with one of the 8 toppings into a grand total of $155 \times 8 = 1240$ different cone/flavor/topping combinations. The implications—as well as the calories—are staggering! ⬭

We will now use the multiplication rule to count sequential coalitions. Specifically, we want to know, how many sequential coalitions with N players are possible?

> #### EXAMPLE 2.16   Counting Sequential Coalitions

Let's start with a warm up: How do we count the number of sequential coalitions with four players? Using the multiplication rule, we can argue as follows: We can choose any one of the *four* players to go first, then choose any one of the remaining *three* players to go second, then choose any one of the remaining *two* players to go third, and finally the *one* player left goes last. Using the multiplication rule we get a total of $4 \times 3 \times 2 \times 1 = 24$ sequential coalitions with four players.

If we have five players, following up on our previous argument, we can count on a total of $5 \times 4 \times 3 \times 2 \times 1 = 120$ sequential coalitions, and with N players the number of sequential coalitions is $N \times (N-1) \times \cdots \times 3 \times 2 \times 1$. The number $N \times (N-1) \times \cdots \times 3 \times 2 \times 1$ is called the **factorial** of N and is written in the shorthand form $N!$.

■ This is the end of our mathematical detour. We will return to factorials in Chapter 6, but for the time being, if you have a business or scientific calculator, you might want to find the factorial key and familiarize yourself with its use. This is also a good time to give Exercises 35–40 a try!

> ┌ **THE NUMBER OF SEQUENTIAL COALITIONS**
>
> $N! = N \times (N-1) \times \cdots \times 3 \times 2 \times 1$ gives the number of sequential coalitions with N players.

Back to the Shapley-Shubik Power Index

We will illustrate the procedure for computing Shapley-Shubik power by revisiting a couple of examples done earlier in this chapter.

> #### EXAMPLE 2.17   Shapley-Shubik Power in [4: 3, 2, 1]

In this example we will compute the Shapley-Shubik power distribution of the weighted voting system $[4 : 3, 2, 1]$. (We computed the Banzhaf power distribution of this weighted voting system in Example 2.9, so at the end we will be able to compare the results and see if there is a difference.)

Steps 1 and 2. Table 2-9 shows the 6 sequential coalitions of three players, with the pivotal player in each sequential coalition underlined. The second column shows the weights as they are being added from left to right until the tally reaches or exceeds the quota $q = 4$.

TABLE 2-9	Sequential Coalitions for [4: 3, 2, 1] (Pivotal Players Underlined)
Sequential coalition	**Weight tallies (from left to right)**
$\langle P_1, \underline{P_2}, P_3 \rangle$	$3; 3 + 2 = \mathbf{5}$
$\langle P_1, \underline{P_3}, P_2 \rangle$	$3; 3 + 1 = \mathbf{4}$
$\langle P_2, \underline{P_1}, P_3 \rangle$	$2; 2 + 3 = \mathbf{5}$
$\langle P_2, P_3, \underline{P_1} \rangle$	$2; 2 + 1 = 3; 2 + 1 + 3 = \mathbf{6}$
$\langle P_3, \underline{P_1}, P_2 \rangle$	$1; 1 + 3 = \mathbf{4}$
$\langle P_3, P_2, \underline{P_1} \rangle$	$1; 1 + 2 = 3; 1 + 2 + 3 = \mathbf{6}$

Step 3. The player's pivotal counts are, respectively, $SS_1 = 4$, $SS_2 = 1$, and $SS_3 = 1$.

Step 4. The Shapley-Shubik power distribution is given by $\sigma_1 = \frac{4}{6} = 66\frac{2}{3}\%$, $\sigma_2 = \frac{1}{6} = 16\frac{2}{3}\%$, and $\sigma_3 = \frac{1}{6} = 16\frac{2}{3}\%$.

When we compare this power distribution with the Banzhaf power distribution found in Example 2.9 ($\beta_1 = \frac{3}{5} = 60\%$, $\beta_2 = \frac{1}{5} = 20\%$, and $\beta_3 = \frac{1}{5} = 20\%$), we see that the two interpretations of power are indeed different. ⬭

(**EXAMPLE 2.18**)   **Shapley-Shubik Power and the NBA Draft**

We will now revisit Example 2.10, the NBA draft example. The weighted voting system in this example is [6: 4, 3, 2, 1], and we will now find its Shapley-Shubik power distribution.

Steps 1 and 2. Table 2-10 shows the 24 sequential coalitions of P_1, P_2, P_3, and P_4. In each sequential coalition the pivotal player is underlined. (Note there is a method to the madness of Table 2-9—each column corresponds to the sequential coalitions with a given first player. You may want to use the same or a similar pattern when you do the exercises on Shapley-Shubik power distributions.)

TABLE 2-10	Sequential Coalitions for [6: 4, 3, 2, 1] (Pivotal Players Underlined)		
$\langle P_1, \underline{P_2}, P_3, P_4 \rangle$	$\langle P_2, \underline{P_1}, P_3, P_4 \rangle$	$\langle P_3, \underline{P_1}, P_2, P_4 \rangle$	$\langle P_4, P_1, \underline{P_2}, P_3 \rangle$
$\langle P_1, \underline{P_2}, P_4, P_3 \rangle$	$\langle P_2, \underline{P_1}, P_4, P_3 \rangle$	$\langle P_3, \underline{P_1}, P_4, P_2 \rangle$	$\langle P_4, P_1, \underline{P_3}, P_2 \rangle$
$\langle P_1, \underline{P_3}, P_2, P_4 \rangle$	$\langle P_2, P_3, \underline{P_1}, P_4 \rangle$	$\langle P_3, P_2, \underline{P_1}, P_4 \rangle$	$\langle P_4, P_2, \underline{P_1}, P_3 \rangle$
$\langle P_1, \underline{P_3}, P_4, P_2 \rangle$	$\langle P_2, P_3, \underline{P_4}, P_1 \rangle$	$\langle P_3, P_2, \underline{P_4}, P_1 \rangle$	$\langle P_4, P_2, \underline{P_3}, P_1 \rangle$
$\langle P_1, P_4, \underline{P_2}, P_3 \rangle$	$\langle P_2, P_4, \underline{P_1}, P_3 \rangle$	$\langle P_3, P_4, \underline{P_1}, P_2 \rangle$	$\langle P_4, P_3, \underline{P_1}, P_2 \rangle$
$\langle P_1, P_4, \underline{P_3}, P_2 \rangle$	$\langle P_2, P_4, \underline{P_3}, P_1 \rangle$	$\langle P_3, P_4, \underline{P_2}, P_1 \rangle$	$\langle P_4, P_3, \underline{P_2}, P_1 \rangle$

Step 3. The pivotal counts are $SS_1 = 10$, $SS_2 = 6$, $SS_3 = 6$, and $SS_4 = 2$.

Step 4. The Shapley-Shubik power distribution is given by $\sigma_1 = \frac{10}{24} = 41\frac{2}{3}\%$, $\sigma_2 = \frac{6}{24} = 25\%$, $\sigma_3 = \frac{6}{24} = 25\%$, and $\sigma_4 = \frac{2}{24} = 8\frac{1}{3}\%$.

If you compare this result with the Banzhaf power distribution obtained in Example 2.10, you will notice that here the two power distributions are the same. If nothing else, this shows that it is not impossible for the Banzhaf and Shapley-Shubik power distributions to agree. In general, however, for randomly chosen real-life situations it is very unlikely that the Banzhaf and Shapley-Shubik methods will give the same answer. ⬭

(**EXAMPLE 2.19**)   **Shapley-Shubik in the City**

In some cities the city council operates under what is known as the "strong-mayor" system. Under this system the city council can pass a motion under simple majority, but the mayor has the power to veto the decision. The mayor's veto can then be overruled by a "supermajority" of the council members. As an illustration of the strong-mayor system we will consider the city of Cleansburg. In Cleansburg the city council has four members plus a strong mayor who has a vote as well as the power to veto motions supported by a simple majority of the council members. On the other hand, the mayor cannot veto motions supported by all four council members. Thus, a motion can pass if the mayor plus two or more council members support it or, alternatively, if the mayor is against it but the four council members support it.

Common sense tells us that under these rules, the four council members have the same amount of power but the mayor has more. We will compute the Shapley-Shubik power distribution of this weighted voting system to figure out exactly how much more.

With five players, this weighted voting system has $5! = 120$ sequential coalitions to consider. Obviously, we would prefer not to have to write them all down, and in fact we can skip that step and just use our imagination. Specifically, we will imagine a table in which the 120 sequential coalitions are listed in 5 columns. The first column has all the sequential coalitions where the mayor (let's call him P_1) appears as the first player in the coalition, the second column has all the sequential coalitions where P_1 is in second place, the third column has all sequential coalitions with P_1 in third place, and so on. Now close your eyes and try to visualize this table in your mind's eye for a moment or two. Most likely, you visualized a table that had the same number of sequential coalitions in each of the columns. This is a key observation, as it allows us to conclude that there are $120 \div 5 = 24$ sequential coalitions in each column.

We are now in a position to find the pivotal player count of the mayor (P_1). We first observe that the mayor is the pivotal player only when he is in the third or fourth positions of a sequential coalition, as illustrated in Fig. 2-4(a) and (b). (When the mayor is in the fifth position [Fig. 2-4(c)] the fourth council member is the pivotal player.) This means that in our imaginary table P_1 is pivotal in every sequential coalition in columns 3 and 4, and no others. So there we have it—the pivotal count for the mayor is $24 + 24 = 48$, and his Shapley-Shubik power index is $48/120 = 2/5 = 40\%$.

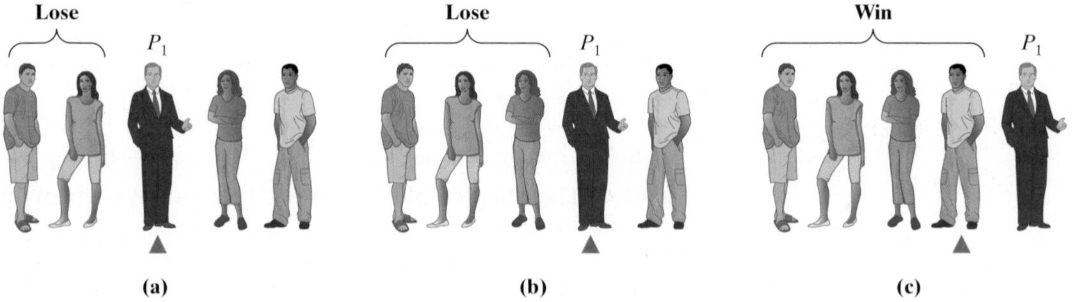

■ See Exercise 68 for an important generalization of these observations.

FIGURE 2-4

Since the Shapley-Shubik power index of the mayor equals 40% and the four council members must share the remaining 60% of the power equally, it follows that each council member has a Shapley-Shubik power index of 15%. In conclusion, the Shapley-Shubik power distribution of the Cleansburg city council is $\sigma_1 = 40\%, \sigma_2 = \sigma_3 = \sigma_4 = \sigma_5 = 15\%$.

In this example it's not so much the conclusion but how we got it that is worth remembering.

For the purposes of comparison, the reader is encouraged to calculate the Banzhaf power distribution of the Cleansburg city council (see Exercise 71).

2.5 Applications of the Shapley-Shubik Power Index

The Electoral College Calculating the Shapley-Shubik power index of the states in the Electoral College is no easy task. There are 51! sequential coalitions, a number so large (67 digits long) that we don't even have a name for it. Individually checking all possible sequential coalitions is out of the question, even for the world's fastest computer. There are, however, some sophisticated mathematical shortcuts that, when coupled with the right kind of software, allow the calculations to be done by an ordinary computer in a matter of seconds (see reference 16 for details).

Appendix A at the end of this book shows both the Banzhaf and the Shapley-Shubik power indexes for each of the 50 states and the District of Columbia. Comparing the Banzhaf and the Shapley-Shubik power indexes shows that there is a very small difference between the two. This example shows that in some situations the Banzhaf and Shapley-Shubik power indexes give essentially the same answer. The United Nations Security Council example below illustrates a very different situation.

The United Nations Security Council As mentioned in Section 2.3, the United Nations Security Council consists of 15 member nations—5 are permanent members and 10 are nonpermanent members appointed on a rotating basis. For a motion to pass it must have a Yes vote from each of the 5 permanent members plus at least 4 of the 10 nonpermanent members. It can be shown that this arrangement is equivalent to giving the permanent members 7 votes each, the nonpermanent members 1 vote each, and making the quota equal to 39 votes [see Exercise 67(e)].

We will sketch a rough outline of how the Shapley-Shubik power distribution of the Security Council can be calculated. The details, while not terribly difficult, go beyond the scope of this book.

1. There are 15! sequential coalitions of 15 players (roughly about *1.3 trillion*).

2. A nonpermanent member can be pivotal only if it is the 9th player in the coalition, preceded by all five of the permanent members and three nonpermanent members. (There are approximately *2.44 billion* sequential coalitions of this type.)

3. From steps 1 and 2 we can conclude that the Shapley-Shubik power index of a nonpermanent member is approximately 2.44 billion/1.3 trillion $\approx$ 0.0019 = 0.19%. (For the purposes of comparison it is worth noting that there is a big difference between this Shapley-Shubik power index and the corresponding Banzhaf power index of 1.65% obtained in Section 2.3.)

4. The 10 nonpermanent members (each with a Shapley-Shubik power index of 0.19%) have together 1.9% of the power pie, leaving the remaining 98.2% to be divided equally among the 5 permanent members. Thus, the Shapley-Shubik power index of each permanent member is approximately 98.2/5 = 19.64%.

This analysis shows the enormous difference between the Shapley-Shubik power of the permanent and nonpermanent members of the Security Council— permanent members have roughly 100 times the Shapley-Shubik power of nonpermanent members!

The European Union We introduced the European Union Council of Ministers in Section 2.3 and observed that the Banzhaf power index of each country is reasonably close to that country's weight when the weight is expressed as a percent of the total number of votes. We will now do a similar analysis for the Shapley-Shubik power distribution.

Computing Shapley-Shubik power in a weighted voting system with 27 players cannot be done using the direct approach we introduced in this chapter—even the world's fastest supercomputer would need many centuries to list the 27! different sequential coalitions that would have to be examined. [If you don't believe that, try Exercise 35(d).] And yet, the last column of Table 2-11 shows the Shapley-Shubik power distribution in the EU Council of Ministers. The calculations took just a couple of seconds using an ordinary desktop computer and some fancy mathematics software. (An applet that will allow you to compute Shapley-Shubik power distributions of moderately large weighted voting systems is available in MyMathLab.*)

When we compare the Banzhaf and Shapley-Shubik power indexes of the various nations in the EU (see Tables 2-8 and 2-11), we can see that there are differences, but the differences are small (less than 1% in all cases). In both cases there is a close match between weights and power but with a twist: In the Shapley-Shubik power distribution the larger countries have a tad more power than they should and the smaller countries have a little less power than they should; with the Banzhaf power distribution this situation is exactly reversed.

TABLE 2-11	The European Union (2008): Shapley-Shubik Power Distribution		
Member nation	Weight	Percentage of votes	Shapley-Shubik Power
France, Germany, Italy, United Kingdom	29	8.41%	8.67%
Spain, Poland	27	7.83%	8.00%
Romania	14	4.06%	3.99%
Netherlands	13	3.77%	3.68%
Belgium, Greece, Czech Republic, Hungary, Portugal	12	3.48%	3.41%
Austria, Bulgaria, Sweden	10	2.90%	2.82%
Denmark, Ireland, Lithuania, Slovakia, Finland	7	2.03%	2.00%
Cyprus, Estonia, Latvia, Luxembourg, Slovenia	4	1.16%	1.10%
Malta	3	0.87%	0.82%

*To use MyMathLab, you need a course ID and a student access code. Contact your instructor for more information.

CONCLUSION

In any society, no matter how democratic, some individuals and groups have more power than others. This is simply a consequence of the fact that individuals and groups are not all equal. Diversity is the inherent reason the concept of power exists.

Power itself comes in many different forms. We often hear clichés such as "In strength lies power" or "Money is power" (and the newer cyber version, "Information is power"). In this chapter we discussed the notion of power as it applies to formal voting situations called *weighted voting systems* and saw how mathematical methods allow us to measure the power of an individual or group by means of a *power index*. In particular, we looked at two different kinds of power indexes: the *Banzhaf power index* and the *Shapley-Shubik power index*.

These indexes provide two different ways to measure power, and, while they occasionally agree, they often differ significantly. Of the two, which one better measures real power?

Unfortunately, there is no simple answer. Both of them are useful, and in some sense the choice is subjective. Perhaps the best way to evaluate them is to think of them as being based on a slightly different set of assumptions. The idea behind the Banzhaf interpretation of power is that players are free to come and go, negotiating their allegiance for power (somewhat like professional athletes since the advent of free agency). Underlying the Shapley-Shubik interpretation of power is the assumption that when a player joins a coalition, he or she is making a commitment to stay. In the latter case a player's power is generated by his ability to be in the right place at the right time.

In practice the choice of which method to use for measuring power is based on which of the assumptions better fits the specifics of the situation. Contrary to what we've often come to expect, mathematics does not give us the answer, just the tools that might help us make an informed decision.

PROFILE: Lloyd S. Shapley (1923–   )

Lloyd Shapley is one of the giants of modern *game theory*—which, in a nutshell, is the mathematical study of *games*. (The word *game* is used here as a metaphor for any situation involving competition and cooperation among individuals or groups each trying to further their own separate goals. This broad interpretation encompasses not only real games like poker and *Monopoly,* but also economic, political, and military "games," thus making game theory one of the most important branches of modern applied mathematics.)

Lloyd Shapley was born in Cambridge, Massachusetts, to a prominent scientific family. His father, Harlow Shapley, was a renowned astronomer at Harvard University and director of the Harvard Observatory. Shapley entered Harvard in 1942, but he interrupted his studies to join the Army dur-

ing World War II. From 1943 to 1945 he served in the Army Air Corps in China, where he was awarded a Bronze Star for breaking the Japanese weather code. In 1945 he returned to Harvard, receiving a bachelor of arts in mathematics in 1948.

Upon graduation from Harvard, Shapley joined the Rand Corporation, a famous think tank in Santa Monica, California, where much of the pioneering research in game theory was being conducted. After a year at Rand, and with a reputation as a budding young star in game theory in tow, Shapley went to Princeton University as a graduate student in mathematics. At Princeton he joined a circle that included some of the most brilliant mathematical minds of his time, including John von Neumann, the father of game theory, and fellow graduate student John Nash, the mathematical genius whose life is portrayed in the Academy Award–winning movie *A Beautiful Mind* (with Russell Crowe playing John Nash).

As a graduate student, Shapley had a reputation as a fierce competitor in everything he took on, including card games and board games, some of his own invention. One of the games Shapley invented with a couple of other graduate students (one of whom was Nash and the other Martin Shubik, who at the time was an economics graduate student as well as Shapley's roommate) was a fiendish board game called *So Long Sucker* in which players formed coalitions to gang up on and double-cross each other. Some of the experiences and observations that Shapley and Shubik (who became lifelong friends) derived from their graduate student game-playing days eventually led to their development of the Shapley-Shubik index for measuring power in a weighted voting system (a weighted voting system, after all, is just a simple *game* in which players form coalitions in order to pass or block a specific action by the group).

After receiving his Ph.D. in mathematics from Princeton in 1953, Shapley returned to the Rand Corporation, where he worked as a research mathematician until 1981. During this time he made many major contributions to game theory, including the invention of *convex* and *stochastic* games, the *Shapley value* of a game, and the *Shapley-Shubik power index*. The citation when he was awarded the Von Neumann Prize in Game Theory in 1981 partly read, "His individual work and his joint research with Martin Shubik has helped build bridges between game theory, economics, political science, and practice."

In 1981 Shapley joined the Mathematics department at UCLA, where he is currently Professor Emeritus of Mathematics and Economics.   ■

KEY CONCEPTS

Banzhaf power distribution, **49**
Banzhaf power index, **49**
coalition, **48**
critical player, **48**
dictator, **46**
dummy, **46**
factorial, **60**
grand coalition, **48**

losing coalition, **48**
motion, **44**
multiplication rule, **59**
pivotal player, **57–8**
player, **44**
quota, **44**
sequential coalition, **57**

Shapley-Shubik power
  distribution, **58**
Shapley-Shubik power index, **58**
veto power, **47**
weight, **44**
weighted voting system, **44**
winning coalition, **48**

EXERCISES

WALKING

A Weighted Voting Systems

1. Consider the weighted voting system [65: 30, 28, 22, 15, 7, 6]. [*Historical footnote*: This weighted voting system describes the voting in the Nassau County Board of Supervisors (N.Y.) in the 1990s.] Find

   (a) the total number of players.

   (b) the total number of votes.

   (c) the weight of P_2.

   (d) the minimum percentage of the votes needed to pass a motion (rounded to the next whole percent).

2. Consider the weighted voting system [62: 10, 10, 10, 10, 8, 5, 5, 5, 5, 4, 4, 3, 3, 3, 2]. [*Historical footnote*: This weighted voting system describes the voting in the European Union Council of Ministers prior to 2004.) Find

   (a) the total number of players.

   (b) the total number of votes.

   (c) the weight of P_6.

   (d) the minimum percentage of the votes needed to pass a motion (rounded to the next whole percent).

3. Consider the weighted voting system [q: 10, 6, 5, 4, 2].

   (a) What is the smallest value that the quota q can take?

   (b) What is the largest value that the quota q can take?

   (c) What is the value of the quota if *at least* two-thirds of the votes are required to pass a motion?

   (d) What is the value of the quota if *more* than two-thirds of the votes are required to pass a motion?

4. Consider the weighted voting system $[q: 6, 4, 3, 3, 2, 2]$.

(a) What is the smallest value that the quota q can take?

(b) What is the largest value that the quota q can take?

(c) What is the value of the quota if *at least* three-fourths of the votes are required to pass a motion?

(d) What is the value of the quota if *more* than three-fourths of the votes are required to pass a motion?

5. A committee has four members (P_1, P_2, P_3 and P_4). In this committee P_1 has twice as many votes as P_2; P_2 has twice as many votes as P_3; P_3 and P_4 have the same number of votes. The quota is $q = 49$. For each of the given definitions of the *quota*, describe the committee using the notation $[q: w_1, w_2, w_3, w_4]$. (*Hint*: Write the weighted voting system as $[49: 4x, 2x, x, x]$, and then solve for x.)

(a) The quota is defined as a *simple majority* of the votes.

(b) The quota is defined as *more than two-thirds* of the votes.

(c) The quota is defined as *more than three-fourths* of the votes.

6. A committee has six members (P_1, P_2, P_3, P_4, P_5, and P_6). In this committee P_1 has twice as many votes as P_2; P_2 and P_3 each has twice as many votes as P_4; P_4 has twice as many votes as P_5; P_5 and P_6 have the same number of votes. The quota is $q = 121$. For each of the given definitions of the *quota*, describe the committee using the notation $[q: w_1, w_2, w_3, w_4, w_5, w_6]$. (*Hint*: Write the weighted voting system as $[121: 8x, 4x, 4x, 2x, x, x]$, and then solve for x.)

(a) The quota is defined as a *simple majority* of the votes.

(b) The quota is defined as *more than two-thirds* of the votes.

(c) The quota is defined as *more than three-fourths* of the votes.

7. In each of the following weighted voting systems, determine which players, if any, (i) are dictators; (ii) have veto power; (iii) are dummies.

(a) $[15: 16, 8, 4, 1]$

(b) $[18: 16, 8, 4, 1]$

(c) $[24: 16, 8, 4, 1]$

8. In each of the following weighted voting systems, determine which players, if any, (i) are dictators; (ii) have veto power; (iii) are dummies.

(a) $[8: 8, 4, 2, 1]$

(b) $[9: 8, 4, 2, 1]$

(c) $[12: 8, 4, 2, 1]$

(d) $[15: 8, 4, 2, 1]$

9. Consider the weighted voting system $[q: 8, 4, 2]$. Find the *smallest* value of q for which

(a) all three players have veto power.

(b) P_2 has veto power but P_3 does not.

(c) P_3 is the only dummy.

10. Consider the weighted voting system $[q: 7, 5, 3]$. Find the *smallest* value of q for which

(a) all three players have veto power.

(b) P_2 has veto power but P_3 does not.

(c) P_3 is the only dummy.

B Banzhaf Power

11. Consider the weighted voting system $[10: 6, 5, 4, 2]$.

(a) What is the weight of the coalition formed by P_1 and P_3?

(b) Write down all winning coalitions.

(c) Which players are critical in the coalition $\{P_1, P_2, P_3\}$?

(d) Find the Banzhaf power distribution of this weighted voting system.

12. Consider the weighted voting system $[5: 3, 2, 1, 1]$.

(a) What is the weight of the coalition formed by P_1 and P_3?

(b) Which players are critical in the coalition $\{P_1, P_2, P_3\}$?

(c) Which players are critical in the coalition $\{P_1, P_3, P_4\}$?

(d) Write down all winning coalitions.

(e) Find the Banzhaf power distribution of this weighted voting system.

13. (a) Find the Banzhaf power distribution of the weighted voting system $[6: 5, 2, 1]$.

(b) Find the Banzhaf power distribution of the weighted voting system $[3: 2, 1, 1]$. Compare your answers in (a) and (b).

14. (a) Find the Banzhaf power distribution of the weighted voting system $[7: 5, 2, 1]$.

(b) Find the Banzhaf power distribution of the weighted voting system $[5: 3, 2, 1]$. Compare your answers in (a) and (b).

15. (a) Find the Banzhaf power distribution of the weighted voting system $[10: 5, 4, 3, 2, 1]$.

(b) Find the Banzhaf power distribution of the weighted voting system $[11: 5, 4, 3, 2, 1]$. [*Hint*: Note that the only change from (a) is in the quota and use this fact to your advantage.]

16. (a) Find the Banzhaf power distribution of the weighted voting system [9: 5, 5, 4, 2, 1].

(b) Find the Banzhaf power distribution of the weighted voting system [9: 5, 5, 3, 2, 1].

17. Consider the weighted voting system $[q: 8, 4, 2, 1]$. Find the Banzhaf power distribution of this weighted voting system when

(a) $q = 8$

(b) $q = 9$

(c) $q = 10$

(d) $q = 12$

(e) $q = 14$

18. Consider the weighted voting system $[q: 16, 8, 4, 1]$. Find the Banzhaf power distribution of this weighted voting system when

(a) $q = 16$

(b) $q = 17$

(c) $q = 18$

(d) $q = 24$

(e) $q = 28$

19. In a weighted voting system with three players the winning coalitions are: $\{P_1, P_2\}, \{P_1, P_3\}$, and $\{P_1, P_2, P_3\}$.

(a) Find the critical players in each winning coalition.

(b) Find the Banzhaf power distribution of the weighted voting system.

20. In a weighted voting system with four players the winning coalitions are: $\{P_1, P_2\}, \{P_1, P_2, P_3\}, \{P_1, P_2, P_4\}$, and $\{P_1, P_2, P_3, P_4\}$.

(a) Find the critical players in each winning coalition.

(b) Find the Banzhaf power distribution of the weighted voting system.

21. The Nassau County (N.Y.) Board of Supervisors (1960s version). In the 1960s, the Nassau County Board of Supervisors operated as the weighted voting system [58: 31, 31, 28, 21, 2, 2]. Assume that the players are P_1 through P_6.

(a) List all the *two-* and *three-player* winning coalitions and find the critical players in each coalition.

(b) List all the winning coalitions that have P_4 as a member and find the critical players in each coalition.

(c) Use the results in (b) to find the Banzhaf power index of P_4.

(d) Use the results in (a) and (c) to find the Banzhaf power distribution of the weighted voting system.

22. The Nassau County Board of Supervisors (1990s version). In the 1990s, after a series of legal challenges, the Nassau County Board of Supervisors was redesigned to operate as the weighted voting system [65: 30, 28, 22, 15, 7, 6].

(a) List all the *three-player* winning coalitions and find the critical players in each coalition.

(b) List all the *four-player* winning coalitions and find the critical players in each coalition. (*Hint*: There are 11 four-player winning coalitions.)

(c) List all the *five-player* winning coalitions and find the critical players in each coalition.

(d) Use the results in (a), (b), and (c) to find the Banzhaf power distribution of the weighted voting system.

23. A law firm is run by four partners (A, B, C, and D). Each partner has one vote and decisions are made by majority rule, but in the case of a 2-2 tie, the coalition with D (the junior partner) loses. (For example, $\{A, B\}$ wins, but $\{A, D\}$ loses.)

(a) List all the winning coalitions in this voting system and find the critical players in each.

(b) Find the Banzhaf power distribution in this law firm.

24. A law firm is run by four partners (A, B, C, and D). Each partner has one vote and decisions are made by majority rule, but in the case of a 2-2 tie, the coalition with A (the senior partner) wins.

(a) List all the winning coalitions in this voting system and the critical players in each.

(b) Find the Banzhaf power index of this law firm.

C Shapley-Shubik Power

25. Consider the weighted voting system [16: 9, 8, 7].

(a) Write down all the sequential coalitions, and in each sequential coalition identify the pivotal player.

(b) Find the Shapley-Shubik power distribution of this weighted voting system.

26. Consider the weighted voting system [8: 7, 6, 2].

(a) Write down all the sequential coalitions, and in each sequential coalition identify the pivotal player.

(b) Find the Shapley-Shubik power distribution of this weighted voting system.

27. Find the Shapley-Shubik power distribution of each of the following weighted voting systems.

(a) [15: 16, 8, 4, 1]

(b) [18: 16, 8, 4, 1]

(c) [24: 16, 8, 4, 1]

(d) [28: 16, 8, 4, 1]

28. Find the Shapley-Shubik power distribution of each of the following weighted voting systems.

   (a) $[8:8,4,2,1]$

   (b) $[9:8,4,2,1]$

   (c) $[12:8,4,2,1]$

   (d) $[14:8,4,2,1]$

29. Find the Shapley-Shubik power distribution of each of the following weighted voting systems.

   (a) $[51:40,30,20,10]$

   (b) $[59:40,30,20,10]$ (*Hint*: Compare this situation with the one in (a).)

   (c) $[60:40,30,20,10]$

30. Find the Shapley-Shubik power distribution of each of the following weighted voting systems.

   (a) $[41:40,10,10,10]$

   (b) $[49:40,10,10,10]$ (*Hint*: Compare this situation with the one in (a).)

   (c) $[50:40,10,10,10]$

31. In a weighted voting system with three players the only winning coalitions are: $\{P_1, P_2\}, \{P_1, P_3\}$, and $\{P_1, P_2, P_3\}$ (see Exercise 19).

   (a) List the sequential coalitions and identify the pivotal player in each one.

   (b) Find the Shapley-Shubik power distribution of the weighted voting system.

32. In a weighted voting system with three players the winning coalitions are: $\{P_1, P_2\}$ and $\{P_1, P_2, P_3\}$.

   (a) List the sequential coalitions and identify the pivotal player in each sequential coalition.

   (b) Find the Shapley-Shubik power distribution of the weighted voting system.

33. A law firm is run by four partners $(A, B, C,$ and $D)$. Each partner has one vote and decisions are made by majority rule, but in the case of a 2-2 tie, the coalition with D (the junior partner) loses. (See Exercise 23.) Find the Shapley-Shubik power distribution in this law firm.

34. A law firm is run by four partners $(A, B, C,$ and $D)$. Each partner has one vote and decisions are made by majority rule, but in the case of a 2-2 tie, the coalition with A (the senior partner) wins. (See Exercise 24.) Find the Shapley-Shubik power distribution in this law firm.

D Miscellaneous

For Exercises 35 and 36 you should use a calculator with a factorial key (typically, it's a key labeled either x! or n!). All scientific calculators and most business calculators have such a key.

35. Use a calculator to compute each of the following.

   (a) 13!

   (b) 18!

   (c) 25!

   (d) Suppose that you have a supercomputer that can list *one trillion* (10^{12}) sequential coalitions per second. Estimate (in years) how long it would take the computer to list all the sequential coalitions of 25 players.

36. Use a calculator to compute each of the following.

   (a) 12!

   (b) 15!

   (c) 20!

   (d) Suppose that you have a supercomputer that can list one billion (10^9) sequential coalitions per second. Estimate (in years) how long it would take the computer to list all the sequential coalitions of 20 players.

The purpose of Exercises 37 through 40 is for you to learn how to numerically manipulate factorials. If you use a calculator to answer these questions, you are defeating the purpose of the exercise. Please answer Exercises 37 through 40 without using a calculator.

37. (a) Given that $10! = 3,628,800$, find 9!

   (b) Find 11!/10!

   (c) Find 11!/9!

   (d) Find 9!/6!

   (e) Find 101!/99!

38. (a) Given that $20! = 2,432,902,008,176,640,000$, find 19!

   (b) Find 20!/19!

   (c) Find 201!/199!

   (d) Find 11!/8!

39. Find the value of each of the following. Give the answer in decimal form.

   (a) $(9! + 11!)/10!$

   (b) $(101! + 99!)/100!$

40. Find the value of each of the following. Give the answer in decimal form.

   (a) $(19! + 21!)/20!$ (*Hint*: $1/20 = 0.05$)

   (b) $(201! + 199!)/200!$

41. Consider a weighted voting system with six players $(P_1$ through $P_6)$.

   (a) Find the total number of coalitions in this weighted voting system.

(b) How many coalitions in this weighted voting system do not include P_1? (*Hint:* Think of all the possible coalitions of the remaining players.)

(c) How many coalitions in this weighted voting system do not include P_3? (*Hint:* Is this really different from (b)?)

(d) How many coalitions in this weighted voting system do not include both P_1 and P_3?

(e) How many coalitions in this weighted voting system include both P_1 and P_3? [*Hint:* Use your answers for (a) and (d).]

42. Consider a weighted voting system with five players (P_1 through P_5).

(a) Find the total number of coalitions in this weighted voting system.

(b) How many coalitions in this weighted voting system do not include P_1? (*Hint:* Think of all the possible coalitions of the remaining players.)

(c) How many coalitions in this weighted voting system do not include P_5? (*Hint:* Is this really different from (b)?)

(d) How many coalitions in this weighted voting system do not include P_1 or P_5?

(e) How many coalitions in this weighted voting system include both P_1 and P_5? [*Hint:* Use your answers of (a) and (d).]

43. Consider a weighted voting system with five players (P_1 through P_5).

(a) Find the number of sequential coalitions in this weighted voting system.

(b) How many sequential coalitions in this weighted voting system have P_1 as the *first* player? (*Hint:* Think of all the possible sequential coalitions of the remaining players.)

(c) How many sequential coalitions in this weighted voting system have P_1 as the *last* player? [*Hint:* Is this really different from (b)?]

(d) How many sequential coalitions in this weighted voting system do not have P_1 as the *first* player? [*Hint:* Use your answers for (a) and (b).]

44. Consider a weighted voting system with six players (P_1 through P_6).

(a) Find the number of sequential coalitions in this weighted voting system.

(b) How many sequential coalitions in this weighted voting system have P_1 as the *first* player? (*Hint:* Think of all the possible sequential coalitions of the remaining players.)

(c) How many sequential coalitions in this weighted voting system have P_1 as the *third* player? [*Hint:* Is this really different from (b)?]

(d) How many sequential coalitions in this weighted voting system do not have P_1 as the *first* player? [*Hint:* Use your answers for (a) and (b).]

45. A law firm has five partners: a senior partner (P_1) with 4 votes and four junior partners (P_2 through P_5) with 1 vote each. The quota is a *simple majority* of the votes. (This law firm operates as the weighted voting system $[5: 4, 1, 1, 1, 1]$.)

(a) In how many sequential coalitions is the senior partner P_1 the pivotal player? (*Hint:* First note that P_1 is the pivotal player in all sequential coalitions except those in which he is the *first* player. Then see Exercise 43.)

(b) Using your answer in (a), find the Shapley-Shubik power index of the senior partner P_1.

(c) Using your answer in (b), find the Shapley-Shubik power distribution in this law firm.

46. A law firm has six partners: a senior partner (P_1) with 5 votes and five junior partners (P_2 through P_6) with 1 vote each. The quota is a *simple majority* of the votes. (This law firm operates as the weighted voting system $[6: 5, 1, 1, 1, 1, 1]$.)

(a) In how many sequential coalitions is the senior partner P_1 the pivotal player? (*Hint:* First note that P_1 is the pivotal player in all sequential coalitions except those in which he is the *first* player. Then see Exercise 44.)

(b) Using your answer in (a), find the Shapley-Shubik power index of the senior partner P_1.

(c) Using your answer in (b), find the Shapley-Shubik power distribution in this law firm.

JOGGING

47. A partnership has four partners (P_1, P_2, P_3 and P_4). In this partnership P_1 has twice as many votes as P_2; P_2 has twice as many votes as P_3; P_3 has twice as many votes as P_4. The quota is a *simple majority* of the votes. Show that P_1 is always a *dictator*. (*Hint:* Write the weighted voting system in the form $[q: 8x, 4x, 2x, x]$, and express q in terms of x. Consider separately the case when x is even and the case when x is odd.)

48. In a weighted voting system with four players the winning coalitions are: $\{P_1, P_2, P_3\}, \{P_1, P_2, P_4\}, \{P_1, P_3, P_4\}$ and $\{P_1, P_2, P_3, P_4\}$.

(a) Find the Banzhaf power distribution of the weighted voting system.

(b) Find the Shapley-Shubik power distribution of the weighted voting system.

49. In a weighted voting system with three players, the six sequential coalitions (each with the pivotal player underlined) are: $\langle P_1, \underline{P_2}, P_3 \rangle$, $\langle P_1, \underline{P_3}, P_2 \rangle$, $\langle P_2, \underline{P_1}, P_3 \rangle$, $\langle P_2, P_3, \underline{P_1} \rangle$, $\langle P_3, \underline{P_1}, P_2 \rangle$, and $\langle P_3, P_2, \underline{P_1} \rangle$. Find the Banzhaf power distribution of the weighted voting system.

50. Four players (P_2, P_3, P_4, and P_5) are seated forming a square as shown in Fig. 2-5, with a fifth player (P_1) seated in the center of the square. The *three-player* winning coalitions consist of P_1 together with two other players as long as the two other players are sitting next to each other. The *four-player* winning coalitions are all the ones that include P_1. The grand coalition is always a winning coalition. Find the Banzhaf power distribution of this weighted voting system.

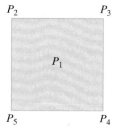

FIGURE 2-5

51. A professional basketball team has four coaches, a head coach (H), and three assistant coaches (A_1, A_2, A_3). Player personnel decisions require at least three yes votes, one of which must be H's.

  (a) If we use $[q: h, a, a, a]$ to describe this weighted voting system, find q, h, and a.

  (b) Find the Shapley-Shubik power distribution of the weighted voting system.

52. Veto power. A player P with weight w is said to have *veto power* if and only if $V - w < q$ (where V denotes the total number of votes in the weighted voting system). Explain why each of the following is true:

  (a) A player has veto power if and only if the player is a member of every winning coalition.

  (b) A player has veto power if and only if the player is a critical player in the *grand* coalition.

53. Dummies. We defined a *dummy* as a player that is never critical. Explain why each of the following is true:

  (a) If P is a dummy, then any winning coalition that contains P would also be a winning coalition without P.

  (b) P is a dummy if and only if the Banzhaf Power Index of P is 0.

  (c) P is a dummy if and only if the Shapley-Shubik Power Index of P is 0.

54. (a) Consider the weighted voting system [22: 10, 10, 10, 10, 1]. Are there any dummies? Explain your answer.

  (b) Without doing any work [but using your answer for (a)], find the Banzhaf and Shapley-Shubik power distributions of this weighted voting system.

  (c) Consider the weighted voting system [q: 10, 10, 10, 10, 1]. Find all the possible values of q for which P_5 is not a dummy.

  (d) Consider the weighted voting system [34: 10, 10, 10, 10, w]. Find all positive integers w which make P_5 a dummy.

55. Consider the weighted voting system $[q: 8, 4, 1]$.

  (a) What are the possible values of q?

  (b) Which values of q result in a dictator? (Who? Why?)

  (c) Which values of q result in exactly one player with veto power? (Who? Why?)

  (d) Which values of q result in more than one player with veto power? (Who? Why?)

  (e) Which values of q result in one or more dummies? (Who? Why?)

56. Consider the weighted voting systems $[9: w, 5, 2, 1]$.

  (a) What are the possible values of w?

  (b) Which values of w result in a dictator? (Who? Why?)

  (c) Which values of w result in a player with veto power? (Who? Why?)

  (d) Which values of w result in one or more dummies? (Who? Why?)

57. (a) Verify that the weighted voting systems [12: 7, 4, 3, 2] and [24: 14, 8, 6, 4] result in exactly the same Banzhaf power distribution. (If you need to make calculations, do them for both systems side by side and look for patterns.)

  (b) Based on your work in (a), explain why the two proportional weighted voting systems $[q: w_1, w_2, \ldots, w_N]$ and $[cq: cw_1, cw_2, \ldots, cw_N]$ always have the same Banzhaf power distribution.

58. (a) Verify that the weighted voting systems [12: 7, 4, 3, 2] and [24: 14, 8, 6, 4] result in exactly the same Shapley-Shubik power distribution. (If you need to make calculations, do them for both systems side by side and look for patterns.)

  (b) Based on your work in (a), explain why the two proportional weighted voting systems $[q: w_1, w_2, \ldots, w_N]$ and $[cq: cw_1, cw_2, \ldots, cw_N]$ always have the same Shapley-Shubik power distribution.

59. A law firm has $N + 1$ partners: the senior partner with N votes, and N junior partners with one vote each. The quota is a simple majority of the votes. Find the Shapley-Shubik power distribution in this weighted voting system. (*Hint:* Try Exercises 45 and 46 first.)

60. Consider the generic weighted voting system $[q: w_1, w_2, \ldots, w_N]$. (Assume $w_1 \geq w_2 \geq \cdots \geq w_N$.)

(a) Find all the possible values of q for which no player has veto power.

(b) Find all the possible values of q for which every player has veto power.

(c) Find all the possible values of q for which P_i has veto power but P_{i+1} does not. (*Hint:* See Exercise 52.)

61. The weighted voting system $[8: 6, 4, 2, 1]$ represents a partnership among four partners (P_1, P_2, P_3 and you!). You are the partner with just one vote, and in this situation you have no power (you dummy!). Not wanting to remain a dummy, you offer to buy one vote. Each of the other four partners is willing to sell you one of their votes, and they are all asking the same price. From which partner should you buy in order to get as much power for your buck as possible? Use the Banzhaf power index for your calculations. Explain your answer.

62. The weighted voting system $[27: 10, 8, 6, 4, 2]$ represents a partnership among five people ($P_1, P_2, P_3, P_4,$ and P_5). You are P_5, the one with two votes. You want to increase your power in the partnership and are prepared to buy one share (one share equals one vote) from any of the other partners. Partners $P_1, P_2,$ and P_3 are each willing to sell cheap ($1000 for one share), but P_4 is not being quite as cooperative—she wants $5000 for one of her shares. Given that you still want to buy one share, from whom should you buy it? Use the Banzhaf power index for your calculations. Explain your answer.

63. The weighted voting system $[18: 10, 8, 6, 4, 2]$ represents a partnership among five people ($P_1, P_2, P_3, P_4,$ and P_5). You are P_5, the one with two votes. You want to increase your power in the partnership and are prepared to buy shares (one share equals one vote) from any of the other partners.

(a) Suppose that each partner is willing to sell one share and that they are all asking the same price. Assuming that you decide to buy only one share, from which partner should you buy? Use the Banzhaf power index for your calculations.

(b) Suppose that each partner is willing to sell two shares and that they are all asking the same price. Assuming that you decide to buy two shares from a single partner, from which partner should you buy? Use the Banzhaf power index for your calculations.

(c) If you have the money and the cost per share is fixed, should you buy one share or two shares (from a single person)? Explain.

64. Mergers. Sometimes in a weighted voting system two or more players decide to merge—that is to say, to combine their votes and always vote the same way. (Note that a merger is different from a coalition—coalitions are temporary, whereas mergers are permanent.) For example, if in the weighted voting system $[7: 5, 3, 1]$ P_2 and P_3 were to merge, the weighted voting system would then become $[7: 5, 4]$. In this exercise we explore the effects of mergers on a player's power.

(a) Consider the weighted voting system $[4: 3, 2, 1]$. In Example 9 we saw that P_2 and P_3 each have a Banzhaf power index of $\frac{1}{5}$. Suppose that P_2 and P_3 merge and become a single player P^*. What is the Banzhaf power index of P^*?

(b) Consider the weighted voting system $[5: 3, 2, 1]$. Find first the Banzhaf power indexes of players P_2 and P_3 and then the Banzhaf power index of P^* (the merger of P_2 and P_3). Compare.

(c) Rework the problem in (b) for the weighted voting system $[6: 3, 2, 1]$.

(d) What are your conclusions from (a), (b), and (c)?

65. Decisive voting systems. A weighted voting system is called **decisive** if for every losing coalition, the coalition consisting of the remaining players (called the *complement*) must be a winning coalition.

(a) Show that the weighted voting system $[5: 4, 3, 2]$ is decisive.

(b) Show that the weighted voting system $[3: 2, 1, 1, 1]$ is decisive.

(c) Explain why any weighted voting system with a dictator is decisive.

(d) Find the number of winning coalitions in a decisive voting system with N players.

66. Equivalent voting systems. Two weighted voting systems are **equivalent** if they have the same number of players and exactly the same winning coalitions.

(a) Show that the weighted voting systems $[8: 5, 3, 2]$ and $[2: 1, 1, 0]$ are equivalent.

(b) Show that the weighted voting systems $[7: 4, 3, 2, 1]$ and $[5: 3, 2, 1, 1]$ are equivalent.

(c) Show that the weighted voting system discussed in Example 2.12 is equivalent to $[3: 1, 1, 1, 1, 1]$.

(d) Explain why equivalent weighted voting systems must have the same Banzhaf power distribution.

(e) Explain why equivalent weighted voting systems must have the same Shapley-Shubik power distribution.

67. The United Nations Security Council. The U.N. Security Council consists of 15 member countries—5 permanent members and 10 nonpermanent members. A motion can pass only if it has the vote of *all* 5 of the permanent members plus at least 4 of the nonpermanent members.

(a) Describe the critical players in a winning coalition. (*Hint*: Consider separately 9-member winning coalitions and winning coalitions with 10 or more members.)

(b) Use your answer in (a), together with the fact that there are 210 coalitions with 9 members and 638 coalitions with 10 or more members to explain why the total number of times all players are critical is 5080.

(c) Using the results of (a) and (b), show that the Banzhaf power index of a permanent member is given by the ratio 848/5040.

(d) Using the results of (a), (b), and (c) show that the Banzhaf power index of a nonpermanent member is given by the ratio 84/5040.

(e) Explain why the U.N. Security Council is equivalent to a weighted voting system in which each nonpermanent member has 1 vote, each permanent member has 7 votes, and the quota is 39 votes.

RUNNING

68. Consider a weighted voting system with N players, P_1 through P_N.

(a) Choose a random player P. In how many sequential coalitions is P in the first place? Second place? Last place?

(b) Choose randomly two players. Call them P and Q. In how many sequential coalitions is P in the first place and Q in the second place? How about P in the last place and Q in the first place? How about P in place j and Q in place k, where j and k are any two distinct numbers between 1 and N?

(c) Choose randomly three players. Call them P, Q, and R. Choose randomly three different numbers between 1 and N. Call them j, k, and m. In how many sequential coalitions is P in the jth place, Q in the kth place, and R in the mth place?

(d) Generalize the results of (a), (b), and (c).

69. Minimal voting systems. A weighted voting system is called **minimal** if there is no equivalent weighted voting system with a smaller quota or with a smaller total number of votes. (For the definition of equivalent weighted voting systems, see Exercise 66.)

(a) Show that the weighted voting system [3: 2, 1, 1] is minimal.

(b) Show that the weighted voting system [4: 2, 2, 1] is not minimal and find an equivalent weighted voting system that is minimal.

(c) Show that the weighted voting system [8: 5, 3, 1] is not minimal and find an equivalent weighted voting system that is minimal.

(d) Given a weighted voting system with N players and a dictator, describe the minimal voting system equivalent to it.

70. The Nassau County Board of Supervisors (1990s). In the mid 1990s, the Nassau County Board operated as the weighted voting system [65: 30, 28, 22, 15, 7, 6] (see Exercise 22). Show that the weighted voting system [15: 7, 6, 5, 4, 2, 1] is

(a) equivalent to [65: 30, 28, 22, 15, 7, 6].

(b) minimal (see Exercise 67). (*Hint*: First show that all six weights must be positive and all must be different. Then examine possible quotas that would give the correct results for the three coalitions $\{P_1, P_2, P_5\}$, $\{P_1, P_2, P_6\}$, and $\{P_2, P_3, P_4\}$. Conclude that the players' weights *cannot* be 6, 5, 4, 3, 2, 1. Use the same coalitions to conclude that $w_3 + w_4 > w_1 + w_6$, and finally that if $w_1 = 7$, then $w_4 = 4$.)

71. The Cleansburg City Council. Find the Banzhaf power distribution in the Cleansburg City Council. (See Example 2.19 for details.)

72. The Fresno City Council. In Fresno, California, the city council consists of seven members (the mayor and six other council members). A motion can be passed by the mayor and at least three other council members, or by at least five of the six ordinary council members.

(a) Describe the Fresno City Council as a weighted voting system.

(b) Find the Shapley-Shubik power distribution for the Fresno City Council. (*Hint*: See Example 2.19 for some useful ideas.)

73. Suppose that in a weighted voting system there is a player A who hates another player P so much that he will always vote the opposite way of P, regardless of the issue. We will call A the **antagonist** of P.

(a) Suppose that in the weighted voting system [8; 5, 4, 3, 2], P is the player with two votes and his antagonist A is the player with five votes. What are the possible coalitions under these circumstances? What is the Banzhaf power distribution under these circumstances?

(b) Suppose that in a generic weighted voting system with N players there is a player P who has an antagonist A. How many coalitions are there under these circumstances?

(c) Give examples of weighted voting systems where a player A can

(i) increase his Banzhaf power index by becoming an antagonist of another player

(ii) decrease his Banzhaf power index by becoming an antagonist of another player

(d) Suppose that the antagonist A has more votes than his enemy P. What is a strategy that P can use to gain power at the expense of A?

74. (a) Give an example of a weighted voting system with four players and such that the Shapley-Shubik power index of P_1 is 3/4.

(b) Show that in any weighted voting system with four players a player cannot have a Shapley-Shubik power index of more than 3/4 unless he or she is a dictator.

(c) Show that in any weighted voting system with N players a player cannot have a Shapley-Shubik power index of more than $(N - 1)/N$ unless he or she is a dictator.

(d) Give an example of a weighted voting system with N players and such that P_1 has a Shapley-Shubik power index of $(N - 1)/N$.

75. (a) Give an example of a weighted voting system with three players and such that the Shapley-Shubik power index of P_3 is 1/6.

(b) Explain why in any weighted voting system with three players any player that is not a dummy must have a Shapley-Shubik power index bigger or equal to 1/6.

(c) Give an example of a weighted voting system with four players and such that the Shapley-Shubik power index of P_4 is 1/12.

(d) Explain why in any weighted voting system with four players any player that is not a dummy must have a Shapley-Shubik power index bigger than or equal to 1/12.

76. (a) Explain why in any weighted voting system with N players a player with veto power must have a Banzhaf power index bigger than or equal to $1/N$.

(b) Explain why in any weighted voting system with N players a player with veto power must have a Shapley-Shubik power index bigger than or equal to $1/N$.

77. A law firm has $N + 1$ partners: the senior partner with N votes, and N junior partners with one vote each. The quota is simple majority of the votes (see Exercise 59). Find the Banzhaf power distribution in this weighted voting system.

78. An alternative way to compute Banzhaf power. As we know, the first step in computing the Banzhaf power index of player P is to compute B, the total number of times P is a critical player. The following formula gives an alternative way to compute B: $B = 2W_P - W$, where W is the number of winning coalitions and W_P is the number of winning coalitions containing P. Explain why the preceding formula is true.

PROJECTS AND PAPERS

A The Johnston Power Index

The Banzhaf and Shapley-Shubik power indexes are not the only two mathematical methods for measuring power. The Johnston power index is a subtle but rarely used variation of the Banzhaf power index in which the power of a player is based not only on how often he or she is critical in a coalition, but also on the number of other players in the coalition. Specifically, being a critical player in a coalition of 2 players contributes 1/2 toward your power score; being critical in a coalition of 3 players contributes 1/3 toward your power score; and being critical in a coalition of 10 contributes only 1/10 toward your power score.

A player's *Johnston power score* is obtained by adding all such fractions over all coalitions in which the player is critical. The player's *Johnston power index* is his or her Johnston power score divided by the sum of all players' power scores. (A more detailed description of how to compute the Johnston power index is given in reference 17.)

Prepare a presentation on the Johnston power index. Include a mathematical description of the procedure for computing Johnston power, give examples, and compare the results with the ones obtained using the Banzhaf method. Include your own personal analysis on the merits of the Johnston method compared with the Banzhaf method.

B The Past, Present, and Future of the Electoral College

Starting with the Constitutional Convention of 1776 and ending with the Bush-Gore presidential election of 2000, give a historical and political analysis of the Electoral College. You should address some or all of the following issues: How did the Electoral College get started? Why did some of the Founding Fathers want it? How did it evolve? What has been its impact over the years in affecting presidential elections? (Pay particular attention to the 2000 presidential election.) What does the future hold for the Electoral College? What are the prospects that it will be reformed or eliminated?

C Mathematical Arguments in Favor of the Electoral College

As a method for electing the president, the Electoral College is widely criticized as being undemocratic. At the same time, different arguments have been made over the years to support the case that the Electoral College is not nearly as bad as it seems. Massachusetts Institute of Technology physicist Alan Natapoff has recently used mathematical ideas (many of which are connected to the material in this chapter) to make the claim that the Electoral College is a better system than a direct presidential election. Summarize and analyze Natapoff's mathematical arguments in support of the Electoral College. (Natapoff's arguments are nicely described in reference 6.)

D Banzhaf Power and the Law

John Banzhaf was a lawyer, and he made his original arguments on behalf of his mathematical method to measure power in court cases, most of which involved the Nassau County Board of Supervisors in New York State. (See the discussion in Section 2.3.) Among the more significant court cases were *Graham v. Board of Supervisors* (1966); *Bechtle v. Board of Supervisors* (1981); *League of Women Voters v. Board of Supervisors* (1983); and *Jackson v. Board of Supervisors* (1991). Other important legal cases based on the Banzhaf method for measuring power but not involving Nassau County were *Ianucci v. Board of Supervisors of Washington County* and *Morris v. Board of Estimate* (U.S. Supreme Court, 1989). Choose one or two of these cases, read their background, arguments, and the court's decision, and write a brief for each. This is a good project for pre-law and political science majors, but it might require access to a good law library.

REFERENCES AND FURTHER READINGS

1. Banzhaf, John F., III, "One Man, 3.312 Votes: A Mathematical Analysis of the Electoral College," *Villanova Law Review*, 13 (1968), 304–332.
2. Banzhaf, John F., III, "Weighted Voting Doesn't Work," *Rutgers Law Review*, 19 (1965), 317–343.
3. Brams, Steven J., *Game Theory and Politics*. New York: Free Press, 1975, chap. 5.
4. Felsenthal, Dan, and Moshe Machover, *The Measurement of Voting Power: Theory and Practice, Problems and Paradoxes*. Cheltenham, England: Edward Elgar, 1998.
5. Grofman, B., "Fair Apportionment and the Banzhaf Index," *American Mathematical Monthly*, 88 (1981), 1–5.
6. Hively, Will, "Math Against Tyranny," *Discover*, November 1996, 74–85.
7. Imrie, Robert W., "The Impact of the Weighted Vote on Representation in Municipal Governing Bodies of New York State," *Annals of the New York Academy of Sciences*, 219 (November 1973), 192–199.
8. Lambert, John P., "Voting Games, Power Indices and Presidential Elections," *UMAP Journal*, 3 (1988), 213–267.
9. Merrill, Samuel, "Approximations to the Banzhaf Index of Voting Power," *American Mathematical Monthly*, 89 (1982), 108–110.
10. Meyerson, Michael I., *Political Numeracy: Mathematical Perspectives on Our Chaotic Constitution*, New York: W. W. Norton, 2002, chap. 2.
11. Riker, William H., and Peter G. Ordeshook, *An Introduction to Positive Political Theory*. Englewood Cliffs, NJ: Prentice-Hall, Inc., 1973, chap. 6.
12. Shapley, Lloyd, and Martin Shubik, "A Method for Evaluating the Distribution of Power in a Committee System," *American Political Science Review*, 48 (1954), 787–792.
13. Sickels, Robert J., "The Power Index and the Electoral College: A Challenge to Banzhaf's Analysis," *Villanova Law Review*, 14 (1968), 92–96.
14. Straffin, Philip D., Jr., "The Power of Voting Blocs: An Example," *Mathematics Magazine*, 50 (1977), 22–24.
15. Straffin, Philip D., Jr., *Topics in the Theory of Voting, UMAP Expository Monograph*. Boston: Birkhäuser, 1980, chap. 1.
16. Tannenbaum, Peter, "Power in Weighted Voting Systems," *The Mathematica Journal*, 7 (1997), 58–63.
17. Taylor, Alan, *Mathematics and Politics: Strategy, Voting, Power and Proof*. New York: Springer-Verlag, 1995, chaps. 4 and 9.

3 The Mathematics of Sharing

Fair-Division Games

If you are a lion, dividing a meal with friends is easy — if you don't like the division, just make some of the friends part of the meal. What are good friends for, anyway? Of course, most of us are not lions, and when we divide food with friends we make an effort to do it so that everyone gets something like a fair share. Fairness is an innate human value, based on empathy, the instinct for cooperation, and, to some extent, self-interest.

But talking fairness is one thing. Accomplishing it is something else. Sharing things fairly is not always easy, and history repeatedly shows that we humans can play the sharing game very poorly. At the root of most wars is a failure to share something of value (land, resources, rights, etc.) in a way that all parties involved might find acceptable. By the same token, if we set our minds to it, we can share, and share really well. A theory for sharing things fairly based on cooperation, reason, and logic is one of the great achievements of social science, and, once again, we can trace the roots of this achievement to simple mathematics.

The Lion, the Fox, and the Ass

The Lion, the Fox, and the Ass entered into an agreement to assist each other in the hunt. Having secured a large booty, the Lion on their return from the forest asked the Ass to allot its due portion to each of the three partners in the treaty.

The Ass carefully divided the spoil into three equal shares and modestly requested the two others to make the first choice.

The Lion, bursting out into a great rage, devoured the Ass. Then he requested the Fox to do him the favor to make the division. The Fox accumulated all that they had killed into one large heap and left to himself the smallest possible morsel.

The Lion said, "Who has taught you, my very excellent fellow, the art of division? You are perfect to a fraction," to which the Fox replied, "I learned it from the Ass, by witnessing his fate."

—Aesop

We take our first mathematical stab at *fair division* somewhere around the third or fourth grade. A typical problem goes like this: There are 20 pieces of candy to be divided among four equally deserving children. What is a fair solution? We know, of course, the standard answer: Give each child five pieces. The problem with this answer is that it may not produce a fair division after all. What if the pieces of candy are not all the same? Say, for example, that the 20 pieces are made up of a wide variety of goodies—Snickers, Hershey's, mints, bubblegum, and so forth—with some pieces clearly more desirable than others and with each child having a different set of preferences. Can we take these diverse opinions into account and still divide the pieces fairly? And by the way, what does *fairly* mean in this situation? These are some of the many thought-provoking questions we will discuss in this chapter.

Why are these questions important, you may wonder. After all, it's only candy and kids. Not exactly. Just like the booty in Aesop's fable, the candy is a metaphor—we could just as well be dividing family heirlooms, land, works of art, and other items of real value, as is often the case in an inheritance or a divorce proceeding. And, on an even grander scale, questions of fair division may involve issues of global significance like dividing an entire country (consider the division of the former Yugoslavia in the 1990s) or dividing the rights to mine the ocean's resources (consider the 1982 Convention of the Law of the Sea) or even dividing the global responsibilities for reducing greenhouse gas emissions (consider the 1997 Kyoto treaty).

In its generic form the problem of *fair division* can be stated in reasonably simple terms: How can something that must be shared by a set of competing parties be divided among them in a way that ensures that each party receives a fair share? It could be argued that a good general answer to this question would go a long way in solving most of the problems of humankind, but unfortunately, good general answers are not possible. Under the right set of circumstances, however, it is possible to solve special types of fair-division problems using basic mathematical ideas. We will explore a few of these ideas in this chapter.

The chapter is organized in a manner very similar to that of Chapters 1 and 2. We will start with an introduction to the concept of a *fair-division game* (Section 3.1), followed by a description and an analysis of several specific *fair-division methods* that have been developed over the years (Sections 3.2 through 3.7). In the conclusion we recap the main ideas of the chapter and take a brief look at the "big picture"—what is fair division all about and why should we really care?

3.1   Fair-Division Games

In this section we will introduce some of the basic concepts and terminology of fair division. Much as we did in Chapter 2 when we studied weighted voting systems, we will think of fair-division problems in terms of games—with players, goals, rules, and strategies.

Basic Elements of a Fair-Division Game

The underlying elements of every *fair-division game* are as follows:

- **The goods (or "booty").** This is the informal name we will give to the item or items being divided. Typically, these items are tangible physical objects with a positive value, such as candy, cake, pizza, jewelry, art, land, and so on. In some situations the items being divided may be intangible things such as rights—water rights, drilling rights, broadcast licenses, and so on—or things with a negative value such as chores, liabilities, obligations, and so on. Regardless of the nature of the items being divided, we will use the symbol S throughout this chapter to denote the booty.

  ■ For more on the fair division of negative items, see Exercises 78 and 79.

- **The players.** In every fair-division game there is a set of parties with the right (or in some cases the duty) to share S. They are the *players* in the game. Most of the time the players in a fair-division game are individuals, but it is worth noting that some of the most significant applications of fair division occur when the players are *institutions* (ethnic groups, political parties, states, and even nations).

- **The value systems.** The fundamental assumption we will make is that each player has an internalized *value system* that gives the player the ability to quantify the value of the booty or any of its parts. Specifically, this means that each player can look at the set S or any subset of S and assign to it a value—either in absolute terms ("to me, that's worth \$147.50") or in relative terms ("to me, that piece is worth 30% of the total value of S").

Basic Assumptions

Like most games, fair-division games are predicated on certain assumptions about the players. For the purposes of our discussion, we will make the following four assumptions:

- **Rationality.** Each of the players is a thinking, rational entity seeking to maximize his or her share of the booty S. We will further assume that in the pursuit of this goal, a player's moves are based on reason alone (we are taking emotion, psychology, mind games, and all other nonrational elements out of the picture.)

- **Cooperation.** The players are willing participants and accept the rules of the game as binding. The rules are such that after a *finite* number of moves by the players, the game terminates with a division of S. (There are no outsiders such as judges or referees involved in these games—just the players and the rules.)

- **Privacy.** Players have *no* useful information on the other players' value systems and thus of what kinds of moves they are going to make in the game. (This assumption does not always hold in real life, especially if the players are siblings or friends.)

- **Symmetry.** Players have *equal* rights in sharing the set S. A consequence of this assumption is that, at a minimum, each player is entitled to a *proportional* share of S—when there are two players, each is entitled to at least one-half of S, with three players each is entitled to at least one-third of S, and so on.

■ Fair-division methods can also be used in situations in which different players are entitled by right to different-sized shares of the booty. For more details on *asymmetric fair division*, see Exercise 83 and Project B.

Fair Shares and Fair-Division Methods

Given the booty S and players $P_1, P_2, P_3, \ldots, P_N$, each with his or her own value system, the ultimate goal of the game is to end up with a **fair division** of S, that is, to divide S into N shares and *assign shares to players in such a way that each player gets a fair share*. This leads us to what is perhaps the most important definition of this section.

> ┌─**FAIR SHARE**───────────────────────────────────
>
> Suppose that s denotes a share of the booty S and that P is one of the players in a fair-division game with N players. We will say that s is a **fair share** to player P if s is worth *at least* 1/Nth of the total value of S *in the opinion of* P. (Such a share is often called a *proportional fair share*, but for simplicity we will refer to it just as a *fair share*.)

■ A fair division in which each player gets a share that he or she considers to be the *best* share is called an *envy-free fair division*. For more on envy-free fair division, see Project A.

For example, suppose that we have four players and that S has been divided into four shares $s_1, s_2, s_3,$ and s_4. Paul (one of the four players) values s_1 at 12% of S, s_2 at 27% of S, s_3 at 18% of S, and s_4 at 43% of S. In this situation s_2 *and* s_4 are both fair shares to Paul, since the threshold for a fair share when dividing among four players is 25% of S. Note that a share can be a fair share without being the most valuable share—clearly Paul would be most happy if he got s_4, but he may very well have to settle for s_2.

In the coming sections of this chapter we will discuss several different **fair-division methods** (also known as fair-division *protocols* or fair-division *schemes*). Essentially, we can think of a *fair-division method* as the *set of rules* that define how the game is to be played. Thus, in a fair-division game we must consider not only the booty S and the players $P_1, P_2, P_3, \ldots, P_N$ (each with his or her own opinions about how S should be divided), but also a specific method by which we plan to accomplish the fair division.

There are many different fair-division methods known, but in this chapter we will only discuss a few of the classic ones. Depending on the nature of the set S, a fair-division game can be classified as one of three types: *continuous*, *discrete*, or *mixed*, and the fair-division methods used depend on which of these types we are facing.

In a **continuous** fair-division game the set S is divisible in infinitely many ways, and shares can be increased or decreased by arbitrarily small amounts. Typical examples of continuous fair-division games involve the division of land, a cake, a pizza, and so forth.

A fair-division game is **discrete** when the set S is made up of objects that are indivisible like paintings, houses, cars, boats, jewelry, and so on. (What about pieces of candy? One might argue that with a sharp enough knife a piece of candy could be chopped up into smaller and smaller pieces, but nobody really does that—it's messy. As a semantic convenience let's agree that candy is indivisible, and therefore dividing candy is a discrete fair-division game.)

A **mixed** fair-division game is one in which some of the components are continuous and some are discrete. Dividing an estate consisting of jewelry, a house, and a parcel of land is a mixed fair-division game.

Fair-division methods are classified according to the nature of the problem involved. Thus, there are *discrete fair-division* methods (used when the set S is made up of indivisible, discrete objects), and there are *continuous fair-division* methods (used when the set S is an infinitely divisible, continuous set). Mixed fair-division games can usually be solved by dividing the continuous and discrete parts separately, so we will not discuss them in this chapter.

3.2   Two Players: The Divider-Chooser Method

The **divider-chooser method** is undoubtedly the best known of all continuous fair-division methods. This method can be used anytime the fair-division game involves two players and a *continuous* set S (a cake, a pizza, a plot of land, etc.). Most of us have unwittingly used it at some time or another, and informally it is best known as the *you cut—I choose method*. As this name suggests, one player, called the *divider*, divides S into two shares, and the second player, called the *chooser*, picks the share he or she wants, leaving the other share to the divider.

Under the *rationality* and *privacy* assumptions we introduced in the previous section, this method guarantees that both divider and chooser will get a fair share (with two players, this means a share worth 50% or more of the total value of S). Not knowing the chooser's likes and dislikes (privacy assumption), the divider can only guarantee himself a 50% share by dividing S into two halves of equal value (rationality assumption); the chooser is guaranteed a 50% or better share by choosing the piece he or she likes best.

EXAMPLE 3.1   **Damian and Cleo Divide a Cheesecake**

On their first date, Damian and Cleo go to the county fair. They buy jointly a raffle ticket, and as luck would have it, they win a half chocolate–half strawberry cheesecake.

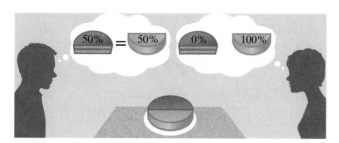
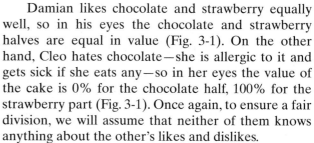

FIGURE 3-1

Damian likes chocolate and strawberry equally well, so in his eyes the chocolate and strawberry halves are equal in value (Fig. 3-1). On the other hand, Cleo hates chocolate—she is allergic to it and gets sick if she eats any—so in her eyes the value of the cake is 0% for the chocolate half, 100% for the strawberry part (Fig. 3-1). Once again, to ensure a fair division, we will assume that neither of them knows anything about the other's likes and dislikes.

Let's now see how Damian and Cleo might divide this cake using the divider-chooser method. Damian volunteers to go first and be the divider. His cut is shown in Fig. 3-2(a). Notice that this is a perfectly rational division of the cake based on Damian's value system—each piece is half of the cake and to him worth one-half of the total value of the cake. It is now Cleo's turn to choose, and her choice is obvious—she will pick the piece having the larger strawberry part [Fig. 3-2(b)].

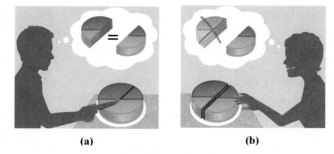

FIGURE 3-2 (a) Damian cuts. (b) Cleo picks.

The final outcome of this division is that Damian gets a piece that in his own eyes is worth exactly half of the cake, but Cleo ends up with a much sweeter deal—a piece that in her own eyes is worth about two-thirds of the cake. This is, nonetheless, a fair division of the cake—both players get pieces worth 50% or more. Mission accomplished!

Example 3.1 illustrates why, given a choice, it is always *better to be the chooser than the divider*—the divider is guaranteed a share worth exactly 50% of the total value of *S*, but with just a little luck the chooser can end up with a share worth more than 50%. Since a fair-division method should treat all players equally, both players should have an equal chance of being the chooser. This is best done by means of a coin toss, with the winner of the coin toss getting the privilege of making the choice.

The idea behind the divider-chooser method goes all the way back to the Old Testament. [When Lot and Abraham argued over grazing rights, Abraham proposed, "Let us divide the land into left and right. If you go left, I will go right; and if you go right, I will go left" (Genesis 13:1–9).] But how do we implement the same idea when the division is among three, or four, or *N* players? Surprisingly, going from two to three or more players is not simple, and generalizations of the divider-chooser method did not come about until the 1940s, primarily as a result of the pioneering work of one man, Polish mathematician Hugo Steinhaus.

In the next few sections we will discuss three methods that can be used to extend the divider-chooser method to *continuous* fair-division games involving *three or more* players.

Hugo Steinhaus (1887–1972). A brief biographical profile of Steinhaus appears at the end of the chapter.

3.3   The Lone-Divider Method

The first important breakthrough in the mathematics of fair division came in 1943, when Steinhaus came up with a clever way to extend some of the ideas in the divider-chooser method to the case of *three* players, one of whom plays the role of the *divider* and the other two who play the role of *choosers*. Steinhaus's approach was subsequently generalized to any number of players N (one divider and $N - 1$ choosers) by Princeton mathematician Harold Kuhn. In either case we will refer to this method as the **lone-divider method**.

We start this section with a description of Steinhaus's *lone-divider method* for the case of $N = 3$ players. We will describe the process in terms of dividing a cake, a commonly used and convenient metaphor for any continuous set S.

The Lone-Divider Method for Three Players

■ **Preliminaries.** One of the three players will be the divider; the other two players will be choosers. Since it is better to be a chooser than a divider, the decision of who is what is made by a random draw (rolling dice, drawing cards from a deck, etc.). We'll call the divider D and the choosers C_1 and C_2.

■ **Step 1 (Division).** The divider D divides the cake into three shares (s_1, s_2, and s_3). D will get one of these shares, but at this point does not know which one. (Not knowing which share will be his is critical—it forces D to divide the cake into three shares of equal value.)

■ **Step 2 (Bidding).** C_1 declares (usually by writing on a slip of paper) which of the three pieces are fair shares to her. Independently, C_2 does the same. These are the *bids*. A chooser's bid *must list every single piece that he or she considers to be a fair share* (i.e., worth one-third or more of the cake)—it may be tempting to bid only for the very best piece, but this is a strategy that can easily backfire (see Exercise 77). To preserve the privacy requirement, it is important that the bids be made independently, without the choosers being privy to each other's bids.

■ **Step 3 (Distribution).** Who gets which piece? The answer, of course, depends on which pieces are listed in the bids. For convenience, we will separate the pieces into two types: *C-pieces* (these are pieces *chosen* by either one or both of the choosers) and *U-pieces* (these are *unwanted* pieces that did not appear in either of the bids). Expressed in terms of value, a *U-piece* is a piece that *both* choosers value at less than $33\frac{1}{3}\%$ of the cake, and a *C-piece* is a piece that *at least* one of the choosers (maybe both) value at $33\frac{1}{3}\%$ or more. Depending on the number of *C-pieces*, there are two separate cases to consider.

Case 1. When there are two or more *C-pieces*, there is always a way to give each chooser a different piece from among the pieces listed in her bid. (The details will be covered in Examples 3.2 and 3.3.) Once each chooser gets her piece, the divider gets the last remaining piece. At this point every player has received a fair share, and a fair division has been accomplished. (Sometimes we might end up in a situation in which C_1 likes C_2's piece better than her own and vice versa. In that case it is perfectly reasonable to add a final, informal step and let them swap pieces—this would make each of them happier than they already were, and who could be against that?)

Case 2. When there is only one C-piece, we have a bit of a problem because it means that both choosers are bidding for the very same piece [Fig. 3-3(a)]. The solution here requires a little more creativity. First, we take care of the divider D—to whom all pieces are equal in value—by giving him one of the pieces that neither chooser wants [Fig. 3-3(b).] (If the two choosers can agree on the least desirable piece, then so much the better; if they can't agree, then the choice of which piece to give the divider can be made randomly.)

After D gets his piece, the two pieces left (the C-piece and the remaining U-piece) are recombined into one piece that we call the B-piece [Fig. 3-3(c)]. Now we have one piece and two players and we can revert to the *divider-chooser method* to finish the fair division: one player cuts the B-piece into two pieces; the other player chooses the piece she likes better [Fig. 3-3(d)].

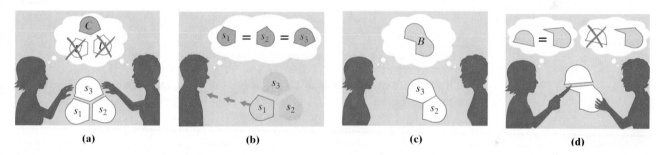

(a)   **(b)**   **(c)**   **(d)**

FIGURE 3-3 Case 2 in the lone-divider method (three players). (a) Both choosers covet the same piece. (b) The divider walks away with one of the U-pieces. (c) The two remaining pieces are recombined into the B-piece. (d) The B-piece is divided by the remaining two players using the divider-chooser method.

This process results in a fair division of the cake because it guarantees fair shares for all players. We know that D ends up with a fair share by the very fact that D did the original division, but what about C_1 and C_2? The key observation is that in the eyes of both C_1 and C_2 the B-piece is worth *more than two-thirds of the value of the original cake* (think of the B-piece as 100% of the original cake minus a U-piece worth less than $33\frac{1}{3}\%$), so when we divide *it* fairly into *two* shares, each party is guaranteed *more than one-third* of the original cake. We will come back to this point in Example 3.4.

We will now illustrate the details of Steinhaus's lone-divider method for three players with several examples. In all these examples we will assume that the divider has already divided the cake into three shares s_1, s_2, and s_3. In each example the values that each of the three players assigns to the shares, expressed as percentages of the total value of the cake, are shown in the form of a table. The reader should remember, however, that this information is never available in full to the players—an individual player only knows the percentages on his or her row.

EXAMPLE 3.2   **Lone Divider with 3 Players (Case 1, Version 1)**

Dale, Cindy, and Cher are dividing a cake using Steinhaus's lone-divider method. They draw cards from a well-shuffled deck of cards, and Dale draws the low card (bad luck!) and has to be the divider.

Step 1 (Division). Dale divides the cake into three pieces s_1, s_2, and s_3. Table 3-1 shows the values of the three pieces in the eyes of each of the players.

Step 2 (Bidding). From Table 3-1 we can assume that Cindy's bid list is $\{s_1, s_3\}$ and Cher's bid list is also $\{s_1, s_3\}$.

Step 3 (Distribution). The C-pieces are s_1 and s_3. There are two possible distributions. One distribution would be: Cindy gets s_1, Cher gets s_3, and Dale gets s_2. An even better distribution (the *optimal* distribution) would be: Cindy gets s_3, Cher gets s_1, and Dale gets s_2. In the case of the first distribution, both Cindy and Cher would benefit by swapping pieces, and there is no rational reason why they would not do so. Thus, using the rationality assumption, we can conclude that in either case the final result will be the same: Cindy gets s_3, Cher gets s_1, and Dale gets s_2.

TABLE 3-1

	s_1	s_2	s_3
Dale	$33\frac{1}{3}\%$	$33\frac{1}{3}\%$	$33\frac{1}{3}\%$
Cindy	35%	10%	55%
Cher	40%	25%	35%

EXAMPLE 3.3   **Lone Divider with 3 Players (Case 1, Version 2)**

We'll use the same setup as in Example 3.2—Dale is the divider, Cindy and Cher are the choosers.

Step 1 (Division). Dale divides the cake into three pieces s_1, s_2, and s_3. Table 3-2 shows the values of the three pieces in the eyes of each of the players.

Step 2 (Bidding). Here Cindy's bid list is $\{s_2\}$ only, and Cher's bid list is $\{s_1\}$ only.

Step 3 (Distribution). This is the simplest of all situations, as there is only one possible distribution of the pieces: Cindy gets s_2, Cher gets s_1, and Dale gets s_3.

TABLE 3-2

	s_1	s_2	s_3
Dale	$33\frac{1}{3}\%$	$33\frac{1}{3}\%$	$33\frac{1}{3}\%$
Cindy	30%	40%	30%
Cher	60%	15%	25%

EXAMPLE 3.4   **Lone Divider with 3 Players (Case 2)**

The gang of Examples 3.2 and 3.3 is back at it again.

Step 1 (Division). Dale divides the cake into three pieces s_1, s_2, and s_3. Table 3-3 shows the values of the three pieces in the eyes of each of the players.

Step 2 (Bidding). Here Cindy's and Cher's bid lists consist of just $\{s_3\}$.

Step 3 (Distribution). The only C-piece is s_3. Cindy and Cher talk it over, and without giving away any other information agree that of the two U-pieces, s_1 is the least desirable, so they all agree that Dale gets s_1. (Dale doesn't care which of the three pieces he gets, so he has no rational objection.) The remaining pieces (s_2 and s_3) are then recombined to form the B-piece, to be divided between Cindy and Cher using the divider-chooser method (one of them divides the B-piece into two shares, the other one chooses the share she likes better). Re-

TABLE 3-3

	s_1	s_2	s_3
Dale	$33\frac{1}{3}\%$	$33\frac{1}{3}\%$	$33\frac{1}{3}\%$
Cindy	20%	30%	50%
Cher	10%	20%	70%

gardless of how this plays out, both of them will get a very healthy share of the cake: Cindy will end up with a piece worth at least 40% of the original cake (the B-piece is worth 80% of the original cake to Cindy), and Cher will end up with a piece worth at least 45% of the original cake (the B-piece is worth 90% of the original cake to Cher).

The Lone-Divider Method for More Than Three Players

In 1967 Harold Kuhn, a mathematician at Princeton University, was able to extend Steinhaus's lone-divider method to any number of players $N > 3$. The first two steps of Kuhn's method are a straightforward generalization of Steinhaus's lone-divider method for three players, but the distribution step requires some fairly sophisticated mathematical ideas and is rather difficult to describe in full generality, so we will only give an outline here and will illustrate the details with a couple of examples for $N = 4$ players. (For a complete description of Kuhn's lone-divider method, see reference 12.)

- **Preliminaries.** One of the players is chosen to be the divider D, and the remaining $N - 1$ players are all going to be choosers. As always, it's better to be a chooser than a divider, so the decision should be made by a random draw.
- **Step 1 (Division).** The divider D divides the set S into N shares $s_1, s_2, s_3, \ldots, s_N$. D is guaranteed of getting one of these shares, but doesn't know which one.
- **Step 2 (Bidding).** Each of the $N - 1$ choosers independently submits a bid list consisting of *every share that he or she considers to be a fair share* (i.e., worth $1/N$th or more of the booty S).
- **Step 3 (Distribution).** The bid lists are opened. Much as we did with three players, we will have to consider two separate cases, depending on how these bid lists turn out.

  Case 1. If there is a way to assign a different share to each of the $N - 1$ choosers, then that should be done. (Needless to say, the share assigned to a chooser should be from his or her bid list.) The divider, to whom all shares are presumed to be of equal value, gets the last unassigned share. At the end, players may choose to swap pieces if they want.

  Case 2. There is a *standoff*—in other words, there are two choosers both bidding for just one share, or three choosers bidding for just two shares, or K choosers bidding for less than K shares. This is a much more complicated case, and what follows is a rough sketch of what to do. To resolve a standoff, we first set aside the shares involved in the standoff from the remaining shares. Likewise, the players involved in the standoff are temporarily separated from the rest. Each of the remaining players (including the divider) can be assigned a fair share from among the remaining shares and sent packing. All the shares left are recombined into a new booty S to be divided among the players involved in the standoff, and the process starts all over again.

The following two examples will illustrate some of the ideas behind the lone-divider method in the case of $N = 4$ players. The first example is one without a standoff; the second example involves a standoff.

EXAMPLE 3.5   **Lone Divider with 4 Players (Case 1)**

We have one divider, Demi, and three choosers, Chan, Chloe, and Chris.

Step 1 (Division). Demi divides the cake into four shares s_1, s_2, s_3, and s_4. Table 3-4 shows how each of the players values each of the four shares. Remember that the information on each row of Table 3-4 is private and known only to that player.

Step 2 (Bidding). Chan's bid list is $\{s_1, s_3\}$; Chloe's bid list is $\{s_3\}$ only; Chris's bid list is $\{s_1, s_4\}$.

Step 3 (Distribution). The bid lists are opened. It is clear that for starters Chloe must get s_3 — there is no other option. This forces the rest of the distribution: s_1 must then go to Chan, and s_4 goes to Chris. Finally, we give the last remaining piece, s_2, to Demi.

This distribution results in a fair division of the cake, although it is not entirely "envy-free" — Chan wishes he had Chloe's piece (35% is better than 30%) but Chloe is not about to trade pieces with him, so he is stuck with s_1. (From a strictly rational point of view, Chan has no reason to gripe — he did not get the best piece, but got a piece worth 30% of the total, better than the 25% he is entitled to.)

TABLE 3-4

	s_1	s_2	s_3	s_4
Demi	25%	25%	25%	25%
Chan	30%	20%	35%	15%
Chloe	20%	20%	40%	20%
Chris	25%	20%	20%	35%

EXAMPLE 3.6   **Lone Divider with 4 Players (Case 2)**

Once again, we will let Demi be the divider and Chan, Chloe, and Chris be the three choosers (same players, different game).

Step 1 (Division). Demi divides the cake into four shares s_1, s_2, s_3, and s_4. Table 3-5 shows how each of the players values each of the four shares.

Step 2 (Bidding). Chan's bid list is $\{s_4\}$; Chloe's bid list is $\{s_2, s_3\}$; Chris's bid list is $\{s_4\}$.

Step 3 (Distribution). The bid lists are opened, and the players can see that there is a standoff brewing on the horizon — Chan and Chris are both bidding for s_4. The first step is to set s_4 aside and assign Chloe and Demi a fair share from s_1, s_2, and s_3. Chloe could be given either s_2 or s_3. (She would rather have s_2, of course, but it's not for her to decide.) A coin toss is used to determine which one. Let's say Chloe ends up with s_3 (bad luck!). Demi could be now given either s_1 or s_2. Another coin toss, and Demi ends up with s_1. The final move is ... you guessed it! — recombine s_2 and s_4 into a single piece to be divided between Chan and Chris using the divider-chooser method. Since $(s_2 + s_4)$ is worth 60% to Chan and 58% to Chris (you can check it out in Table 3-5), regardless of how this final division plays out they are both guaranteed a final share worth more than 25% of the cake.

Mission accomplished! We have produced a fair division of the cake.

TABLE 3-5

	s_1	s_2	s_3	s_4
Demi	25%	25%	25%	25%
Chan	20%	20%	20%	40%
Chloe	15%	35%	30%	20%
Chris	22%	23%	20%	35%

3.4 The Lone-Chooser Method

A completely different approach for extending the divider-chooser method was proposed in 1964 by A. M. Fink, a mathematician at Iowa State University. In this method one player plays the role of chooser, all the other players start out playing the role of dividers. For this reason, the method is known as the **lone-chooser method**. Once again, we will start with a description of the method for the case of three players.

The Lone-Chooser Method for Three Players

- **Preliminaries.** We have one chooser and two dividers. Let's call the chooser C and the dividers D_1 and D_2. As usual, we decide who is what by a random draw.
- **Step 1 (Division).** D_1 and D_2 divide S [Fig. 3-4(a)] between themselves into *two* fair shares. To do this, they use the divider-chooser method. Let's say that D_1 ends up with s_1 and D_2 ends up with s_2 [Fig. 3-4(b)].
- **Step 2 (Subdivision).** Each divider divides his share into three subshares. Thus, D_1 divides s_1 into three subshares, which we will call s_{1a}, s_{1b}, and s_{1c}. Likewise, D_2 divides s_2 into three subshares, which we will call s_{2a}, s_{2b}, and s_{2c} [Fig. 3-4(c)].
- **Step 3 (Selection).** The chooser C now selects one of D_1's three subshares and one of D_2's three subshares (whichever she likes best). These two subshares make up C's final share. D_1 then keeps the remaining two subshares from s_1, and D_2 keeps the remaining two subshares from s_2 [Fig. 3-4(d)].

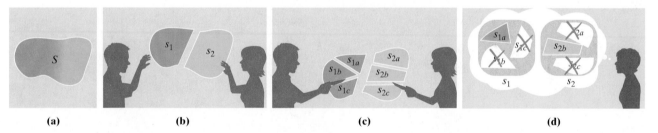

FIGURE 3-4 (a) The original S, (b) first division, (c) second division, and (d) selection.

Why is this a fair division of S? D_1 ends up with two-thirds of s_1. To D_1, s_1 is worth at least one-half of the total value of S, so two-thirds of s_1 is at least one-third—a fair share. The same argument applies to D_2. What about the chooser's share? We don't know what s_1 and s_2 are each worth to C, but it really doesn't matter—a one-third or better share of s_1 plus a one-third or better share of s_2 equals a one-third or better share of $(s_1 + s_2)$ and thus a fair share of the cake.

The following example illustrates in detail how the lone-chooser method works with three players.

EXAMPLE 3.7 **Lone Chooser with 3 Players**

David, Dinah, and Cher are dividing the orange-pineapple cake shown in Fig. 3-5 using the lone-chooser method. The cake is valued by each of them at $27, so each of them expects to end up with a share worth at least $9.

Their individual value systems (not known to one another, but available to us as outside observers) are as follows:

- David likes pineapple and orange the same. [In his eyes the cake looks like Fig. 3-5(a).]
- Dinah likes orange but hates pineapple. [In her eyes the cake looks like Fig. 3-5(b).]
- Cher likes pineapple twice as much as she likes orange. [In her eyes the cake looks like Fig. 3-5(c).]

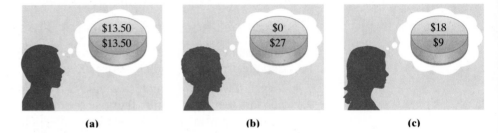

FIGURE 3-5 The values of the pineapple and orange parts in the eyes of each player.

(a)   (b)   (c)

After a random selection, Cher gets to be the chooser and thus gets to sit out Steps 1 and 2.

Step 1 (Division). David and Dinah start by dividing the cake between themselves using the divider-chooser method. After a coin flip, David cuts the cake into two pieces, as shown in Fig. 3-6(a). Since Dinah doesn't like pineapple, she will take the share with the most orange [Fig. 3-6(c)].

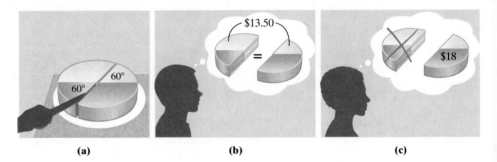

FIGURE 3-6 The first cut and the values of the shares in the eyes of each player.

(a)   (b)   (c)

Step 2 (Subdivision). David divides his share into three subshares that in his opinion are of equal value. Notice that the subshares [Fig. 3-7(a)] are all the same size. Dinah also divides her share into three smaller subshares that in her opinion are of equal value [Fig. 3-7(b)]. (Remember that Dinah hates pineapple. Thus, she has made her cuts in such a way as to have one-third of the orange in each of the subshares.)

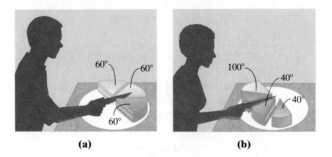

FIGURE 3-7 (a) David cuts his share. (b) Dinah cuts her share.

(a)   (b)

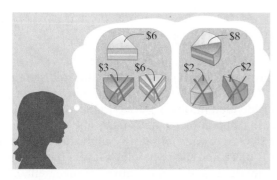

FIGURE 3-8 The values of the subshares in Cher's eyes.

Step 3 (Selection). It's now Cher's turn to choose one sub-share from David's three and one subshare from Dinah's three. Figure 3-8 shows the values of the subshares in Cher's eyes. It's clear what her choices will be: She will choose one of the two pineapple wedges from David's subshares and the big orange-pineapple wedge from Dinah's subshares.

The final fair division of the cake is shown in Fig. 3-9: David gets a final share worth $9 [Fig. 3-9(a)], Dinah gets a final share worth $12 [Fig. 3-9(b)], and Cher gets a final share worth $14 [Fig. 3-9(c)]. David is satisfied, Dinah is happy, and Cher is ecstatic.

FIGURE 3-9 Each player's final fair share.

(a)   (b)   (c)

The Lone-Chooser Method for *N* Players

In the general case of N players, the lone-chooser method involves one chooser C and $N - 1$ dividers $D_1, D_2, \ldots, D_{N-1}$. As always, it is preferable to be a chooser than a divider, so the chooser is determined by a random draw. The method is based on an inductive strategy—if you can do it for three players, then you can do it for four players; if you can do it for four, then you can do it for five; and so on. Thus, when we get to N players, we can assume that we can use the lone-chooser method with $N - 1$ players.

- **Step 1 (Division).** $D_1, D_2, \ldots, D_{N-1}$ divide fairly the set S among themselves, as if C didn't exist. This is a fair division among $N - 1$ players, so each one gets a share he or she considers worth at least $1/(N - 1)$th of S.

- **Step 2 (Subdivision).** Each divider subdivides his or her share into N sub-shares.

- **Step 3 (Selection).** The chooser C finally gets to play. C selects one sub-share from each divider—one subshare from D_1, one from D_2, and so on. At the end, C ends up with $N - 1$ subshares, which make up C's final share, and each divider gets to keep the remaining $N - 1$ subshares in his or her subdivision.

When properly played, the lone-chooser method guarantees that everyone, dividers and chooser alike, ends up with a fair share (see Exercise 84).

3.5 The Last-Diminisher Method

The **last-diminisher method** was proposed by Polish mathematicians Stefan Banach and Bronislaw Knaster in the 1940s. The basic idea behind this method is that throughout the game, the set S is divided into two pieces—a piece currently "owned" by one of the players (we will call that piece the C-piece and the player claiming it the "claimant") and the rest of S, "owned" jointly by all the other players. We will call this latter piece the R-piece and the remaining players the "nonclaimants." The tricky part of this method is that the entire arrangement is temporary—as the game progresses, each nonclaimant has a chance to become the current claimant (and bump the old claimant back into the nonclaimant group) by making changes to the C-piece and consequently to the R-piece. Thus, the claimant, the nonclaimants, the C-piece, and the R-piece all keep changing throughout the game.

Here are the specific details of how the last-diminisher method works.

- **Preliminaries.** Before the game starts the players are randomly assigned an order of play (like in a game of Monopoly this can be done by rolling a pair of dice). We will assume that P_1 plays first, P_2 second, . . . , P_N last, and the players will play in this order throughout the game. The game is played in rounds, and at the end of each round there is one fewer player and a smaller S to be divided.

- **Round 1.**

  - P_1 kicks the game off by "cutting" for herself a $1/N$th share of S (i.e., a share whose value equals $1/N$th of the value of S). This will be the current C-piece, and P_1 is its claimant. P_1 does not know whether or not she will end up with this share, so she must be careful that her claim is neither too small (in case she does) nor too large (in case someone else does).

  - P_2 comes next and has a choice: *pass* (remain a nonclaimant) or *diminish* the C-piece into a share that is a $1/N$th share of S. Obviously, P_2 can be a diminisher only if he thinks that the value of the current C-piece is *more than $1/N$th the value of S*. If P_2 *diminishes*, several changes take place: P_2 becomes the new claimant; P_1 is bumped back into the nonclaimant group; the diminished C-piece becomes the new current C-piece; and the "trimmed" piece is added to the old R-piece to form a new, larger R-piece (there is a lot going on here and the best way to visualize it is by taking a close look at Fig. 3-10).

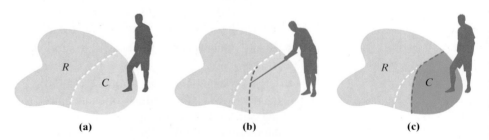

FIGURE 3-10 (a) C-piece claimed by claimant. (b) Diminisher "trims" C-piece. (c) Diminisher becomes new claimant.

- P_3 comes next and has exactly the same opportunity as P_2: *pass* or *diminish* the current *C*-piece. If P_3 passes, then there are no changes and we move on to the next player. If P_3 diminishes (only because in her value system the current *C*-piece is worth more than $1/N$th of *S*), she does so by trimming the *C*-piece to a $1/N$th share of *S*. The rest of the routine is always the same: The trimmed piece is added to the *R*-piece, and the previous claimant (P_1 or P_2) is bumped back into the nonclaimant group.

- The round continues this way, each player in turn having an opportunity to pass or diminish. The last player P_N can also pass or diminish, but if he chooses to diminish, he has a certain advantage over the previous players—knowing that there is no player behind him who could further diminish his claim. In this situation if P_N chooses to be a diminisher, the logical move would be to trim the tiniest possible sliver from the current *C*-piece—a sliver so small that for all practical purposes it has zero value. (Remember that a player's goal is to maximize the size of his or her share.) We will describe this move as "trimming by 0%," although in practice there has to be something trimmed, however small it may be. At the end of Round 1, the current claimant, or *last diminisher*, gets to keep the *C*- piece (it's her fair share) and is out of the game. The remaining players (the nonclaimants) move on to the next round, to divide the *R*-piece among themselves. At this point everyone is happy—the last diminisher got his or her claimed piece, and the nonclaimants are happy to move to the next round, where they will have a chance to divide the *R*-piece among themselves.

- **Round 2.** The *R*-piece becomes the new *S*, and a new version of the game is played with the new *S* and the $N - 1$ remaining players [this means that the new standard for a fair share is a value of $1/(N - 1)$th or more of the new *S*]. At the end of this round, the last diminisher gets to keep the current *C*-piece and is out of the game.

- **Rounds 3, 4, and so on.** Repeat the process, each time with one fewer player and a smaller *S*, until there are just two players left. At this point, divide the remaining piece between the final two players using the *divider-chooser method*.

(EXAMPLE 3.8)   **The Castaways**

A new reality TV show called *The Castaways* is making its debut this season. In the show five contestants (let's call them P_1, P_2, P_3, P_4, and P_5) are dropped off on a deserted tropical island in the middle of nowhere and left there for a year to manage on their own. After one year, the player who has succeeded and prospered the most wins the million-dollar prize. (The producers are counting on the quarreling, double-crossing, and backbiting among the players to make for great reality TV!) In the first episode of the show, the players are instructed to divide up the island among themselves any way they see fit. Instead of quarreling and double-dealing as the producers were hoping for, these five players choose to divide the island using the last-diminisher method. This being reality TV, pictures speak louder than words, and the whole episode unfolds in Figs. 3-11 through 3-15.

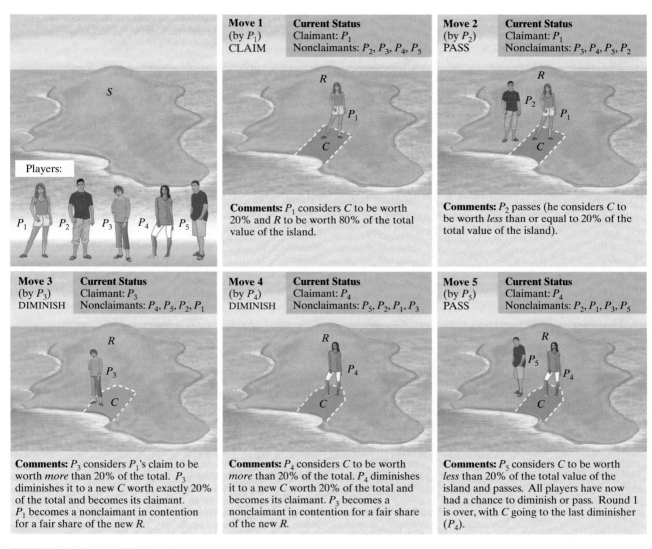

Move 1
(by P_1)
CLAIM

Current Status
Claimant: P_1
Nonclaimants: P_2, P_3, P_4, P_5

Comments: P_1 considers C to be worth 20% and R to be worth 80% of the total value of the island.

Move 2
(by P_2)
PASS

Current Status
Claimant: P_1
Nonclaimants: P_3, P_4, P_5, P_2

Comments: P_2 passes (he considers C to be worth *less* than or equal to 20% of the total value of the island).

Move 3
(by P_3)
DIMINISH

Current Status
Claimant: P_3
Nonclaimants: P_4, P_5, P_2, P_1

Comments: P_3 considers P_1's claim to be worth *more* than 20% of the total. P_3 diminishes it to a new C worth exactly 20% of the total and becomes its claimant. P_1 becomes a nonclaimant in contention for a fair share of the new R.

Move 4
(by P_4)
DIMINISH

Current Status
Claimant: P_4
Nonclaimants: P_5, P_2, P_1, P_3

Comments: P_4 considers C to be worth *more* than 20% of the total. P_4 diminishes it to a new C worth 20% of the total and becomes its claimant. P_3 becomes a nonclaimant in contention for a fair share of the new R.

Move 5
(by P_5)
PASS

Current Status
Claimant: P_4
Nonclaimants: P_2, P_1, P_3, P_5

Comments: P_5 considers C to be worth *less* than 20% of the total value of the island and passes. All players have now had a chance to diminish or pass. Round 1 is over, with C going to the last diminisher (P_4).

FIGURE 3-11 Example 3.8, Round 1.

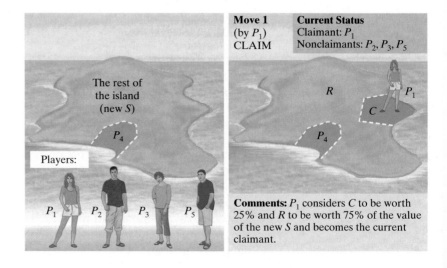

Move 1
(by P_1)
CLAIM

Current Status
Claimant: P_1
Nonclaimants: P_2, P_3, P_5

Comments: P_1 considers C to be worth 25% and R to be worth 75% of the value of the new S and becomes the current claimant.

FIGURE 3-12 Example 3.8, Round 2 (four players left).

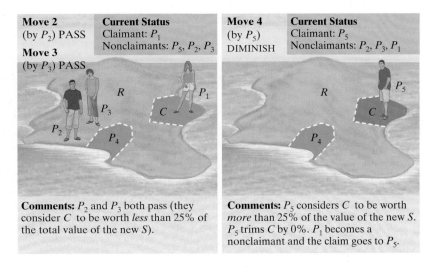

Move 2
(by P_2) PASS

Move 3
(by P_3) PASS

Current Status
Claimant: P_1
Nonclaimants: P_5, P_2, P_3

Comments: P_2 and P_3 both pass (they consider C to be worth *less* than 25% of the total value of the new S).

Move 4
(by P_5)
DIMINISH

Current Status
Claimant: P_5
Nonclaimants: P_2, P_3, P_1

Comments: P_5 considers C to be worth *more* than 25% of the value of the new S. P_5 trims C by 0%. P_1 becomes a nonclaimant and the claim goes to P_5.

FIGURE 3-12 Example 3.8, Round 2 (*continued*).

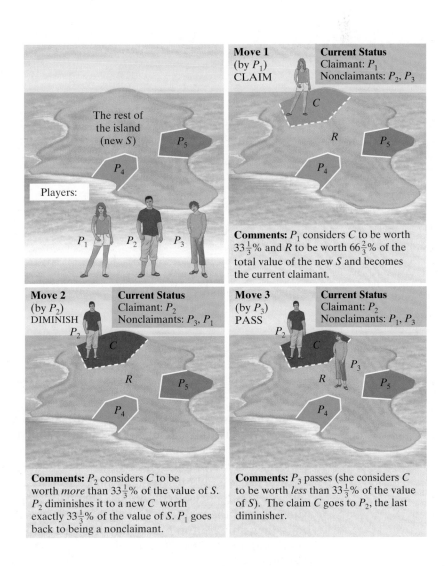

The rest of the island (new S)

Players:

Move 1
(by P_1)
CLAIM

Current Status
Claimant: P_1
Nonclaimants: P_2, P_3

Comments: P_1 considers C to be worth $33\frac{1}{3}$% and R to be worth $66\frac{2}{3}$% of the total value of the new S and becomes the current claimant.

Move 2
(by P_2)
DIMINISH

Current Status
Claimant: P_2
Nonclaimants: P_3, P_1

Comments: P_2 considers C to be worth *more* than $33\frac{1}{3}$% of the value of S. P_2 diminishes it to a new C worth exactly $33\frac{1}{3}$% of the value of S. P_1 goes back to being a nonclaimant.

Move 3
(by P_3)
PASS

Current Status
Claimant: P_2
Nonclaimants: P_1, P_3

Comments: P_3 passes (she considers C to be worth *less* than $33\frac{1}{3}$% of the value of S). The claim C goes to P_2, the last diminisher.

FIGURE 3-13 Example 3.8, Round 3 (three players left).

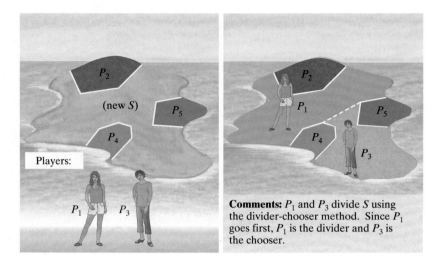

FIGURE 3-14 Example 3.8, last round (divider-chooser method).

Comments: P_1 and P_3 divide S using the divider-chooser method. Since P_1 goes first, P_1 is the divider and P_3 is the chooser.

FIGURE 3-15 The final division of the island.

In the next two sections we will discuss *discrete* fair-division methods—methods for dividing a booty S consisting of indivisible objects such as art, jewels, or candy. As a general rule of thumb, discrete fair division is harder to achieve than continuous fair division because there is a lot less flexibility in the division process, and discrete fair divisions that are truly fair are only possible under a limited set of conditions. Thus, it is important to keep in mind that while both of the methods we will discuss in the next two sections have limitations, they still are the best methods we have available. Moreover, when they work, both methods work remarkably well and produce surprisingly good fair divisions.

3.6　The Method of Sealed Bids

The **method of sealed bids** was originally proposed by Hugo Steinhaus and Bronislaw Knaster around 1948. The best way to illustrate how this method works is by means of an example.

> **EXAMPLE 3.9**　Settling Grandma's Estate

In her last will and testament, Grandma plays a little joke on her four grandchildren (Art, Betty, Carla, and Dave) by leaving just three items—a cabin in the mountains, a vintage 1955 Rolls Royce, and a Picasso painting—with the stipulation that the items must remain with the grandchildren (not sold to outsiders) and must be divided fairly in equal shares among them. How can we possibly resolve this conundrum? The method of sealed bids will give an ingenious and elegant solution.

- **Step 1　(Bidding).** Each of the players makes a bid (in dollars) for each of the items in the estate, giving his or her honest assessment of the actual value of each item. To satisfy the privacy assumption, it is important that the bids are done independently, and no player should be privy to another player's bids before making his or her own. The easiest way to accomplish this is for each player to submit his or her bid in a sealed envelope. When all the bids are in, they are opened. Table 3-6 shows each player's bid on each item in the estate.

TABLE 3-6　The Players' Bids

	Art	Betty	Carla	Dave
Cabin	220,000	250,000	211,000	198,000
Rolls Royce	40,000	30,000	47,000	52,000
Picasso	280,000	240,000	234,000	190,000

- **Step 2　(Allocation).** Each item will go to the highest bidder for that item. (If there is a tie, the tie can be broken with a coin flip.) In this example the cabin goes to Betty, the vintage Rolls Royce goes to Dave, and the Picasso painting goes to Art. Notice that Carla gets nothing. Not to worry—it all works out at the end! (In this method it is possible for one player to get none of the items and another player to get many or all of the items. Much like in a silent auction, it's a matter of who bids the highest.)

- **Step 3　(First Settlement).** It's now time to settle things up. Depending on what items (if any) a player gets in Step 2, he or she will owe money to or be owed money by the estate. To determine how much a player owes or is owed, we first calculate each player's *fair-dollar share* of the estate. A player's *fair-dollar share* is found by adding that player's bids and dividing the total by the number of players. For example, Art's bids on the three items add up to $540,000. Since there are four equal heirs, Art realizes he is only entitled to one-fourth of that—his fair-dollar share is therefore $135,000. The last row of Table 3-7 shows the fair-dollar share of each player.

TABLE 3-7

	Art	Betty	Carla	Dave
Cabin	220,000	250,000	211,000	198,000
Rolls Royce	40,000	30,000	47,000	52,000
Picasso	280,000	240,000	234,000	190,000
Total	540,000	520,000	492,000	440,000
Fair-dollar share	135,000	130,000	123,000	110,000

The fair-dollar shares are the baseline for the settlements—if the total value of the items that the player gets in Step 2 is more than his or her fair-dollar share, then the player *pays* the estate the difference. If the total value of the items that the player gets is *less* than his or her fair-dollar share, then the player *gets* the difference in cash. Here are the details of how the settlement works out for each of our four players.

Art: As we have seen, Art's fair dollar share is $135,000. At the same time, Art is getting a Picasso painting worth (to him) $280,000, so Art must *pay* the estate the difference of $145,000 ($280,000 − $135,000). The combination of getting the $280,000 Picasso painting but paying $145,000 for it in cash results in Art getting his fair share of the estate.

Betty: Betty's fair-dollar share is $130,000. Since she is getting the cabin, which she values at $250,000, she must pay the estate the difference of $120,000. By virtue of getting the $250,000 house for $120,000, Betty ends up with her fair share of the estate.

Carla: Carla's fair-dollar share is $123,000. Since she is getting no items from the estate, she receives her full $123,000 in cash. Clearly, she is getting her fair share of the estate.

Dave: Dave's fair-dollar share is $110,000. Dave is getting the vintage Rolls, which he values at $52,000, so he has an additional $58,000 coming to him in cash. The Rolls plus the cash constitute Dave's fair share of the estate.

At this point each of the four heirs has received a fair share, and we might consider our job done, but this is not the end of the story—there is more to come (good news mostly!). If we add Art and Betty's payments to the estate and subtract the payments made by the estate to Carla and Dave, we discover that there is a *surplus* of $84,000! ($145,000 and $120,000 came in from Art and Betty; $123,000 and $58,000 went out to Carla and Dave.)

- **Step 4 (Division of the Surplus).** The surplus is common money that belongs to the estate, and thus to be divided equally among the players. In our example each player's share of the $84,000 surplus is $21,000.
- **Step 5 (Final Settlement).** The final settlement is obtained by adding the surplus money to the first settlement obtained in Step 3.

Art: Gets the Picasso painting and pays the estate $124,000—the original $135,000 he had to pay in Step 3 minus the $21,000 he gets from his share of the surplus. (Everything done up to this point could be done on paper, but now, finally, real money needs to change hands!)

Betty: Gets the cabin and has to pay the estate only $99,000 ($120,000 − $21,000).

Carla: Gets $144,000 in cash ($123,000 + $21,000).

Dave: Gets the vintage Rolls Royce plus $79,000 ($58,000 + $21,000).

The method of sealed bids works so well because it tweaks a basic idea in economics. In most ordinary transactions there is a buyer and a seller, and the buyer knows the other party is the seller and vice versa. In the method of sealed bids, each player is simultaneously a buyer and a seller, without actually knowing which one until all the bids are opened. This keeps the players honest and, in the long run, works out to everyone's advantage. At the same time, the method of sealed bids will not work unless the following two important conditions are satisfied.

- Each player must have enough money to play the game. If a player is going to make honest bids on the items, he or she must be prepared to buy some or all of them, which means that he or she may have to pay the estate certain sums of money. If the player does not have this money available, he or she is at a definite disadvantage in playing the game.

- Each player must accept money (if it is a sufficiently large amount) as a substitute for any item. This means that no player can consider any of the items priceless.

The method of sealed bids takes a particularly simple form in the case of two players and one item. Consider the following example.

EXAMPLE 3.10   Splitting Up the House

Al and Betty are getting a divorce. The only joint property of any value is their house. Rather than hiring attorneys and going to court to figure out how to split up the house, they agree to give the method of sealed bids a try.

Al's bid on the house is $340,000; Betty's bid is $364,000. Their fair-dollar shares of the "estate" are $170,000 and $182,000, respectively. Since Betty is the higher bidder, she gets to keep the house and must pay Al cash for his share. The computation of how much cash Betty pays Al can be done in two steps. In the first settlement, Betty owes the estate $182,000. Of this money, $170,000 pays for Al's fair share, leaving a surplus of $12,000 to be split equally between them. The bottom line is that Betty ends up paying $176,000 to Al for his share of the house, and both come out $6000 ahead.

The method of sealed bids can provide an excellent solution not only to settlements of property in a divorce but also to the equally difficult and often contentious issue of splitting up a partnership. The catch is that in these kinds of splits we can rarely count on the rationality assumption to hold. A divorce or partnership split devoid of emotion, spite, and hard feelings is a rare thing indeed!

3.7   The Method of Markers

The **method of markers** is a discrete fair-division method proposed in 1975 by William F. Lucas, a mathematician at the Claremont Graduate School. The method has the great virtue that it does not require the players to put up any of their own money. On the other hand, unlike the method of sealed bids, this method cannot be used effectively unless (1) there are many more items to be divided than there are players in the game and (2) the items are reasonably close in value.

In this method we start with the items lined up in a random but fixed sequence called an *array*. Each of the players then gets to make an independent *bid* on the items in the array. A player's bid consists of dividing the array into segments of consecutive items (as many segments as there are players) so that each of the segments represents a fair share of the entire set of items.

For convenience, we might think of the array as a string. Each player then "cuts" the string into N segments, each of which he or she considers an acceptable share. (Notice that to cut a string into N sections, we need $N - 1$ cuts.) In practice, one way to make the "cuts" is to lay markers in the places where the cuts are made. Thus, each player can make his or her bids by placing $N - 1$ markers so that they divide the array into N segments. To ensure privacy, no player should see the markers of another player before laying down his or her own.

The final step is to give to each player one of the segments in his or her bid. The easiest way to explain how this can be done is with an example.

FIGURE 3-16 The Halloween leftovers.

EXAMPLE 3.11   **Dividing the Halloween Leftovers**

Alice, Bianca, Carla, and Dana want to divide the Halloween leftovers shown in Fig. 3-16 among themselves. There are 20 pieces, but having each randomly choose 5 pieces is not likely to work well—the pieces are too varied for that. Their teacher, Mrs. Jones, offers to divide the candy for them, but the children reply that they just learned about a cool fair-division game they want to try, and they can do it themselves, thank you.

As a preliminary step, the 20 pieces are arranged in an array (Fig. 3-17). For convenience, we will label the pieces of candy 1 through 20. The order in which the pieces are lined up should be random (the easiest way to do this is to dump the pieces into a paper bag, shake the bag, and take the pieces out of the bag one at a time).

FIGURE 3-17 The items lined up in an array.

- **Step 1 (Bidding).** Each child writes down independently on a piece of paper exactly where she wants to place her three markers. (Three markers divide the array into four sections.) The bids are opened, and the results are shown in Fig. 3-18. The A-labels indicate the position of Alice's markers (A_1 denotes her first marker, A_2 her second marker, and A_3 her third and last marker). Alice's bid means that she is willing to accept one of the following as a fair share of the candy: (1) pieces 1 through 5 (first segment), (2) pieces 6 through 11 (second segment), (3) pieces 12 through 16 (third segment), or (4) pieces 17 through 20 (last segment). Bianca's bid is shown by the B-markers and indicates how she would break up the array into four segments that are fair shares; ditto for Carla's bid (shown by the C-markers) and Dana's bid (shown by the D-markers).

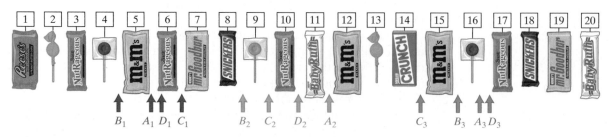

FIGURE 3-18 The bids.

- **Step 2 (Allocations).** This is the tricky part, where we are going to give to each child one of the segments in her bid. Here is how to do it: Scan the array from left to right until the first *first marker* comes up. Here the first *first marker* is Bianca's B_1. This means that Bianca will be the first player to get her fair share consisting of the first segment in her bid (pieces 1 through 4, Fig. 3-19).

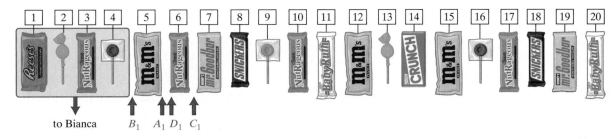

FIGURE 3-19 B_1 is the first 1-marker. Bianca goes first, gets her first segment.

Bianca is done now, and her markers can be removed since they are no longer needed. Continue scanning from left to right looking for the first *second marker*. Here the first second marker is Carla's C_2, so Carla will be the second player taken care of. Carla gets the second segment in her bid (pieces 7 through 9, Fig. 3-20). Carla's remaining markers can now be removed.

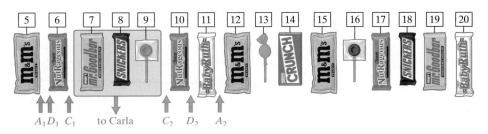

FIGURE 3-20 C_2 is the first 2-marker (among A's, C's, and D's). Carla goes second, gets her second segment.

Continue scanning from left to right looking for the first *third marker*. Here there is a tie between Alice's A_3 and Dana's D_3. As usual, a coin toss is used to break the tie and Alice will be the third player to go—she will get the third segment in her bid (pieces 12 through 16, Fig. 3-21). Dana is the last player

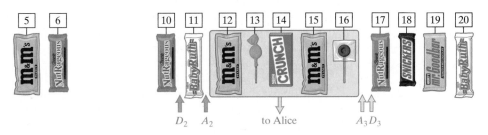

FIGURE 3-21 A_3 and D_3 are tied as the first 3-marker. After a coin toss, Alice gets her third segment.

and gets the last segment in her bid (pieces 17 through 20, Fig. 3-22). At this point each player has gotten a fair share of the 20 pieces of candy. The amazing part is that there is *leftover candy*!

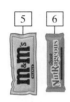

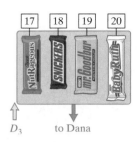

D_3 to Dana

FIGURE 3-22 Dana is last, gets her last segment.

FIGURE 3-23 The surplus (to be given randomly to the players one at a time) are a bonus.

- **Step 3 (Dividing the Surplus).** The easiest way to divide the surplus is to randomly draw lots and let the players take turns choosing one piece at a time until there are no more pieces left. Here the leftover pieces are 5, 6, 10, and 11 (Fig. 3-23). The players now draw lots; Carla gets to choose first and takes piece 11. Dana chooses next and takes piece 5. Bianca and Alice receive pieces 6 and 10, respectively.

The ideas behind Example 3.11 can be easily generalized to any number of players. We now give the general description of the method of markers with N players and M discrete items.

- **Preliminaries.** The items are arranged randomly into an array. For convenience, label the items 1 through M, going from left to right.
- **Step 1 (Bidding).** Each player independently divides the array into N segments (segments 1, 2, . . . , N) by placing $N - 1$ markers along the array. These segments are assumed to represent the fair shares of the array in the opinion of that player.
- **Step 2 (Allocations).** Scan the array from left to right until the first *first marker* is located. The player owning that marker (let's call him P_1) goes first and gets the first segment in his bid. (In case of a tie, break the tie randomly.) P_1's markers are removed, and we continue scanning from left to right, looking for the first *second marker*. The player owning that marker (let's call her P_2) goes second and gets the second segment in her bid. Continue this process, assigning to each player in turn one of the segments in her bid. The last player gets the last segment in her bid.
- **Step 3 (Dividing the Surplus).** The players get to go in some random order and pick one item at a time until all the surplus items are given out.

Despite its simple elegance, the method of markers can be used only under some fairly restrictive conditions. In particular, the method assumes that every player is able to divide the array of items into segments in such a way that each of the segments has approximately equal value. This is usually possible when the items are of small and homogeneous value, but almost impossible to accomplish when there is a combination of expensive and inexpensive items (good luck trying to divide fairly 19 candy bars plus an iPod using the method of markers!).

CONCLUSION

The Judgement of Solomon by
Nicolas Poussin (1649)

Problems of fair division are as old as humankind. One of the best-known and best-loved biblical stories is built around one such problem. Two women, both claiming to be mothers of the same baby, make their case to King Solomon. As a solution King Solomon proposes to cut the baby in two and give each woman a share. (Basically, King Solomon was proposing a continuous solution to a discrete fair-division problem!) This solution is totally unacceptable to the true mother, who would rather see the baby go to the other woman than be "divided" into two shares. The final settlement, of course, is that the baby is returned to its rightful mother.

The problem of dividing an object or a set of objects among the members of a group is a practical problem that comes up regularly in our daily lives. When the object is a pizza, a cake, or a bunch of candy, we don't always pay a great deal of attention to the issue of fairness, but when the object is an estate, land, jewelry, or some other valuable asset, dividing things fairly becomes a critical issue.

In this chapter we looked at fair division from a mathematical perspective. Within mathematics, problems of fair division are considered part of *game theory* (a branch of mathematics that provides the tools to analyze "games" of competition or cooperation between individuals or groups), and we approached fair-division problems within this context. We found that, when certain conditions are satisfied (especially assumptions about the behavior and rationality of the players), mathematics can provide solutions that guarantee that each player will always receive a fair share.

Over the years, mathematicians have developed different *methods* for solving fair-division problems. Each of these methods works under some restricted set of circumstances (there is no single method that will work in every possible situation), so it is important not only to know how the method works but also to understand under what circumstances it can and cannot be used.

In our analysis we classified fair-division problems into *continuous* and *discrete*, and we discussed different fair-division methods for each type of problem. In the case of a continuous fair-division problem with two players, the *divider-chooser method* is always the method of choice. For continuous fair-division problems with three or more players, there are many possible methods that can be considered, and we discussed three of these: the *lone-divider method*, the *lone-chooser method*, and the *last-diminisher method*. There are several other lesser-known methods that we did not discuss.

> **❝ Grief can take care of itself, but to get the full value of joy you must have someone to divide it with. ❞**
>
> —Mark Twain

Discrete fair-division problems are inherently harder than continuous problems, and the number of methods available to tackle them is very limited. The best-known discrete fair-division method is the *method of sealed bids*, and this method has many useful applications in real life. An alternative is the *method of markers*, which works well but only for very limited types of objects.

This chapter gave an overview of what is an interesting and a surprising application of mathematics to one of the fundamental questions of social science—how to get humans to share in a reasonable and fair way. Mathematics can provide only some answers, and these answers work under a fairly limited set of circumstances, but when they do work, they can work remarkably well. Remember this next time you must divide an inheritance, a piece of real estate, or even an old-fashioned joy. It may serve you well.

P ROFILE: Hugo Steinhaus (1887–1972)

Dividing things, be it the spoils of war or the fruits of peace, is an issue as old as man. Throughout most of human history, fairness in dividing things was seen as a problem far removed from the mathematical arena—the province of kings, priests, judges, and politicians. It wasn't until the 1940s that the revolutionary notion that mathematical methods could be used to tackle successfully many types of fair-division problems came about. This breakthrough can be traced to the inspiration and creativity of one man—the great Polish mathematician Hugo Steinhaus.

Hugo Steinhaus was born in Jaslo, Poland. After completing his secondary education in his homeland, Steinhaus went to Germany to study mathematics at the University of Göttingen, which in the early 1900s was the most prestigious center of mathematical research in the world. At Göttingen, Steinhaus studied under many famous mathematicians, including David Hilbert, his doctoral

supervisor and arguably the most famous mathematician of his time. Steinhaus was awarded a doctorate in mathematics (with distinction) in 1911.

After completing four years of military service in World War I, Steinhaus returned to Poland, where he began a distinguished academic career as a professor of mathematics, first in Kraków and then in Lvov, where he founded the famous Lvov school of mathematics. Steinhaus was a prolific and versatile mathematician who made important contributions to many different branches of mathematics, including functional analysis, trigonometric series, and probability theory. But Steinhaus's interests went beyond conventional mathematical research—he loved to discuss and expound on simple but interesting real-life applications of mathematics and to foster an appreciation for the power and beauty of mathematics among young people. In 1937 Steinhaus wrote a classic book called *Mathematical Snapshots*, a collection of vignettes intended to appeal, in Steinhaus's own words, to the "scientist in the child and the child in the scientist." After many editions, *Mathematical Snapshots* is still in print and widely quoted.

Sometime in the late 1930s, Steinhaus made one of his more famous mathematical discoveries, a theorem informally known as the *ham sandwich theorem*. Essentially, the theorem can be paraphrased as follows: If you have a sandwich consisting of *three* ingredients (say bread, ham, and cheese), then there is a way to slice the sandwich (using a single straight cut) into two parts each of which has exactly half of the bread, half of the ham, and half of the cheese—you can share the sandwich with your friend and each of you gets exactly half of each of the three ingredients. (You can do this with even the funkiest of sandwiches as long as you stick to just three ingredients. Unfortunately, if you want to divide *n* ingredients equally, you need to live in the *n*th dimension!)

The ham sandwich theorem is a beautiful theoretical result but has little practical value—it says that the desired cut is possible in theory, but it doesn't give even the slightest hint as to how to do it. It promises a fair division, but it cannot deliver it. The limitations of the ham sandwich theorem forced Steinhaus to think about a different, truly practical approach to the problem of dividing things *equally*, and this led to the foundations of the theory of fair division as we know it today. By the late 1940s, Steinhaus, together with Stefan Banach and Bronislaw Knaster, two of his students and collaborators, had developed many of the most important fair-division methods we know today, including the *last-diminisher method*, the *method of sealed bids*, and a continuous version of the *method of markers*.

Hugo Steinhaus died in 1972, at the age of 85. His mathematical legacy includes more than 170 articles, five books, and the invaluable insight that mathematics can help humankind solve some of its disputes. ▪

KEY CONCEPTS

continuous fair-division game, **80**
discrete fair-division game, **80**
divider-chooser method, **80**
fair division, **79**

fair-division method, **80**
fair share, **79**
last-diminisher method, **90**
lone-chooser method, **87**

lone-divider method, **82**
method of markers, **97**
method of sealed bids, **95**

EXERCISES

WALKING

A Shares, Fair Shares, and Fair Divisions

(a)	(b)	(c)

FIGURE 3-24

1. Angelina and Brad jointly buy the chocolate-strawberry mousse cake shown in Fig. 3-24(a) for $24. Suppose that Angelina values chocolate cake *three* times as much as she values strawberry cake. Find the dollar value to Angelina of each of the following pieces of cake:

(a) the strawberry half of the cake

(b) the chocolate half of the cake

(c) the slice of strawberry cake shown in Fig. 3-24(b)

(d) the slice of chocolate cake shown in Fig. 3-24(c)

2. Brad and Angelina jointly buy the chocolate-strawberry mousse cake shown in Fig. 3-24(a) for $24. Suppose that Brad values strawberry cake *four* times as much as he values chocolate cake. Find the dollar value to Brad of each of the following pieces of cake:

(a) the strawberry half of the cake

(b) the chocolate half of the cake

(c) the slice of strawberry cake shown in Fig. 3-24(b)

(d) the slice of chocolate cake shown in Fig. 3-24(c)

3. Homer and Marge jointly buy the half pepperoni–half mushroom pizza shown in Fig. 3-25(a) for $18. Suppose that Homer values pepperoni pizza *four* times as much as he values mushroom pizza. Find the dollar value to Homer of each of the following pieces of pizza:

   (a) the mushroom half of the pizza

   (b) the pepperoni half of the pizza

   (c) the slice of mushroom pizza shown in Fig. 3-25(b)

   (d) the slice of pepperoni pizza shown in Fig. 3-25(c)

   (e) the half pepperoni-half mushroom slice shown in Fig. 3-25(d)

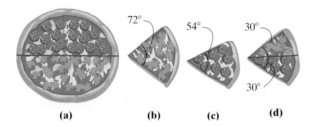

(a)   (b)   (c)   (d)

FIGURE 3-25

4. Homer and Marge jointly buy the half pepperoni–half mushroom pizza shown in Fig. 3-25(a) for $18. Suppose that Marge values mushroom pizza *three* times as much as she values pepperoni pizza. Find the dollar value to Marge of each of the following pieces of pizza:

   (a) the mushroom half of the pizza

   (b) the pepperoni half of the pizza

   (c) the slice of mushroom pizza shown in Fig. 3-25(b)

   (d) the slice of pepperoni pizza shown in Fig. 3-25(c)

   (e) the half pepperoni–half mushroom slice shown in Fig. 3-25(d)

5. Karla and five other friends jointly buy the chocolate-strawberry-vanilla cake shown in Fig. 3-26(a) for $30 and plan to divide the cake fairly among themselves. After much discussion, the cake is divided into the six equal-sized slices $s_1, s_2, \ldots, s_6$ shown in Fig. 3-26(b). Suppose that Karla values strawberry cake *twice* as

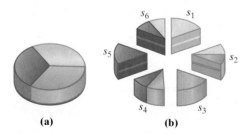

(a)   (b)

FIGURE 3-26

much as vanilla cake and chocolate cake *three* times as much as vanilla cake.

   (a) Find the dollar value to Karla of each of the slices s_1 through s_6.

   (b) Which of the slices s_1 through s_6 are fair shares to Karla?

6. Marla and five other friends jointly buy the chocolate-strawberry-vanilla cake shown in Fig. 3-26(a) for $30 and plan to divide the cake fairly among themselves. After much discussion, the cake is divided into the six equal-sized slices $s_1, s_2, \ldots, s_6$ shown in Fig. 3-26(b). Suppose that Marla values vanilla cake *twice* as much as chocolate cake and chocolate cake *three* times as much as strawberry cake.

   (a) Find the dollar value to Marla of each of the slices s_1 through s_6.

   (b) Which of the slices s_1 through s_6 are fair shares to Marla?

7. Three players (Ana, Ben, and Cara) are sharing a cake. Suppose that the cake is divided into three slices (s_1, s_2, and s_3). The following table gives the value of each slice in the eyes of each of the players.

	s_1	s_2	s_3
Ana	$3.00	$5.00	$4.00
Ben	$4.00	$4.50	$6.50
Cara	$4.50	$4.50	$4.50

   (a) Which of the three slices are fair shares to Ana?

   (b) Which of the three slices are fair shares to Ben?

   (c) Which of the three slices are fair shares to Cara?

   (d) Find a fair division of the cake using s_1, s_2, and s_3 as fair shares. Explain why there is only one such fair division possible.

8. Three players (Amy, Bella, and Chia) are sharing a cake. Suppose that the cake is divided into three slices (s_1, s_2, and s_3). The following table gives the value of each slice in the eyes of each of the players.

	s_1	s_2	s_3
Amy	$6.00	$7.00	$8.00
Bella	$6.00	$6.50	$5.50
Chia	$8.00	$6.25	$6.75

   (a) Which of the three slices are fair shares to Amy?

   (b) Which of the three slices are fair shares to Bella?

(c) Which of the three slices are fair shares to Chia?

(d) Find a fair division of the cake using s_1, s_2, and s_3 as fair shares. Explain why there is only one such fair division possible.

9. Three players (Alex, Betty, and Cindy) are sharing a cake. Suppose that the cake is divided into three slices (s_1, s_2, and s_3). The following table shows the value of s_1 and of s_2 to each of the players. The values of s_3 are missing. (The percentages represent the value of the slice as a percent of the value of the entire cake.)

	s_1	s_2
Alex	30%	40%
Betty	31%	35%
Cindy	30%	35%

(a) Which of the three slices are fair shares to Alex?

(b) Which of the three slices are fair shares to Betty?

(c) Which of the three slices are fair shares to Cindy?

(d) If possible, find a fair division of the cake using s_1, s_2, and s_3 as fair shares. If this is not possible, explain why not.

10. Three players (Alex, Betty, and Cindy) are sharing a cake. Suppose that the cake is divided into three slices (s_1, s_2, and s_3). The following table shows the value of s_1 and of s_2 to each of the players. The values of s_3 are missing. (The percentages represent the value of the slice as a percent of the value of the entire cake.)

	s_1	s_2
Alex	30%	34%
Betty	28%	36%
Cindy	30%	$33\frac{1}{3}\%$

(a) Which of the three slices are fair shares to Alex?

(b) Which of the three slices are fair shares to Betty?

(c) Which of the three slices are fair shares to Cindy?

(d) If possible, find a fair division of the cake using s_1, s_2, and s_3 as fair shares. If this is not possible, explain why not.

11. Four players (Abe, Betty, Cory, and Dana) are sharing a cake. Suppose that the cake is divided into four slices s_1, s_2, s_3, and s_4. The following table gives the value of each slice in the eyes of each of the players.

	s_1	s_2	s_3	s_4
Abe	$3.00	$5.00	$5.00	$2.00
Betty	$4.50	$4.50	$4.50	$4.50
Cory	$4.00	$3.50	$2.00	$2.50
Dana	$2.75	$2.40	$2.45	$2.40

(a) Which of the four slices are fair shares to Abe?

(b) Which of the four slices are fair shares to Betty?

(c) Which of the four slices are fair shares to Cory?

(d) Which of the four slices are fair shares to Dana?

(e) Explain why there is only one possible fair division of the cake using s_1, s_2, s_3, and s_4 as fair shares.

12. Four partners (Adams, Benson, Cagle, and Duncan) jointly own a piece of land with a market value of $400,000. Suppose that the land is subdivided into four parcels s_1, s_2, s_3, and s_4. The partners are planning to split up, with each partner getting one of the four parcels. The following table gives the value of some of the parcels in the eyes of each of the partners.

	s_1	s_2	s_3	s_4
Adams	$90,000		$110,000	$104,000
Benson	$98,000	$100,000		$98,000
Cagle	$106,000	$102,000	$94,000	
Duncan		$96,000	$112,000	$96,000

(a) Which of the four parcels are fair shares to Adams?

(b) Which of the four parcels are fair shares to Benson?

(c) Which of the four parcels are fair shares to Cagle?

(d) Which of the four parcels are fair shares to Duncan?

(e) Explain why there is only one possible fair division of the land using the parcels s_1, s_2, s_3, and s_4 as fair shares.

13. Four partners (Adams, Benson, Cagle, and Duncan) jointly own a piece of land with a market value of $400,000. Suppose that the land is subdivided into four parcels s_1, s_2, s_3, and s_4. The partners are planning to split up, with each partner getting one of the four parcels.

(a) To Adams, s_1 is worth $40,000 more than s_2, s_2 and s_3 are equal in value, and s_4 is worth $20,000 more than s_1. Determine which of the four parcels are fair shares to Adams.

(b) To Benson, s_1 is worth $40,000 more than s_2, s_4 is $8,000 more than s_3, and together s_4 and s_3 have a combined value equal to 40% of the value of the land. Determine which of the four parcels are fair shares to Benson.

(c) To Cagle, s_1 is worth $40,000 more than s_2 and $20,000 more than s_4, and s_3 is worth twice as much as s_4. Determine which of the four parcels are fair shares to Cagle.

(d) To Duncan, s_1 is worth $4,000 more than s_2; s_2 and s_3 have equal value; and s_1, s_2, and s_3 have a combined value equal to 70% of the value of the land. Determine which of the four parcels are fair shares to Duncan.

(e) Find a fair division of the land using the parcels s_1, s_2, s_3, and s_4 as fair shares.

14. Four players (Abe, Betty, Cory, and Dana) are sharing a cake. Suppose that the cake is divided into four slices s_1, s_2, s_3, and s_4.

(a) To Abe, s_1 is worth $3.60, s_4 is worth $3.50, s_2 and s_3 have equal value, and the entire cake is worth $15.00. Determine which of the four slices are fair shares to Abe.

(b) To Betty, s_2 is worth twice as much as s_1, s_3 is worth three times as much as s_1, and s_4 is worth four times as much as s_1. Determine which of the four slices are fair shares to Betty.

(c) To Cory, s_1, s_2 and s_4 have equal value, and s_3 is worth as much as s_1, s_2 and s_4 combined. Determine which of the four slices are fair shares to Cory.

(d) To Dana, s_1 is worth $1.00 more than s_2, s_3 is worth $1.00 more than s_1, s_4 is worth $3.00, and the entire cake is worth $18.00. Determine which of the four slices are fair shares to Dana.

(e) Find a fair division of the cake using s_1, s_2, s_3 and s_4 as fair shares.

B The Divider-Chooser Method

Exercises 15 and 16 refer to the following situation: Jared and Karla jointly bought the half meatball–half vegetarian foot-long sub shown in Fig. 3-27 for $8.00. They plan to divide the sandwich fairly using the divider-chooser method. Jared likes meatball subs three times as much as vegetarian subs; Karla is a strict vegetarian and does not eat meat at all. Assume that Jared just met Karla and has no idea that she is a vegetarian. Assume also that when the sandwich is cut, the cut is made perpendicular to the length of the sandwich. (You can describe

different shares of the sandwich using the ruler and interval notation. For example, [0, 6] describes the vegetarian half, [6, 8] describes one-third of the meatball half, etc.).

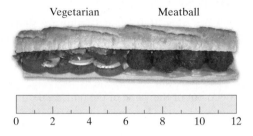

FIGURE 3-27

15. Suppose that they flip a coin and Jared ends up being the divider.

(a) Describe how Jared should cut the sandwich into two shares s_1 and s_2.

(b) After Jared cuts, Karla gets to choose. Specify which of the two shares Karla should choose and give the value of the share to Karla.

16. Suppose they flip a coin and Karla ends up being the divider.

(a) Describe how Karla should cut the sandwich into two shares s_1 and s_2.

(b) After Karla cuts, Jared gets to choose. Specify which of the two shares Jared should choose and give the value of the share to Jared.

Exercises 17 and 18 refer to the following situation: Martha and Nick jointly bought the giant 28-in. sub sandwich shown in Fig. 3-28 for $9. They plan to divide the sandwich fairly using the divider-chooser method. Martha likes ham subs twice as much as she likes turkey subs, and she likes turkey and roast beef subs the same. Nick likes roast beef subs twice as much as he likes ham subs, and he likes ham and turkey subs the same. Assume that Nick and Martha just met and know nothing of each other's likes and dislikes. Assume also that when the sandwich is cut, the cut is made perpendicular to the length of the sandwich. (You can describe different shares of the sandwich using the ruler and interval notation. For example, [0, 8] describes the ham part, [8, 12] describes one-third of the turkey part, etc.).

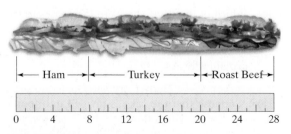

FIGURE 3-28

17. Suppose that they flip a coin and Martha ends up being the divider.

    (a) Describe how Martha would cut the sandwich into two shares s_1 and s_2.

    (b) After Martha cuts, Nick gets to choose. Specify which of the two shares Nick should choose, and give the value of the share to Nick.

18. Suppose that they flip a coin and Nick ends up being the divider.

    (a) Describe how Nick would cut the sandwich into two shares s_1 and s_2.

    (b) After Nick cuts, Martha gets to choose. Specify which of the two shares Martha should choose and give the value of the share to Martha.

Exercises 19 and 20 refer to the following situation: David and Paula are planning to divide the pizza shown in Fig. 3-29(a) using the divider-chooser method. David likes pepperoni and sausage pizza equally well, and he likes sausage pizza twice as much as mushroom pizza. Paula likes sausage and mushroom pizza equally well, but she hates pepperoni pizza.

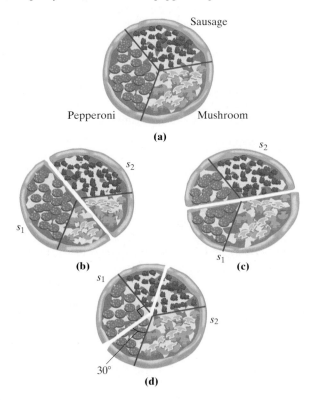

FIGURE 3-29

19. Suppose that they flip a coin and David ends up being the divider. (Assume that David is playing the game by the rules and knows nothing about Paula's likes and dislikes.)

    (a) Is the cut shown in Fig. 3-29(b) a possible 50-50 cut that David might have made as the divider? If so, describe the share Paula should choose and give the value (as a percent) of that share to Paula. If the cut is not a 50-50 cut, give the values of the two shares to David.

    (b) Is the cut shown in Fig. 3-29(c) a possible 50-50 cut that David might have made as the divider? If so, describe the share Paula should choose and give the value (as a percent) of that share to Paula. If the cut is not a 50-50 cut, give the values of the two shares to David.

    (c) Is the cut shown in Fig. 3-29(d) a possible 50-50 cut that David might have made as the divider? If so, describe the share Paula should choose and give the value (as a percent) of that share to Paula. If the cut is not a 50-50 cut, give the values of the two shares to David.

20. Suppose that they flip a coin and Paula ends up being the divider. (Assume that Paula is playing the game by the rules and knows nothing about David's likes and dislikes.)

    (a) Is the cut shown in Fig. 3-29(b) a possible 50-50 cut that Paula might have made as the divider? If so, describe the share David should choose and give the value (as a percent) of that share to David. If the cut is not a 50-50 cut, give the values of the two shares to Paula.

    (b) Is the cut shown in Fig. 3-29(c) a possible 50-50 cut that Paula might have made as the divider? If so, describe the share David should choose and give the value (as a percent) of that share to David. If the cut is not a 50-50 cut, give the values of the two shares to Paula.

    (c) Is the cut shown in Fig. 3-29(d) a possible 50-50 cut that Paula might have made as the divider? If so, describe the share David should choose and give the value (as a percent) of that share to David. If the cut is not a 50-50 cut, give the values of the two shares to Paula.

C The Lone-Divider Method

21. Three partners are dividing a plot of land among themselves using the *lone-divider* method. After the divider D divides the land into three shares s_1, s_2, and s_3, the choosers C_1 and C_2 submit their bids for these shares.

    (a) Suppose that the chooser's bids are C_1: $\{s_2, s_3\}$; C_2: $\{s_2, s_3\}$. Describe *two* different fair divisions of the land.

    (b) Suppose that the chooser's bids are C_1: $\{s_2, s_3\}$; C_2: $\{s_1, s_3\}$. Describe *three* different fair divisions of the land.

22. Three partners are dividing a plot of land among themselves using the *lone-divider* method. After the divider D divides the land into three shares s_1, s_2, and s_3, the choosers C_1 and C_2 submit their bids for these shares.

(a) Suppose that the chooser's bids are C_1: $\{s_2\}$; C_2: $\{s_1, s_3\}$. Describe *two* different fair divisions of the land.

(b) Suppose that the chooser's bids are C_1: $\{s_1, s_2\}$; C_2: $\{s_2, s_3\}$. Describe *three* different fair divisions of the land.

23. Four partners are dividing a plot of land among themselves using the *lone-divider* method. After the divider D divides the land into four shares s_1, s_2, s_3, and s_4, the choosers C_1, C_2, and C_3 submit their bids for these shares.

(a) Suppose that the chooser's bids are C_1: $\{s_2\}$; C_2: $\{s_1, s_3\}$; C_3: $\{s_2, s_3\}$. Find a fair division of the land. Explain why this is the only possible fair division.

(b) Suppose that the chooser's bids are C_1: $\{s_2, s_3\}$; C_2: $\{s_1, s_3\}$; C_3: $\{s_1, s_2\}$. Describe *two* different fair divisions of the land.

(c) Suppose that the chooser's bids are C_1: $\{s_2\}$; C_2: $\{s_1, s_3\}$; C_3: $\{s_1, s_4\}$. Describe *three* different fair divisions of the land.

24. Four partners are dividing a plot of land among themselves using the *lone-divider* method. After the divider D divides the land into four shares s_1, s_2, s_3, and s_4, the choosers C_1, C_2, and C_3 submit their bids for these shares.

(a) Suppose that the chooser's bids are C_1: $\{s_2\}$; C_2: $\{s_1, s_3\}$; C_3: $\{s_2, s_3\}$. Find a fair division of the land. Explain why this is the only possible fair division.

(b) Suppose that the chooser's bids are C_1: $\{s_2\}$; C_2: $\{s_1, s_3\}$; C_3: $\{s_1, s_4\}$. Describe *three* different fair divisions of the land.

(c) Suppose that the chooser's bids are C_1: $\{s_2\}$; C_2: $\{s_1, s_2, s_3\}$; C_3: $\{s_2, s_3, s_4\}$. Describe *three* different fair divisions of the land.

25. Four partners are dividing a plot of land among themselves using the *lone-divider* method. After the divider D divides the land into four shares s_1, s_2, s_3, and s_4, the choosers C_1, C_2, and C_3 submit the following bids: C_1: $\{s_2\}$; C_2: $\{s_1, s_2\}$; C_3: $\{s_1, s_2\}$. For each of the following possible divisions, determine if it is a fair division or not. If not, explain why not.

(a) D gets s_3; s_1, s_2, and s_4 are recombined into a single piece that is then divided fairly among C_1, C_2, and C_3 using the lone-divider method for three players.

(b) D gets s_1; s_2, s_3, and s_4 are recombined into a single piece that is then divided fairly among C_1, C_2, and C_3 using the lone-divider method for three players.

(c) D gets s_4; s_1, s_2, and s_3 are recombined into a single piece that is then divided fairly among C_1, C_2, and C_3 using the lone-divider method for three players.

(d) D gets s_3; C_1 gets s_2; and s_1, s_4 are recombined into a single piece that is then divided fairly between C_2 and C_3 using the divider-chooser method.

26. Four partners are dividing a plot of land among themselves using the *lone-divider* method. After the divider D divides the land into four shares s_1, s_2, s_3, and s_4, the choosers C_1, C_2, and C_3 submit the following bids: C_1: $\{s_3, s_4\}$; C_2: $\{s_4\}$; C_3: $\{s_3\}$. For each of the following possible divisions, determine if it is a fair division or not. If not, explain why not.

(a) D gets s_1; s_2, s_3, and s_4 are recombined into a single piece that is then divided fairly among C_1, C_2, and C_3 using the lone-divider method for three players.

(b) D gets s_3; s_1, s_2, and s_4 are recombined into a single piece that is then divided fairly among C_1, C_2, and C_3 using the lone-divider method for three players.

(c) D gets s_2; s_1, s_3, and s_4 are recombined into a single piece that is then divided fairly among C_1, C_2, and C_3 using the lone-divider method for three players.

(d) C_2 gets s_4; C_3 gets s_3; s_1, s_2 are recombined into a single piece that is then divided fairly between C_1 and D using the divider-chooser method.

27. Five players are dividing a cake among themselves using the *lone-divider* method. After the divider D cuts the cake into five slices (s_1, s_2, s_3, s_4, s_5), the choosers C_1, C_2, C_3, and C_4 submit their bids for these shares.

(a) Suppose that the chooser's bids are C_1: $\{s_2, s_4\}$; C_2: $\{s_2, s_4\}$; C_3: $\{s_2, s_3, s_5\}$; C_4: $\{s_2, s_3, s_4\}$. Describe *two* different fair divisions of the cake. Explain why that's it—why there are no others.

(b) Suppose that the chooser's bids are C_1: $\{s_2\}$; C_2: $\{s_2, s_4\}$; C_3: $\{s_2, s_3, s_5\}$; C_4: $\{s_2, s_3, s_4\}$. Find a fair division of the cake. Explain why that's it—there are no others.

28. Five players are dividing a cake among themselves using the *lone-divider* method. After the divider D cuts the cake into five slices (s_1, s_2, s_3, s_4, s_5), the choosers C_1, C_2, C_3, and C_4 submit their bids for these shares.

(a) Suppose that the chooser's bids are C_1: $\{s_2, s_3\}$; C_2: $\{s_2, s_4\}$; C_3: $\{s_1, s_2\}$; C_4: $\{s_1, s_3, s_4\}$. Describe *three* different fair divisions of the land. Explain why that's it—why there are no others.

(b) Suppose that the chooser's bids are C_1: $\{s_1, s_4\}$; C_2: $\{s_2, s_4\}$; C_3: $\{s_2, s_4, s_5\}$; C_4: $\{s_2\}$. Find a fair division of the land. Explain why that's it—why there are no others.

29. Four partners (Egan, Fine, Gong, and Hart) jointly own a piece of land with a market value of \$480,000. The partnership is breaking up, and the partners decide to divide the land among themselves using the *lone-divider* method. Using a map, the divider divides the property into four parcels s_1, s_2, s_3, and s_4. The following table shows the value of the four parcels in the eyes of each partner, but some of the entries in the table are missing.

	s_1	s_2	s_3	s_4
Egan	$80,000	$85,000		$195,000
Fine		$100,000	$135,000	$120,000
Gong			$120,000	
Hart	$95,000	$100,000		$110,000

(a) Who was the divider? Explain.

(b) Determine each chooser's bid.

(c) Find a fair division of the property.

30. Four players (Abe, Betty, Cory, and Dana) are dividing a pizza worth $18.00 among themselves using the *lone-divider* method. The divider divides into four shares s_1, s_2, s_3, and s_4. The following table shows the value of the four shares in the eyes of each player, but some of the entries in the table are missing.

	s_1	s_2	s_3	s_4
Abe	$5.00	$5.00	$3.50	
Betty		$4.50		
Cory	$4.80	$4.20	$4.00	
Dana	$4.00	$3.75	$4.25	

(a) Who was the divider? Explain.

(b) Determine each chooser's bid.

(c) Find a fair division of the pizza.

Exercises 31 and 32 refer to the following situation: Jared, Karla, and Lori are planning to divide the half vegetarian–half meatball foot-long sub sandwich shown in Fig. 3-27 among themselves using the lone-divider method. Jared likes the meatball and vegetarian parts equally well; Karla is a strict vegetarian and does not eat meat at all; Lori likes the meatball part twice as much as the vegetarian part. (Assume that when the sandwich is cut, the cuts are always made perpendicular to the length of the sandwich. You can describe different shares of the sandwich using the ruler and interval notation—for example, [0, 6] describes the vegetarian half, [6, 8] describes one-third of the meatball half, etc.).

31. Suppose that Jared is the divider.

(a) Describe how Jared should cut the sandwich into three shares. Label the three shares s_1 for the leftmost piece, s_2 for the middle piece, and s_3 for the rightmost piece. Use the ruler and interval notation to describe the three shares. (Assume that Jared knows nothing about Karla and Lori's likes and dislikes.)

(b) Which of the three shares are fair shares to Karla?

(c) Which of the three shares are fair shares to Lori?

(d) Find three different fair divisions of the sandwich.

32. Suppose that Lori ends up being the divider.

(a) Describe how Lori should cut the sandwich into three shares. Label the three shares s_1 for the leftmost piece, s_2 for the middle piece, and s_3 for the rightmost piece. Use the ruler and interval notation to describe the three shares. (Assume that Lori knows nothing about Karla and Jared's likes and dislikes.)

(b) Which of the three shares are fair shares to Jared?

(c) Which of the three shares are fair shares to Karla?

(d) Suppose that Lori gets s_3. Describe how to proceed to find a fair division of the sandwich.

D The Lone-Chooser Method

Exercises 33 through 36 refer to the following situation: Angela, Boris, and Carlos are dividing the vanilla-strawberry cake shown in Fig. 3-30(a) using the lone-chooser method. Figure 3-30(b) shows how each player values each half of the cake. In your answers assume that all cuts are normal "cake cuts" from the center to the edge of the cake. You can describe each piece of cake by giving the angles of the vanilla and strawberry parts, as in "15° strawberry–40° vanilla" or "60° vanilla only."

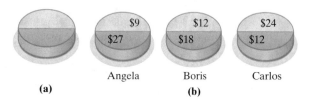

	Angela	Boris	Carlos
	$9 / $27	$12 / $18	$24 / $12

(a)   (b)

FIGURE 3-30

33. Suppose that Angela and Boris are the dividers and Carlos is the chooser. In the first division, Boris cuts the cake vertically through the center as shown in Fig. 3-31, with Angela choosing s_1 (the left half) and Boris s_2 (the right half). In the second division, Angela subdivides s_1 into three pieces and Boris subdivides s_2 into three pieces.

FIGURE 3-31

(a) Describe how Angela would subdivide s_1 into three pieces.

(b) Describe how Boris would subdivide s_2 into three pieces.

(c) Based on the subdivisions in (a) and (b), describe a possible final fair division of the cake.

(d) For the final fair division you described in (c), find the value (in dollars and cents) of each share in the eyes of the player receiving it.

34. Suppose that Carlos and Angela are the dividers and Boris is the chooser. In the first division, Carlos cuts the cake vertically through the center as shown in Fig. 3-31, with Angela choosing s_1 (the left half) and Carlos s_2 (the right half). In the second division, Angela subdivides s_1 into three pieces and Carlos subdivides s_2 into three pieces.

(a) Describe how Carlos would subdivide s_2 into three pieces.

(b) Describe how Angela would subdivide s_1 into three pieces.

(c) Based on the subdivisions in (a) and (b), describe a possible final fair division of the cake.

(d) For the final fair division you described in (c), find the value (in dollars and cents) of each share in the eyes of the player receiving it.

35. Suppose that Angela and Boris are the dividers and Carlos is the chooser. In the first division, Angela cuts the cake into two shares: s_1 (a 120° strawberry-only piece) and s_2 (a 60° strawberry–180° vanilla piece) as shown in Fig. 3-32. Boris picks the share he likes best, and Angela gets the other share. In the second division, Angela subdivides her share of the cake into three pieces and Boris subdivides his share of the cake into three pieces.

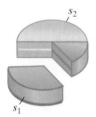

FIGURE 3-32

(a) Describe which share (s_1 or s_2) Boris picks and how he might subdivide it.

(b) Describe how Angela would subdivide her share of the cake.

(c) Based on the subdivisions in (a) and (b), describe a possible final fair division of the cake.

(d) For the final fair division you described in (c), find the value (in dollars and cents) of each share in the eyes of the player receiving it.

36. Suppose that Carlos and Angela are the dividers and Boris is the chooser. In the first division, Carlos cuts the cake into two shares: s_1 (a 135° vanilla-only piece) and s_2 (a 45° vanilla–180° strawberry piece) as shown in Fig. 3-33. Angela picks the share she likes better and Carlos gets the other share. In the second division, Angela subdivides her share of the cake into three pieces and Carlos subdivides his share of the cake into three pieces.

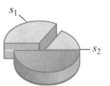

FIGURE 3-33

(a) Describe which share (s_1 or s_2) Angela picks and how she might subdivide it.

(b) Describe how Carlos might subdivide his share of the cake.

(c) Based on the subdivisions in (a) and (b), describe a possible final fair division of the cake.

(d) For the final fair division you described in (c), find the value (in dollars and cents) of each share in the eyes of the player receiving it.

Exercises 37 through 40 refer to the following: Arthur, Brian, and Carl are dividing the cake shown in Fig. 3-34 using the lone-chooser method. Arthur loves chocolate cake and orange cake equally but hates strawberry cake and vanilla cake. Brian loves chocolate cake and strawberry cake equally but hates orange cake and vanilla cake. Carl loves chocolate cake and vanilla cake equally but hates orange cake and strawberry cake. In your answers, assume all cuts are normal "cake cuts" from the center to the edge of the cake. You can describe each piece of cake by giving the angles of its parts, as in "15° strawberry–40° chocolate" or "60° orange only."

FIGURE 3-34

37. Suppose that Arthur and Brian are the dividers and Carl is the chooser. In the first division, Arthur cuts the cake vertically through the center as shown in Fig. 3-35 and Brian picks the share he likes better. In the second division, Brian subdivides the share he chose into three pieces and Arthur subdivides the other share into three pieces.

FIGURE 3-35

(a) Describe which share (s_1 or s_2) Brian picks and how he might subdivide it.

(b) Describe how Arthur might subdivide the other share.

(c) Based on the subdivisions in (a) and (b), describe a possible final fair division of the cake.

(d) For the final fair division you described in (c), find the value of each share (as a percentage of the total value of the cake) in the eyes of the player receiving it.

38. Suppose that Carl and Arthur are the dividers and Brian is the chooser. In the first division, Carl makes the cut shown in Fig. 3-36 and Arthur picks the share he likes better. In the second division, Arthur subdivides the share he chose into three pieces and Carl subdivides the other share into three pieces.

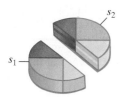

FIGURE 3-36

(a) Describe which share (s_1 or s_2) Arthur picks and how he might subdivide it.

(b) Describe how Carl might subdivide the other share.

(c) Based on the subdivisions in (a) and (b), describe a possible final fair division of the cake.

(d) For the final fair division you described in (c), find the value of each share (as a percentage of the total value of the cake) in the eyes of the player receiving it.

39. Suppose that Brian and Carl are the dividers and Arthur is the chooser. In the first division, Brian cuts the cake horizontally as shown in Fig. 3-37 and Carl picks the share he likes better. In the second division, Carl subdivides the share he chose into three pieces and Brian subdivides the other share into three pieces.

(a) Describe which share (s_1 or s_2) Carl picks and how he might subdivide it.

(b) Describe how Brian might subdivide the other share.

FIGURE 3-37

(c) Based on the subdivisions in (a) and (b), describe a possible final fair division of the cake.

(d) For the final fair division you described in (c), find the value of each share (as a percentage of the total value of the cake) in the eyes of the player receiving it.

40. Suppose that Arthur and Carl are the dividers and Brian is the chooser. In the first division, Arthur makes a vertical cut as shown in Fig. 3-38 and Carl picks the share he likes better. In the second division, Carl subdivides the share he chose into three pieces and Arthur subdivides the other share into three pieces.

FIGURE 3-38

(a) Describe which share (s_1 or s_2) Carl picks and how he might subdivide it.

(b) Describe how Arthur might subdivide the other share.

(c) Based on the subdivisions in (a) and (b), describe a possible final fair division of the cake.

(d) For the final fair division you described in (c), find the value of each share (as a percentage of the total value of the cake) in the eyes of the player receiving it.

41. Jared, Karla, and Lori are dividing the foot-long half meatball–half vegetarian sub shown in Fig. 3-39 using the *lone-chooser* method. Jared likes the vegetarian and meatball parts equally well, Karla is a strict vegetarian and does not eat meat at all, and Lori likes the meatball part twice as much as she likes the vegetarian part. Suppose that Karla and Jared are the dividers and Lori is the chooser. In the first division, Karla divides the sub into two shares (a left share s_1 and a right share s_2) and Jared picks the share he likes better. In the second division, Jared subdivides the share he picks into three pieces (a "left" piece J_1, a "middle" piece J_2, and a "right" piece J_3) and Karla subdivides the other share into three pieces (a "left" piece K_1, a "middle" piece K_2, and a "right" piece K_3). Assume that all cuts are perpendicular to the length of the sub. (You can describe the pieces of sub using the ruler and interval notation, as in $[3, 7]$ for the piece that starts at inch 3 and ends at inch 7.)

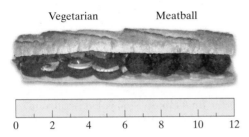

Vegetarian   Meatball

0   2   4   6   8   10   12

FIGURE 3-39

(a) Describe Karla's first division into s_1 and s_2.

(b) Describe which share (s_1 or s_2) Jared picks and how he would then subdivide it into the three pieces J_1, J_2, and J_3.

(c) Describe how Karla would subdivide her share into three pieces K_1, K_2, and K_3.

(d) Based on the subdivisions in (a), (b), and (c), describe the final fair division of the sub and give the value of each player's share (as a percentage of the total value of the sub) in the eyes of the player receiving it.

42. Jared, Karla, and Lori are dividing the foot-long half meatball–half vegetarian sub shown in Fig. 3-39 using the *lone-chooser* method. Jared likes the vegetarian and meatball parts equally well, Karla is a strict vegetarian and does not eat meat at all, and Lori likes the meatball part twice as much as she likes the vegetarian part. Suppose that Karla and Lori are the dividers and Jared is the chooser. In the first division, Lori divides the sub into two shares (a left share s_1 and a right share s_2) and Karla picks the share she likes better. In the second division, Karla subdivides the share she picks into three pieces (a "left" piece K_1, a "middle" piece K_2, and a "right" piece K_3) and Lori subdivides the other share into three pieces (a "left" piece L_1, a "middle" piece L_2, and a "right" piece L_3). Assume that all cuts are perpendicular to the length of the sub. (You can describe the pieces of sub using the ruler and interval notation, as in $[3, 7]$ for the piece that starts at inch 3 and ends at inch 7.)

(a) Describe Lori's first division into s_1 and s_2.

(b) Describe which share (s_1 or s_2) Karla picks and how she would then subdivide it into the three pieces K_1, K_2, and K_3.

(c) Describe how Lori would subdivide her share into three pieces L_1, L_2, and L_3.

(d) Based on the subdivisions in (a), (b), and (c), describe the final fair division of the sub and give the value of each player's share (as a percentage of the total value of the sub) in the eyes of the player receiving it.

E   The Last-Diminisher Method

43. A cake valued at \$30 is divided among five players (P_1, P_2, P_3, P_4, and P_5) using the *last-diminisher* method. The order in which the players play is P_1 first, P_2 second, and so on. Round 1: P_1 makes the first cut and makes a claim on a C-piece. Each of the other players, in turn, has a chance to evaluate the current C-piece (which may or may not be the same as the previous player's). The following table gives the value of the *current C-piece* in the eyes of a player at the time it is that player's turn to play:

	P_2	P_3	P_4	P_5
Value of C-piece	\$7.00	\$4.50	\$6.50	\$4.00

(a) Determine which players are *diminishers* in round 1.

(b) Determine which player gets a fair share at the end of round 1 and the value of that share to the player.

(c) (Fill in the blanks.) Round 2: _____ makes the first cut. The value of the R-piece to each of the remaining players is _____ \$ _____ (enter = or ≥ and the dollar amount).

44. A cake valued at \$30 is divided among four players (P_1, P_2, P_3, and P_4) using the *last-diminisher* method. The order in which the players play is P_1 first, P_2 second, and so on. Round 1: P_1 makes the first cut and makes a claim on a C-piece. Each of the other players, in turn, has a chance to evaluate the current C-piece (which may or may not be the same as the previous player's). The following table gives the value of the *current C-piece* in the eyes of a player at the time it is that player's turn to play:

	P_2	P_3	P_4
Value of C-piece	\$9.00	\$8.50	\$6.50

(a) Determine which players are *diminishers* in round 1.

(b) Determine which player gets a fair share at the end of round 1 and the value of that share to the player.

(c) (Fill in the blanks.) Round 2:_____ makes the first cut. The value of the R-piece to each of the remaining players is _____ \$ _____ (enter = or ≥ and the dollar amount).

45. A piece of land with a market value of $300,000 is divided among four players ($P_1, P_2, P_3,$ and P_4) using the *last-diminisher* method. The order in which the players play is P_1 first, P_2 second, and so on. Round 1: P_1 makes the first cut and makes a claim on a *C*-piece. Each of the other players, in turn, has a chance to evaluate the current *C*-piece (which may or may not be the same as the previous player's). The following table gives the value of the *current C-piece* in the eyes of a player at the time it is that player's turn to play:

	P_2	P_3	P_4
Value of the C-piece	$65,000	$85,000	$80,000

(a) Determine which players are *diminishers* in round 1.

(b) Determine which player gets a fair share at the end of round 1 and the value of that share to the player.

(c) (Fill in the blanks.) Round 2: _____ makes the first cut. The value of the *R*-piece to each of the remaining players is _____ $ _____ (enter = or ≥ and the dollar amount).

46. A piece of land with a market value of $300,000 is divided among five players ($P_1, P_2, P_3, P_4,$ and P_5) using the *last-diminisher* method. The order in which the players play is P_1 first, P_2 second, and so on. Round 1: P_1 makes the first cut and makes a claim on a *C*-piece. Each of the other players, in turn, has a chance to evaluate the current *C*-piece (which may or may not be the same as the previous player's). The following table gives the value of the *current C-piece* in the eyes of a player at the time it is that player's turn to play:

	P_2	P_3	P_4	P_5
Value of the C-piece	$50,000	$55,000	$85,000	$70,000

(a) Determine which players are *diminishers* in round 1.

(b) Determine which player gets a fair share at the end of round 1 and the value of that share to the player.

(c) (Fill in the blanks.) Round 2: _____ makes the first cut. The value of the *R*-piece to each of the remaining players is _____ $ _____ (enter = or ≥ and the dollar amount).

47. A piece of land is divided among 12 players ($P_1, P_2, P_3, \ldots, P_{12}$) using the *last-diminisher* method. The order in which the players play is P_1 first, P_2 second, and so on.

Round 1: P_1 makes the first cut; P_2 passes; P_3 diminishes; $P_4, P_5,$ and P_6 pass; P_7 diminishes; P_8 passes; P_9 diminishes; $P_{10}, P_{11},$ and P_{12} pass. Round 2: After the first claim is made, P_5 is the only diminisher. Round 3: After the first claim is made, all the remaining players pass.

(a) Which player gets his or her share at the end of round 1?

(b) Which player makes the first cut to start round 2?

(c) Which player gets his or her share at the end of round 2?

(d) Which player gets his or her share at the end of round 3?

(e) Which player makes the first cut to start round 4?

(f) What is the total number of rounds needed to play this game through to the end?

48. A cake is divided among six players ($P_1, P_2, P_3, P_4, P_5, P_6$) using the *last-diminisher* method. The order in which the players play is P_1 first, P_2 second, and so on. Round 1: After P_1 makes the first cut, $P_2, P_5,$ and P_6 are the only diminishers. Round 2: After the first claim is made, there are no diminishers. Round 3: After the first claim is made, every player is a diminisher.

(a) Which player gets his or her share at the end of round 1?

(b) Which player makes the first cut to start round 2?

(c) Which player gets his or her share at the end of round 2?

(d) Which player gets his or her share at the end of round 3?

(e) Which player makes the first cut to start round 4?

(f) What is the total number of rounds needed to play this game through to the end?

Exercises 49 and 50 refer to the following situation (see Exercises 33 through 36): Angela, Boris, and Carlos are dividing the vanilla-strawberry cake shown in Fig. 3-40(a) using the last-diminisher method. Figure 3-40(b) shows how each player values each half of the cake. In your answers, assume all cuts are normal "cake cuts" from the center to the edge of the cake. You can describe each piece of cake by giving the angles of the vanilla and strawberry parts, as in "15° strawberry–40° vanilla" or "60° vanilla only."

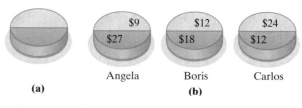

Angela   Boris   Carlos

(a)   **(b)**

FIGURE 3-40

49. Suppose that the order of play is Carlos first, Boris second, and Angela last. Round 1: Carlos claims the entire strawberry half of the cake. Round 2: The first player to cut claims a vanilla only–zero strawberry share.

(a) Who was the last diminisher in round 1? Describe his or her share.

(b) Who was the first player to cut in round 2?

(c) Who was the last diminisher in round 2? Describe his or her share.

(d) Describe the value of each player's share (as a fraction of the total value of the cake) in the eyes of the player receiving it. (*Hint*: Each of these fractions should have a different denominator.)

50. Suppose that the order of play is Angela first, Boris second, and Carlos last. Round 1: Angela claims a piece that is all strawberry. Round 2: The first player to cut claims a vanilla only–zero strawberry share.

(a) Who was the last diminisher in round 1? Describe his or her share.

(b) Who was the first player to cut in round 2?

(c) Who was the last diminisher in round 2? Describe his or her share.

(d) Describe the value of each player's share (as a fraction of the total value of the cake) in the eyes of the player receiving it. (*Hint*: Each of these fractions should have a different denominator.)

Exercises 51 and 52 refer to the following (see Exercises 41 and 42): Jared, Karla, and Lori are dividing the foot-long half meatball–half vegetarian sub shown in Fig. 3-41 using the last-diminisher method. Jared likes the vegetarian and meatball parts equally well, Karla is a strict vegetarian and does not eat meat at all, and Lori likes the vegetarian part three times as much as the meatball part. Assume that all players make their cuts perpendicular to the length of the sub. (You can describe the pieces of sub using the ruler and interval notation, as in [3, 7] for the piece that starts at inch 3 and ends at inch 7.)

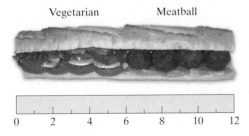

Vegetarian   Meatball

0   2   4   6   8   10   12

FIGURE 3-41

51. Suppose that the order of play is Lori first, Karla second, and Jared last. Round 1: Lori starts by claiming the C-piece $[0, x]$.

(a) Find the value of x in Lori's claim in round 1.

(b) Who was the last diminisher in round 1? Describe his or her share.

(c) Describe the final fair division of the sub.

52. Suppose that the order of play is Jared first, Lori second, and Karla last. Round 1: Jared starts by claiming the C-piece $[0, x]$.

(a) Find the value of x in Jared's claim in round 1.

(b) Who was the last diminisher in round 1? Describe his or her share.

(c) Describe the final fair division of the sub.

F The Method of Sealed Bids

53. Ana, Belle, and Chloe are dividing four pieces of furniture using the method of sealed bids. Their bids on each of the items are given in the following table.

	Ana	Belle	Chloe
Dresser	$150	$300	$275
Desk	$180	$150	$165
Vanity	$170	$200	$260
Tapestry	$400	$250	$500

(a) Find the value of each player's fair share.

(b) Describe the first settlement (who gets which item and how much do they pay or get in cash).

(c) Find the surplus after the first settlement is over.

(d) Describe the final settlement (who gets which item and how much do they pay or get in cash)

54. Andre, Bea, and Chad are dividing an estate consisting of a house, a small farm, and a painting using the method of sealed bids. Their bids on each of the items are given in the following table.

	Andre	Bea	Chad
House	$150,000	$146,000	$175,000
Farm	$430,000	$425,000	$428,000
Painting	$50,000	$59,000	$57,000

(a) Describe the first settlement of this fair division and compute the surplus.

(b) Describe the final settlement of this fair-division problem.

55. Five heirs (A, B, C, D, and E) are dividing an estate consisting of six items using the method of sealed bids. The heirs' bids on each of the items are given in the following table.

	A	B	C	D	E
Item 1	$352	$295	$395	$368	$324
Item 2	$98	$102	$98	$95	$105
Item 3	$460	$449	$510	$501	$476
Item 4	$852	$825	$832	$817	$843
Item 5	$513	$501	$505	$505	$491
Item 6	$725	$738	$750	$744	$761

(a) Find the value of each player's fair share.

(b) Describe the first settlement (who gets which item and how much do they pay or get in cash).

(c) Find the surplus after the first settlement is over.

(d) Describe the final settlement (who gets which item and how much do they pay or get in cash)

56. Alan, Bly, and Claire are dividing five items using the method of sealed bids. Their bids on each of the items are given in the following table.

	Alan	Bly	Claire
Item 1	$14,000	$12,000	$22,000
Item 2	$24,000	$15,000	$33,000
Item 3	$16,000	$18,000	$14,000
Item 4	$16,000	$16,000	$18,000
Item 5	$18,000	$24,000	$20,000

(a) Find the value of each player's fair share. (Round your answers to the nearest cent.)

(b) Describe the first settlement (who gets which item and how much do they pay or get in cash).

(c) Find the surplus after the first settlement is over.

(d) Describe the final settlement (who gets which item and how much do they pay or get in cash)

57. After breaking off their engagement, Angelina and Brad agree to divide their common assets (a plasma TV, a digital camera, a laptop computer, and an MP3 player) using the method of sealed bids. Their bids on the items are shown in the following table, but Angelina's bid on the laptop computer is missing.

	Angelina	Brad
Plasma TV	$2100	$2200
Camera	$500	$580
Laptop	?	$1600
MP3 Player	$300	$260

In the final settlement Angelina got the laptop computer, the MP3 player, and $355 in cash. Determine Angelina's bid on the laptop.

58. Alan, Bly, and Claire are dividing three items using the method of sealed bids. Their bids on each of the items are given in the following table, but Claire's bid on item 2 is missing.

	Alan	Bly	Claire
Item 1	$3400	$4000	$3600
Item 2	$500	$580	?
Item 3	$1800	$1600	$1500

In the final settlement Claire got item 2 and $1480 in cash. Determine Claire's bid on item 2.

59. Anne, Bette, and Chia jointly own a flower shop. They can't get along anymore and decide to break up the partnership using the method of sealed bids, with the understanding that one of them will get the flower shop and the other two will get cash. Anne bids $210,000, Bette bids $240,000, and Chia bids $225,000. How much money do Anne and Chia each get from Bette for their third share of the flower shop?

60. Al, Ben, and Cal jointly own a fruit stand. They can't get along anymore and decide to break up the partnership using the method of sealed bids, with the understanding that one of them will get the fruit stand and the other two will get cash. Al bids $156,000, Ben bids $150,000, and Cal bids $171,000. How much money do Al and Ben each get from Cal for their one-third share of the fruit stand?

G The Method of Markers

61. Three players (A, B, and C) are dividing the array of 13 items shown in Fig. 3-42 using the *method of markers*. The players' bids are as indicated in the figure.

(a) Which items go to A?

(b) Which items go to B?

(c) Which items go to C?

(d) Which items are left over?

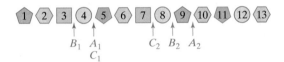

FIGURE 3-42

62. Three players (A, B, and C) are dividing the array of 13 items shown in Fig. 3-43 using the *method of markers*. The players' bids are as indicated in the figure.

FIGURE 3-43

(a) Which items go to A?

(b) Which items go to B?

(c) Which items go to C?

(d) Which items are left over?

63. Three players (A, B, and C) are dividing the array of 12 items shown in Fig. 3-44 using the *method of markers*. The players' bids are as indicated in the figure.

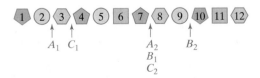

FIGURE 3-44

(a) Which items go to A?

(b) Which items go to B?

(c) Which items go to C?

(d) Which items are left over?

64. Three players (A, B, and C) are dividing the array of 12 items shown in Fig. 3-45 using the *method of markers*. The players' bids are indicated in the figure.

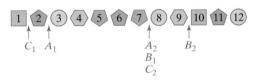

FIGURE 3-45

(a) Which items go to A?

(b) Which items go to B?

(c) Which items go to C?

(d) Which items are left over?

65. Five players ($A, B, C, D,$ and E) are dividing the array of 20 items shown in Fig. 3-46 using the *method of markers*. The players' bids are as indicated in the figure.

(a) Describe the allocation of items to each player.

(b) Which items are left over?

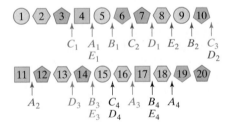

FIGURE 3-46

66. Four players (A, B, C, and D) are dividing the array of 15 items shown in Fig. 3-47 using the method of markers. The players' bids are as indicated in the figure.

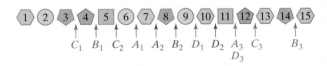

FIGURE 3-47

(a) Describe the allocation of items to each player.

(b) Which items are left over?

67. Quintin, Ramon, Stephone, and Tim are dividing a collection of 18 classic superhero comic books using the method of markers. The comic books are randomly lined up in the array shown below. (The W's are Wonder Woman comic books, the S's are Spider-Man comic books, the G's are Green Lantern comic books, and the B's are Batman comic books.)

W S S G S W W B G G G S G S G S B B

The value of the comic books in the eyes of each player is shown in the following table.

	Quintin	Ramon	Stephone	Tim
Each W is worth	$12	$9	$8	$5
Each S is worth	$7	$5	$7	$4
Each G is worth	$4	$5	$6	$4
Each B is worth	$6	$11	$14	$7

(a) Describe the placement of each player's markers. (Use $Q_1, Q_2,$ and Q_3 for Quintin's markers, $R_1, R_2,$ and R_3 for Ramon's markers, and so on.)

(b) Describe the allocation of comic books to each player and describe what comic books are left over.

68. Queenie, Roxy, and Sophie are dividing a set of 15 CDs—6 Beach Boys CDs, 6 Grateful Dead CDs, and 3 opera CDs using the method of markers. Queenie loves the Beach Boys, but hates the Grateful Dead and opera. Roxy loves the Grateful Dead and the Beach Boys equally well, but hates opera. Sophie loves the Grateful Dead and opera equally well, but hates the Beach Boys. The CDs are lined up in an array as follows:

O O O GD GD GD BB BB BB GD GD GD O O O

(O represents the opera CDs, GD the Grateful Dead CDs, and BB the Beach Boys CDs.)

(a) Describe the placement of each player's markers. (Use Q_1 and Q_2 for Queenie's markers, R_1 and R_2 for Roxy's markers, etc.)

(b) Describe the allocation of CDs to each player and describe what CDs are left over.

(c) Suppose that the players agree that each one gets to pick an extra CD from the leftover CDs. Suppose

that Queenie picks first, Sophie picks second, and Roxy picks third. Describe which leftover CDs each one would pick.

69. Ana, Belle, and Chloe are dividing 3 Snickers bars, 3 Nestle Crunch bars, and 3 Reese's Peanut Butter Cups among themselves. The easy solution would be for each of them to get one of each type of candy bar, but that's boring, so they decide to give the method of markers a try. The players' value systems are as follows: (1) Ana likes all candy bars the same; (2) Belle loves Nestle Crunch bars but hates Snickers bars and Reese's Peanut Butter Cups; (3) Chloe likes Reese's Peanut Butter Cups twice as much as she likes Snickers or Nestle Crunch bars. Suppose that the candy is lined up exactly as shown in Fig. 3-48.

FIGURE 3-48

(a) Describe the placement of each player's markers. (Use A_1 and A_2 for Ana's markers, B_1 and B_2 for Belle's markers, and C_1 and C_2 for Chloe's markers). (*Hint:* For each player, compute the value of each piece as a fraction of the value of the booty first. This will help you figure out where the players would place their markers.)

(b) Describe the allocation of candy bars to each player and which candy bars are left over.

(c) Suppose that the players decide to divide the leftover pieces by a random lottery in which each player gets to choose one piece. Suppose that Belle gets to choose first, Chloe second, and Ana last. Describe the division of the leftover pieces.

70. Arne, Bruno, Chloe, and Daphne are dividing 3 Snickers bars, 3 Nestle Crunch bars, 3 Reese's Peanut Butter Cups, and 6 Baby Ruths among themselves using the method of markers. Arne hates Snickers Bars, but likes Nestle Crunch bars, Reese's Peanut Butter Cups, and Baby Ruth equally well. Bruno hates Nestle Crunch bars, but likes Snickers bars, Reese's Peanut Butter Cups, and Baby Ruths equally well. Chloe hates Reese's Peanut Butter Cups and Baby Ruths, and likes Snickers three times as much as Nestle Crunch bars. Daphne hates Snickers and Nestle Crunch bars, and values a Reese's Peanut Butter Cup as equal to two-thirds the value of a Baby Ruth (i.e., 2 Baby Ruth bars equal

3 Reese's Cups). Suppose the candy is lined up exactly as shown in Fig. 3-49.

FIGURE 3-49

(a) Describe the placement of each player's markers. (*Hint*: For each player, compute the value of each piece as a fraction of the value of the booty first. This will help you figure out where the players would place their markers.)

(b) Describe the allocation of candy bars to each player and which candy bars are left over.

(c) After the allocation, each player is allowed to pick one candy bar. Will there be any arguments?

JOGGING

71. Three players (P_1, P_2, and P_3) agree to divide the property shown in Fig. 3-50 using the *last-diminisher method*. The players play in the order P_1, P_2, P_3. In round 1, P_1 claims the parcel marked C in the figure. Assume that P_2 and P_3 both value the land uniformly ($\$x$ per square meter throughout).

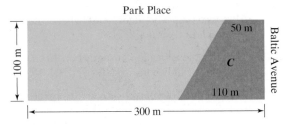

FIGURE 3-50

(a) Explain why P_2 and P_3 both pass in round 1 and P_1 ends up with C.

(b) Suppose that in round 2 the cut must be made parallel to Park Place. Describe the cut made by P_2 in this case.

(c) Suppose that in round 2 the cut must be made parallel to Baltic Avenue. Describe the cut made by P_2 in this case.

72. Three players (P_1, P_2, and P_3) agree to divide the property shown in Fig. 3-51 using the *last-diminisher method*. The players play in the order P_1, P_2, P_3. In round 1, P_1 claims the parcel marked C in the figure. Assume that P_2 and P_3 both value the land uniformly ($\$x$ per square meter throughout).

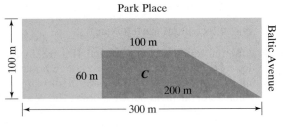

FIGURE 3-51

(a) Explain why P_2 and P_3 both pass in round 1 and P_1 ends up with C.

(b) Suppose that in round 2 the cut must be made parallel to Baltic Avenue. Describe the cut made by P_2 in this case.

73. Two partners (A and B) jointly own a business but wish to dissolve the partnership using the *method of sealed bids*. One of the partners will keep the business; the other will get cash for his half of the business. Suppose that A bids $\$x$ and B bids $\$y$. Assume that B is the high bidder. Describe the final settlement of this fair division in terms of x and y. (*Hint*: See Example 3.10 first.)

74. Three partners (A, B, and C) jointly own a business but wish to dissolve the partnership using the *method of sealed bids*. One of the partners will keep the business; the other two will each get cash for their one-third share of the business. Suppose that A bids $\$x$, B bids $\$y$, and C bids $\$z$. Assume that C is the high bidder. Describe the final settlement of this fair division in terms of x, y and z. (*Hint*: Try Exercises 59 and 60 first.)

75. Three players (A, B, and C) are sharing the chocolate-strawberry-vanilla cake shown in Fig. 3-52(a). Figure 3-52(b) shows the relative value that each player gives to each of the three parts of the cake. There is a way to divide this cake into three pieces (using just three cuts) so that each player ends up with a piece that he or she will value at exactly 50% of the value of the cake. Find such a fair division.

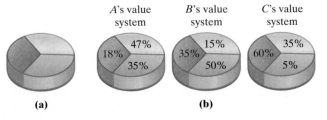

FIGURE 3-52

76. Angelina and Brad are planning to divide the chocolate-strawberry cake shown in Fig. 3-53(a) using the *divider-chooser* method, with Angelina being the divider. Suppose that Angelina values chocolate cake *three* times as much as she values strawberry cake. Figure 3-53(b) shows a generic cut made by Angelina dividing the cake into two shares s_1 and s_2 of equal value to her. Think of s_1 as a "$x°$ chocolate–$y°$ strawberry" share. For each given measure of the angle x, find the corresponding measure of the angle y.

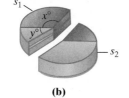

(a)   (b)

FIGURE 3-53

(a) $x = 60°$.

(b) $x = 72°$.

(c) $x = 108°$.

(d) $x = 120°$.

77. The benefits of honesty. Every fair-division method has a built-in disincentive for dishonest play. This exercise illustrates the disincentive for dishonest bidding in the *lone-divider method*. Four partners (Burly, Curly, Greedy, and Dandy) are dividing a million dollar property using the *lone-divider method*. Using a map, Dandy divides the property into four parcels s_1, s_2, s_3, and s_4. The table shows the value of the four parcels in the eyes of each partner.

	s_1	s_2	s_3	s_4
Dandy	$250,000	$250,000	$250,000	$250,000
Burly	$400,000	$200,000	$200,000	$200,000
Curly	$280,000	$320,000	$200,000	$200,000
Greedy	$320,000	$280,000	$280,000	$120,000

(a)   Describe the outcome of the fair division assuming that all players make honest bids.

(b)   Suppose that Burly and Curly both bid honestly, but Greedy decides to cheat and bid only for s_1 (figuring that he will then get that parcel). Under the right set of circumstances, Greedy could end up with a share worth only $220,000. Describe how this could happen.

*Exercises 78 and 79 show how the method of sealed bids can be used for the fair division of chores. We use the term **chore** here to describe any item that has a negative value (i.e., some-*

thing that you would rather not do or own, but if paid the right amount of money, you are willing to do or assume ownership of). To "bid" on a chore we use negative numbers: a bid of $–x$ means that you are willing to do that chore for x, but not for less. The highest bidder for a chore (i.e., the player willing to do the chore for the smallest payment) gets the chore.

78. Fair division of just chores. Three roommates share a house and are planning to divide the routine household chores among themselves. The following table shows each player's bids. Use the *method of sealed bids* to divide the chores fairly among the roommates. Who does which chores? Who gets paid, and how much? Who pays, and how much?

	Anne	Bess	Cindy
Clean bathrooms	$ −20	$ −30	$ −40
Cook dinners	$ −50	$ −10	$ −25
Wash dishes	$ −30	$ −20	$ −15
Mow the lawn	$ −30	$ −20	$ −10
Vacuum and dust	$ −20	$ −40	$ −15

79. Fair division of positive and negative items. Four college roommates are going their separate ways after graduation and want to divide up their jointly owned goods (furniture, stereo, etc.) as well as the chores (cleanup and repair of apartment) using the method of sealed bids. Their bids on the items are shown in the following table. Use the method of sealed bids to divide the items (good and bad) fairly among the roommates. Who gets what? Who does what? Who gets paid, and how much? Who pays, and how much?

	Arne	Brent	Carlos	Dale
Stereo	300	250	200	280
Couch	200	350	300	100
Table	250	200	240	80
Desk	150	150	200	220
Cleaning rugs	−80	−70	−100	−60
Patching holes in wall	−60	−30	−60	−40
Fixing broken window	−60	−50	−80	−80

80. **Standoffs in the lone-divider method.** In the lone-divider method, a *standoff* occurs when a set of k choosers are bidding for less than k items. The types of *standoffs* possible depend on the number of players. With $N = 3$ players, there is only *one* type of standoff—the two choosers are bidding for the same item. With $N = 4$ players, there are *three* possible types of standoffs: two choosers are bidding on the same item, or three choosers are bidding on the same item, or three choosers are bidding on just two items. With $N = 5$ players, there are *six* possible types of standoffs, and the number of possible types of standoffs increases rapidly as the number of players increases.

   (a) List the six possible types of *standoffs* with $N = 5$ players.

   (b) What is the number of possible types of *standoffs* with $N = 6$ players?

   (c) What is the number of possible types of *standoffs* with N players? Give your answer in terms of N. (*Hint:* You will need to use the formula given in Chapter 1, Section 5, for computing the sum of consecutive integers.)

RUNNING

81. Suppose that N players bid on M items using the method of sealed bids. Let T denote the table with M rows (one for each item) and N columns (one for each player) containing all the player's bids (i.e., the entry in column j, row k represents player j's bid for item k). Let $c_1, c_2, \ldots, c_N$ denote, respectively, the sum of the entries in column 1, column 2, ..., column N of T, and let $r_1, r_2, \ldots, r_M$ denote, respectively, the sum of the entries in row 1, row 2, ..., row M of T. Let $w_1, w_2, \ldots, w_M$ denote the winning bids for items $1, 2, \ldots, M$, respectively (i.e., w_1 is the largest entry in row 1 of T, w_2 is the largest entry in row 2, etc.). Let S denote the surplus money left after the first settlement.

   (a) Show that
   $$S = (w_1 + w_2 + \cdots + w_M) - \frac{(c_1 + c_2 + \cdots + c_N)}{N}.$$

   (b) Using (a), show that
   $$S = \left(w_1 - \frac{r_1}{N}\right) + \left(w_2 - \frac{r_2}{N}\right) + \cdots + \left(w_M - \frac{r_M}{N}\right).$$

   (c) Using (b), show that $S \geq 0$.

   (d) Describe the conditions under which $S = 0$.

82. **Efficient and envy-free fair divisions.** A fair division is called **efficient** if there is no other fair division that gives *every* player a share that is as good or better (i.e.,

any other fair division that gives some players a better share must give some other players a worse share). A fair division is called **envy-free** if every player ends up with a share that he or she feels is *as good as or better than* that of any other player.

Suppose that three partners (A, B, and C) jointly own a piece of land that has been subdivided into six parcels ($s_1, s_2, \ldots, s_6$). The partnership is splitting up, and the partners are going to divide fairly the six parcels among themselves. The following table shows the value of each parcel in the eyes of each partner.

	A	B	C
s_1	\$20,000	\$16,000	\$19,000
s_2	\$19,000	\$18,000	\$18,000
s_3	\$18,000	\$19,000	\$15,000
s_4	\$16,000	\$20,000	\$12,000
s_5	\$15,000	\$15,000	\$20,000
s_6	\$12,000	\$12,000	\$16,000

   (a) Find a fair division of the six parcels among the three partners that is *efficient*.

   (b) Find a fair division of the six parcels among the three partners that is *envy-free*.

   (c) Find a fair division of the six parcels among the three partners that is *efficient but not envy-free*.

   (d) Find a fair division of the six parcels among the three partners that is *envy-free but not efficient*.

83. **Asymmetric method of sealed bids.** Suppose that an estate consisting of M indivisible items is to be divided among N heirs ($P_1, P_2, \ldots, P_N$) using the *method of sealed bids*. Suppose that Grandma's will stipulates that P_1 is entitled to $x_1\%$ of the estate, P_2 is entitled to $x_2\%$ of the estate, ..., P_N is entitled to $x_N\%$ of the estate. The percentages add up to 100%, but they are not all equal (Grandma loved some grandchildren more than others). Describe a variation of the method of sealed bids that ensures that each player receives a "fair share" (i.e., P_1 receives a share that she considers to be worth at least $x_1\%$ of the estate, P_2 receives a share that he considers to be worth at least $x_2\%$ of the estate, etc.).

84. **Lone-chooser is a fair-division method.** Suppose that N players divide a cake using the lone-chooser method. The chooser is C and the dividers are $D_1, D_2, \ldots, D_{N-1}$. Explain why, when properly played, the method guarantees to each player a fair share. (You will need one argument for the dividers and a different argument for the chooser.)

PROJECTS AND PAPERS

A   Envy-Free Fair Division

An *envy-free fair division* is a fair division in which each player ends up with a share that he or she feels is as good as or better than that of any other player. Thus, in an envy-free fair division a player would never envy or covet another player's share. Over the last twenty years several envy-free fair-division methods have been developed. (Some of these are discussed in references 7 and 15.)

Write a paper discussing the topic of envy-free fair division.

Notes: Some ideas of topics for your paper: (1) Discuss how envy-free fair division differs from the (proportional) type of fair division discussed in this chapter. (2) Describe a continuous envy-free fair-division method for $N = 3$ players. (3) Give an outline of the Brams-Taylor method for continuous envy-free fair division for any number of players.

B   Fair Divisions with Unequal Shares

All the fair-division problems we discussed in this chapter were based on the assumption of *symmetry* (i.e., all players have equal rights in the division). Sometimes, players are not all equal and are entitled to larger or smaller shares than other players. This type of fair-division problem is called an *asymmetric* fair division (*asymmetric* means that the players are not all equal in their rights.) For example, Grandma's will may stipulate that her estate is to be divided as follows: Art is entitled to 25%, Betty is entitled to 35%, Carla is entitled to 30%, and Dave is entitled to 10%. (After all, it is her will, and if she wants to be difficult, she can!)

Write a paper discussing how some of the fair-division methods discussed in this chapter can be adapted for the case of asymmetric fair division. Discuss at least one discrete and one continuous asymmetric fair-division method.

REFERENCES AND FURTHER READINGS

1. Ariely, Dan, *Predictably Irrational: The Hidden Forces That Shape Our Decisions*. New York: HarperCollins, 2008.

2. Brams, Steven, Paul H. Edelman, and Peter C. Fishburn, "Paradoxes of Fair Division," *Journal of Philosophy*, 98 (2001), 300–314.

3. Brams, Steven, and Alan Taylor, *Fair Division: From Cake-Cutting to Dispute Resolution*. Cambridge, England: Cambridge University Press, 1996.

4. Fink, A. M., "A Note on the Fair Division Problem," *Mathematics Magazine*, 37 (1964), 341–342.

5. Gardner, Martin, *aha! Insight*. New York: W. H. Freeman, 1978.

6. Hill, Theodore, "Mathematical Devices for Getting a Fair Share," *American Scientist*, 88 (2000), 325–331.

7. Hively, Will, "Dividing the Spoils," *Discover*, 16 (1995), 49–57.

8. Jones, Martin, "A Note on a Cake Cutting Algorithm of Banach and Knaster," *American Mathematical Monthly*, 104 (1997), 353–355.

9. Kuhn, Harold W., "On Games of Fair Division," *Essays in Mathematical Economics*, Martin Shubik, ed. Princeton, NJ: Princeton University Press, 1967, 29–37.

10. Peterson, Elisha, and Francis E. Su, "Four-Person Envy-Free Chore Division." *Mathematics Magazine*, 75 (2002), 117–122.

11. Robertson, Jack, and William Webb, *Cake Cutting Algorithms: Be Fair If You Can*. Natick, MA: A. K. Peters, 1998.

12. Steinhaus, Hugo, *Mathematical Snapshots*, 3rd ed. New York: Dover, 1999.

13. Steinhaus, Hugo, "The Problem of Fair Division," *Econometrica*, 16 (1948), 101–104.

14. Stewart, Ian, "Fair Shares for All," *New Scientist*, 146 (June, 1995), 42–46.

15. Stewart, Ian, "Mathematical Recreations: Division without Envy," *Scientific American*, (1999), 110–111.

4 The Mathematics of Apportionment

Making the Rounds

In the stifling heat of the Philadelphia summer of 1787, delegates from the 13 states met to draft a Constitution for a new nation. Except for Thomas Jefferson (then minister to France) and Patrick Henry (who refused to participate), all the main names of the American Revolution were there — George Washington, Ben Franklin, Alexander Hamilton, John Adams.

Without a doubt, the most important and heated debate at the Constitutional Convention concerned the makeup of the legislature. The small states wanted all states to have the same number of representatives; the larger states wanted some form of proportional representation. The Constitutional compromise, known as the Connecticut Plan, was a Senate, in which every state has two senators, and a House of Representatives, in which each state has a number of representatives that is a function of its population.

Signing the Constitution, by Hy Hintermeister (1897–1792).

W hile the Constitution makes clear that seats in the House of Representatives are to be allocated to the states based on their populations
("according to their respective Numbers"), it does not prescribe a specific
method for the calculations. Undoubtedly, the Founding Fathers felt
that this was a relatively minor detail—a matter of simple arithmetic that could be easily figured out and agreed upon by reasonable people. Certainly it was not the kind of thing to clutter
a Constitution with or spend time arguing over in the heat of
the summer. What the Founding Fathers did not realize is that
Article I, Section 2, set the Constitution of the United States
into a collision course with a mathematical iceberg known
today (but certainly not then) as *the apportionment problem.*

This chapter is an excursion into the mathematics behind the apportionment problem, and there are many interesting and surprising twists and turns to this excursion. We will start the chapter
with an introduction to the basic concepts of apportionment: What is an apportionment problem? What is
the nature of the problem? What are the mathematical issues we must deal with? Why
should we care? In Sections 4.2 through 4.6
we will discuss four of the best-known apportionment methods that have been proposed over the years to solve apportionment
problems. Surprisingly, the names associated
with these methods are not names you
would expect to find attached to mathematical
procedures—Alexander Hamilton, Thomas
Jefferson, John Quincy Adams, and Daniel
Webster. This is a testament to the historical
importance of the topic of this chapter.
While our primary focus will be the mathe-
matical aspects of apportionment, as we navigate the mathematics we will also
get a glimpse of a little known but important chapter in U.S. history.

4.1 Apportionment Problems

■ **ap·pôr′·tion**: to *divide and assign* in due and proper *proportion* or according to some *plan*.
          —*Webster's New Twentieth Century Dictionary*

Obviously, the word *apportion* is the key word in this chapter. There are two critical elements in the dictionary definition of the word: (1) We are dividing and assigning things, and (2) we are doing this on a proportional basis and in a planned, organized fashion.

We will start this section with a pair of examples that illustrate the nature of the problem we are dealing with. These examples will raise several questions but not give any answers. Most of the answers will come later.

EXAMPLE 4.1   **Kitchen Capitalism**

Mom has a total of 50 identical pieces of candy (let's say caramels), which she is planning to divide among her five children (this is the *division* part). Like any good mom, she is intent on doing this fairly. Of course, the easiest thing to do would be to give each child 10 caramels—by most standards, that would be fair. Mom, however, is thinking of the long-term picture—she wants to teach her children about the value of work and about the relationship between work and reward. This leads her to the following idea: She announces to the kids that the candy is going to be divided at the end of the week in proportion to the amount of time each of them spends helping with the weekly kitchen chores—if you worked twice as long as your brother you get twice as much candy, and so on (this is the *due and proper proportion* part). Unwittingly, mom has turned this division problem into an apportionment problem.

At the end of the week, the numbers are in. Table 4-1 shows the amount of work done by each child during the week. (Yes, mom did keep up-to-the-minute records!)

TABLE 4-1   **Amount of Work (in Minutes) per Child**

Child	Alan	Betty	Connie	Doug	Ellie	**Total**
Minutes worked	150	78	173	204	295	900

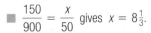

$\frac{150}{900} = \frac{x}{50}$ gives $x = 8\frac{1}{3}$.

According to the ground rules, Alan, who worked 150 out of a total of 900 minutes, should get $8\frac{1}{3}$ pieces. Here comes the problem: Since the pieces of candy are indivisible, it is impossible for Alan to get his $8\frac{1}{3}$ pieces—he can get 8 pieces (and get shorted) or he can get 9 pieces (and someone else will get shorted). A similar problem occurs with each of the other children. Betty's exact fair share should be $4\frac{1}{3}$ pieces; Connie's should be $9\frac{11}{18}$ pieces; Doug's, $11\frac{1}{3}$ pieces; and Ellie's, $16\frac{7}{18}$ pieces. (Be sure to double-check these figures!) Because none of these shares can be realized, an absolutely fair apportionment of the candy is going to be impossible. What should mom do?

Example 4.1 shows all the elements of an **apportionment problem**—there are objects to be divided (the pieces of candy), and there is a proportionality criterion for the division (number of minutes worked during the week). We will say that the pieces of candy are *apportioned* to the kids, and we will describe the final solution as an *apportionment* (Alan's *apportionment* is *x* pieces, Betty's *apportionment* is *y* pieces, etc.).

Let's now consider a seemingly different apportionment problem.

EXAMPLE 4.2   **The Intergalactic Congress of Utopia**

It is the year 2525, and all the planets in the Utopia galaxy have finally signed a peace treaty. Five of the planets (Alanos, Betta, Conii, Dugos, and Ellisium) decide to join forces and form an Intergalactic Federation. The Federation will be ruled by an Intergalactic Congress consisting of 50 "elders," and the 50 seats in the Intergalactic Congress are to be *apportioned* among the planets according to their respective populations.

The population data for each of the planets (in billions) are shown in Table 4-2. Based on these population figures, what is the correct *apportionment* of seats to each planet? (Good question. We will come back to it in Section 4.3.)

TABLE 4-2	Intergalactic Federation: Population Figures (in Billions) for 2525					
Planet	Alanos	Betta	Conii	Dugos	Ellisium	**Total**
Population	150	78	173	204	295	900

Source: Intergalactic Census Bureau.

Example 4.2 is another example of an apportionment problem—here the objects being divided are the 50 seats in the Intergalactic Congress, and the proportionality criterion is population. But there is more to it than that. The observant reader may have noticed that the numbers in Examples 4.1 and 4.2 are identical—it is only the framing of the problem that has changed. Mathematically speaking, Examples 4.1 and 4.2 are one and the same apportionment problem—solve either one and you have solved both!

Between the extremes of apportioning the seats in the Intergalactic Congress of Utopia (important, but too far away!) and apportioning the candy among the children in the house (closer to home, but the galaxy will not come to an end if the apportionments aren't done right!) fall many other real-life apportionment problems that are both important and relevant: apportioning nurses to shifts in a hospital, apportioning telephone calls to switchboards in a network, apportioning classrooms to departments in a university, and so on.

We will now introduce the most important apportionment example of this chapter. We will return to this example many times. While the story behind the example is obviously fiction, the issues it raises are real.

EXAMPLE 4.3   **The Congress of Parador**

Parador is a small republic located in Central America and consists of six states: Azucar, Bahia, Cafe, Diamante, Esmeralda, and Felicidad (*A*, *B*, *C*, *D*, *E*, and *F* for short). There are 250 seats in the Congress, which, according to the laws of Parador, are to be apportioned among the six states in proportion to their respective populations. What is the "correct" apportionment?

Table 4-3 shows the population figures for the six states according to the most recent census.

TABLE 4-3	Republic of Parador (Populations by State)						
State	*A*	*B*	*C*	*D*	*E*	*F*	**Total**
Population	1,646,000	6,936,000	154,000	2,091,000	685,000	988,000	12,500,000

The first step we will take to tackle this apportionment problem is to compute the *population to seats* ratio, called the **standard divisor (SD)**. In the case of Parador, the standard divisor is 12,500,000/250 = 50,000.

The standard divisor $SD = 50,000$ tells us that in Parador, each seat in the Congress corresponds to 50,000 people. We can now use this yardstick to find the number of seats that each state *should get by the proportionality criterion* — all we have to do is divide the state's population by 50,000. For example, take state A. If we divide the population of A by the standard divisor, we get 1,646,000/50,000 = 32.92. This number is called the **standard quota** (sometimes also known as the *exact* or *fair quota*) of state A. If seats in the Congress could be apportioned in fractional parts, then the fair and exact apportionment to A would be 32.92 seats, but, of course, this is impossible — seats in the Congress are indivisible objects! Thus, the standard divisor of 32.92 is a standard that we cannot meet, although in this case we might be able to come close to it: The next best solution might be to apportion 33 seats to state A. Hold that thought!

Using the standard divisor $SD = 50,000$, we can quickly find the standard quotas of each of the other states. These are shown in Table 4-4 (rounded to two decimal places). Notice that the sum of the standard quotas equals 250, the number of seats being apportioned.

■ Typically, the standard divisor is not going to turn out to be such a nice whole number, but it is always going to be a rational number.

■ Standard quota = (state population) ÷ 50,000

TABLE 4-4   Republic of Parador: Standard Quotas for Each State ($SD = 50,000$)

State	A	B	C	D	E	F	**Total**
Population	1,646,000	6,936,000	154,000	2,091,000	685,000	988,000	12,500,000
Standard quota	32.92	138.72	3.08	41.82	13.70	19.76	250

Once we have computed the standard quotas, we get to the heart of the apportionment problem: How should we round these quotas into whole numbers? At first glance, this seems like a dumb question. After all, we all learned in school how to round decimals to whole numbers — round down if the fractional part is less than 0.5, round up otherwise. This kind of rounding is called *rounding to the nearest integer*, or simply *conventional rounding*. Unfortunately, conventional rounding will not work in this example, and Table 4-5 shows why not — we would end up giving out 251 seats in the Congress, and there are only 250 seats to give out! (Add an extra seat, you say? Sorry, this is not a banquet. The number of seats is fixed — we don't have the luxury of adding or subtracting seats!)

TABLE 4-5   Conventional Rounding of the Standard Quotas

State	A	B	C	D	E	F	**Total**
Standard quota	32.92	138.72	3.08	41.82	13.70	19.76	250.00
Nearest integer	33	139	3	42	14	20	**251**

So now that we know that conventional rounding of the standard quotas is not going to give us the solution to the apportionment of Parador's Congress, what do we try next? For now, we leave this example — a simple apportionment problem looking for a simple solution — with no answers (but, we hope, with some gained insight!). Our search for a good solution to this problem will be the theme of our journey through the rest of this chapter.

Apportionment: Basic Concepts and Terminology

Apportionment problems can arise in a variety of real-life applications—dividing candy among children (see Example 4.1), assigning nurses to shifts (see Exercise 4.4), assigning buses to routes (see Exercise 4.3), and so on. But the gold standard for apportionment applications is the allocation of seats in a legislature, and thus it is standard practice to borrow the terminology of legislative apportionment and apply it to apportionment problems in general.

The basic elements of every apportionment problem are as follows:

- **The "states."** This is the term we will use to describe the parties having a stake in the apportionment. Unless they have specific names (Azucar, Bahia, etc.), we will use $A_1, A_2, \ldots, A_N$ to denote the states.

- **The "seats."** This term describes the set of M *identical, indivisible objects* that are being divided among the N states. For convenience, we will assume that there are more seats than there are states, thus ensuring that every state can potentially get a seat. (This assumption does not imply that every state *must* get a seat! Such a requirement is not part of the general apportionment problem, although it is part of the constitutional requirements for the apportionment of the U.S. House of Representatives.)

- **The "populations."** This is a set of N positive numbers (for simplicity we will assume that they are whole numbers) that are used as the basis for the apportionment of the seats to the states. We will use $p_1, p_2, \ldots, p_N$ to denote the state's respective populations and P to denote the total population ($P = p_1 + p_2 + \cdots + p_N$).

Two of the most important concepts of the chapter are the *standard divisor* and the *standard quotas*. We can now formally define these concepts using our new terminology and notation.

$$SD = \frac{\text{Population}}{M}$$

- **The standard divisor (SD).** This is the ratio of population to seats. It gives us a unit of measurement (SD people $= 1$ seat) for our apportionment calculations.

$$\text{standard quota} = \frac{\text{state population}}{SD}$$

- **The standard quotas.** The standard quota of a state is the exact fractional number of seats that the state would get if fractional seats were allowed. We will use the notation $q_1, q_2, \ldots, q_N$ to denote the standard quotas of the respective states. To find a state's standard quota, we divide the state's population by the standard divisor. (In general, the standard quotas can be expressed as decimals or fractions—it would be almost a miracle if one of them turned out to be a whole number—and when using the decimal form, it's customary to round them to two or three decimal places.)

- **Upper and lower quotas.** Associated with each standard quota are two other important numbers—the **lower quota** (the standard quota rounded down) and the **upper quota** (the standard quota rounded up). In the unlikely event that the standard quota is a whole number, the lower and upper quotas are the same. We will use L's to denote lower quotas and U's to denote upper quotas. For example, the standard quota $q_1 = 32.92$ has lower quota $L_1 = 32$ and upper quota $U_1 = 33$.

One of the important morals of Example 4.3 is that the intuitively obvious approach to solving an apportionment problem (take the standard quotas and round them to either the lower or upper quotas using conventional rounding) is not a strategy we can rely on, since the final apportionment may not add up to M seats. In Example 4.3 we ended up apportioning more seats than we were supposed to. In other apportionment problems we could end up apportioning fewer seats than we are supposed to. Sometimes the stars are aligned and we end up apportioning the right number of seats. As a general method, this will not do.

Our main goal in this chapter is to discover a "good" **apportionment method**—a reliable procedure that (1) will always produce a valid apportionment (exactly *M* seats are apportioned) and (2) will always produce a "fair" apportionment. In this quest we will discuss several different methods and find out what is good and bad about each one.

4.2   Hamilton's Method and the Quota Rule

Hamilton's method can be described quite briefly: *Every state gets at least its lower quota. As many states as possible get their upper quota, with the one with highest residue (i.e., fractional part) having first priority, the one with second highest residue second priority, and so on.* A little more formally, it goes like this:

Alexander Hamilton (1757–1804). Hamilton's method (also known as *Vinton's method* or the *method of largest remainders*) was used in the United States only between 1850 and 1900, but it is still used today to apportion the legislatures of Costa Rica, Namibia, and Sweden.

HAMILTON'S METHOD

- **Step 1.** Calculate each state's standard quota.
- **Step 2.** Give to each state its *lower quota*.
- **Step 3.** Give the surplus seats (one at a time) to the states with the largest *residues* (fractional parts) until there are no more surplus seats.

(**EXAMPLE 4.4**)   **Parador's Congress (Hamilton's Method)**

As promised, we are revisiting the Parador Congress apportionment problem. We are now going to find our first real solution to this problem—the solution given by Hamilton's method (sometimes, for the sake of brevity, we will call it the *Hamilton apportionment*). Table 4-6 shows all the details and speaks for itself. (Reminder to the reader: The standard quotas in the second row of the table were computed in Example 4.3.)

TABLE 4-6	Parador's Congress: Hamilton Apportionment						
State	*A*	*B*	*C*	*D*	*E*	*F*	**Total**
Standard quota	32.92	138.72	3.08	41.82	13.70	19.76	250.00
Lower quota	32	138	3	41	13	19	246
Residue	0.92	0.72	0.08	0.82	0.70	0.76	4.00
Order of Surplus	First	Last		Second		Third	
Apportionment	33	139	3	42	13	20	250

At first glance, Hamilton's method appears to be quite fair. However, a careful look at Example 4.4 already shows some hints of possible unfairness. Compare the fates of state B, with a residue of 0.72, and state E, with a residue of 0.70. State B gets the last surplus seat; state E gets nothing! Sure enough, 0.72 is more than 0.70, so following the rules of Hamilton's method, we give priority to B. By the same token, B is a huge state, and as a percentage of its population the 0.72 represents an insignificant amount, whereas state E is a relatively small state, and its residue of 0.70 (for which it gets nothing) represents more than 5% of its population ($0.70/13.70 \approx 0.051 = 5.1\%$).

■ The idea of using the relative size of the residues to determine the order of priority for the surplus seats is the basis of an apportionment method known as *Lowndes's method*. For more on Lowndes's method, see Exercises 62 and 63.

It could be reasonably argued that Hamilton's method has a major flaw in the way it relies entirely on the size of the residues without consideration of what those residues represent as a percent of the state's population. In so doing, Hamilton's method creates a systematic bias in favor of larger states over smaller ones. This is bad—a good apportionment method should be *population neutral*, meaning that it should not be biased in favor of large states over small ones or vice versa. In the next section we will see that bias in favor of large states is only one of the many serious flaws of Hamilton's method.

To be totally fair, Hamilton's method has two important things going for it: (1) It is very easy to understand, and (2) it satisfies an extremely important requirement for fairness called the *quota rule*.

The Quota Rule

Suppose that the standard quota of state A is $q = 38.59$. This number represents the benchmark for what a fair apportionment of seats to A ought to be. Since giving A exactly 38.59 seats is impossible, it is reasonable to argue that A should get either its upper quota ($U = 39$ seats) or, in the worst of cases, its lower quota ($L = 38$ seats). Thus, no reasonable or fair apportionment should give A, say, 37 seats, or, for that matter, 40 seats. Either of these outcomes would be a clear violation of what seems like a fundamental requirement for fairness—*a state should not be apportioned a number of seats smaller than its lower quota or larger than its upper quota*. This rule is known as the *quota rule*.

> **QUOTA RULE**
>
> No state should be apportioned a number of seats smaller than its lower quota or larger than its upper quota. (When a state is apportioned *a number smaller than its lower quota*, we call it a **lower-quota violation**; when a state is apportioned *a number larger than its upper quota*, we call it an **upper-quota violation**.)

An apportionment method that guarantees that every state will be apportioned either its lower quota or its upper quota is said to *satisfy the quota rule*. It is not hard to see that Hamilton's method satisfies the quota rule: Step 2 of Hamilton's method hands out to each state its lower quota. Right off the bat this guarantees that there will be no lower-quota violations. In Step 3 some states get one extra seat, some get none; no state can get more than one. This guarantees that there will be no upper-quota violations.

4.3   The Alabama and Other Paradoxes

The most serious (in fact, the fatal) flaw of Hamilton's method is commonly known as the **Alabama paradox**. In essence, the Alabama paradox occurs when an *increase in the total number of seats being apportioned, in and of itself, forces a state to lose one of its seats*. The best way to understand what this means is to look carefully at the following example.

(**EXAMPLE 4.5**)   **More Seats Means Fewer Seats**

The small country of Calavos consists of three states: Bama, Tecos, and Ilnos. With a total population of 20,000 and 200 seats in the House of Representatives, the apportionment of the 200 seats under Hamilton's method is shown in Table 4-7.

TABLE 4-7	Hamilton Apportionment for $M = 200$ ($SD = 100$)				
State	Population	Quota	Lower quota	Surplus	Apportionment
Bama	940	9.4	9	First	**10**
Tecos	9030	90.3	90	0	90
Ilnos	10,030	100.3	100	0	100
Total	20,000	200.0	199	1	**200**

Now imagine that overnight the number of seats is increased to 201, but nothing else changes. Since there is one more seat to give out, the apportionment has to be recomputed. Table 4-8 shows the new apportionment (still using Hamilton's method) for a House with 201 seats. (You are strongly encouraged to verify these short calculations. Notice that for $M = 200$, the SD is 100; for $M = 201$, the SD drops to 99.5.)

TABLE 4-8	Hamilton Apportionment for $M = 201$ ($SD \approx 99.5$)				
State	Population	Quota	Lower quota	Surplus	Apportionment
Bama	940	9.45	9	0	**9**
Tecos	9030	90.75	90	Second	91
Ilnos	10,030	100.80	100	First	101
Total	20,000	201.00	199	2	**201**

The shocking part of this story is the fate of Bama, the "little guy." When the House of Representatives had 200 seats, Bama got 10 seats, but when the number of seats to be divided increased to 201, Bama's apportionment went down to 9 seats. How did this paradox occur? Notice the effect of the increase in M on the size of the residues: In a House with 200 seats, Bama is at the head of the priority line for surplus seats, but when the number of seats goes up to 201, Bama gets shuffled to the back of the line. Welcome to the wacky arithmetic of Hamilton's method!

Example 4.5 illustrates the quirk of arithmetic behind the Alabama paradox: When we increase the number of seats to be apportioned, each state's standard quota goes up, but not by the same amount. As the residues change, some states can move ahead of others in the priority order for the surplus seats. This can result in some state or states losing seats they already had.

Historical note. The origin of the name *Alabama paradox* can be traced to the events surrounding the 1882 apportionment of the U.S. House of Representatives. As different apportionment bills for the House were being debated, it came to light that in a House with 299 seats, Alabama's apportionment would be 8 seats, whereas in a House with 300 seats, Alabama's apportionment would be 7 seats. The effect of adding that extra seat would be to give one more seat to Texas and one more seat to Illinois. To do this, a seat would have to be taken away from Alabama! The details are shown in Table 4-9.

TABLE 4-9	The Real Alabama Paradox (1882)				
	With $M = 299$			With $M = 300$	
State	Quota	Apportionment		Quota	Apportionment
Alabama	7.646	**8**		7.671	**7**
Texas	9.64	9		9.672	10
Illinois	18.64	18		18.702	19

The final straw for Hamilton's method came during the 1901 apportionment debate. As part of the debate, House sizes of anywhere between 350 and 400 seats were considered. Within that range, the apportionment for the state of Maine went up and down like a roller coaster: For M between 350 and 356, Maine would get 4 seats, but for $M = 357$, Maine's apportionment goes down to 3 seats. Then it would go up again to 4 seats for M between 358 and 381, then back down to 3 for $M = 382$, and so on. When a bill with $M = 357$ was proposed, all hell broke loose on the House floor. Fortunately, cooler heads prevailed and the bill never passed. Hamilton's method was never to be used again. (For a brief history of the apportionment controversies plaguing the United States House of Representatives over the years, see the Historical Note at the end of this chapter.)

> "In Maine comes and out Maine goes — God help the State of Maine when mathematics reach for her to strike her down."
>
> —Charles E. Littlefield
> *Maine Congressional Representative, 1901*

If the Alabama paradox weren't bad enough, Hamilton's method can fall victim to two other paradoxes called the *population paradox* and the *new-states paradox*. We will discuss these two paradoxes very briefly in the remainder of this section.

The Population Paradox

Sometime in the early 1900s it was discovered that under Hamilton's method a state could potentially lose some seats because its population got too big! This phenomenon is known as the *population paradox*. To be more precise, the **population paradox** occurs when state *A loses* a seat to state *B* even though the population of *A grew at a higher rate* than the population of *B*. Too weird to be true, you say? Check out the next example.

(**EXAMPLE 4.6**)   **A Tale of Two Planets**

In the year 2525 the five planets in the Utopia galaxy finally signed a peace treaty and agreed to form an Intergalactic Federation governed by an Intergalactic Congress. This is the story of the two apportionments that broke up the Federation.

Part I. The Apportionment of 2525. The first Intergalactic Congress was apportioned using Hamilton's method, based on the population figures (in billions) shown in the second column of Table 4-10. (This is the apportionment problem discussed in Example 4.2.) There were 50 seats apportioned.

TABLE 4-10   Intergalactic Congress: Apportionment of 2525

Planet	Population	St. quota	Lower quota	Surplus	Apportionment
Alanos	150	$8.\overline{3}$	8	0	8
Betta	**78**	$4.\overline{3}$	4	0	**4**
Conii	173	$9.6\overline{1}$	9	First	10
Dugos	204	$11.\overline{3}$	11	0	11
Ellisium	**295**	$16.3\overline{8}$	16	Second	**17**
Total	900	50.00	48	2	50

Source: Intergalactic Census Bureau.

Let's go over the calculations. Since the total population of the galaxy is 900 billion, the standard divisor is $SD = 900/50 = 18$ billion. Dividing the planet populations by this standard divisor gives the standard quotas shown in the third column of Table 4-10. After the lower quotas are handed out (column 4), there are two surplus seats. The first surplus seat goes to Conii and the other one to Ellisium. The last column shows the apportionments. (Keep an eye on the apportionments of Betta and Ellisium—they are central to how this story unfolds.)

Part II. The Apportionment of 2535. After 10 years of peace, all was well in the Intergalactic Federation. The Intergalactic Census of 2535 showed only a few changes in the planet's populations—an 8 billion increase in the population of

TABLE 4-11   Intergalactic Congress: Apportionment of 2535

Planet	Population	St. quota	Lower quota	Surplus	Apportionment
Alanos	150	8.25	8	0	8
Betta	**78**	4.29	4	Second	**5**
Conii	181	9.96	9	First	10
Dugos	204	11.22	11	0	11
Ellisium	**296**	16.28	16	0	**16**
Total	909	50.00	48	2	50

Source: Intergalactic Census Bureau.

Conii, and a 1 billion increase in the population of Ellisium. The populations of the other planets remained unchanged from 2525. Nonetheless, a new apportionment was required. Table 4-11 shows the details of the 2535 apportionment under Hamilton's method. Notice that the total population increased to 909 billion, so the standard divisor for this apportionment was $SD = 909/50 = 18.18$. (You might want to check all the calculations in Table 4-11 for yourself!)

The one remarkable thing about the 2535 apportionment is that Ellisium *lost a seat* while its *population went up* and Betta *gained that seat* while its *population remained unchanged*! The epilogue to this story is that Ellisium used the 2535 apportionment as an excuse to invade Betta, and the years of peace and prosperity in Utopia were over.

⊂⊃

Example 4.6 is a rather trite story illustrating a fundamental paradox: *Under Hamilton's method, it is possible for a state with a positive population growth rate to lose one (or more) of its seats to another state with a smaller (or zero) population growth rate.* How can this population paradox happen? Go back to Example 4.6 and look at the effect that a small change of the standard divisor has on the relative sizes of the residues. Once again, Hamilton's reliance on the residues to allocate the surplus seats is its undoing. But wait, there is one more!

The New-States Paradox

In 1907 Oklahoma joined the Union. Prior to Oklahoma becoming a state, there were 386 seats in the House of Representatives. At the time the fair apportionment to Oklahoma was five seats, so the size of the House of Representatives was changed from 386 to 391. The point of adding these five seats was to give Oklahoma its fair share of seats and leave the apportionments of the other states unchanged. However, when the new apportionments were calculated, another paradox surfaced: Maine's apportionment went up from three to four seats and New York's went down from 38 to 37 seats. The perplexing fact that *the addition of a new state with its fair share of seats can, in and of itself, affect the apportionments of other states* is called the **new-states paradox**.

The following example gives a simple illustration of the new-states paradox. For a change of pace, we will discuss something other than legislatures.

(**EXAMPLE 4.7**) **Garbage Time**

The Metro Garbage Company has a contract to provide garbage collection and recycling services in two districts of Metropolis, Northtown (with 10,450 homes) and the much larger Southtown (89,550 homes). The company runs 100 garbage trucks, which are apportioned under Hamilton's method according to the number of homes in the district. A quick calculation shows that the standard divisor is $SD = 1000$ homes, a nice, round number which makes the rest of the calculations (shown in Table 4-12) easy. As a result of the apportionment, 10 garbage trucks are assigned to service Northtown and 90 garbage trucks to service Southtown.

TABLE 4-12	Metro Garbage Truck Apportionments		
District	Homes	Quota	Apportionment
Northtown	10,450	10.45	**10**
Southtown	89,550	89.55	**90**
Total	100,000	100.00	100

Now imagine that the Metro Garbage Company is bidding to expand its territory by adding the district of Newtown (5250 homes) to its service area. In its bid to the City Council the company promises to buy five additional garbage trucks for the Newtown run so that its service to the other two districts is not affected. But when the new calculations (shown in Table 4-13) are carried out, there is a surprise: One of the garbage trucks assigned to Southtown has to be reassigned to Northtown! (You should check these calculations for yourself. Notice that the standard divisor has gone up a little and is now approximately 1002.38.)

TABLE 4-13	Revised Metro Garbage Truck Apportionments		
District	Homes	Quota	Apportionment
Northtown	10,450	10.42	**11**
Southtown	89,550	89.34	**89**
Newtown	5,250	5.24	5
Total	105,250	105.00	105

There are two key lessons we should take from this section: (1) In terms of fairness, Hamilton's method leaves a lot to be desired; and (2) the critical flaw in Hamilton's method is the way it handles the surplus seats. Clearly, there must be a better apportionment method, so let's move on!

4.4 Jefferson's Method

Jefferson's method is based on an approach very different from Hamilton's method, so there is some irony in the fact that we will explain the idea behind Jefferson's method by taking one more look at Hamilton's method.

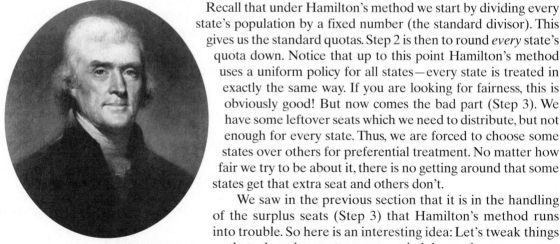

Thomas Jefferson (1743–1826). Jefferson's method (also known as the *method of greatest divisors*) was used in the United States from 1792 to 1840 and is still used to apportion the legislature in many countries, including Austria, Brazil, Finland, Germany, and the Netherlands.

Recall that under Hamilton's method we start by dividing every state's population by a fixed number (the standard divisor). This gives us the standard quotas. Step 2 is then to round *every* state's quota down. Notice that up to this point Hamilton's method uses a uniform policy for all states—every state is treated in exactly the same way. If you are looking for fairness, this is obviously good! But now comes the bad part (Step 3). We have some leftover seats which we need to distribute, but not enough for every state. Thus, we are forced to choose some states over others for preferential treatment. No matter how fair we try to be about it, there is no getting around that some states get that extra seat and others don't.

We saw in the previous section that it is in the handling of the surplus seats (Step 3) that Hamilton's method runs into trouble. So here is an interesting idea: Let's tweak things so that when the quotas are rounded down, *there are no surplus seats*! (The idea, in other words, is to get rid of Step 3.) But under Hamilton's method there is always going to be at least one surplus seat, so how do we work this bit of magic? The answer is by changing the divisor, which then changes the quotas. The idea is that by using a *smaller* divisor, we make the quotas bigger. If we hit it just right, when we round the slightly bigger quotas down . . . poof! the surplus seats are gone. When we pull that bit of numerical magic, we have in fact implemented *Jefferson's method*.

The best way to get a good feel for the inner workings of Jefferson's method is through an example, so we will do that next.

EXAMPLE 4.8   **Parador's Congress (Jefferson's Method)**

As promised we are returning to the apportionment of the Parador Congress. First, a quick review of some the calculations we already did in Example 4.4. When we divide the populations by the standard divisor $SD = 50,000$, we get the standard quotas (third row of Table 4-14). When these quotas are rounded down, we end up with four surplus seats (fourth row of Table 4-14). Now here are some new calculations: When we divide the populations by the slightly smaller divisor $D = 49,500$, we get a slightly bigger set of quotas (fifth row of Table 4-14). When these *modified quotas* are rounded down (last row of Table 4-14), the surplus seats are gone and we have a valid apportionment. It's nothing short of magical! And, of course, you are wondering, Where did that 49,500 come from? Let's call it a lucky guess for now.

TABLE 4-14	Parador's Congress: Jefferson Apportionment						
State	A	B	C	D	E	F	**Total**
Population	1,646,000	6,936,000	154,000	2,091,000	685,000	988,000	12,500,000
Standard quota ($SD = 50,000$)	32.92	138.72	3.08	41.82	13.70	19.76	250.00
Lower quota	32	138	3	41	13	19	246
Modified quota ($D = 49,500$)	33.25	140.12	3.11	42.24	13.84	19.96	
Lower quota	**33**	**140**	**3**	**42**	**13**	**19**	250

Example 4.8 illustrates a key point: *Apportionments don't have to be based exclusively on the standard divisor.* If our intent is to *round the quotas down*, then the standard divisor $SD = 50,000$ is a bad choice (it forces us to deal with surplus seats), whereas the modified divisor $D = 49,500$ is a good choice (no surplus seats to worry about). One useful way to think about this change in divisors is as a *change in the units of measurement*. Instead of using the standard unit 1 seat = 50,000 people, we chose to use the unit 1 seat = 49,500 people. There is nothing that says we can't do that!

Jefferson's method is but one of a group of apportionment methods based on the principle that the standard yardstick of 1 seat = SD people is not set in concrete and that, if necessary, we can change to a different yardstick: 1 seat = D people, where D is a suitably chosen number. The number D is called a **divisor** (sometimes we use the term **modified divisor** to emphasize the point that it is not the standard divisor), and apportionment methods that use modified divisors are called **divisor methods**. Different divisor methods are based on different philosophies of how the **modified quotas** *should be rounded to whole numbers*, but they all follow one script: When you are done rounding the modified quotas, all M seats have been apportioned (no more, and no less). To be able to do this, you just need to find a suitable divisor D. For Jefferson's method, the rounding rule is to always round the modified quotas down. We will soon discuss two other divisor methods that use different rounding rules.

The formal description of Jefferson's method is surprisingly short. The devil is in the details.

> **JEFFERSON'S METHOD**
>
> - **Step 1.** Find a "suitable" divisor D.
> - **Step 2.** Using D as the divisor, compute each state's modified quota (modified quota = state population/D).
> - **Step 3.** Each state is apportioned its modified *lower quota*.

The hard part of implementing Jefferson's method is Step 1. How does one find a suitable divisor D? There are different ways to do it, and some are quite sophisticated (see, for example, Project D), but we will use a basic, blue-collar approach: educated trial and error. Our target is a set of modified lower quotas whose sum is M. For the sum of the modified *lower quotas* to equal M, we need to make the modified quotas *somewhat bigger than the standard quotas*. This can only be accomplished by choosing a divisor *somewhat smaller than SD*. (As the divisor goes down, the quota goes up, and vice versa.) We start by making an educated guess and choose a divisor D smaller than SD. If our first guess works, then we are lucky (or smart), and we are done. If we don't hit the target on our first shot, then we adjust our aim. If the sum of the lower quotas is less than M, then we know that the modified quotas are not quite big enough, and we need to make them bigger. This we do by *choosing an even smaller value for D*. On the other hand, if the sum of the lower quotas is more than M, then we need to bring the modified quotas down by *choosing a bigger value for D*. Using this educated trial-and-error approach, each guess is closer to the target than the previous one, and usually in a matter of a few guesses we can find a divisor D that works. (In most problems there is a full range of values of D that work, so it's not like we have to hit a bull's-eye!) A flowchart outlining the trial-and-error implementation of Jefferson's method is shown in Fig. 4-1.

■ In some very rare cases, there is no divisor D that works — see Exercise 59.

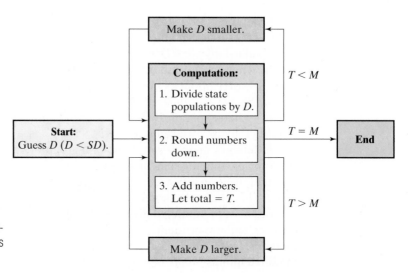

FIGURE 4-1 Flowchart for trial-and-error implementation of Jefferson's method.

To illustrate the details of how to use the flowchart shown in Fig. 4-1, let's go back to Example 4.8. The first guess should be a divisor somewhat smaller than

$SD = 50,000$. Suppose that we start with $D = 49,000$. Using this divisor, we calculate the quotas, round them down, and add. We get a total of $T = 252$ seats. We overshot our target by two seats! We now refine our guess by choosing a *larger* divisor D (the point is to make the quotas *smaller*). A reasonable next guess (halfway between 50,000 and 49,000) is $D = 49,500$. We go through the computation, and it works! (Notice that $D = 49,450$ also works, as do many other divisors.)

Jefferson's Method and the Quota Rule

By design Jefferson's method is consistent in its approach to apportionment—the same formula applies to every state (find a suitable divisor, find the quota, round it down). By doing this, Jefferson's method is able to avoid the major pitfalls of Hamilton's method. There is, however, a serious problem with Jefferson's method that can surface when we least expect it—*it can produce upper-quota violations!* In fact, take another look at the last row of Table 4-14 in Example 4.8 and look specifically at B's apportionment (140 seats) and standard quota (138.72). This is bad! To make matters worse, the upper-quota violations tend to consistently favor the larger states.

Historical Note. One of the most heated and contentious apportionment debates in U.S. history took place in 1832. The apportionment bill of 1832, based on Jefferson's method, gave New York 40 seats in the House of Representatives. When it came to light that New York's standard quota was only 38.59 seats, many delegates were horrified by such an unexpected violation of the quota rule, and the constitutionality of Jefferson's method came into question. Leading the move to dump Jefferson's method was Daniel Webster. The following is a brief excerpt from one of Daniel Webster's most famous speeches to the House of Representatives.

> The House is to consist of 240 members. Now, the precise portion of power, out of the whole mass presented by the number of 240, to which New York would be entitled according to her population, is 38.59; that is to say, she would be entitled to thirty-eight members, and would have a residuum or fraction; and even if a member were given her for that fraction, she would still have but thirty-nine. But the bill gives her forty . . . for what is such a fortieth member given? Not for her absolute numbers, for her absolute numbers do not entitle her to thirty-nine. Not for the sake of apportioning her members to her numbers as near as may be because thirty-nine is a nearer apportionment of members to numbers than forty. But it is given, say the advocates of the bill, because the process [Jefferson's method] which has been adopted gives it. The answer is, no such process is enjoined by the Constitution.[1]

The 1832 apportionment bill passed nonetheless, but it was to be the last time the House of Representatives would use Jefferson's method. Two alternative methods were being proposed at the time as improvements over Jefferson's method—one by John Quincy Adams, the other by Daniel Webster himself. We will discuss them next.

---

[1]Daniel Webster, *The Writings and Speeches of Daniel Webster*, Vol. VI (Boston: Little, Brown and Company, 1903).

John Quincy Adams (1767–1848).
Sixth President of the United States.
Son of John Adams (second President)
and distant cousin of Samuel Adams
(political activist and brewer). Adams's
method is also known as the *method
of smallest divisors.*

4.5 Adams's Method

Like Jefferson's method, Adams's method is a divisor method, but instead of rounding the quotas down, it rounds them up. For this to work the modified quotas have to be made *smaller*, and this requires the use of a divisor D *larger* than the standard divisor SD. In essence, Adams's method is a mirror image of Jefferson's method but based on a "round 'em up" philosophy instead of a "round 'em down" philosophy.

ADAMS'S METHOD

- **Step 1.** Find a "suitable" divisor D.
- **Step 2.** Using D as the divisor, compute each state's modified quota (modified quota = state population/D).
- **Step 3.** Each state is apportioned its modified *upper quota*.

As before, we will illustrate Adams's method with the apportionment of Parador's Congress.

(**EXAMPLE 4.9**) **Parador's Congress (Adams's Method)**

By now you know the background story by heart. The populations are shown once again in the second row of Table 4-15. The challenge, as is the case with any divisor method, is to find a suitable divisor D. We know that in Adams's method, the modified divisor D will have to be bigger than the standard divisor of 50,000. We start with the guess $D = 50,500$. (After all, $D = 49,500$ worked for Jefferson's method!) When the computations are carried out, the total is $T = 251$. (See rows 3 and 4 of Table 4-15 for the details.) This total is just one seat above our target of 250, so we need to make the quotas just a tad smaller. To do this, we must increase the divisor a little bit. We try $D = 50,700$. The new computations are shown in the last two rows of Table 4-15. This divisor works! The apportionment under Adams's method is shown in the last row of Table 4-15.

TABLE 4-15	Parador's Congress: Adams' Apportionment						
State	A	B	C	D	E	F	**Total**
Population	1,646,000	6,936,000	154,000	2,091,000	685,000	988,000	12,500,000
Modified quota 1 (D = 50,500)	32.59	137.35	3.05	41.41	13.56	19.56	
Upper quota	33	138	4	42	14	20	251
Modified quota 2 (D = 50,700)	32.47	136.80	3.04	41.24	13.51	19.49	
Upper quota	**33**	**137**	**4**	**42**	**14**	**20**	**250**

Beyond illustrating how Adams's method can be implemented, Example 4.9 also highlights a serious weakness of this method—it can produce *lower-quota violations*! Once again, look at the apportionment of *B* in Table 4-15 (137 seats) and compare it with its standard quota (138.72). This is a different kind of violation, but just as serious as the one in Example 4.8—state *B* got 1.72 fewer seats than what it rightfully deserves! We can reasonably conclude that Adams's method is no better (or worse) than Jefferson's method—just different. (After considerable debate, the apportionment bill based on Adams's method never passed, and Adams's method was never used to apportion the House of Representatives.)

■ Whereas under Jefferson's method all violations of the quota rule are upper-quota violations, under Adams's method the violations of the quota rule are always lower-quota violations — see Exercise 56.

4.6 Webster's Method

Daniel Webster (1782–1852). Lawyer, statesman, and senator from Massachusetts. First proposed the apportionment method that bears his name in 1832. The method is also known as the *Webster-Willcox method* and *method of major fractions*.

What is the obvious compromise between rounding all the quotas down (Jefferson's method) and rounding all the quotas up (Adams's method)? What about conventional rounding (Round the quotas down when the fractional part is less than 0.5 and up otherwise.)?

It makes a lot of sense, but haven't we been there before and found it didn't quite work? Yes and no. Yes, we have tried conventional rounding with the standard quotas and found it didn't work (see Example 4.3, Table 4-5). No, we haven't considered this approach using *modified quotas*. Now that we know that we can use modified divisors to manipulate the quotas, it is always possible to find a suitable divisor that will make conventional rounding work. This is the idea behind Webster's method.

> **WEBSTER'S METHOD**
>
> ■ **Step 1.** Find a "suitable" divisor *D*.
> ■ **Step 2.** Using *D* as the divisor, compute each state's modified quota (modified quota = state population/*D*).
> ■ **Step 3.** Find the apportionments by rounding each modified quota the conventional way.

EXAMPLE 4.10 **Parador's Congress (Webster's Method)**

We will now apportion Parador's Congress using Webster's method. Our first decision is to make a guess at the divisor *D*: Should it be more than the standard divisor (50,000), or should it be less? Here we will use the standard quotas as a starting point. When we round off the standard quotas to the nearest integer, we get a total of 251 (row 4 of Table 4-16). This number is too high (just by one seat), which tells us that we should try a divisor *D* a tad larger than the standard divisor. We try *D* = 50,100. Row 5 of Table 4-16 shows the respective modified quotas, and the last row shows these quotas rounded to the nearest integer. Now we have a valid apportionment! The last row of the table shows the final apportionment under Webster's method.

TABLE 4-16	Parador's Congress: Webster's Apportionment						
State	*A*	*B*	*C*	*D*	*E*	*F*	**Total**
Population	1,646,000	6,936,000	154,000	2,091,000	685,000	988,000	12,500,000
Standard quota (*SD* = 50,000)	32.92	138.72	3.08	41.82	13.70	19.76	250.00
Nearest integer	33	139	3	42	14	20	251
Modified quota (*D* = 50,100)	32.85	138.44	3.07	41.74	13.67	19.72	
Nearest integer	**33**	**138**	**3**	**42**	**14**	**20**	**250**

A flowchart illustrating how to find a suitable divisor *D* for Webster's method using educated trial and error is shown in Fig. 4-2. The most significant difference when we use trial and error to implement Webster's method as opposed to Jefferson's method (compare Figs. 4-1 and 4-2) is the choice of the starting value of *D*. With Webster's method *we always start with the standard divisor SD*. If we are lucky and *SD* happens to work, we are done! If the standard divisor *SD* doesn't quite do the job, we proceed with the trial-and-error approach as before until we find a divisor that works.

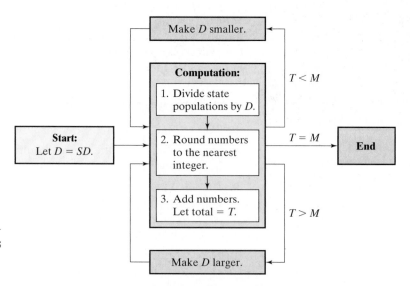

FIGURE 4-2 Flowchart for trial-and-error implementation of Webster's method.

When the standard divisor works as a suitable divisor for Webster's method, every state gets an apportionment that is within 0.5 of its standard quota. This is as good an apportionment as one can hope for. If the standard divisor doesn't quite do the job, there will be at least one state with an apportionment that differs by more than 0.5 from its standard quota. In general, Webster's method tends to produce apportionments that don't stray too far from the standard quotas, although occasional violations of the quota rule (both lower- and upper-quota violations) are possible.

Webster's method has a lot going for it—it does not suffer from any paradoxes, and it shows no bias between small and large states. Its one flaw is that it can violate the quota rule, but such violations are rare in real-life apportionments. (Had Webster's method been used for every apportionment between 1790 and 2000, not a single violation of the quota rule would have shown up.) Surprisingly, Webster's

■ For a detailed discussion of the Huntington-Hill method, see Mini-Excursion 1.

method had a rather short tenure in the U. S. House of Representatives. It was used for the apportionment of 1842, then replaced by Hamilton's method, then reintroduced for the apportionments of 1901, 1911, and 1931, and then replaced again by the Huntington-Hill method, the apportionment method we currently use.

CONCLUSION

Apportionment is a problem of surprising mathematical depth—to say nothing of its political and historical complexities. Many of our Founding Fathers were deeply concerned with the issue, and the names of the apportionment methods we discussed bear witness to this fact.

In this chapter we covered apportionment methods named after Alexander Hamilton, Thomas Jefferson, John Quincy Adams, and Daniel Webster. These are some, but not all, of the apportionment methods that have been proposed and used over the years to apportion the U.S. House of Representatives. (Of the four, Adams's method was the only one never adopted.) Hamilton's method is known as a *quota method* and makes apportionment decisions sticking strictly to the use of the *standard divisor* and the *standard quotas*. (A different quota method called *Lowndes's method* is discussed in Exercises 62 and 63.) On the other hand, the methods proposed by Jefferson, Adams, and Webster are examples of *divisor methods*—methods based on the notion that divisors and quotas can be modified to work under different rounding philosophies. The only difference between divisor methods is in the rule used for rounding the quotas.

For the purposes of comparison, it is useful to apply the different methods to the same apportionment problem, and this we did with the apportionment of Parador's Congress (Examples 4.4, 4.8, 4.9, and 4.10). The four solutions are summarized in Table 4-17. It should be noted that in this example each of the four methods produced a different solution, but this is not always going to happen. Often, two different methods applied to the same apportionment problem will produce identical apportionments.

TABLE 4-17	Parador's Congress: A Tale of Four Methods						
State	*A*	*B*	*C*	*D*	*E*	*F*	**Total**
Population	1,646,000	6,936,000	154,000	2,091,000	685,000	988,000	12,500,000
Standard quota	32.92	138.72	3.08	41.82	13.70	19.76	250.00
Hamilton	33	139	3	42	13	20	250
Jefferson	33	140	3	42	13	19	250
Adams	33	137	4	42	14	20	250
Webster	33	138	3	42	14	20	250

While some of the methods are clearly better than others, none of them is perfect. Some are subject to paradoxes (*Alabama paradox*, *population paradox*, and *new-states paradox*); others produce violations of the quota rule (*upper-quota* violations, *lower-quota* violations, or both). In addition, some methods show a systematic bias in favor of large states, others in favor of small states. Table 4-18 summarizes the potential problems with each method.

TABLE 4-18	Summary of the Four Apportionment Methods		
Method	**Violates Quota Rule?**	**Paradoxes*?**	**Favors**
Hamilton	No	Yes (all three)	Large states
Jefferson	Yes (upper quota only)	None	Large states
Adams	Yes (lower quota only)	None	Small states
Webster	Yes (upper and lower quota)	None	Neutral

*Alabama paradox, population paradox, and new-states paradox.

From a purely mathematical point of view, an *ideal* apportionment method would be one that will not violate the quota rule, will not produce any paradoxes, and will not show bias between small and large states. For many years, the goal of mathematicians, social scientists, and politicians interested in the apportionment problem was to find such a method. In 1980 two mathematicians—Michel L. Balinski of the State University of New York at Stony Brook and H. Peyton Young of Johns Hopkins University—made a surprising discovery: *An apportionment method that does not violate the quota rule and does not produce any paradoxes is a mathematical impossibility.* This fact is known as **Balinski and Young's impossibility theorem**. In essence, the theorem tells us that there are no perfect solutions to the riddle of apportionment—any apportionment method that satisfies the quota rule will produce paradoxes, and any apportionment method that does not produce paradoxes will produce violations of the quota rule.

Much like Arrow's impossibility theorem in Chapter 1, Balinski and Young's impossibility theorem teaches us that there are inherent limitations to our democratic ideals—we believe in proportional representation and we believe in fairness, but we cannot consistently have both. Some problems do not have perfect solutions, and it is good to know that. We can then focus on the more realistic goal of finding, among all the imperfect solutions, the better ones.

ISTORICAL NOTE: A Brief History of Apportionment in the United States

The apportionment of the U.S. House of Representatives has an intriguing and convoluted history. As mandated by the United States Constitution, apportionments of the House are to take place every 10 years, following each census of the population. The real problem was that the Constitution left the method of apportionment and the number of seats to be apportioned essentially up to Congress. The only two restrictions stated in the constitution were that (1) "each State shall have at least one Representative," and (2) "The number of Representatives shall not exceed one for every thirty thousand" (Article I, Section 2).

Following the 1790 Census, and after considerable and heated debate, Congress passed the first "act of apportionment" in 1792. The bill, sponsored by Alexander Hamilton (then Secretary of the Treasury), established a House of Representatives with $M = 120$ seats and apportioned under the method we now call *Hamilton's method*. In April of 1792, at the urging of then Secretary of State Thomas Jefferson, President George Washington vetoed the bill. (This was the first presidential veto in U.S. history.) Jefferson convinced Washington to support a different apportionment bill, based on a House of Representatives with $M = 105$ seats apportioned under the method we now call *Jefferson's method.* Unable to override the president's veto and facing a damaging political stalemate, Congress finally adopted Jefferson's bill. This is how the first House of Representatives came to be constituted.

Jefferson's method remained in use for five decades, up to and including the apportionment of 1832. The great controversy during the 1832 apportionment debate centered on New York's apportionment. Under Jefferson's method, New York would get 40 seats even though its *standard quota* was only 38.59. This apportionment exposed, in a very dramatic way, the critical weakness of Jefferson's method—it produces *upper-quota violations*. Two alternative apportionment bills were considered during the 1832 apportionment debate, one proposed by John Quincy Adams that would apportion the House using the method we now call *Adams's method* and a second one, sponsored by Daniel Webster, that would do the same using the method we now call *Webster's method*. Both of these proposals were defeated and the original apportionment bill passed, but the 1832 apportionment was the last gasp for Jefferson's method.

Webster's method was adopted for the 1842 apportionment, but in 1852 Congress passed a law making Hamilton's method the "official" apportionment method for the House of Representatives. Since it is not unusual for Hamilton's method and Webster's method to produce exactly the same apportionment, an "unofficial" compromise was also adopted in 1852: Choose the number of seats in the House so that the apportionment is the same under either method. This was done again with the apportionment bills of 1852 and 1862.

In 1872, as a result of a power grab among states, an apportionment bill was passed that can only be described as a total mess—it was based on no particular method and produced an apportionment that was inconsistent with both Hamilton's method and Webster's method. The apportionment of 1872 was in violation of both the constitution (which requires that some method be used) and the 1852 law (which designated Hamilton's method as the method of choice).

In 1876 Rutherford B. Hayes defeated Samuel L. Tilden in one of the most controversial and disputed presidential elections in U.S. history. Hayes won in the Electoral College (despite having lost the popular vote) after Congress awarded him the disputed electoral votes from three southern states—Florida, Louisiana, and South Carolina. One of the many dark sidebars of the 1876 election was that, had the House of Representatives been legally apportioned, Tilden would have been the clear-cut winner in the Electoral College.

In 1882 the *Alabama paradox* first surfaced. In looking at possible apportionments for different House sizes, it was discovered that for $M = 299$ Alabama would get 8 seats, but if the House size were increased to $M = 300$, then Alabama's apportionment would decrease to 7 seats. So how did Congress deal with this disturbing discovery? It essentially glossed it over, choosing a House with $M = 325$ seats, a number for which Hamilton's method and Webster's method would give the same apportionment. The same strategy was adopted in the apportionment bill of 1892.

In 1901 the *Alabama paradox* finally caught up with Congress. When the Census Bureau presented to Congress tables showing the possible apportionments under Hamilton's method for all House sizes between 350 and 400 seats, it was pointed out that two states—Maine and Colorado—were impacted by the Alabama paradox: For most values of M starting with $M = 350$, Maine would get 4 seats, but for $M = 357, 382, 386, 389$, and 390, Maine's apportionment would go down to 3 seats. Colorado would get 3 seats for all possible values of M except $M = 357$, for which it would only get 2 seats. For $M = 357$, both Maine and Colorado lose, and, coincidentally, this just happened to be the House size that was proposed for the 1901 bill. Faster than you can say "we are being robbed," the debate in Congress escalated into a frenzy of name-calling and accusations, with the end result being that the bill was defeated and Hamilton's method was scratched for good. The final apportionment of 1901 used Webster's method and a House with $M = 386$ seats.

Webster's method remained in use for the apportionments of 1901, 1911, and 1931. (No apportionment bill was passed following the 1920 Census, in direct violation of the Constitution.)

In 1941 Congress passed a law that established a fixed size for the House of Representatives (435 seats) and a permanent method of apportionment known as the *method of equal proportions*, or *Huntington-Hill method*. (For a discussion of the Huntington-Hill method see Mini-Excursion 1.) The 1941 law (*Public Law 291, H.R. 2665, 55 Stat 761: An Act to Provide for Apportioning Representatives in Congress among the Several States by the Equal Proportions Method*) represented a realization by Congress that politics should be taken out of the apportionment debate and that the apportionment of the House of Representatives should be purely a mathematical issue.

Are apportionment controversies then over? Not a chance. With a fixed-size House, one state's gain has to be another state's loss. In the 1990 apportionment Montana was facing the prospect of losing one of its two seats—seats it had held in the House for 80 years. Not liking the message, Montana tried to kill the messenger. In 1991 Montana filed a lawsuit in federal District Court (*Montana v. United States Department of Commerce*) in which it argued that the Huntington-Hill method is unconstitutional and that either *Adams's method* or *Dean's method* (see Project A) should be used. (Under either of these methods, Montana would keep its two seats.) A panel of three federal judges ruled by a 2-to-1 vote in favor of Montana. The case then went on appeal to the Supreme Court, which overturned the decision of the lower federal court and upheld the constitutionality of the Huntington-Hill method. But it is just a matter of time before the issue will surface again. As long as there are winners and losers in the apportionment game, the debate over what is the "right" apportionment method is not going to go away.

KEY CONCEPTS

EXERCISES

WALKING

A Standard Divisors and Standard Quotas

1. The Bandana Republic is a small country consisting of four states: Apure (population 3,310,000), Barinas (population 2,670,000), Carabobo (population 1,330,000), and Dolores (population 690,000). Suppose that there are $M = 160$ seats in the Bandana Congress, to be apportioned among the four states based on their respective populations.

    (a) Find the standard divisor.

    (b) Find each state's standard quota.

2. The Bandana Republic is a small country consisting of four states: Apure (population 3,310,000), Barinas (population 2,670,000), Carabobo (population 1,330,000), and Dolores (population 690,000). Suppose that there are $M = 200$ seats in the Bandana Congress, to be apportioned among the four states based on their respective populations.

    (a) Find the standard divisor.

    (b) Find each state's standard quota.

3. The Scotia Metropolitan Area Rapid Transit Service (SMARTS) operates six bus routes (A, B, C, D, E, and F) and 130 buses. The number of buses apportioned to each route is based on the number of passengers riding that route. The following table shows the daily average ridership (per thousand) on each route.

Route	A	B	C	D	E	F
Ridership*	45.3	31.07	20.49	14.16	10.26	8.72

*in thousands

    (a) Find the standard divisor.

    (b) Explain what the standard divisor represents in this problem.

    (c) Find the standard quotas.

4. The Placerville General Hospital has a nursing staff of 225 nurses working in four shifts: A (7:00 A.M. to 1:00 P.M.), B (1:00 P.M. to 7:00 P.M.), C (7:00 P.M. to 1:00 A.M.), and D (1:00 A.M. to 7:00 A.M.). The number of nurses apportioned to each shift is based on the average number of patients treated in that shift, given in the following table:

Shift	A	B	C	D
Patients	871	1029	610	190

    (a) Find the standard divisor.

    (b) Explain what the standard divisor represents in this problem.

    (c) Find the standard quotas.

5. The Republic of Tropicana is a small country consisting of five states (A, B, C, D, and E). The total population of Tropicana is 23.8 million. According to the Tropicana constitution, the seats in the legislature are apportioned to the states according to their populations. The following table shows each state's standard quota:

State	A	B	C	D	E
Standard quota	40.50	29.70	23.65	14.60	10.55

   (a) Find the number of seats in the Tropicana legislature.

   (b) Find the standard divisor.

   (c) Find the population of each state.

6. Tasmania State University is made up of five different schools: Agriculture, Business, Education, Humanities, and Science (A, B, E, H, and S for short). The total number of students at TSU is 12,500. The faculty positions at TSU are apportioned to the various schools based on the schools' respective enrollments. The following table shows each school's standard quota:

School	A	B	E	H	S
Standard quota	32.92	15.24	41.62	21.32	138.90

   (a) Find the number of faculty positions at TSU.

   (b) Find the standard divisor.

   (c) Find the number of students enrolled in each school.

7. According to the 2000 U.S. Census, 7.43% of the U.S. population lived in Texas. Compute Texas's standard quota in 2000 (rounded to two decimal places). (*Hint*: There are 435 seats in the U.S. House of Representatives.)

8. At the time of the 2000 U.S. Census, California had a standard quota of 52.45. Estimate what percent of the U.S. population lived in California in 2000. Give your answer to the nearest tenth of a percent. (*Hint*: There are 435 seats in the U.S. House of Representatives.)

9. The Interplanetary Federation of Fraternia consists of six planets: Alpha Kappa, Beta Theta, Chi Omega, Delta Gamma, Epsilon Tau, and Phi Sigma (A, B, C, D, E, and F for short). The federation is governed by the Inter-Fraternia Congress, consisting of 200 seats apportioned among the planets according to their populations. The following table gives the planet populations as percentages of the total population of Fraternia:

Planet	A	B	C	D	E	F
Population percentage	11.37	8.07	38.62	14.98	10.42	16.54

   (a) Find the standard divisor (expressed as a percent of the total population).

   (b) Find the standard quota for each planet.

10. The small island nation of Margarita is made up of four islands: Aleta, Bonita, Corona, and Doritos (A, B, C, and D for short). There are 125 seats in the Margarita Congress, which are apportioned among the islands according to their populations. The following table gives the island populations as percentages of the total population of Margarita:

Island	A	B	C	D
Population percentage	6.24	26.16	28.48	39.12

   (a) Find the standard divisor (expressed as a percent of the total population).

   (b) Find the standard quota for each island.

B Hamilton's Method

11. Find the apportionment of the Bandana Republic Congress as described in Exercise 1 ($M = 160$) under Hamilton's method.

12. Find the apportionment of the Bandana Republic Congress as described in Exercise 2 ($M = 200$) under Hamilton's method.

13. Find the apportionment of the SMARTS buses described in Exercise 3 under Hamilton's method.

14. Find the apportionment of the Placerville General Hospital nurses described in Exercise 4 under Hamilton's method.

15. Find the apportionment of the Republic of Tropicana legislature described in Exercise 5 under Hamilton's method.

16. Find the apportionment of the faculty at Tasmania State University described in Exercise 6 under Hamilton's method.

17. Find the apportionment of the Inter-Fraternia Congress described in Exercise 9 under Hamilton's method.

18. Find the apportionment of the Margarita Congress described in Exercise 10 under Hamilton's method.

Exercises 19 through 21 are based on the following story: Mom found an open box of her children's favorite candy bars. She decides to apportion the candy bars among her three youngest children according to the number of minutes each child spent doing homework during the week.

19. (a) Suppose that there were 11 candy bars in the box. Given that Bob did homework for a total of 54 minutes, Peter did homework for a total of 243 minutes, and Ron did homework for a total of 703 minutes, apportion the 11 candy bars among the children using Hamilton's method.

(b) Suppose that before mom hands out the candy bars, the children decide to spend a "little" extra time on homework. Bob puts in an extra 2 minutes (for a total of 56 minutes), Peter an extra 12 minutes (for a total of 255 minutes), and Ron an extra 86 minutes (for a total of 789 minutes). Using these new totals, apportion the 11 candy bars among the children using Hamilton's method.

(c) The results of (a) and (b) illustrate one of the paradoxes of Hamilton's method. Which one? Explain.

20. (a) Suppose that there were 10 candy bars in the box. Given that Bob did homework for a total of 54 minutes, Peter did homework for a total of 243 minutes, and Ron did homework for a total of 703 minutes, apportion the 10 candy bars among the children using Hamilton's method.

(b) Suppose that just before she hands out the candy bars, mom finds one extra candy bar. Using the same total minutes as in (a), apportion now the 11 candy bars among the children using Hamilton's method.

(c) The results of (a) and (b) illustrate one of the paradoxes of Hamilton's method. Which one? Explain.

21. (a) Suppose that there were 11 candy bars in the box. Given that Bob did homework for a total of 54 minutes, Peter did homework for a total of 243 minutes, and Ron did homework for a total of 703 minutes, apportion the 11 candy bars among the children using Hamilton's method. [This is a repeat of Exercise 19(a).]

(b) Suppose that before the 11 candy bars are given out, cousin Jim shows up and wants to be included in the game. Jim did homework for 580 minutes during the previous week [the other three children's totals remain the same as in (a)]. To be fair, Mom adds six more candy bars to the original 11. Apportion now the 17 candy bars among the four children using Hamilton's method.

(c) The results of (a) and (b) illustrate one of the paradoxes of Hamilton's method. Which one? Explain.

22. A mother wishes to apportion 15 pieces of candy among her three daughters, Katie, Lilly, and Jaime, based on the number of minutes each child spent studying. The only information we have is that mom will use Hamilton's method and that Katie's standard quota is 6.53.

(a) Explain why it is impossible for all three daughters to end up with five pieces of candy each.

(b) Explain why it is impossible for Katie to end up with nine pieces of candy.

(c) Explain why it is impossible for Lilly to end up with nine pieces of candy.

C Jefferson's Method

23. Find the apportionment of the Bandana Republic Congress as described in Exercise 1 under Jefferson's method.

24. Find the apportionment of the Bandana Republic Congress as described in Exercise 2 under Jefferson's method.

25. Find the apportionment of the SMARTS buses described in Exercise 3 under Jefferson's method.

26. Find the apportionment of the Placerville General Hospital nurses described in Exercise 4 under Jefferson's method.

27. Find the apportionment of the Republic of Tropicana legislature described in Exercise 5 under Jefferson's method.

28. Find the apportionment of the faculty at Tasmania State University described in Exercise 6 under Jefferson's method.

29. Find the apportionment of the Inter-Fraternia Congress described in Exercise 9 under Jefferson's method. (*Hint*: Express the modified divisors in terms of percents of the total population.)

30. Find the apportionment of the Margarita Congress described in Exercise 10 under Jefferson's method. (*Hint*: Express the modified divisors in terms of percents of the total population.)

31. Suppose that you just took a test on the apportionment chapter. One of the questions in the test was to calculate what Texas's apportionment on the House of Representatives would be for the 2000 Census under Jefferson's method. Your friend claims that he did all the calculations and the answer is 31. It's immediately clear to you that your friend has the wrong answer. Explain how you know. (*Hint*: Do Exercise 7 first.)

32. Under Jefferson's method California's apportionment in the 2000 House of Representatives would have been 55 seats. What does this fact illustrate about Jefferson's method? (*Hint*: Do Exercise 8 first.)

D Adams's Method

33. Find the apportionment of the Bandana Republic Congress as described in Exercise 1 under Adams's method.

34. Find the apportionment of the Bandana Republic Congress as described in Exercise 2 under Adams's method.

35. Find the apportionment of the SMARTS buses described in Exercise 3 under Adams's method.

36. Find the apportionment of the Placerville General Hospital nurses described in Exercise 4 under Adams's method.

37. Find the apportionment of the Republic of Tropicana legislature described in Exercise 5 under Adams's method.

38. Find the apportionment of the faculty at Tasmania State University described in Exercise 6 under Adams's method.

39. Find the apportionment of the Inter Fraternia Congress described in Exercise 9 under Adams's method. (*Hint*: Express the modified divisors in terms of percents of the total population.)

40. Find the apportionment of the Margarita Congress described in Exercise 10 under Adams's method. (*Hint*: Express the modified divisors in terms of percents of the total population.)

41. Under Adams's method, California's apportionment in the 2000 House of Representatives would have been 50 seats even though California's standard quota is 52.45. What does this fact illustrate about Adam's method? (*Hint*: Do Exercise 8 first.)

42. Suppose that you just took a test on the apportionment chapter. One of the questions on the test was to calculate what Texas's apportionment in the House of Representatives would be for the 2000 Census under Adams's method. Your friend claims that he did all the calculations and the answer is 34. It's immediately clear to you that your friend has the wrong answer. Explain how you know. (*Hint*: Do Exercise 7 first.)

E Webster's Method

43. Find the apportionment of the Bandana Republic Congress as described in Exercise 1 under Webster's method.

44. Find the apportionment of the Bandana Republic Congress as described in Exercise 2 under Webster's method.

45. Find the apportionment of the SMARTS buses described in Exercise 3 under Webster's method.

46. Find the apportionment of the Placerville General Hospital nurses described in Exercise 4 under Webster's method.

47. Find the apportionment of the Republic of Tropicana legislature discussed in Exercise 5 under Webster's method.

48. Find the apportionment of the faculty at Tasmania State University discussed in Exercise 6 under Webster's method.

49. Find the apportionment of the Inter Fraternia Congress discussed in Exercise 9 under Webster's method

(*Hint*: Express the modified divisors in terms of percents of the total population.)

50. Find the apportionment of the Margarita Congress discussed in Exercise 10 under Webster's method. (*Hint*: Express the modified divisors in terms of percents of the total population.)

JOGGING

51. Consider an apportionment problem with N states. The populations of the states are given by $p_1, p_2, \ldots, p_N$, and the standard quotas are $q_1, q_2, \ldots, q_N$, respectively. Describe in words what each of the following quantities represents.

(a) $q_1 + q_2 + \cdots + q_N$

(b) $\dfrac{p_1 + p_2 + \cdots + p_N}{q_1 + q_2 + \cdots + q_N}$

(c) $\left(\dfrac{p_N}{p_1 + p_2 + \cdots + p_N} \right) \times 100$

52. For an arbitrary state X, let q represent its standard quota and s represent the number of seats apportioned to X under some unspecified apportionment method. Interpret in words the meaning of each of the following mathematical statements.

(a) $s - q \geq 1$

(b) $q - s \geq 1$

(c) $|s - q| \leq 0.5$

(d) $0.5 < |s - q| < 1$

53. Consider the problem of apportioning $M = 3$ seats between two states, A and B, using Jefferson's method. Let p_A and p_B denote the populations of A and B, respectively. Show that if the apportionment under Jefferson's method gives all three seats to A and none to B, then more than 75% of the country's population must live in state A. (*Hint*: Show that a Jefferson apportionment of 3 and 0 seats implies that $p_A > 3p_B$.)

54. Consider the problem of apportioning $M = 3$ seats between two states, A and B, using Webster's method. Let p_A and p_B denote the populations of A and B, respectively. Show that if the apportionment under Webster's method gives all three seats to A and none to B, then more than $83\frac{1}{3}\%$ of the country's population must live in state A. (*Hint*: Show that a Webster apportionment of 3 and 0 seats implies that $p_A > 5p_B$.)

55. Consider the problem of apportioning M seats between two states, A and B. Let q_A and q_B denote the standard quotas of A and B, respectively, and assume that these quotas have decimal parts that are not equal to 0.5. Explain why in this case

(a) Hamilton's and Webster's methods must give the same apportionment.

(b) the Alabama or population paradoxes cannot occur under Hamilton's method. [*Hint*: Use the result of (a).]

(c) violations of the quota rule cannot occur under Webster's method. [*Hint*: Use the result of (a).]

56. (a) Explain why, when Jefferson's method is used, any violations of the quota rule must be upper-quota violations.

(b) Explain why, when Adams's method is used, any violations of the quota rule must be lower-quota violations.

(c) Explain why, in the case of an apportionment problem with two states, violations of the quota rule cannot occur under either Jefferson's or Adams's method. [Hint: Use the results of (a) and (b).]

57. This exercise is based on actual data taken from the 1880 census. In 1880 the population of Alabama was given at 1,262,505. With a House of Representatives consisting of $M = 300$ seats, the standard quota for Alabama was 7.671.

(a) Find the 1880 census population for the United States (rounded to the nearest person).

(b) Given that the standard quota for Texas was 9.672, find the population of Texas (to the nearest person).

58. According to the 2000 U. S. Census, the population of the U.S. was 281,424,177. If Hamilton's method had been used to apportion the House of Representatives, Minnesota would have been apportioned eight seats. Use this information (and that $M = 435$) to determine lower and upper bounds on the 2000 population of Minnesota. (*Hint*: Hamilton's method satisfies the quota rule.)

59. The purpose of this exercise is to show that, under rare circumstances, Jefferson's method may not work. A small country consists of four states with populations given in the following table, and there are $M = 49$ seats to be apportioned.

State	E	F	G	H
Population	500	1000	1500	2000

(a) Find the apportionment under Adams's method.

(b) Attempt to apportion the seats under Jefferson's method using the divisor $D = 100$. What happens if $D < 100$? What happens if $D > 100$?

(c) Explain why Jefferson's method will not work for this example.

60. The purpose of this exercise is to show that, under rare circumstances, Adams's method may not work. A small

country consists of four states with populations given in the following table, and there are $M = 51$ seats to be apportioned.

State	E	F	G	H
Population	500	1000	1500	2000

(a) Find the apportionment under Jefferson's method.

(b) Attempt to apportion the seats under Adams's method using the divisor $D = 100$. What happens if $D < 100$? What happens if $D > 100$?

(c) Explain why Adams's method will not work for this example.

61. **Alternate version of Hamilton's method.** Consider the following description of an apportionment method:

■ **Step 1.** Find each state's standard quota.

■ **Step 2.** Give each state (temporarily) its *upper quota* of seats. (You have now given away more seats than the number of seats available.)

■ **Step 3.** Let K denote the number of extra seats you have given away in Step 2. Take away the K extra seats from the K states with the smallest fractional parts in their standard quotas.

Explain why this method produces exactly the same apportionment as Hamilton's method.

Lowndes's Method. Exercises 62 and 63 refer to a variation of Hamilton's method known as Lowndes's method, first proposed in 1822 by South Carolina Representative William Lowndes. The basic difference between Hamilton's and Lowndes's methods is that in Lowndes's method, after each state is assigned the lower quota, the surplus seats are handed out in order of **relative fractional parts**. *(The relative fractional part of a number is the fractional part divided by the integer part. For example, the relative fractional part of 41.82 is 0.82/41 = 0.02, and the relative fractional part of 3.08 is 0.08/3 = 0.027. Notice that while* 41.82 *would have priority over* 3.08 *under Hamilton's method,* 3.08 *has priority over* 41.82 *under Lowndes's method because 0.027 is greater than 0.02.)*

62. (a) Find the apportionment of Parador's Congress (Example 4.3) under Lowndes's method.

(b) Verify that the resulting apportionment is different from each of the apportionments shown in Table 4-17. In particular, list which states do better under Lowndes's method than under Hamilton's method.

63. Consider an apportionment problem with two states, A and B. Suppose that state A has standard quota q_1 and state B has standard quota q_2, neither of which is a whole number. (Of course, $q_1 + q_2 = M$ must be a whole number.) Let f_1 represent the fractional part of q_1 and f_2 the fractional part of q_2.

(a) Find values q_1 and q_2 such that Lowndes's method and Hamilton's method result in the same apportionment.

(b) Find values q_1 and q_2 such that Lowndes's method and Hamilton's method result in different apportionments.

(c) Write an inequality involving q_1, q_2, f_1, and f_2 that would guarantee that Lowndes' method and Hamilton's method result in different apportionments.

RUNNING

64. The Hamilton-Jefferson hybrid method. The Hamilton-Jefferson hybrid method starts by giving each state its lower quota (as per Hamilton's method) and then apportioning the surplus seats using Jefferson's method.

(a) Use the Hamilton-Jefferson hybrid method to apportion $M = 22$ seats among four states according to the following populations: A (population 18,000), B (population 18,179), C (population 40,950), and D (population 122,871).

(b) Explain why the Hamilton-Jefferson hybrid method can produce apportionments that are different from both Hamilton and Jefferson apportionments.

(c) Explain why the Hamilton-Jefferson hybrid method can violate the quota rule.

65. Consider an apportionment problem with N states and M seats ($M \geq N$). Let k denote the number of surplus seats in Step 3 of Hamilton's method.

(a) What is the *smallest* possible value of k? Give an example of an apportionment problem in which k has this value.

(b) What is the *largest* possible value of k? Give an example of an apportionment problem in which k has this value.

66. Explain why Jefferson's method cannot produce

(a) the Alabama paradox

(b) the new-states paradox

67. Explain why Adams's method cannot produce

(a) the Alabama paradox

(b) the new-states paradox

68. Explain why Webster's method cannot produce

(a) the Alabama paradox

(b) the new-states paradox

PROJECTS AND PAPERS

A Dean's Method

Dean's method (also known as the *method of harmonic means*) was first proposed in 1832 by James Dean. [This is a different James Dean. Dean (1776–1849) was a Professor of Astronomy and Mathematics at Dartmouth College and the University of Vermont.] Dean's method is a divisor method similar to Webster's method but with a slightly different set of rules for rounding the quotas. In this project you are to prepare a report on Dean's method. Your report should include (1) a discussion of the *harmonic mean*, including some of its properties and applications; (2) a description of Dean's method; (3) a discussion of how Dean's method can be implemented using trial and error; and (4) a comparison of Dean's method with Webster's method with an example showing that the two methods can produce different apportionments.

You can find a good description of Dean's method in reference 3.

B Apportionment Methods and the 2000 Presidential Election

After tremendous controversy over hanging chads and missed votes, the 2000 presidential election was decided in the United States Supreme Court, with George W. Bush get-

ting the disputed electoral votes from Florida and thus beating Al Gore by a margin of four electoral votes. Ignored in all the controversy was the significant role that the choice of apportionment method plays in a close presidential election. Would things have turned out differently if the House of Representatives had been apportioned under a different method?

In this project, you are asked to analyze and speculate how the election would have turned out had the House of Representatives (and thus the Electoral College) been apportioned under (1) Hamilton's method, (2) Jefferson's method, and (3) Webster's method.

Notes: (1) The 2000 Electoral College was based on the 1990 U.S. Census. (Remember to add 2 to the seats in the House to get each state's Electoral College votes!) You can go to *www.census.gov/main/www//cen1990.html* for the 1990 state population figures. (2) You can use the Apportionment applets in MyMathLab* to do the calculations.

*To use MyMathLab, you need a course ID and a student access code. Contact your instructor for more information.

C Apportionment Calculations via Spreadsheets

The power of a spreadsheet is its ability to be *dynamic*: Any and all calculations will automatically update after a change is made to the value in any given cell. This is particularly useful in implementing apportionment calculations that use educated trial and error as well as when trying to rig up examples of apportionment problems that meet specific requirements.

This project has two parts.

Part 1. Implement each of the four apportionment methods discussed in the chapter using a spreadsheet. (Jefferson's, Adams's, and Webster's methods all fall under a similar template, so once you can implement one, the others are quite similar. Hamilton's method is a little different and, ironically, a bit harder than the other three to implement with a spreadsheet.) To check that everything works, do the different apportionments of Parador's Congress via the spreadsheets.

Part 2. Use your spreadsheets to create examples of the following situations:

(1) An apportionment problem in which Hamilton's method and Jefferson's method produce exactly the same apportionments

(2) An apportionment problem in which Hamilton's method and Adams's method produce exactly the same apportionments

(3) An apportionment problem in which Hamilton's method and Webster's method produce exactly the same apportionments

(4) An apportionment problem in which Hamilton's method, Jefferson's method, Adams's method, and Webster's method all produce different apportionments

(5) An apportionment problem in which Webster's method violates the quota rule

To do this project you should be reasonably familiar with Excel or a similar spreadsheet program.

D Rank Index Implementations of Divisor Methods

From a computational point of view, the divisor methods of Jefferson, Adams, and Webster can be implemented using a priority system (called a **rank index**) that gives away seats one by one according to a table of values specifically constructed for this purpose. Each of the methods uses a slightly different set of rules to construct the rank index for the seats. In this project you are to describe how to construct the rank index for each of the three divisor methods studied in this chapter (Jefferson, Adams, and Webster). You should use a simple example to illustrate the procedure for each method.

A good starting point for this project is the American Mathematical Society apportionment Web site (reference 1). A more technical account can be found in reference 10.

E The First Apportionment of the House of Representatives

The following table shows the United States population figures from the 1790 U.S. Census.

State	Population
Connecticut	236,841
Delaware	55,540
Georgia	70,835
Kentucky	68,705
Maryland	278,514
Massachusetts	475,327
New Hampshire	141,822
New Jersey	179,570
New York	331,589
North Carolina	353,523
Pennsylvania	432,879
Rhode Island	68,446
South Carolina	206,236
Vermont	85,533
Virginia	630,560
Total	3,615,920

For the first apportionment of 1792, two competing apportionment bills were considered: a bill to apportion 120 seats using Hamilton's method (sponsored by Hamilton), and a bill to apportion 105 seats using Jefferson's method (sponsored by Jefferson). Originally, Congress passed Hamilton's bill, but Washington was persuaded by Jefferson to veto it. Eventually, Congress approved Jefferson's bill, which became the basis for the first mathematical apportionment of the House of Representatives.

This project has three parts.

Part 1. Historical Research. Summarize the arguments presented by Hamilton and Jefferson on behalf of their respective proposals. [Suggested sources: *The Papers of Alexander Hamilton, Vol. XI*, Harold C. Syrett (editor), and *The Works of Thomas Jefferson, Vol. VI*, Paul Leicester Ford (editor)].

Part 2. Mathematics. Calculate the apportionments for the House of Representatives under both proposals.

Part 3. Analysis. It has been argued by some scholars that there was more than mathematical merit behind Jefferson's thinking and Washington's support of it. Looking at the apportionments obtained in Part 2, explain why one could be suspicious of Jefferson and Washington's motives.

(*Hint*: Jefferson and Washington were both from the same state.)

REFERENCES AND FURTHER READINGS

1. American Mathematical Society, "Apportionment Systems," *http://www.ams.org/featurecolumn/archive/apportionII1.html*.

2. Balinski, Michel L., and H. Peyton Young, "The Apportionment of Representation," *Fair Allocation: Proceedings of Symposia on Applied Mathematics*, 33 (1985), 1–29.

3. Balinski, Michel L., and H. Peyton Young, *Fair Representation; Meeting the Ideal of One Man, One Vote*. New Haven, CT: Yale University Press, 1982.

4. Balinski, Michel L., and H. Peyton Young, "The Quota Method of Apportionment," *American Mathematical Monthly*, 82 (1975), 701–730.

5. Brams, Steven, and Philip Straffin, Sr., "The Apportionment Problem," *Science*, 217 (1982), 437–438.

6. Census Bureau Web site, *http://www.census.gov*.

7. Eisner, Milton, *Methods of Congressional Apportionment*, COMAP Module no. 620.

8. Hoffman, Paul, *Archimedes' Revenge: The Joys and Perils of Mathematics*. New York: W. W. Norton & Co., 1988, chap. 13.

9. Huntington, E. V., "The Apportionment of Representatives in Congress." *Transactions of the American Mathematical Society*, 30 (1928), 85–110.

10. Huntington, E. V., "The Mathematical Theory of the Apportionment of Representatives," *Proceedings of the National Academy of Sciences, U.S.A.*, 7 (1921), 123–127.

11. Meder, Albert E., Jr., *Legislative Apportionment*. Boston: Houghton Mifflin Co., 1966.

12. Saari, D. G., "Apportionment Methods and the House of Representatives," *American Mathematical Monthly*, 85 (1978), 792–802.

13. Schmeckebier, L. F., *Congressional Apportionment*. Washington, DC: The Brookings Institution, 1941.

14. Steen, Lynn A., "The Arithmetic of Apportionment," *Science News*, 121 (May 8, 1982), 317–318.

15. Woodall, D. R., "How Proportional Is Proportional Representation?," *The Mathematical Intelligencer*, 8 (1986), 36–46.

16. Young, H. Peyton, "Dividing the House: Why Congress Should Reinstate an Old Reapportionment Formula," Brookings Institution, Policy Brief no. 88, 2001.

Apportionment Today

The Huntington-Hill Method

Chapter 4 made for a rugged but illuminating mathematical and historical excursion. But no apportionment excursion would be complete (at least from an American perspective) without an additional trip into the present. The purpose of this mini-excursion is to discuss the modern method of apportionment of the U.S. House of Representatives. In so doing, we will venture into some new mathematical territory.

Historical Background

By 1929 Congress was fed up with the controversies surrounding the decennial apportionment of the House of Representatives, and there was a strong push within Congress to place the issue on a firm mathematical foundation. At the formal request of the Speaker of the House of Representatives, Nicholas Longworth, the National Academy of Sciences commissioned a panel of distinguished mathematicians to investigate the different apportionment methods that had been proposed over the years and to recommend, in the words of Representative Gibson, the "correct mathematical formula." After deliberate consideration and discussion, the panel endorsed a method then called the **method of equal proportions**, a method originally proposed sometime around 1911 by Joseph A. Hill, the Chief Statistician of the Bureau of the Census, and later improved and refined by Edward V. Huntington, a Professor of Mechanics and Mathematics at Harvard University. The method is now known as the **Huntington-Hill method** (or some variation of this name, such as the *Hill method* or the *Hill-Huntington method*).

> "The apportionment of Representatives . . . is a mathematical problem. Then why not use a method that will stand the test . . . under a correct mathematical formula? "
>
> —Representative Ernest Gibson of Vermont (1929)

It took a while, but eventually the politicians paid heed to the mathematicians. In 1941 Congress passed, and President Franklin D. Roosevelt signed, "An Act to Provide for Apportioning Representatives in Congress among the Several States by the Equal Proportions Method," generally known as *the 1941 apportionment act*. This act had three key elements: (1) It set the Huntington-Hill method as the *permanent* method for apportionment of the House of Representatives, (2) it made the decennial apportionments *self-executing* (once the official population figures are in, the apportionment formula is applied automatically and changes take effect without the need for Congressional approval), and (3) it permanently fixed the size of the House of Representatives at 435 seats (with an exception made if a new state were to join the Union).

With a "fixed-size" House, one state's gain must be another state's loss, and, typically, the states losing those seats tend to put up a fight. These fights are fought in the federal courts (including the Supreme Court—see the Project at the end of this mini-excursion), and ultimately the legal arguments always boil down to a core mathematical question—what is the "correct mathematical formula" for apportionment? States losing seats under Huntington-Hill always try to make the argument that the Huntington-Hill method is not it and that some other method (whichever one lets them keep their seat) is the correct mathematical formula. Ultimately, the courts have ruled against these challenges, and the Huntington-Hill method remains the method of choice—at least until the next lawsuit.

Before we embark on a full-fledged description of the Huntington-Hill method, we will take a brief mathematical detour.

The Arithmetic and Geometric Means

The standard way of "averaging" two numbers (we are taught sometime in elementary school) is to take the number halfway between them. This kind of average is called the **arithmetic mean** of the two numbers. Less known or understood is that there are other important and useful ways to "average" numbers. To avoid confusion with the conventional use of the word *average*, these other types of "averages" are referred to as **means**. Examples of important means beyond the arithmetic mean are the *geometric mean*, the *harmonic mean*, and the *quadratic mean* (see Exercises 21 and 22). Of these three, the *geometric mean* is of particular relevance to our discussion.

■ The requirement that the numbers be positive ensures that we don't have to deal with negative square roots.

┌─ **GEOMETRIC MEAN** ────────────────────────────

The *geometric mean* of two positive numbers a and b is the number $G = \sqrt{a \cdot b}$.

We are going to do a lot of comparing between the geometric and arithmetic means of the same numbers, so for good measure we are including a formal definition of the arithmetic mean as well.

■ Here there is no requirement that the numbers be positive, but in our examples and applications they will turn out to be so.

┌─ **ARITHMETIC MEAN** ────────────────────────────

The *arithmetic mean* of two numbers a and b is the number $A = \dfrac{a + b}{2}$.

One useful way to think about the difference between the arithmetic and geometric means is that the arithmetic mean is an average under *addition*, while the geometric mean is an average under *multiplication*.

(**EXAMPLE ME1.1**)   **Averaging Growth Rates**

Over the last two years, the student body at Tasmania State University grew at a rate of 5% during the first year and 75% during the second year. What was the *average annual growth rate* over the two-year period?

Tempting as it may seem to say that the average annual growth rate was 40% $[(5 + 75)/2 = 40]$, the correct average growth rate was roughly 35.55%. To explain where this number came from, we will work our way backwards: we will first look at the calculation and then give a brief explanation of why we do it this way.

To compute the average growth rate, we first find the *geometric mean* of 1.05 and 1.75. This is given by $\sqrt{(1.05)(1.75)} \approx 1.3555$. From this number we deduce that the average annual growth rate is approximately 35.55%. Let's first deal with the 1's that mysteriously showed up. The 1.05 and the 1.75 are *multipliers* corresponding to the *growth rates* of 5% and 75%, respectively. [A growth rate of 5% means that in the first year a student population of P students grew to $(1.05)P$ students. A growth rate of 75% in the second year means that a student population of $(1.05)P$ students grew to $(1.75)(1.05)P$ students.] Likewise, the 1.3555 represents a growth rate of approximately 35.55%.

But why did we take the geometric mean of 1.05 and 1.75? The answer is that growth rates are not added, but rather compounded multiplicatively, and *when quantities are multiplied, the correct way to average them is to use the geometric mean*. In this case, what we really mean when we claim that the average growth rate is 35.55% is that a 5% growth rate followed by a 75% growth rate are together equivalent to two years at a 35.55% growth rate. The basis for this claim is that $(1.3555)(1.3555) \approx (1.05)(1.75)$.

▪ A more detailed discussion of growth rates as percentages can be found in Chapter 10, Section 1.

The next pair of examples will highlight some important numerical differences between the arithmetic and geometric means.

(**EXAMPLE ME1.2**)   **The Arithmetic-Geometric Mean Inequality**

Table ME1-1 shows geometric means in the third column (rounded to two decimal places when necessary) and arithmetic means in the last column. These values can be easily verified as long as you have a calculator handy (which you should).

What is Table ME1-1 telling us? Think of a and b as starting as the same number (10 and 10). In this case, the arithmetic and geometric means are equal ($G = A = a$). Now think of a staying put but b separating away from a and steadily getting bigger. As b moves away from a, both the geometric mean G and the arithmetic mean A fall somewhere in between a and b, but while A is always exactly halfway between a and b, G is closer to a than to b. (Think of G and A as two runners moving away from a, with A moving at a steady rate but G moving at a slower and slower rate as time goes on.)

TABLE ME1-1	Geometric Means Versus Arithmetic Means		
a	b	$G = \sqrt{a \cdot b}$	$A = (a + b)/2$
10	10	10	10
10	20	14.14	15
10	40	20	25
10	100	31.62	55
10	1000	100	505
10	9000	300	4505

The most important implication of our observation in Example ME1.2 is that for any two positive numbers a and b, *the arithmetic mean is bigger than the geometric mean* except when $a = b$, in which case the two means are equal. This fact is known as the *arithmetic-geometric mean inequality*.

There are many ways to prove the arithmetic-geometric mean inequality — see Exercise 10. For a nice geometric proof using the Pythagorean theorem, see Exercise 11.

┌─ **ARITHMETIC-GEOMETRIC MEAN INEQUALITY** ─────────┐

Let a and b be any two positive numbers (assume $a \le b$), and let A and G denote their arithmetic and geometric means, respectively. Then,

$$a \le G \le A \le b$$

(In the special case when $a = b$, we end up with $G = A$.)

└───┘

EXAMPLE ME1.3   **Scaling Geometric and Arithmetic Means**

See Exercise 9(a).

Table ME1-2 illustrates another simple but useful property of the geometric and arithmetic means: When a and b are multiplied by a common positive scaling factor k, their geometric and arithmetic means are also multiplied by the same scaling factor k. This fact can be proved using basic high school algebra.

TABLE ME1-2

a	b	Geometric Mean	Arithmetic Mean
1	4	2	2.5
10	40	20	25
0.25	1	0.5	0.625
1.5	6	3	3.75
k	$4k$	$2k$	$2.5k$

EXAMPLE ME1.4   **Geometric Means of Consecutive Integers**

In this example we will examine the behavior of the geometric mean when the numbers a and b are consecutive integers. The geometric means here are rounded to three decimal places.

See Exercise 18.

The last column of Table ME1-3 shows the difference between the arithmetic and the geometric means. It's hard to miss the gist of the message of Table ME1-3: *The difference between the arithmetic and the geometric means of two consecutive integers is very small, and as the integers get larger, this difference becomes even smaller.*

TABLE ME1-3

a, b	G	A	$A - G$	a, b	G	A	$A - G$
1, 2	1.414	1.5	0.086	7, 8	7.483	7.5	0.017
2, 3	2.449	2.5	0.051	8, 9	8.485	8.5	0.015
3, 4	3.464	3.5	0.036	9, 10	9.487	9.5	0.013
4, 5	4.472	4.5	0.028	20, 21	20.494	20.5	0.006
5, 6	5.477	5.5	0.023	100, 101	100.499	100.5	0.001
6, 7	6.481	6.5	0.019	200, 201	200.499	200.5	0.001

This is the end of our mathematical detour into the geometric mean, but obviously not the end of our dealings with the concept.

The Huntington-Hill Method

There are several ways to describe the Huntington-Hill method, but by far the easiest way to do it is by analogy to Webster's method. In fact, Huntington-Hill's method is a very close cousin of Webster's method, and we can say that the difference between the two methods boils down to *the difference between the arithmetic mean and the geometric mean*. We will start our discussion by briefly revisiting Webster's method.

Webster's Method Revisited

Without going into all the details (for details, see Section 4.6), the defining philosophy of Webster's method is that the state's *quotas are to be rounded to the nearest integer* (down if the decimal part is less than 0.5, up otherwise). The term *quota* here refers to either the *standard quota* (if that works then so much the better) or some *modified quota*.

A different way to describe the rounding rule for Webster's method rule is in terms of *cutoff points*—the cutoff point for rounding a quota is the *arithmetic mean* of the *lower* and *upper* quotas. If the quota is less than this cutoff point, then we round it down; otherwise, we round it up.

The following formal description of Webster's method may seem like a bit of overkill, but it makes for a good precedent.

> **WEBSTER'S ROUNDING RULE**
>
> For a quota q, let L denote its lower quota, U its upper quota, and A the *arithmetic mean* of L and U. If $q < A$, then round q *down to* L; otherwise, *round q up to U.*

> **WEBSTER'S METHOD**
>
> - **Step 1.** Find a "suitable" divisor D. [Here a suitable divisor means a divisor that produces an apportionment of exactly M seats when the *quotas* (populations divided by D) are rounded using *Webster's rounding rule*.]
> - **Step 2.** Find the apportionment of each state by rounding its quota using Webster's rounding rule.

The description of the Huntington-Hill method follows exactly the same script but uses the *geometric mean* rather than the arithmetic mean to define the cutoff points.

> **HUNTINGTON-HILL ROUNDING RULE**
>
> For a quota q, let L denote its lower quota, U its upper quota, and G the *geometric mean* of L and U. If $q < G$, then round q *down to* L; otherwise, *round q up to U.*

HUNTINGTON-HILL METHOD

- **Step 1.** Find a "suitable" divisor D. [Here a suitable divisor means a divisor that produces an apportionment of exactly M seats when the quotas (populations divided by D) are rounded using the *Huntington-Hill rounding rule*.]

- **Step 2.** Find the apportionment of each state by rounding its quota using the Huntington-Hill rounding rule.

When doing apportionments under the Huntington-Hill method, it is helpful to have a table of cutoff values handy. Table ME1-4 gives cutoff values (rounded to three decimal places) for quotas between 1 and 15.

TABLE ME1-4

Lower quota L	Upper quota U	Huntington-Hill cutoff G	Webster cutoff A
1	2	1.414	1.5
2	3	2.449	2.5
3	4	3.464	3.5
4	5	4.472	4.5
5	6	5.477	5.5
6	7	6.481	6.5
7	8	7.483	7.5
8	9	8.485	8.5
9	10	9.487	9.5
10	11	10.488	10.5
11	12	11.489	11.5
12	13	12.490	12.5
13	14	13.491	13.5
14	15	14.491	14.5

Table ME1-4 illustrates how and where the two sets of rounding rules differ. For example, a quota of 3.48 would be *rounded up under the Huntington-Hill rule* (it's above the cutoff point of 3.464), but rounded down under Webster's rule. On the other hand, a quota of 14.48 would be rounded down under both rules. The impression we get, looking carefully at Table ME1-4, is that there is only a very small "window of opportunity" for the two rounding rules to act differently and that, as the quotas get larger, this small window gets even smaller. This is indeed

■ See Exercise 18. what happens.

We will now illustrate the difference between Webster's method and the Huntington-Hill method with a pair of examples.

EXAMPLE ME1.5   **Huntington-Hill Can Be Easier Than Webster**

A small country consists of five states A, B, C, E, and F, with populations of 34,800 (A); 104,800 (B); 64,800 (C); 140,800 (E); and 54,800 (F). There are $M = 40$ seats in the legislature. We will apportion the seats first using the Huntington-Hill method and then, for comparison purposes, using Webster's method.

Table ME1-5 shows the first set of calculations, using the standard divisor and standard quotas. The third row of the table gives the standard quotas, based on the standard divisor $SD = 400,000/40 = 10,000$. The last row shows the results of rounding the standard quotas using the Huntington-Hill rounding rules. Notice that under the Huntington-Hill rounding rules the quotas 3.48 and 5.48 get *rounded up* (they are both above their cutoff points).

TABLE ME1-5   Huntington-Hill Apportionment

State	A	B	C	E	F	Total
Population	34,800	104,800	64,800	140,800	54,800	400,000
Standard quota (SD = 10,000)	3.48	10.48	6.48	14.08	5.48	40
Huntington-Hill	4	10	6	14	6	40

The nice surprise is that using the Huntington-Hill rounding rules we get an apportionment—the standard divisor is just the right divisor to apportion under the Huntington-Hill method! On the other hand, we can see from the third row that the standard quotas would not work for Webster's method, as 3.48 and 5.48 would have to be rounded down.

Table ME1-6 shows a set of modified quotas that work for Webster's method. They were obtained using the modified divisor $D = 9965$. (It takes a few educated guesses to hit on this divisor. Other divisors between 9964 and 9969 also work.) The last row shows the apportionment under Webster's method.

TABLE ME1-6

State	A	B	C	E	F	Total
Population	34,800	104,800	64,800	140,800	54,800	400,000
Modified quota (D = 9965)	3.492	10.517	6.503	14.13	5.499	
Webster	3	11	7	14	5	40

The next example is, in a sense, the mirror image of Example ME1.5. The background story is the same, with only a few minor changes in the population figures.

EXAMPLE ME1.6   **Huntington-Hill Can Be Harder Than Webster**

A small country consists of five states A, B, C, E, and F, with populations of 34,800 (A); 105,100 (B); 65,100 (C); 140,200 (E), and 54,800 (F). There are $M = 40$ seats in the legislature. We will apportion the seats using both the Huntington-Hill method and Webster's method.

Table ME 1-7 shows the first set of calculations based on the standard divisor $SD = 400{,}000/40 = 10{,}000$. The third row of the table gives the standard quotas. The fourth row shows the results of rounding the standard quotas using the Huntington-Hill rounding rules, and the last row shows the results of rounding the quotas using Webster's rounding rules. We can see that in this case the standard divisor $SD = 10{,}000$ is a suitable divisor to apportion under Webster's method but is not a suitable divisor for the Huntington-Hill method!

TABLE ME1-7

State	A	B	C	E	F	Total
Population	34,800	105,100	65,100	140,200	54,800	400,000
Standard quota (SD = 10,000)	3.48	10.51	6.51	14.02	5.48	40
Huntington-Hill	4	11	7	14	6	42
Webster	3	11	7	14	5	40

A suitable divisor for the Huntington-Hill method will have to be a number bigger than 10,000 (we need to bring the standard quotas down a bit). A little nip here, a little tuck there, and we find the modified divisor $D = 10{,}030$, a divisor that works! Table ME1-8 shows the quotas when we use the modified divisor $D = 10{,}030$. When we carefully round these quotas using the Huntington-Hill rounding rules (good idea to have Table ME1-4 handy!) we get the Huntington-Hill apportionment shown in the last row of Table ME1-8.

TABLE ME1-8

State	A	B	C	E	F	Total
Population	34,800	105,100	65,100	140,200	54,800	400,000
Modified quota (D = 10,030)	3.47	10.479	6.491	13.978	5.464	
Huntington-Hill	4	10	7	14	5	40

■ Since the implementation of the Huntington-Hill method in 1941, every apportionment of the House of Representatives would have turned out exactly the same had Webster's method been used.

We conclude this discussion with a note of caution. Looking at the results of Examples ME1.5 and ME1.6, we might get the impression that the Huntington-Hill method and Webster's method are likely to produce different apportionments, but this is far from being the case. The numbers in Examples ME1.5 and ME1.6 were carefully rigged to show that the two methods *can* give different results—we should hardly conclude that this happens all the time. In fact, in a real-life example the odds are pretty high that the same divisor will work for both Webster and Huntington-Hill apportionments.

CONCLUSION

The Huntington-Hill method is of interest primarily for a practical reason—it is, by law, the way the United States has chosen (after trying several other methods) to apportion its House of Representatives. The method was chosen as the "best" of all possible apportionment methods based on mathematical considerations that go beyond the scope of our discussion, but, as a practical matter, the method is almost identical to Webster's method (and most of the time produces exactly the same results).

As educated citizens, it behooves us to have some basic understanding of how our political institutions work, and every once in a while this may involve a little mathematics. True, the mathematics behind the Huntington-Hill method introduced in this mini-excursion go a bit above and beyond the level that most people are comfortable with, but to us, veterans of the Chapter 4 expedition, this mini-excursion was just a natural extension of that trip. You should derive some sense of satisfaction and pat yourself on the back for having completed the journey.

■ First, try some of the exercises.

EXERCISES

A The Geometric Mean

1. Without using a calculator, find the geometric mean of each pair of numbers.

   (a) 10 and 1000

   (b) 20 and 2000

   (c) 1/20 and 1/2000

2. Without using a calculator, find the geometric mean of each pair of numbers.

   (a) 2 and 8

   (b) 20 and 80

   (c) 1/20 and 1/80

3. Without using a calculator, find the geometric mean of $2 \cdot 3^4 \cdot 7^3 \cdot 11$ and $2^7 \cdot 7^5 \cdot 11$.

   (*Hint*: These are very big numbers, but you don't need a calculator—work with their prime factorizations.)

4. Without using a calculator, find the geometric mean of $11^5 \cdot 13^8 \cdot 17^3 \cdot 19$ and $11^7 \cdot 17^5 \cdot 19$.

   (*Hint*: These are very big numbers, but you don't need a calculator—work with their prime factorizations.)

5. Home prices in Royaltown increased by 11.7% in 2005 and 25.9% in 2006. Find the average annual increase in home prices over the two-year period (rounded to the nearest tenth of a percent).

6. Over a two-year period a company's shares went up by 7.25% the first year and 11.45% the second year. Find the average annual increase in the value of the shares over the two-year period (rounded to the nearest tenth of a percent).

7. Using a calculator, complete the following table. Round the answers to three decimal places when necessary.

Consecutive integers	Geometric mean (G)	Arithmetic mean (A)	Difference (A − G)
15, 16			
16, 17			
17, 18			
18, 19			
19, 20			
29, 30			
39, 40			
49, 50			

8. Let a and b be positive numbers, and let G be their geometric mean. Express the geometric mean of each of the following pairs of numbers in terms of G.

(a) $3.75a$ and $3.75b$

(b) $10a$ and $1000b$

(c) $1/a$ and $1/b$

(d) a^2 and b^2

9. Let a and b be two positive numbers, let G be their geometric mean, and let k be a positive constant. Show that

(a) The geometric mean of $k \cdot a$ and $k \cdot b$ is $k \cdot G$.

(b) The geometric mean of k/a and k/b is k/G.

10. The arithmetic-geometric mean inequality. Let a and b be positive numbers, let A be their arithmetic mean, and let G be their geometric mean. Show that

(a) If $b = a$, then $A = G$.

(b) If $b > a$, then $A > G$.
(*Hint:* Show that $a^2 + b^2 \geq 2ab$, and use this to show that $\left(\dfrac{a+b}{2}\right)^2 \geq ab$.)

11. Pythagoras and the arithmetic-geometric mean inequality. Let a and b be positive numbers, and let $b > a$.

(a) In the right triangle ABC shown in Fig. ME1-1, find the length of BC.

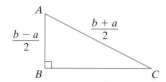

FIGURE ME1-1

(b) Explain why the result of (a) proves the arithmetic-geometric mean inequality.

B The Huntington-Hill Method

12. Round each quota using the Huntington-Hill rounding rules.

(a) $q = 4.46$

(b) $q = 4.48$

(c) $q = 50.498$

(d) $q = \sqrt{12.01}$

13. A small country consists of five states: A, B, C, D, and E. The standard quotas of each state are given in the following table. Find the apportionment under the Huntington-Hill method.

State	A	B	C	D	E
Standard quota	25.26	18.32	2.58	37.16	40.68

14. (a) Use the Huntington-Hill method to apportion Parador's Congress (Chapter 4, Example 4.3).

(b) Compare your answer in (a) with the apportionment produced by Webster's method. What's your conclusion?

15. A country consists of six states, with the state's populations given in the following table. The number of seats to be apportioned is $M = 200$.

State	A	B	C	D	E	F
Population	344,970	408,700	219,200	587,210	154,920	285,000

(a) Find the apportionment under Webster's method.

(b) Find the apportionment under the Huntington-Hill method.

(c) Compare the apportionments found in (a) and (b).

16. A country consists of six states, with the state's populations given in the following table. The number of seats to be apportioned is $M = 200$.

State	A	B	C	D	E	F
Population	344,970	204,950	515,100	84,860	154,960	695,160

(a) Find the apportionment under Webster's method.

(b) Find the apportionment under the Huntington-Hill method.

(c) Compare the apportionments found in (a) and (b).

17. The small island nation of Margarita consists of four islands: Aleta, Bonita, Corona, and Doritos. The state's population of each island is given in the following table. The number of seats to be apportioned is $M = 100$.

State	A	B	C	D
Population	86,915	4,325	5,400	3,360

(a) Find the apportionment under the Huntington-Hill method.

(b) Describe any possible violations that occurred under the apportionment in (a).

C Miscellaneous

18. Show that the difference between the arithmetic and geometric means of consecutive positive integers gets smaller as the integers get bigger.

(*Hint:* You should show that for any positive integer N, the difference between the arithmetic and geometric means of N and $N + 1$ is bigger than the difference between the arithmetic and geometric means of $N + 1$ and $N + 2$.)

19. Let a, b, and c be positive numbers such that $a < b < c$. Let A denote the arithmetic mean of a and b, and let A' denote the arithmetic mean of a and c. Likewise, let G denote the geometric mean of a and b, and let G' denote the geometric mean of a and c. Show that $A' - G' > A - G$.

(*Hint:* Start by showing that $\sqrt{c} - \sqrt{a} > \sqrt{c} - \sqrt{b}$, and then square both sides of the inequality.)

20. The purpose of this exercise is to show that under rare circumstances, there is no suitable divisor for the Huntington-Hill method and thus the method will not work. (Imagine the headache this scenario would create in the U.S. Congress!) A small country consists of three states, with populations given in the following table. The number of seats to be apportioned is $M = 12$.

State	A	B	C
Population	7290	1495	1215

(a) Show that using the Huntington-Hill rounding rules with the divisor $D = \dfrac{1215}{\sqrt{2}}$ does not work. Explain why this divisor is too small!

(*Hint:* Do not convert the numbers to decimal form—do all your computations using radicals.)

(b) Show that using the Huntington-Hill rounding rules any divisor $D > \dfrac{1215}{\sqrt{2}}$ does not work. Explain why any divisor of this size is too big!

(c) Explain why (a) and (b) imply that the Huntington-Hill method will not work in this apportionment problem.

21. **The quadratic mean.** The *quadratic mean* of two numbers a and b is defined as $Q = \sqrt{\dfrac{a^2 + b^2}{2}}$.

(a) Using a calculator, complete the following table. Round the answer to three decimal places.

a	b	Q	a	b	Q
1	2		9	10	
2	3		19	20	
3	4		29	30	
4	5		99	100	
5	6		10	20	
6	7		10	90	
7	8		10	190	
8	9		10	990	

(b) Let a and b be positive numbers such that $a < b$. Show that $a < Q < b$.

(c) Let a and b be positive numbers such that $a < b$. Show that the quadratic mean of a and b is bigger than the arithmetic mean of a and b ($Q > A$).

22. **The harmonic mean.** The *harmonic mean* of two numbers a and b is defined as $H = \dfrac{2ab}{a + b}$.

(a) Using a calculator, complete the following table. Round the answer to three decimal places.

a	b	H	a	b	H
1	2		9	10	
2	3		19	20	
3	4		29	30	
4	5		99	100	
5	6		10	20	
6	7		10	90	
7	8		10	190	
8	9		10	990	

(b) Let a and b be positive numbers such that $a < b$. Show that $a < H < b$.

(c) Let a and b be positive numbers such that $a < b$. Show that the harmonic mean of a and b is smaller than the geometric mean of a and b ($H < G$).

KEY CONCEPTS

arithmetic mean, **153**
Huntington-Hill method, **152**
mean, **153**
method of equal proportions, **152**

PROJECTS AND PAPERS

A Dean's Method

Dean's method is a method almost identical to the Huntington-Hill method (and thus to Webster's method) that uses the *harmonic means* of the lower and upper quotas (see Exercise 22) as the cutoff points for rounding. In this project you are to discuss Dean's method by comparing it to the Huntington-Hill method (in a manner similar to how the Huntington-Hill method was discussed in the mini-excursion by comparison to Webster's method). Give examples that illustrate the difference between Dean's method and the Huntington-Hill method. You should also include a brief history of Dean's method.

B Montana v. U.S. Department of Commerce (U.S. District Court, 1991); U.S. Department of Commerce v. Montana (U.S. Supreme Court, 1992)

Write a paper discussing these two important legal cases concerning the apportionment of the U.S. House of Representatives. In this paper you should (1) present the background preceding Montana's challenge to the constitutionality of the Huntington-Hill method, (2) summarize the arguments presented by Montana and the government in both cases, (3) summarize the arguments given by the District Court in ruling for Montana, and (4) summarize the arguments given by the Supreme Court in unanimously overturning the District Court ruling.

REFERENCES AND FURTHER READINGS

1. Balinski, Michel L., and H. Peyton Young, *Fair Representation; Meeting the Ideal of One Man, One Vote.* New Haven, CT: Yale University Press, 1982.

2. Census Bureau Web site, *http://www.census.gov.*

3. Huntington, E. V., "The Apportionment of Representatives in Congress," *Transactions of the American Mathematical Society,* 30 (1928), 85–110.

4. Huntington, E. V., "The Mathematical Theory of the Apportionment of Representatives," *Proceedings of the National Academy of Sciences, U.S.A.,* 7 (1921), 123–127.

5. Neubauer, Michael G., and Joel Zeitlin, "Apportionment and the 2000 Election," *The College Mathematics Journal,* 34 (Jan. 2003), 2–10.

6. Saari, D. G., "Apportionment Methods and the House of Representatives," *American Mathematical Monthly,* 85 (1978), 792–802.

7. Schmeckebier, L. F., *Congressional Apportionment.* Washington, DC: The Brookings Institution, 1941.

part 2

Management Science

5 The Mathematics of Getting Around

Euler Paths and Circuits

Sometimes great discoveries arise from the humblest and most unexpected of origins. Such is the case with the theme we will explore in the next four chapters — the mathematical study of how things are interconnected. Our story begins in the 1700s in the medieval town of Königsberg, in Eastern Europe. At the time, Königsberg was divided by a river into four separate sections, which were connected to one another by seven bridges. The old map of Königsberg shown here gives the layout of the city in 1735, the year a brilliant young mathematician named Leonhard Euler came passing through.

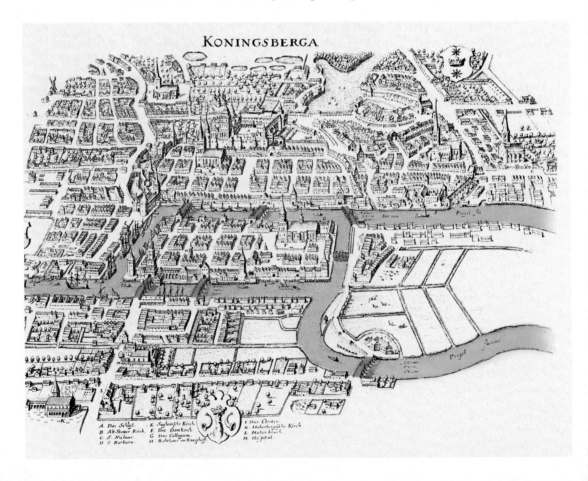

KONINGSBERGA

A. Das Schloß.   E. Sugbeingsche Kirch.   I. Das Closter.
B. Alt Stetter Kirch. F. Die Domkirch.   K. Haberbergische Kirch.
C. S. Niclaus.   G. Das Collegium.   L. Haber kirch.
D. S. Barbara.   H. Rathhaus im Kneiphoff. M. Hospital.

> **❝** The question is whether a journey can be arranged that will pass over all the bridges but not over any of them more than once. **❞**
>
> —Leonhard Euler

While visiting Königsberg, Euler was told of an innocent little puzzle of disarming simplicity: Is it possible for a person to take a walk around town in such a way that each of the seven bridges is crossed *once, but only once*? He heard the locals had tried, repeatedly and without success. Could he prove mathematically that this could not be done?

Euler, perhaps sensing that something important lay behind the frivolity of the puzzle, proceeded to solve it by demonstrating that indeed such a walk was impossible. But he actually did much more. In solving the puzzle of the Königsberg bridges, Euler laid the foundations for what was at the time a totally new type of geometry, which he called *geometris situs* ("the geometry of location"). From these modest beginnings, the basic ideas set forth by Euler eventually developed and matured into one of the most important and practical branches of modern mathematics, now known as *graph theory*. Modern applications of graph theory span practically every area of science and technology—from chemistry, biology, and computer science to psychology, sociology, and business management.

Leonhard Euler (1707–1783). For more on Euler, see the biographical profile at the end of the chapter.

Over the next four chapters we will discuss the uses of graph theory as a tool for solving *management science* problems. These are problems in which the ultimate goal is to find efficient ways to organize and carry out complex tasks, usually tasks that involve a large number of variables and in which it is not at all obvious how to make optimal decisions.

The theme of this chapter is the question of how to create *efficient routes* for the delivery of goods and services—such as mail delivery, garbage collection, police patrols, newspaper deliveries, and, most important, late-night pizza deliveries—along the streets of a city, town, or neighborhood. These types of management science problems are known as *Euler circuit problems* (named, of course, after the founder of the field). Section 5.1 sets the table with a description of *routing problems* in general and the introduction of several examples of *Euler circuit problems* (which are solved at the end of the chapter). Sections 5.2 and 5.3 give an introduction to the *language and concepts* of *graph theory*—the mathematical toolkit that will help us tackle routing problems like the ones introduced in Section 5.1. Section 5.4 introduces the concept of *graph modeling*—the process by which a complicated real-life problem can be *translated* into the clean and precise language of graph theory. In Sections 5.5 and 5.6 we will learn about *Euler's theorems* and *Fleury's algorithm*, which provide the theoretical framework that will help us understand the structural aspects of Euler circuit problems. In Section 5.7 we will learn how to combine all the preceding ideas to develop a strategy that allows us to solve Euler circuit problems.

5.1   Euler Circuit Problems

We will start this section with a brief discussion of routing problems. What is a **routing problem**? To put it in the most general way, routing problems are concerned with finding ways to route the delivery of *goods* and/or *services* to an assortment of *destinations*. The goods or services in question could be packages, mail, newspapers, pizzas, garbage collection, bus service, and so on. The delivery destinations could be homes, warehouses, distribution centers, terminals, and the like.

There are two basic questions that we are typically interested in when dealing with a routing problem. The first is called the *existence* question. The existence question is simple: Is an actual route possible? For most routing problems, the existence question is easy to answer, and the answer takes the form of a simple yes or no. When the answer to the existence question is yes, then a second question—the *optimization question*—comes into play. Of all the possible routes, which one is the *optimal route*? (*Optimal* here means "the best" when measured against some predetermined variable such as *cost*, *distance*, or *time*.) In most management science problems, the optimization question is where the action is.

In this chapter we will learn how to answer both the existence and optimization questions for a special class of routing problems known as **Euler circuit problems**. The common thread in all Euler circuit problems is what we might call, for lack of a better term, the *exhaustion requirement*—the requirement that the route must wind its way through . . . *everywhere*. Thus, in an Euler circuit problem, by definition every single one of the streets (or bridges, or lanes, or highways) within a defined area (be it a town, an area of town, or a subdivision) must be covered by the route. We will refer to these types of routes as *exhaustive routes*. The most common services that typically require exhaustive routing are mail delivery, police patrols, garbage collection, street sweeping, and snow removal. More exotic examples can be census taking, precinct walking, electric meter reading, routing parades, tour buses, and so on.

> 66 **ex-haust:** To treat completely; cover thoroughly. 99
>
> — *American Heritage Dictionary*

To clarify some of the ideas we will introduce several examples of Euler circuit problems (just the problems for now—their solutions will come later in the chapter).

(**EXAMPLE 5.1**)   **Walking the 'Hood'**

After a rash of burglaries, a private security guard is hired to patrol the streets of the Sunnyside neighborhood shown in Fig. 5-1. The security guard's assignment is to make an exhaustive patrol, on foot, through the entire neighborhood. Obvi-

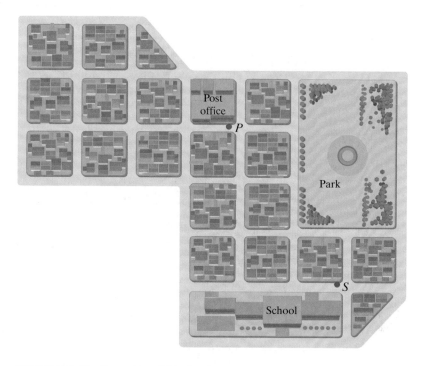

FIGURE 5-1 The Sunnyside neighborhood.

ously, he doesn't want to walk any more than what is necessary. His starting point is the southeast corner across from the school (*S* in Fig. 5-1)—that's where he parks his car. (This is relevant because at the end of his patrol he needs to come back to *S* to pick up his car.) Being a practical person, the security guard would like the answers to two questions. (1) Is it possible to start and end at *S*, cover every block of the neighborhood, and pass through each block *just once*? (2) If some of the blocks will have to be covered more than once, what is an *optimal* route that covers the entire neighborhood? (*Optimal* here means "with the minimal amount of walking.")

EXAMPLE 5.2   **Delivering the Mail**

A mail carrier has to deliver mail in the same Sunnyside neighborhood (Fig. 5-1). The difference between the mail carrier's route and the security guard's route is that the mail carrier must make *two* passes through blocks with houses on both sides of the street and only one pass through blocks with houses on only one side of the street; and where there are no homes on either side of the street, the mail carrier does not have to walk at all. In addition, the mail carrier has no choice as to her starting and ending points—she has to start and end her route at the local post office (*P* in Fig. 5-1). Much like the security guard, the mail carrier wants to find the optimal route that would allow her to cover the neighborhood with the least amount of walking. (Put yourself in her shoes and you would do the same—good weather or bad, she walks this route 300 days a year!)

EXAMPLE 5.3)   **The Seven Bridges of Königsberg**

Figure 5-2(a) shows an old map of the city of Königsberg and its seven bridges; Fig. 5-2(b) shows a modernized version of the very same layout. We opened the chapter with this question: Can a walker take a stroll and cross each of the seven bridges of Königsberg without crossing any of them more than once?

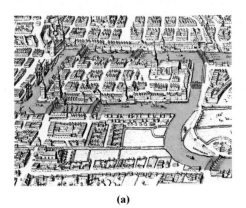
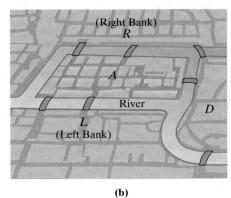

FIGURE 5-2

(a)   (b)

EXAMPLE 5.4)   **The Bridges of Madison County**

This is a more modern version of Example 5.3. Madison County is a quaint old place, famous for its quaint old bridges. A beautiful river runs through the county, and there are four islands (A, B, C, and D) and 11 bridges joining the islands to both banks of the river (R and L) and one another (Fig. 5-3). A famous photographer is hired to take pictures of each of the 11 bridges for a national magazine. The photographer needs to drive across each bridge once for the photo shoot. Moreover, since there is a $25 toll (the locals call it a "maintenance tax") every time an out-of-town visitor drives across a bridge, the photographer wants to minimize the total cost of his trip and to recross bridges only if it is absolutely necessary. What is the optimal (cheapest) route for him to follow?

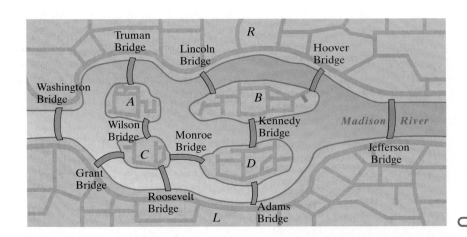

FIGURE 5-3

EXAMPLE 5.5 Child's Play

Figure 5-4 shows a few simple line drawings. The name of the game is to trace each drawing *without lifting the pencil or retracing any of the lines*. These kinds of tracings are called **unicursal tracings**. (When we end in the same place we started, we call it a *closed* unicursal tracing; when we start and end in different places, we call it an *open* unicursal tracing.) Which of the drawings in Fig. 5-4 can be traced with closed unicursal tracings? Which with only open ones? Which can't be traced (without cheating)? Some of us played such games in our childhood (in the good old days before X-Boxes and PlayStations) and may be able quickly to figure out the answers in the case of Figs. 5-4(a), (b), and (c). But what about slightly more complicated shapes, such as the one in Fig. 5-4(d)? How can we tell if a unicursal tracing (open or closed) is possible? Good question. We will answer it in Section 5.5.

> " u'ni*cur'sal: That can be passed over in a single course. "
>
> —*Webster's Revised Unabridged Dictionary*

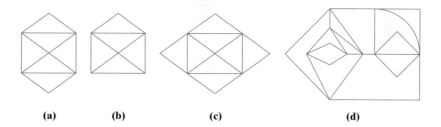

FIGURE 5-4
(a) (b) (c) (d)

5.2 What Is a Graph?

■ A note of warning: The graphs we will be discussing here have no relation to the graphs of functions you may have studied in algebra or calculus.

As Examples 5.1 through 5.5 illustrate, Euler circuit problems can come in many forms. Fortunately, they can all be tackled by means of a single unifying mathematical concept—the concept of a **graph**.

The most common way to describe a *graph* is by means of a picture. The basic elements of such a picture are a set of "dots" called the **vertices** of the graph and a collection of "lines" called the **edges** of the graph. On the surface, that's all there is to it—lines connecting dots! Below the surface there is a surprisingly rich theory. Let's explore a few basic concepts first.

EXAMPLE 5.6 Connect the Dots

Figure 5-5 shows the picture of a graph. This graph has six vertices called A, B, C, D, E, and F. Each edge can be described by listing (in any order) the pair of vertices that are connected by the edge. Thus, the edges of this graph, listed in random order, are AB, BC, CD, AD, DE, EB, CD, and BB.

Notice several important things about the edges of the graph:

■ It is possible for an edge to connect a vertex back to itself, as is the case with BB. These type of edges are called **loops**.

■ It is possible for two edges to connect the same pair of vertices, as is the case with CD, which is a "double edge." In general, we refer to such edges as **multiple edges**.

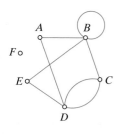

FIGURE 5-5

- Sometimes edges "cross" each other at incidental crossing points that are not themselves vertices of the graph. Such is the case with the crossing point created by edges AD and BE in Fig. 5-5.
- Edges do not have a direction; thus, there is no right or wrong order to write an edge—AB or BA are both acceptable.

■ To avoid confusion with vertices, we use a special "chancery" font for the $\mathcal{V}$ and $\mathcal{E}$ that represent the vertex and edge sets.

A convenient way to describe the vertices and edges of a graph is by using the notation of sets. For the graph shown in Fig. 5-5 the **vertex set** is $\mathcal{V} = \{A, B, C, D, E, F\}$, and the **edge set** is $\mathcal{E} = \{AB, AD, BB, BC, BE, CD, CD, \text{and } DE\}$. Notice that CD appears twice in the edge set, indicating that there are two edges connecting C and D.

One reason graphs are so useful is that they can tell a story in ways that simple words often can't.

(**EXAMPLE 5.7**)   **Relationship Graphs**

Imagine that as part of a sociology study we want to describe the network of "friendships" that develops among a group of students through their Facebook sites. We can illustrate this very nicely with a graph such as the one in Fig. 5-6. In this graph the vertices represent people (the students), and an edge connecting vertex X to vertex Y implies that X and Y are "Facebook friends."

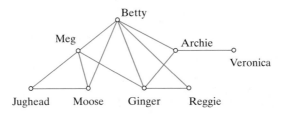

FIGURE 5-6

The graph in Fig. 5-6 provides a simple illustration of an important idea—the vertices of the graph as objects and the edges of the graph describing a specific type of relationship that may exist between pairs of objects. In this particular case we used a small group to illustrate the idea, but the same idea on a much grander scale is used frequently by social scientists to describe and analyze complex social networks (see Fig. 5-7).

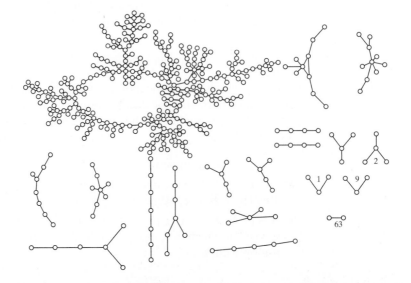

FIGURE 5-7 A graph showing the romantic relationships among the students in an unnamed high school in the Midwest over an eighteen-month period. (*Source:* Adapted from Peter Bearman, James Moody, and Katherine Stovel, "Chains of Affection: The Structure of Adolescent Romantic and Sexual Networks," *American Journal of Sociology*, 110, July 2004, 49–91.)

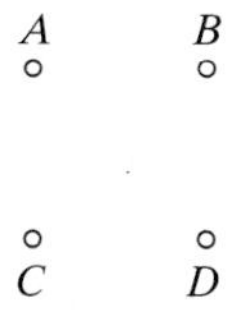

A   B
o   o

o   o
C   D

FIGURE 5-8

EXAMPLE 5.8   **Pure Isolation**

Figure 5-8 shows a graph with four **isolated vertices** and having no edges. We won't be seeing graphs like this too often, but it's important to know that graphs with no edges are allowed. The edge set of a graph with no edges is the *empty* set (we can write it as $\mathcal{E} = \{\ \ \}$ or $\mathcal{E} = \phi$).

EXAMPLE 5.9   **Pictures Optional**

Suppose you are given the following information about a graph: The vertex set is $V = \{A, D, L,$ and $R\}$, and the edge set is $\mathcal{E} = \{AD, AL, AL, AR, AR, DL, DR\}$. But where is the picture?

You are told that if you really want a picture, you can make up your own. That's fine, but where should you place the vertices? What should the edges look like? Good news: These issues are irrelevant! You have total freedom to place the vertices anywhere you please, and as long as you connect the right pairs of vertices, you can connect them any way you like! Figure 5-9 shows two pictures of the same graph. While they may look like two totally different pictures to the casual observer, they both describe the same vertex set $V = \{A, D, L,$ and $R\}$ and the same edge set $\mathcal{E} = \{AD, AL, AL, AR, AR, DL, DR\}$. [From a visual point of view, however, one could argue that the graph in Fig. 5-9(a) looks nice, while the graph in Fig. 5-9(b) looks gnarly.]

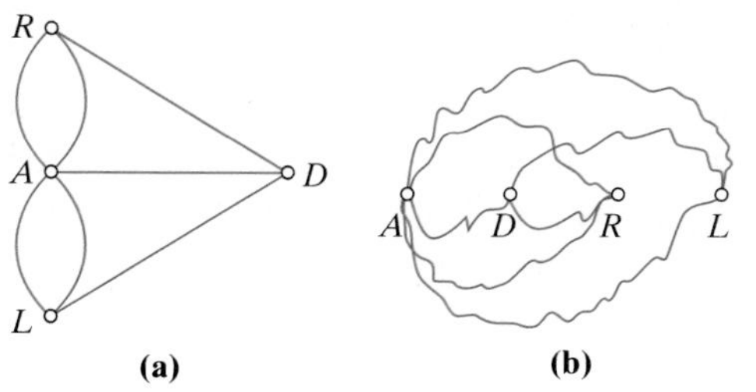

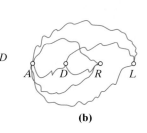

(a)   **(b)**

FIGURE 5-9

Example 5.9 illustrates a very important point. We think of a graph as *a picture* consisting of dots and lines, but the picture is just a visual representation of a more abstract idea: a set of objects (represented by the vertices of the graph) and a relationship among pairs of objects (represented by the edges of the graph.) All one needs to describe these two things are a vertex set V and an edge set $\mathcal{E}$. A picture is nice but not essential. Thus, if we simply give the vertex set $V = \{A, D, L,$ and $R\}$ and the edge set $\mathcal{E} = \{AD, AL, AL, AR, AR, DL, DR\}$, we have defined a graph.

So, if a graph is not just a set of points connected by a bunch of lines, what is it? The following is a somewhat formal, but proper, definition.

GRAPH

A **graph** is a structure consisting of a set of objects (the vertex set) and a list describing how pairs of objects are related (the edge set). Relations among objects include the possibility of an object being related to itself (a loop) as well as multiple relations between the same pair of objects (multiple edges).

5.3 Graph Concepts and Terminology

Every branch of mathematics has its own peculiar jargon, and graph theory has more than its share. In this section we will introduce some important concepts and terminology that we will need in the next three chapters.

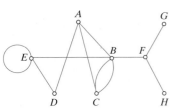

FIGURE 5-10

(**EXAMPLE 5.10**)   **Adjacency**

We say that two vertices in a graph are **adjacent** if they are joined by an edge. In the graph shown in Fig. 5-10, vertices A and B are adjacent and so are vertices B and C. Vertices C and D are not adjacent, and neither are vertices A and E. Because of the loop EE, we say that vertex E is adjacent to itself.

We can also speak of edges being adjacent. Two edges are said to be **adjacent** if they share a common vertex. In the graph shown in Fig. 5-10, AB and BF are adjacent edges and so are AB and AD. On the other hand, AB and DE are not adjacent.

(**EXAMPLE 5.11**)   **Degree of a Vertex**

The **degree** of a vertex is the number of edges meeting at that vertex. A loop counts twice toward the degree. We will use the notation $\deg(V)$ to denote the degree of vertex V. In Fig. 5-10 the degrees of the vertices are as follows: $\deg(A) = 3$, $\deg(B) = 5$, $\deg(C) = 3$, $\deg(D) = 2$, $\deg(E) = 4$, $\deg(F) = 3$, $\deg(G) = 1$, and $\deg(H) = 1$.

Because distinguishing vertices with an even degree from vertices with an odd degree is going to be critical later on, we will often refer to vertices as **even vertices** or **odd vertices**, depending on their degree. The graph in Fig. 5-10 has two even vertices (D and E) and six odd vertices (all the others).

(**EXAMPLE 5.12**)   **Paths and Circuits**

Paths and circuits both describe "trips" along the edges of a graph. The only real difference between a path and a circuit is that a **circuit** is a "closed" trip (the trip ends back at the starting point), whereas a **path** is an "open" trip (the starting and ending points are different). In this context, by a "trip" (be it a path or a circuit), we mean a sequence of adjacent edges with the property that *an edge can be traveled just once.*

The standard way to describe a path or a circuit is by listing the vertices in order of travel. Here are a few examples of paths and circuits using the graph in Fig. 5-10:

- A, B, E, D is a *path* from vertex A to vertex D. The edges of this path in order of travel are AB, BE, and ED. The **length** of the path (i.e., the number of edges in the path) is 3.
- A, B, C, A, D, E is a *path* of length 5 from A to E. This path visits vertex A twice (that's fine), but no edge is repeated.
- A, B, C, B, E is another *path* from A to E. This path is only possible because there are two edges connecting B and C.
- A, C, B, E, E, D is a *path* of length 5 from A to D. One of the edges in this path is the loop EE.
- A, B, C, B, A, D is *not a path* because the edge AB is traveled twice.
- $A, B, C, B, E, E, D, A, C, B$ is *not a path* because the edge BC is traveled three times. (The first two passes are fine, since there are two edges connecting B and C, but a third pass requires that we travel through one of those edges a second time.)
- A, B, C, A is a *circuit* of length 3. The edges of this circuit are AB, BC, and CA. Unlike a path, a circuit has no specific starting or ending point. We arbitrarily have to choose a starting point to describe the circuit on paper, but the circuit A, B, C, A can also be written as B, C, A, B or C, A, B, C.

- B, C, B is a *circuit* of length 2. Circuits of length 2 are possible when there are multiple edges.
- The *EE* loop is considered to be a *circuit* of length 1.

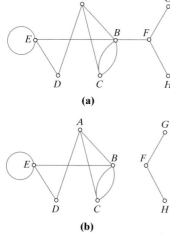

EXAMPLE 5.13   **Connectedness and Bridges**

A graph is **connected** if you can get from any vertex to any other vertex along a path. Essentially, this means that the graph is all in one piece. The graph shown in Fig. 5-11(a) is a connected graph, whereas the graphs shown in Fig. 5-11(b) and (c) are **disconnected** graphs. A disconnected graph is made up of separate connected **components**. Figure 5-11(b) shows a disconnected graph with two components, and Fig. 5-11(c) shows a disconnected graph with three components (an isolated vertex is a component in and of itself).

Notice that Fig. 5-11(b) is the graph we get when we remove the edge *BF* from the graph in Fig. 5-11(a). This illustrates how the removal of just one edge from a connected graph can sometimes make the graph disconnected. An edge whose removal makes a connected graph disconnected is called a **bridge**. Thus, we say that *BF* is a bridge in the graph in Fig. 5-11(a). The graph has two other bridges—*FG* and *FH*.

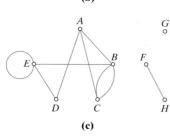

(b)

EXAMPLE 5.14   **Euler Paths and Euler Circuits**

An **Euler path** in a connected graph is a path that travels through *all* the edges of the graph. Being a path, edges can only be traveled once, so in an Euler path *every* edge of the graph is traveled *once and only once*. By definition, only a connected graph can have Euler paths, but, of course, just being connected is not enough.

Much like Euler paths, we can also define Euler circuits. An **Euler circuit** is a circuit that travels through *every* edge of a connected graph. Being a circuit, the trip must end where it started and travel along *every* edge of the graph *once and only once*.

A connected graph cannot have both an Euler path *and* an Euler circuit—it can have one or the other or neither. In the graph shown in Fig. 5-12(a), the path $C, A, B, E, A, D, B, C, D$ travels along each of the eight edges of the graph and is therefore an Euler path. This graph has several other Euler paths—you may want to try to find one that does not start at C.

(c)

FIGURE 5-11

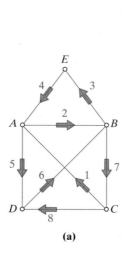

(a)

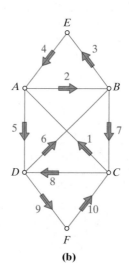

(b)

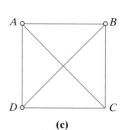

(c)

FIGURE 5-12

In the graph shown in Fig. 5-12(b), the circuit $C, A, B, E, A, D, B, C, D, F, C$ is an Euler circuit (one of many). You may want to find a different one. (Remember that traveling the edges in the same sequence but using a different starting point is cheating—you are rewriting the same circuit.)

The graph shown in Fig. 5-12(c) has neither an Euler path nor an Euler circuit. (We will learn how to tell in Section 5.5.)

5.4 Graph Models

One of Euler's most important insights was the observation that certain types of problems can be conveniently rephrased as graph problems and that, in fact, graphs are just the right tool for describing many real-life situations. The notion of using a mathematical concept to describe and solve a real-life problem is one of the oldest and grandest traditions in mathematics. It is called *modeling*. Unwittingly, we have all done simple forms of modeling before, all the way back to elementary school. Every time we turn a word problem into an arithmetic calculation, an algebraic equation, or a geometric picture, we are modeling. We can now add to our repertoire one more tool for modeling: graph models.

In the next set of examples we are going to illustrate how we can use graphs to *model* some of the problems introduced in Section 5.1.

(**EXAMPLE 5.15**)   **The Seven Bridges of Königsberg: Part 2**

The Königsberg bridges question discussed in Example 5.3 asked whether it was possible to take a stroll through the old city of Königsberg and cross each of the seven bridges once and only once. To answer this question one obviously needs to take a look at the layout of the old city. A stylized map of the city of Königsberg is shown in Fig. 5-13(a). This map is not entirely accurate—the drawing is not to scale and the exact positions and angles of some of the bridges are changed. Does it matter?

A moment's reflection should convince us that many details on the original map are irrelevant to the question in point. The shape and size of the islands, the width of the river, the lengths of the bridges—none of these things really matter. So, then, what is it that does matter? Surprisingly little. *The only thing that truly matters to the solution of this problem is the relationship between land masses (islands and banks) and bridges.* Which land masses are connected to each other and by how many bridges? This information is captured by the red edges in Fig. 5-13(b). Thus, when we strip the map of all its superfluous information, we end

FIGURE 5-13

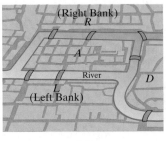

(a)

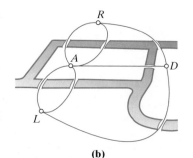

(b)

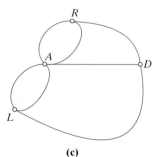

(c)

up with the graph model shown in Fig. 5-13(c). The four vertices of the graph represent each of the four land masses; the edges represent the seven bridges. In this graph an *Euler circuit* would represent a stroll around the town that crosses each bridge once and ends back at the starting point; an *Euler path* would represent a stroll that crosses each bridge once but does not return to the starting point.

As big moments go this one may not seem like much, but Euler's idea to turn a puzzle about walking across bridges in a quaint medieval city into an abstract question about graphs was a "eureka" moment in the history of mathematics. (Here is a thought: Before you read on, take a break now, get yourself a cup of coffee, and try to mull over the significance of Example 5.15.)

(**EXAMPLE 5.16**) **Walking the 'Hood': Part 2**

In Example 5.1 we were introduced to the problem of the security guard who needs to walk the streets of the Sunnyside neighborhood [Fig. 5-14(a)]. The graph in Fig. 5-14(b)—where each edge represents a block of the neighborhood and each vertex an intersection—is a graph model of this problem. Does the graph have an Euler circuit? An Euler path? Neither? (These are relevant questions that we will learn how to answer in the next section.)

(**EXAMPLE 5.17**) **Delivering the Mail: Part 2**

Recall that unlike the security guard, the mail carrier (see Example 5.2) must make two passes through every block that has homes on both sides of the street (she has to physically place the mail in the mailboxes), must make one pass through blocks that have homes on only one side of the street, and does not have to walk along blocks where there are no houses. In this situation an appropriate graph model requires two edges on the blocks that have homes on both sides of the street, one edge for the blocks that have homes on only one side of the street, and no edges for blocks having no homes on either side of the street. The graph that models this situation is shown in Fig. 5-14(c).

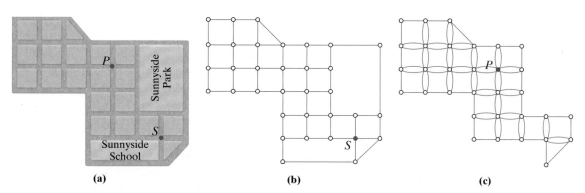

FIGURE 5-14 (a) The Sunnyside neighborhood. (b) A graph model for the security guard. (c) A graph model for the mail carrier.

5.5   Euler's Theorems

In this section we are going to develop the basic theory that will allow us to determine if a graph has an Euler circuit, an Euler path, or neither. This is important because, as we saw in the previous section, what are Euler circuit or Euler path questions in theory are real-life routing questions in practice. The three theorems we are going to see next (all due to Euler) are surprisingly simple and yet tremendously useful.

> **EULER'S CIRCUIT THEOREM**
>
> - If a graph is *connected* and *every vertex is even*, then it has an Euler circuit (at least one, usually more).
> - If a graph has *any odd vertices*, then it does not have an Euler circuit.

In practice, if we want to know if a graph has an Euler circuit or not, here is how we can use Euler's circuit theorem. First we make sure the graph is connected. (If it isn't, then no matter what else, an Euler circuit is impossible.) If the graph is connected, then we start checking the degrees of the vertices, one by one. As soon as we hit an odd vertex, we know that an Euler circuit is out of the question. If there are no odd vertices, then we know that the answer is yes—the graph does have an Euler circuit! (The theorem doesn't tell us how to find it—that will come soon.) Figure 5-15 illustrates the three possible scenarios. The graph in Fig. 5-15(a) cannot have an Euler circuit for the simple reason that it is disconnected. The graph in Fig. 5-15(b) is connected, but we can quickly spot odd vertices (*C* is one of them; there are others). This graph has no Euler circuits either. But the graph in Fig. 5-15(c) is connected and all the vertices are even. This graph does have Euler circuits.

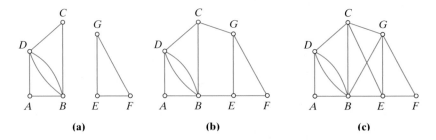

FIGURE 5-15 (a) Not connected. (b) Some vertices are odd. (c) All vertices are even.

The basic idea behind Euler's circuit theorem is that as we travel along an Euler circuit, every time we go through a vertex we use up two different edges at that vertex—one to come in and one to go out. We can keep doing this as long as the vertices are even. A single odd vertex means that at some point we are going to come into it and not be able to get out. An analogous theorem will work with Euler paths, but now we do need odd vertices for the starting and ending points of the path. All the other vertices have to be even. Thus, we have the following theorem.

EULER'S PATH THEOREM

- If a graph is *connected* and has *exactly two odd vertices*, then it has an Euler path (at least one, usually more). Any such path must start at one of the odd vertices and end at the other one.

- If a graph has *more than two* odd vertices, then it cannot have an Euler path.

(**EXAMPLE 5.18**)   **The Seven Bridges of Königsberg: Part 3**

Back to the Königsberg bridges problem. In Example 5.15 we saw that the layout of the bridges in the old city can be modeled by the graph in Fig. 5-16(a). This graph has four odd vertices; thus, neither an Euler circuit nor an Euler path can exist. We now have an unequivocal answer to the puzzle: *There is no possible way anyone can walk across all the bridges without having to recross some of them!* How many bridges will need to be recrossed? It depends. If we want to start and end in the same place, we must recross at least two of the bridges. One of the many possible routes is shown in Fig. 5-16(b). In this route the bridge connecting *L* and *D* is crossed twice, and so is one of the two bridges connecting *A* and *R*. If we are allowed to start and end in different places, we can do it by recrossing just one of the bridges. One possible route starting at *A*, crossing bridge *LD* twice, and ending at *R* is shown in Fig. 5-16(c).

FIGURE 5-16 (a) The original graph with four odd vertices. (b) A walk recrossing bridges *DL* and *RA*. (c) A walk recrossing bridge *DL* only.

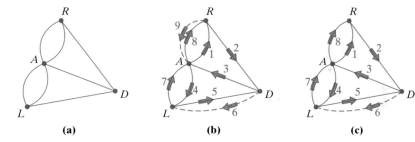

(a)   (b)   (c)

(**EXAMPLE 5.19**)   **Child's Play: Part 2**

Figure 5-17 shows four graphs. These graphs correspond to the line drawings in Example 5.5. Using Euler's theorems, we can answer all questions about unicursal tracings.

The graph in Fig. 5-17(a) is connected, and the vertices are all even. By Euler's circuit theorem we know that the graph has an Euler circuit, which implies that the original line drawing has a closed unicursal tracing.

The graph in Fig. 5-17(b) is connected and has exactly two odd vertices (*C* and *D*). By Euler's path theorem, the graph has an Euler path (open unicursal

FIGURE 5-17 Odd vertices shown in red.

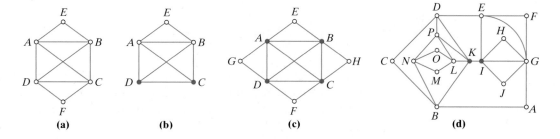

(a)   (b)   (c)   (d)

tracing). Moreover, we now know that the path has to start at C and end at D, or vice versa.

The graph in Fig. 5-17(c) has four odd vertices (A, B, C, and D), so it has neither an Euler path nor an Euler circuit.

The full power of Euler's theorems is best appreciated when the graphs get bigger. The graph in Fig. 5-17(d) is not extremely big, but we can no longer "eyeball" an Euler circuit or path. On the other hand, a quick check of the degrees of the vertices shows that K and I are odd vertices and all the rest are even. We are now in business—an open unicursal tracing is possible as long as we start it at K or I (and end it at the other one). Starting anyplace else will lead to a dead end. Good to know!

Euler's circuit theorem deals with graphs with zero odd vertices, whereas Euler's Path Theorem deals with graphs with two or more odd vertices. The only scenario not covered by the two theorems is that of graphs with just *one* odd vertex. Euler's third theorem rules out this possibility—a graph cannot have just one odd vertex. In fact, Euler's third theorem says much more.

EULER'S SUM OF DEGREES THEOREM

- The sum of the degrees of all the vertices of a graph equals twice the number of edges (and therefore is an even number).
- A graph always has an even number of *odd* vertices.

Euler's sum of degrees theorem is based on the following basic observation: Take any edge—let's call it XY. The edge contributes once to the degree of vertex X and once to the degree of vertex Y, so, in all, that edge makes a total contribution of 2 to the sum of the degrees. Thus, when the degrees of all the vertices of a graph are added, the total is twice the number of edges. Since the total sum is an even number, it is impossible to have just one odd vertex, or three odd vertices, or five odd vertices, and so on. To put it in a slightly different way, *the odd vertices of a graph always come in twos*.

Table 5-1 is a summary of Euler's three theorems. It shows the relationship between the number of odd vertices in a connected graph G and the existence of Euler paths or Euler circuits. (The assumption that G is connected is essential—a disconnected graph cannot have Euler paths or circuits regardless of what else is going on.)

TABLE 5-1	Euler's Theorems (Summary)
Number of odd vertices	**Conclusion**
0	G has Euler circuit
2	G has Euler path
$4, 6, 8, \ldots$	G has neither
$1, 3, 5, \ldots$	Better go back and double check! This is impossible!

5.6 Fleury's Algorithm

Euler's theorems help us answer the following existence question: Does the graph have an Euler circuit, an Euler path, or neither? But when the graph has an Euler circuit or path, how do we find it? For small graphs, simple trial-and-error usually works fine, but real-life applications sometimes involve graphs with hundreds, or even thousands, of vertices. In these cases a trial-and-error approach is out of the question, and what is needed is a systematic strategy that tells us how to create an Euler circuit or path. In other words, we need an *algorithm*.

Algorithms

There are many types of problems that can be solved by simply following a set of procedural rules—very specific rules like *when you get to this point, do this, . . . after you finish this, do that*, and so on. Given a specific problem *X*, an **algorithm** for solving *X* is a set of *procedural rules* that, when followed, always lead to some sort of "solution" to *X*. *X* need not be a mathematics problem—algorithms are used, sometimes unwittingly, in all walks of life: directions to find someone's house, the instructions for assembling a new bike, or a recipe for baking an apple pie are all examples of real-life algorithms. A useful analogy is to think of the problem as a *dish* we want to prepare and the algorithm as a *recipe* for preparing that dish.

In many cases, there are several different algorithms for solving the same problem (there is more than one way to bake an apple pie); in other cases, the problem does not lend itself to an algorithmic solution. In mathematics, algorithms are either *formula* driven (you just apply the formula or formulas to the appropriate inputs) or *directive* driven (you must follow a specific set of directives). In this part of the book (Chapters 5 through 8) we will discuss many important algorithms of the latter type.

Algorithms may be complicated but are rarely difficult. (There is a world of difference between complicated and difficult—accounting is complicated, calculus is difficult!) You don't have to be a brilliant and creative thinker to implement most algorithms—you just have to learn how to follow instructions carefully and methodically. For most of the algorithms we will discuss in this and the next three chapters, the key to success is simple: practice, practice, and more practice!

Fleury's Algorithm

■ For a completely different algorithm, known as **Hierholzer's algorithm**, see Exercise 68.

We will now turn our attention to an algorithm that finds an *Euler circuit* or an *Euler path* in a connected graph. Technically speaking, these are two separate algorithms, but in essence they are identical, so they can be described as one. (The algorithm we will give here is attributed to a Frenchman by the name of M. Fleury, who is alleged to have published a description of the algorithm in 1885. Other than his connection to this algorithm, little else is known about Monsieur Fleury.)

The idea behind Fleury's algorithm can be paraphrased by that old piece of folk wisdom: *Don't burn your bridges behind you*. In graph theory the word *bridge* has a very specific meaning—it is the only edge connecting two separate sections (call them *A* and *B*) of a graph, as illustrated in Fig. 5-18. This means that if you are in *A*, you can only get to *B* by crossing the bridge. If you do that and then want to get back to *A*, you will need to recross that same

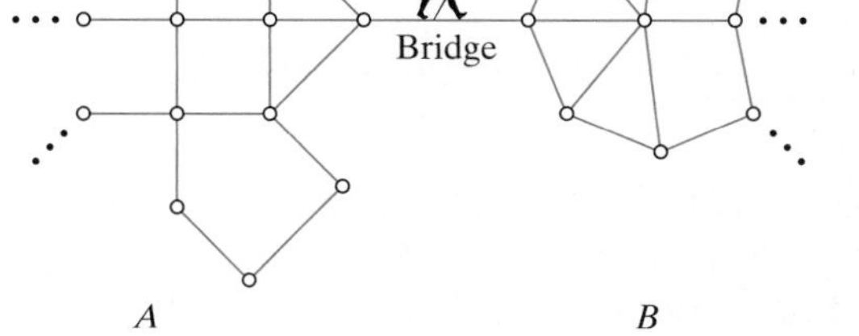

FIGURE 5-18 The bridge separates the two sections. Once you cross from *A* to *B*, the only way to get back to *A* is by recrossing the bridge.

bridge. It follows that if you don't want to recross bridges, you better finish your business at *A* before you move on to *B*.

Thus, Fleury's algorithm is based on a simple principle: To find an Euler circuit or an Euler path, *bridges are the last edges you want to cross*. Only do it if you have no choice! Simple enough, but there is a rub: The graph whose bridges we are supposed to avoid is not necessarily the original graph of the problem. Instead, it is that part of the original graph that has yet to be traveled. The point is this: Once we travel along an edge, we are done with it! We will never cross it again, so from that point on, as far as we are concerned, it is as if that edge never existed. Our concerns lie only on how we are going to get around the *yet-to-be-traveled* part of the graph. Thus, when we talk about bridges that we want to leave as a last resort, we are really referring to *bridges of the to-be-traveled part of the graph*.

┌─ **FLEURY'S ALGORITHM FOR FINDING AN EULER CIRCUIT (PATH)** ─┐

- **Preliminaries.** Make sure that the graph is connected and either (1) has no odd vertices (circuit) or (2) has just two odd vertices (path).
- **Start.** Choose a starting vertex. [In case (1) this can be any vertex; in case (2) it must be one of the two *odd* vertices.]
- **Intermediate steps.** At each step, if you have a choice, *don't choose a bridge of the yet-to-be-traveled part* of the graph. However, if you have only one choice, take it.
- **End.** When you can't travel any more, the circuit (path) is complete. [In case (1) you will be back at the starting vertex; in case (2) you will end at the other odd vertex.]

└──┘

The only complicated aspect of Fleury's algorithm is the bookkeeping. With each new step, the untraveled part of the graph changes and there may be new bridges formed. Thus, in implementing Fleury's algorithm it is critical to separate the *past* (the part of the graph that has already been traveled) from the *future* (the part of the graph that still needs to be traveled). While there are many different ways to accomplish this (you are certainly encouraged to come up with one of your own), a fairly reliable way goes like this: Start with *two* copies of the graph. Copy 1 is to keep track of the "future"; copy 2 is to keep track of the "past." Every time you travel along an edge, *erase* the edge from copy 1, but mark it (say in red) and label it with the appropriate number on copy 2. As you move forward, copy 1 gets smaller and copy 2 gets redder. At the end, copy 1 has disappeared; copy 2 shows the actual Euler circuit or path.

It's time to look at a couple of examples.

(**EXAMPLE 5.20**) **Implementing Fleury's Algorithm**

The graph in Fig. 5-19(a) is a very simple graph—it would be easier to find an Euler circuit just by trial-and-error than by using Fleury's algorithm. Nonetheless, we will do it using Fleury's algorithm. The real purpose of the example is to see the algorithm at work. Each step of the algorithm is explained in Figs. 5-19(b) through (h).

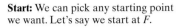

(a)

Start: We can pick any starting point we want. Let's say we start at *F*.

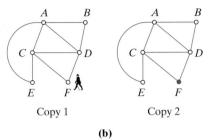

Copy 1   Copy 2

(b)

Step 1: Travel from *F* to *C*. (Could have also gone from *F* to *D*.)

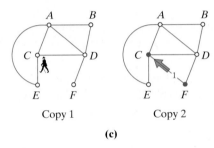

Copy 1   Copy 2

(c)

Step 2: Travel from *C* to *D*. (Could have also gone to *A* or to *E*.)

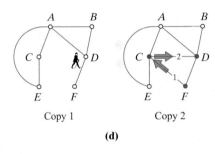

Copy 1   Copy 2

(d)

Step 3: Travel from *D* to *A*. (Could have also gone to *B* but not to *F*—*DF* is a bridge!)

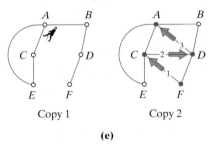

Copy 1   Copy 2

(e)

Step 4: Travel from *A* to *C*. (Could have also gone to *E* but not to *B*—*AB* is a bridge!)

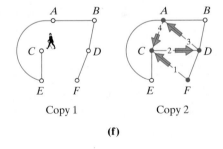

Copy 1   Copy 2

(f)

Step 5: Travel from *C* to *E*. (There is no choice!)

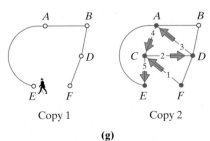

Copy 1   Copy 2

(g)

Steps 6, 7, 8, and 9: Only one way to go at each step.

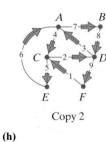

Copy 1   Copy 2

(h)

FIGURE 5-19

(**EXAMPLE 5.21**)   **Fleury's Algorithm for Euler Paths**

We will apply Fleury's algorithm to the graph in Fig. 5-20. Since it would be a little impractical to show each step of the algorithm with a separate picture as we did in Example 5.20, you are going to have to do some of the work. Start by making two copies of the graph. (If you haven't already done so, get some paper, a pencil, and an eraser.)

- **Start.** This graph has two odd vertices, E and J. We can pick either one as the starting vertex. Let's start at J.

- **Step 1.** From J we have five choices, all of which are OK. We'll randomly pick K. (Erase JK on copy 1, and mark and label JK with a 1 on copy 2.)

FIGURE 5-20

- **Step 2.** From K we have three choices (B, L, or H). Any of these choices is OK. Say we choose B. (Now erase KB from copy 1 and mark and label KB with a 2 on copy 2.)

- **Step 3.** From B we have three choices (A, C, or J). Any of these choices is OK. Say we choose C. (Now erase BC from copy 1 and mark and label BC with a 3 on copy 2.)

- **Step 4.** From C we have three choices (D, E, or L). Any of these choices is OK. Say we choose L. (EML—that's shorthand for erase, mark, and label.)

- **Step 5.** From L we have three choices (E, G, or K). Any of these choices is OK. Say we choose K. (EML.)

- **Step 6.** From K we have only one choice—to H. Without further ado, we choose H. (EML.)

- **Step 7.** From H we have three choices (G, I, or J). But for the first time, one of the choices is a bad choice. We should not choose G, as HG is a bridge of the yet-to-be-traveled part of the graph (see Fig. 5-21). Either of the other two choices is OK. Say we choose J. (EML.)

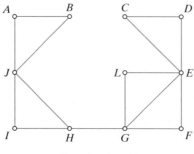

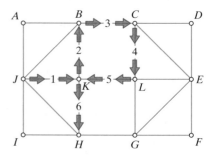

FIGURE 5-21   Copy 1 at Step 7   Copy 2 at Step 7

- **Step 8.** From J we have three choices (A, B, or I), but we should not choose I, as JI has just become a bridge. Either of the other two choices is OK. Say we choose B. (EML)

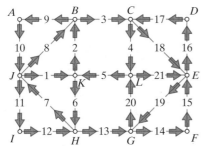

FIGURE 5-22

- **Steps 9 through 13.** Each time we have only one choice. From B we have to go to A, then to J, I, H, and G.
- **Steps 14 through 21.** Not to belabor the point, let's just cut to the chase. The rest of the path is given by $G, F, E, D, C, E, G, L, E$. There are many possible endings, and you should find a different one by yourself.

The completed Euler path (one of hundreds of possible ones) is shown in Fig. 5-22.

5.7 Eulerizing Graphs

In this section we will finally answer some of the routing problems raised at the beginning of the chapter. Their common thread is the need to find *optimal exhaustive routes* in a connected graph. How is this done? Let's first refresh our memories of what this means. We will use the term **exhaustive route** to describe a route that travels along the edges of a graph and passes through *each and every edge* of the graph *at least once*. Such a route could be an Euler circuit (if the graph has no odd vertices) or an Euler path (if the graph has two odd vertices), but for graphs with more than two odd vertices, an exhaustive route will have to *recross* some of the edges. This follows from Euler's theorems.

We are interested in finding exhaustive routes that *recross* the fewest number of edges. Why? In many applications, each edge represents a unit of cost. The more edges along the route, the higher the cost of the route. In an exhaustive route, the first pass along an edge is a necessary expense, part of the requirements of the job. Any additional pass along that edge represents a wasted expense (these extra passes are often described as *deadhead* travel). Thus, an exhaustive route that minimizes cost (*optimal exhaustive route*) is one with the fewest number of deadhead edges. (This is only true under the assumption that each edge equals one unit of cost.)

We are now going to see how the theory developed in the preceding sections will help us design optimal exhaustive routes for graphs with many (more than two) odd vertices. The key idea is that we can turn odd vertices into even vertices by adding "duplicate" edges in strategic places. This process is called **eulerizing** the graph.

> **EXAMPLE 5.22**   **Covering a 3 by 3 Street Grid**

The graph in Fig. 5-23(a) (shown on p. 186) represents a 3 block by 3 block street grid consisting of 24 blocks (count them if you don't believe it!). How can we find an optimal route that covers all the edges of the graph and ends back at the starting vertex?

Our first step is to identify the odd vertices. This graph has eight odd vertices ($B, C, E, F, H, I, K,$ and L), shown in red. When we add a duplicate copy of edges $BC, EF, HI,$ and KL, we get the graph in Fig. 5-23(b). This is a *eulerized* version of the original graph—its vertices are all even, so we know it has an Euler circuit. Moreover, it's clear we couldn't have done this with fewer than four duplicate edges. Figure 5-23(c) shows one of the many possible Euler

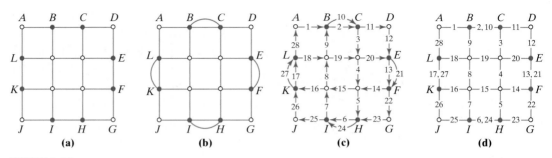

FIGURE 5-23

circuits, with the edges numbered in the order they are traveled. The Euler circuit described in Fig. 5-23(c) represents an exhaustive closed route along the edges of the original graph [Fig. 5-23(d)], with the four duplicate edges (*BC*, *EF*, *HI*, and *KL*) indicating the deadhead blocks where a second pass is required. The total length of this route is 28 blocks (24 blocks in the grid plus 4 deadhead blocks), and this route is optimal—no matter how clever you are or how hard you try, if you want to travel along each block of the grid and start and end at the same vertex, you will have to pass through a minimum of 28 blocks! (There are many other ways to do it using just 28 blocks, but none with fewer than 28.)

EXAMPLE 5.23   **Covering a 4 by 4 Street Grid**

The graph in Fig. 5-24(a) represents a 4 block by 4 block street grid consisting of 40 blocks. The 12 odd vertices in the graph are shown in red. We want to eulerize the graph by adding the least number of edges. Figure 5-24(b) shows how *not to do it!* This graph violates the cardinal rule of eulerization—you can only duplicate edges that are part of the original graph. Edges *DF* and *NL* are new edges, not duplicates, so Fig. 5-24(b) is out! Figure 5-24(c) shows a legal eulerization, but it is not optimal, as it is obvious that we could have accomplished the same thing by adding fewer duplicate edges. Figure 5-24(d) shows an *optimal eulerization* of the original graph—one of several possible. Once we have an optimal eulerization, we have the blueprint for an optimal exhaustive closed route on the original graph. Regardless of the specific details, we now know that the route will travel along 48 blocks—the 40 original blocks in the grid plus 8 deadhead blocks.

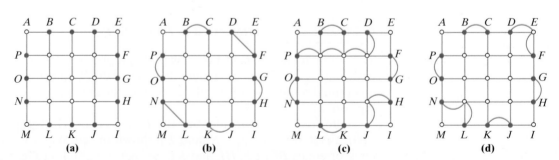

FIGURE 5-24 (a) The original graph. Odd vertices shown in red. (b) Bad move — *DF* and *NL* were not edges of the original graph! (c) This is a eulerization, but not an efficient one! (d) This is one of the many possible optimal eulerizations.

In some situations we need to find an exhaustive route, but there is no requirement that it be closed—the route may start and end at different points. In these cases we want to leave two odd vertices on the graph unchanged and change the other odd vertices into even vertices by duplicating appropriate edges of the graph. This process is called a **semi-eulerization** of the graph. When we strategically choose how to do this so that the number of duplicate edges is as small as possible, we can obtain an *optimal exhaustive open route*. In this case the route will start at one of the two odd vertices and end at the other one.

(**EXAMPLE 5.24**)   **Parade Routes**

Let's consider once again the 4 by 4 street grid shown in Fig. 5-24(a). Imagine that your job is to design a good route for a Fourth of July parade that must pass through each of the 40 blocks of the street grid. The key difference between this example and Example 5.23 is that when routing a parade, you do not want the parade to start and end in the same place. (In fact, for traffic control it is usually desirable to keep the starting and ending points of a parade as far from each other as possible.) The fire department has added one additional requirement to the parade route: The parade has to start at B. Your task, then, is to find a *semi-eulerization* of the graph that leaves B and one more odd vertex unchanged (preferably a vertex far from B) and that changes all the other odd vertices into even vertices. (Why don't you give it a try before you read on?)

Figure 5-25 shows two different semi-eulerizations. The semi-eulerization in Fig. 5-25(a) is *optimal* because it required only six duplicate edges, and this is as good as one can do. The optimal parade route could be found by finding an Euler path on the graph in Fig. 5-25(a). The only bad thing about this route is that the parade would end at P, a point a bit too close to the starting point, and the traffic control people are unhappy about that. Back to the drawing board.

A different semi-eulerization is shown in Fig. 5-25(b). The parade route in this case would not be optimal (it has seven deadhead blocks), but because it ends at K, it better satisfies the requirement that the starting and ending points be far apart. The traffic control folks are happy now.

FIGURE 5-25 (a) An optimal semi-eulerization. (b) A good semi-eulerization (not optimal) with an ending point far from the starting point.

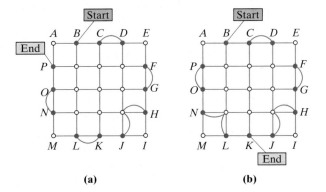

(a)   (b)

(**EXAMPLE 5.25**)   **The Bridges of Madison County: Part 2**

This is the conclusion of the routing problem first introduced in Example 5.4. A photographer needs to take photos of each of the 11 bridges in Madison County [Fig. 5-26(a)]. A graph model of the layout (vertices represent land masses, edges

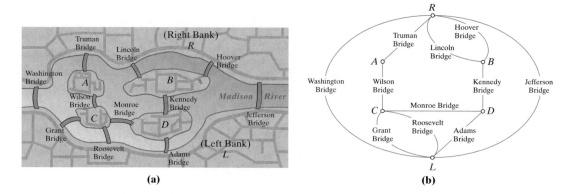

FIGURE 5-26   **(a)**   **(b)**

represent bridges) is shown in Fig. 5-26(b). The graph has four odd vertices (R, L, B, and D), so some bridges are definitely going to have to be recrossed. How many and which ones depends on the other parameters of the problem. (Recall that it costs $25 in toll fees to cross a bridge, so the baseline cost for crossing the 11 bridges is $275. Each recrossing is at an additional cost of $25.)

The following are a few of the possible scenarios one might have to consider. Each one requires a different eulerization and will result in a different route.

- The photographer needs to start and end his trip in the same place. This scenario requires an optimal eulerization of the graph in Fig. 5-26(b). This is not hard to do, and an optimal route can be found for a cost of $325. (See Exercise 59.)

- The photographer has the freedom to choose any starting and ending points for his trip. In this case we can find an optimal semi-eulerization of the graph, requiring only one duplicate edge. Now an optimal route is possible at a cost of $300. [See Exercise 58(a).]

- The photographer has to start his trip at B and end the trip at L. In this case we must find a semi-eulerization of the graph where B and L remain as odd vertices and R and D become even vertices. It is possible to do this with just two duplicate edges and thus find an optimal route that will cost $325. [See Exercise 58(b).]

Even more complicated scenarios are possible, but we have to move on.

EXAMPLE 5.26   **The Exhausted Patrol and the Grateful No Deadhead**

This example brings us full circle to the first couple of examples of this chapter. In Example 5.1 we raised the question of finding an optimal exhaustive closed route for a security guard hired to patrol the streets of the Sunnyside subdivision. In Example 5.16 we created the graph model for this problem shown in Fig. 5-27(a). The graph has 18 odd vertices, shown in red. We now know that the name of the game is to find an optimal eulerization of this graph. In this case the odd vertices pair up beautifully, and the optimal eulerization requires only nine duplicate edges, shown in Fig. 5-27(b). All the answers to the security guards questions can now be answered: An optimal route will require nine deadhead blocks. The actual route can be found using trial and error or Fleury's algorithm.

A slightly different problem is the one facing the mail carrier delivering mail along the streets of the Sunnyside subdivision. Much to the mail carrier's pleasant surprise, in the graph that models her situation [Fig. 5-27(c)] all the vertices are even (you just go ahead and check it out!). This means that the optimal route is an Euler circuit, which can be found once again using Fleury's algorithm (or trial

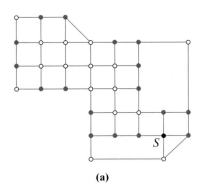
(a)

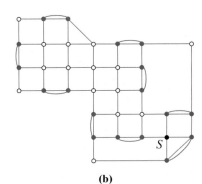
(b)

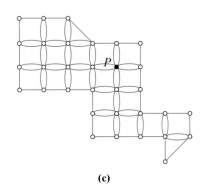
(c)

FIGURE 5-27 Odd vertices shown in red.

and error if you prefer). Thanks to the lessons of this chapter, this mail carrier will not have to deadhead, for which she is extremely grateful!

CONCLUSION

In this chapter we got our first introduction to three fundamental ideas. First, we learned about a simple but powerful concept for describing relationships within a set of objects—the concept of a *graph*. This idea can be traced back to Euler, some 270 years ago. Since then, the study of graphs has grown into one of the most important and useful branches of modern mathematics.

The second important idea of this chapter is the concept of a *graph model*. Every time we take a real-life problem and turn it into a mathematical problem, we are, in effect, modeling. Unwittingly, we have all done some form of mathematical modeling at one time or another—first using arithmetic and later using equations and functions to describe real-life situations. In this chapter we learned about a new type of modeling called graph modeling, in which we use graphs and the mathematical theory of graphs to solve certain types of routing problems.

By necessity, the routing problems that we solved in this chapter were fairly simplistic—crossing a few bridges, patrolling a small neighborhood, designing a parade route—what's all the fuss about? We should not be deceived by the simplicity of these examples—larger-scale variations on these themes have significant practical importance. In many big cities, where the efficient routing of municipal services (police patrols, garbage collection, etc.) is a major issue, the very theory that we developed in this chapter is being used on a large scale, the only difference being that many of the more tedious details are mechanized and carried out by a computer. (In New York City, for example, garbage collection, curb sweeping, snow removal, and other municipal services have been scheduled and organized using graph models since the 1970s, and the improved efficiency has yielded savings estimated in the tens of millions of dollars a year.)

The third important concept introduced in this chapter is that of an *algorithm*—a set of procedural rules that, when followed, provide solutions to certain types of problems. Perhaps without even realizing it, we had our first exposure to algorithms in elementary school, when we learned how to add, multiply, and divide numbers following precise and exacting procedural rules. In this chapter we learned about *Fleury's algorithm*, which helps us find an Euler circuit or an Euler path in a graph. In the next few chapters we will learn many other *graph algorithms*, some quite simple, others a bit more complicated. When it comes to algorithms of any kind, be they for doing arithmetic calculations or for finding circuits in graphs, there is one standard piece of advice that always applies: *Practice makes perfect*. (As is often the case, Yogi Berra said it better.)

P ROFILE: Leonhard Euler (1707–1783)

Leonhard Euler is universally recognized as one of the great, if not the greatest, mathematical geniuses in history. In the words of one of his biographers, "Euler was the Shakespeare of mathematics—universal, richly detailed and inexhaustible." In terms of sheer volume, Euler's mathematical production is staggering—his *Opera Omnia* (collected works) fills more than 80 volumes and runs over 25,000 pages. No other mathematician—in fact, no other scientist—in the history of humankind can come close to matching Euler in terms of creative output. Euler produced groundbreaking discoveries in practically every area of mathematics—analysis, number theory, algebra, geometry, and topology (which he started). Euler's theorems in graph theory (see Section 5.5) and his solution of the Königsberg bridge problem are just one note in the mathematical symphony that is his work.

Leonhard Euler was born in Basel, Switzerland, the son of a Protestant minister. From an early age, Euler exhibited two traits that set him apart from mere mortals—a prodigious memory and the ability to perform incredibly complicated calculations in his head. At the age of 14 Euler entered the University of Basel to study theology and follow in the footsteps of his father. The young man's incredible mathematical gifts soon came to the attention of Johann Bernoulli, professor of mathematics at Basel and one of the most famous and influential mathematicians of his generation. Under Bernoulli's mentorship Euler eventually dropped theology and decided to concentrate on the study of mathematics. At the age of 19, Euler graduated from the University of Basel, having completed a doctoral thesis in which he analyzed the works of Descartes and Newton.

By this time Euler was already a mathematician of some fame and was offered an academic position at the St. Petersburg Academy of Sciences in Russia—a remarkable and an unprecedented offer for someone his age. For the next 14 years (1727 to 1741), Euler worked at the St. Petersburg Academy under the patronage of Catherine I. In addition to his groundbreaking work in pure mathematics, Euler was a practical man who worked on many important applied problems in physics, engineering, astronomy, and cartography. In 1733, at the age of 27, he was promoted to Professor of Mathematics at the St. Petersburg Academy. His first book, *Mechanica*, a two-volume text published in 1736, became the definitive work in mechanics for the next 50 years.

In 1738, due to an infection, Euler lost the sight in his right eye. Twenty-eight years later, he would lose the sight in his left eye, leaving him totally blind. Remarkably, neither event had a significant impact on his ability to produce copious amounts of mathematical work—he just dictated to his assistants and did most calculations in his head. In 1741 Euler accepted an offer from Frederick the Great of Prussia to become mathematics director of the Berlin Academy of Sciences. Euler worked in Berlin for the next 25 years (1741–1766), where he produced some of his greatest work, including his most famous book—a multivolume textbook in elementary science called *Letters to a German Princess*. In 1766, mostly for political reasons, Euler returned to the St. Petersburg Academy of Sciences, where, despite his progressive blindness, he carried on with his prolific research for another 17 years.

The day he died, September 18, 1783, Euler worked in the morning on mathematical questions related to balloon flights, and in the afternoon he performed important calculations concerning the orbit of the planet Uranus—all just another day's work for him. In the late afternoon, while playing with one of his grandchildren, he suffered a massive stroke and died. His death marked the passing of one of the most creative men in the history of science.

KEY CONCEPTS

adjacent edges, **174**
adjacent vertices, **174**
algorithm, **181**
bridge (of a graph), **175**
circuit, **174**
component (of a graph), **175**
connected graph, **175**
degree (of a vertex), **174**
disconnected graph, **175**
edge, **171**
edge set, **172**

Euler circuit, **175**
Euler circuit problem, **168**
eulerization (of a graph), **185**
Euler path, **175**
Euler's theorems, **178**
even vertex, **174**
exhaustive route, **185**
Fleury's algorithm, **182**
graph, **171**
graph model, **176**
isolated vertex, **173**

length (of a path or a circuit), **174**
loop, **171**
multiple edges, **171**
odd vertex, **174**
path, **174**
routing problems, **168**
semi-eulerization, **187**
unicursal tracing, **171**
vertex, **171**
vertex set, **172**

EXERCISES

WALKING

A Graphs: Basic Concepts

1. For the graph shown in Fig. 5-28,

    (a) give the vertex set $\mathcal{V}$.

    (b) give the edge set $\mathcal{E}$.

    (c) list the degree of each vertex.

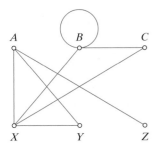

FIGURE 5-28

2. For the graph shown in Fig. 5-29,

    (a) give the vertex set $\mathcal{V}$.

    (b) give the edge set $\mathcal{E}$.

    (c) list the degree of each vertex.

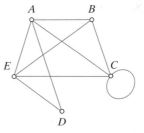

FIGURE 5-29

3. For the graph shown in Fig. 5-30,

    (a) give the vertex set $\mathcal{V}$.

    (b) give the edge set $\mathcal{E}$.

    (c) list the degree of each vertex.

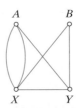

FIGURE 5-30

4. For the graph shown in Fig. 5-31,

    (a) give the vertex set V.

    (b) give the edge set E.

    (c) list the degree of each vertex.

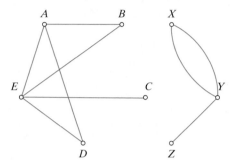

FIGURE 5-31

5. Consider the graph with $V = \{K, R, S, T, W\}$ and $E = \{RS, RT, TT, TS, SW, WW, WS\}$. Draw two different pictures of the graph.

6. Consider the graph with $V = \{A, B, C, D, E\}$ and $E = \{AC, AE, BD, BE, CA, CD, CE, DE\}$. Draw two different pictures of the graph.

7. Consider the graph with $V = \{A, B, C, D, E\}$ and $E = \{AD, AE, BC, BD, DD, DE\}$. Without drawing a picture of the graph,

    (a) list all the vertices adjacent to D.

    (b) list all the edges adjacent to BD.

    (c) find the degree of D.

    (d) find the sum of the degrees of the vertices.

8. Consider the graph with $V = \{A, B, C, X, Y, Z\}$ and $E = \{AX, AY, AZ, BB, CX, CY, CZ, YY\}$. Without drawing a picture of the graph,

    (a) list all the vertices adjacent to Y.

    (b) list all the edges adjacent to AY.

    (c) find the degree of Y.

    (d) find the sum of the degrees of the vertices.

9. (a) Give an example of a connected graph with six vertices such that each vertex has degree 2.

    (b) Give an example of a disconnected graph with six vertices such that each vertex has degree 2.

    (c) Give an example of a graph with six vertices such that each vertex has degree 1.

10. (a) Give an example of a connected graph with eight vertices such that each vertex has degree 3.

    (b) Give an example of a disconnected graph with eight vertices such that each vertex has degree 3.

(c) Give an example of a graph with eight vertices such that each vertex has degree 1.

11. Consider the graph in Fig. 5-32.

    (a) Find a path from C to F passing through vertex B but not through vertex D.

    (b) Find a path from C to F passing through both vertex B and vertex D.

    (c) Find a path of length 4 from C to F.

    (d) Find a path of length 7 from C to F.

    (e) How many paths are there from C to A?

    (f) How many paths are there from H to F?

    (g) How many paths are there from C to F?

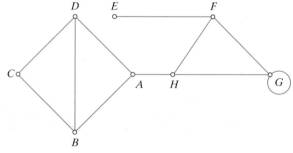

FIGURE 5-32

12. Consider the graph in Fig. 5-32.

    (a) Find a path from D to E passing through vertex G only once.

    (b) Find a path from D to E passing through vertex G twice.

    (c) Find a path of length 4 from D to E.

    (d) Find a path of length 8 from D to E.

    (e) How many paths are there from D to A?

    (f) How many paths are there from H to E?

    (g) How many paths are there from D to E?

13. Consider the graph in Fig. 5-32.

    (a) Find all circuits of length 1.

    (b) Find all circuits of length 2.

    (c) Find all circuits of length 3.

    (d) Find all circuits of length 4.

    (e) What is the total number of circuits in the graph?

14. Consider the graph in Fig. 5-33.

    (a) Find all circuits of length 1.

    (b) Find all circuits of length 2.

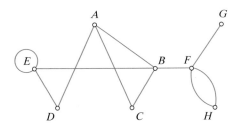

FIGURE 5-33

(c) Find all circuits of length 3.

(d) Find all circuits of length 4.

(e) Find all circuits of length 5.

(f) What is the total number of circuits in the graph?

15. List all the bridges in each of the following graphs:

(a) the graph in Fig. 5-32

(b) the graph with $V = \{A, B, C, D, E\}$ and $E = \{AB, AE, BC, CD, DE\}$

(c) the graph with $V = \{A, B, C, D, E\}$ and $E = \{AB, BC, BE, CD\}$

16. List all the bridges in each of the following graphs:

(a) the graph in Fig. 5-33

(b) the graph with $V = \{A, B, C, D, E\}$ and $E = \{AB, AD, AE, BC, CE, DE\}$

(c) the graph with $V = \{A, B, C, D, E\}$ and $E = \{AB, BC, CD, DE\}$

B Graph Models

17. Figure 5-34 shows a map of the downtown area of the picturesque hamlet of Kingsburg, with the Kings River running through the downtown area and the three islands (A, B, and C) connected to each other and both banks by seven bridges. You have been hired by the Kingsburg Chamber of Commerce to organize the annual downtown

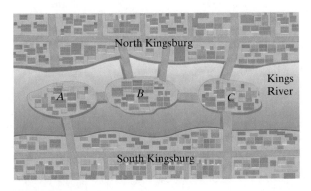

FIGURE 5-34

parade. Part of your job is to plan the route for the parade. Draw a graph that models the layout of Kingsburg.

18. Figure 5-35 is a map of downtown Royalton, showing the Royalton River running through the downtown area and the three islands (A, B, and C) connected to each other and both banks by eight bridges. The Downtown Athletic Club wants to design the route for a marathon through the downtown area. Draw a graph that models the layout of Royalton.

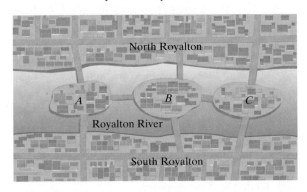

FIGURE 5-35

19. A night watchman must walk the streets of the Green Hills subdivision shown in Fig. 5-36. The night watchman needs to walk only once along each block. Draw a graph that models this situation.

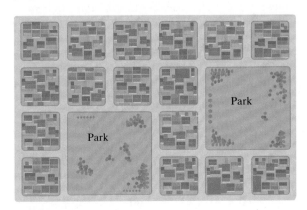

FIGURE 5-36

20. A mail carrier must deliver mail on foot along the streets of the Green Hills subdivision shown in Fig. 5-36. The mail carrier must make two passes on every block that has houses on both sides of the street (once for each side of the street), but only one pass on blocks that have houses on only one side of the street. Draw a graph that models this situation.

21. Six teams ($A, B, C, D, E,$ and F) are entered in a softball tournament. The top two seeded teams (A and B) only have to play three games; the other teams have to play four games each. The tournament pairings are A plays against C, E, and F; B plays against C, D, and F; C plays against every team except *F*; *D* plays against every team except *A*; *E* plays against every team except *B*; and *F* plays against every team except *C*. Draw a graph that models the tournament.

22. The Kangaroo Lodge of Madison County has 10 members (*A, B, C, D, E, F, G, H, I*, and *J*). The club has five working committees: the Rules Committee (*A, C, D, E, I*, and *J*), the Public Relations Committee (*B, C, D, H, I*, and *J*), the Guest Speaker Committee (*A, D, E, F*, and *H*), the New Year's Eve Party Committee (*D, F, G, H*, and *I*), and the Fund Raising Committee (*B, D, F, H*, and *J*).

(a) Suppose we are interested in knowing which pairs of members are on the same committee. Draw a graph that models this situation. (*Hint*: Let the vertices of the graph represent the members.)

(b) Suppose we are interested in knowing which committees have members in common. Draw a graph that models this situation. (*Hint*: Let the vertices of the graph represent the committees.)

C Euler's Theorems

In Exercises 23 through 28 only one of the following answers is correct: (i) the graph has an Euler circuit, (ii) the graph has an Euler path, (iii) the graph has neither an Euler circuit nor an Euler path, (iv) the graph may or may not have an Euler circuit, or (v) the graph may or may not have an Euler path. Choose the correct answer—(i), (ii), (iii), (iv), or (v)—followed with an explanation. You do not have to show an actual path or circuit.

23. (a) Fig. 5-37(a)

(b) Fig. 5-37(b)

(c) A graph with six vertices, all of degree 2

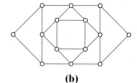

(a)   (b)

FIGURE 5-37

24. (a) Fig. 5-38(a)

(b) Fig. 5-38(b)

(c) A graph with eight vertices: six vertices of degree 2 and two vertices of degree 3

(a)   (b)

FIGURE 5-38

25. (a) Fig. 5-39(a)

(b) Fig. 5-39(b)

(c) A disconnected graph with six vertices, all of degree 2

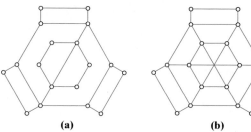

(a)   (b)

FIGURE 5-39

26. (a) Fig. 5-40(a)

(b) Fig. 5-40(b)

(c) A disconnected graph with eight vertices: six vertices of degree 2 and two vertices of degree 3

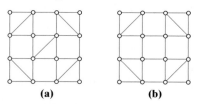

(a)   (b)

FIGURE 5-40

27. (a) Fig. 5-41(a)

(b) Fig. 5-41(b)

(c) A graph with six vertices, all of degree 1. [*Hint*: Try Exercise 9(c) first.]

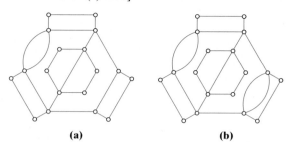

(a)   (b)

FIGURE 5-41

28. (a) Fig. 5-42(a)

(b) Fig. 5-42(b)

(c) A graph with eight vertices, all of degree 1. [*Hint*: Try Exercise 10(c) first.]

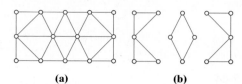

(a)   (b)

FIGURE 5-42

D   Finding Euler Circuits and Euler Paths

29. Find an Euler circuit for the graph in Fig. 5-43. Show your answer by labeling the edges 1, 2, 3, and so on in the order in which they are traveled.

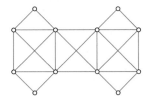

FIGURE 5-43

30. Find an Euler circuit for the graph in Fig. 5-44. Show your answer by labeling the edges 1, 2, 3, and so on in the order in which they can be traveled.

FIGURE 5-44

31. Jack set out to find an Euler circuit for the graph in Fig. 5-45. He started labeling the edges 1, 2, 3, and so on as shown in the figure, but did not finish the job. Finish the job for Jack.

FIGURE 5-45

32. Jill set out to find an Euler path for the graph in Fig. 5-46. She started labeling the edges 1, 2, 3, and so on as shown in the figure, but did not finish the job. Finish the job for Jill.

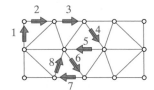

FIGURE 5-46

33. Find an Euler path for the graph in Fig. 5-47. Show your answer by labeling the edges 1, 2, 3, and so on in the order in which they are traveled.

FIGURE 5-47

34. Find an Euler path for the graph in Fig. 5-48. Show your answer by labeling the edges 1, 2, 3, and so on in the order in which they can be traveled.

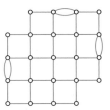

FIGURE 5-48

35. Find an Euler circuit for the graph in Fig. 5-49. Use B as the starting and ending point of the circuit. Show your answer by labeling the edges 1, 2, 3, and so on in the order in which they are traveled.

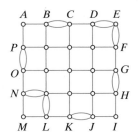

FIGURE 5-49

36. Find an Euler circuit for the graph in Fig. 5-50. Use S as the starting and ending point of the circuit. Show your answer by labeling the edges 1, 2, 3, and so on in the order in which they are traveled.

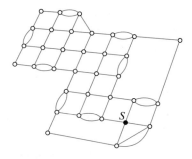

FIGURE 5-50

E   Eulerizations and Semi-eulerizations

37. Find an optimal eulerization for the graph in Fig. 5-51.

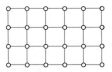

FIGURE 5-51

38. Find an optimal eulerization for the graph in Fig. 5-52.

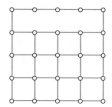

FIGURE 5-52

39. Find an optimal eulerization for the graph in Fig. 5-53.

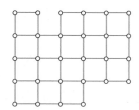

FIGURE 5-53

40. Find an optimal eulerization for the graph in Fig. 5-54.

FIGURE 5-54

41. Find an optimal semi-eulerization for the graph in Fig. 5-53.

42. Find an optimal semi-eulerization for the graph in Fig. 5-54.

F Miscellaneous

43. Give an example of a connected graph with six vertices such that each edge of the graph is a *bridge* and

(a) the graph has exactly two vertices of degree 1.

(b) the graph has exactly three vertices of degree 1.

(c) the graph has exactly four vertices of degree 1.

(d) the graph has exactly five vertices of degree 1.

44. Give an example of a connected graph with five vertices such that each edge of the graph is a *bridge* and

(a) the graph has exactly two vertices of degree 1.

(b) the graph has exactly three vertices of degree 1.

(c) the graph has exactly four vertices of degree 1.

45. A security guard must patrol on foot the streets of the Green Hills subdivision shown in Fig. 5-55. The security guard wants to start and end his walk at the corner labeled *A*, and he needs to cover each block of the subdivision at least once. Find an optimal route for the security guard. Describe the route by labeling the edges 1, 2, 3, and so on in the order in which they are traveled. (*Hint*: You should do Exercise 19 first.)

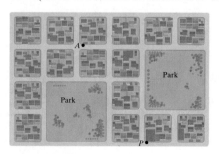

FIGURE 5-55

46. A mail carrier must deliver mail on foot along the streets of the Green Hills subdivision shown in Fig. 5-55. His route must start and end at the Post Office, labeled *P* in the figure. The mail carrier must walk along each block twice if there are houses on both sides of the street and once along blocks where there are houses on only one side of the street. Find an optimal route for the mail carrier. Describe the route by labeling the edges 1, 2, 3, and so on in the order in which they are traveled. (*Hint*: You should do Exercise 20 first.)

47. Assume you want to trace the tennis court diagram shown in Fig. 5-56 without retracing any lines. How many times would you have to lift your pencil to do it? Explain. (*Note*: This is the problem facing a groundskeeper trying to mark the chalk lines of a tennis court.)

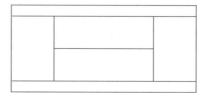

FIGURE 5-56

48. Assume you want to trace the diagram of a basketball court shown in Fig. 5-57 without retracing any lines. How many times would you have to lift your pencil to do it? Explain.

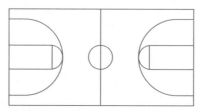

FIGURE 5-57

JOGGING

49. (a) Explain why in every graph the sum of the degrees of all the vertices equals twice the number of edges.

(b) Explain why every graph must have an even number of odd vertices.

50. Suppose a connected graph G has k odd vertices and you want to trace all its edges. Assuming you would not trace over any edges twice, what is the least number of times you would have to lift your pencil? Explain.

51. If G is a connected graph with no bridges, how many vertices of degree 1 can G have? Explain your answer.

52. Regular graphs. A graph is called *regular* if every vertex has the same degree. Let G be a connected regular graph with N vertices.

(a) Explain why if N is odd, then G must have an Euler circuit.

(b) When N is even, then G may or may not have an Euler circuit. Give examples of both situations.

53. Complete bipartite graphs. A complete bipartite graph is a graph having the property that the vertices of the graph can be divided into two groups A and B and each vertex in A is adjacent to each vertex in B, as shown in Fig. 5-58. Two vertices in A are never adjacent, and neither are two vertices in B. Let m and n denote the number of vertices in A and B, respectively, and assume $m \le n$.

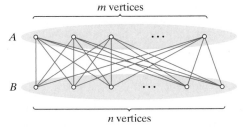

m vertices

A

B

n vertices

FIGURE 5-58

(a) Describe all the possible values of m and n for which the complete bipartite graph has an Euler circuit. (*Hint*: There are infinitely many values of m and n.)

(b) Describe all the possible values of m and n for which the complete bipartite graph has an Euler path.

54. Consider the following game. You are given N vertices and are required to build a graph by adding edges connecting these vertices. Each time you add an edge you must pay $1. You can stop when the graph is connected.

(a) Describe the strategy that will cost you the least amount of money.

(b) What is the minimum amount of money needed to build the graph? (Give your answer in terms of N.)

55. Consider the following game. You are given N vertices and allowed to build a graph by adding edges connecting these vertices. For each edge you can add, you make $1. You are not allowed to add loops or multiple edges, and you must stop before the graph is connected (i.e., the graph you end up with must be disconnected).

(a) Describe the strategy that will give you the greatest amount of money.

(b) What is the maximum amount of money you can make building the graph? (Give your answer in terms of N.)

56. Figure 5-59 shows a map of the downtown area of the picturesque hamlet of Kingsburg. You have been hired by the Kingsburg Chamber of Commerce to organize the annual downtown parade. Part of your job is to plan the route for the parade. An *optimal* parade route is one that keeps the bridge crossings to a minimum and yet crosses each of the seven bridges in the downtown area at least once.

(a) Find an optimal parade route if the parade is supposed to start in North Kingsburg but can end anywhere.

(b) Find an optimal parade route if the parade is supposed to start in North Kingsburg and end in South Kingsburg.

(c) Find an optimal parade route if the parade is supposed to start in North Kingsburg and end on island B.

(d) Find an optimal parade route if the parade is supposed to start in North Kingsburg and end on island A.

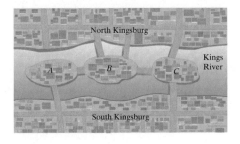

FIGURE 5-59

57. A policeman has to patrol on foot the streets of the subdivision shown in Fig. 5-60. The policeman needs to start his route at the police station, located at X, and end the route at the local coffee shop, located at Y. He needs to cover each block of the subdivision at least once, but he wants to make his route as efficient as possible and duplicate the fewest possible number of blocks.

(a) How many blocks will he have to duplicate in an optimal trip through the subdivision?

(b) Describe an optimal trip through the subdivision. Label the edges 1, 2, 3, and so on in the order the policeman would travel them.

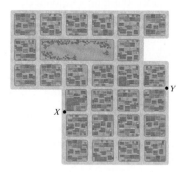

FIGURE 5-60

Exercises 58 and 59 refer to Example 5.25. In this example, the problem is to find an optimal route (i.e., a route with the fewest bridge crossings) for a photographer who needs to cross each of the 11 bridges of Madison County for a photo shoot. The layout of the 11 bridges is shown in Fig. 5-61. You may find it helpful to review Example 5.25 before trying these two exercises.

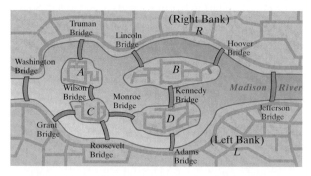

FIGURE 5-61

58. Describe an optimal route for the photographer if

(a) there are no restrictions on the starting and ending vertices.

(b) the route must start at B and end at L.

59. Describe an optimal route for the photographer if

(a) the route must start and end in D and the first bridge crossed must be the Adams Bridge.

(b) the route must start and end in the same place, the first bridge crossed must be the Adams bridge, and the last bridge crossed must be the Grant Bridge.

60. This exercise comes to you courtesy of Euler himself. Here is the question in Euler's own words, accompanied by the diagram shown in Fig. 5-62. (See reference 7.)

> *Let us take an example of two islands with four rivers forming the surrounding water. There are fifteen bridges marked a, b, c, d, etc., across the water around the islands and the adjoining rivers. The question is whether a journey can be arranged that will pass over all the bridges but not over any of them more than once.*

What is the answer to Euler's question? If the "journey" is possible, describe it. If it isn't, explain why not.

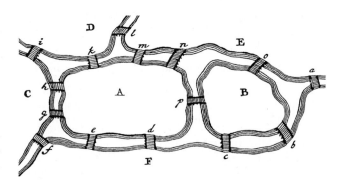

FIGURE 5-62

RUNNING

61. Most of the time, the edge set $\mathcal{E}$ of a graph provides all the information necessary to describe the graph, and in these cases the vertex set is *redundant*. Every once in a while, however, the vertex set is not redundant—it carries information that is not provided by the edge set.

(a) Give an example of a graph such that its vertex set is *not redundant*.

(b) Give a description (in words) of all graphs for which the vertex set is *not redundant*.

62. Suppose G is a connected graph with N vertices, all of even degree. Let k denote the number of bridges in G. Find the value(s) of k. Explain your answer.

63. Suppose G is a connected graph with $N - 2$ even vertices and two odd vertices. Let k denote the number of bridges in G. Find all the possible values of k. Explain your answer.

64. Suppose G is a disconnected graph with exactly two odd vertices. Explain why the two odd vertices must be in the same component of the graph.

65. Suppose G is a graph with N vertices ($N \geq 2$) with no loops and no multiple edges. Explain why G must have at least two vertices of the same degree.

66. Complements. The *complement* of a graph G is a graph having the same vertex set as G but consisting of exactly those edges that are not edges of G. (If AB *is* an edge in G, then it *is not* an edge in the complement; if AB *is not* an edge in G, then it *is* an edge in the complement.) Let N denote the number of vertices in G.

(a) Suppose N is even ($N \geq 4$) and G has an Euler circuit. Explain why the complement of G cannot have an Euler circuit.

(b) Suppose N is odd ($N \geq 5$) and G has an Euler circuit. Explain why the complement of G may or may not have an Euler circuit.

67. Kissing circuits. When two circuits in a graph have no edges in common but share the common vertex v, they are said to be *kissing at v*.

(a) For the graph shown in Fig. 5-63, find a circuit kissing the circuit A, D, C, A (there is only one), and find two different circuits kissing the circuit A, B, D, A.

(b) Suppose G is a connected graph and every vertex in G is even. Explain why the following statement is true: *If a circuit in G has no kissing circuits, then that circuit must be an Euler circuit.*

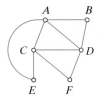

FIGURE 5-63

68. Hierholzer's algorithm. *Hierholzer's algorithm* is another algorithm for finding an Euler circuit in a graph. The basic idea behind Hierholzer's algorithm is to start with an arbitrary circuit and then enlarge it by patching to it a *kissing circuit*, continuing this way and making larger and larger circuits until the circuit cannot be enlarged any far-

ther. (For the definition of kissing circuits, see Exercise 67.) More formally, Hierholzer's algorithm is as follows:

Step 1. Start with an arbitrary circuit C_0.

Step 2. Find a kissing circuit to C_0. If there are no kissing circuits to C_0, then you are finished—C_0 is itself an Euler circuit of the graph [see Exercise 67(b)]. If there is a kissing circuit to C_0, let's call it K_0, and let V denote the vertex at which the two circuits kiss. Go to Step 3.

Step 3. Let C_1 denote the circuit obtained by "patching" K_0 to C_0 at vertex V (i.e., start at V, travel along C_0 back to V, and then travel along K_0 back again to V). Now find a kissing circuit to C_1. (If there are no kissing circuits to C_1, then you are finished—C_1 is your Euler circuit.) If there is a kissing circuit to C_1, let's call it K_1, and let W denote the vertex at which the two circuits kiss. Go to Step 4.

Step 4, 5, and so on. Continue this way until there are no more kissing circuits available.

(a) Use Hierholzer's algorithm to find an Euler circuit for the graph shown in Fig. 5-64 (this is the graph in Example 5-17).

(b) Describe a modification of Hierholzer's algorithm that allows you to find an Euler path in a connected graph having exactly two vertices of odd degree. (*Hint:* A path can also have a kissing circuit.)

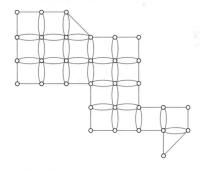

FIGURE 5-64

PROJECTS AND PAPERS

A Original Sources

Whenever possible, it is instructive to read about a great discovery from the original source. Euler's landmark paper in graph theory with his solution to the Königsberg bridge problem was published in 1736. Luckily, the paper was written in a very readable style and the full English translation is quite accessible—it appears as an article in *Scientific American*

(reference 7) and in Newman's book The *World of Mathematics* (reference 9).

Write a summary/analysis of Euler's original paper. Include (1) a description of how Euler originally tackled the Königsberg bridge problem, (2) a discussion of Euler's general conclusions, and (3) a discussion of Euler's approach toward finding an Euler circuit/path.

B Computer Representation of Graphs

In many real-life routing problems, we have to deal with very large graphs—a graph could have thousands of vertices and tens of thousands of edges. In these cases algorithms such as Fleury's algorithm (as well as others we will study in later chapters) are done by computer. Unlike humans, computers are not very good at interpreting pictures, so the first step in using computers to perform computations with graphs is to describe the graph in a way the computer can understand it. The two most common ways to do so are by means of matrices.

In this project you are asked to write a short research paper describing the use of matrices to represent graphs. Explain (1) what is a **matrix**, (2) what is the **adjacency matrix** of a graph, and (3) what is the **incidence matrix** of a graph. Illustrate some of the graph concepts from this chapter (degrees of vertices, multiple edges, loops, etc.) in matrix terms. Include plenty of examples. You can find definitions and information on adjacency and incidence matrices of graphs in many graph theory books (see, for example, references 5 and 14).

C The Chinese Postman Problem

A *weighted graph* is a graph in which the edges are assigned positive numbers called weights. The weights represent distances, times, or costs. Finding optimal routes that cover *all* the edges of a *weighted graph* is a problem known as the *Chinese postman problem*. (Chinese postman problems are a generalization of *Euler circuit* problems and, as a general rule, are much harder to solve, but most of the concepts developed in this chapter still apply.)

In this project you are asked to prepare a presentation on the Chinese postman problem for your class.

Some suggestions: (1) Give several examples to illustrate Chinese postman problems and how they differ from corresponding Euler circuit problems. (2) Describe some possible real-life applications of Chinese postman problems. (3) Discuss how to solve a Chinese postman problem in the simplest cases when the weighted graph has no odd vertices or has only two odd vertices (these cases can be solved using techniques learned in this chapter). (4) Give a rough outline of how one might attempt to solve a Chinese postman problem for a graph with four odd vertices.

REFERENCES AND FURTHER READINGS

1. Beltrami, E., *Models for Public Systems Analysis*. New York: Academic Press, Inc., 1977.
2. Beltrami, E., and L. Bodin, "Networks and Vehicle Routing for Municipal Waste Collection," *Networks*, 4 (1974), 65–94.
3. Bogomolny, A., "Graphs," *http://www.cut-the-knot.org/do_you_know/graphs.shtml.*
4. Chartrand, Gary, *Graphs as Mathematical Models*. Belmont, CA: Wadsworth Publishing Co., 1977.
5. Chartrand, Gary, and Ortud R. Oellerman, *Applied and Algorithmic Graph Theory*. New York: McGraw-Hill, 1993.
6. Dunham, William, *Euler: The Master of Us All*. Washington, DC: The Mathematical Association of America, 1999.
7. Euler, Leonhard, "The Königsberg Bridges," trans. James Newman, *Scientific American*, 189 (1953), 66–70.
8. Minieka, E., *Optimization Algorithms for Networks and Graphs*. New York: Marcel Dekker, Inc., 1978.
9. Newman, James R., *The World of Mathematics*, vol. 1. New York: Simon & Schuster, 1956.
10. Roberts, Fred S., "Graph Theory and Its Applications to Problems of Society," *CBMS-NSF Monograph No. 29*. Philadelphia: Society for Industrial and Applied Mathematics, 1978, chap. 8.
11. Stein, S. K., *Mathematics: The Man-Made Universe* (3rd ed.). New York: Dover, 2000.

12. Steinhaus, H., *Mathematical Snapshots* (3rd ed.). New York: Dover, 1999.

13. Tucker, A. C., and L. Bodin, "A Model for Municipal Street-Sweeping Operations," in *Modules in Applied Mathematics*, Vol. 3, eds. W. Lucas, F. Roberts, and R. M. Thrall. New York: Springer-Verlag, 1983, 76–111.

14. West, Douglas, *Introduction to Graph Theory*. Upper Saddle River, NJ: Prentice Hall, 1996.

6 The Mathematics of Touring

The Traveling Salesman Problem

To those of us who get a thrill out of planning a big trip, the next decade holds a truly special treat: We humans are on our way to Mars. Why are we going to Mars, and what are we doing there?

For starters, Mars is the next great frontier for human exploration. Its promise lies in the fact that it is the most earthlike of all the planets in the solar system, the Martian soil is rich in chemicals and minerals, and, most important of all, there is strong evidence that there may be water just below the surface. The ultimate attraction, however, is the hope of finding life on Mars. Of all the planets in our solar system, Mars is the most likely place to show some evidence of life — probably primitive bacterial forms buried inside Martian rocks or under the Martian surface. But finding life on Mars — assuming there is any — raises many technical and logistical questions. What are the best places on Mars to explore? How do we get the equipment there? How long will it take to do the job? And, above all, how much will it cost? Once again, lurking behind the complexities of Mars exploration is an interesting and important mathematics problem.

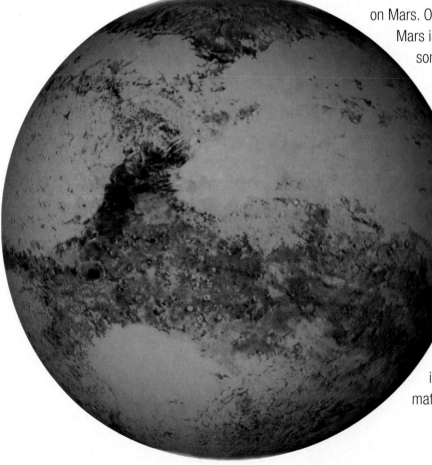

H ere are the details in a nutshell. Based on geological data already collected from earlier orbiting missions, NASA has identified a set of sites on the surface of Mars where the likelihood of finding evidence of past life is the highest (see Fig. 6-1). In January 2004 two rovers named *Spirit* and *Opportunity* successfully landed on Mars, returning incredible pictures and a lot of scientific data (so far, no evidence of life). But *Spirit* and *Opportunity* have limited mobility and were designed to only explore the area near their respective landing sites. A main goal for the next stage of Mars exploration is a *robotic sample-return mission*, in which a lander would land at one of the designated sites and release an unmanned rover controlled from Earth. The rover would then travel to each of the other sites, collecting

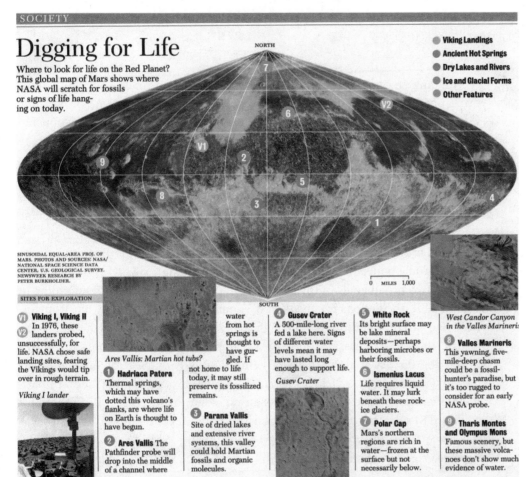

SOCIETY

Digging for Life

Where to look for life on the Red Planet? This global map of Mars shows where NASA will scratch for fossils or signs of life hanging on today.

● Viking Landings
● Ancient Hot Springs
◗ Dry Lakes and Rivers
● Ice and Glacial Forms
● Other Features

SINUSOIDAL EQUAL-AREA PROJ. OF MARS. PHOTOS AND SOURCES: NASA/ NATIONAL SPACE SCIENCE DATA CENTER, U.S. GEOLOGICAL SURVEY. NEWSWEEK RESEARCH BY PETER BURKHOLDER.

0 MILES 1,000

SITES FOR EXPLORATION

V1 V2 Viking I, Viking II
In 1976, these landers probed, unsuccessfully, for life. NASA chose safe landing sites, fearing the Vikings would tip over in rough terrain.

Viking I lander

1 Hadriaca Patera
Thermal springs, which may have dotted this volcano's flanks, are where life on Earth is thought to have begun.

2 Ares Vallis The Pathfinder probe will drop into the middle of a channel where

Ares Vallis: Martian hot tubs?

water from hot springs is thought to have gurgled. If not home to life today, it may still preserve its fossilized remains.

3 Parana Vallis
Site of dried lakes and extensive river systems, this valley could hold Martian fossils and organic molecules.

4 Gusev Crater
A 500-mile-long river fed a lake here. Signs of different water levels mean it may have lasted long enough to support life.

Gusev Crater

5 White Rock
Its bright surface may be mineral deposits—perhaps harboring microbes or their fossils.

6 Ismenius Lacus
Life requires liquid water. It may lurk beneath these rock-ice glaciers.

7 Polar Cap
Mars's northern regions are rich in water—frozen at the surface but not necessarily below.

West Candor Canyon in the Valles Marineris

8 Valles Marineris
This yawning, five-mile-deep chasm could be a fossil-hunter's paradise, but it's too rugged to consider for an early NASA probe.

9 Tharis Montes and Olympus Mons
Famous scenery, but these massive volcanoes don't show much evidence of water.

FIGURE 6-1
Source: NASA/National Space Science Data Center, U.S. Geological Survey.

soil samples and performing experiments. After all the sites have been visited, the rover would return to the landing site where a return rocket would bring the samples back to Earth. The first of these sample-return missions is scheduled to be launched in 2014.

Figure 6-2 shows one of the many possible routes that the rover might take as it travels to each of the sites to look for life and collect soil samples. And there are hundreds of other possible ways to route the rover. Which one is the best?

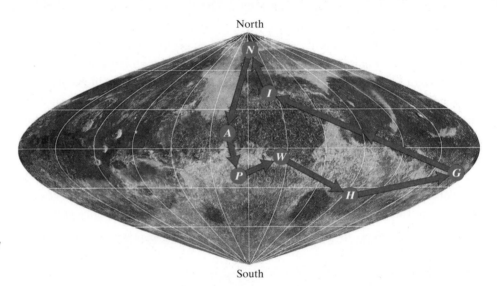

North

South

FIGURE 6-2
A: Ares Vallis (starting and ending point); *P*: Parana Vallis; *W*: White Rock; *H*: Hadriaca Patera; *G*: Gusev Crater; *I*: Ismenius Lacus; *N*: North Polar Cap.

Finding an optimal route for an unmanned Martian rover is an exotic illustration of a class of routing problems that go by the general name of *traveling salesman problems*, or TSPs. The unusual name comes from what has by now become the standard version of the problem: a traveling salesman trying to find the cheapest route that takes him to each of several cities, returning at the end of the trip to the starting city. The name has stuck as a generic name given to all sorts of equivalent problems, even if they have nothing to do with traveling salesmen. It is best to think of the "traveling salesman" as just a metaphor—a rover searching for the best route on Mars, a UPS driver trying to find the best way to deliver packages around town, or any ordinary Joe trying to plan the fastest route by which to run a bunch of errands on a Saturday morning.

The two broad goals of this chapter are (1) to develop an understanding of what a TSP is (Sections 6.1 through 6.3) and (2) to understand what it means to "solve" a TSP (Sections 6.4 through 6.8). As usual, to accomplish these goals a little theory is necessary, and several important graph concepts are introduced and discussed along the way—*Hamilton circuits* (Section 6.1), *complete graphs* (Section 6.2), *efficient algorithms* (Section 6.5), *optimal* and *approximate algorithms* (Section 6.6). In addition, some of the basic algorithms for solving TSPs are introduced and discussed in the second half of the chapter—the *brute-force* and *nearest-neighbor* algorithms in Section 6.5, the *repetitive nearest-neighbor* algorithm in Section 6.7, and the *cheapest-link* algorithm in Section 6.8.

6.1 Hamilton Paths and Hamilton Circuits

■ Hamilton paths and circuits are named after Irish mathematician Sir William Rowan Hamilton (1805–1865). For more on Hamilton, see the biographical profile at the end of this chapter.

In Chaper 5 we discussed Euler paths and Euler circuits. There, the name of the game was to find paths or circuits that include every *edge* of the gaph once (and only once). We are now going to discuss a seemingly related game: finding paths and circuits that include every *vertex* of the graph once and only once. Paths and circuits having this property are called **Hamilton paths** and **Hamilton circuits**.

┌─ **HAMILTON PATHS AND CIRCUITS** ──────────────────────────

- A Hamilton path in a graph is a path that includes each vertex of the graph once and only once.

- A Hamilton circuit is a circuit that includes each vertex of the graph once and only once. (At the end, of course, the circuit must return to the staring vertex.)

On the surface, there is a one-word difference between Euler paths/circuits and Hamilton paths/circuits: The former covers all *edges*; the latter covers all *vertices*. but oh my, what a difference that one word makes!

(**EXAMPLE 6.1**)   **Hamilton Versus Euler**

The various graphs shown in Fig. 6-3 will help us highlight the differences between Hamilton circuits/paths and Euler circuits/paths.

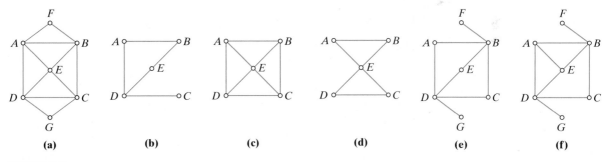

FIGURE 6-3

- Figure 6-3(a) shows a graph that (1) has Euler circuits (the vertices are all even) and (2) has Hamilton circuits. One such Hamilton circuit is A, F, B, C, G, D, E, A—there are plenty more. Note that *if a graph has a Hamilton circuit, then it automatically has a Hamilton path*—the Hamilton circuit can always be truncated into a Hamilton path by dropping the last vertex of the circuit. (For example, the Hamilton circuit A, F, B, C, G, D, E, A can be truncated into the Hamilton path A, F, B, C, G, D, E.) Contrast this with the mutually exclusive relationship between Euler circuits and paths: If a graph has an Euler circuit it cannot have an Euler path and vice versa.

- Figure 6-3(b) shows a graph that (1) has no Euler circuits but does have Euler paths (for example C, D, E, B, A, D) and (2) has no Hamilton circuits (sooner or later you have to go to C, and then you are stuck) but does have Hamilton paths (for example, A, B, E, D, C). This illustrates that *a graph can have a Hamilton path but no Hamilton circuit!*

- Figure 6-3(c) shows a graph that (1) has neither Euler circuits nor paths (it has four odd vertices) and (2) has Hamilton circuits (for example A, B, C, D, E, A—there are plenty more) and consequently has Hamilton paths (for example, A, B, C, D, E).

- Figure 6-3(d) shows a graph that (1) has Euler circuits (the vertices are all even) and (2) has no Hamilton circuits (no matter what, you are going to have to go through E more than once!) but has Hamilton paths (for example, A, B, E, D, C).

- Figure 6-3(e) shows a graph that (1) has no Euler circuits but has Euler paths (F and G are the two odd vertices) and (2) has neither Hamilton circuits nor Hamilton paths (see Exercise 9).

- Figure 6-3(f) shows a graph that (1) has neither Euler circuits nor Euler paths (too many odd vertices) and (2) has neither Hamilton circuits nor Hamilton paths.

Table 6-1 summarizes the preceding observations.

TABLE 6-1	Existence of Euler and/or Hamilton Circuits/Paths			
	Euler circuit	Euler path	Hamilton circuit	Hamilton path
Fig. 6-3(a)	Yes	No	Yes	Yes
Fig. 6-3(b)	No	Yes	No	Yes
Fig. 6-3(c)	No	No	Yes	Yes
Fig. 6-3(d)	Yes	No	No	Yes
Fig. 6-3(e)	No	Yes	No	No
Fig. 6-3(f)	No	No	No	No

The lesson of Example 6.1 is that the existence of an Euler path or circuit in a graph tells us nothing about the existence of a Hamilton path or circuit in that graph. This is important because it implies that Euler's circuit and path theorems from Chapter 5 are useless when it comes to identifying Hamilton circuits and paths. But surely, there must be analogous "Hamilton circuit and path theorems" that we could use to determine if a graph has a Hamilton circuit, a Hamilton path, or neither. Surprisingly, no such theorems exist. Determining when a given graph does or does not have a Hamilton circuit or path can be very easy, but it also can be very hard—it all depends on the graph.

■ For a theorem that gives a *sufficient condition* for a graph to have a Hamilton circuit, see Exercise 77.

6.2   Complete Graphs

Sometimes the question, *Does the graph have a Hamilton circuit?* has an obvious *yes* answer, and the more relevant question turns out to be, *How many different Hamilton circuits does it have?* In this section we will answer this question for an important family of graphs called complete graphs.

A graph with N vertices in which *every* pair of distinct vertices is joined by an edge is called a **complete graph** on N vertices and denoted by the symbol K_N. Figure 6-4 shows the complete graphs on three, four, five, and six vertices, respectively.

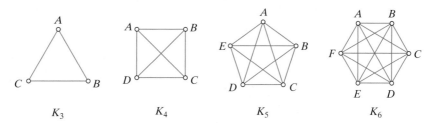

FIGURE 6-4
K_3 K_4 K_5 K_6

One of the key properties of K_N is that every vertex has degree $N - 1$. This implies that the sum of the degrees of all the vertices is $N(N - 1)$, and it follows from Euler's sum of degrees theorem (see page 180) that the number of edges in K_N is $N(N - 1)/2$. For a graph with N vertices and no multiple edges or loops, $N(N - 1)/2$ is the maximum number of edges possible, and this maximum can only occur when the graph is K_N.

NUMBER OF EDGES IN K_N

- K_N has $N(N - 1)/2$ edges.
- Of all graphs with N vertices and no multiple edges or loops, K_N has the most edges.

Because K_N has a complete set of edges (every vertex is connected to every other vertex), it also has a complete set of Hamilton circuits—you can travel the vertices in *any* sequence you choose and you will not get stuck.

EXAMPLE 6.2   **Hamilton Circuits in K_4**

If we travel the four vertices of K_4 in an arbitrary order, we get a Hamilton path. For example, C, A, D, B is a Hamilton path [Fig. 6-5(a)]; D, C, A, B is another one [Fig. 6-5(b)]; and so on. Each of these Hamilton paths can be closed into a Hamilton circuit—the path C, A, D, B begets the circuit C, A, D, B, C [Fig. 6-5(c)]; the path D, C, A, B begets the circuit D, C, A, B, D [Fig. 6-5(d)]; and so on. It looks like we have an abundance of Hamilton circuits, but it is important to remember that the same Hamilton circuit can be written in many ways. For example, C, A, D, B, C is the same circuit as A, D, B, C, A [Fig. 6-5(c) describes either one]—the only difference is that in the first case we used C as the *reference point*; in the second case we used A as the reference point. There are two additional sequences that describe this same Hamilton circuit: D, B, C, A, D (with reference point D) and B, C, A, D, B (with reference point B). Taking all this into account, there are six

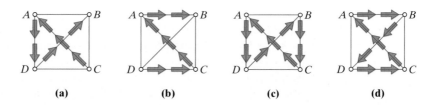

FIGURE 6-5   **(a)**   **(b)**   **(c)**   **(d)**

different Hamilton circuits in K_4, as shown in Table 6-2 (the table also shows the four different ways each circuit can be written).

TABLE 6-2	The Six Hamilton Circuits in K_4			
	Reference point is A	Reference point is B	Reference point is C	Reference point is D
1	A, B, C, D, A	B, C, D, A, B	C, D, A, B, C	D, A, B, C, D
2	A, B, D, C, A	B, D, C, A, B	C, A, B, D, C	D, C, A, B, D
3	A, C, B, D, A	B, D, A, C, B	C, B, D, A, C	D, A, C, B, D
4	A, C, D, B, A	B, A, C, D, B	C, D, B, A, C	D, B, A, C, D
5	A, D, B, C, A	B, C, A, D, B	C, A, D, B, C	D, B, C, A, D
6	A, D, C, B, A	B, A, D, C, B	C, B, A, D, C	D, C, B, A, D

EXAMPLE 6.3   **Hamilton Circuits in K_5**

Let's try to list all the Hamilton circuits in K_5. For simplicity, we will write each circuit just once, using a common reference point—say A. (As long as we are consistent, it doesn't really matter which reference point we pick.) Each of the Hamilton circuits will be described by a sequence that starts and ends with A, with the letters B, C, D, and E sandwiched in between in some order. There are $4 \times 3 \times 2 \times 1 = 24$ different ways to shuffle the letters B, C, D, and E, each producing a different Hamilton circuit. The complete list of the 24 Hamilton circuits in K_5 is shown in Table 6-3. The table is laid out so that each of the circuits in the table is directly opposite its *mirror-image circuit* (the circuit with vertices listed in reverse order). Although they are close relatives, a circuit and its mirror image are not considered the same circuit.

TABLE 6-3	The 24 Hamilton Circuits in K_5		
1	A, B, C, D, E, A	13	A, E, D, C, B, A
2	A, B, C, E, D, A	14	A, D, E, C, B, A
3	A, B, D, C, E, A	15	A, E, C, D, B, A
4	A, B, D, E, C, A	16	A, C, E, D, B, A
5	A, B, E, C, D, A	17	A, D, C, E, B, A
6	A, B, E, D, C, A	18	A, C, D, E, B, A
7	A, C, B, D, E, A	19	A, E, D, B, C, A
8	A, C, B, E, D, A	20	A, D, E, B, C, A
9	A, C, D, B, E, A	21	A, E, B, D, C, A
10	A, C, E, B, D, A	22	A, D, B, E, C, A
11	A, D, B, C, E, A	23	A, E, C, B, D, A
12	A, D, C, B, E, A	24	A, E, B, C, D, A

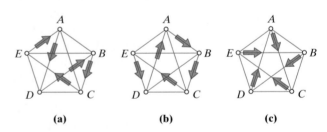

FIGURE 6-6 Three of the 24 Hamilton circuits in K_5. (Try to find them in Table 6.3.)

We can now generalize our observations in Examples 6.2 and 6.3. Suppose we have K_N, the complete graph with N vertices. We pick one of the vertices—call it A if you please—and make it the designated reference point for all the Hamilton circuits. Every Hamilton circuit will then take the form $A, *, *, \ldots, *, A$ (the *'s are wildcards that denote the remaining $N - 1$ vertices), and with each

■ If this is your first exposure to factorials (or you need a refresher), it might be a good idea to stop here, read the brief discussion of factorials in Chapter 2, and then take a stab at Exercises 17 through 24.

reordering of the *'s we get a different Hamilton circuit. This means that the question, "What is the number of Hamilton circuits in K_N?" boils down to the equivalent question, "How many different ways are there to rearrange the $(N - 1)$ vertices represented by the *'s?" The answer, as you may recall from Chapter 2, is given by the number $1 \times 2 \times 3 \times \cdots \times (N - 1)$, called the **factorial** of $(N - 1)$ and written as $(N - 1)!$ for short.

> **NUMBER OF HAMILTON CIRCUITS IN K_N**
>
> There are $(N - 1)!$ distinct Hamilton circuits in K_N.

Table 6-4 shows the number of Hamilton circuits in complete graphs with up to $N = 20$ vertices. The main point of Table 6-4 is to convince you that as we increase the number of vertices, the number of Hamilton circuits in the complete graph goes through the roof. Even a relatively small graph such as K_8 has more than five thousand Hamilton circuits. Double the number of vertices to K_{16} and the number of Hamilton circuits exceeds 1.3 *trillion*. Double the number of vertices again to K_{32} and the number of Hamilton circuits—about *eight billion trillion trillion*—is so large that it defies ordinary human comprehension. (This incredible growth in the number of Hamilton circuits in K_N is not just an idle observation, and it will become very relevant to us soon.)

TABLE 6-4	Number of Distinct Hamilton Circuits in K_N		
N	**(N − 1)!**	**N**	**(N − 1)!**
3	2	12	39,916,800
4	6	13	479,001,600
5	24	14	6,227,020,800
6	120	15	87,178,291,200
7	720	16	1,307,674,368,000
8	5040	17	20,922,789,888,000
9	40,320	18	355,687,428,096,000
10	362,880	19	6,402,373,705,728,000
11	3,628,800	20	121,645,100,408,832,000

6.3 Traveling Salesman Problems

The title *Traveling Salesman Problems* is catchy but a bit misleading, since most of the time the problems that fall under this heading have nothing to do with salespeople living out of a suitcase. The "traveling salesman" is a convenient metaphor for many different real-life applications. The next few examples illustrate a few of the many possible settings for a "traveling salesman" problem, starting, of course, with the traveling salesman's "traveling salesman" problem.

(From here on we are following custom and dropping the quotes around *traveling salesman*, but remember—don't take the phrase too literally.)

> **EXAMPLE 6.4**   **A Tour of Five Cities**

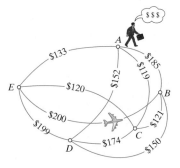

FIGURE 6-7

Meet Willy, the traveling salesman. Willy has customers in five cities, which for the sake of brevity we will call A, B, C, D, and E. Willy needs to schedule a sales trip that will start and end at A (that's Willy's hometown) and goes to each of the other four cities once. We will call the trip "Willy's sales tour." Other than starting and ending at A, there are no restrictions as to the sequence in which Willy's sales tour visits the other four cities.

The graph in Fig. 6-7 shows the cost of a *one-way* airline ticket between each pair of cities. (For simplicity we are making the assumption here that a one-way ticket costs the same regardless of whether you go from city X to city Y or vice versa. This is not always true—today's airline prices have a logic of their own.) Like most people, Willy hates to waste money. Thus, among the many possibilities for his sales tour, Willy wants to find the optimal (cheapest) one. How? We will return to this question soon.

> **EXAMPLE 6.5**   **Touring the Outer Moons**

It is the year 2020. An expedition to explore the outer planetary moons in our solar system is about to be launched from planet Earth. The expedition is scheduled to visit Callisto, Ganymede, Io, Mimas, and Titan (the first three are moons of Jupiter; the last two, of Saturn), collect rock samples at each, and then return to Earth with the loot.

Figure 6-8 shows the mission time (in years) between any two moons. An important goal of the mission planners is to complete the mission in the least amount of time. What is the optimal (shortest) tour of the outer moons?

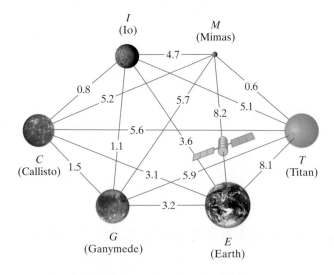

FIGURE 6-8

> **EXAMPLE 6.6**   **Roving the Red Planet**

Figure 6-9 shows seven locations on Mars where NASA scientists believe there is a good chance of finding evidence of life (see the chapter opener for details). Imagine that you are in charge of planning a *sample-return* mission. First, you must land an unmanned rover in the Ares Vallis (A). Then you must direct the

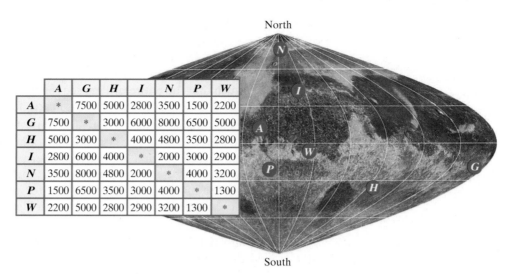

North

	A	*G*	*H*	*I*	*N*	*P*	*W*
A	*	7500	5000	2800	3500	1500	2200
G	7500	*	3000	6000	8000	6500	5000
H	5000	3000	*	4000	4800	3500	2800
I	2800	6000	4000	*	2000	3000	2900
N	3500	8000	4800	2000	*	4000	3200
P	1500	6500	3500	3000	4000	*	1300
W	2200	5000	2800	2900	3200	1300	*

South

FIGURE 6-9

rover to travel to each site and collect and analyze soil samples. Finally, you must instruct the rover to return to the Ares Vallis landing site, where a return rocket will bring the best samples back to Earth. A Mars tour like this will take several years and cost several billion dollars, so good planning is critical.

Figure 6-9 shows the estimated distances (in miles) that a rover would have to travel to get from one Martian site to another. What is the optimal (shortest) tour for the Mars rover?

While representing very different real-life problems, Examples 6.4 through 6.6 have much in common. In all three examples we have a *traveler* (Willy, a spaceship, an unmanned rover) and a set of *sites* the traveler must visit (cities, moons in the solar system, locations on Mars). In addition, there is a known *cost* (money, time, distance) associated with each leg of the trip. In each case the problem is to find a **tour** of the sites (i.e., a *trip* that starts and ends at a designated site and visits each of the other sites once) and has the property of being **optimal** (i.e., has the least total cost). Any problem that shares these common elements (a *traveler*, a set of *sites*, a *cost* function for travel between pairs of sites, a need to *tour all* the sites, and a desire to *minimize* the total cost of the tour) is known as a **traveling salesman problem**, or **TSP**.

The following are just a few examples of real-life TSPs.

- **Routing school buses.** A school bus (the *traveler*) picks up children in the morning and drops them off at the end of the day at designated stops (the *sites*). On a typical school bus route there may be 20 to 30 such stops. With school buses, total time on the bus is always the most important variable (students have to get to school on time), and there is a known time of travel (the *cost*) between any two bus stops. Since children must be picked up at every bus stop, a *tour* of all the sites (starting and ending at the school) is required. Since the bus repeats its route every day during the school year, finding an *optimal tour* is crucial.

- **Delivering packages.** Package delivery companies such as UPS and FedEx deal with TSPs on a daily basis. Each truck is a *traveler* that must deliver packages to a specific list of delivery destinations (the *sites*). The travel time between any two delivery sites (the *cost*) is known or can be estimated. Each day the truck must deliver to all the sites on its list (that's why sometimes you see a UPS truck delivering at 8 P.M.), so a *tour* is an implied part of the requirements. Since one can assume that the driver would rather be home than out delivering packages, an *optimal* tour is a highly desirable goal.

- **Fabricating circuit boards.** In the process of fabricating integrated-circuit boards, tens of thousands of tiny holes (the *sites*) must be drilled in each board. This is done by using a stationary laser beam and moving the board (the *traveler*). To do this efficiently, the order in which the holes are drilled should be such that the entire drilling sequence (the *tour*) is completed in the least amount of time (*optimal cost*). This makes for a very high tech TSP.

- **Running errands around town.** On a typical Saturday morning, an average Joe or Jane (the *traveler*) sets out to run a bunch of errands around town, visiting various sites (grocery store, hair salon, bakery, post office). When gas was cheap, *time* used to be the key *cost* variable, but with the cost of gas these days, people are more likely to be looking for the tour that minimizes the total *distance* traveled (see Exercise 53 for an illustration of this TSP).

Regardless of the original real-life application, every TSP can be modeled by a **weighted graph**, that is, a graph such that there is a number associated with each edge (called the **weight** of the edge). The beauty of this approach is that the model always has the same structure: The vertices of the graph are the *sites* of the TSP, and there is an edge between X and Y if there is a direct link for the *traveler* to travel from site X to site Y. Moreover, the weight of the edge XY is the cost of travel between X and Y. In this setting a *tour* is a Hamilton circuit of the graph, and an *optimal tour* is the Hamilton circuit of least total weight.

In all the applications and examples we will be considering in this chapter, we will make the assumption that there is an edge connecting every pair of sites, which implies that the underlying graph model is always a **complete weighted graph**. (Figs. 6-7 and 6-8 are typical examples of complete weighted graphs. Figure 6-9 is not in and of itself a complete weighted graph, but all the necessary information is there, and if we have to, we can create the corresponding graph.)

The following is a summary of the preceding observations.

GRAPH MODEL OF A TRAVELING SALESMAN PROBLEM

- Sites → vertices of the graph
- Costs → weights of the edges
- Tour → Hamilton circuit
- Optimal tour → Hamilton circuit of least total weight

6.4 Simple Strategies for Solving TSPs

> **EXAMPLE 6.7**   **A Tour of Five Cities: Part 2**

In Example 6.4 we met Willy the traveling salesman pondering his upcoming sales trip, dollar signs running through his head. (The one-way airfares between any two cities are shown again in Fig. 6-10.) Imagine now that Willy, unwilling or unable to work out the problem for himself, decides to offer a reward of $50 to anyone who can find an optimal tour. Would it be worth $50 to you to work out this problem? (Yes.) So how would you do it? (At this point you should take a break from your reading, get a pencil and a piece of paper, and try to work out the problem on your own. Give yourself about 15 to 20 minutes.)

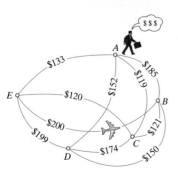

FIGURE 6-10

. . .

Welcome back. If you are like most people, you probably followed one of two strategies in working through this problem.

> **STRATEGY 1 (EXHAUSTIVE SEARCH)**
>
> Make a list of all possible Hamilton circuits. For each circuit in the list, calculate the weight of the circuit. From all the circuits, choose a circuit with least total weight. This is your optimal tour.

Table 6-5 shows a detailed implementation of this strategy. The 24 Hamilton circuits are split into two columns consisting of circuits and their mirror images. (Notice that since the one-way airfares are the same in either direction, a circuit and its mirror-image circuit will always cost the same. This is a useful shortcut that cuts the number of calculations in half.) The total costs for the circuits are shown in the middle column of the table. There are two optimal tours, with a total cost of $676—$A$, D, B, C, E, A and its mirror image A, E, C, B, D, A. The two tours are shown in Fig. 6-11. (There are always going to be at least two optimal tours, since the mirror image of an optimal tour is also optimal. Either one of them can be used for a solution.)

TABLE 6-5	Willy's Possible Tours and Their Costs	
	Hamilton circuit	**Total cost**	**Mirror-image circuit**
1	A, B, C, D, E, A	$185 + 121 + 174 + 199 + 133 = 812$	A, E, D, C, B, A
2	A, B, C, E, D, A	$185 + 121 + 120 + 199 + 152 = 777$	A, D, E, C, B, A
3	A, B, D, C, E, A	$185 + 150 + 174 + 120 + 133 = 762$	A, E, C, D, B, A
4	A, B, D, E, C, A	$185 + 150 + 199 + 120 + 119 = 773$	A, C, E, D, B, A
5	A, B, E, C, D, A	$185 + 200 + 120 + 174 + 152 = 831$	A, D, C, E, B, A
6	A, B, E, D, C, A	$185 + 200 + 199 + 174 + 119 = 877$	A, C, D, E, B, A
7	A, C, B, D, E, A	$119 + 121 + 150 + 199 + 133 = 722$	A, E, D, B, C, A
8	A, C, B, E, D, A	$119 + 121 + 200 + 199 + 152 = 791$	A, D, E, B, C, A
9	A, C, D, B, E, A	$119 + 174 + 150 + 200 + 133 = 776$	A, E, B, D, C, A
10	A, C, E, B, D, A	$119 + 120 + 200 + 150 + 152 = 741$	A, D, B, E, C, A
11	A, D, B, C, E, A	$152 + 150 + 121 + 120 + 133 = 676$	A, E, C, B, D, A
12	A, D, C, B, E, A	$152 + 174 + 121 + 200 + 133 = 780$	A, E, B, C, D, A

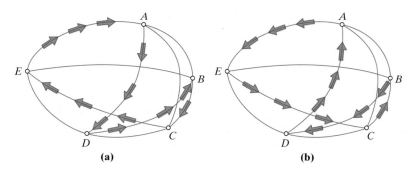

FIGURE 6-11 The two optimal tours for Willy's trip. (Total travel cost = $676.)

All the preceding could be reasonably done in somewhere between 10 and 20 minutes. Not a bad gig for $50!

STRATEGY 2 (GO CHEAP)

Start from the home city. From there go to the city that is the cheapest to get to. From each new city go to the next new city that is cheapest to get to. When there are no more new cities to go to, go back home.

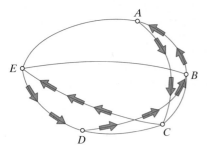

FIGURE 6-12

In Willy's case, this strategy works like this: Start at A. From A Willy goes to C (the cheapest place he can fly to from A). From C Willy goes to E (the cheapest city to fly to other than A). From E Willy goes to D, and from D he has little choice—the last new city to go to is B. From B Willy has to return home to A. The total cost of this tour (shown in Fig. 6-12) is $773 ($119 + $120 + $199 + $150 + $185 = $773).

The "go cheap" strategy takes a lot less work than the "exhaustive search" strategy, but there is a hitch—the cost of the tour we get is $773, which is $97 more than the optimal tour found under the exhaustive search strategy. Willy says this is not a solution and refuses to pay the $50. The "go cheap" strategy looks like a bust, but be patient—there is more to come.

EXAMPLE 6.8   **A Tour of (Gasp!) 10 Cities**

Let's imagine now that Willy, who has done very well with his business, has expanded his sales territory to 10 cities (let's call them A through K). Willy wants us to help him once again find an optimal tour of his sales territory. Flush with success and generosity, he is offering a whopping $200 as a reward for a solution to this problem. Should we accept the challenge?

The graph in Fig. 6-13 represents the 10 cities, and the fares table shows the one-way fares between them. (With this many edges the graph would get pretty cluttered if we tried to show the weights on the graph itself. It is better to put them in a table.) We now know that a foolproof method for tackling this problem is the "exhaustive search" strategy. But before we plunge into it, let's think of what we are getting into. With 10 cities we have 9! = 362,880 Hamilton circuits. Assuming we could compute the cost of a new circuit every 30 seconds—and that's working fast—it would take about 3000 hours to do all 362,880 possible circuits. Shortcuts? Say you cut the work in half by skipping the mirror-image circuits. That's 1500 hours of work. If you worked nonstop, 24 hours a day, 7 days a week it would still take a couple of months!

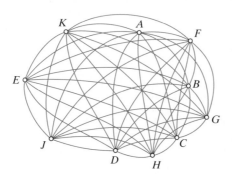

FIGURE 6-13

	A	B	C	D	E	F	G	H	J	K
A	*	185	119	152	133	321	297	277	412	381
B	185	*	121	150	200	404	458	492	379	427
C	119	121	*	174	120	332	439	348	245	443
D	152	150	174	*	199	495	480	500	454	489
E	133	200	120	199	*	315	463	204	396	487
F	321	404	332	495	315	*	356	211	369	222
G	297	458	439	480	463	356	*	471	241	235
H	277	492	348	500	204	211	471	*	283	478
J	412	379	245	454	396	369	241	283	*	304
K	381	427	443	489	487	222	235	478	304	*

Let's now try the "go cheap" strategy, the one that didn't work out so well in Example 6.7. This time we have to implement the strategy using the fare table in Fig. 6-13. The trip would start at A. Starting at the A-row of the table, we search for the smallest number in the row. It is 119, located under the C-column. This means that the trip should start with a flight from A to C. Now we go to the C-row in the table searching for the smallest number other than the 119 (A is already taken). That number is 120, located under the E-column. This means that the next leg of the trip is from C to E. Now we go to the E-row and locate the smallest number other than the numbers under the A and C columns (we don't want to go back to either of those cities). That number is 199, located under the D-column. The process can be continued, one city at a time, and you are encouraged to finish this on your own—it shouldn't take more than a few minutes. The final result is the tour $A, C, E, D, B, J, G, K, F, H, A$, with a total cost of $2153.

■ A good way to keep track of the cities that have been visited is to cross out their columns as we move on.

So we now have a tour, but is it the optimal tour? If it isn't—is it at least close? Willy refuses to pay the $200 unless these questions can be answered to his satisfaction. We will settle this issue later in the chapter.

6.5 The Brute-Force and Nearest-Neighbor Algorithms

In this section we will look at the two strategies we informally developed in connection with Willy's sales trips and recast them in the language of algorithms. The "exhaustive search" strategy can be formalized into an algorithm generally known as the **brute-force algorithm**; the "go cheap" strategy can be formalized into an algorithm known as the **nearest-neighbor algorithm**.

> ### ALGORITHM 1: THE BRUTE-FORCE ALGORITHM
>
> - **Step 1.** Make a list of *all* the possible Hamilton circuits of the graph. Each of these circuits represents a tour of the vertices of the graph.
> - **Step 2.** For each tour calculate its *weight* (i.e., add the weights of all the edges in the circuit).
> - **Step 3.** Choose an *optimal* tour (there is always more than one optimal tour to choose from!).

> ### ALGORITHM 2: THE NEAREST-NEIGHBOR ALGORITHM
>
> - **Start:** Start at the designated starting vertex. If there is no designated starting vertex pick any vertex.
> - **First step:** From the starting vertex go to its *nearest neighbor* (i.e., the vertex for which the corresponding edge has the smallest weight).
> - **Middle steps:** From each vertex go to its *nearest neighbor*, choosing only among *the vertices that haven't been yet visited*. (If there is more than one nearest neighbor choose among them at random.) Keep doing this until all the vertices have been visited.
> - **Last step:** From the last vertex return to the starting vertex.

Both of the preceding algorithms have some pros and some cons, which we will discuss next. The positive aspect of the brute-force algorithm is that it is an optimal algorithm. (An **optimal algorithm** is an algorithm that, when correctly implemented, is guaranteed to produce an optimal solution.) In the case of the brute-force algorithm, we know we are getting an optimal solution because we are choosing from among *all* possible tours.

The negative aspect of the brute-force algorithm is the amount of effort that goes into implementing the algorithm, which is (roughly) *proportional to the number of Hamilton circuits* in the graph. As we first saw in Table 6-4, as the number of vertices grows just a little, the number of Hamilton circuits in a complete graph grows at an incredibly fast rate.

What does this mean in practical terms? Let's start with human computation. To solve a TSP for a graph with 10 vertices (as real-life problems go, that's puny), the brute-force algorithm requires checking 362,880 Hamilton circuits. To do this by hand, even for a fast and clever human, it would take over 1000 hours. Thus, at $N = 10$ we are already beyond the limit of what can be considered reasonable human effort.

A possible way to get around this difficulty, we might think, is to recruit a fast helper, such as a powerful computer. (In principle, this is a good idea, since a computer is the perfect tool to implement the brute-force algorithm—the algorithm is essentially a mindless exercise in arithmetic with a little bookkeeping thrown in, and both of these are things that computers are very good at.) Let's imagine, for the sake of argument, that we have unlimited free access to SUPER-HERO—the fastest supercomputer on the planet—which can compute *a quadrillion* tours per second. (One quadrillion equals a million billion. This is much faster than even the fastest current supercomputers can perform, but since we are just fantasizing, let's think big!) For *N* less than 20, SUPERHERO can run through all possible tours in a matter of seconds (or less). Things get more interesting when we start considering what happens beyond $N = 20$. Table 6-6 shows the approximate computation time required by SUPERHERO when implementing the brute-force algorithm for values of *N* ranging between 20 and 30.

The surprising lesson of Table 6-6 is that even with the world's best technology on our side, we very quickly reach the point beyond which using the brute-force algorithm is completely unrealistic (clearly for $N \geq 25$ and some might argue even for $N = 24$).

The brute-force algorithm is a classic example of what is formally known as an **inefficient algorithm**—an algorithm for which the number of steps needed to carry it out grows disproportionately with the size of the problem. The trouble with inefficient algorithms is that they are of limited practical use—they can realistically be carried out only when the problem is small.

The extraordinary computational effort required by the brute-force algorithm is a direct consequence of the way factorials grow—each time we increase the number of vertices of the graph from *N* to $N + 1$, *the amount of work required to carry out the brute-force algorithm increases by a factor of N.* For example, it takes 5 times as much work to go from 5 vertices to 6, 10 times as much work to go from 10 vertices to 11, and 100 times as much work to go from 100 vertices to 101.

Fortunately, not all algorithms are inefficient. Let's discuss now the nearest-neighbor algorithm, in which we hop from vertex to vertex using a simple criterion: Choose the next available "nearest" vertex and go for it. Let's call the process of checking among the available vertices and finding the nearest one a single computation. Then, for a TSP with $N = 5$, we need to perform 5 computations. (Actually, the last step—go back to the starting vertex—is automatic, but for the sake of simplicity let's still call it a computation.) What happens when we double

TABLE 6-6

	SUPERHERO
N	computation time
20	2 minutes
21	40 minutes
22	14 hours
23	13 days
24	10 months
25	20 years
26	500 years
27	13,000 years
28	350,000 years
29	9.8 million years
30	284 million years

the number of vertices to $N = 10$? We now have to perform 10 computations. What if we jump to $N = 30$? Now we have to perform roughly 30 computations. Think about that—even if we are conservative and assume a couple of minutes per computation—a human being without a computer can implement the nearest-neighbor algorithm on a TSP with $N = 30$ vertices in an hour or less! Compare that with the last row of Table 6-6.

We can summarize the above observations by saying that the nearest-neighbor algorithm is an *efficient algorithm*. Roughly speaking, an **efficient algorithm** is an algorithm for which the amount of computational effort required to implement the algorithm grows in some reasonable proportion with the size of the input to the problem. (Granted, this is a pretty vague definition, with nettlesome fudge words such as *computational effort* and *reasonable proportion*. Unfortunately, a precise definition requires a more technical background that is beyond the scope of this book.)

The main problem with the nearest-neighbor algorithm is that it is *not an optimal algorithm*. If you go back to Example 6.7, you'll find that the **nearest-neighbor tour** (*nearest-neighbor tour* is short for "the tour obtained using the nearest-neighbor algorithm") had a cost of $773, whereas the *optimal tour* had a cost of $676. In absolute terms the nearest-neighbor tour is off by $773 − $676 = $97 (this is called the *absolute error*). A better way to describe how far "off" this tour is from the optimal tour is to use the concept of *relative error*. In this example the absolute error is $97 out of $676, giving a relative error of $97/$676 = 0.1434911 ... ≈ 14.35%.

In general, for any tour that might be proposed as a "solution" to a TSP, we can find its **relative error** ε as follows:

RELATIVE ERROR OF A TOUR (ε)

$$\varepsilon = \frac{(\text{cost of tour} - \text{cost of optimal tour})}{\text{cost of optimal tour}}$$

It is customary to express the relative error as a percentage (usually rounded to two decimal places). By using the notion of relative error, we can characterize the optimal tours as those with relative error of 0%. All other tours give "approximate solutions," the relative merits of which we can judge by their respective relative errors: tours with "small" relative errors are good, and tours with "large" relative errors are not good. Of course, the interpretation of what is "small" or "large" is subjective and depends on one's tolerance for error and the circumstances surrounding the problem. When people use the expression "it's good enough for government work," there is the unspoken implication that "it" (whatever "it" is) may not be good enough under different circumstances. A similar principle often applies to judging the fitness or lack thereof of an approximate solution to a TSP.

6.6 Approximate Algorithms

A really good algorithm for solving TSPs in general would have to be both *efficient* (like the nearest-neighbor algorithm) and *optimal* (like the brute-force algorithm). In practice this would ensure that we could find the optimal solution to any TSP. Unfortunately, nobody knows of such an algorithm. Moreover, *we don't even know why we don't know*. Is it because such an algorithm is actually a mathematical impossibility? Or is it because no one has yet been clever enough to find one?

Despite the efforts of some of the best mathematicians of our time, the answers to these questions have remained quite elusive. So far, no one has been able to come up with an efficient and optimal algorithm for solving TSPs or, alternatively, to prove that such an algorithm does not exist. Because this question has profound implications in an area of computer science called *complexity theory*, it has become one of the most famous unsolved problems in modern mathematics.

In the meantime, we are faced with a quandary. In many real-world applications, it is necessary to find some sort of "solution" for TSPs involving graphs with hundreds and even thousands of vertices, and to do so in a reasonable time (using computers as helpers, for sure). Since the brute-force algorithm is out of the question, and since no efficient algorithm that guarantees an optimal tour is known, the only practical fallback strategy is to compromise—we give up on the expectation of an optimal tour and accept as a solution a tour that may be less than optimal. In exchange, we ask for quick results. Nowadays, this is the way that most real-life applications of TSPs are "solved."

We will use the term **approximate algorithm** to describe any algorithm that produces solutions that are, most of the time, reasonably close to the optimal solution. Sounds good, but what does "most of the time" mean? And how about "reasonably close"? Unfortunately, to answer these questions properly would take us beyond the scope of this book. We will have to accept that in the area of analyzing algorithms we will be dealing with informal ideas rather than precise definitions. (A good introduction to the analysis of algorithms can be found in references 8 and 10.)

Let's return to the unfinished saga of Willy, the traveling salesman trying to find an optimal tour for his 10-city sales trip (Example 6.8).

(**EXAMPLE 6.9**)   **A Tour of 10 Cities: Part 2**

When we left Willy a few pages back (Example 6.8), we had (1) decided not to try the brute-force algorithm (way too much work for just a $200 reward) and (2) used the nearest-neighbor algorithm to find the tour $A, C, E, D, B, J, G, K, F, H, A$, with a total cost of $2153 (it took about 10 minutes). Is this an optimal tour? If not, how close is it to the optimal tour?

To answer these questions we would have to know the cost of the optimal tour. It turns out that this cost is $1914. (This value was found using a computer program in just a few seconds, but this is deceptive. Somebody had to spend a lot of time writing the program, and someone—me—had to spend a fair amount of time entering the city-to-city costs. Even with a computer, it's a hassle.) We can conclude, with hindsight, that the nearest-neighbor tour is off by a relative error of about $[(2153 - 1914)/1914] \approx 0.1249 = 12.49\%$.   ⊂⊃

The key point of Example 6.9 is that choosing an *approximate but efficient* algorithm over an *optimal but inefficient* algorithm is often a good tradeoff. As they say, "time is money." (This is a big departure from the conventional wisdom that math problems have either "right" or "wrong" answers. When dealing with algorithmic problems we need to consider not only the quality of the answer but also the amount of effort it takes to find it.)

In the next two sections we will discuss two other approximate but efficient algorithms for "solving" TSPs: the *repetitive nearest-neighbor algorithm* and the *cheapest-link algorithm*.

6.7 The Repetitive Nearest-Neighbor Algorithm

As one might guess, the *repetitive nearest-neighbor algorithm* is a variation of the nearest-neighbor algorithm in which we repeat several times the entire nearest-neighbor circuit-building process. Why would we want to do this? The reason is that the nearest-neighbor tour depends on the choice of the starting vertex. If we change the starting vertex, the nearest-neighbor tour is likely to be different, and, if we are lucky, better. Since finding a nearest-neighbor tour is an efficient process, it is not unreasonable to repeat the process several times, each time starting at a different vertex of the graph. In this way, we can obtain several different "nearest-neighbor solutions," from which we can then pick the best.

But what do we do with a tour that starts somewhere other than the vertex we really want to start at? That's not a problem. Remember that once we have a circuit, we can start the circuit anywhere we want. In fact, in an abstract sense, a circuit has no starting or ending point.

To illustrate how the repetitive nearest-neighbor algorithm works, let's return to the original five-city problem we last discussed in Example 6.5.

> **EXAMPLE 6.10**   **A Tour of Five Cities: Part 3**

Once again, we are going to look at Willy's original five-city TSP introduced in Example 6-4 [Fig. 6-14(a)] but this time use the repetitive nearest-neighbor algorithm to find a tour.

In Example 6.7 we computed the nearest-neighbor tour with A as the starting vertex, and we got A, C, E, D, B, A with a total cost of $773 [Fig. 6-14(b)]. But if we use B as the starting vertex, the nearest-neighbor tour takes us from B to C, then to A, E, D, and back to B, with a total cost of $722, as shown in Fig. 6-14(c). Well, that is certainly an improvement! As for Willy—who must start and end his trip at A—this very same tour would take the form A, E, D, B, C, A.

The process is once again repeated using C, D, and E as the starting vertices. When the starting point is C, we get the nearest-neighbor tour C, A, E, D, B, C (total cost is $722); when the starting point is D, we get the nearest-neighbor tour D, B, C, A, E, D (total cost is $722); and when the starting point is E, we get the nearest-neighbor tour E, C, A, D, B, E (total cost is $741). Among all the nearest-neighbor tours we pick the cheapest one, shown in Fig. 6-14(c).

■ To verify these claims, try Exercise 31.

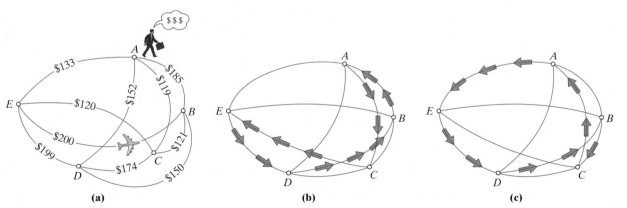

FIGURE 6-14

(a)   (b)   (c)

A formal description of the **repetitive nearest-neighbor algorithm** is as follows.

ALGORITHM 3: THE REPETITIVE NEAREST-NEIGHBOR ALGORITHM

- Let X be any vertex. Find the nearest-neighbor tour with X as the starting vertex, and calculate the cost of this tour.
- Repeat the process with each of the other vertices of the graph as the starting vertex.
- Of the nearest-neighbor tours thus obtained, choose one with least cost. If necessary, rewrite the tour so that it starts at the designated starting vertex. We will call this tour the **repetitive nearest-neighbor tour**.

6.8 The Cheapest-Link Algorithm

The last—but not least—of the algorithms we will consider is known as the **cheapest-link algorithm**. The idea behind the cheapest-link algorithm is to piece together a tour by picking the separate "links" (i.e., legs) of the tour on the basis of cost. It doesn't matter if in the intermediate stages the "links" are all over the place—if we are careful at the end, they will all come together and form a tour.

This is how it works: Look at the entire graph and choose the cheapest edge of the graph, wherever that edge may be. Once this is done, choose the next cheapest edge of the graph, wherever that edge may be. (Don't worry if that edge is not adjacent to the first edge.) Continue this way, each time choosing the cheapest edge available but following two rules: (1) Do not allow a circuit to form except at the very end, and (2) do not allow three edges to come together at a vertex. A violation of either of these two rules will prevent forming a Hamilton circuit. Conversely, following these two rules guarantees that the end result will be a Hamilton circuit.

(**EXAMPLE 6.11**)   **A Tour of Five Cities: Part 4**

What better example to illustrate the cheapest-link algorithm than Willy's five-city TSP? Each step of the algorithm is illustrated in Fig. 6-15.

Among all the edges of the graph, the "cheapest link" is edge AC, with a cost of $119. For the purposes of recordkeeping, we will tag this edge as taken by marking it in red as shown in Fig. 6-15(a). (Nothing magical about red—you can choose any color you want.) The next step is to scan the entire graph again and choose the cheapest link available, in this case edge CE with a cost of $120. Again, we mark it in red, to indicate it is taken [Fig. 6-15(b)]. The next cheapest link available is edge BC ($121), but we should not choose BC—we would have three edges coming out of vertex C and this would prevent us from forming a circuit. This time we put a "do not use" mark on edge BC [Fig. 6-15(c)]. The next cheapest link available is AE ($133). But we can't take AE either—the vertices A, C, and E would form a small circuit, making it impossible to form a Hamilton circuit at the end. So we put a "do not use" mark on AE [Fig. 6-15(d)]. The next cheapest link available is BD ($150). Choosing BD would not violate either of the two rules, so we can add it to our budding circuit and mark it in red [Fig. 6-15(e)]. The next cheapest link available is AD ($152), and it works just fine [Fig. 6-15(f)]. At this point, we have only one way to close up the Hamilton circuit, edge BE, as shown in Fig. 6-15(g).

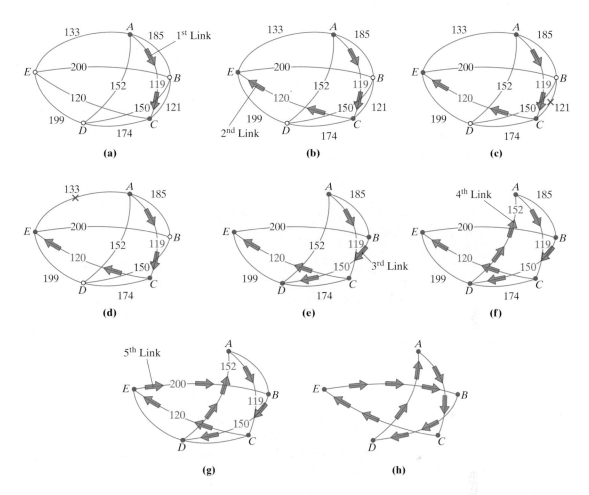

FIGURE 6-15

The cheapest-link tour shown in Fig 6-15(h) could have any starting vertex we want. Since Willy lives at A, we describe it as A, C, E, B, D, A. The total cost of this tour is \$741, which is a little better than the nearest-neighbor tour (see Example 6.7) but not as good as the repetitive nearest-neighbor tour (see Example 6.10).

A formal description of the cheapest-link algorithm is as follows.

ALGORITHM 4: THE CHEAPEST-LINK ALGORITHM

- **Step 1.** Pick the *cheapest link* (i.e., edge with smallest weight) available. (In case of a tie, pick one at random.) Mark it (say in red).
- **Step 2.** Pick the next cheapest link available and mark it.
- **Steps 3, 4, ..., $N - 1$.** Continue picking and marking the cheapest unmarked link available that does not
  (a) close a circuit or
  (b) create three edges coming out of a single vertex
- **Step N.** Connect the last two vertices to close the red circuit. This circuit gives us the **cheapest-link tour**.

For the last example of this chapter, we will return to the problem first described in the chapter opener and Example 6.6—that of finding an optimal tour for a rover exploring Mars.

(**EXAMPLE 6.12**)   **Roving the Red Planet: Part 2**

Figure 6-16(a) shows seven sites on Mars identified as particularly interesting sites for geological exploration. Our job is to find an optimal tour for a rover that will land at A, visit all the sites to collect rock samples, and at the end return to A. The distances between sites (in miles) are given in the graph shown in Fig. 6-16(b).

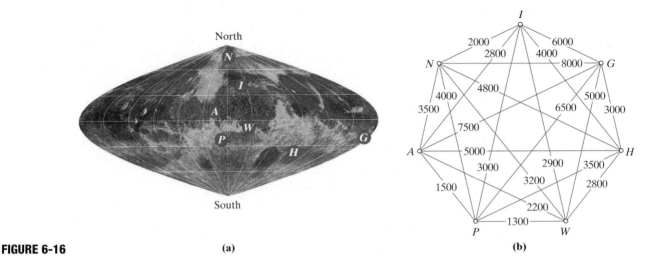

FIGURE 6-16   (a)   (b)

Let's look at some of the approaches one might use to tackle this problem.

Brute force: The brute-force algorithm would require us to check and compute $6! = 720$ different tours. We will pass on that idea for now.

Cheapest link: The cheapest-link algorithm is a reasonable algorithm to use— not trivial but not too hard either. A summary of the steps is shown in Table 6-7.

TABLE 6-7			
Step	Cheapest edge available	Weight	Add to circuit?
1	PW	1300	Yes
2	AP	1500	Yes
3	IN	2000	Yes
4	AW	2200	No ⟨△⟩
5 ⎫ 6 ⎭	HW ⎫ AI ⎭ tie	2800 2800	Yes Yes
7	IW	2900	No ⟨◇ & ⋎⟩
8 ⎫ 9 ⎭	IP ⎫ GH ⎭ tie	3000 3000	No ⟨△ & ⋎⟩ Yes
Last	GN only way to close circuit	8000	Yes

The cheapest-link tour $(A, P, W, H, G, N, I, A$ with a total length of 21,400 miles) is shown in Fig. 6-17.

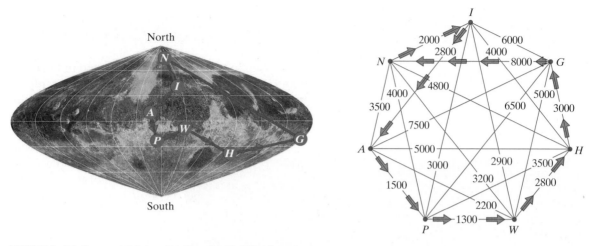

FIGURE 6-17 Cheapest-link tour. Total length: 21,400 miles.

Nearest neighbor: The nearest-neighbor algorithm is the simplest of all the algorithms we learned. Starting from A we go to P, then to W, then to H, then to G, then to I, then to N, and finally back to A. The nearest-neighbor tour (A, P, W, H, G, I, N, A with a total length of 20,100 miles) is shown in Fig. 6-18. (We know that we can repeat this method using different starting points, but we won't bother with that at this time.)

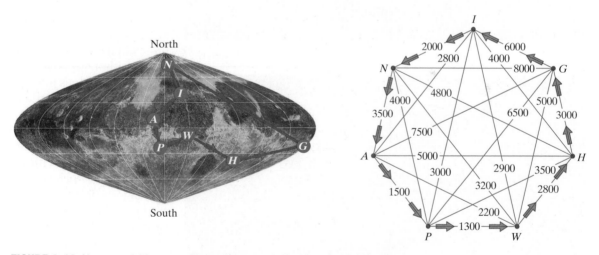

FIGURE 6-18 Nearest-neighbor tour with starting vertex A. Total length: 20,100 miles.

The first surprise in Example 6.12 is that the nearest-neighbor algorithm gave us a better tour than the cheapest-link algorithm. Sometimes the cheapest-link algorithm produces a better tour than the nearest-neighbor algorithm, but just as often, it's the other way around. The two algorithms are different but of equal standing—neither one can be said to be superior to the other one in terms of the quality of the tours it produces.

The second surprise is that the nearest-neighbor tour A, P, W, H, G, I, N, A turns out to be an *optimal tour*. (This can be verified using a computer and the brute-force algorithm.) Essentially, this means that in this particular example, the simplest of all methods happens to produce the optimal answer—a nice turn of events. Too bad we can't count on this happening on a more consistent basis!

Well, maybe next chapter.

CONCLUSION

The basic problem discussed in this chapter can be stated in fairly simple terms: How does one find an *optimal tour* in a *complete weighted graph*? (For historical reasons, these types of problems are known as *traveling salesman problems*, or TSPs, even though most of the real-life applications of the problem have nothing to do with traveling salesmen.) Once the meaning of the technical terms *optimal tour* and *complete weighted graph* is clear, the problem does not sound all that difficult. Thus, that this happens to be one of the most difficult problems in modern mathematics is quite remarkable.

How is this problem so difficult? After all, in this chapter we looked at several examples of TSPs and in most cases were able to find the optimal tour without too much trouble. But we haven't actually "solved" a general class of problems unless we have a method that produces a solution for any instance of the problem. In addition, when the method of solution is an algorithm, we have to worry about the effort that goes into finding a solution. A general algorithm for solving TSPs that (1) always finds an optimal tour and (2) does it efficiently, has so far eluded the efforts of some of the world's best mathematicians.

The *nearest-neighbor* and *cheapest-link* algorithms are two fairly simple strategies for attacking TSPs. We have seen that both algorithms are *approximate algorithms*. This means that they are not likely to give us an optimal solution, although, as we saw in Example 6.12, on a lucky day, even that is possible. In most typical problems, however, we can expect either of these algorithms to give an approximate solution that is within a reasonable margin of error. With some problems the cheapest-link algorithm gives a better solution than the nearest-neighbor algorithm; with other problems it's the other way around. Thus, while they are not the same, neither is superior to the other.

(a)

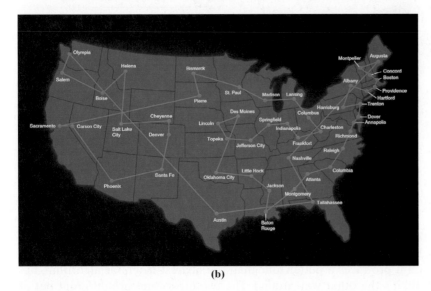

(b)

FIGURE 6-19 What is the shortest tour that visits all 48 state capitals in the contiguous United States? The *optimal* solution shown in (a) has total length of approximately 12,000 miles. Even with a fast computer and special software, it could take hundreds of computer hours to find this solution using an *inefficient algorithm*. For the same 48-city TSP, *approximate solutions* can be found quickly using *efficient algorithms*. The tour shown in (b) was found in a matter of seconds using the nearest-neighbor algorithm (starting at Olympia, Washington). The total length of the tour is 14,500 miles (a relative error of about 20%).

Solving a TSP involving a large number of cities requires a combination of smart strategy and raw computational power—it is the perfect marriage of mathematics and computer science. (See Fig. 6-19.) Nowadays, research on TSPs—typically done by teams that include mathematicians and computer scientists—is carried out on several fronts. One major area of research interest is the quest for finding *optimal solutions* to larger and larger TSPs. (As of the writing of this edition the record solution is an optimal 89,500-city tour found in 2006 by a team of researchers from AT&T Labs. For details, visit "The Traveling Salesman Problem Homepage" at *www.tsp.gatech.edu*.)

A second area that is attracting tremendous interest is that of finding improved and more efficient *approximate algorithms* for "solving" very large TSPs. Here progress is much faster, and many sophisticated approximate algorithms have been developed, including algorithms based on *recombinant DNA* techniques (see Project C and reference 1) as well as algorithms based on the collective intelligence of *ant colonies* (yes, that's ants, as in insects—see Project D and reference 5). Today, the best approximate algorithms can tackle TSPs with hundreds of thousands of vertices and produce solutions with relative errors *guaranteed* to be less than 1%.

On a purely theoretical level, the holy grail of TSP research still is the search for an *optimal and efficient general algorithm* that can solve all TSPs, or, alternatively, prove that such an algorithm does not exist. This conundrum is waiting for another Euler to come along.

PROFILE: Sir William Rowan Hamilton (1805–1865)

Hamilton is one of the greatest and yet most tragic figures in the history of mathematics. A prodigy in languages, science, and mathematics at a very young age, he suffered from bouts of depression and fought the demon of alcohol for much of his adult life. In spite of many important mathematical discoveries, the story of his life is one of what might have been had the circumstances of his personal life turned out differently.

William Rowan Hamilton was born in Dublin, Ireland, exactly at midnight on August 3, 1805 (causing some confusion as to whether his birthday was August 3 or August 4). From a very young age he was taught by his uncle the Reverend James Hamilton, an eccentric clergyman and brilliant linguist. Under his uncle's tutelage, Hamilton learned Greek, Latin, and Hebrew by the age of 6, and by the time he was 13 he knew a dozen languages, including Persian, Arabic, and Sanskrit. At the age of 10 Hamilton became interested in mathematics, and by the age of 15 he was reading the works of Newton and Laplace. When he was just 17, Hamilton wrote his first two mathematical research papers, *On Contacts be-*

tween Algebraic Curves and Surfaces and *Developments*. After reading the papers, John Brinkley, the Astronomer Royal of Ireland at that time, remarked, "This young man, I do not say will be, but is, the first [foremost] mathematician of his age."

In 1823 Hamilton entered Trinity College in Dublin, where he excelled in all subjects while continuing to publish groundbreaking research in mathematical physics. A year after entering college, Hamilton met and fell madly in love with Catherine Disney, a young lady from an upper-class family. Although Catherine might have reciprocated his affections, there was a problem. Hamilton was just a 19-year-old university student and did not come from a particularly wealthy family—according to the social norms of the day he was not a suitable match for Catherine. On the other hand, Hamilton was one of those mathematical geniuses who come around once in a generation. Not lacking for confidence, he was banking on his status and fame as a scientist to bring Catherine's parents around to eventually accepting him.

The following year Hamilton found out that Catherine's family had arranged for her marriage to another man. This was devastating news to him and he even considered suicide. Ironically, soon after Catherine was married, Hamilton's social status and financial position took a quantum leap forward. In 1827, not quite yet 22 years old and still an undergraduate, Hamilton won the appointment to the prestigious position of Astronomer Royal of Ireland as well as to a Professorship of Astronomy at Trinity College, an incredible personal triumph. Unfortunately, for him and Catherine it all happened a couple of years too late. By now Catherine was married and had a child.

Over the next 20 years, Hamilton's life was a checkered mixture of professional successes and personal setbacks. On the rebound, he soon entered into an ill-advised and unhappy marriage that bore him three children. He fought intermittent bouts of depression and started having serious problems with alcohol. At the same time, he pursued his research in mathematics and optics, where he produced many significant discoveries, including the formulas for conic refraction, which gained him much fame. In recognition of his many scientific accomplishments, Hamilton was knighted in 1835.

In 1843 Hamilton produced the mathematical discovery he is best known for, the *calculus of quaternions*. (Quaternions are essentially four-dimensional numbers that generalize the complex numbers and are ideally suited to describe mathematically many complex phenomena in physics.) Hamilton predicted that quaternions would revolutionize nineteenth-century physics, but he was a century ahead of his time—quaternions revolutionized *twentieth*-century physics, as they are a key element in quantum mechanics and Einstein's theory of relativity.

Hamilton's connection with graph theory and *Hamilton circuits* came late in his life and in a rather roundabout way. In 1857, Hamilton invented a board game he called the *Icosian game*, the purpose of which was to create a trip passing through each of 20 European cities once and only once. The cities were connected in the form of a *dodecahedral graph* (see Exercise 59). Hamilton sold the rights to the game to a London game dealer for 25 pounds, and the game was actually marketed throughout Europe, a precursor to some of the "connect-the-dots" games that we know today. ■

KEY CONCEPTS

EXERCISES

WALKING

A Hamilton Circuits and Hamilton Paths

1. For the graph shown in Fig. 6-20,

   (a) find three different Hamilton circuits.

   (b) find a Hamilton path that starts at A and ends at B.

   (c) find a Hamilton path that starts at D and ends at F.

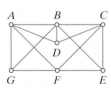

FIGURE 6-20

2. For the graph shown in Fig. 6-21,

(a) find three different Hamilton circuits.

(b) find a Hamilton path that starts at *A* and ends at *B*.

(c) find a Hamilton path that starts at *F* and ends at *I*.

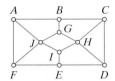

FIGURE 6-21

3. Find all possible Hamilton circuits in the given graph. (Use *A* as your reference point.)

(a) Figure 6-22(a)

(b) Figure 6-22(b)

(c) Figure 6-22(c)

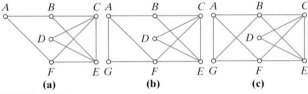

(a)   **(b)**   **(c)**

FIGURE 6-22

4. Find all possible Hamilton circuits in the given graph. (Use *A* as your reference point.)

(a) Figure 6-23(a)

(b) Figure 6-23(b)

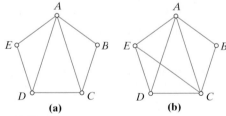

(a)   **(b)**

FIGURE 6-23

5. For the graph shown in Fig. 6-24,

(a) find a Hamilton path that starts at *A* and ends at *E*.

(b) find a Hamilton circuit that starts at *A* and ends with the edge *EA*.

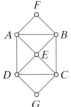

FIGURE 6-24

(c) find a Hamilton path that starts at *A* and ends at *C*.

(d) find a Hamilton path that starts at *F* and ends at *G*.

6. For the graph shown in Fig. 6-25,

(a) find a Hamilton path that starts at *A* and ends at *E*.

(b) find a Hamilton circuit that starts at *A* and ends with the edge *EA*.

(c) find a Hamilton path that starts at *A* and ends at *G*.

(d) find a Hamilton path that starts at *F* and ends at *G*.

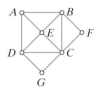

FIGURE 6-25

7. For the graph shown in Fig. 6-26,

(a) list all Hamilton circuits using *A* as the reference point.

(b) list all Hamilton circuits using *D* as the reference point.

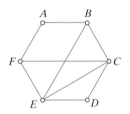

FIGURE 6-26

8. For the graph shown in Fig. 6-27,

(a) list all Hamilton circuits using *A* as the reference point.

(b) list all Hamilton circuits using *D* as the reference point.

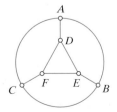

FIGURE 6-27

9. Explain why the graph shown in Fig. 6-28 has neither Hamilton circuits nor Hamilton paths.

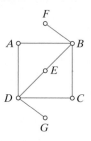

FIGURE 6-28

10. Explain why the graph shown in Fig. 6-29 has no Hamilton circuit but does have a Hamilton path.

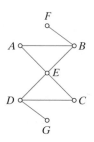

FIGURE 6-29

11. For the graph shown in Fig. 6-30,

    (a) find a Hamilton path that starts at *A* and ends at *D*.

    (b) find a Hamilton path that starts at *G* and ends at *H*.

    (c) explain why the graph has no Hamilton path that starts at *B*.

    (d) explain why the graph has no Hamilton circuit.

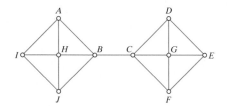

FIGURE 6-30

12. For the graph shown in Fig. 6-31,

    (a) find a Hamilton path that starts at *D*.

    (b) find a Hamilton path that starts at *B*.

(c) explain why the graph has no Hamilton path that starts at *A* or *C*.

(d) explain why the graph has no Hamilton circuit.

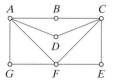

FIGURE 6-31

13. For the weighted graph shown in Fig. 6-32,

    (a) find the weight of edge *BD*.

    (b) find a Hamilton circuit that starts with edge *BD*, and give its weight.

    (c) find a Hamilton circuit that ends with edge *DB*, and give its weight.

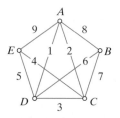

FIGURE 6-32

14. For the weighted graph shown in Fig. 6-33,

    (a) find the weight of edge *AD*.

    (b) find a Hamilton circuit that starts with edge *AD*, and give its weight.

    (c) find a Hamilton circuit that ends with edge *DA*, and give its weight.

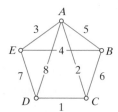

FIGURE 6-33

15. For the weighted graph shown in Fig. 6-34,

    (a) find a Hamilton path that starts at *A* and ends at *C*, and give its weight.

    (b) find a second Hamilton path that starts at *A* and ends at *C*, and give its weight.

(c) find the optimal (least weight) Hamilton path that starts at A and ends at C, and give its weight.

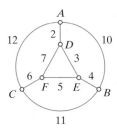

FIGURE 6-34

16. For the weighted graph shown in Fig. 6-35,

(a) find a Hamilton path that starts at B and ends at D, and give its weight.

(b) find a second Hamilton path that starts at B and ends at D, and give its weight.

(c) find the optimal (least weight) Hamilton path that starts at B and ends at D, and give its weight.

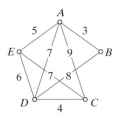

FIGURE 6-35

B Factorials and Complete Graphs

Exercises 17 through 20 are a repeat of Exercises 37 through 40 in Chapter 2. Please answer these exercises without using a calculator—the point of the exercises is for you to practice how to manipulate factorials numerically.

17. (a) Given that $10! = 3,628,800$, find $9!$.

(b) Find $11!/10!$.

(c) Find $11!/9!$.

(d) Find $9!/6!$.

(e) Find $101!/99!$.

18. (a) Given that $20! = 2,432,902,008,176,640,000$, find $19!$.

(b) Find $20!/19!$.

(c) Find $201!/199!$.

19. Find the value of each of the following. Give the answer in decimal form.

(a) $(9! + 11!)/10!$

(b) $(101! + 99!)/100!$

20. Find the value of each of the following. Give the answer in decimal form.

(a) $(19! + 21!)/20!$ (*Hint*: $1/20 = 0.05$.)

(b) $(201! + 199!)/200!$

For Exercises 21 through 24, you will need a calculator with a factorial key. (Any scientific or graphing calculator should do.)

21. Using a calculator with a factorial key,

(a) compute $20!$.

(b) compute $22!$.

(c) compute the number of Hamilton circuits in K_{21}.

22. Using a calculator with a factorial key,

(a) compute $25!$.

(b) compute $27!$.

(c) compute the number of Hamilton circuits in K_{26}.

23. Suppose you have a supercomputer that can generate one *billion* Hamilton circuits per second.

(a) Estimate (in years) how long it would take the supercomputer to generate all the Hamilton circuits in K_{21}.

(b) Estimate (in years) how long it would take the supercomputer to generate all the Hamilton circuits in K_{22}.

24. Suppose you have a supercomputer that can generate one *trillion* Hamilton circuits per second.

(a) Estimate (in years) how long it would take the supercomputer to generate all the Hamilton circuits in K_{26}.

(b) Estimate (in years) how long it would take the supercomputer to generate all the Hamilton circuits in K_{27}.

25. (a) How many edges are there in K_{20}?

(b) How many edges are there in K_{21}?

(c) If the number of edges in K_{50} is x, and the number of edges in K_{51} is y, what is the value of $y - x$?

26. (a) How many edges are there in K_{200}?

(b) How many edges are there in K_{201}?

(c) If the number of edges in K_{500} is x, and the number of edges in K_{501} is y, what is the value of $y - x$?

27. In each case, find the value of N.

(a) K_N has 120 distinct Hamilton circuits.

(b) K_N has 45 edges.

(c) K_N has 20,100 edges.

28. In each case, find the value of N.

   (a) K_N has 720 distinct Hamilton circuits.

   (b) K_N has 66 edges.

   (c) K_N has 80,200 edges.

C Brute-Force and Nearest-Neighbor Algorithms

29. For the weighted graph shown in Fig. 6-36, (i) find the indicated tour, and (ii) give its cost.

   (a) An optimal tour (use the brute-force algorithm)

   (b) The nearest-neighbor tour with starting vertex A

   (c) The nearest-neighbor tour with starting vertex B

   (d) The nearest-neighbor tour with starting vertex C

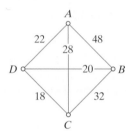

FIGURE 6-36

30. For the weighted graph shown in Fig. 6-37, (i) find the indicated tour, and (ii) give its cost.

   (a) An optimal tour (use the brute-force algorithm)

   (b) The nearest-neighbor tour with starting vertex A

   (c) The nearest-neighbor tour with starting vertex C

   (d) The nearest-neighbor tour with starting vertex D

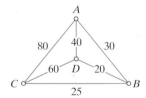

FIGURE 6-37

31. For the weighted graph shown in Fig. 6-38, (i) find the indicated tour, and (ii) give its cost. (*Note*: This is the graph discussed in Example 6.7.)

   (a) The nearest-neighbor tour with starting vertex B

   (b) The nearest-neighbor tour with starting vertex C

   (c) The nearest-neighbor tour with starting vertex D

   (d) The nearest-neighbor tour with starting vertex E

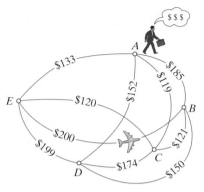

FIGURE 6-38

32. A delivery service must deliver packages at Buckman (B), Chatfield (C), Dayton (D), and Evansville (E), and then return to Arlington (A), the home base. Figure 6-39 shows a graph of the estimated travel times (in minutes) between the cities.

   (a) Find the nearest-neighbor tour with starting vertex A. What is the total travel time of this trip?

   (b) Find the nearest-neighbor tour with starting vertex D. Write the tour as it would be traveled if starting and ending at A.

   (c) Suppose that the delivery truck's last stop before returning to A has to be at D. Find the optimal tour that satisfies this requirement. What is the total travel time of this tour?

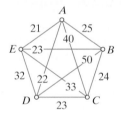

FIGURE 6-39

33. The Brute Force Bandits is a rock band planning a five-city concert tour. The cities and the distances (in miles) between them are given in the weighted graph shown in Fig. 6-40. The tour must start and end at A. The cost of the chartered bus the band is traveling in is $8 per mile.

   (a) Find the nearest-neighbor tour with starting vertex A. What is the bus cost of this tour?

   (b) Find the nearest-neighbor tour with starting vertex B. Write the tour as it would be traveled by the band, starting and ending the trip at A. What is the bus cost of this tour?

(c) Suppose that the band decides to make *B* the first stop after *A*. Find the optimal tour that starts with the pair of cities *A*, *B*. What is the bus cost of this tour?

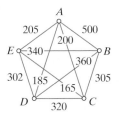

FIGURE 6-40

34. A space expedition is scheduled to visit the moons Callisto (*C*), Ganymede (*G*), Io (*I*), Mimas (*M*), and Titan (*T*) to collect rock samples at each and then return to Earth (*E*). The travel times (in years) are given in the weighted graph shown in Fig. 6-41. (*Note:* This is the graph discussed in Example 6.5.)

(a) Find the nearest-neighbor tour with starting vertex *E*. What is the total travel time of this tour?

(b) Find the nearest-neighbor tour with starting vertex *T*. Write the tour as it would be traveled by an expedition starting and ending at *E*.

(c) Suppose that scientific considerations require that the *last two* moons visited before the expedition returns to *E* should be *T* and *C* (in either order). Find the optimal tour that satisfies this requirement. What is the total travel time of this tour?

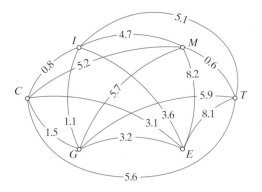

FIGURE 6-41

In Exercises 35 and 36, you are given the distance between the cities in a mileage chart, rather than in a weighted graph. The purpose of the exercises is for you to try to implement the nearest-neighbor algorithm using the mileage chart and without drawing a weighted graph. (Nobody can stop you from drawing the graph, but you would be doing a lot more work than is really necessary.)

35. Darren is a sales rep whose territory consists of the six cities in the mileage chart shown in Fig. 6-42. Darren wants to visit customers at each of the cities, starting and ending his trip in his home city of Atlanta. His travel costs (gas, insurance, etc.) average $0.75 per mile.

(a) Find the nearest-neighbor tour with Atlanta as the starting city. What is the total cost of this tour?

(b) Find the nearest-neighbor tour using Kansas City as the starting city. Write the tour as it would be traveled by Darren, who must start and end the trip in Atlanta. What is the total cost of this tour?

Mileage Chart

	Atlanta	Columbus	Kansas City	Minneapolis	Pierre	Tulsa
Atlanta	*	533	798	1068	1361	772
Columbus	533	*	656	713	1071	802
Kansas City	798	656	*	447	592	248
Minneapolis	1068	713	447	*	394	695
Pierre	1361	1071	592	394	*	760
Tulsa	772	802	248	695	760	*

FIGURE 6-42

36. The Platonic Cowboys is a country western band based in Nashville. The Cowboys are planning a concert tour to the seven cities in the mileage chart shown in Fig. 6-43.

(a) Find the nearest-neighbor tour with Nashville as the starting city. What is the total length of this tour?

(b) Find the nearest-neighbor tour using St. Louis as the starting city. Write the tour as it would be traveled by the band, which must start and end the tour in Nashville. What is the total length of this tour?

Mileage Chart

	Boston	Dallas	Houston	Louisville	Nashville	Pittsburgh	St. Louis
Boston	*	1748	1804	941	1088	561	1141
Dallas	1748	*	243	819	660	1204	630
Houston	1804	243	*	928	769	1313	779
Louisville	941	819	928	*	168	388	263
Nashville	1088	660	769	168	*	553	299
Pittsburgh	561	1204	1313	388	553	*	588
St. Louis	1141	630	779	263	299	588	*

FIGURE 6-43

D Repetitive Nearest-Neighbor Algorithm

37. For the weighted graph shown in Fig. 6-44, find the repetitive nearest-neighbor tour. Write the tour using B as the starting vertex.

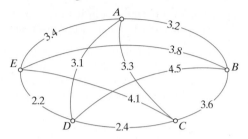

FIGURE 6-44

38. For the weighted graph shown in Fig. 6-45, find the repetitive nearest-neighbor tour. Write the tour using D as the starting vertex.

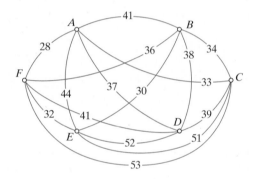

FIGURE 6-45

39. This exercise is a continuation of the Brute Force Bandits concert tour (Exercise 33). Find the repetitive nearest-neighbor tour, and give the bus cost for this tour.

40. This exercise is a continuation of the space expedition problem (Exercise 34). Find the repetitive nearest-neighbor tour, and give the total travel time for this tour.

41. This exercise is a continuation of Darren's sales trip problem (Exercise 35). Find the repetitive nearest-neighbor tour, and give the total cost for this tour.

42. This exercise is a continuation of The Platonic Cowboys concert tour (Exercise 36). Find the repetitive nearest-neighbor tour, and give the total mileage for this tour.

E Cheapest-Link Algorithm

43. For the weighted graph shown in Fig. 6-46, find the cheapest-link tour. Write the tour using B as the starting vertex. (*Note*: This is the graph in Exercise 37.)

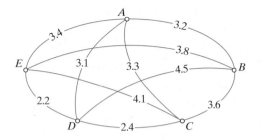

FIGURE 6-46

44. For the weighted graph shown in Fig. 6-47, find the cheapest-link tour, and give the total mileage for this tour. Write the tour using B as the starting vertex. (*Note*: This is the graph in Exercise 38.)

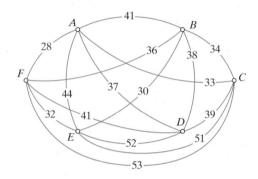

FIGURE 6-47

45. For the Brute Force Bandits concert tour discussed in Exercise 33, find the cheapest-link tour, and give the bus cost for this tour.

46. For the space expedition problem discussed in Exercise 34, find the cheapest-link tour, and give the total travel time for this tour.

47. For Darren's sales trip problem discussed in Exercise 35, find the cheapest-link tour, and give the total cost for this tour.

48. For the Platonic Cowboys concert tour discussed in Exercise 36, find the cheapest-link tour, and give the total mileage for this tour.

F Miscellaneous

49. Suppose that in solving a TSP you use the nearest-neighbor algorithm and find a nearest-neighbor tour with a total cost of $13,500. Suppose that you later find out that the cost of an optimal tour is $12,000. What was the relative error of your nearest-neighbor tour? Express your answer as a percentage, rounded to the nearest tenth of a percent.

50. Suppose that in solving a TSP you use the cheapest-link algorithm and find a cheapest-link tour with a total length of 21,400 miles. Suppose that you later find out that the length of an optimal tour is 20,100 miles. What was the relative error of your cheapest-link tour? Express your answer as a percentage, rounded to the nearest tenth of a percent.

51. Suppose that in solving a TSP you find an approximate solution with a cost of $1614, and suppose that you later find out that the relative error of your solution was 7.6%. What was the cost of the optimal solution?

52. Suppose that in solving a TSP you find an approximate solution with a cost of $2508, and suppose that you later find out that the relative error of your solution was 4.5%. What was the cost of the optimal solution?

53. You have a busy day ahead of you. You must run the following errands (in no particular order): Go to the post office, deposit a check at the bank, pick up some French bread at the deli, visit a friend at the hospital, and get a haircut at Karl's Beauty Salon. You must start and end at home. Each block on the map shown in Fig. 6-48 is exactly 1 mile.

(a) Draw a weighted graph corresponding to this problem.

(b) Find an optimal tour for running all the errands. (Use any algorithm you think is appropriate.)

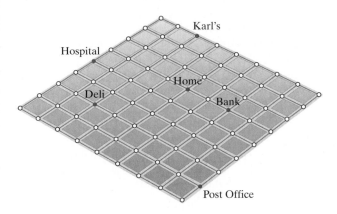

FIGURE 6-48

54. Rosa's Floral must deliver flowers to each of the five locations A, B, C, D, and E labeled on the map shown in Fig. 6-49. The trip must start and end at the flower shop, located at F. Each block on the map is exactly 1 mile.

(a) Draw a weighted graph corresponding to this problem.

(b) Find an optimal tour for making all the deliveries. (Use any algorithm you think is appropriate.)

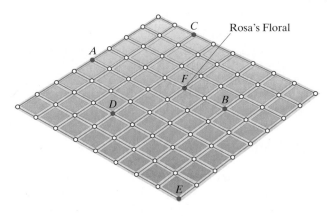

FIGURE 6-49

In Exercises 55 through 58, you are scheduling a dinner party for six people ($A, B, C, D, E,$ and F). The guests are to be seated around a circular table, and you want to arrange the seating so that each guest is seated between two friends (i.e., the guests to the left and to the right are friends of the guest in between). You can assume that all friendships are mutual (when X is a friend of Y, Y is also a friend of X).

55. Suppose that you are told that all possible friendships can be deduced from the following information:

A is friends with B and F; B is friends with $A, C,$ and E; C is friends with $B, D, E,$ and F; E is friends with $B, C, D,$ and F.

(a) Draw a "friendship graph" for the dinner guests.

(b) Find a possible seating arrangement for the party.

(c) Is there a possible seating arrangement in which B and E are seated next to each other? If there is, find it. If there isn't, explain why not.

56. Suppose that you are told that all possible friendships can be deduced from the following information:

A is friends with $B, C,$ and D; B is friends with $A, C,$ and E; D is friends with $A, E,$ and F; F is friends with $C, D,$ and E.

(a) Draw a "friendship graph" for the dinner guests.

(b) Find a possible seating arrangement for the party.

(c) Is there a possible seating arrangement in which B and E are seated next to each other? If there is, find it. If there isn't, explain why not.

57. Suppose that you are told that all possible friendships can be deduced from the following information:

A is friends with $C, D, E,$ and F; B is friends with $C, D,$ and E; C is friends with $A, B,$ and E; D is friends with $A, B,$ and E.

Explain why it is impossible to have a seating arrangement in which each guest is seated between friends.

58. Suppose that you are told that all possible friendships can be deduced from the following information:

A is friends with *B*, *D*, and *F*; *C* is friends with *B*, *D*, and *F*; *E* is friends with *C* and *F*.

Explain why it is impossible to have a seating arrangement in which each guest is seated between friends.

JOGGING

59. The graph shown in Fig. 6-50 is called the *dodecahedral graph*, because it describes the relationship between the vertices and edges of a *dodecahedron*, a regular three-dimensional solid consisting of 12 faces, all of which are regular pentagons. Find a Hamilton circuit in the graph. Indicate your answer by labeling the vertices in the order of travel.

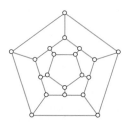

FIGURE 6-50

60. The graph shown in Fig. 6-51 is called the *icosahedral graph*, because it describes the relationship between the vertices and edges of an *icosahedron*, a regular three-dimensional solid consisting of twenty faces all of which are equilateral triangles. Find a Hamilton circuit in the graph. Show your answer by labeling the vertices in the order of travel.

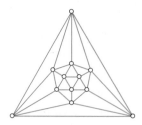

FIGURE 6-51

61. (a) If the number of edges in K_{500} is x and the number of edges in K_{502} is y, what is the value of $y - x$? (*Hint*: Try Exercise 26 first.)

(b) If the number of edges in K_{500} is x and the number of edges in K_{503} is y, what is the value of $y - x$?

62. A 2 by 2 grid graph. The graph shown in Fig. 6-52 represents a street grid that is 2 blocks by 2 blocks. (Such a graph is called a *2 by 2 grid graph*.) For convenience, the

vertices are labeled by type: corner vertices C_1, C_2, C_3, and C_4, boundary vertices B_1, B_2, B_3, and B_4, and the interior vertex I.

(a) Find a Hamilton path in the graph that starts at I.

(b) Find a Hamilton path in the graph that starts at one of the corner vertices and ends at a different corner vertex.

(c) Find a Hamilton path that starts at one of the corner vertices and ends at I.

(d) Find (if you can) a Hamilton path that starts at one of the corner vertices and ends at one of the boundary vertices. If this is impossible, explain why.

FIGURE 6-52

63. Find (if you can) a Hamilton circuit in the 2 by 2 grid graph discussed in Exercise 62. If this is impossible, explain why.

64. A 3 by 3 grid graph. The graph shown in Fig. 6-53 represents a street grid that is 3 blocks by 3 blocks. The graph has four corner vertices (C_1, C_2, C_3, and C_4), eight boundary vertices (B_1 through B_8), and four interior vertices (I_1, I_2, I_3, and I_4).

(a) Find a Hamilton circuit in the graph.

(b) Find a Hamilton path in the graph that starts at one of the corner vertices and ends at a different corner vertex.

(c) Find (if you can) a Hamilton path that starts at one of the corner vertices and ends at one of the interior vertices. If this is impossible, explain why.

(d) Given any two adjacent vertices of the graph, explain why there always is a Hamilton path that starts at one and ends at the other one.

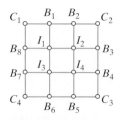

FIGURE 6-53

65. **A 3 by 4 grid graph.** The graph shown in Fig. 6-54 represents a street grid that is 3 blocks by 4 blocks.

   (a) Find a Hamilton circuit in the graph.

   (b) Find a Hamilton path in the graph that starts at C_1 and ends at C_3.

   (c) Find (if you can) a Hamilton path in the graph that starts at C_1 and ends at C_2. If this is impossible, explain why.

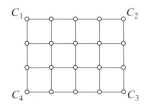

FIGURE 6-54

66. Explain why the cheapest edge in any graph is always part of the Hamilton circuit obtained using the nearest-neighbor algorithm.

67. Give an example of a six-city TSP such that a nearest-neighbor tour and the cheapest-link tour are both optimal. Choose the weights of the edges to be all different.

68. **(a)** Give an example of a graph with four vertices in which the same circuit can be both an Euler circuit and a Hamilton circuit.

   (b) Give an example of a graph with N vertices in which the same circuit can be both an Euler circuit and a Hamilton circuit. Explain why there is only one kind of graph for which this is possible.

69. **(a)** Explain why a graph that has a bridge cannot have a Hamilton circuit.

   (b) Give an example of a graph with bridges that has a Hamilton path.

70. Explain why 21! is more than 100 billion times bigger than 10! (i.e., show that $21! > 10^{11} \times 10!$).

Exercises 71 and 72 refer to the following situation. Nick is a traveling salesman. His territory consists of the 11 cities shown on the mileage chart in Fig. 6-55. Nick must organize a round trip that starts and ends in Dallas (that's his home) and visits each of the other 10 cities exactly once.

71. Working directly from the mileage chart, find a nearest-neighbor tour that starts at Dallas. Explain the procedure you used to find the answer.

72. Working directly from the mileage chart, find the a cheapest-link tour. Explain the procedure you used to find the answer.

73. Julie is the marketing manager for a small software company based in Boston. She is planning a sales trip to Michigan to visit customers in each of the nine cities shown on the mileage chart in Fig. 6-56 (see p. 236). She can fly from Boston to any one of the cities and fly out of any one of the cities back to Boston for the same price (call the arrival city A and the departure city D). Her plan is to pick up a rental car at A, drive to each of the other cities, and drop off the rental car at the last city D. Slightly complicating the situation is that Michigan has two separate peninsulas—an upper peninsula and a lower peninsula—and the only way to get from one to the other is through the Mackinaw Bridge connecting Cheboygan to Sault Ste. Marie. (There is a $3 toll to cross the bridge in either direction.)

Mileage Chart

	Atlanta	Boston	Buffalo	Chicago	Columbus	Dallas	Denver	Houston	Kansas City	Louisville	Memphis
Atlanta	*	1037	859	674	533	795	1398	789	798	382	371
Boston	1037	*	446	963	735	1748	1949	1804	1391	941	1293
Buffalo	859	446	*	522	326	1346	1508	1460	966	532	899
Chicago	674	963	522	*	308	917	996	1067	499	292	530
Columbus	533	735	326	308	*	1028	1229	1137	656	209	576
Dallas	795	1748	1346	917	1028	*	781	243	489	819	452
Denver	1398	1949	1508	996	1229	781	*	1019	600	1120	1040
Houston	789	1804	1460	1067	1137	243	1019	*	710	928	561
Kansas City	798	1391	966	499	656	489	600	710	*	520	451
Louisville	382	941	532	292	209	819	1120	928	520	*	367
Memphis	371	1293	899	530	576	452	1040	561	451	367	*

FIGURE 6-55

(a) Suppose that the rental car company charges 39 cents per mile plus a drop off fee of $250 if A and D are different cities (there is no charge if $A = D$). Find the optimal (cheapest) route and give the total cost.

(b) Suppose that the rental car company charges 49 cents per mile but the car can be returned to any city without a drop off fee. Find the optimal route and give the total cost.

Mileage Chart

	Detroit	Lansing	Grand Rapids	Flint	Cheboygan	Sault Ste. Marie	Marquette	Escanaba	Menominee
Detroit	*	90	158	68	280				
Lansing	90	*	68	56	221				
Grand Rapids	158	68	*	114	233				
Flint	68	56	114	*	215				
Cheboygan	280	221	233	215	*	78			
Sault Ste. Marie					78	*	164	174	227
Marquette						164	*	67	120
Escanaba						174	67	*	55
Menominee						227	120	55	*

FIGURE 6-56

RUNNING

74. Complements. Let G be a graph with N vertices. The complement of G is the graph with the same vertices and the complementary set of edges. (Two distinct vertices that are adjacent in G are not adjacent in the complement, and, conversely, two distinct vertices not adjacent in G are adjacent in the complement.)

(a) If the number of edges in G is k, find the number of edges in the complement of G. (Assume no loops or multiple edges in either graph.)

(b) Explain why for $N = 4$ there is no graph such that both the graph and its complement have Hamilton circuits.

(c) Explain why for all $N \geq 5$ there are graphs such that both the graph and its complement have Hamilton circuits.

75. Complete bipartite graphs. A complete bipartite graph is a graph with the property that the vertices can be divided into two sets A and B and each vertex in set A is adjacent to each of the vertices in set B. There are no other edges! If there are m vertices in set A and n vertices in set B, the complete bipartite graph is written as $K_{m,n}$. Figure 6-57 shows a generic bipartite graph.

(a) For $n > 1$, the complete bipartite graphs of the form $K_{n,n}$ all have Hamilton circuits. Explain why.

(b) If the difference between m and n is exactly 1 (i.e., $|m - n| = 1$), the complete bipartite graph $K_{m,n}$ has a Hamilton path. Explain why.

(c) When the difference between m and n is more than 1, then the complete bipartite graph $K_{m,n}$ has neither a Hamilton circuit nor a Hamilton path. Explain why.

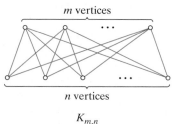

$K_{m,n}$

FIGURE 6-57

76. m by n grid graphs. An m by n grid graph represents a rectangular street grid that is m blocks by n blocks, as indicated in Fig. 6-58. (You should try Exercises 62 through 65 before you try this one.)

(a) If m and n are both odd, then the m by n grid graph has a Hamilton circuit. Describe the circuit by drawing it on a generic graph.

(b) If either m or n is even and the other one is odd, then the m by n grid graph has a Hamilton circuit. Describe the circuit by drawing it on a generic graph.

(c) If m and n are both even, then the m by n grid graph does not have a Hamilton circuit. Explain why a Hamilton circuit is impossible.

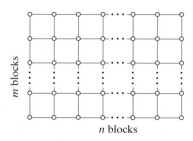

FIGURE 6-58

77. Ore's theorem. A connected graph with N vertices is said to satisfy *Ore's condition* if $\deg(X) + \deg(Y) \geq N$ for every pair of vertices X and Y of the graph. Ore's theorem states that *if a graph satisfies Ore's condition, then it has a Hamilton circuit.*

(a) Explain why the complete bipartite graph $K_{n,n}$ (see Exercise 75) satisfies Ore's condition.

(b) Explain why for $m \neq n$, the complete bipartite graph $K_{m,n}$ (see Exercise 75) does *not* satisfy Ore's condition.

(c) Ore's condition is sufficient to guarantee that a connected graph has a Hamilton circuit, but is not a necessary condition. Give an example of a graph that has a Hamilton circuit but does not satisfy Ore's condition.

(d) *Dirac's Theorem* states that if $\deg(X) \geq N/2$ for every vertex X of a connected graph, then G has a Hamilton circuit. Explain why Dirac's theorem is a direct consequence of Ore's theorem.

PROJECTS AND PAPERS

A The Great Kaliningrad Circus

The Great Kaliningrad Circus has been signed for an extended 21-city tour of the United States, starting and ending in Miami, Florida. The 21 cities and the travel distances between cities are shown in Fig. 6-59 on page 238. The cost of transporting an entire circus the size of the Great Kaliningrad can be estimated to be about $1000 per mile, so determining the best, or near best, route for the circus tour is clearly a top priority for the tour manager, who, by the way, is you!

In this project, your job is to find the best route you can for the circus tour. (You are not asked to find the optimal route, just to find the best route you can!) You should prepare a presentation to the board of directors of the circus explaining what strategies you used to come up with your route, and why you think your route is a good one.

Note: The tools at your disposal to carry out this project are everything you learned in this chapter, your ingenuity, and your time (a wall-sized map of the United States and some pins may come in handy too!). This should be a low-tech project—you should not go to the Web to try to find some software that will do this job for you!

B The Nearest-Insertion Algorithm

The *nearest-insertion algorithm* is another approximate algorithm used for tackling TSPs. The basic idea of the algorithm is to start with a subcircuit (a circuit that includes some, but not all, of the vertices) and enlarge it, one step at a time, by adding an extra vertex—the one that is closest to some vertex in the circuit. By the time we have added all of the vertices, we have a full-fledged Hamilton circuit.

In this project, you should prepare a class presentation on the nearest-insertion algorithm. Your presentation should include a detailed description of the algorithm, at least two carefully worked-out examples, and a comparison of the nearest-insertion and the nearest-neighbor algorithms.

C Computing with DNA

DNA is the basic molecule of life—it encodes the genetic information that characterizes all living organisms. Due to the recent great advances in biochemistry, scientists can now snip, splice, and recombine segments of DNA almost at will. In 1994, Leonard Adleman, a professor of computer science at the University of Southern California, was able to encode a graph representing seven cities into a set of DNA segments and to use the chemical reactions of the DNA fragments to uncover the existence of a Hamilton path in the graph. Basically, he was able to use the biochemistry of DNA to solve a graph theory problem. While the actual problem solved was insignificant, the idea was revolutionary, as it opened the door for the possibility of someday using DNA computers to solve problems beyond the reach of even the most powerful of today's electronic computers.

Write a research paper telling the story of Adleman's landmark discovery. How did he encode the graph into DNA? How did he extract the mathematical solution (Hamilton path) from the chemical solution? What other kinds of problems might be solved using DNA computing? What are the implications of Adleman's discovery for the future of computing?

D Ant Colony Optimization

An individual ant is, by most standards, a dumb little creature, but collectively an entire ant colony can perform surprising feats of teamwork (such as lifting and carrying large leaves or branches) and self-organize to solve remarkably complex problems (finding the shortest route to a food source, optimizing foraging strategies, managing a smooth and steady traffic flow in congested ant highways). The ability of ants and other social insects to perform sophisticated group tasks goes by the name of *swarm intelligence*. Since ants don't talk to each other and don't have bosses telling them what to do, swarm intelligence is a decentralized, spontaneous type of intelligence that has many potential applications at the human scale. In recent years, computer scientists have been able to approach many difficult optimization problems (including TSPs) using *ant colony optimization* software (i.e., computer programs that use *virtual ants* to imitate the problem-solving strategies of real ants).

Write a research paper describing the concept of swarm intelligence and some of the recent developments in ant colony optimization methods and, in particular, the use of virtual ant algorithms for solving TSPs.

Mileage Chart

	Atlanta	Boston	Buffalo	Chicago	Columbus	Dallas	Denver	Houston	Kansas City	Louisville	Memphis	Miami	Minneapolis	Nashville	New York	Omaha	Pierre	Pittsburgh	Raleigh	St. Louis	Tulsa
Atlanta	*	1037	859	674	533	795	1398	789	798	382	371	655	1068	242	841	986	1361	687	372	541	772
Boston	1037	*	446	963	735	1748	1949	1804	1391	941	1293	1504	1368	1088	206	1412	1726	561	685	1141	1537
Buffalo	859	446	*	522	326	1346	1508	1460	966	532	899	1409	927	700	372	971	1285	216	605	716	1112
Chicago	674	963	522	*	308	917	996	1067	499	292	530	1329	405	446	802	459	763	452	784	289	683
Columbus	533	735	326	308	*	1028	1229	1137	656	209	576	1160	713	377	542	750	1071	182	491	406	802
Dallas	795	1748	1346	917	1028	*	781	243	489	819	452	1300	936	660	1552	644	943	1204	1166	630	257
Denver	1398	1949	1508	996	1229	781	*	1019	600	1120	1040	2037	841	1156	1771	537	518	1411	1661	857	681
Houston	789	1804	1460	1067	1137	243	1019	*	710	928	561	1190	1157	769	1608	865	1186	1313	1160	779	478
Kansas City	798	1391	966	499	656	489	600	710	*	520	451	1448	447	556	1198	201	592	838	1061	257	248
Louisville	382	941	532	292	209	819	1120	928	520	*	367	1037	697	168	748	687	1055	388	541	263	659
Memphis	371	1293	899	530	576	452	1040	561	451	367	*	997	826	208	1100	652	1043	752	728	285	401
Miami	655	1504	1409	1329	1160	1300	2037	1190	1448	1037	997	*	1723	897	1308	1641	2016	1200	819	1196	1398
Minneapolis	1068	1368	927	405	713	936	841	1157	447	697	826	1723	*	826	1207	357	394	857	1189	552	695
Nashville	242	1088	700	446	377	660	1156	769	556	168	208	897	826	*	892	744	1119	553	521	299	609
New York	841	206	372	802	542	1552	1771	1608	1198	748	1100	1308	1207	892	*	1251	1565	368	489	948	1344
Omaha	986	1412	971	459	750	644	537	865	201	687	652	1641	357	744	1251	*	391	895	1214	449	387
Pierre	1361	1726	1285	763	1071	943	518	1186	592	1055	1043	2016	394	1119	1565	391	*	1215	1547	824	760
Pittsburgh	687	561	216	452	182	1204	1411	1313	838	388	752	1200	857	553	368	895	1215	*	445	588	984
Raleigh	372	685	605	784	491	1166	1661	1160	1061	541	728	819	1189	521	489	1214	1547	445	*	804	1129
St. Louis	541	1141	716	289	406	630	857	779	257	263	285	1196	552	299	948	449	824	588	804	*	396
Tulsa	772	1537	1112	683	802	257	681	478	248	659	401	1398	695	609	1344	387	760	984	1129	396	*

FIGURE 6-59

E The Knight's Tour

The knight's tour is a mathematical problem with a long history and closely tied to the topic of this chapter. A legal move for a *knight* on a "chessboard" allows the knight to move another square that is two squares away in one direction (Left/Right or Up/Down) and one square away in the other direction. By a "chessboard" here we mean an array of squares such as the ones in an 8 by 8 traditional chessboard, but of arbitrary dimensions (*m* rows by *n* columns). A knight's tour is a trip by a knight starting at some specified square of the chessboard that passes through each of the other squares exactly once. If from the last square the knight can return to the starting square by one more move, then we have a closed knight's tour. Otherwise, it's an open knight's tour. Depending on the dimensions of the chessboard and the starting position, a knight's tour may or may not be possible. (For example, in a 3 by 4 chessboard, closed tours are impossible but open tours are possible; in a 4 by 4 chessboard, neither open nor closed tours are possible; in a 4 by 5 chessboard, both open and closed tours are possible.) In this project, you are asked to prepare a presentation/paper discussing knight tours on chessboards of different sizes and the connection between knight tours and Hamilton circuits/paths on a graph.

You should have no trouble finding plenty of really good information on knight tours on the Web. This is a well-traveled topic.

REFERENCES AND FURTHER READINGS

1. Adleman, Leonard, "Computing with DNA," *Scientific American*, 279 (August 1998), 54–61.
2. Bell, E. T., "An Irish Tragedy: Hamilton," in *Men of Mathematics*. New York: Simon and Schuster, 1986, chap. 19.
3. Bellman, R., K. L. Cooke, and J. A. Lockett, *Algorithms, Graphs and Computers*. New York: Academic Press, Inc., 1970, chap. 8.
4. Berge, C., *The Theory of Graphs and Its Applications*. New York: Wiley, 1962.
5. Bonabeau, Eric, and G. Théraulaz, "Swarm Smarts," *Scientific American*, 282 (March 2000), 73–79.
6. Chartrand, Gary, *Graphs as Mathematical Models*. Belmont, CA: Wadsworth Publishing Co., Inc., 1977, chap. 3.
7. Devlin, Keith, *Mathematics: The New Golden Age*. London: Penguin Books, 1988, chap. 11.
8. Kolata, Gina, "Analysis of Algorithms: Coping with Hard Problems," *Science*, 186 (November 1974), 520–521.
9. Lawler, E. L., Lenstra, J. K., Rinooy Kan, A. H. G., and D. B. Shmoys, *The Traveling Salesman Problem*. New York: John Wiley & Sons, Inc., 1985.
10. Lewis, H. R., and C. H. Papadimitriou, "The Efficiency of Algorithms," *Scientific American*, 238 (January 1978), 96–109.
11. Michalewicz, Z., and D. B. Fogel, *How to Solve It: Modern Heuristics*. New York: Springer-Verlag, 2000.
12. Papadimitriou, Christos H., and Kenneth Steiglitz, *Combinatorial Optimization: Algorithms and Complexity*. New York: Dover, 1998, chap. 17.
13. Peterson, Ivars, *Islands of Truth: A Mathematical Mystery Cruise*. New York: W. H. Freeman & Co., 1990, chap. 6.
14. Wilson, Robin, and John J. Watkins, *Graphs: An Introductory Approach*. New York: John Wiley & Sons, Inc., 1990.

7 The Mathematics of Networks

The Cost of Being Connected

Social networks, communication networks, transportation networks, the Internet. Nowadays networks are everywhere, which is probably why it's so hard to find a good definition of the concept in the dictionary. Dictionary.com lists 14 different definitions of the word *network*, which run the gamut, from "a group of transmitting stations linked by wire or microwave relay" to "any netlike combination of filaments, lines, veins, passages, or the like." Huh?

Fortunately, our definition of a network is going to be really simple—essentially, *a network is a graph that is connected*. In this context the term is most commonly used when the graph models a real-life "network." Typically, the vertices of a network (sometimes called *nodes* or *terminals*) are "objects"—transmitting stations, computer servers, places, cell phones, people, and so on. The edges of a network (which in this context are often called *links*) indicate *connections* among the objects—wires, cables, roads, Internet connections, social connections, and so on.

Undoubtedly, one of the largest, most complex, and most unpredictable networks around is the World Wide Web. The World Wide Web is a classic example of an *evolutionary* network—it follows no predetermined master plan and essentially evolves on its own without structure or centralized direction (yes, there are some rules of behavior but, as we all know, not that many). Most social animals, including humans, belong to many different overlapping evolutionary networks. (Think of how many different such networks you belong to—your family, your friends, facebook.com, and so on.)

At the opposite end of the spectrum from evolutionary networks are networks that are centrally planned and carefully designed to meet certain goals and objectives. Often these types of networks are very expensive to build, and one of the primary considerations when designing such networks is minimizing their cost. This certainly applies to networks of roads, fiber-optic cable lines, rail lines, power lines, and so on. In most of these cases the cost of the network is roughly proportional to its length (*x* dollars per mile), which means that the cheapest networks are going to be the shortest ones. Of course, there are other situations in which shortest may not necessarily be cheapest (freeways, for example—a bridge or an overpass can eat up more of the budget than many miles of roadbed).

241

The general theme of this chapter is the problem of finding *optimal networks* connecting a set of points. Once again, as was the case in Chapter 6, *optimal* means shortest, cheapest, or fastest, depending on whether the cost variable is distance, money, or time. Thus, the design of an optimal network involves two basic goals: (1) to make sure that all the vertices (stations, places, people, etc.) end up connected to the network and (2) to minimize the total cost of the network. For obvious reasons, problems of this type are known as *minimum network problems*.

The backbone of a minimum network is a special type of graph called a *tree*. This chapter starts with a discussion of the properties of *trees* (Section 7.1) and *minimum spanning trees* (Section 7.2). Section 7.3 describes *Kruskal's algorithm*, an algorithm for finding the minimum spanning tree of any network. Sections 7.4 and 7.5 give an introduction to the geometry of *shortest networks*. (In contrast to the general theory of minimum spanning trees, which is very simple, the general theory of shortest networks is quite complicated, and Sections 7.4 and 7.5 are intended to offer just a brief glimpse of the overall picture.)

7.1   Trees

(EXAMPLE 7.1)   **The Amazonian Cable Network**

The Amazonia Telephone Company is contracted to provide telephone, cable, and Internet service to the seven small mining towns shown in Fig. 7-1(a). These towns are located deep in the heart of the Amazon jungle, which makes the project particularly difficult and expensive. In this environment the most practical and environmentally friendly option is to create a network of underground fiber-optic cable lines connecting the towns. In addition, it makes sense to bury the underground cable lines along the already existing roads connecting the towns. (How would you maintain and repair the lines otherwise?) The existing network of roads is also shown in Fig. 7-1(a). The weighted graph in Fig. 7-1(b) is a graph model describing the situation. The vertices of the graph represent the towns, the edges of the graph represent the existing roads, and the weight of each edge represents the cost (in millions of dollars) of creating a fiber-optic cable connection along that particular road.

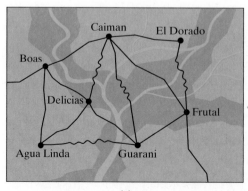

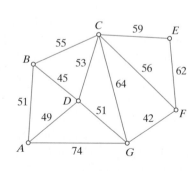

FIGURE 7-1   (a)   (b)

The problem facing the engineers and planners at the Amazonia Telephone Company is to build a cable network that (1) utilizes the existing network of roads, (2) connects all the towns, and (3) has the least cost. The challenge, of course, is in meeting the last requirement. ⬭

When we reformulate the three requirements of Example 7.1 in the language of graphs, this is what we get:

66 **span:** to extend, reach, or stretch across. **99**

—Webster's 20th Century Dictionary

1. The network must be a **subgraph** of the original graph (in other words, its edges must come from the original graph).

2. The network must **span** the original graph (in other words, it must include *all* the vertices of the original graph).

3. The network must be **minimal** (in other words, the total weight of the network should be as small as possible).

The last requirement has an important corollary—a minimal network *cannot have any circuits*. Why not? Imagine that the solid edges in Fig. 7-2 represent already existing links in a minimal network. Why would you then build the link between X and Y and close the circuit? The edge XY would be a *redundant* link of the network. (In many real-life applications, other factors beyond cost—such as reliability, and convenience—come into play. In these situations redundant links are quite important.)

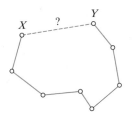

FIGURE 7-2

It is now time for a few formal definitions.

- A **network** is just another name for a *connected* graph. (This terminology is most commonly used when the graph models a real-life situation.) When the network has weights associated to the edges, we call it a **weighted network**.

- A network with no circuits is called a **tree**.

- A **spanning tree** of a network is a *subgraph* that connects *all* the vertices of the network and has no circuits.

- Among all spanning trees of a weighted network, one with least total weight is called a **minimum spanning tree** (MST) of the network.

(**EXAMPLE 7.2**) **Networks, Trees, and Spanning Trees**

The six graphs shown in Fig. 7-3 all have the same set of vertices (A through L). Let's imagine, for the purposes of illustration, that these vertices represent computer labs at a university, and that the edges are Ethernet connections between pairs of labs.

In Fig. 7-3(a) there is no network—the graph is disconnected.

Figure 7-3(b) is the same as Fig. 7-3(a) plus an additional link connecting labs E and K. The graph is now connected and we have a network. However, in this network there are several circuits (for example, K, H, I, J, K) that create redundant connections. In other words, this network is not a tree.

In Fig. 7-3(c) there is a partial tree connecting some of, but not all, the labs. G and I are left out, so once again, we have no network.

Figure 7-3(d) shows a tree that *spans* (i.e., reaches) all the vertices. We now have a network connecting all the labs and without any redundant connections. Here, the tree is the network.

Figure 7-3(e) shows (in red) the same tree as in Fig. 7-3(d) but now highlighted inside of a larger network [in fact, the network in Fig. 7-3(b)]. In this case we describe the red tree as a *spanning tree* of the larger network. Figure 7-3(e) shows a different spanning tree for the same network.

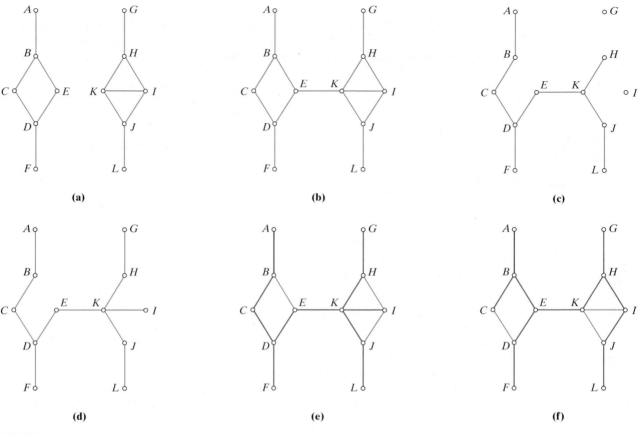

FIGURE 7-3

Properties of Trees

As graphs go, trees occupy an important niche between disconnected graphs and "overconnected" graphs. A tree is special by virtue of the fact that it is *barely connected*. This means several things:

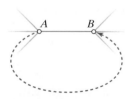

FIGURE 7-4 Two different paths joining *X* and *Y* make a circuit.

- For *any two* vertices *X* and *Y* of a tree, there is one and *only one* path joining *X* to *Y*. (If there were *two different* paths joining *X* and *Y*, then these two paths would form a circuit, as shown in Fig. 7-4.)

- Every edge of a tree is a *bridge*. (Suppose that some edge *AB* is not a bridge. Then without *AB* the graph is still connected, so there must be an alternative path from *A* to *B*. This would imply that the edge *AB* is part of a circuit as illustrated in Fig. 7-5.)

- Among all networks with *N* vertices, a tree is the one with the *fewest* number of edges.

The next example illustrates the special role that trees play in the creation of a network.

FIGURE 7-5 If *AB* is not a bridge, then it must be part of a circuit.

(**EXAMPLE 7.3**)   **Connect the Dots (and Then Stop)**

Imagine the following "connect-the-dots" game: Start with eight isolated vertices. The object of the game is to create a network connecting the vertices by adding edges, one at a time. You are free to create any network you want. In this game, bridges are good and circuits are bad. (Imagine, for example, that for each bridge in your network you get a $10 reward, but for each circuit in your network you pay a $10 penalty.)

So grab a marker and start playing. We will let M denote the number of edges you have added at any point in time. In the early stages of the game ($M = 1, 2, \ldots, 6$) the graph is disconnected [Figs. 7-6(a) through (d)].

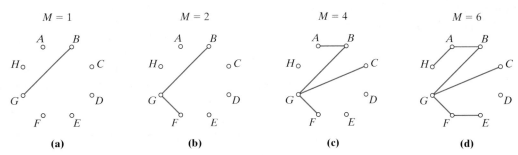

FIGURE 7-6 For small values of M, the graph is disconnected.

As soon as you get to $M = 7$ (and if you stayed away from forming any circuits) the graph becomes connected. Some of the possible configurations are shown in Fig. 7-7. Each of these networks is a tree, and thus each of the seven edges is a bridge. Stop here and you will come out $70 richer.

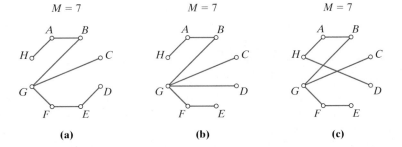

FIGURE 7-7 When $M = 7$, we have a tree — a network with no circuits.

Interestingly, this is as good as it will get. When $M = 8$, the graph will have a circuit—it just can't be avoided. In addition, none of the edges in that circuit can be bridges of the graph. As a consequence, the larger the circuit that we create, the fewer the bridges left in the graph. [The graph in Fig. 7-8(a) has a circuit and five bridges, the graph in Fig. 7-8(b) has a circuit and two bridges, and the graph in Fig. 7-8(c) has a circuit and only one bridge.] As the value of M increases, the number of circuits goes up (very quickly) and the number of bridges goes down [Figs. 7-9(a), (b), and (c)].

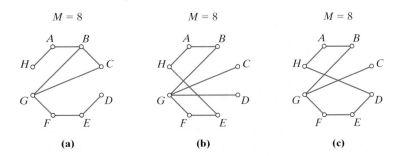

FIGURE 7-8 When $M = 8$, we have a network with a circuit.

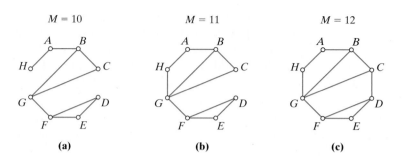

FIGURE 7-9 For larger values of M, we have a network with lots of circuits.

Our discussion of trees can be summarized into the following three key properties.

> **PROPERTY 1**
> - In a tree, there is one and only one path joining any two vertices.
> - If there is one and only one path joining any two vertices of a graph, then the graph must be a tree.

> **PROPERTY 2**
> - In a tree, every edge is a bridge.
> - If every edge of a graph is a bridge, then the graph must be a tree.

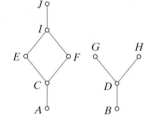

FIGURE 7-10 A disconnected graph with 10 vertices and 9 edges.

> **PROPERTY 3**
> - A tree with N vertices has $N - 1$ edges.
> - If a *network* has N vertices and $N - 1$ edges, then it must be a tree.

Notice that a disconnected graph (i.e., not a network) can have N vertices and $N - 1$ edges, as shown in Fig. 7-10.

7.2   Spanning Trees

When we reformulate Property 3 in a slightly more general context, we get the following useful property:

> **PROPERTY 4**
> - If a network has N vertices and M edges, then $M \geq N - 1$. [For convenience, we will refer to the difference $R = M - (N - 1)$ as the **redundancy** of the network.]
> - If $M = N - 1$, the network is a tree; if $M > N - 1$, the network has circuits and is not a tree. (In other words, a tree is a network with zero redundancy, and a network with positive redundancy is not a tree.)

In the case of a network with positive redundancy, there are many trees *within* the network that connect its vertices—these are the *spanning trees* of the network.

(EXAMPLE 7.4)   **Counting Spanning Trees**

The network in Fig. 7-11(a) has $N = 8$ vertices and $M = 8$ edges. The redundancy of the network is $R = M - (N - 1) = 1$, so to find a spanning tree we will have to "discard" one edge. Five of these edges are bridges of the network, and they *will have to be part of any spanning tree.* The other three edges (BC, CG, and GB) form a circuit of length 3, and if we exclude any one of the three edges, then we will have a spanning tree. Thus, the network has three different spanning trees [Figs. 7-11(b), (c), and (d)].

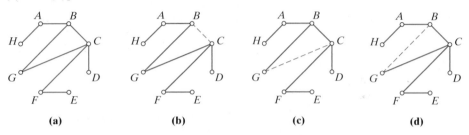

FIGURE 7-11   (a)   (b)   (c)   (d)

The network in Fig. 7-12(a) has $M = 9$ edges and $N = 8$ vertices. The redundancy of the network is $R = 2$, so to find a spanning tree we will have to "discard" two edges. Edges AB and AH are bridges of the network, so they will have to be part of any spanning tree. The other seven edges are split into two separate circuits (B, C, G, B of length 3 and C, D, E, F, C of length 4). A spanning tree can be found by "busting" each of the two circuits. This means excluding any one of the three edges of circuit B, C, G, B and any one of the four edges of circuit C, D, E, F, C. For example, if we exclude BC and CD, we get the spanning tree shown in Fig. 7-12(b). We could also exclude BC and DE and get the spanning tree shown in Fig. 7-12(c), and so on. Given that there are $3 \times 4 = 12$ different ways to choose an edge from the circuit of length 3 and an edge from the circuit of length 4, we will not show all 12 spanning trees. [Figs. 7-12(b) through (e) show some examples.]

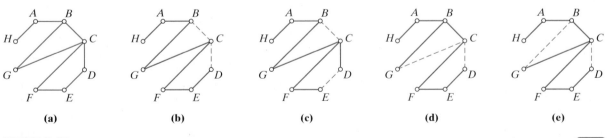

    (a)   (b)   (c)   (d)   (e)

FIGURE 7-12

(EXAMPLE 7.5)   **More Counting of Spanning Trees**

The network in Fig. 7-13(a) is another network with $M = 9$ edges and $N = 8$ vertices. The difference between this network and the one in Fig. 7-12(a) is that here the circuits B, C, G, B and C, D, E, G, C share a common edge CG. Determining which pairs of edges can be excluded in this case is a bit more complicated.

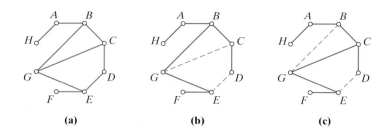

FIGURE 7-13   (a)   (b)   (c)

If one of the excluded edges is the common edge CG, then the other excluded edge can be any other edge in a circuit. There are five choices (BC, CD, DE, EG, and GB), and each choice will result in a different spanning tree [one of these is shown in Fig. 7-13(b)]. The alternative scenario is to exclude two edges neither of which is the common edge CG. In this case one excluded edge has to be either BC or BG (to "bust" circuit B, C, G, B), and the other excluded edge has to be either CD, DE, or EG (to "bust" circuit C, D, E, G, C). There are $2 \times 3 = 6$ possible spanning trees that can be formed this way [one of these is shown in Fig. 7-13(c)]. Combining the two scenarios gives a total of 11 possible spanning trees for the network in Fig. 7-13(a).

■ For a little practice counting spanning trees, see Exercises 17, 18, and 51.

(**EXAMPLE 7.6**)   **Minimum Spanning Trees**

This example builds on the ideas introduced in Example 7.2. The network shown in Fig. 7-14(a) is the same network as in Fig. 7-3(b), now with weights added to the edges. As before, the vertices represent computer labs at a university, and the edges are potential Ethernet connections among the various labs. The weights represent the cost in K's (1K = \$1000) of installing the Ethernet connections.

Using the terminology we introduced in this section, the weighted network in Fig. 7-14(a) has a redundancy of $R = 3$ (the network has $N = 12$ vertices and $M = 14$ edges). The network has many possible spanning trees, and our job is to find one with least weight, that is, a minimum spanning tree (MST) for the network.

From Examples 7.4 and 7.5, we know that we can find a spanning tree by *excluding* certain edges of the network (those that close circuits) from the spanning tree and that there are many different ways in which this can be done. Given that the edges now have weights, a reasonable strategy for sorting through the many options would be to always try to exclude from the network the most expensive edges.

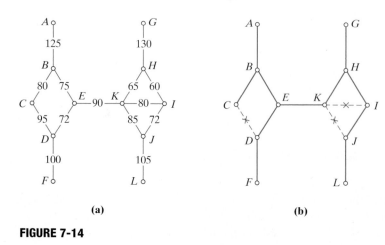

(a)   **(b)**

FIGURE 7-14

To illustrate the point, let's look at circuit B, C, D, E, B on the left side of the network in Fig. 7-14(a). We can "bust" the circuit by excluding any one of the four edges from the circuit, and it makes sense to exclude CD (which costs \$95,000 to build) rather than one of the other three (which cost less). Likewise, on the right side of the network we have a configuration of two circuits K, H, I, K and K, J, I, K that share the common edge KI. Once again, we exclude the two most expensive edges from these two circuits (KJ and KI). At this point, all the circuits of the original network are "busted," and we end up with the red spanning tree shown in Fig. 7-14(b).

Is the red spanning tree obtained in Example 7.6 the MST of the original network? We would like to think it is, but how can we be sure, especially after the lessons learned in Chapter 6? And even if it is, what assurances do we have that our simple strategy will work in more complicated graphs? These are the questions we will answer next.

7.3 Kruskal's Algorithm

There are several well-known algorithms for finding minimum spanning trees. In this section we will discuss one of the nicest of these, called **Kruskal's algorithm** (after American mathematician Joseph Kruskal, who first proposed the algorithm in 1956).

Kruskal's algorithm is wickedly simple and almost identical to the *cheapest-link algorithm* we discussed in Chapter 6: We build the minimum spanning tree one edge at a time, choosing at each step the cheapest available edge. The only restriction to our choice of edges is that we should never choose an edge that creates a circuit. (Having three or more edges coming out of a vertex, however, is now OK.) What is truly remarkable about Kruskal's algorithm is that—unlike the cheapest-link algorithm—*it always gives an optimal solution.*

Before giving a formal description of Kruskal's algorithm, let's look at a simple example.

(**EXAMPLE 7.7**)   **The Amazonian Cable Network: Part 2**

In Example 7.1 we raised the following question: What is the optimal fiber-optic cable network connecting the seven towns shown in Fig. 7-15(a)? The weighted graph in Fig. 7-15(b) shows the costs (in millions of dollars) of laying the cable lines along each of the potential links of the network.

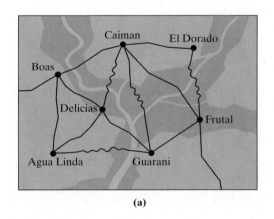
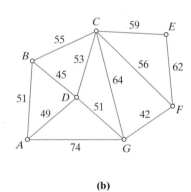

FIGURE 7-15   (a)   (b)

The answer, as we now know, is to find the MST of the graph in Fig. 7-15(b). We will use Kruskal's algorithm to do it. Here are the details:

■ **Step 1.** Among all the possible links, we choose the cheapest one, in this particular case *GF* (at a cost of $42 million). This link is going to be part of the MST, and we mark it in red (or any other color) as shown in Fig. 7-16(a). (Note that this does not have to be the first link actually built—we are putting the network together on paper only. On the ground, the schedule of construction is a different story, and there are many other factors that need to be considered.)

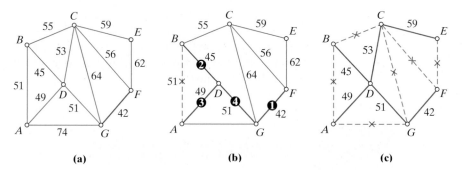

FIGURE 7-16   (a)   (b)   (c)

- **Step 2.** The next cheapest link available is *BD* at $45 million. We choose it for the MST and mark it in red.
- **Step 3.** The next cheapest link available is *AD* at $49 million. Again, we choose it for the MST and mark it in red.
- **Step 4.** For the next cheapest link there is a tie between *AB* and *DG*, both at $51 million. But we can rule out *AB*—it would create a circuit in the MST, and we can't have that! (For bookkeeping purposes it is a good idea to erase or cross out the link.) The link *DG*, on the other hand, is just fine, so we mark it in red and make it part of the MST.
    Figure 7-16(b) shows how things look at this point.
- **Step 5.** The next cheapest link available is *CD* at $53 million. No problems here, so again, we mark it in red and make it part of the MST.
- **Step 6.** The next cheapest link available is *BC* at $55 million, but this link would create a circuit, so we cross it out. The next possible choice is *CF* at $56 million, but once again, this choice creates a circuit so we must cross it out. The next possible choice is *CE* at $59 million, and this is one we do choose. We mark it in red and make it part of the MST.
- **Step ...** Wait a second—we are finished! Even without looking at a picture, we can tell we are done—six links is exactly what is needed for an MST on seven vertices.

    Figure 7-16(c) shows the MST in red. The total cost of the network is $299 million.

The following is a formal description of Kruskal's algorithm. (For simplicity, we use "cheapest" to denote "of least weight.")

> **KRUSKAL'S ALGORITHM**
>
> - **Step 1.** Pick the *cheapest link* available. (In case of a tie, pick one at random.) Mark it (say in red).
> - **Step 2.** Pick the next cheapest link available and mark it.
> - **Steps 3, 4, ..., *N* − 1.** Continue picking and marking the cheapest unmarked link available that does not create a circuit.

    As algorithms go, Kruskal's algorithm is as good as it gets. First, it is easy to implement. Second, it is an *efficient* algorithm. As we increase the number of vertices and edges of the network, the amount of work grows more or less proportionally (roughly speaking, if finding the MST of a 30-city network takes you, say, 30 minutes, finding the MST of a 60-city network might take you 60 minutes).

Last, but not least, Kruskal's algorithm is an *optimal* algorithm—it will always find a minimum spanning tree.

Thus, we have reached a surprisingly satisfying end to the MST story: No matter how complicated the network, we can find its minimum spanning tree by means of an algorithm that is easy to understand and implement, is *efficient*, and is also *optimal*. We couldn't ask for better karma.

7.4 The Shortest Network Connecting Three Points

Minimum spanning trees represent the optimal way to connect a set of points based on one key assumption—that all the connections should be along the links of the network. For example, in the Amazonian cable network problem, the optimal network was an MST because it didn't make sense to bury the fiber-optic lines through uncleared jungle—the only realistic placement of the underground fiber-optic lines was along the already existing roads (the links of the network). But what if, in a manner of speaking, we don't have to *follow the road*? What if we are free to create new vertices and links "outside" the original network? To clarify the distinction, let's look at a new type of cable network problem.

EXAMPLE 7.8 **The Outback Cable Network**

The Amazonia Telephone Company has relocated to Australia and now calls itself the Outback Cable Network. The company has a new contract to create an underground fiber-optic network connecting the towns of Alcie Springs, Booker Creek, and Camoorea—three isolated towns located in the heart of the Australian outback [Fig. 7-17(a)]. By freakish coincidence (or good planning) the towns are equidistant, forming an equilateral triangle 500 miles on each side. The three towns are connected by three unpaved straight roads [Fig. 7-17(b)]. The question, once again, is, What is the cheapest underground fiber-optic cable network connecting the three towns?

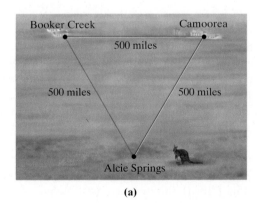
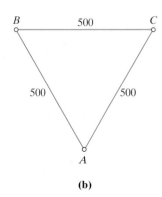

FIGURE 7-17
(a)   (b)

Obviously, the story has a lot of similarities to the Amazonian cable network problem we solved in Example 7.7, but there is one crucial difference, and this difference is based on where we can and cannot put the underground cable lines. The Amazon is heavy jungle, big rivers, and dangerous critters—you want to stay on the road (i.e., no shortcuts!). The Australian outback is flat scrub with few major roads to speak of—there is little or no disadvantage to going offroad (i.e., if you can find a shortcut, you take it!). Based on the flat and homogeneous nature of the terrain, we will assume that in the outback cable lines can be laid anywhere (along the road or offroad) and that there is a fixed cost of $100,000 per mile. Here, *cheapest* means "shortest," so the name of the game is to design a network that is as short as possible. We shall call such a network the **shortest network** (SN).

■ A shortest network by definition is shorter than the MST or, in the worst of cases, equal to the MST— to have a shortest network longer than the MST is a contradiction in terms.

The search for the shortest network often starts with a look at the minimum spanning tree. The MST can always be found using Kruskal's algorithm, and it gives us a ceiling on the length of the shortest network. In this example the MST consists of two (any two) of the three sides of the equilateral triangle [Fig. 7-18(a)], and its length is 1000 miles.

It is not hard to find a network connecting the three towns shorter than the MST. The T-network in Fig. 7-18(b) is clearly shorter. The length of the height AJ of the triangle can be computed using properties of 30-60-90 triangles and is approximately 433 miles. It follows that the length of this network is approximately 933 miles.

■ See Exercise 49. (For a primer on 30-60-90 triangles, see first Exercises 43 and 44.)

We can do better yet. The Y-network shown in Fig. 7-18(c) is shorter than the T-network in Fig. 7-18(b). In this network there is a "Y"-junction at S, with three equal branches connecting S to each of A, B, and C. This network is approximately 866 miles long. A key feature of this network is the way the three branches come together at the junction point S, forming equal 120° angles.

■ See Exercise 50.

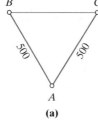  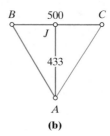 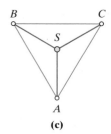

FIGURE 7-18 (a) The MST. (b) A shorter network with a T-junction. (c) The SN, with an interior Y-junction.

Even without a formal mathematical argument, it is not hard to convince oneself that the Y-network in Fig. 7-18(c) is the shortest network connecting the three towns. An informal argument goes like this: Since the original layout of the towns is completely symmetric (it looks the same from each of the three towns), we would expect that the shortest network should also be completely symmetric. The only network that looks the same from each of the three towns is a "perfect" Y where all three branches of the Y have equal length.

Our discussion of shortest networks will rely heavily on a careful analysis of the *junction points* of the network we create. In some cases, as in the red network shown in Fig. 7-18(a), the junction point is one of the vertices of the original network—we will refer to such a junction point as a *native junction point*. In other cases, as in the red networks shown in Figs. 7-18(b) and (c), the junction points are points that were not in the original network—they are new points introduced in the process of creating the network. We call such points *interior junction points* of the network. In addition, the junction point S in Fig. 7-18(c) has the special property that it makes for a "perfect Y" junction—three branches coming together at equal 120° angles. We will call such a junction point a *Steiner point* of the network.

The following summarizes the new terminology we introduced concerning junction points of a network:

Steiner points are named after Swiss mathematician Jakob Steiner (1796–1863).

> ┌─ **JUNCTION POINTS** ─────────────────────────────────
>
> - A **junction point** of a network is a point where two or more branches of the network come together.
> - A **native junction point** is a junction point that is also one of the original vertices of the graph.
> - A junction point that is not a native junction point is called an **interior junction point** of the network.
> - An interior junction point formed by three branches coming together at equal 120° angles is called a **Steiner point** of the network.

(**EXAMPLE 7.9**)   **The Third Trans-Pacific Cable Network**

This is a true story. In 1989 a consortium of several of the world's biggest telephone companies (among them AT&T, MCI, Sprint, and British Telephone) completed the Third Trans-Pacific Cable (TPC-3) line, a network of submarine fiber-optic lines linking Japan and Guam to the United States (via Hawaii). The approximate straight-line distances (in miles) between the three terminals of the network (Chikura, Japan; Tanguisson Point, Guam; and Oahu, Hawaii) are shown in Fig. 7-19. By and large, laying submarine cable has a fixed cost per mile (somewhere between $50,000 and $70,000), so *cheapest* means "shortest" and the problem is once again to find the shortest network connecting the three terminals.

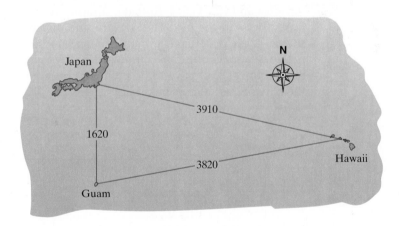

FIGURE 7-19

Given what we learned in Example 7.8, a reasonable guess is that the shortest network is going to require an interior junction point, somewhere inside the Japan-Guam-Hawaii triangle. But where? Figure 7-20(a) shows the answer: The junction point *S* is located in such a way that three branches of the network coming out of *S* form equal 120° angles—in other words, *S* is a *Steiner point of the network*.

Figure 7-20(b) shows a stamp issued by the Japanese Postal Service to commemorate the completion of TCP-3. The length of the shortest network is 5180 miles, but this is a theoretical length only, based on straight-line distances. With submarine cable one has to add as much as 10% to the straight-line lengths because of the uneven nature of the ocean floor. (The actual length of submarine cable used in TPC-3 is about 5690 miles.)

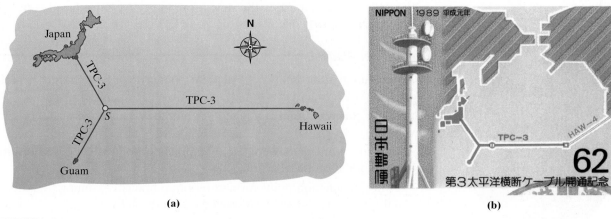

FIGURE 7-20

From Examples 7.8 and 7.9 we are tempted to conclude that the key to finding the shortest network connecting three points (cities) *A*, *B*, and *C* is to find a *Steiner point S* inside triangle *ABC* and make this point the junction point of the network (i.e., the network consists of the three segments *AS*, *BS*, and *CS* forming equal 120° angles at *S*). This is true as long as we can find a Steiner point inside the triangle, but, as we will see in the next example, *not every triangle has a Steiner point*!

> **EXAMPLE 7.10**   **A High-Speed Rail Network**

Off and on, there is talk of building a high-speed rail connection between Los Angeles and Las Vegas. To make it more interesting, let's add a third city—Salt Lake City—to the mix. The straight-line distances between the three cities are shown in Fig. 7-21(a). The mathematical question again is, What is the shortest network connecting these three cities? (High-speed trains run on expensive special tracks that have to be built expressly for them, and it is critical to make the network as short as possible.)

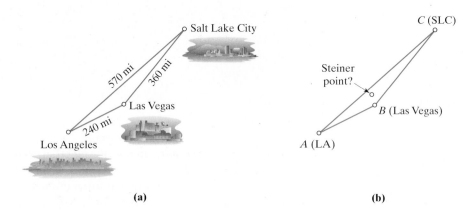

FIGURE 7-21   (a)   (b)

The first idea that comes to mind is that we should locate the Steiner point of the triangle *ABC* [Fig. 7-21(b)]. Imagine that there is such a point somewhere inside the triangle *ABC* and call it *S*. The first observation is a simple but important general property of triangles illustrated in Fig. 7-22: *For any triangle ABC*

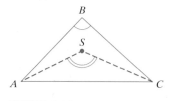

FIGURE 7-22

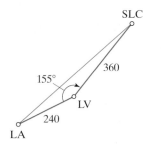

FIGURE 7-23 The SN connecting LA, LV (Las Vegas), and SLC (Salt Lake City).

and interior point S, angle ASC must be bigger than angle ABC. It follows that for angle *ASC* to measure 120°, the measure of angle *ABC* must be *less* than 120°.

Unfortunately (or fortunately—depends on how you look at it), the angle *ABC* in this example measures about 155°; therefore, there is no Steiner junction point inside the triangle. Without a Steiner junction point, how do we find the shortest network? The answer turns out to be surprisingly simple: In this situation the shortest network consists of the two shortest sides of the triangle, as shown in Fig. 7-23. Please notice that this shortest network happens to be the *minimum spanning tree* as well!

Two important lessons are to be drawn from Example 7.10. First, if a triangle has an angle whose measure is bigger than or equal to 120°, then the triangle has no Steiner point. Second, when a triangle has no Steiner point, the shortest network connecting the vertices of the triangle consists of the two shorter sides of the triangle, and this happens to also be the *minimum spanning tree* of the triangle.

When we put it all together, combining the lessons of Examples 7.8, 7.9, and 7.10, the shortest network connecting three points is given by the following general solution.

THE SHORTEST NETWORK CONNECTING THREE POINTS

- If one of the angles of the triangle is 120° or more, the shortest network linking the three vertices consists of the two shortest sides of the triangle [Fig. 7-24(a)]. In this situation, *the shortest network coincides with the minimum spanning tree.*

- If all three angles of the triangle are less than 120°, the shortest network is obtained by finding a *Steiner point S* inside the triangle and joining *S* to each of the vertices [Fig. 7-24(b)].

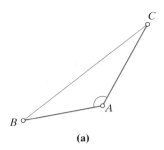

(a)

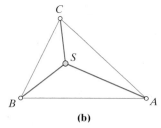

(b)

FIGURE 7-24 The SN connecting *A*, *B*, and *C*: (a) One angle is greater than or equal to 120°. (b) All three angles are less than 120°.

Finding the Steiner Point: Torricelli's Construction

Finding the shortest network connecting three points is a problem with a long and an interesting history going back some 400 years. In the early 1600s Italian Evangelista Torricelli came up with a remarkably simple and elegant method for locating a Steiner junction point inside a triangle. All you need to carry out Torricelli's construction is a straightedge and a compass; all you need to understand why it works is a few facts from high school geometry. Here it goes:

TORRICELLI'S CONSTRUCTION

Suppose A, B, and C form a triangle such that all three angles of the triangle are less than 120° [Fig. 7-25(a)].

■ **Step 1.** Choose any of the three sides of the triangle (say BC) and construct an equilateral triangle BCX so that X and A are on opposite sides of BC [Fig. 7-25(b)].

■ **Step 2.** Circumscribe a circle around equilateral triangle BCX [Fig. 7-25(c)].

■ **Step 3.** Join X to A with a straight line [Fig. 7-25(d)]. The point of intersection of the line segment XA with the circle is the Steiner point!

Why does the construction work? It is sufficient to prove that under this construction, angles BSA and CSA both measure 120°. Then angle BSC will also measure 120°, proving that S is indeed the Steiner point.

Fact 1: Angle BSX measures 60°. *Reasons:* (a) Angle BCX measures 60° since BCX is an equilateral triangle by construction. (b) Angles BSX and BCX have equal measures because the two angles are inscribed in the same circle and are opposite to the same chord BX.

Fact 2: Angle CSX measures 60°. *Reason:* Use an analogous argument as for Fact 1.

Fact 3: Angle BSA measures 120°. *Reason:* Fact 1.

Fact 4: Angle CSA measures 120°. *Reason:* Fact 2.

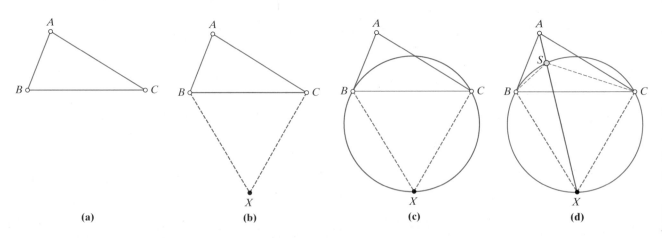

FIGURE 7-25 Torricelli's construction: (a) Triangle ABC. (b) Construct equilateral triangle BXC. (c) Circumscribe triangle BXC in a circle. (d) S is the point of intersection of the circle and the line segment joining X and A.

7.5 Shortest Networks for Four or More Points

When it comes to finding shortest networks, things get really interesting when we have to connect four points.

(**EXAMPLE 7.11**) **Four-City Networks**

Imagine four cities (A, B, C, and D) that need to be connected by an underground fiber-optic cable network. Suppose that the cities sit on the vertices of a square 500 miles on each side, as shown in Fig. 7-26(a). What does the optimal network connecting these cities look like? It depends on the situation.

If we *don't want to create any interior junction points in the network* (either because we don't want to venture off the prescribed paths—as in the jungle scenario—or because the cost of creating a new junction is too high), then the answer is a *minimum spanning tree*, such as the one shown in Fig. 7-26(b). The length of the MST is 1500 miles.

On the other hand, if *interior junction points* are allowed, somewhat shorter networks are possible. One obvious improvement is the network shown in Fig. 7-26(c), with an X-junction located at O, the center of the square. The length of this network is approximately 1414 miles.

We can shorten the network even more if we place not one but two interior junction points inside the square. It's not immediately obvious that this is a good move, but once we recognize that two junction points might be better than one, then it's not hard to see that the best option is to make the two interior junction points Steiner points. There are two different networks possible with two Steiner points inside the square, and they are shown in Figs. 7-26(d) and (e). These two networks are essentially equal (one is a rotated version of the other) and clearly have the same length—approximately 1366 miles. It is impossible to shorten these any further—the two networks in Figs. 7-26(d) and (e) are the shortest networks connecting the four cities.

■ See Exercise 62.

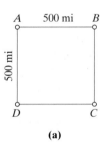

(a)

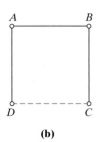

(b)

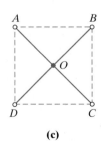

(c)

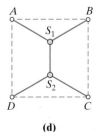

(d)

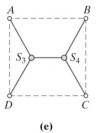

(e)

FIGURE 7-26 (a) Four cities located at the vertices of a square. (b) An MST (1500 miles). (c) A shorter network (approximately 1414 miles). (d) An SN (approximately 1366 miles). Junction points S_1 and S_2 are both Steiner points. (e) An equivalent SN, with Steiner points S_3 and S_4.

EXAMPLE 7.12 **Four-City Networks: Part 2**

Let's repeat what we did in Example 7.11, but this time imagine that the four cities are located at the vertices of a rectangle, as shown in Fig. 7-27(a). By now, we have some experience on our side, so we can cut to the chase. We know that the MST is 1000 miles long [Fig. 7-27(b)]. That's the easy part.

See Exercise 63.

For the shortest network, let's think about what happened in the previous example — a square, after all, is just a special case of a rectangle. An obvious candidate would be a network with two interior Steiner junction points. There are two such networks, shown in Figs. 7-27(c) and 7-27(d), but this time they are not equivalent. The length of the network in Fig. 7-27(c) is approximately 993 miles, while the length of the network in Fig. 7-27(d) is approximately 920 miles — a pretty significant difference. Given these lengths, it is obvious that the network in Fig. 7-27(c) cannot be the shortest network, but what about Fig. 7-27(d)? If there is any justice then this network fits the pattern and ought to be the shortest. In fact, it is!

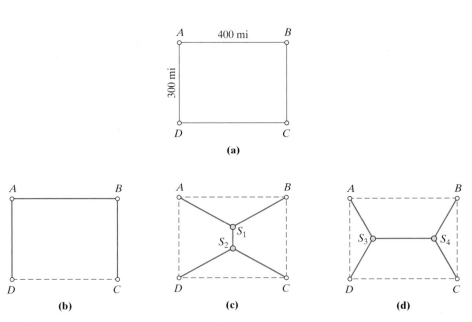

FIGURE 7-27 (a) Four cities located at the vertices of a rectangle. (b) An MST (1000 miles). (c) A network with two Steiner junction points (approximately 993 miles.) (d) The SN (approximately 920 miles).

EXAMPLE 7.13 **Four-City Networks: Part 3**

Let's look at four cities once more. This time, imagine that the cities are located at the vertices of a skinny trapezoid, as shown in Fig. 7-28(a). The minimum spanning tree is shown in Fig. 7-28(b), and it is 600 miles long.

What about the shortest network? Based on our experience with Examples 7.11 and 7.12, we are fairly certain that we should be looking for a network with a couple of interior Steiner junction points. After a little trial and error, however, we realize that such a layout is impossible! The trapezoid is too skinny, or, to put it in a slightly more formal way, the angles at A and B are greater than 120°. Since no Steiner points can be placed inside the trapezoid, the shortest network, whatever it is, will have to be one without Steiner junction points.

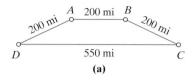

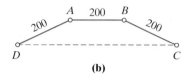

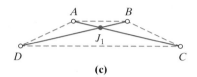

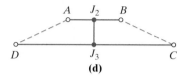

FIGURE 7-28 (a) Four cities located at the vertices of a trapezoid. (b) The MST (600 miles). (c) A network with an X-junction at J_1 is longer than the MST (774.6 miles). (d) A network with a couple of T-junction points at J_2 and J_3 is even worse (846.8 miles).

Well, we say to ourselves, if not Steiner junction points, then how about other kinds of interior junction points? How about X-junctions [Fig. 7-28(c)], or T-junctions [Fig. 7-28(d)], or Y-junctions where the angles are not all 120°? As reasonable as this idea sounds, a remarkable thing happens: *The only possible interior junction points in a shortest network are Steiner points.* For convenience, we will call this the **interior junction rule** for shortest networks.   ⊂⊃

> **THE INTERIOR JUNCTION RULE FOR SHORTEST NETWORKS**
>
> In a shortest network the *interior junction points* are all *Steiner points*.

The interior junction rule is an important and a powerful piece of information in building shortest networks, and we will come back to it soon. Meanwhile, what does it tell us about the situation of Example 7.13? It tells us that the shortest network cannot have any interior junction points. Steiner junction points are impossible because of the geometry; other types of junction points do not work because of the interior junction rule. But we also know that the shortest network without interior junction points is the minimum spanning tree! Conclusion: For the four cities of Example 7.13, *the shortest network is the minimum spanning tree!*

(**EXAMPLE 7.14**)   **Four-City Networks: Part 4**

For the last time, let's look at four cities *A*, *B*, *C*, and *D*. This time, the cities sit as shown in Fig. 7-29(a). The MST is shown in Fig. 7-29(b), and its length is 1000 miles.

We know by now that in searching for the shortest network, the MST is a good starting point—it might even be the shortest network. We also know that if the shortest network is not the MST, then it will have interior junction points, which by the interior junction rule will have to be Steiner points. Because of the layout of these cities, it is geometrically impossible to build a network with two interior Steiner points. On the other hand, there are three possible networks with a single interior Steiner point [Figs. 7-29(c), (d), and (e)].

We have now narrowed the list of contenders for the shortest network title to four: the MST or one of the networks shown in Figs. 7-29(c), (d), or (e). All we have

■ See Exercise 71.

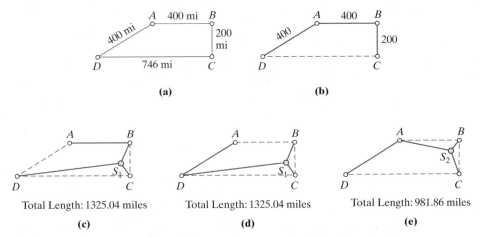

FIGURE 7-29 (a) The four cities. (b) First contender: the MST (1000 miles). (b) Second contender (1325 miles). (c) Third contender (1325 miles). (e) Fourth contender (982 miles) is the SN.

to do now to figure which one is the SN is to compute their lengths. The length of the MST is 1000 miles (we knew that!). The networks shown in Figs. 7-29(c) and (d) are both approximately 1325 miles long—that certainly rules them out! The network shown in Fig. 7-29(e) is approximately 982 miles long. This last network is the shortest in our list and thus the shortest network connecting, the four cities.

There are several lessons that we can draw from the four examples we discussed in this section. The first and clearest lesson is that finding the shortest network is far from easy—even for four cities it's a tough nut to crack. The second lesson is that we should not use the first lesson as an excuse and give up—we can learn a lot about a problem by just poking around. The third lesson is—don't forget the MST! The MST is always a candidate for the title of shortest network, and even when it isn't it gives us a good baseline for the SN. The fourth lesson is that the shortest networks seem to have an affinity for Steiner points—whenever the SN has interior points, it likes them to be Steiner points.

What happens when the number of cities grows—five, six, one hundred? How do we find the shortest network? Here, mathematicians face a situation analogous to the one discussed in Chapter 6—no optimal and efficient algorithm is currently known. At this point, the best we can do is to take advantage of the following rule, which we will informally call the **shortest network rule**.

THE SHORTEST NETWORK RULE

The shortest network connecting a set of points is either

- a *minimum spanning tree* (no interior junction points) or
- a *Steiner tree*. [A **Steiner tree** is a network with no circuits (i.e., a tree) such that all interior junction points are Steiner points.]

Based on the shortest network rule, we have the following algorithm for finding a shortest network connecting any set of points (cities, terminals, etc.) in the plane:

1. Make a list of all possible Steiner trees connecting the points, and find the shortest network in that list.

2. Using Kruskal's algorithm, find the minimum spanning tree connecting the points.

3. Compare the trees found in Steps 1 and 2. The shorter of these two is the desired shortest network.

The above algorithm is in essence the strategy we successfully used to find the SN in Example 7.14, and although it may seem reasonable in theory, it is totally impractical. The problem is with Step 1. With as few as 10 points, we might have to compute over a million possible Steiner trees; with 20 cities, the number of possible Steiner trees is in the billions. Using the terminology introduced in Chapter 6, we describe this algorithm as *optimal* (it guarantees the shortest network), but *inefficient* (as the number of points grows, the amount of effort needed to find the solution grows too fast for the algorithm to be practical). At the moment, no optimal and efficient algorithm for finding the shortest network is known.

What's the alternative? Just as in Chapter 6, if we are willing to settle for *approximate* solutions (in other words, if we are willing to accept a short network—not necessarily *the shortest*), then we can do very well, even when the number of points is large. Recently developed sophisticated *approximate algorithms* for finding short networks can tackle problems involving hundreds of points and efficiently produce short networks with lengths that are no more than 1% off the shortest network.

Even an old workhorse like Kruskal's algorithm can be used as a reasonably good approximate algorithm for finding short networks. The reason is that it has been recently demonstrated (see the *New York Times* article on the following page) that for any set of points—no matter how many and how they are laid out—the MST is never that much longer than the shortest network: 13.4% longer in the worst case scenario, but usually 3% or less (see, for example, Fig. 7-30).

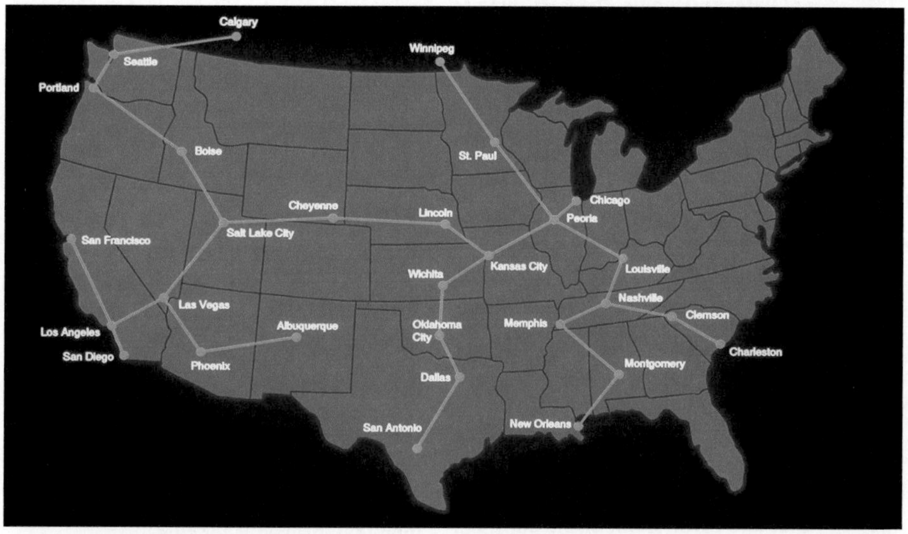

(a)

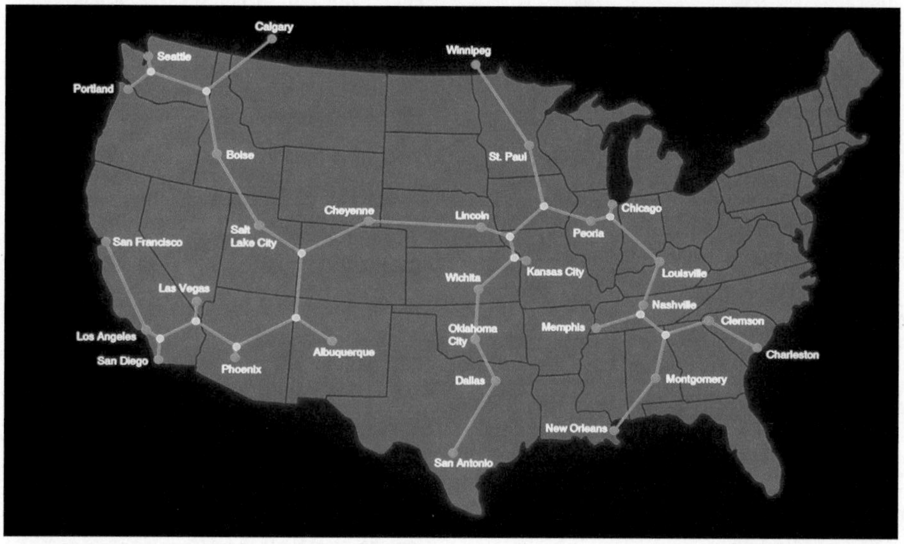

(b)

FIGURE 7-30 (a) The minimum spanning tree connecting 29 cities in the United States and Canada. Total length: approximately 7600 miles. This MST was found using Kruskal's algorithm in a matter of a few seconds. (b) The shortest network connecting the same 29 cities. (The network has 13 interior Steiner points, shown in yellow.) Total length: approximately 7400 miles (a savings of about 3%). This SN was found by a computer using a sophisticated algorithm developed by E. J. Cockayne and D. E. Hewgill of the University of Victoria.

Solution to Old Puzzle: How Short a Shortcut?

By GINA KOLATA

Two mathematicians have solved an old problem in the design of networks that has enormous practical importance but has baffled some of the sharpest minds in the business.

Dr. Frank Hwang of A.T.&T. Bell Laboratories in Murray Hill, N.J., and Dr. Ding Xhu Du, a postdoctoral student at Princeton University, announced at a meeting of theoretical computer scientists last week that they had found a precise limit to the design of paths connecting three or more points.

Designers of things like computer circuits, long-distance telephone lines, or mail routings place great stake in finding the shortest path to connect a set of points, like the cities an airline will travel to or the switching stations for long-distance telephone lines. Dr. Hwang and Dr. Du proved, without using any calculations, that an old trick of adding extra points to a network to make the path shorter can reduce the length of the path by no more than 13.4 percent.

"This problem has been open for 22 years," said Dr. Ronald L. Graham, a research director at A.T.&T. Bell Laboratories who spent years trying in vain to solve it. "The problem is of tremendous interest at Bell Laboratories, for obvious reasons."

In 1968, two other mathematicians guessed the proper answer, but until now no one had proved or disproved their conjecture.

In the tradition of Dr. Paul Erdos, an eccentric Hungarian mathematician who offers prizes for solutions to hard problems that interest him, Dr. Graham offered $500 for the solution to this one. He said he is about to mail his check.

Add Point, Reduce Path

The problem has its roots in an observation made by mathematicians in 1640. They noticed that if you want to find a path connecting three points that form the vertices of an equilateral triangle, the best thing to do is to add an extra point in the middle of the set. By connecting these four points, you can get a shorter route around the original three than if you had simply drawn lines between those three points in the first place.

Later, mathematicians discovered many more examples of this paradoxical phenomenon. It turned out to be generally true that if you want to find the shortest path connecting a set of points, you can shorten the path by adding an extra point or two.

But it was not always clear where to add the extra points, and even more important, it was not always clear how much you would gain. In 1977, Dr. Graham and his colleagues, Dr. Michael Garey and Dr. David Johnson of Bell Laboratories, proved that there was no feasible way to find where to place the extra points, and since there are so many possibilities, trial and error is out of the question.

The Shorter Route

The length of string needed to join the three corners of a triangle is shorter if a point is added in the middle, as in the figure at right.

Source: F.K. Hwang

These researchers and others found ways to guess where an extra point or so might be beneficial, but they were always left with the nagging question of whether this was the best they could do.

The conjecture, made in 1968 by Dr. Henry Pollack, who was formerly a research director at Bell Laboratories, and Dr. Edgar Gilbert, a Bell mathematician, was that the best you could ever do by adding points was to reduce the length of the path by the square root of three divided by two, or about 13.4 percent. This was exact-

Relief for designers of computer and telephone networks.

ly the amount the path was reduced in the original problem with three vertices of an equilateral triangle.

Knowing Your Limits

In the years since, researchers proved that the conjecture was true if you wanted to connect four points, then they proved it for five.

"Everyone who contributed devised some special trick," Dr. Hwang said. "But still, if you wanted to generalize to the next number of points, you couldn't do it."

Dr. Du and Dr. Hwang's proof relied on converting the problem into one involving mathematics of continuously moving variables. The idea was to suppose that there is an arrangement for which added points make you do worse than the square root of three over two. Then, Dr. Du and Dr. Hwang showed, you can slide the points in your network around and see what happens to your attempt to make the connecting path shorter. They proved that there was no way you could slide points around to do any better than a reduction in the path length of the square root of three over two.

Dr. Graham said the proof should allow network designers to relax as they search for the shortest routes. "If you can never save more than the square route of three over two, it may not be worth all the effort it takes to look for the best solution," he said.

CONCLUSION

In this chapter we discussed the problem of creating an optimal network connecting a set of points, where *optimal* usually means "cheapest" or "shortest" (sometimes both). In practice, the points represent geographical locations (cities, telephone centrals, pumping stations, etc.), and the links can be fiber-optic cable lines, rail lines, power lines, roads, and so on. Depending on the circumstances, we considered two different ways of doing this.

The MST version. In the first half of the chapter, we were required to build the network in such a way that no junction points other than the original locations were allowed, in which case the optimal network turned out to be a *minimum spanning tree*. Finding a minimum spanning tree was the great success story of the chapter—Kruskal's algorithm is a simple *optimal* and *efficient* algorithm for finding MSTs.

The SN version. In the second half of the chapter, we considered what on the surface appears to be a minor change—allowing the network to have interior junction points. In this case, the problem became one of finding the *shortest network* connecting the points. Here the situation got considerably more complicated, but we did learn a few things:

- Sometimes the minimum spanning tree and the shortest network are one and the same, but more often than not they are different. In the latter case, the shortest network is obviously shorter than the minimum spanning tree, but not by all that much. The percentage difference between the minimum spanning tree and the shortest network is always under 13.4%, and in most problems it is a lot less than that.

- In a shortest network, the interior junction points must always be *Steiner points*—junction points formed by three lines joining at 120° angles. It follows that *if a shortest network has interior junction points, then it must be a Steiner tree* (a tree whose interior junction points are all Steiner points) *and if it doesn't, then it must be the minimum spanning tree.*

- In general, the number of minimum spanning trees for a given set of points is small (usually just one) and easy to find using Kruskal's algorithm, but the number of possible Steiner trees for the same set of points is usually very large and difficult to find. With 7 points, for example, the possible number of Steiner trees is in the thousands, and with 10, it's in the millions.

Finding the shortest network connecting a set of points is, in general, a very hard problem, similar in many ways to the traveling salesman problems discussed in Chapter 6. While no optimal and efficient algorithms for finding shortest networks are currently known, there are approximate algorithms that do amazingly well, giving solutions with relative errors of less than 1%. In most cases this is good enough. In a few cases it isn't, and mathematicians are constantly striving to find even better algorithms.

P ROFILE: Evangelista Torricelli (1608–1647)

We usually first hear about Torricelli in a physics class. He was, after all, the man who invented the barometer (originally called *Torricelli's tube*), the man who discovered the laws of atmospheric pressure (called *Torricelli's laws*), and the man who gave us the equation describing the flow of a liquid discharging through an opening in the bottom of a container (known as *Torricelli's equation*). Somewhat overshadowed by his fame as a physicist is that Torricelli was also one of the most gifted mathematicians of his time. In spite of a short life span, he produced many important contributions to geometry, including the elegant method described earlier in the chapter for finding the shortest network connecting the vertices of a triangle (*Torricelli's construction*).

Evangelista Torricelli was born in Faenza, Italy, the son of a textile artisan of modest means. As a young man he was educated at the Jesuit school in Faenza and tutored by his uncle, a Catholic monk. At the age of 18 he was sent by his uncle to continue his studies in Rome. His tutor in Rome was Father Benedetto Castelli, a well-known scholar who had studied under Galileo. Castelli recognized that he had a brilliant student on his hands and tutored Torricelli in most of the mathematics and physics that were known at that time. Torricelli completed his studies by writing a treatise on the path of projectiles using Galileo's recently discovered laws of motion. Castelli forwarded Torricelli's completed manuscript on projectiles to Galileo, who was so impressed with the mathematical sophistication used by Torricelli in this work that he offered him a position as his secretary and personal assistant.

By this time, Galileo was in serious trouble with the Catholic Church for writing a book supporting the theory that it's the Earth that moves around the Sun, and not the other way around. This was considered heresy, and Galileo was tried and placed under house arrest at his own estate near Florence. Torricelli, as much as he admired Galileo, was afraid of getting into trouble himself (guilt by association was common practice in those days), so he postponed accepting Galileo's offer for some time. For the next few years, Torricelli supported himself by working as an assistant to several professors at the University of Rome and by grinding lenses for telescopes, a delicate craft at which he was a master. (Torricelli became known as the best telescope maker of his time and also developed an ingenious new type of microscope using tiny drops of crystal.)

In 1641 Torricelli finally decided to join Galileo and work for him, but by this time Galileo was blind due to an eye infection and in pretty poor health. Galileo died three months after Torricelli joined him. Upon Galileo's death, Torricelli was invited by the Grand Duke Ferdinand II of Tuscany to succeed Galileo as court mathematician and professor of mathematics at the University of Florence. Torricelli remained in this position and lived in the ducal palace in Florence until his death in 1647.

Having a prestigious academic post and being financially secure allowed Torricelli to concentrate on his scientific work. During this period, Torricelli conducted the experiment he is best remembered for, filling a long glass tube with mercury and turning it upside down (carefully keeping any air from getting in) into a dish also containing mercury. The level of mercury always fell a certain amount, creating a vacuum inside the tube. This single experiment secured Torricelli's name in history, as it disproved the prevailing doctrine of the time that vacuums cannot be sustained in nature. The experiment also paved the way to his discovery of the barometer.

Torricelli died in Florence at the age of 39 due to an illness of undetermined nature. A man best remembered for his practical contributions to our understanding of our physical world, he was most proud of his theoretical work in mathematics. Once, when challenged because of some inaccuracies in his calculations on the paths of cannonballs, he replied, "I write for philosophers, not bombardiers." ■

KEY CONCEPTS

EXERCISES

WALKING

A Trees

1. For each graph, determine whether the graph is a tree or not. If the graph is not a tree, give the reason(s) why not.

(a) The graph shown in Fig. 7-31(a)

(b) The graph shown in Fig. 7-31(b)

(c) The graph shown in Fig. 7-31(c)

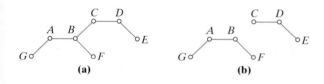

(a)   (b)

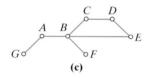

(c)

FIGURE 7-31

2. For each graph, determine whether the graph is a tree or not. If the graph is not a tree, give the reason(s) why not.

(a) The graph shown in Fig. 7-32(a)

(b) The graph shown in Fig. 7-32(b)

(c) The graph shown in Fig. 7-32(c)

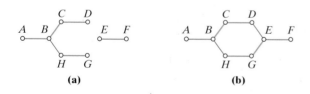

(a)   (b)

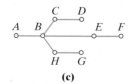

(c)

FIGURE 7-32

3. Let T be a tree with eight vertices.

(a) How many edges does T have?

(b) How many bridges does T have?

4. Let T be a tree with nine edges.

(a) How many vertices does T have?

(b) How many bridges does T have?

In Exercises 5 through 8, assume that G is a graph with no loops or multiple edges, and choose the option that best applies: (I) G is definitely a tree (explain why); (II) G is definitely not a tree (explain why); or (III) G may or may not be a tree (in this case, give two examples of graphs that fit the description—one a tree and the other one not).

5. (a) G has 8 vertices and 10 edges.

(b) G has 8 vertices and 7 edges.

(c) G has 8 vertices and is connected, and every edge in G is a bridge.

(d) G has 8 vertices, and there is exactly one path from any vertex to any other vertex.

(e) G has 8 vertices and 7 bridges.

6. (a) G has 10 vertices and 11 edges and is a connected graph.

(b) G has 10 vertices and 9 edges.

(c) G has 10 vertices and for some pair of vertices X and Y in G there are two paths from X to Y.

(d) G has 10 vertices and for every pair of vertices X and Y in G there is at least one path from X to Y.

(e) G has 10 vertices and for every pair of vertices X and Y in G there is exactly one path from X to Y.

(f) G has 10 vertices and 9 bridges.

7. (a) G has 8 vertices and no circuits.

(b) G has 8 vertices, 7 edges, and exactly one circuit.

(c) G has 8 vertices, 7 edges, and no circuits.

8. (a) G has 10 vertices, and there is a Hamilton circuit in G.

(b) G is connected and has 10 vertices. Every vertex has degree 9.

(c) G is connected and has 10 vertices. One of the vertices has degree 9, and all other vertices have degree less than 9.

(d) G is connected and has 10 vertices, and every vertex has degree 2.

To answer Exercises 9 and 10, you need to be familiar with the concepts of an Euler path and an Euler circuit discussed in Chapter 5.

9. Explain why if every vertex of a graph has even degree, then the graph is not a tree.

10. (a) Give an example of a graph that has an Euler path and is not a tree.

  (b) Give an example of a graph that has an Euler path and is a tree.

B Spanning Trees

11. Consider the network shown in Fig. 7-33.

  (a) Find a spanning tree of the network.

  (b) Calculate the redundancy of the network.

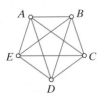

FIGURE 7-33

12. Consider the network shown in Fig. 7-34.

  (a) Find a spanning tree of the network.

  (b) Calculate the redundancy of the network.

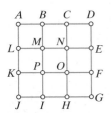

FIGURE 7-34

13. Consider the network shown in Fig. 7-35.

  (a) Find a spanning tree of the network.

  (b) Calculate the redundancy of the network.

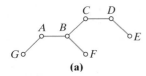

(a)

FIGURE 7-35

14. Consider the network shown in Fig. 7-36.

  (a) Find a spanning tree of the network.

  (b) Calculate the redundancy of the network.

C D
A B   E F
      H G

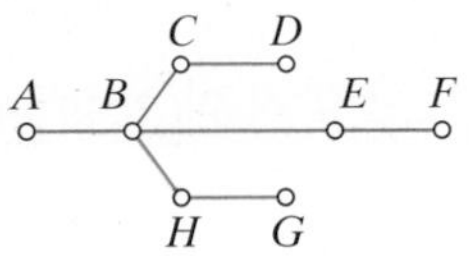

FIGURE 7-36

15. (a) Find all the spanning trees of the network shown in Fig. 7-37(a).

  (b) Find all the spanning trees of the network shown in Fig. 7-37(b).

  (c) How many different spanning trees does the network shown in Fig. 7-37(c) have?

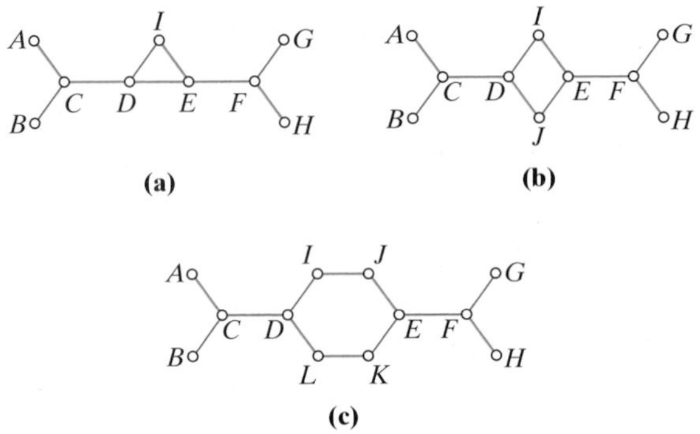

(a)   (b)

(c)

FIGURE 7-37

16. (a) Find all the spanning trees of the network shown in Fig. 7-38(a).

  (b) Find all the spanning trees of the network shown in Fig. 7-38(b).

  (c) How many different spanning trees does the network shown in Fig. 7-38(c) have?

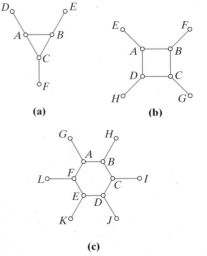

(a)   (b)

(c)

FIGURE 7-38

17. (a) How many different spanning trees does the network shown in Fig. 7-39(a) have?

(b) How many different spanning trees does the network shown in Fig. 7-39(b) have?

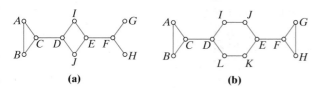

(a)   **(b)**

FIGURE 7-39

18. (a) How many different spanning trees does the network shown in Fig. 7-40(a) have?

(b) How many different spanning trees does the network shown in Fig. 7-40(b) have?

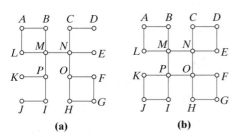

(a)   **(b)**

FIGURE 7-40

C Minimum Spanning Trees and Kruskal's Algorithm

19. For the network shown in Fig. 7-41,

(a) find the MST of the network using Kruskal's algorithm.

(b) give the weight of the MST found in (a).

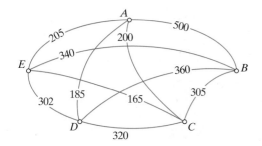

FIGURE 7-41

20. For the network shown in Fig. 7-42,

(a) find the MST of the network using Kruskal's algorithm.

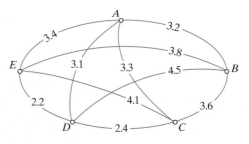

FIGURE 7-42

(b) give the weight of the MST found in (a).

21. For the network shown in Fig. 7-43,

(a) find the MST of the network using Kruskal's algorithm.

(b) give the weight of the MST found in (a).

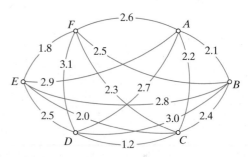

FIGURE 7-43

22. For the network shown in Fig. 7-44,

(a) find the MST of the network using Kruskal's algorithm.

(b) give the weight of the MST found in (a).

Table layout for Figure 7-44:

```
      A    B    C    D
      o-30-o-15-o-30-o
      30   10   10   30
   L  o-20-o-10-o-20-o E
      15   M    N    15
      30   10   10   30
   K  o-35-o-20-o-35-o F
      30   P    O    30
      30   25   25   30
      o-30-o-15-o-30-o G
      J    I    H
```

FIGURE 7-44

23. The mileage chart in Fig. 7-45 shows the distances between Atlanta, Columbus, Kansas City, Minneapolis, Pierre, and Tulsa. Working directly from the mileage chart use Kruskal's algorithm to find the MST connecting the six cities.

Mileage Chart

	Atlanta	Columbus	Kansas City	Minneapolis	Pierre	Tulsa
Atlanta	*	533	798	1068	1361	772
Columbus	533	*	656	713	1071	802
Kansas City	798	656	*	447	592	248
Minneapolis	1068	713	447	*	394	695
Pierre	1361	1071	592	394	*	760
Tulsa	772	802	248	695	760	*

FIGURE 7-45

24. The mileage chart in Fig. 7-46 shows the distances between Boston, Dallas, Houston, Louisville, Nashville, Pittsburgh, and St. Louis. Working directly from the mileage chart use Kruskal's algorithm to find the MST connecting the seven cities.

Mileage Chart

	Boston	Dallas	Houston	Louisville	Nashville	Pittsburgh	St. Louis
Boston	*	1748	1804	941	1088	561	1141
Dallas	1748	*	243	819	660	1204	630
Houston	1804	243	*	928	769	1313	779
Louisville	941	819	928	*	168	388	263
Nashville	1088	660	769	168	*	553	299
Pittsburgh	561	1204	1313	388	553	*	588
St. Louis	1141	630	779	263	299	588	*

FIGURE 7-46

25. The 3 by 4 grid shown in Fig. 7-47 represents a network of streets (3 blocks by 4 blocks) in a small subdivision. For landscaping purposes, it is necessary to get water to each of the corners by laying down a system of pipes along the streets. The cost of laying down the pipes is $40,000 per mile, and each block of the grid is exactly half a mile long. Find the cost of the cheapest network of pipes connecting all the corners of the subdivision. Explain your answer. (*Hint*: First determine the number of blocks in the MST.)

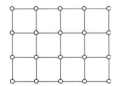

FIGURE 7-47

26. The 4 by 5 grid shown in Fig. 7-48 represents a network of streets (4 blocks by 5 blocks) in a small subdivision. For landscaping purposes, it is necessary to get water to

each of the corners by laying down a system of pipes along the streets. The cost of laying down the pipes is $40,000 per mile, and each block of the grid is exactly half a mile long. Find the cost of the cheapest network of pipes connecting all the corners of the subdivision. Explain your answer. (*Hint*: First determine the number of blocks in the MST.)

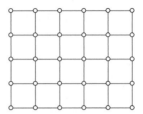

FIGURE 7-48

D Steiner Points and Shortest Networks

27. Find the length of the shortest network connecting the indicated cities. All distances are rounded to the nearest mile. (Figures are not drawn to scale.)

(a) The three cities A, B, and C shown in Fig. 7-49(a)

(b) The three cities A, B, and C shown in Fig. 7-49(b)

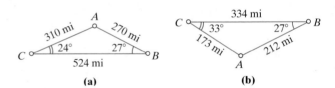

FIGURE 7-49

28. Find the length of the shortest network connecting the indicated cities. All distances are rounded to the nearest mile. (Figures are not drawn to scale.)

(a) The three cities A, B, and C shown in Fig. 7-50(a)

(b) The three cities A, B, and C shown in Fig. 7-50(b)

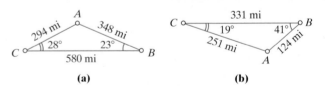

FIGURE 7-50

29. Find the length of the shortest network connecting the indicated cities. All distances are rounded to the nearest kilometer. (Figures are not drawn to scale.)

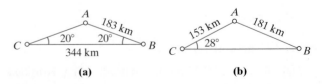

FIGURE 7-51

(a) The three cities A, B, and C shown in Fig. 7-51(a)

(b) The three cities A, B, and C shown in Fig. 7-51(b)

30. Find the length of the shortest network connecting the indicated cities. All distances are rounded to the nearest kilometer. (Figures are not drawn to scale.)

(a) The three cities A, B, and C shown in Fig. 7-52(a)

(b) The three cities A, B, and C shown in Fig. 7-52(b)

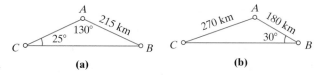

(a)   **(b)**

FIGURE 7-52

31. Find the length of the shortest network (rounded to the nearest mile) connecting the three cities A, B, and C shown in Fig. 7-53. (Figure is not drawn to scale.) (*Hint*: Use the Pythagorean Theorem.)

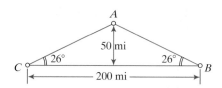

FIGURE 7-53

32. Find the length of the shortest network (rounded to the nearest mile) connecting the three cities A, B, and C shown in Fig. 7-54. (Figure is not drawn to scale.) (*Hint*: Use the Pythagorean Theorem.)

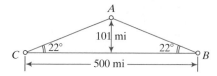

FIGURE 7-54

33. Assume that one of three points—X, Y, or Z—is a Steiner point of triangle ABC. The table in Fig. 7-55 shows the distances (rounded to the nearest mile) of each of these three points to the vertices of the triangle. Determine which of the three points is the Steiner point. Explain your answer.

Distance in miles

	X	Y	Z
A	54	72	41
B	61	94	87
C	125	77	104

FIGURE 7-55

34. Assume that one of three points—X, Y, or Z—is a Steiner point of triangle ABC. The table in Fig. 7-56 shows the distances (rounded to the nearest mile) of each of these three points to the vertices of the triangle. Determine which of the three points is the Steiner point. Explain your answer.

Distance in miles

	X	Y	Z
A	380	390	700
B	620	680	260
C	300	190	550

FIGURE 7-56

Exercises 35 and 36 refer to five cities (A, B, C, D, and E) located as shown in Fig. 7-57. Assume that $AC = AB$, that $DC = DB$, and that angle CDB measures $120°$.

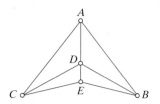

FIGURE 7-57

35. (a) Which is larger, $CD + DB$ or $CE + ED + EB$? Explain.

(b) What is the shortest network connecting cities C, D, and B? Explain.

(c) What is the shortest network connecting cities C, E, and B? Explain.

36. (a) Which is larger, $CA + AB$ or $DC + DA + DB$? Explain.

(b) Which is larger, $EC + EA + EB$ or $DC + DA + DB$? Explain.

(c) What is the shortest network connecting cities A, B, and C? Explain.

E **Miscellaneous**

37. Consider a network with M edges. Let k denote the number of bridges in the network.

(a) If $M = 5$, list all the possible values of k.

(b) If $M = 123$, describe the set of all possible values of k.

38. Consider a network with M edges. Let k denote the number of bridges in the network.

(a) If $M = 7$, list all the possible values of k.

(b) If $M = 2481$, describe the set of all possible values of k.

39. Consider a network with N vertices and M edges.

   (a) If $N = 123$ and $M = 122$, find the number of circuits in the network.

   (b) If $N = 123$ and $M = 123$, find the number of circuits in the network.

   (c) If $N = 123$, $M = 123$, and there is a circuit of length 5, find the number of bridges in the network.

40. Consider a network with N vertices and M edges.

   (a) If $N = 2481$ and $M = 2480$, find the number of circuits in the network.

   (b) If $N = 2481$ and $M = 2481$, find the number of circuits in the network.

   (c) If $N = 2481$, $M = 2481$, and there is a circuit of length 10, find the number of bridges in the network.

41. Give an example of a graph with $N = 11$ vertices and $M = 10$ edges having

   (a) exactly one circuit.

   (b) exactly two circuits.

   (c) exactly three circuits.

42. Give an example of a graph with $N = 14$ vertices and $M = 13$ edges having

   (a) exactly one circuit.

   (b) exactly two circuits.

   (c) exactly three circuits.

30-60-90 Triangles. *30-60-90 triangles are typically covered in high school geometry. The purpose of Exercises 43 and 44 is to review the relationship between the lengths of the three sides of a 30-60-90 triangle. Figure 7-58 shows a generic 30-60-90 triangle. As shown in the figure, we denote the lengths of the three sides by s (for short leg), l (for long leg), and h (for hypotenuse).* (Use the approximation $\sqrt{3} \approx 1.73$ in your calculations.)

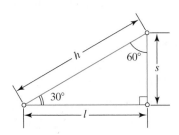

    FIGURE 7-58

43. (a) Express h in terms of s. (*Hint*: A 30-60-90 triangle is half of an equilateral triangle.)

   (b) Express l in terms of s. [*Hint*: Use (a) and the Pythagorean theorem.]

(c) If $s = 21$, find h and l (rounded to two decimal places).

(d) If $h = 30.4$, find s and l (rounded to two decimal places).

44. (a) Express s in terms of h. [*Hint*: Try Exercise 45(a) first.]

   (b) Express l in terms of h. [*Hint*: Try Exercise 45(b) first.]

   (c) Express h in terms of l.

   (d) If $l = 173$, find s and h (rounded to the nearest integer).

45. Three cities A, B, and C are located as shown in Fig. 7-59, forming an isosceles triangle ($AB = AC$). Suppose that S is the Steiner point of the triangle. (Figure is not drawn to scale.) Using the information given in the figure, find the length of the shortest network connecting A, B, and C. (*Hint*: Look for some 30-60-90 triangles.)

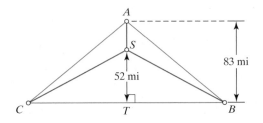

FIGURE 7-59

46. Three cities A, B, and C are located as shown in Fig. 7-60, forming an isosceles triangle ($AB = AC$). Suppose that S is the Steiner point of the triangle. (Figure is not drawn to scale.) Using the information given in the figure, find the length of the shortest network connecting A, B, and C. (*Hint*: Look for some 30-60-90 triangles.)

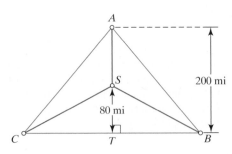

FIGURE 7-60

47. Three cities A, B, and C are located as shown in Fig. 7-61, forming an isosceles triangle ($AB = AC$). Suppose that S is the Steiner point of the triangle. (Figure is not drawn to scale.) Using the information given in the figure, find the length of the shortest network connecting A, B, and C. (Use $\sqrt{3} \approx 1.73$ *in your calculations*, and round your answer to the nearest mile. (*Hint*: Look for some 30-60-90 triangles.)

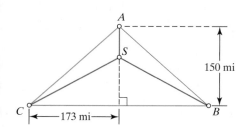

FIGURE 7-61

48. Three cities $A, B,$ and C are located as shown in Fig. 7-62, forming an isosceles triangle ($AB = AC$). Suppose that S is the Steiner point of the triangle. (Figure is not drawn to scale.) Using the information given in the figure, find the length of the shortest network connecting $A, B,$ and C. (Use $\sqrt{3} \approx 1.73$ *in your calculations*, and round your answer to the nearest mile. (*Hint*: Look for some 30-60-90 triangles.)

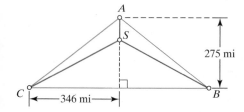

FIGURE 7-62

Exercises 49 and 50 refer to the Outback Cable Network discussed in Example 7.8. The three cities A, B, and C form an equilateral triangle with sides of length equal to 500 miles.

49. Show that the length of the T-network shown in Fig. 7-63 (rounded to the nearest mile) is 933 miles. (*Hint*: Look for some 30-60-90 triangles.)

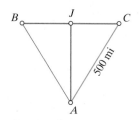

FIGURE 7-63

50. Show that the length of the shortest network shown in Fig. 7-64 (rounded to the nearest mile) is 866 miles. (*Hint*: Look for some 30-60-90 triangles.)

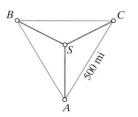

FIGURE 7-64

JOGGING

51. (a) How many spanning trees does the network shown in Fig. 7-65(a) have?

(b) How many different spanning trees does the network shown in Fig. 7-65(b) have?

(c) How many different spanning trees does the network shown in Fig. 7-65(c) have?

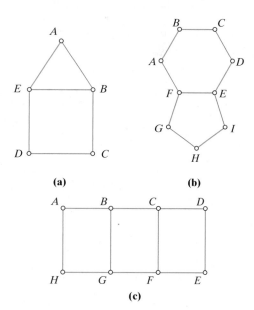

(a) **(b)**

(c)

FIGURE 7-65

52. (a) Let G be a tree with N vertices. Find the sum of the degrees of all the vertices in G.

(b) Explain why a tree must have at least two vertices of degree 1. (A vertex of degree 1 in a tree is called a *leaf*.)

(c) Explain why in a tree with three or more vertices the degrees of the vertices cannot all be the same.

53. (a) Give an example of a tree with six vertices such that the degrees of the vertices are $1, 1, 2, 2, 2, 2$.

(b) Give an example of a tree with N vertices such that the degrees of the vertices are $1, 1, 2, 2, 2, \ldots, 2$.

(c) Give an example of a tree with five vertices such that the degrees of the vertices are $1, 1, 1, 1, 4$.

(d) Give an example of a tree with N vertices such that the degrees of the vertices are $1, 1, 1, \ldots, 1,$ $N - 1$.

54. Redundancy. Recall that the *redundancy* of a network with N vertices and M edges is given by $R = M - (N - 1)$.

(a) Describe all networks with $R = 0$.

(b) Explain why a network with $R = 1$ has exactly one circuit.

(c) Explain why if there are no loops or multiple edges, then the maximum value of the redundancy of a network is given by $R = (N^2 - 3N + 2)/2$.

55. Suppose that G is a network with N vertices and M edges, with $M \geq N$. Explain why the number of bridges in G is smaller than or equal to $M - 3$.

56. A highway system connecting nine cities—$C_1, C_2,$ $C_3, \ldots, C_9$—is to be built. The accompanying table shows the projected cost (in millions of dollars) of a highway link between any two cities. Using Kruskal's algorithm, find the minimum spanning tree connecting the nine cities.

	C_1	C_2	C_3	C_4	C_5	C_6	C_7	C_8	C_9
C_1	*	1.3	3.4	6.6	2.6	3.5	5.7	1.1	3.8
C_2	1.3	*	2.4	7.9	1.7	2.3	7.0	2.4	3.9
C_3	3.4	2.4	*	9.9	3.4	1.0	9.1	4.4	6.5
C_4	6.6	7.9	9.9	*	8.2	9.7	0.9	5.5	4.9
C_5	2.6	1.7	3.4	8.2	*	4.8	7.4	3.7	3.5
C_6	3.5	2.3	1.0	9.7	4.8	*	8.9	4.4	5.8
C_7	5.7	7.0	9.1	0.9	7.4	8.9	*	4.7	3.9
C_8	1.1	2.4	4.4	5.5	3.7	4.4	4.7	*	2.8
C_9	3.8	3.9	6.5	4.9	3.5	5.8	3.9	2.8	*

57. The four cities (A, B, C, and D), shown in Fig. 7-66 must be connected into a telephone network. The cost

of laying down telephone cable connecting any two of the cities is given (in millions of dollars) by the weights of the edges in the graph. In addition, in any city that serves as a junction point of the network, expensive switching equipment must be installed. The cost of installing this equipment is given (in millions of dollars) by the numbers inside the circles. Find the minimum-cost telephone network connecting these four cities.

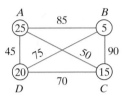

FIGURE 7-66

58. Figure 7-67(a) shows a network of roads connecting cities A through G. The weights of the edges represent the cost (in millions of dollars) of putting underground fiber-optic lines along the roads, and the MST of the network is shown in red. Figure 7-67(b) shows the same network except that one additional road (connecting E and G) has been added. Let x be the cost (in millions) of putting fiber-optic lines along this new road.

(a) Describe the MST of the network in Fig. 7-67(b) in the case $x > 59$. Explain your answer.

(b) Describe the MST of the network in Fig. 7-67(b) in the case $x < 59$. Explain your answer.

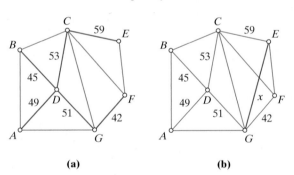

(a)   **(b)**

FIGURE 7-67

59. Middletown is designing a new high-speed rail system to connect Downtown, the Airport, and the Coliseum. The major avenues of Middletown form the grid shown in Fig. 7-68, and the elevated tracks for the rail system will have to run along the these major avenues. In addition, a switching station (junction point) will be needed somewhere on the grid.

(a) Determine the optimal location for the switching station.

(b) Suppose it costs $1,000,000 per mile to lay the elevated track and $500,000 for the switching station.

Find an optimal network connecting the three locations, and give the cost of this network.

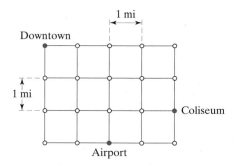

FIGURE 7-68

60. Happyville is building a new high-speed rail system to connect the Airport, the Midtown Mall, the University, and Slugger's Ballpark. The major avenues of Happyville form the grid shown in Fig. 7-69, and the elevated tracks for the rail system will have to run along these major avenues. In addition, two switching stations (junction points) will be placed at two different locations on the grid.

(a) Determine the optimal location for the two switching stations.

(b) Suppose that it costs $2,000,000 per mile of elevated track plus $500,000 for each switching station. Find the optimal network connecting the four locations, and give the cost of the network.

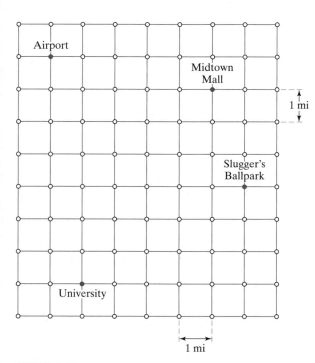

FIGURE 7-69

61. Consider triangle ABC having an equilateral triangle EFG inside, as shown in Fig. 7-70.

(a) Find the measure of angles BFA, AEC, and CGB.

(b) Explain why triangle ABC has a Steiner point S.

(c) Explain why the Steiner point of triangle ABC lies inside triangle EFG.

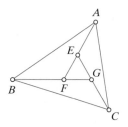

FIGURE 7-70

62. (a) Compute the length of the red network shown in Fig. 7-71(a). Round your answer to the nearest mile.

(b) Show that the length of the red network shown in Fig. 7-71(b) (rounded to the nearest mile) is 1366 miles. (S_1 and S_2 are Steiner points.) (*Hint:* Look for hidden 30-60-90 triangles.)

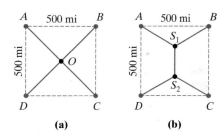

FIGURE 7-71

63. (a) Show that the length of the red network shown in Fig. 7-72(a) (rounded to the nearest mile) is 993 miles. (S_1 and S_2 are Steiner points.) [*Hint:* Do Exercise 62(b) first.]

(b) Show that the length of the red network shown in Fig. 7-72(b) (rounded to the nearest tenth of a mile) is 919.6 miles. (S_1 and S_2 are Steiner points.)

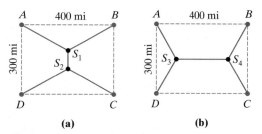

FIGURE 7-72

64. Viviani's theorem. Let P be an arbitrary point inside an equilateral triangle ABC. Let h denote the height of the triangle, and let x, y, and z denote the distances from P to the three sides of the triangle, as shown in Fig. 7-73. Show that $x + y + z = h$.

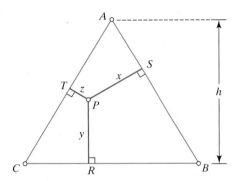

FIGURE 7-73

65. Boruvka's algorithm. In 1926 Czech mathematician Otakar Boruvka gave the following algorithm for finding the minimum spanning tree of a network.

Start: Start with a "blank" copy of the network (no edges, just the vertices).

Step 1. For each vertex, choose an edge incident to that vertex that (a) has least weight and (b) does not create any cycles. Add these edges (to the blank copy). At this point, you will have a collection of disconnected subtrees.

Main Step. For each subtree formed in the previous step, choose an edge incident to that subtree that (a) has least weight and (b) does not create any cycles. Continue doing this until the individual subtrees are all connected into a network. This will be the MST.

Consider the network shown in Fig. 7-74 (this is the same network as in Exercise 19).

(a) Show the subtrees formed at Step 1 of Boruvka's algorithm.

(b) Show the subtrees formed at Step 2 of Boruvka's algorithm.

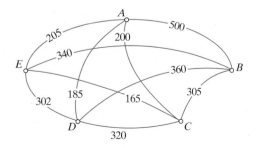

FIGURE 7-74

66. Cayley's theorem. Cayley's theorem says that the number of spanning trees in a complete graph with N vertices is given by N^{N-2}.

(a) Verify this result for the cases $N = 3$ and $N = 4$ by finding all spanning trees for complete graphs with three and four vertices.

(b) Which is larger, the number of Hamilton circuits or the number of spanning trees in a complete graph with N vertices? Explain.

RUNNING

67. Show that if a tree has a vertex of degree K, then there are at least K vertices in the tree of degree 1.

68. A *bipartite graph* is a graph with the property that the vertices of the graph can be divided into two sets A and B so that every edge of the graph joins a vertex from A to a vertex from B (Fig. 7-75). Explain why trees are always bipartite graphs.

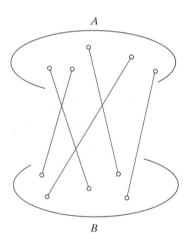

FIGURE 7-75

69. (a) Suppose that there is an edge in a network that must be included in any spanning tree. Give an algorithm for finding the minimum spanning tree that includes a given edge. (*Hint*: Modify Kruskal's algorithm.)

(b) Consider the mileage chart given in Exercise 56. Suppose that C_3 and C_4 are the two largest cities in the area and that the Chamber of Commerce insists that a section of highway directly connecting them must be built. Find the minimum spanning tree that includes the section of highway between C_3 and C_4.

70. In Fig. 7-76 A, B, C, and D are the vertices of a square. In Fig. 7-76(a) S_1 and S_2 are Steiner points. In Fig. 7-76(b) X and Y are two arbitrary points that are not Steiner points. Show that the red network in (a) is shorter than the red network in (b). (You can assume any facts concerning shortest networks for three points.)

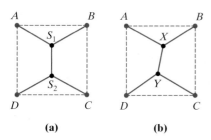

FIGURE 7-76

71. Suppose that A, B, C, and D are four towns located as shown in Fig. 7-77. Show that it is impossible to have a network with two Steiner points inside the trapezoid connecting the four cities.

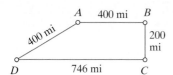

FIGURE 7-77

72. Suppose that A, B, C, and D are four towns located at the vertices of a rectangle with length b and height a as shown in Fig. 7-78(a). Assume that $a < b$. Determine the conditions that must be satisfied by a and b so that the length of the MST shown in Fig. 7-78(b) is less than the length of the Steiner tree shown in Fig. 7-78(c).

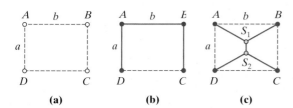

FIGURE 7-78

73. This example describes a method for constructing a Steiner tree inside a quadrilateral when the Steiner tree has two interior Steiner points. Consider the quadrilateral $ABCD$ shown in Fig. 7-79(a).

Step 1: Use two opposite sides of the quadrilateral (say AB and CD) to construct equilateral triangles ABX and CDY, as illustrated in Fig. 7-79(b).

Step 2: Circumscribe a circle around each of the two equilateral triangles constructed in Step 1.

Step 3: Let S_1 and S_2 denote the points where the line segment XY intersects the circles constructed in Step 2 [Fig. 7-79(b)].

Explain why the network shown in red in Fig. 7-79(c) is a Steiner tree connecting the points A, B, C, and D. (*Hint*: Look carefully at the arguments used to justify Torricelli's construction in Section 7.4.)

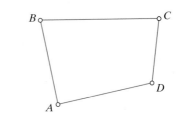

(a)

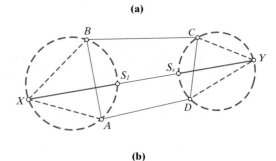

(b)

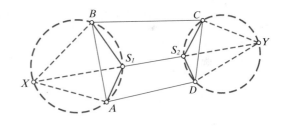

(c)

FIGURE 7-79

74. Consider the 1 by 3 square grid shown in Fig. 7-80.

(a) Find the shortest network connecting the eight vertices of the grid.

(b) Compute the length of the shortest network found in (a). (Assume that each square in the grid is a 1 by 1 square.)

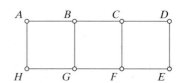

FIGURE 7-80

(c) Compare the length of the shortest network found in (b) with the length of the minimum spanning tree for the same 1 by 3 grid. What is the percentage difference between the length of the MST and the length of the shortest network?

75. Consider the 3 by 3 square grid shown in Fig. 7-81.

(a) Find the shortest network connecting the 16 vertices in the grid.

(b) Compute the length of the shortest network found in (a). (Assume that each square in the grid is a 1 by 1 square.)

(c) Compare the length of the shortest network found in (b) with the length of the minimum spanning tree for the same 3 by 3 grid. What is the percentage difference between the length of the MST and the length of the shortest network?

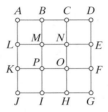

FIGURE 7-81

PROJECTS AND PAPERS

A The Kruskal-Steiner Fiber-Optic Cable Network Project

Kruskal-Steiner Corporation is a new telecommunications company providing phone/Internet service to selected cities in the United States. Kruskal-Steiner plans to build a major fiber-optic cable network connecting the 21 cities shown in the mileage chart on the opposite page. The network construction costs are estimated at $100,000 per mile, so this is serious stuff.

You have been hired as a consultant for this project. Your job is to design a network connecting the 21 cities and sell your design to the board of directors. Your contract states that your consulting fee of $10,000 will be paid only if your network is better (shorter) than the *minimum spanning tree* (MST) connecting the 21 cities. In addition, you will get a bonus of $1000 for every mile that your network saves over the MST, so it's in your best interest to try to improve on the MST as much as you can.

For this project you are asked to prepare a presentation to the board. In your presentation you should (1) describe the MST and compute the total cost of its construction, (2) describe your proposed network and show where it provides improvements over the MST, and (3) estimate the length of your network and the savings it generates over the MST.

Notes: (1) You should start by getting two accurate, good-quality maps of the contiguous United States. (2) Use Kruskal's algorithm to first find the MST, drawing it on one of the maps. (To implement Kruskal's algorithm directly on the map, you may find it convenient to first sort the distances in the mileage chart from smallest to biggest, and use this sorted list to build the MST.) (3) Looking at the angles

formed by the branches of the MST in the map, try to identify places where shortcuts are possible by introducing an interior Steiner junction point. (4) Estimate the length of each of the shortcuts and then the total length of your network. You will need the following formula: The length of the shortest network connecting the vertices of a triangle having angles that measure less than 120° each and sides of lengths a, b, and c is given by

$$\sqrt{\frac{a^2 + b^2 + c^2}{2} + \frac{\sqrt{3}}{2}\sqrt{2a^2b^2 + 2a^2c^2 + 2b^2c^2 - (a^4 + b^4 + c^4)}}$$

(5) The last, and most important step: Calculate how much money you made for your efforts ($10,000 plus $1000 for every mile you saved over the MST).

B Validating Torricelli's Construction

In this chapter we introduced Torricelli's method for finding a Steiner point inside a triangle but did not give a justification as to why the Steiner point gives the shortest network.

Torricelli proved the following three facts about a triangle ABC having all angles less than 120°.

Fact 1. There is a unique point S inside the triangle with the property that the angles ASB, ASC, and BSC are all 120° (the *Steiner point* of the triangle), as shown in Fig. 7-82(a).

Fact 2. For any point P other than the Steiner point S, $\overline{PA} + \overline{PB} + \overline{PC} > \overline{SA} + \overline{SB} + \overline{SC}$. (This implies that the network obtained using S as the junction point is the shortest network connecting the three vertices of the triangle.) [Fig. 7-82(b).]

Mileage Chart

	Atlanta	Boston	Buffalo	Chicago	Columbus	Dallas	Denver	Houston	Kansas City	Louisville	Memphis	Miami	Minneapolis	Nashville	New York	Omaha	Pierre	Pittsburgh	Raleigh	St. Louis	Tulsa
Atlanta	*	1037	859	674	533	795	1398	789	798	382	371	655	1068	242	841	986	1361	687	372	541	772
Boston	1037	*	446	963	735	1748	1949	1804	1391	941	1293	1504	1368	1088	206	1412	1726	561	685	1141	1537
Buffalo	859	446	*	522	326	1346	1508	1460	966	532	899	1409	927	700	372	971	1285	216	605	716	1112
Chicago	674	963	522	*	308	917	996	1067	499	292	530	1329	405	446	802	459	763	452	784	289	683
Columbus	533	735	326	308	*	1028	1229	1137	656	209	576	1160	713	377	542	750	1071	182	491	406	802
Dallas	795	1748	1346	917	1028	*	781	243	489	819	452	1300	936	660	1552	644	943	1204	1166	630	257
Denver	1398	1949	1508	996	1229	781	*	1019	600	1120	1040	2037	841	1156	1771	537	518	1411	1661	857	681
Houston	789	1804	1460	1067	1137	243	1019	*	710	928	561	1190	1157	769	1608	865	1186	1313	1160	779	478
Kansas City	798	1391	966	499	656	489	600	710	*	520	451	1448	447	556	1198	201	592	838	1061	257	248
Louisville	382	941	532	292	209	819	1120	928	520	*	367	1037	697	168	748	687	1055	388	541	263	659
Memphis	371	1293	899	530	576	452	1040	561	451	367	*	997	826	208	1100	652	1043	752	728	285	401
Miami	655	1504	1409	1329	1160	1300	2037	1190	1448	1037	997	*	1723	897	1308	1641	2016	1200	819	1196	1398
Minneapolis	1068	1368	927	405	713	936	841	1157	447	697	826	1723	*	826	1207	357	394	857	1189	552	695
Nashville	242	1088	700	446	377	660	1156	769	556	168	208	897	826	*	892	744	1119	553	521	299	609
New York	841	206	372	802	542	1552	1771	1608	1198	748	1100	1308	1207	892	*	1251	1565	368	489	948	1344
Omaha	986	1412	971	459	750	644	537	865	201	687	652	1641	357	744	1251	*	391	895	1214	449	387
Pierre	1361	1726	1285	763	1071	943	518	1186	592	1055	1043	2016	394	1119	1565	391	*	1215	1547	824	760
Pittsburgh	687	561	216	452	182	1204	1411	1313	838	388	752	1200	857	553	368	895	1215	*	445	588	984
Raleigh	372	685	605	784	491	1166	1661	1160	1061	541	728	819	1189	521	489	1214	1547	445	*	804	1129
St. Louis	541	1141	716	289	406	630	857	779	257	263	285	1196	552	299	948	449	824	588	804	*	396
Tulsa	772	1537	1112	683	802	257	681	478	248	659	401	1398	695	609	1344	387	760	984	1129	396	*

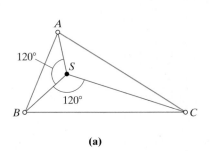

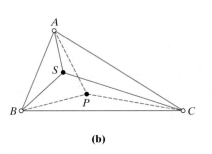

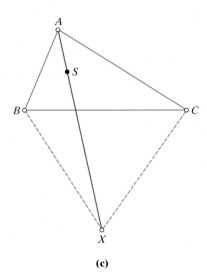

FIGURE 7-82   (a) S is unique. (b) $\overline{PA} + \overline{PB} + \overline{PC} > \overline{SA} + \overline{SB} + \overline{SC}$. (c) $\overline{SA} + \overline{SB} + \overline{PC} = \overline{AX}$.

Fact 3. $\overline{SA} + \overline{SB} + \overline{SC} = \overline{AX}$, where X is the vertex of the equilateral triangle built on side BC and on the opposite side of A, as shown in the Figure. 7-82(c).

In this project you are asked to give a complete mathematical proof of each of the facts.

Notes: (1) All the foregoing can be proved using standard facts from Euclidean geometry. (2) There are several different ways you can approach these proofs, and you can use other well-known theorems in Euclidean geometry [such as *Viviani's theorem* (see Exercise 64) and *Ptolemy's theorem*] in your proofs.

C   Prim's Algorithm

Prim's algorithm is another well-known algorithm for finding the minimum spanning tree of a weighted graph. Like Kruskal's algorithm, Prim's algorithm is an optimal and efficient algorithm for finding MSTs.

Prepare a classroom presentation on Prim's algorithm. (1) Describe the algorithm. (2) Illustrate how the algorithm works using the following two examples: (a) the weighted network in Exercise 22 and (b) the mileage chart in Exercise 24. (3) Compare Prim's algorithm to Kruskal's algorithm and discuss their differences.

D   Minimizing with Soap Film

Occasionally, we can enlist nature's aid to help us solve reasonably interesting math problems. One of the better-known examples of this is the use of soap-film solutions to solve *minimal surface* and *shortest distance* problems.

For this project, prepare a classroom presentation on soap-film solutions and how they can be used to solve certain optimization problems. Discuss (1) the types of optimization problems in which soap-film solutions can be used and (2) the basic laws of physics that explain why and how soap-film solutions work in these problems.

REFERENCES AND FURTHER READINGS

1. Almgren, Fred J., Jr., and Jean E. Taylor, "The Geometry of Soap Films and Soap Bubbles," *Scientific American*, 235 (July 1976), 82–93.

2. Barabási, Albert-László, *Linked: The New Science of Networks.* Cambridge, MA: Perseus Publishing, 2002.

3. Barabási, Albert-László, and Eric Bonabeau, "Scale-Free Networks," *Scientific American*, 288 (May 2003), 60–69.

4. Bern, M., and R. L. Graham, "The Shortest Network Problem," *Scientific American*, 260 (January 1989), 84–89.

5. Buchanan, Mark, *Nexus: Small Worlds and the Groundbreaking Theory of Networks.* New York: W. W. Norton, 2003.

6. Chung, F., M. Gardner, and R. Graham, "Steiner Trees on a Checkerboard," *Mathematics Magazine*, 62 (April 1984), 83–96.

7. Cockayne, E. J., and D. E. Hewgill, "Exact Computation of Steiner Minimal Trees in the Plane," *Information Processing Letters*, 22 (1986), 151–156.

8. Du, D.-Z., and F. K. Hwang, "The Steiner Ratio Conjecture of Gilbert and Pollack is True," *Proceedings of the National Academy of Sciences, U.S.A.*, 87 (December 1990), 9464–9466.

9. Gardner, Martin, "Mathematical Games: Casting a Net on a Checkerboard and Other Puzzles of the Forest," *Scientific American,* 254 (June 1986), 16–23.

10. Gilbert, E. N., and H. O. Pollack, "Steiner Minimal Trees," *SIAM Journal of Applied Mathematics*, 16 (1968), 1–29.

11. Graham, R. L., and P. Hell, "On the History of the Minimum Spanning Tree Problem," *Annals of the History of Computing*, 7 (January 1985), 43–57.

12. Gross, Jonathan, and Jay Yellen, *Graph Theory*. Boca Raton, FL: CRC Press, 1999, chap. 4.

13. Hayes, Brian, "Graph Theory in Practice: Part I," *American Scientist*, 88 (January–February 2000), 9–13.

14. Hayes, Brian, "Graph Theory in Practice: Part II," *American Scientist*, 88 (March–April 2000), 104–109.

15. Hwang, F. K., D. S. Richards, and P., Winter, *The Steiner Tree Problem*. Amsterdam: North Holland, 1992.

16. Robinson, P., "Evangelista Torricelli," *Mathematical Gazette*, 78 (1994), 37–47.

17. Wilson, Robin J., and John J. Watkins, *Graphs: An Introductory Approach*. New York: John Wiley & Sons, 1990, chap 10.

8 The Mathematics of Scheduling

Chasing the Critical Path

How long does it take to build a house? Here is a deceptively simple question that defies an easy answer. Some of the factors involved are obvious: the size of the house, the type of construction, the number of workers, the kinds of tools and machinery used. Less obvious, but equally important, is another variable: the ability to organize and coordinate the timing of people, equipment, and work so that things get done in a timely way. For better or for worse, this last issue boils down to a mathematics problem, just one example in a large family of problems that fall under an area of graph theory known as *combinatorial scheduling*. Discussing some of the basic ideas behind combinatorial scheduling will be the theme of this chapter.

Let's get back to our original question. According to the National Association of Home Builders, it takes 1092 work-hours to build the average American house. Since we are not being picky about details, we'll just call it 1100 hours. Essentially this means that it would take a single construction worker (assuming that this worker can do every single job required for building a house) about 1100 hours of labor to finish this hypothetical average American house. Let's now turn the question on its head. If we had 1100 equally capable workers, could we get the same house built in one hour? Of course not! In fact, we could put tens of thousands of workers on the job and we still could not get the house built in one hour. Some inherent physical limitations to the speed with which a house can be built are outside of the builder's control. Some jobs cannot be speeded up beyond a certain point, regardless of how many workers one puts on that job. Even more significantly, some jobs can be started only after certain other jobs have been completed. (Roofing, for example, can be started only after framing has been completed.)

Combinatorial scheduling involves such questions as: How fast could a house be built if one cared only about speed? (To the best of our knowledge, the record is a bit under 24 hours.) How fast could we build the house if we had 10 equally able construction workers at our disposal at all times? What if we had only three workers? What if we needed to finish the entire project within a given time frame—say, 30 days? How many workers should we hire then? The same questions could equally well be asked if we replaced "building a house" with many other types of projects—from preparing a banquet to launching a space shuttle.

This chapter starts with an introduction to the *key concepts* and terminology of combinatorial scheduling (Section 8.1). Typically, scheduling problems are modeled using a special type of graph called a *directed graph* (or digraph for

short)—we discuss digraphs and some of their basic properties in Section 8.2. In Section 8.3 we will discuss the general rules for creating schedules—the key concept here is that of a *priority list*. The two most commonly used algorithms for "solving" a scheduling problem are the *decreasing-time* and the *critical-path algorithms*—we discuss these in Sections 8.4, 8.5, and 8.6. In Section 8.7 we briefly discuss scheduling tasks that have no precedence relations among them (*independent tasks*).

8.1   The Basic Elements of Scheduling

We will now introduce the principal characters in any scheduling story.

- **The processors.** Every job requires workers. We will use the term **processors** to describe the "workers" who carry out the work. While the word *processor* may sound a little cold and impersonal, it does underscore an important point: Processors need not be human beings. In scheduling, a processor could just as well be a robot, a computer, an automated teller machine, and so on. For the purposes of our discussion, we will use N to represent the number of processors and $P_1, P_2, P_3, \ldots, P_N$ to denote the processors themselves. We will assume throughout the chapter that $N \geq 2$ (for $N = 1$ scheduling is trivial and not very interesting).

- **The tasks.** In every complex project there are individual pieces of work, often called "jobs" or "tasks." We will need to be a little more precise than that, however. We will define a **task** as an indivisible unit of work that (either by nature or by choice) cannot be broken up into smaller units. Thus, by definition a task cannot be shared—it is always *carried out by a single processor*. In general, we will use capital letters $A, B, C, \ldots$, to represent the tasks, although in specific situations it is convenient to use abbreviations (such as WE for "wiring the electrical system," or PL for "plumbing").

    At a particular moment in time a task can be in one of four possible states: (1) *ineligible* (the task cannot be started because some of the prerequisites for the task have not yet been completed), (2) *ready* (the task has not been started but could be started at this time), (3) *in execution* (the task is being carried out by one of the processors), (4) *completed*.

- **The processing times.** The **processing time** of a given task X is the amount of time, without interruption, required by *one processor* to execute X. When dealing with human processors, there are many variables (ability, attitude, work ethic, etc.) that can affect the processing time of a task, and this adds another layer of complexity to an already complex situation. On the other hand, if we assume a "robotic" interpretation of the processors (either because they are indeed machines or because they are human beings trained to work in a very standardized and uniform way), then scheduling becomes somewhat more manageable.

    To keep things simple we will work under the following three assumptions:

- Any processor can execute any task (call this the *versatility* assumption).

- The processing time for a task is the same regardless of which processor is executing the task (call this the *uniformity* assumption).

- Once a processor starts a task, it will complete it without interruption (call this the *perseverance* assumption).

Under the preceding assumptions, the concept of *processing time* for a task (we will sometimes call it the *P*-time) makes good sense—and we can conveniently incorporate this information by including it inside parentheses next to the name of the task. Thus, the notation $X(5)$ tells us that the task called X has a processing time of 5 units (be it minutes, hours, days, or any other unit of time) *regardless of which processor is assigned to execute the task.*

- **The precedence relations.** Precedence relations are formal restrictions on the order in which the tasks can be executed, much like those course prerequisites in the school catalog that tell you that you can't take course Y until you have completed course X. In the case of tasks, these prerequisites are called **precedence relations.** A typical precedence relation is of the form *task X precedes task Y* (we also say X is *precedent* to Y), and it means that *task Y cannot be started until task X has been completed.* A precedence relation can be conveniently abbreviated by writing $X \rightarrow Y$, or described graphically as shown in Fig. 8-1(a). A single scheduling problem can have hundreds or even thousands of precedence relations, each adding another restriction on the scheduler's freedom.

  At the same time, it also happens fairly often that there are no restrictions on the order of execution between two tasks in a project. When a pair of tasks X and Y have no precedence requirements between them (neither $X \rightarrow Y$ nor $Y \rightarrow X$), we say that the tasks are **independent.** When two tasks are independent, either one can be started before the other one, or they can both be started at the same time (some people put their shoes on before their shirt, others put their shirt on before their shoes, and, occasionally, some of us have been known to put our shoes and shirt on at the same time). Graphically, we can tell that two tasks are independent if there are no arrows connecting them [Fig. 8-1(b)].

  Two final comments about precedence relations are in order. First, precedence relations are *transitive:* If $X \rightarrow Y$ and $Y \rightarrow Z$, then it must be true that $X \rightarrow Z$. In a sense, the last precedence relation is implied by the first two, and it is really unnecessary to mention it [Fig. 8-1(c)]. Thus, we will make a distinction between two types of precedence relations: *basic* and *implicit.* Basic precedence relations are the ones that come with the problem and that we must follow in the process of creating a schedule. If we do this, the implicit precedence relations will be taken care of automatically.

  The second observation is that *we cannot have a set of precedence relations that form a cycle!* Imagine having to schedule the tasks shown in Fig. 8-1(d): X precedes Y, which precedes Z, which precedes W, which in turn precedes X. Clearly, this is logically impossible. From here on, we will assume that there are no cycles of precedence relations among the tasks.

FIGURE 8-1 (a) X is precedent to Y. (b) X and Y are independent tasks. (c) When $X \rightarrow Y$ and $Y \rightarrow Z$, then $X \rightarrow Z$ is implied. (d) These tasks cannot be scheduled because of the cyclical nature of the precedence relations.

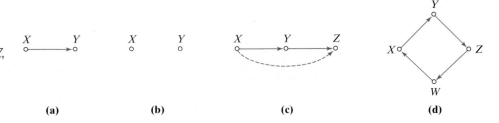

(a)   (b)   (c)   (d)

Processors, *tasks*, *processing times*, and *precedence relations* are the basic ingredients that make up a scheduling problem. They constitute, in a manner of speaking, the hand that is dealt to us. But how do we play such a hand? To get a small inkling of what's to come, let's look at the following simple example.

EXAMPLE 8.1   **Repairing a Wreck**

Imagine that you just wrecked your car, but thank heavens you are OK, and the insurance company will pick up the tab. You take the car to the best garage in town, operated by the Tappet brothers Click and Clack (we'll just call them P_1 and P_2). The repairs on the car can be broken into four different tasks: (A) exterior body work (4 hours), (B) engine repairs (5 hours), (C) painting and exterior finish work (7 hours), and (D) transmission repair (3 hours). The only precedence relation for this set of tasks is that the painting and exterior finish work cannot be started until the exterior body work has been completed ($A \rightarrow C$). The two brothers always work together on a repair project, but each takes on a different task (so they won't argue with each other). Under these assumptions, how should the different tasks be scheduled? Who should do what and when?

Even in this simple situation, many different schedules are possible. Figure 8-2 shows several possibilities, each one illustrated by means of a timeline. Figure 8-2(a) shows a schedule that is very inefficient. All the short tasks are assigned to one processor (P_1) and all the long tasks to the other processor (P_2)—obviously not a very clever strategy. Under this schedule, the project **finishing time** (the duration of the project from the start of the first task to the completion of the last task) is 12 hours. (We will use *Fin* to denote the project finishing time, so for this project we can write *Fin* = 12 hours.)

Figure 8-2(b) shows what looks like a much better schedule, but it violates the precedence relation $A \rightarrow C$ (as much as we would love to, we cannot start task C until task A is completed). On the other hand, if we force P_2 to be idle for one hour,

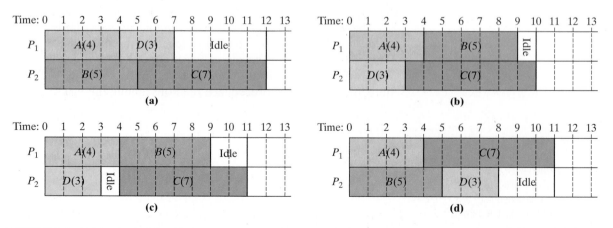

FIGURE 8-2 (a) A legal schedule with *Fin* = 12 hours. (b) An *illegal* schedule (the precedence relation $A \rightarrow C$ is violated). (c) An optimal schedule (*Opt* = 11 hours). (d) A different optimal schedule.

waiting for the green light to start task C, we get a perfectly good schedule, shown in Fig. 8-2(c). Under this schedule the finishing time of the project is $Fin = 11$ hours.

The schedule shown in Fig. 8-2(c) is an improvement over the first schedule. Can we do even better? No! No matter how clever we are and no matter how many processors we have at our disposal, the precedence relation $A(4) \rightarrow C(7)$ implies that 11 hours is a minimum barrier that we cannot break—it takes 4 hours to complete A, 7 hours to complete C, and *we cannot start C until A is completed*! Thus, the schedule shown in Fig. 8-2(c) is an **optimal schedule** and the finishing time of $Fin = 11$ hours is the **optimal finishing time**. (From now on we will use Opt instead of Fin when we are referring to the optimal finishing time.) Figure 8-2(d) shows a different optimal schedule with finishing time $Opt = 11$ hours.   ⬭

As scheduling problems go, Example 8.1 was a fairly simple one. But even from this simple example, we can draw some useful lessons. First, notice that even though we had only four tasks and two processors, we were able to create several different schedules. The four we looked at were just a sampler—there are other possible schedules that we didn't bother to discuss. Imagine what would happen if we had hundreds of tasks and dozens of processors—the number of possible schedules to consider would be overwhelming. In looking for a good, or even an optimal, schedule, we are going to need a systematic way to sort through the many possibilities. In other words, we are going to need some good *scheduling algorithms*.

The second useful thing we learned in Example 8.1 is that when it comes to the finishing time of a project, there is an *absolute minimum* time that no schedule can break, no matter how good an algorithm we use or how many processors we put to work. In Example 8.1 this absolute minimum was 11 hours, and, as luck would have it, we easily found a schedule [actually two—Figs. 8-2(c) and (d)] with a finishing time to match it. Every project, no matter how simple or complicated, has such an absolute minimum (called the *critical time*) that depends on the processing times and precedence relations for the tasks and not on the number of processors used. We will return to the concept of critical time in Section 8.5.

To set the stage for a more formal discussion of scheduling algorithms, we will introduce the most important example of this chapter. While couched in what seems like science fiction terms, the situation it describes is not totally farfetched.

(**EXAMPLE 8.2**)   **Building That Dream Home on Mars**

It is the year 2050, and several human colonies have already been established on Mars. Imagine that you accept a job offer to work in one of these colonies. What will you do about housing?

Like everyone else on Mars, you will be provided with a living pod called a Martian Habitat Unit (MHU). MHUs are shipped to Mars in the form of prefabricated kits that have to be assembled on the spot—an elaborate and unpleasant job if you are going to do it yourself. A better option is to use specialized "construction" robots that can do all the assembly tasks much more efficiently than human beings can. These construction robots can be rented by the hour at the local Rent-a-Robot outlet.

The assembly of an MHU consists of 15 separate tasks, and there are 17 different precedence relations among these tasks that must be followed. The tasks, their respective processing times, and their precedent tasks are all shown in Table 8-1.

TABLE 8-1

Task	Label (*P*-time)	Precedent tasks
Assemble pad	AP(7)	
Assemble flooring	AF(5)	
Assemble wall units	AW(6)	
Assemble dome frame	AD(8)	
Install floors	IF(5)	AP, AF
Install interior walls	IW(7)	IF, AW
Install dome frame	ID(5)	AD, IW
Install plumbing	PL(4)	IF
Install atomic power plant	IP(4)	IW
Install pressurization unit	PU(3)	IP, ID
Install heating units	HU(4)	IP
Install commode	IC(1)	PL, HU
Complete interior finish work	FW(6)	IC
Pressurize dome	PD(3)	HU
Install entertainment unit	EU(2)	PU, HU

Here are some of the basic questions we will want to address: How can we get your MHU built quickly? How many robots should you rent to do the job? How do we create a suitable work schedule that will get the job done? (A robot will do whatever it is told, but someone has to tell it what to do and when.) These are all tough questions to answer, but we will be able to do it later in the chapter.

8.2 Directed Graphs (Digraphs)

A directed graph, or **digraph** for short, is a graph in which the edges have a direction associated with them, typically indicated by an arrowhead. Digraphs are particularly useful when we want to describe *asymmetric* relationships (*X* related to *Y* does not imply that *Y* must be related to *X*).

The classic example of an asymmetric relationship is romantic love: Just because *X* is in love with *Y*, there is no guarantee that *Y* reciprocates that love. Given two individuals *X* and *Y* and some asymmetric relationship (say love), we have four possible scenarios: Neither loves the other [Fig. 8-3(a)], *X* loves *Y* but *Y* does not love *X* [Fig. 8-3(b)], *Y* loves *X* but *X* does not love *Y* [Fig. 8-3(c)], and they love each other [Fig. 8-3(d)].

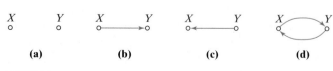

(a)   (b)   (c)   (d)

FIGURE 8-3

To distinguish digraphs from ordinary graphs, we use slightly different terminology.

- In a digraph, instead of talking about edges we talk about **arcs**. Every arc is defined by its *starting vertex* and its *ending vertex*, and we respect that order when we write the arc. Thus, if we write XY, we are describing the arc in Fig. 8-3(b) as opposed to the arc YX shown in Fig. 8-3(c).

- A list of all the arcs in a digraph is called the **arc-set** of the digraph. The digraph in Fig. 8-3(d) has arc-set $\mathcal{A} = \{XY, YX\}$.

- If XY is an arc in the digraph, we say that vertex X is **incident to** vertex Y, or, equivalently, that Y is **incident from** X).

- The arc YZ is said to be **adjacent** to the arc XY if the starting point of YZ is the ending point of XY. (Essentially, this means one can go from X to Z by way of Y).

- In a digraph, a **path** from vertex X to vertex W consists of a sequence of arcs $XY, YZ, ZU, \ldots, VW$ such that each arc is adjacent to the one before it and no arc appears more than once in the sequence—it is essentially a trip from X to W along the arcs in the digraph. The best way to describe the path is by listing the vertices in the order of travel: X, Y, Z, and so on.

- When the path starts and ends at the same vertex, we call it a **cycle** of the digraph. Just like circuits in a regular graph, cycles in digraphs can be written in more than one way—the cycle X, Y, Z, X is the same as the cycles Y, Z, X, Y and Z, X, Y, Z.

- In a digraph, the notion of the degree of a vertex is replaced by the concepts of *indegree* and *outdegree*. The **outdegree** of X is the number of arcs that have X as their *starting point* (outgoing arcs); the **indegree** of X is the number of arcs that have X as their *ending point* (incoming arcs).

The following example illustrates some of the above concepts.

EXAMPLE 8.3   **Digraph Basics**

The digraph in Fig. 8-4 has vertex-set $\mathcal{V} = \{A, B, C, D, E\}$ and arc-set $\mathcal{A} = \{AB, AC, BD, CA, CD, CE, EA, ED\}$. In this digraph, A is *incident to B* and C, but not to E. By the same token, A is *incident from E* as well as from C. The indegree of vertex A is 2, and so is the outdegree. The indegree of vertex D is 3, and the outdegree is 0. We leave it to the reader to find the indegrees and outdegrees of each of the other vertices of the graph.

In this digraph, there are several paths from A to D, such as $A, C, D; A, C, E,$ $D; A, B, D$; and even A, C, A, B, D. On the other hand, A, E, D is not a path from A to D (because you can't travel from A to E). There are two cycles in this digraph: A, C, E, A (which can also be written as C, E, A, C and E, A, C, E), and A, C, A. Notice that there is no cycle passing through D because D has outdegree 0 and thus is a "dead-end," and there is no cycle passing through B because from B you can only go to D.

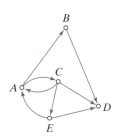

FIGURE 8-4

While love is not to be minimized as a subject of study, there are many other equally important applications of digraphs:

- **Traffic flow.** In most cities some streets are one-way streets and others are two-way streets. In this situation, digraphs allow us to visualize the flow of traffic through the city's streets. The vertices are intersections, and the *arcs* represent one-way streets. (To represent a two-way street, we use two arcs, one for each direction.)

- **Telephone traffic.** To track and analyze the traffic of telephone calls through their network, telephone companies use "call digraphs." In these digraphs the vertices are telephone numbers, and an arc from X to Y indicates that a call was initiated from telephone number X to telephone number Y.

- **Tournaments.** Digraphs are frequently used to describe certain types of tournaments, with the vertices representing the teams (or individual players) and the arcs representing the outcomes of the games played in the tournament (the arc XY indicates that X defeated Y). Tournament digraphs can be used in any sport in which the games cannot end in a tie (basketball, tennis, etc.).

- **Organization charts.** In any large organization (a corporation, the military, a university, etc.) it is important to have a well-defined chain of command. The best way to describe the chain of command is by means of a digraph often called an *organization chart*. In this digraph the *vertices* are the individuals in the organization, and an *arc* from X to Y indicates that X is Y's immediate boss (i.e., Y takes orders directly from X).

As you probably guessed by now, digraphs are also used in scheduling. There is no better way to visualize the tasks, processing times, and precedence relations in a project than by means of a digraph in which the vertices represent the tasks (with their processing times indicated in parentheses) and the arcs represent the precedence relations.

(**EXAMPLE 8.4**)   **Building That Dream Home on Mars: Part 2**

Let's return to the scheduling problem first discussed in Example 8.2. We can take the tasks and precedence relations given in Table 8-1 and create a digraph like the one shown in Fig. 8-5(a). It is helpful to try to place the vertices of the digraph so that the arcs point from left to right, and this can usually be done with a little trial-and-error. After this is done, it is customary to add two fictitious tasks called START and END, where START indicates the imaginary task of getting the project started (cutting the red ribbon, so to speak), and END indicates the

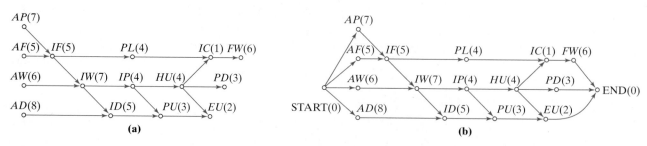

FIGURE 8-5

imaginary task of declaring the project complete (pop the champagne). By giving these fictitious tasks zero processing time, we avoid affecting the time calculations for the project. This modified digraph, shown in Fig. 8-5(b), is called the **project digraph**. The project digraph allows us to better visualize the execution of the project as a flow, moving from left to right.

8.3   Scheduling with Priority Lists

The project digraph is the basic graph model used to package all the information in a scheduling problem, but there is nothing in the project digraph itself that specifically tells us how to create a schedule. We are going to need something else, some set of instructions that indicates the order in which tasks should be executed. We can accomplish this by the simple act of prioritizing the tasks in some specified order, called a *priority list*.

A **priority list** is a list of all the tasks prioritized in the order we prefer to execute them. If task X is ahead of task Y in the priority list, then X gets priority over Y. This means that when it comes to a choice between the two, *X is executed ahead of Y*. However, if X is not yet *ready* for execution, then *we skip over it and move on to the first ready task after X in the priority list*. If there are no ready tasks after X in the priority list, the processors must sit idle and wait until a task becomes ready.

The process of scheduling tasks using a priority list and following these basic rules is known as the *priority-list model* for scheduling. The priority-list model is a completely general model for scheduling—every priority list produces a schedule, and any schedule can be created from some (usually more than one) "parent" priority list. The trick is going to be to figure out *which* priority lists give us good schedules and which don't. We will come back to this topic in Sections 8.4 and 8.5.

Since each time we change the order of the tasks we get a different priority list, there are as many priority lists as there are ways to order the tasks. For three tasks, there are six possible priority lists; for 4 tasks, there are 24 priority lists; for 10 tasks, there are more than 3 million priority lists; and for 100 tasks, there are more priority lists than there are molecules in the universe.

Clearly, a shortage of priority lists is not going to be our problem. If this sounds familiar, it's because we have seen the concept behind this type of explosive growth before. Like sequential coalitions (Chapter 2) and Hamilton circuits (Chapter 6), the number of priority lists is given by a factorial.

■ For a review of factorials, see Section 2.4.

NUMBER OF PRIORITY LISTS

The number of possible priority lists in a project consisting of M tasks, is
$$M! = M \times (M - 1) \times \cdots \times 2 \times 1$$

Before we proceed, we will illustrate how the priority-list model for scheduling works with a couple of small but important examples. Even with such small examples, there is a lot to keep track of, and you are well advised to have pencil and paper in front of you as you follow the details.

> **EXAMPLE 8.5   Preparing for Launch**

Immediately preceding the launch of a satellite into space, last-minute system checks need to be performed by the on-board computers, and it is important to complete these system checks as quickly as possible—for both cost and safety reasons. Suppose that there are five system checks required: $A(6)$, $B(5)$, $C(7)$, $D(2)$, and $E(5)$, with the numbers in parentheses representing the hours it takes one computer to perform that system check. In addition, there are precedence relations: D cannot be started until both A and B have been finished, and E cannot be started until C has been finished. The project digraph is shown in Fig. 8-6.

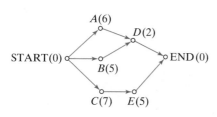

FIGURE 8-6

Let's assume that there are two identical computers on board (P_1 and P_2) that will carry out the individual system checks. How do we use the priority-list model to create a schedule for these two processors? For starters, we will need a priority list. Suppose that the priority list is given by listing the system checks in alphabetical order.

Priority list: $A(6)$, $B(5)$, $C(7)$, $D(2)$, $E(5)$

Let's look at the evolution of the project under the priority-list model. (To keep track of things you will find it helpful to refer to the timeline in Fig. 8-7.) We will use T to indicate the elapsed time in hours.

- **$T = 0$ (START).** $A(6)$, $B(5)$, and $C(7)$ are the only *ready* tasks. Following the priority list, we assign $A(6)$ to P_1 and $B(5)$ to P_2.

- **$T = 5$.** P_1 is still *busy* with $A(6)$; P_2 has just *completed* $B(5)$. $C(7)$ is the only available *ready* task. We assign $C(7)$ to P_2.

- **$T = 6$.** P_1 has just *completed* $A(6)$; P_2 is *busy* with $C(7)$. $D(2)$ has just become a *ready* task (A and B have been completed). We assign $D(2)$ to P_1.

- **$T = 8$.** P_1 has just *completed* $D(2)$; P_2 is still *busy* with $C(7)$. There are no *ready* tasks at this time for P_1, so P_1 has to sit *idle*.

- **$T = 12$.** P_1 is *idle*; P_2 has just *completed* $C(7)$. Both processors are *ready* for work. $E(5)$ is the only ready task, so we assign $E(5)$ to P_1, P_2 sits *idle*. (Note that in this situation, we could have just as well assigned $E(5)$ to P_2 and let P_1 sit idle. The processors don't get tired and don't care if they are working or idle, so the choice is random.)

- **$T = 17$ (END).** P_1 has just *completed* $E(5)$, and therefore the project is completed.

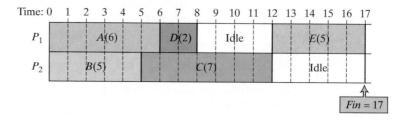

FIGURE 8-7

The evolution of the entire project together with the project finishing time ($Fin = 17$) are captured in the timeline shown in Fig. 8-7. Is this a good schedule? Given the excessive amount of idle time (a total of 9 hours), one might suspect this is a rather bad schedule. How could we improve it? We might try changing the priority list.

EXAMPLE 8.6   **Preparing for Launch: Part 2**

We are going to schedule the same project with the same processors, but with a different priority list. (The project digraph is shown again in Fig. 8-8. When scheduling, it's really useful to have the project digraph right in front of you.)

A(6)
D(2)
START(0)
B(5)
END(0)
C(7)   E(5)

FIGURE 8-8

Let's try this time a reverse alphabetical order for the priority list. Why? Why not—at this point we are just shooting in the dark! (Don't worry, we are just feeling our way around—we will become a lot more enlightened later in the chapter.)

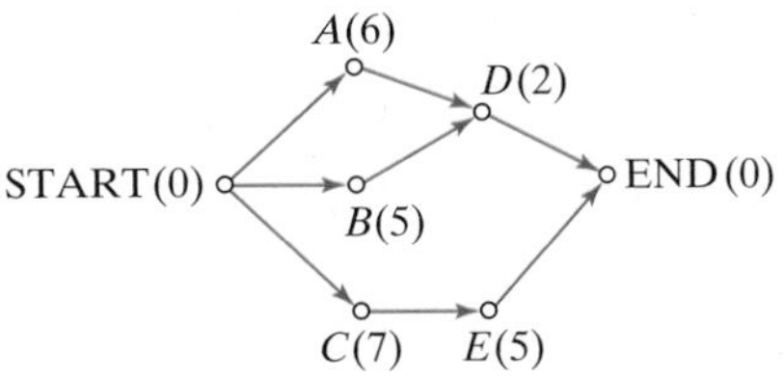

Priority list: $E(5), D(2), C(7), B(5), A(6)$

- **$T = 0$ (START).** $C(7), B(5)$, and $A(6)$ are the only *ready* tasks. Following the priority list, we assign $C(7)$ to P_1 and $B(5)$ to P_2.
- **$T = 5$.** P_1 is still *busy* with $C(7)$; P_2 has just *completed* $B(5)$. $A(6)$ is the only available ready task. We assign $A(6)$ to P_2.
- **$T = 7$.** P_1 has just *completed* $C(7)$; P_2 is *busy* with $A(6)$. $E(5)$ has just become a *ready* task, and we assign it to P_1.
- **$T = 11$.** P_2 has just *completed* $A(6)$; P_1 is *busy* with $E(5)$. $D(2)$ has just become a *ready* task, and we assign it to P_2.
- **$T = 12$.** P_1 has just *completed* $E(5)$; P_2 is *busy* with $D(2)$. There are no tasks left, so P_1 sits idle.
- **$T = 13$ (END).** P_2 has just *completed* the last task, $D(2)$. Project is completed.

The timeline for this schedule is shown in Fig. 8-9. The project finishing time is *Fin* = 13 hours.

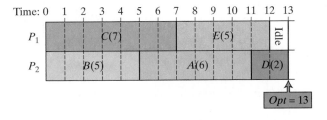

FIGURE 8-9

It's easy to see that this schedule is a lot better than the one obtained in Example 8.5. In fact, we were pretty lucky—this schedule turns out to be optimal for two processors. Two processors cannot finish this project in less than 13 hours because there is a total of 25 hours worth of work (the sum of all processing times), which implies that in the best of cases it would take 12.5 hours to finish the project. But since the processing times are all whole numbers and tasks cannot be split, the finishing time cannot be less than 13 hours! Thus, *Opt* = 13 hours.

Thirteen hours is still a long time for the computers to go over their system checks. Since that's the best we can do with two computers, the only way to speed things up is to add a third computer to the "workforce." Adding another computer to the satellite can be quite expensive, but perhaps it will speed things up enough to make it worth it. Let's see.

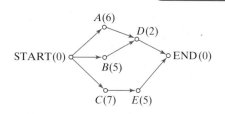

FIGURE 8-10

EXAMPLE 8.7   **Preparing for Launch: Part 3**

We will now schedule the same project using $N = 3$ processors (P_1, P_2, P_3). For the reader's convenience the project digraph is shown again in Fig. 8-10. We will use the "good" priority list we found in the previous example.

Priority list: $E(5), D(2), C(7), B(5), A(6)$

- **$T = 0$ (START).** $C(7)$, $B(5)$, and $A(6)$ are the *ready* tasks. We assign $C(7)$ to P_1, $B(5)$ to P_2, and $A(6)$ to P_3.

- **$T = 5$.** P_1 is *busy* with $C(7)$; P_2 has just *completed* $B(5)$; and P_3 is *busy* with $A(6)$. There are no available ready tasks for P_2 [$E(5)$ can't be started until $C(7)$ is done, and $D(2)$ can't be started until $A(6)$ is done], so P_2 sits idle until further notice.

- **$T = 6$.** P_3 has just *completed* $A(6)$; P_2 is *idle*; and P_1 is still *busy* with $C(7)$. $D(2)$ has just become a *ready* task. We randomly assign $D(2)$ to P_2 and let P_3 be idle, since there are no other ready tasks. [Note that we could have just as well assigned $D(2)$ to P_3 and let P_2 be idle.]

- **$T = 7$.** P_1 has just *completed* $C(7)$ and $E(5)$ has just become a *ready* task, so we assign it to P_1. There are no other tasks to assign, so P_3 continues to sit idle.

- **$T = 8$.** P_2 has just *completed* $D(2)$. There are no other tasks to assign, so P_2 and P_3 both sit *idle*.

- **$T = 12$ (END).** P_1 has just *completed* the last task, $E(5)$, so the project is completed.

■ Strange things can happen when an additional processor is added to a project — see, for example, Exercise 69.

The timeline for this schedule is shown in Fig. 8-11. The project finishing time is $Fin = 12$ hours, a pathetically small improvement over the two-processor schedule found in Example 8.6. The cost of adding a third processor doesn't seem to justify the benefit.

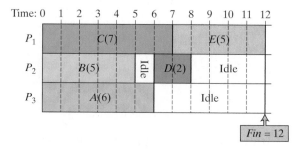

FIGURE 8-11

The Priority-List Model

The previous three examples give us a general sense of how to create a schedule from a project digraph *and* a priority list. We will now formalize the ground rules of the **priority-list model** for scheduling.

At any particular moment in time throughout a project, a processor can be either *busy* or *idle* and a task can be *ineligible, ready, in execution,* or *completed*. Depending on the various combinations of these, there are three different scenarios to consider:

- *All processors are busy.* In this case, there is nothing we can do but wait.

- *One processor is free.* In this case, we scan the priority list from left to right, looking for the first *ready* task in the priority list, which we assign to that processor. (Remember that for a task to be *ready*, all the tasks that are precedent to it must have been completed.) If there are no ready tasks at that moment, the processor must stay idle until things change.

- *More than one processor is free.* In this case, the *first ready* task on the priority list is given to one free processor, the second ready task is given to another free processor, and so on. If there are more free processors than ready tasks, some of the processors will remain idle until one or more tasks become ready. Since the processors are identical and tireless, the choice of which processor is assigned which task is completely arbitrary.

It's fair to say that the basic idea behind the priority-list model is not difficult, but there is a lot of bookkeeping involved, and that becomes critical when the number of tasks is large. At each stage of the schedule we need to keep track of the status of each task—which tasks are *ready* for processing, which tasks are *in execution*, which tasks have been *completed*, which tasks are still *ineligible*. One convenient recordkeeping strategy goes like this: On the priority list itself *ready* tasks are circled in red [Fig. 8-12(a)]. When a ready task is picked up by a processor and goes into *execution*, put a single red slash through the red circle [Fig. 8-12(b)]. When a task that has been in execution is completed, put a second red slash through the circle [Fig. 8-12(c)]. At this point, it is also important to check the project digraph to see if any new tasks have all of a sudden become eligible. Tasks that are *ineligible* remain unmarked [Fig. 8-12(d)].

(a)   **(b)**   **(c)**   **(d)**

FIGURE 8-12 "Road" signs on a priority list. (a) Task *X* is *ready*. (b) Task *X* is *in execution*. (c) Task *X* is *completed*. (d) Task *X* is *ineligible*.

As they say, the devil is in the details, so a slightly more substantive example will help us put everything together—the project digraph, the priority list model, and the bookkeeping strategy.

EXAMPLE 8.8   **Building That Dream Home on Mars: Part 3**

This is the third act of the Martian Habitat Unit building project. We are finally ready to start the project of assembling that MHU, and, like any good scheduler, we will first work the entire schedule out with pencil and paper. Let's start with the assumption that maybe we can get by with just two robots (P_1 and P_2). For the reader's convenience, the project digraph is shown again in Fig. 8-13.

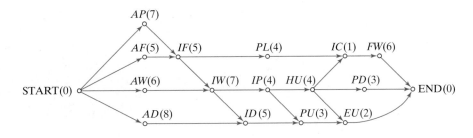

FIGURE 8-13

Let's start with a random priority list.

■ Ready tasks are circled in red.

Priority list: $\widehat{AD(8)}$, $\widehat{AW(6)}$, $\widehat{AF(5)}$, $IF(5)$, $\widehat{AP(7)}$, $IW(7)$, $ID(5)$, $IP(4)$, $PL(4)$, $PU(3)$, $HU(4)$, $IC(1)$, $PD(3)$, $EU(2)$, $FW(6)$

- **$T = 0$ (START).** Status of processors: P_1 starts AD; P_2 starts AW. We put a single red slash through AD and AW.

  Priority list: $\widehat{AD}$, $\widehat{AW}$, $\widehat{AF}$, IF, $\widehat{AP}$, IW, ID, IP, PL, PU, HU, IC, PD, EU, FW.

- **$T = 6$.** P_1 busy (executing AD); P_2 completed AW (put a second slash through AW) and starts AF (put a slash through AF).
  Priority list: ⦶AD⦶, ⦶AW⦶, ⦶AF⦶, IF, ⦶AP⦶, $IW, ID, IP, PL, PU, HU, IC, PD, EU, FW$.

- **$T = 8$.** P_1 completed AD and starts AP; P_2 is busy (executing AF).
  Priority list: ⦶AD⦶, ⦶AW⦶, ⦶AF⦶, IF, ⦶AP⦶, $IW, ID, IP, PL, PU, HU, IC, PD, EU, FW$.

- **$T = 11$.** P_1 busy (executing AP); P_2 completed AF, but since there are no ready tasks, it remains idle.
  Priority list: ⦶AD⦶, ⦶AW⦶, ⦶AF⦶, IF, ⦶AP⦶, $IW, ID, PL, PU, HU, IC, PD, EU, FW$.

- **$T = 15$.** P_1 completed AP. IF becomes a ready task and goes to P_1; P_2 stays idle.
  Priority list: ⦶AD⦶, ⦶AW⦶, ⦶AF⦶, ⦶IF⦶, ⦶AP⦶, $IW, ID, IP, PL, PU, HU, IC, PD, EU, FW$.

■ See Exercise 63.

At this point, we will let you take over and finish the schedule. (Remember—the object is to learn how to keep track of the status of each task, and the only way to do this is with practice.)

After a fair amount of work, we obtain the final schedule shown in Fig. 8-14, with project finishing time $Fin = 44$ hours.

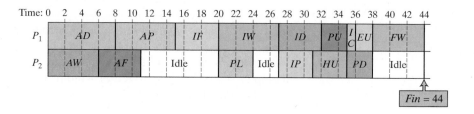

FIGURE 8-14

Scheduling with priority lists is a two-part process: (1) choose a priority list and (2) use the priority list and the ground rules of the priority-list model to come up with a schedule (Fig. 8-15). As we saw in the previous example, the second part is long and tedious, but purely mechanical—it can be done by anyone (or anything) that is able to follow a set of instructions, be it a meticulous student or a properly programmed computer. We will use the term *scheduler* to describe the entity (be it student or machine) that takes a priority list as input and produces the schedule as output.

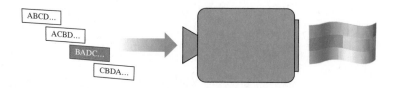

FIGURE 8-15

Ironically, it is the seemingly easiest part of this process—choosing a priority list—that is actually the most interesting. Among all the priority lists there is one or more that give a schedule with the optimal finishing time (we will call these *optimal priority lists*). How do we find an optimal priority list? Short of that, how do we find "good" priority lists, that is, priority lists that give schedules with finishing times reasonably close to the optimal? These are both important questions, and some answers are coming up next.

8.4 The Decreasing-Time Algorithm

Our first attempt to find a good priority list is to formalize what is a seemingly sensible and often used strategy: *Do the longer jobs first and leave the shorter jobs for last*. In terms of priority lists, this strategy is implemented by creating a priority list in which the tasks are listed in decreasing order of processing times—longest first, second longest next, and so on. (When there are two or more tasks with equal processing times, we order them randomly.)

A priority list in which the tasks are listed in decreasing order of processing times is called, not surprisingly, a **decreasing-time priority list**, and the process of creating a schedule using a decreasing-time priority list is called the **decreasing-time algorithm**.

(**EXAMPLE 8.9**)   **Building That Dream Home on Mars: Part 4**

Figure 8-16 shows, once again, the project digraph for the Martian Habitat Unit building project. To use the decreasing-time algorithm we first prioritize the 15 tasks in a decreasing-time priority list.

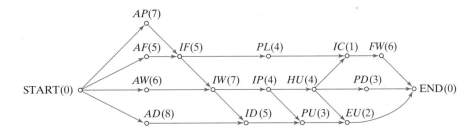

FIGURE 8-16

Decreasing-time priority list: $AD(8)$, $AP(7)$, $IW(7)$, $AW(6)$, $FW(6)$, $AF(5)$, $IF(5)$, $ID(5)$, $IP(4)$, $PL(4)$, $HU(4)$, $PU(3)$, $PD(3)$, $EU(2)$, $IC(1)$

Using the decreasing-time algorithm with $N = 2$ processors, we get the schedule shown in Fig. 8-17, with project finishing time $Fin = 42$ hours.

■ The details of the process are shown in Table 8-2, but you may want to try re-creating the schedule on your own — see Exercise 64.

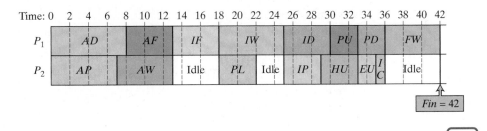

FIGURE 8-17

When looking at the finishing time under the decreasing-time algorithm, one can't help but feel disappointed. The sensible idea of prioritizing the longer jobs ahead of the shorter jobs turned out to be somewhat of a dud—at least in this example! What went wrong? If we work our way backward from the end, we can see that we made a bad choice at $T = 33$ hours. At this point there were three ready tasks [$PD(3)$, $EU(2)$, and $IC(1)$], and both processors were available.

TABLE 8-2

Step	Time	Priority-List Status	Schedule Status
1	$T = 0$	(AD(8)) (AP(7)) IW(7) (AW(6)) FW(6) (AF(5)) IF(5) ID(5) IP(4) PL(4) HU(4) PU(3) PD(3) EU(2) IC(1)	Time: 0 2 4 6 8 10 12 14 16 18 20 22 24 26 28 30 32 34 36 38 40 42 P_1: AD P_2: AP
2	$T = 7$	(AD(8)) (AP(7)) IW(7) (AW(6)) FW(6) (AF(5)) IF(5) ID(5) IP(4) PL(4) HU(4) PU(3) PD(3) EU(2) IC(1)	P_1: AD P_2: AP AW
3	$T = 8$	(AD(8)) (AP(7)) IW(7) (AW(6)) FW(6) (AF(5)) IF(5) ID(5) IP(4) PL(4) HU(4) PU(3) PD(3) EU(2) IC(1)	P_1: AD AF P_2: AP AW
4	$T = 13$	(AD(8)) (AP(7)) IW(7) (AW(6)) FW(6) (AF(5)) (IF(5)) ID(5) IP(4) PL(4) HU(4) PU(3) PD(3) EU(2) IC(1)	P_1: AD AF IF P_2: AP AW
5	$T = 18$	(AD(8)) (AP(7)) (IW(7)) (AW(6)) FW(6) (AF(5)) (IF(5)) ID(5) IP(4) (PL(4)) HU(4) PU(3) PD(3) EU(2) IC(1)	P_1: AD AF IF IW P_2: AP AW Idle PL
6	$T = 22$	(AD(8)) (AP(7)) (IW(7)) (AW(6)) FW(6) (AF(5)) (IF(5)) ID(5) IP(4) (PL(4)) HU(4) PU(3) PD(3) EU(2) IC(1)	P_1: AD AF IF IW P_2: AP AW Idle PL
7	$T = 25$	(AD(8)) (AP(7)) (IW(7)) (AW(6)) FW(6) (AF(5)) (IF(5)) (ID(5)) (IP(4)) (PL(4)) HU(4) PU(3) PD(3) EU(2) IC(1)	P_1: AD AF IF IW ID P_2: AP AW Idle PL Idle IP
8	$T = 29$	(AD(8)) (AP(7)) (IW(7)) (AW(6)) FW(6) (AF(5)) (IF(5)) (ID(5)) (IP(4)) (PL(4)) (HU(4)) PU(3) PD(3) EU(2) IC(1)	P_1: AD AF IF IW ID P_2: AP AW Idle PL Idle IP HU
9	$T = 30$	(AD(8)) (AP(7)) (IW(7)) (AW(6)) FW(6) (AF(5)) (IF(5)) (ID(5)) (IP(4)) (PL(4)) (HU(4)) (PU(3)) PD(3) EU(2) IC(1)	P_1: AD AF IF IW ID PU P_2: AP AW Idle PL Idle IP HU
10	$T = 33$	(AD(8)) (AP(7)) (IW(7)) (AW(6)) FW(6) (AF(5)) (IF(5)) (ID(5)) (IP(4)) (PL(4)) (HU(4)) (PU(3)) (PD(3)) (EU(2)) (IC(1))	P_1: AD AF IF IW ID PU PD P_2: AP AW Idle PL Idle IP HU EU
11	$T = 35$	(AD(8)) (AP(7)) (IW(7)) (AW(6)) FW(6) (AF(5)) (IF(5)) (ID(5)) (IP(4)) (PL(4)) (HU(4)) (PU(3)) (PD(3)) (EU(2)) (IC(1))	P_1: AD AF IF IW ID PU PD P_2: AP AW Idle PL Idle IP HU EU IC
12	$T = 36$	(AD(8)) (AP(7)) (IW(7)) (AW(6)) (FW(6)) (AF(5)) (IF(5)) (ID(5)) (IP(4)) (PL(4)) (HU(4)) (PU(3)) (PD(3)) (EU(2)) (IC(1))	P_1: AD AF IF IW ID PU PD FW P_2: AP AW Idle PL Idle IP HU EU IC
13	$T = 42$	(AD(8)) (AP(7)) (IW(7)) (AW(6)) (FW(6)) (AF(5)) (IF(5)) (ID(5)) (IP(4)) (PL(4)) (HU(4)) (PU(3)) (PD(3)) (EU(2)) (IC(1))	P_1: AD AF IF IW ID PU PD FW P_2: AP AW Idle PL Idle IP HU EU IC Idle

Based on the decreasing-time priority list, we chose the two "longer" tasks, $PD(3)$ and $EU(2)$, ahead of the short task, $IC(1)$. This was a bad move! $IC(1)$ is a much more *critical* task than the other two because we can't start task $FW(6)$ until we finish task $IC(1)$. Had we looked ahead at some of the tasks for which PD, EU, and IC are precedent tasks, we might have noticed this.

An even more blatant example of the weakness of the decreasing-time algorithm occurs at the very start of this schedule when the algorithm fails to take into account that task $AF(5)$ should have a very high priority. Why? $AF(5)$ is one of the two tasks that must be finished before $IF(5)$ can be started, and $IF(5)$ must be finished before $IW(7)$ can be started, which must be finished before $IP(4)$ and $ID(5)$ can be started, and so on down the line.

8.5   Critical Paths

When there is a long path of tasks in the project digraph, it seems clear that the first task along that path should be started as early as possible. This idea leads to the following informal rule: *The greater the total amount of work that lies ahead of a task, the sooner that task should be started.*

To formalize these ideas, we will introduce the concepts of *critical paths* and *critical times*.

> **CRITICAL PATHS AND CRITICAL TIMES**
>
> - For a given vertex X of a project digraph, the **critical path for** X is the path *from X to END* with *longest* processing time. (The *processing time* of a path is defined to be the sum of the processing times of all the vertices in the path.) When we add the processing times of all the tasks along the critical path for a vertex X, we get the **critical time for** X. (By definition, the critical time of END is 0.)
>
> - The path with longest processing time from START to END is called the **critical path** for the project, and the total processing time for this critical path is called the **critical time** for the project.

(**EXAMPLE 8.10**)   **Building That Dream Home on Mars: Part 5**

Figure 8-18 shows the project digraph for the Martian Habitat Unit building project. We will find critical paths and critical times for several vertices of the project digraph.

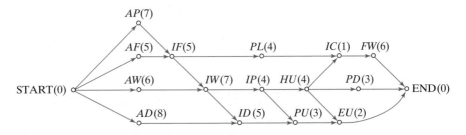

FIGURE 8-18

Let's start with a relatively easy case—the vertex HU. A quick look at Fig. 8-18 should convince you that there are only three paths from HU to END, namely,

- HU, IC, FW, END, with processing time $4 + 1 + 6 = 11$ hours,
- HU, PD, END, with processing time $4 + 3 = 7$ hours, and
- HU, EU, END, with processing time $4 + 2 = 6$ hours.

Of the three paths, the first one has the longest processing time, so HU, IC, FW, END is the *critical path* for vertex HU. The *critical time* for HU is 11 hours.

Next, let's find the critical path for vertex AD. There is only one path from AD to END, namely AD, ID, PU, EU, END, which makes the decision especially easy. Since this is the only path, it is automatically the longest path and therefore the *critical path for AD*. The *critical time* for AD is $8 + 5 + 3 + 2 = 18$ hours.

To find the critical path for the project, we need to find the path from START to END with longest processing time. Since there are dozens of paths from START to END, let's just eyeball the project digraph for a few seconds and take our best guess. . . .

OK, if you guessed START, $AP, IF, IW, IP, HU, IC, FW$, END, you have good eyes. This is indeed the *critical path*. It follows that the *critical time* for the Martian Habitat Unit building project is 34 hours.

We will soon discuss the special role that the critical time and the critical path play in scheduling, but before we do so, let's address the issue of how to find critical paths. In a large project digraph there may be thousands of paths from START to END, and the "eyeballing" approach we used in the preceding example is not likely to work. What we need here is an efficient algorithm, and fortunately there is one—it is called the **backflow algorithm**.

THE BACKFLOW ALGORITHM

- **Step 1.** Find the critical time for every vertex of the project digraph. This is done by starting at END and working backward toward START according to the following rule: *the critical time for a task X equals the processing time of X plus the largest critical time among the vertices incident from X.* The general idea is illustrated in Fig. 8-19. [To help with the recordkeeping, it is suggested that you write the critical time of the vertex in square brackets [ ] to distinguish it from the processing time in parentheses ( ).]

- **Step 2.** Once we have the critical time for every vertex in the project digraph, critical paths are found by just following the *path along largest critical times.* In other words, the critical path for any vertex X (and that includes START) is obtained by starting at X and moving to the adjacent vertex with largest critical time, and from there to the adjacent vertex with largest critical time, and so on.

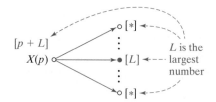

FIGURE 8-19

While the backflow algorithm sounds a little complicated when described in words, it is actually pretty easy to implement in practice, as we will show in the next example.

(**EXAMPLE 8.11**)   **Building That Dream Home on Mars: Part 6**

We are now going to use the backflow algorithm to find the critical time for each of the vertices of the Martian Habitat Unit project digraph (Fig. 8-20).

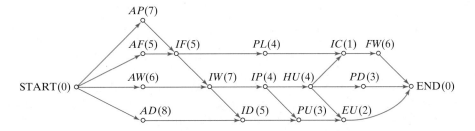

FIGURE 8-20

- **Step 1**
  - Start at END. The critical time of END is 0, so we add a [0] next to END(0).
  - The backflow now moves to the three vertices that are incident to END, namely, $FW(6)$, $PD(3)$, and $EU(2)$. In each case the critical time is the processing time plus 0, so the critical times are $FW[6]$, $PD[3]$, and $EU[2]$. We add this information to the project digraph.
  - From $FW[6]$, the backflow moves to $IC(1)$. The vertex $IC(1)$ is incident only to $FW[6]$, so the critical time for IC is $1 + 6 = 7$. We add a [7] next to IC in the project digraph.
  - The backflow now moves to $HU(4)$, $PL(4)$, and $PU(3)$. There are three vertices $HU(4)$ is incident to ($IC[7]$, $PD[3]$, and $EU[2]$). Of the three, the one with the largest critical time is $IC[7]$. This means that the critical time for HU is $4 + 7 = 11$. $PL(4)$ is only incident to $IC[7]$, so its critical time is $4 + 7 = 11$. $PU(3)$ is only incident to $EU[2]$, so its critical time is $3 + 2 = 5$. Add [11], [11], and [5] next to HU, PL, and PU, respectively.
  - The backflow now moves to $IP(4)$ and $ID(5)$. $IP(4)$ is incident to $HU[11]$ and $PU[5]$, so the critical time for IP is $4 + 11 = 15$. $ID(5)$ is only incident to $PU(5)$, so its critical time is $5 + 5 = 10$. We add [15] next to IP, and [10] next to ID.
  - The backflow now moves to $IW(7)$. The critical time for IW is $7 + 15 = 22$. (Please verify that this is correct!)
  - The backflow now moves to $IF(5)$. The critical time for IF is $5 + 22 = 27$. (Ditto.)
  - The backflow now moves to $AP(7)$, $AF(5)$, $AW(6)$, and $AD(8)$. Their respective critical times are $7 + 27 = 34, 5 + 27 = 32, 6 + 22 = 28,$ and $8 + 10 = 18$.
  - Finally, the backflow reaches START(0). We still follow the same rule—the critical time is $0 + 34 = 34$. This is the critical time for the project!
- **Step 2** The critical time for every vertex of the project digraph is shown in Fig. 8-21. We can now find the critical path by following the trail of largest critical times: START, $AP, IF, IW, IP, HU, IC, FW$, END.

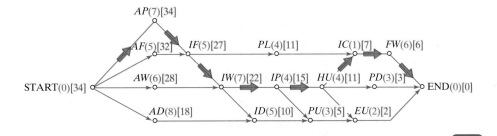

FIGURE 8-21

Why are the critical path and critical time of a project of special significance? We saw earlier in the chapter that for every project there is a theoretical time barrier below which a project cannot be completed, regardless of how clever the scheduler is or how many processors are used. Well, guess what? This theoretical barrier *is the project's critical time.*

If a project is to be completed in the optimal completion time, it is absolutely essential that all the tasks in the critical path be done at the earliest possible time. Any delay in starting up one of the tasks in the critical path will necessarily delay the finishing time of the entire project. (By the way, this is why this path is called *critical.*)

Unfortunately, it is not always possible to schedule the tasks on the critical path one after the other, bang, bang, bang without delay. For one thing, processors are not always free when we need them. (Remember that a processor cannot stop in the middle of one task to start a new task.) Another reason is the problem of uncompleted precedent tasks. We cannot concern ourselves only with tasks along the critical path and disregard other tasks that might affect them through precedence relations. There is a whole web of interrelationships that we need to worry about. Optimal scheduling is extremely complex.

■ A word of caution: The project's critical time is not necessarily the same as the project's *optimal completion time.* The optimal completion time depends on how many processors are working on the project; the critical time has nothing to do with the number of processors at work.

8.6   The Critical-Path Algorithm

The concept of critical paths can be used to create very good (although not necessarily optimal) schedules. The idea is to use *critical times* rather than processing times to prioritize the tasks. The priority list we obtain when we write the tasks in decreasing order of *critical times* (with ties broken randomly) is called the **critical-time priority list**, and the process of creating a schedule using the critical-time priority list is called the **critical-path algorithm**.

CRITICAL-PATH ALGORITHM

- **Step 1 (Find critical times).** Using the backflow algorithm, find the *critical time* for every task in the project.
- **Step 2 (Create priority list).** Using the critical times obtained in Step 1, create a *priority list* with the tasks listed in decreasing order of critical times (i.e., a critical-time priority list).
- **Step 3 (Create schedule).** Using the critical-time priority list obtained in Step 2, create the *schedule.*

There are, of course, plenty of small details that need to be attended to when carrying out the critical-path algorithm, especially in Steps 1 and 3. Fortunately, everything that needs to be done we now know how to do.

(**EXAMPLE 8.12**)   **Building That Dream Home on Mars: Part 7**

We will now schedule the Martian Habitat Unit building project with $N = 2$ processors using the critical-path algorithm.

We took care of Step 1 in Example 8.11. The critical times for each task are shown in red in Fig. 8-22.

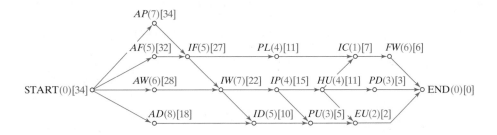

FIGURE 8-22

Step 2 follows directly from Step 1. The critical-time priority list for the project is $AP[34]$, $AF[32]$, $AW[28]$, $IF[27]$, $IW[22]$, $AD[18]$, $IP[15]$, $PL[11]$, $HU[11]$, $ID[10]$, $IC[7]$, $FW[6]$, $PU[5]$, $PD[3]$, $EU[2]$.

■ See Exercise 65.

Step 3 is a lot of busywork—the details are left to the reader.

The timeline for the resulting schedule is given in Fig. 8-23. The project finishing time is $Fin = 36$ hours. This is a very good schedule, but it is not an optimal schedule. (Figure 8-24 shows the timeline for an optimal schedule with finishing time $Opt = 35$ hours.)

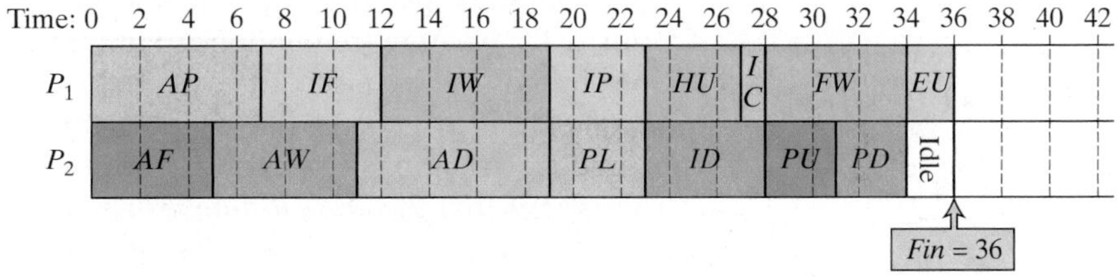

FIGURE 8-23 Timeline for the MHU building project under the critical-path algorithm ($N = 2$).

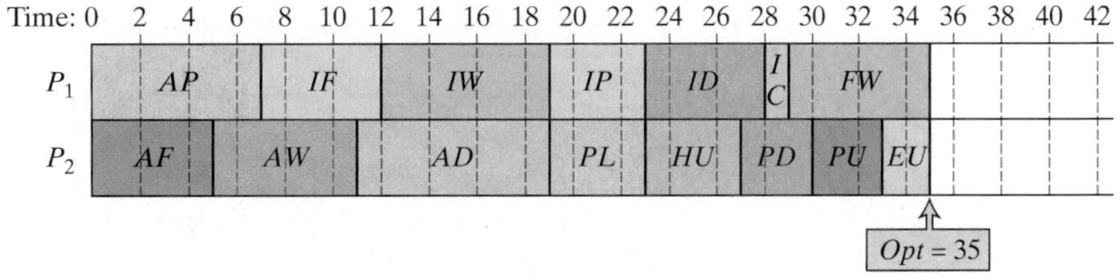

FIGURE 8-24 Timeline for an optimal schedule for the MHU building project ($N = 2$).

The critical-path algorithm is an excellent *approximate* algorithm for scheduling a project, but as Example 8.12 shows, it does not always give an optimal schedule. In this regard, scheduling problems are like traveling salesman problems (Chapter 6) and shortest network problems (Chapter 7)—there are *efficient approximate* algorithms for scheduling, but no *efficient optimal* algorithm is currently known. Of the standard scheduling algorithms, the critical-path algorithm is by far the most commonly used. Other, more sophisticated algorithms have been developed in the last 40 years and under specialized circumstances they can outperform the critical-path algorithm, but as an all-purpose algorithm for scheduling, the critical-path algorithm is hard to beat.

8.7   Scheduling with Independent Tasks

In this section we will briefly discuss what happens to scheduling problems in the special case when there are no precedence relations to worry about. This situation arises whenever we are scheduling tasks that are all independent.

It is tempting to think that without precedence relations hanging over one's head, scheduling becomes a simple problem and one should be able to find optimal schedules without difficulty, but appearances are deceiving. *There are no efficient optimal algorithms known for scheduling, even when the tasks are all independent.*

While, in a theoretical sense, we are not much better able to schedule independent tasks than to schedule tasks with precedence relations, from a purely practical point of view, there are a few differences. For one thing, there is no getting around the fact that the nuts-and-bolts details of creating a schedule using a priority list become tremendously simplified when there are no precedence relations to mess with. In this case, we just assign the tasks to the processors as they become free in exactly the order given by the priority list. Second, without precedence relations, the critical-path time of a task equals its processing time. This means that the *critical-time list* and *decreasing-time list* are exactly the same list, and, therefore, the *decreasing-time algorithm and the critical-path algorithm become one and the same.* Before we go on, let's look at a couple of examples of scheduling with independent tasks.

(**EXAMPLE 8.13**)   **Preparing for Lunch**

Imagine that you and your two best friends are cooking a nine-course luncheon as part of a charity event. We will consider each of the nine courses an independent task, to be done by just one of the chefs (P_1, P_2, or P_3). The nine courses are $A(70)$, $B(90)$, $C(100)$, $D(70)$, $E(80)$, $F(20)$, $G(20)$, $H(80)$, and $I(10)$, with their processing times given in minutes. Let's first use an alphabetical priority list.

Priority list: $A(70)$, $B(90)$, $C(100)$, $D(70)$, $E(80)$, $F(20)$, $G(20)$, $H(80)$, $I(10)$

Since there are no precedence relations, there are no ineligible tasks, and all tasks start out as ready tasks. As soon as a processor is free, it picks up the next available task in the priority list. From the bookkeeping point of view, this is a piece of cake. We leave it to the reader to verify that the resulting schedule is the one in Fig. 8-25, with finishing time *Fin* = 220 minutes. It is obvious from the figure that this is not a very good schedule.

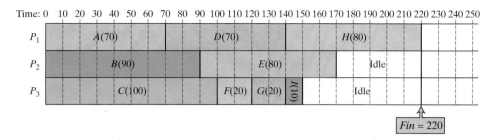

FIGURE 8-25

If we use the decreasing-time priority list (which in this case is also the critical-time priority list), we are bound to get a much better schedule.

Decreasing-time list: $C(100)$, $B(90)$, $E(80)$, $H(80)$, $A(70)$, $D(70)$, $F(20)$, $G(20)$, $I(10)$

The resulting schedule is shown in Fig. 8-26 and it is clearly an optimal schedule, since there is no idle time for any of the processors throughout the project. The optimal finishing time for the project is $Opt = 180$ minutes.

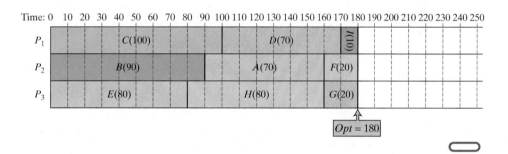

FIGURE 8-26

In Example 8.13, the critical-path algorithm gave us the optimal schedule, but, unfortunately, this need not always be the case.

EXAMPLE 8.14 **Preparing for Lunch: Part 2**

After the success of your last banquet, you and your two friends are asked to prepare another banquet. This time it will be a seven-course meal. The courses are all independent tasks, and their processing times (in minutes) are $A(50)$, $B(30)$, $C(40)$, $D(30)$, $E(50)$, $F(30)$, and $G(40)$.

The decreasing-time priority list is $A(50)$, $E(50)$, $C(40)$, $G(40)$, $B(30)$, $D(30)$, and $F(30)$. The resulting schedule, shown in Fig. 8-27, has project finishing time $Fin = 110$ minutes.

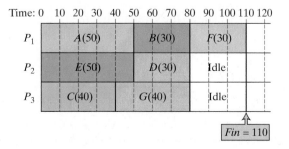

FIGURE 8-27

An optimal schedule (found using old-fashioned trial and error) with finishing time $Opt = 90$ minutes is shown in Fig. 8-28. Knowing the optimal finishing time $Opt = 90$ allows us to measure how much we were off when we used the decreasing-time priority list. As we did with earlier approximate solutions, we use the *relative error*. In this case, the relative error is $(110 - 90)/90 = 20/90 \approx 0.2222 = 22.22\%$.

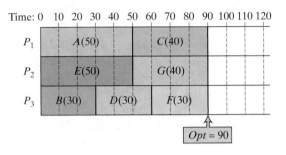

FIGURE 8-28

In general, if we know the optimal finishing time (Opt) of a project, we can compute the relative error of any schedule.

> **RELATIVE ERROR OF A SCHEDULE**
>
> For a schedule with finishing time *Fin*, the **relative error** (denoted by ε) is given by $\varepsilon = \dfrac{Fin - Opt}{Opt}$.

In the last example, we found that the relative error for the schedule obtained using the critical-path algorithm is $\varepsilon \approx 22.22\%$. Interestingly enough, we couldn't do any worse than this—even if we tried.

■ For more on Graham, see the biographical profile at the end of this chapter.

In 1969 American mathematician Ron Graham showed that when scheduling independent tasks using the critical-path algorithm with N processors, *the relative error is at most* $(N - 1)/3N$. We will call this upper bound for the relative error the **Graham bound**.

> **GRAHAM BOUND**
>
> The **Graham bound** for the relative error ε when scheduling a set of independent tasks with N processors is $\varepsilon \leq \dfrac{N - 1}{3N}$.

Table 8-3 shows the Graham bound for the relative error ε for a few small values of N.

TABLE 8-3	Graham Bound for Relative Error ε						
N	2	3	4	5	6	$\cdots$	100
(N − 1)/3N	$1/6 \approx 16.66\%$	$2/9 \approx 22.22\%$	$3/12 = 25\%$	$4/15 \approx 26.66\%$	$5/18 \approx 27.78\%$	$\cdots$	$99/300 = 33\%$

■ See Exercise 67.

Table 8-3 gives us a good sense of what is happening: As N grows, so does the Graham bound, but the Graham bound tapers off very quickly and will never go past $33\frac{1}{3}\%$. Graham's bound essentially implies that with independent tasks we can use the critical-path algorithm with the assurance that the relative error is bounded—no matter how many tasks need to be scheduled or how many processors are available to carry them out the finishing time of the project will never be more than $33\frac{1}{3}\%$ more than the optimal finishing time.

CONCLUSION

In one form or another, the scheduling of human (and nonhuman) activity is a pervasive and fundamental problem of modern life. At its most informal, it is part and parcel of the way we organize our everyday living (so much so that we are often scheduling things without realizing we are doing so). In its more formal incarnation, the systematic scheduling of a set of activities for the purposes of saving either time or money is a critical issue in management science. Business, industry, government, education—wherever there is a big project, there is a schedule behind it.

By now, it should not surprise us that at their very core, scheduling problems are mathematical in nature and that the mathematics of scheduling can range from the simple to the extremely complex. In this chapter we focused on a very specific type of scheduling problem in which we are given a set of *tasks*, a set of *precedence relations* among the tasks, and a set of identical *processors*. The objective is to schedule the tasks by properly assigning tasks to processors so that the *project finishing time* is as small as possible.

To tackle these scheduling problems systematically, we first developed a graph model of the problem, called the *project digraph*, and a general framework by means of which we can create, compare, and analyze schedules, called the *priority-list model*. Within the priority-list model, many strategies can be followed (with each strategy leading to the creation of a specific priority list). In the chapter, we considered two basic strategies for creating schedules. The first was the *decreasing-time algorithm*, a strategy that intuitively makes a lot of sense but that in practice often results in inefficient schedules. The second strategy, called the *critical-path algorithm*, is generally a big improvement over the decreasing-time algorithm, but it falls short of the ideal goal of guaranteeing an optimal schedule. The critical-path algorithm is by far the best known and most widely used algorithm for scheduling in business and industry.

When scheduling with independent tasks, the decreasing-time algorithm and the critical-path algorithm become one and the same, and the project finishing times cannot be too far off the optimal finishing time: The relative error cannot exceed the Graham bound $(N - 1)/3N$, where N is the number of processors.

Although several other, more sophisticated strategies for scheduling have been discovered by mathematicians in the last 40 years, no optimal, efficient all-purpose scheduling algorithm is presently known, and the general feeling among the experts is that there is little likelihood that such an algorithm actually exists. Finding ever-faster and more accurate scheduling algorithms remains a challenging open-ended mathematical problem—another task to execute in the grand cosmic project we call organized life.

P ROFILE: Ronald L. Graham (1935–)

Ron Graham is, by all accounts, the consummate juggler (in every sense of the word). For a period of almost 40 years, as a mathematician, director, vice president, and head scientist at Bell Labs, AT&T's legendary research arm, Graham was famous for his ability to keep an eclectic array of activities all going at once—doing research in several areas of mathematics, administering a major research institute, lecturing and teaching all over the world, teaching himself Chinese, and *really* juggling. (Graham is a past president of the International Jugglers Association, the inventor of many new juggling routines, and an expert on the mathematics of juggling.)

Ron Graham was born in 1935 in Taft, California. Graham discovered his natural talent and love of mathematics at an early age, and by the time he was 15, without even finishing high school, he won a full scholarship to the University of Chicago. After brief stints at Chicago and the University of California, Berkeley, Graham joined the Air Force and was stationed in Fairbanks, Alaska, where he served as a communication specialist while completing his undergraduate studies at the University of Alaska. After his tour of duty with the Air Force, Graham returned to UC Berkeley, where he completed a Ph.D. in mathematics in 1962. While a student at Berkeley, Graham, an accomplished gymnast and trampolinist, supported himself by performing with a small circus troupe called the Bouncing Bears, which performed acrobatic stunts at schools and fairs.

Graham joined Bell Labs in 1962. For the next 37 years he held a long list of scientific and administrative roles: researcher, head of the Mathematics Division, director of the Mathematics Center, vice president of research, and head scientist. Soon after joining Bell Labs, Graham was approached by engineers working on the computer system that was being developed to manage the antiballistic missile (ABM) defense program. One of the major issues engineers were facing was how to schedule the computers to identify and lock onto incoming enemy missiles in an efficient way. This turned out to be a classic scheduling problem of the kind discussed in this chapter (the computers are the *processors*; identifying and destroying the incoming missiles are the *tasks*). To work on these problems, Graham developed much of the theory of scheduling as we know it today. In so doing he also created a new field in the study of algorithms known as *worst-case analysis* (see Project A).

Graham retired from Bell Labs in 1999. Not one to sit idle, he promptly accepted an endowed chair as the Irwin and Joan Jacobs Professor of Computer Science and Information Science at the University of California, San Diego. Graham continues to teach, carry on with his mathematics research, lecture around the world, juggle, and, in general, keep going through life at a pace that amazes everyone else. When asked how he manages to keep so many different things going, Graham's pat answer is, "There are 168 hours in a week." ∎

KEY CONCEPTS

EXERCISES

WALKING

A Directed Graphs

1. For the digraph shown in Fig. 8-29, give

   (a) the vertex-set and arc-set.

   (b) the indegree of each vertex.

   (c) the outdegree of each vertex.

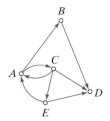

FIGURE 8-29

2. For the digraph shown in Fig. 8-30, give

   (a) the vertex-set and arc-set.

   (b) the indegree of each vertex.

   (c) the outdegree of each vertex.

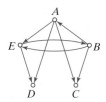

FIGURE 8-30

3. For the digraph shown in Fig. 8-29, find

   (a) the number of arcs in the digraph.

   (b) the sum of the indegrees.

   (c) the sum of the outdegrees.

4. For the digraph shown in Fig. 8-30, find

   (a) the number of arcs in the digraph.

   (b) the sum of the indegrees.

   (c) the sum of the outdegrees.

5. For the digraph shown in Fig. 8-29,

   (a) find all vertices that are incident *to* A.

   (b) find all vertices that are incident *from* A.

   (c) find all vertices that are incident *to* D.

   (d) find all vertices that are incident *from* D.

   (e) find all the arcs adjacent to AC.

   (f) find all the arcs adjacent to CD.

6. For the digraph shown in Fig. 8-30,

   (a) find all vertices that are incident *to* A.

   (b) find all vertices that are incident *from* A.

   (c) find all vertices that are incident *to* D.

   (d) find all vertices that are incident *from* D.

   (e) find all the arcs adjacent to BA.

   (f) find all the arcs adjacent to BC.

7. (a) Draw a digraph with vertex-set $V = \{A, B, C, D\}$ and arc-set $A = \{AB, AC, AD, BD, DB\}$.

   (b) Draw a digraph with vertex-set $V = \{A, B, C, D, E\}$ and arc-set $A = \{AC, AE, BD, BE, CD, DC, ED\}$.

   (c) Draw a digraph with vertex-set $V = \{W, X, Y, Z\}$ and such that W is incident to X and Y, X is incident to Y and Z, Y is incident to Z and W, and Z is incident to W and X.

8. (a) Draw a digraph with vertex-set $V = \{A, B, C, D\}$ and arc-set $A = \{AB, AC, AD, BC, BD, DB, DC\}$.

   (b) Draw a digraph with vertex-set $V = \{V, W, X, Y, Z\}$ and arc-set $A = \{VW, VZ, WZ, XV, XY, XZ, YW, ZY, ZW\}$.

   (c) Draw a digraph with vertex-set $V = \{W, X, Y, Z\}$ and such that every vertex is incident to every other vertex.

9. Consider the digraph with vertex-set $V = \{A, B, C, D, E\}$ and arc-set $A = \{AB, AE, CB, CE, DB, EA, EB, EC\}$. Without drawing the digraph, determine each of the following.

   (a) The outdegree of A

   (b) The indegree of A

   (c) The outdegree of D

   (d) The indegree of D

10. Consider the digraph with vertex-set $V = \{V, W, X, Y, Z\}$ and arc-set $A = \{VW, VZ, WZ, XY, XZ, YW, ZY, ZW\}$. Without drawing the digraph, determine each of the following.

    (a) The outdegree of V

    (b) The indegree of V

(c) The outdegree of Z

(d) The indegree of Z

11. Consider the digraph with vertex-set $V = \{A, B, C, D, E, F\}$ and arc-set $A = \{AB, BD, CF, DE, EB, EC, EF\}$.

(a) Find a path from vertex A to vertex F.

(b) Find a Hamilton path from vertex A to vertex F. (*Note*: A Hamilton path is a path that passes through every vertex of the graph once.)

(c) Find a cycle in the digraph.

(d) Explain why vertex F cannot be part of any cycle.

(e) Explain why vertex A cannot be part of any cycle.

(f) Find all the cycles in this digraph.

12. Consider the digraph with vertex-set $V = \{A, B, C, D, E\}$ and arc-set $A = \{AB, AE, CB, CD, DB, DE EB, EC\}$.

(a) Find a path from vertex A to vertex D.

(b) Explain why the path you found in (a) is the only possible path from vertex A to vertex D.

(c) Find a cycle in the digraph.

(d) Explain why vertex A cannot be part of a cycle.

(e) Explain why vertex B cannot be part of a cycle.

(f) Find all the cycles in this digraph.

13. The White Pine subdivision is a rectangular area six blocks long and two blocks wide. Streets alternate between one way and two way as shown in Fig. 8-31. Draw a digraph that represents the traffic flow in this neighborhood.

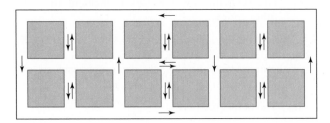

FIGURE 8-31

14. A mathematics textbook consists of 10 chapters. Although many of the chapters are independent of the others, some chapters require that previous chapters be covered first. The following list describes all the chapter dependences: Chapter 1 is a prerequisite to Chapters 3 and 5; Chapters 2 and 9 are both prerequisites to Chapter 10; Chapter 3 is a prerequisite to Chapter 6;

and Chapter 4 is a prerequisite to Chapter 7, which in turn is a prerequisite to Chapter 8. Draw a digraph that describes the dependences among the chapters in the book.

15. The digraph in Fig. 8-32 is a *respect* digraph. That is, the vertices of the digraph represent members of a group and an arc XY represents that X respects Y.

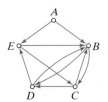

FIGURE 8-32

(a) If you had to choose one person to be the leader of the group, whom would you pick? Explain.

(b) Who would be the worst choice to be the leader of the group? Explain.

(c) Assume that you know the respect digraph of a group of individuals, and that is the only information available to you. Which individual would be the most reasonable choice for leader of the group?

16. The digraph in Fig. 8-33 is an example of a *tournament* digraph. In this example the vertices of the digraph represent five volleyball teams in a round-robin tournament (i.e., every team plays every other team). An arc XY represents that X defeated Y in the tournament. (*Note*: There are no ties in volleyball.)

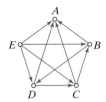

FIGURE 8-33

(a) Which team won the tournament? Explain.

(b) Which team came in last in the tournament? Explain.

(c) Suppose that you are given the tournament digraph of some tournament. What does the *indegree* of a vertex represent? What does the *outdegree* of a vertex represent?

(d) If T denotes the tournament digraph for a round-robin tournament with N teams, then for any vertex X in T, the indegree of X plus the outdegree of $X = N - 1$. Explain why.

B Project Digraphs

17. A project consists of eight tasks labeled *A* through *H*. The processing times (*P*-time) and precedence relations for each task are given in the following table.

(a) Give the vertex-set and arc-set for the project digraph.

(b) Draw the project digraph.

Task	*P*-time	Precedent Tasks
A	3	
B	10	*C, F, G*
C	2	*A*
D	4	*G*
E	5	*C*
F	8	*A, H*
G	7	*H*
H	5	

18. A project consists of eight tasks labeled *A* through *H*. The processing times (*P*-time) and precedence relations for each task are given in the following table.

(a) Give the vertex-set and arc-set for the project digraph.

(b) Draw the project digraph.

Task	*P*-time	Precedent Tasks
A	5	*C*
B	5	*C, D*
C	5	
D	2	*G*
E	15	*A, B*
F	6	*D*
G	2	
H	2	*G*

19. Eight experiments need to be carried out. One of the experiments requires 10 hours to complete, two of the experiments require 7 hours each to complete, two more require 12 hours each to complete, and three of the experiments require 20 hours each to complete. In addition, none of the 20-hour experiments can be started until both of the 7-hour experiments have been completed, and the 10-hour experiment cannot be started until both of the 12-hour experiments have been completed. Draw a project digraph for this scheduling problem.

20. Every fifty thousand miles an aircraft undergoes a complete inspection that involves 10 different diagnostic "checks." Three of the checks require 4 hours each to complete, three more require 7 hours each to complete, and four of the checks require 15 hours each to complete. Moreover, none of the 15-hour checks can be started until all the 4-hour checks have been completed. Draw a project digraph for this scheduling problem.

21. Apartments Unlimited is an apartment maintenance company that refurbishes apartments before new tenants move in. The following table shows the tasks, their processing times, and their precedent tasks performed when refurbishing a one-bedroom apartment (the processing times are given in 15-minute units). Draw a project digraph for the project of refurbishing a one-bedroom apartment.

Tasks	Label (*P*-time)	Precedent tasks
Bathrooms (clean)	*B*(8)	*P*
Carpets (shampoo)	*C*(4)	*S, W*
Filters (replace)	*F*(1)	
General cleaning	*G*(8)	*B, F, K*
Kitchen (clean)	*K*(12)	*P*
Lights (replace bulbs)	*L*(1)	
Paint	*P*(32)	*L*
Smoke detectors (battery)	*S*(1)	*G*
Windows (wash)	*W*(4)	*G*

22. A ballroom is to be set up for a large wedding reception. The following table shows the tasks to be carried out, their processing times (in hours) based on one person doing that task, and their precedent tasks. Draw a project digraph for setting up the wedding reception.

Tasks	Label (*P*-time)	Precedent tasks
Set up tables and chairs	*TC*(1.5)	
Set tablecloths and napkins	*TN*(0.5)	*TC*
Make flower arrangements	*FA*(2.2)	
Unpack crystal, china, and flatware	*CF*(1.2)	
Put place settings on table	*PT*(1.8)	*TN, CF*
Arrange table decorations	*TD*(0.7)	*FA, PT*
Set up the sound system	*SS*(1.4)	
Set up the bar	*SB*(0.8)	*TC*

C Schedules and Priority Lists

Exercises 23 through 26 refer to a project consisting of 11 tasks (A through K) with the following processing times (in hours): A(10), B(7), C(11), D(8), E(9), F(5), G(3), H(6), I(4), J(7), K(5).

23. **(a)** A schedule with $N = 3$ processors produces finishing time $Fin = 31$ hours. What is the total idle time for all the processors?

   (b) Explain why a schedule with $N = 3$ processors must have finishing time $Fin \geq 25$ hours.

24. **(a)** A schedule with $N = 5$ processors has finishing time $Fin = 19$ hours. What is the total idle time for all the processors?

   (b) Explain why a schedule with $N = 5$ processors must have finishing time $Fin \geq 15$ hours.

25. Explain why a schedule with $N = 6$ processors must have finishing time $Fin \geq 13$ hours.

26. **(a)** Explain why a schedule with $N = 10$ processors must have finishing time $Fin \geq 11$ hours.

   (b) Explain why it doesn't make sense to put more than 10 processors on this project.

Exercises 27 and 28 refer to the Martian Habitat Unit scheduling project with $N = 2$ processors discussed in Example 8.8. The purpose of these exercises is to fill in some of the details left out in Example 8.8.

27. For the priority list in Example 8.8, show the priority-list status at time $T = 26$.

28. For the priority list in Example 8.8, show the priority-list status at time $T = 32$.

Exercises 29 through 34 refer to the project described by the project digraph shown in Fig. 8-34.

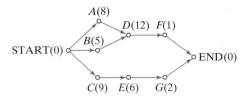

FIGURE 8-34

29. Using the priority list D, C, A, E, B, G, F, schedule the project with $N = 2$ processors. Show the project timeline and give its finishing time.

30. Using the priority list G, F, E, D, C, B, A, schedule the project with $N = 2$ processors. Show the project timeline and give its finishing time.

31. Using the priority list D, C, A, E, B, G, F, schedule the project with $N = 3$ processors. Show the project timeline and give its finishing time.

32. Using the priority list G, F, E, D, C, B, A, schedule the project with $N = 3$ processors. Show the project timeline and give its finishing time.

33. Explain why the priority lists A, B, D, F, C, E, G and F, D, A, B, G, E, C produce the same schedule.

34. Explain why the priority lists E, G, C, B, A, D, F and G, C, E, F, D, B, A produce the same schedule.

Exercises 35 and 36 refer to the Apartments Unlimited scheduling problem given in Exercise 21.

35. Using the priority list $B, C, F, G, K, L, P, S, W$, create a schedule for refurbishing an apartment with two workers. Show the project timeline, and give its finishing time.

36. Using the priority list $W, C, G, S, K, B, L, P, F$, create a schedule for refurbishing an apartment with two workers. Show the project timeline, and give its finishing time.

D The Decreasing-Time Algorithm

Exercises 37 and 38 refer to the project digraph shown in Fig. 8-35.

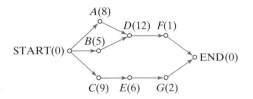

FIGURE 8-35

37. Use the decreasing-time algorithm to schedule the project with $N = 2$ processors. Show the timeline for the project, and give the project finishing time.

38. Use the decreasing-time algorithm to schedule the project with $N = 3$ processors. Show the timeline for the project, and give the project finishing time.

Exercises 39 and 40 refer to the project digraph shown in Fig. 8-36:

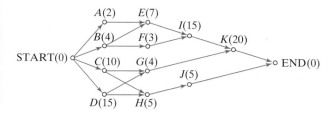

FIGURE 8-36

39. Use the decreasing-time algorithm to schedule the project with $N = 3$ processors. Show the timeline for the project, and give the project finishing time.

40. Use the decreasing-time algorithm to schedule the project with $N = 2$ processors. Show the timeline for the project, and give the project finishing time.

41. Consider the project described by the project digraph shown in Fig. 8-37, and assume that you are to schedule this project using $N = 2$ processors.

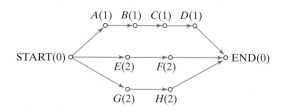

FIGURE 8-37

(a) Use the decreasing-time algorithm to schedule the project. Show the timeline for the project and the finishing time *Fin*.

(b) Find an optimal schedule and the optimal finishing time *Opt*.

(c) Using your answers to (a) and (b), find the relative error (expressed as a percent) of the schedule given by the decreasing-time algorithm.

42. Consider the project described by the project digraph shown in Fig. 8-38, and assume that you are to schedule this project using $N = 2$ processors.

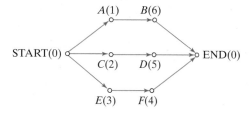

FIGURE 8-38

(a) Use the decreasing-time algorithm to schedule the project. Show the timeline for the project and the finishing time *Fin*.

(b) Find an optimal schedule and the optimal finishing time *Opt*.

(c) Using your answers to (a) and (b), find the relative error (expressed as a percent) of the schedule given by the decreasing-time algorithm.

43. Consider the project described by the project digraph shown in Fig. 8-39, and assume that you are to schedule this project using $N = 3$ processors.

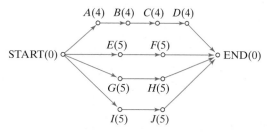

FIGURE 8-39

(a) Use the decreasing-time algorithm to schedule the project. Show the timeline for the project and the finishing time *Fin*.

(b) Find an optimal schedule and the optimal finishing time *Opt*.

(c) Using your answers to (a) and (b), find the relative error (expressed as a percent) of the schedule given by the decreasing-time algorithm.

44. Consider the project described by the project digraph shown in Fig. 8-40, and assume that you are to schedule this project using $N = 4$ processors.

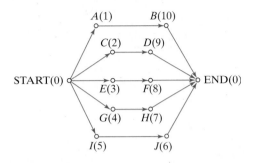

FIGURE 8-40

(a) Use the decreasing-time algorithm to schedule the project. Show the timeline for the project and the finishing time *Fin*.

(b) Find an optimal schedule. (*Hint: Opt* = 17.)

(c) Find the relative error (expressed as a percent) of the schedule given by the decreasing-time algorithm.

E | Critical Paths and the Critical-Path Algorithm

Exercises 45 and 46 refer to the project described by the project digraph shown in Fig. 8-41.

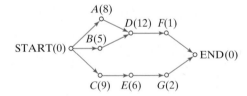

FIGURE 8-41

45. (a) Use the backflow algorithm to find the critical time for each vertex.

(b) Find the critical path for the project.

(c) Schedule the project with $N = 2$ processors using the critical-path algorithm. Show the timeline, and give the project finishing time.

(d) Explain why the schedule obtained in (c) is optimal.

46. (a) Schedule the project with $N = 3$ processors using the critical-path algorithm. Show the timeline, and give the project finishing time.

(b) Explain why the schedule obtained in (a) is optimal.

Exercises 47 and 48 refer to the project described by the project digraph shown in Fig. 8-42.

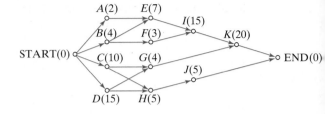

FIGURE 8-42

47. (a) Find the critical path for the project.

(b) Schedule the project with $N = 3$ processors using the critical-path algorithm. Show the timeline, and give the project finishing time.

48. (a) Use the backflow algorithm to find the critical time for each vertex.

(b) Schedule the project with $N = 2$ processors using the critical-path algorithm. Show the timeline, and give the project finishing time.

49. Schedule the Apartments Unlimited project given in Exercise 21 with $N = 2$ processors using the critical-path algorithm. (The tasks, processing times, and precedent tasks are shown again in the table below.) Show the timeline, and give the project finishing time. (Remember that processing times are given in 15-minute units.)

Tasks	Symbol/*P*-time	Precedent tasks
Bathrooms (clean)	$B(8)$	P
Carpets (shampoo)	$C(4)$	S, W
Filters (replace)	$F(1)$	
General cleaning	$G(8)$	B, F, K
Kitchen (clean)	$K(12)$	P
Lights (replace bulbs)	$L(1)$	
Paint	$P(32)$	L
Smoke detectors (battery)	$S(1)$	G
Windows (wash)	$W(4)$	G

50. Schedule the project given in Exercise 22 (setting up a ballroom for a wedding) with $N = 3$ processors using the critical-path algorithm. Show the timeline, and give the project finishing time.

F Scheduling with Independent Tasks

51. Consider eight independent tasks with processing times (in hours) given by 1, 2, 3, 4, 5, 6, 7, and 8.

  (a) Schedule these tasks with $N = 2$ processors using the critical-path algorithm. Show the timeline, and give the project finishing time F.

  (b) Find the optimal finishing time Opt for $N = 2$ processors.

  (c) Compute the relative error of the critical-path schedule found in (a) expressed as a percent.

52. Consider 12 independent tasks with processing times (in hours) given by 5, 6, 7, ..., 15, 16.

  (a) Schedule these tasks with $N = 2$ processors using the critical-path algorithm. Show the timeline, and give the project finishing time.

  (b) Find the optimal finishing time Opt for $N = 2$ processors.

  (c) Compute the relative error of the critical-path schedule found in (a) expressed as a percent.

53. Consider seven independent tasks with processing times (in hours) given by 1, 3, 5, 7, 9, 11, 13.

  (a) Find the finishing time Fin for $N = 3$ processors using the critical-path algorithm.

  (b) Find the optimal finishing time Opt for $N = 3$ processors.

  (c) Compute the relative error of the critical-path schedule found in (a) expressed as a percent.

54. Consider seven independent tasks with processing times (in hours) given by 4, 7, 10, 13, 16, 19, 22.

  (a) Find the finishing time Fin for $N = 3$ processors using the critical-path algorithm.

  (b) Find the optimal finishing time Opt for $N = 3$ processors.

  (c) Compute the relative error of the critical-path schedule found in (a) expressed as a percent.

55. Consider the following set of independent tasks: $A(4)$, $B(4)$, $C(5)$, $D(6)$, $E(7)$, $F(4)$, $G(5)$, $H(6)$, $I(7)$.

  (a) Schedule these tasks with $N = 4$ processors using the critical-path algorithm. Show the timeline, and give the project finishing time Fin.

  (b) Find an optimal schedule for $N = 4$ processors. Show the timeline, and give the optimal finishing time Opt.

  (c) Compute the relative error of the critical-path schedule found in (a) expressed as a percent.

56. Consider the following set of independent tasks: $A(4)$, $B(3)$, $C(2)$, $D(8)$, $E(5)$, $F(3)$, $G(5)$.

  (a) Schedule these tasks with $N = 3$ processors using the critical-path algorithm. Show the timeline, and give the project finishing time Fin.

  (b) Find an optimal schedule for $N = 3$ processors. Show the timeline, and give the optimal finishing time Opt.

  (c) Compute the relative error of the critical-path schedule found in (a) expressed as a percent.

57. (a) Consider seven independent tasks with processing times (in hours) given by 1, 2, 4, 8, 16, 32, and 64. Schedule these tasks with $N = 2$ processors using the critical-path algorithm. Show the timeline, and give the project finishing time.

  (b) Consider seven independent tasks with processing times (in hours) given by 1, $\frac{1}{2}$, $\frac{1}{4}$, $\frac{1}{8}$, $\frac{1}{16}$, $\frac{1}{32}$, and $\frac{1}{64}$. Schedule these tasks with $N = 2$ processors using the critical-path algorithm. Show the timeline, and give the project finishing time.

58. (a) Consider nine independent tasks with processing times (in hours) given by 1, 1, 2, 3, 5, 8, 13, 21, and 34. Schedule these tasks with $N = 2$ processors using the critical-path algorithm. Show the timeline, and give the project finishing time.

  (b) Consider ten independent tasks with processing times (in hours) given by 1, 1, 2, 3, 5, 8, 13, 21, 34, and 55. Schedule these tasks using the critical-path algorithm with two processors. Show the timeline, and give the project finishing time.

G Miscellaneous

59. Explain why, in any digraph, the sum of all the indegrees must equal the sum of all the outdegrees.

60. Symmetric and totally asymmetric digraphs. A digraph is called *symmetric* if, whenever there is an arc from vertex X to vertex Y, there is *also* an arc from vertex Y to vertex X. A digraph is called *totally asymmetric* if, whenever there is an arc from vertex X to vertex Y, there *is not* an arc from vertex Y to vertex X. For each of the following, state whether the digraph is symmetric, totally asymmetric, or neither.

  (a) A digraph representing the streets of a town in which all streets are one-way streets

  (b) A digraph representing the streets of a town in which all streets are two-way streets

  (c) A digraph representing the streets of a town in which there are both one-way and two-way streets

(d) A digraph in which the vertices represent a group of men, and there is an arc from vertex X to vertex Y if X is a brother of Y

(e) A digraph in which the vertices represent a group of men, and there is an arc from vertex X to vertex Y if X is the father of Y

61. For the schedule shown in Fig. 8-43, find a priority list that generates the schedule. (*Note*: There is more than one possible answer.)

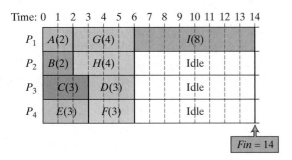

FIGURE 8-43

62. For the schedule shown in Fig. 8-44, find a priority list that generates the schedule. (*Note*: There is more than one possible answer.)

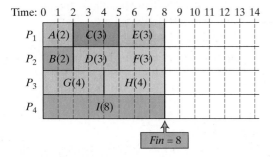

FIGURE 8-44

Exercises 63 through 66 refer to the Martian Habitat Unit building project as discussed in Examples 8.8, 8.9, and 8.12.

63. Find a schedule for building a Martian Habitat Unit with $N = 2$ processors using the priority list $AD(8)$, $AW(6)$, $AF(5)$, $IF(5)$, $AP(7)$, $IW(7)$, $ID(5)$, $IP(4)$, $PL(4)$, $PU(3)$, $HU(4)$, $IC(1)$, $PD(3)$, $EU(2)$, $FW(6)$. Show the timeline, and give the project finishing time. (See Example 8.8.)

64. Find a schedule for building a Martian Habitat Unit with $N = 2$ processors using the decreasing-time algorithm. (See Example 8.9. Do the work on your own, and then compare with the step-by-step details shown in Table 8-2.)

65. Find a schedule for building a Martian Habitat Unit with $N = 2$ processors using the critical-path algorithm.

Show the timeline, and give the project finishing time. (See Example 8.12.)

66. Find a schedule for building a Martian Habitat Unit with $N = 3$ processors using the critical-path algorithm. Show the timeline, and give the project finishing time.

JOGGING

67. When scheduling independent tasks with N processors using the critical-path algorithm, the relative error ε is guaranteed to be between 0 and the Graham bound $(N - 1)/(3N)$. (See the discussion in Section 8.7.)

(a) Calculate the range of values of the relative error ε when scheduling independent tasks with $N = 7$, $N = 8$, $N = 9$, and $N = 10$ processors.

(b) Explain why for any positive integer N, the Graham bound $(N - 1)/(3N)$ is less than $1/3 = 33\frac{1}{3}\%$.

(c) Explain why as N gets larger and larger, the Graham bound $(N - 1)/(3N)$ gets closer and closer to $\frac{1}{3}$.

68. The purpose of this exercise is to illustrate that the relative error can equal the Graham bound. In each case, for the given set of independent tasks and the given N, (i) find a critical-path schedule, (ii) find an optimal schedule, and (iii) show that the relative error of the critical-path schedule equals the Graham bound.

(a) Tasks: $A(3), B(3), C(2), D(2), E(2)$; $N = 2$ processors

(b) Tasks: $A(5)$, $B(5)$, $C(4)$, $D(4)$, $E(3)$, $F(3)$, $G(3)$; $N = 3$ processors

(c) Tasks: $A(7)$, $B(7)$, $C(6)$, $D(6)$, $E(5)$, $F(5)$, $G(4)$, $H(4), I(4)$; $N = 4$ processors

Exercises 69 through 71 introduce three paradoxes that can occur in scheduling. In all three exercises we will use the project described by the project digraph shown in Fig. 8-45. Assume that processing times are given in hours. (This example, together with the three paradoxes, comes from reference 6.)

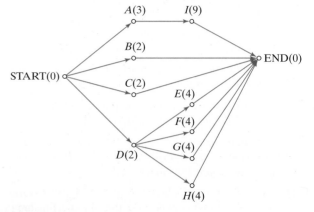

FIGURE 8-45

69. **(a)** Using the priority list $A, B, C, D, E, F, G, H, I,$ schedule the project with $N = 3$ processors.

(b) Using the same priority list as in (a), schedule the project with $N = 4$ processors.

(c) There is a paradox in the results obtained in (a) and (b). Describe the paradox.

70. **(a)** Using the priority list $A, B, C, D, E, F, G, H, I,$ schedule the project with $N = 3$ processors. [This is the same question as in 69(a). If you already answered 69(a), you can recycle the answer.]

(b) Imagine that the processors are fine-tuned so that the processing times for all the tasks are shortened by 1 hour [i.e., the tasks now are $A(2), B(1), \ldots, H(3)$]. Using the same priority list as in (a), schedule this project with $N = 3$ processors.

(c) There is a paradox in the results obtained in (a) and (b). Describe the paradox.

71. **(a)** Using the priority list $A, B, C, D, E, F, G, H, I,$ schedule the project with $N = 3$ processors. [Same question again as 69(a) and 70(a). You're getting your money's worth on this one!]

(b) Suppose that the precedence relations $D \rightarrow E$ and $D \rightarrow F$ are no longer required. All other precedence relations and processing times are as in the original project. Using the same priority list as in (a), schedule this project with $N = 3$ processors.

(c) There is a paradox in the results obtained in (a) and (b). Describe the paradox.

72. In 1961 T. C. Hu of the University of California showed that in any scheduling problem in which all the tasks have equal processing times and in which the original project digraph (without the START and END vertices) is a tree, the critical-path algorithm will give an optimal schedule. Using this result, find an optimal schedule for the project given by the project digraph shown in Fig. 8-46. Schedule the project using $N = 3$ processors. Assume that each task takes three days. (Notice that we have omitted the START and END vertices.)

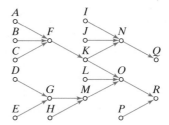

FIGURE 8-46

73. Let W represent the sum of the processing times of all the tasks, N be the number of processors, and Fin be the finishing time for a project.

(a) Explain the meaning of the inequality $Fin \geq W/N$ and why it is true for any schedule.

(b) Under what circumstances is $Fin = W/N$?

(c) What does the value $N \times Fin - W$ represent?

RUNNING

74. You have $N = 2$ processors to process M independent tasks with processing times $1, 2, 3, \ldots, M$. Find the optimal schedule, and give the optimal completion time Opt in terms of M. (*Hint*: Consider four separate cases based on the remainder when N is divided by 4.)

75. You have $N = 3$ processors to process M independent tasks with processing times $1, 2, 3, \ldots, M$. Find the optimal schedule, and give the optimal completion time Opt in terms of M. (*Hint*: Consider six separate cases based on the remainder when N is divided by 6.)

76. You have $N = 2$ processors to process $M + 1$ independent tasks with processing times $1, 2, 4, 8, 16, \ldots, 2^M$. Find the optimal schedule, and give the optimal completion time Opt in terms of M.

PROJECTS AND PAPERS

A Worst-Case Analysis of Scheduling Algorithms

We saw in this chapter that when creating schedules there is a wide range of possibilities, from *very bad schedules* to *optimal* (best possible) schedules. Our focus was on creating good (if possible optimal) schedules, but the study of bad schedules is also of considerable importance in many applications. In the late 1960s and early 1970s, Ronald Graham of AT&T Bell Labs pioneered the study of worst-case

scenarios in scheduling, a field known as *worst-case analysis*. As the name suggests, the issue in worst-case analysis is to analyze how bad a schedule can get. This research was motivated by a deadly serious question—how would the performance of the antiballistic missile defense system of the United States be affected by a failure in the computer programs that run the system?

In his research on worst-case analysis in scheduling, Graham made the critical discovery that there is a limit on how bad a schedule can be—no matter how stupidly a schedule is put together, the project finishing time *Fin* must satisfy the inequality $Fin \leq Opt\,(2 - 1/N)$, where N is the number of processors and Opt is the optimal completion time of the project.

Prepare a presentation explaining Graham's worst-case analysis result and its implications. Give examples comparing optimal schedules and the worst possible schedules for projects involving $N = 2$ and $N = 3$ processors.

B Dijkstra's Shortest-Path Algorithm

Just like there are weighted graphs, there are also *weighted digraphs*. A weighted digraph is a digraph in which each arc has a weight (distance, cost, time) associated with it. Weighted digraphs are used to model many important optimization problems in vehicle routing, pipeline flows, and so on. (Note that a weighted digraph is not the same as a project digraph — in the project digraph the vertices rather than the edges are the ones that have weights.) A fundamental question when working with weighted digraphs is finding the shortest path from a given vertex X to a given vertex Y. (Of course, there may be no path at all from X to Y, in which case the question is moot.) The classic algorithm for finding the shortest path between two vertices in a digraph is

known as *Dijkstra's algorithm*, named after the Dutch computer scientist Edsger Dijkstra, who first proposed the algorithm in 1959.

Prepare a presentation describing Dijkstra's algorithm. Describe the algorithm carefully and illustrate how the algorithm works with a couple of examples. Discuss why Dijkstra's algorithm is important. Is it efficient? Optimal? Discuss some possible applications of the algorithm.

C Tournaments

In the language of graph theory, a *tournament* is a digraph whose underlying graph is a complete graph. In other words, to create a tournament you can start with K_N (the complete graph on N vertices) and then change each edge into an arc by putting an arbitrary direction on it. The reason these graphs are called tournaments is that they can be used to describe the results of a tournament in which every player plays against every other player (no ties allowed).

Write a paper on the mathematics of tournaments (as defined previously). If you studied Chapter 1, you should touch on the connection between tournaments and the results of elections decided under the *method of pairwise comparisons*.

REFERENCES AND FURTHER READINGS

1. Baker, K. R., *Introduction to Sequencing and Scheduling*. New York: John Wiley & Sons, 1974.

2. Brucker, Peter, *Scheduling Algorithms*. New York: Springer-Verlag, 2001.

3. Coffman, E. G., *Computer and Jobshop Scheduling Theory*. New York: John Wiley & Sons, 1976, chaps. 2 and 5.

4. Conway, R. W., W. L. Maxwell, and L. W. Miller, *Theory of Scheduling*. New York: Dover, 2003.

5. Dieffenbach, R. M., "Combinatorial Scheduling," *Mathematics Teacher*, 83 (1990), 269–273.

6. Garey, M. R., R. L. Graham, and D. S. Johnson, "Performance Guarantees for Scheduling Algorithms," *Operations Research*, 26 (1978), 3–21.

7. Graham, R. L., "Bounds on Multiprocessing Timing Anomalies," *SIAM Journal of Applied Mathematics*, 17 (1969), 263–269.

8. Graham, R. L., "The Combinatorial Mathematics of Scheduling," *Scientific American*, 238 (1978), 124–132.

9. Graham, R. L., "Combinatorial Scheduling Theory," in *Mathematics Today*, ed. L. Steen. New York: Springer-Verlag, 1978, 183–211.

10. Graham, R. L., E. L. Lawler, J. K. Lenstra, and A. H. G. Rinnooy Kan, "Optimization and Approximation in Deterministic Sequencing and Scheduling: A Survey," *Annals of Discrete Mathematics*, 5 (1979), 287–326.

11. Gross, Jonathan, and Jay Yellen, *Graph Theory and Its Applications*. Boca Raton, FL: CRC Press, 1999, chap. 11.

12. Hillier, F. S., and G. J. Lieberman, *Introduction to Operations Research*, 3rd ed. San Francisco: Holden-Day, 1980, chap. 6.

13. Kerzner, Harold, *Project Management: A Systems Approach to Planning, Scheduling, and Controlling*, 7th ed. New York: John Wiley & Sons, 2001, chap. 12.

14. Roberts, Fred S., *Graph Theory and Its Applications to Problems of Society*, CBMS/NSF Monograph No. 29. Philadelphia: Society for Industrial and Applied Mathematics, 1978.

15. Wilson, Robin, *Introduction to Graph Theory*, 4th ed. Harlow, England: Addison-Wesley Longman Ltd., 1996.

mini-excursion 2

A Touch of Color

Graph Coloring

FIGURE ME2-1

The map shown in Fig. ME2-1 is a bare bones, no-frills map of the 48 contiguous states of the United States, as bland as a map can be. Maps are supposed to be pleasing to the eye, and the typical way to make them so is to add a touch of color. The standard rules for coloring a map are to use a single color for a state (we will use "state" as a generic term to describe the geographical units in the map, which in reality could be countries, provinces, etc.) and never use the same color for two states that share a common border. (On the other hand, two states whose common border is just one point—for example, Arizona and Colorado—can be colored, if we so choose, with the same color.)

The map shown in Fig. ME2-2(a) is at the opposite extreme from Fig. ME2-1: 48 different colors were used, one for each state. First, it's a little jarring to see that many colors—a classic case of too much of a good thing. Second, let's imagine that there is a production cost incurred each time we add another color to the map, so a map that uses 48 colors is more expensive than a map that uses 47 colors, which in turn is more expensive than a map that uses 46 colors, and so on. (In the old days this was probably true, but with modern laser and inkjet printers the argument doesn't hold up. On the other hand, the cost argument gives us a convenient way to think about the issues in map coloring, so we will pretend it's still true.) A more typical, "less expensive" colored map is shown in Fig. ME2-2(b). Here a total of five different colors were used. Is it possible to cut down the number of colors used to four? How about three? You are encouraged to try answering these questions before you read on. (For a convenient way to play with coloring a U.S. map, go to *http://www.sailor.lib.md.us/MD_topics/kid/ col_applet/color.html.* If you choose from the drop-down menu "Flag," choose "US Map" and then click on "Load Image." You will get a blank map of the lower 48 states and a palette of colors for coloring the map.)

■ See Exercise 16.

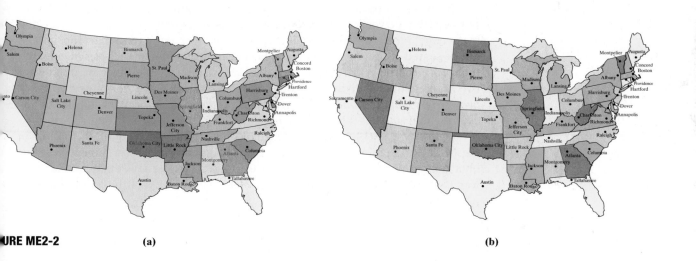

URE ME2-2   **(a)**   **(b)**

Much has been written about the mathematics of map coloring and its connection with graph theory. In this mini-excursion we will explore some of the ideas behind the fascinating topic of graph coloring and some of its more colorful applications. Some familiarity with the basic terminology and concepts in Chapters 5 and 6 is strongly recommended.

Graph Coloring and Chromatic Numbers

(**EXAMPLE ME2.1**)   **Can't We All Get Along?**

Sometimes people just don't get along, and you are caught in the middle. Imagine that you are a wedding planner organizing the rehearsal dinner before a big wedding. There are a total of 16 people attending the rehearsal dinner: $A, B, C,$ $D, E, F, G,$ and H are relatives of the bride and groom; $I, J, K, L, M, N, O,$ and P

are members of the wedding party. If things weren't stressful enough, you are told that some of these people have serious issues:

- *A* doesn't get along with *F, G,* or *H*
- *B* doesn't get along with *C, D,* or *H*
- *C* doesn't get along with *B, D, E, G,* or *H*
- *D* doesn't get along with *B, C,* or *E*
- *E* doesn't get along with *C, D, F,* or *G*
- *F* doesn't get along with *A, E,* or *G*
- *G* doesn't get along with *A, C, E,* or *F*
- *H* doesn't get along with *A, B,* or *C*

To make the rehearsal dinner go smoothly you are instructed to find a way to seat these people so that people who don't get along must be seated at different tables. (*I* through *P* get along with everyone, so they are not a concern.) How are you going to set up the seating arrangements with so many incompatibility issues to worry about? What is the minimum number of tables you will need? You can answer both of these questions using a little graph theory.

We will start by creating the "incompatibility" graph shown in Fig. ME2-3(a). In this graph the vertices represent the individuals, and two vertices are connected by an edge if the corresponding individuals don't get along (and therefore should not be seated at the same table). To make seating assignments we assign colors to the vertices of the graph, with each color representing a table. Since we don't want two incompatible individuals seated at the same table, *we don't want to color two vertices that are connected by an edge with the same color.* We will refer to any coloring that satisfies this rule as a *legal coloring* of the graph.

Figure ME2-3(b) shows a legal coloring of the vertices of the graph in Fig. ME2-3(a) that uses four different colors. This coloring tells us how to seat this obnoxious group using four tables. Can we do better (i.e., use fewer than four tables)? Yes. Figure ME2-3(c) shows a legal coloring of the vertices of the graph that uses just three colors.

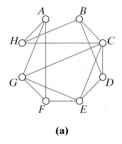
(a)

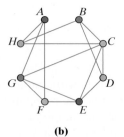
(b)

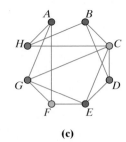
(c)

FIGURE ME2-3

Could we possibly come up with a legal coloring of the vertices of the graph that uses just two colors? No. The reason is clear if we look at *A, F,* and *G* (or any other set of three vertices that form a *triangle*). Since each is adjacent to the other two, *A, F,* and *G* will have to be colored with different colors.

The conclusions to our analysis are (1) the minimum number of tables needed to sit the wedding party is three, and (2) the seating assignment should put *A, B,* and *E* at one table (red), *C* and *F* at a second table (blue), and *D, G,* and *H* at the third table (green). The remaining members of the wedding party can be arbitrarily assigned to fill up the remaining seats at the three tables.

■ This is, in fact, the unique solution to the problem—see Exercise 12.

Example ME2.1 illustrates the two key ideas that we will discuss in this mini-excursion: (1) the idea of coloring the vertices of a graph so that *adjacent vertices don't get the same color* and (2) the idea of trying to do this using as *few colors as possible*. Here are the formal definitions of these concepts.

Note that the actual choice of colors used in coloring the graph is irrelevant. In fact, the colors are just a convenient way to classify the vertices, and we could just as well use numbers or letters to do the same. We use colors primarily because humans are better able to perceive patterns of colors than patterns of symbols.

---

k-COLORING

A **k-coloring** of a graph G is a coloring of the vertices of G using k colors and satisfying the requirement that *adjacent vertices are colored with different colors*.

CHROMATIC NUMBER

The **chromatic number** of a graph G is the *smallest* number k for which a k-coloring of the vertices of G is possible. We will use the notation $\chi(G)$ to denote the chromatic number of G.

Using the above notation, for the graph G in Fig. ME2-3(a) we have $\chi(G) = 3$. This follows from the observation that a 3-coloring of G is possible, as shown in Fig. ME2-3(c), but a 2-coloring of G is not.

EXAMPLE ME2.2   Coloring Complete Graphs

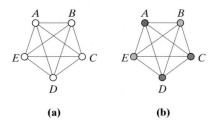

FIGURE ME2-4

(a)   (b)

The graph in Fig. ME2-4(a) is K_5, the complete graph on 5 vertices. In this graph every vertex is adjacent to every other vertex, so no two vertices can have the same color. The only possible way to color K_5 is to use a different color for each vertex, as in Fig. ME2-4(b). Thus, we can conclude that $\chi(K_5) = 5$.

The argument used in Example ME2.2 can be generalized to show that $\chi(K_n) = n$. This illustrates that it is possible to create graphs with arbitrarily large chromatic numbers. You need a graph G with $\chi(G) = 100$? No problem—choose G to be K_{100}.

EXAMPLE ME2.3   Coloring Circuits

A graph consisting of a single circuit with n vertices is denoted by C_n. The graph shown in Fig. ME2-5(a) is C_6. Figure ME2-5(b) shows a 2-coloring of C_6. Since a 1-coloring is clearly out of the question, we can conclude that $\chi(C_6) = 2$.

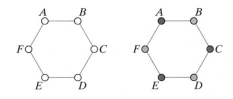

FIGURE ME2-5   (a)   (b)

The graph shown in Fig. ME2-6(a) is C_5. Figure ME2-6(b) shows a 3-coloring of C_5. A little reflection should be enough to convince ourselves that we will not be able to color C_5 with just two colors, as we did with C_6. The problem is that we can alternate two colors around the circuit until we get to the last vertex, but since the number of vertices is odd, the last vertex will be adjacent to two vertices of different colors, and a third color will be needed. It follows that $\chi(C_5) = 3$.

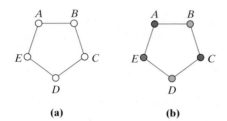

See Exercise 9.

See Exercise 8.

FIGURE ME2-6   (a)   (b)

We can generalize the preceding observations to any C_n ($n \geq 3$): If n is even, then $\chi(C_n) = 2$; if n is odd, then $\chi(C_n) = 3$. Thus, we now know that it is possible to have graphs with a large number of vertices and small chromatic number (2 or 3). Graphs with a large number of vertices and chromatic number 1 are also possible, but are extremely uninteresting—they have no edges.

The Greedy Algorithm for Graph Coloring

Like some of the other graph problems discussed in Chapters 5 through 8 (Euler circuit problems, traveling salesman problems, shortest network problems, scheduling problems) graph coloring can be thought of as an *optimization* problem: How can we color a graph using the fewest possible number of colors? We will call such a coloring an *optimal coloring* of the graph, and the general problem of finding optimal colorings is known as the *coloring problem*.

> **OPTIMAL COLORING**
>
> An **optimal coloring** of a graph G is a coloring of the vertices of G using the fewest possible number of colors. To put it in slightly more formal terminology, an optimal coloring of G is an $\chi(G)$-coloring of G. [Too many G's in the last sentence, but all it says is that when we color a graph using $\chi(G)$ colors by definition we have an optimal coloring of the graph.]

The truly interesting question in graph coloring is the following: Given an arbitrary graph G, how do we find an optimal coloring of G? If you are a veteran of Chapters 5 through 8, you may not be entirely surprised to find out that *no efficient general algorithm is known for finding optimal colorings of graphs*. For small graphs we can use trial and error, and for some families of graphs optimal coloring is reasonably easy (for example, complete graphs or circuits), but those are just special cases. In general, the best we can do is to use *approximate algorithms* that may get us close to an optimal coloring.

In this section we will discuss a classic approximate algorithm for graph coloring known as the *greedy algorithm* (for reasons that will become clear soon). To illustrate the ideas behind the greedy algorithm we will start with a couple of simple examples.

EXAMPLE ME2.4 **Greed Is Good (Sometimes)**

Let's try to color the graph in Fig. ME2-7(a). The strategy we will use is simple. We will start with vertex A and color it with some color, say blue. We will think of blue as the *first* color in some arbitrary priority list of colors (blue goes first, red second, green third, yellow fourth, and so on.) We now move to the next vertex, B, and try to color it with the first color in our list (blue), but we can't do it because B is adjacent to A, which is already blue. OK, fine—we'll just go to the next available color on the list (red) and use it for B. If we go next to vertex C, how should we color it? Other than red (C is adjacent to B) we can use any color we want, but why introduce a third color when we can use blue? We are, after all, trying to cut down on the number of colors used. So we color C with blue.

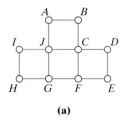

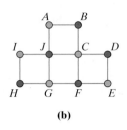

(a)

(b)

FIGURE ME2-7

Using the same philosophy (try the colors in the designated order) we color D with red, and so on. Everything lines up just right and we can alternate blue and red and get the 2-coloring shown in Fig. ME2-7(b). Since we know that the graph cannot be colored with one color, Fig. ME2-7(b) gives an optimal 2-coloring of the graph.

The strategy we used in Example ME2.4 has two key components: (1) we color the vertices of the graph following some designated order (in Example ME2.4 we went through the vertices in alphabetical sequence, but it can be done in any order we choose) and (2) we have a priority list for assigning the colors, and we limit the colors used by always starting at the top of the list and using previously used colors as much as possible. For obvious reasons, we call this a *greedy strategy*.

In Example ME2.4 the greedy strategy gave us an optimal coloring, but there was a bit of good karma involved. By pure luck, the order in which we colored the vertices just worked out perfectly. This is not always the case, and our next example shows how bad luck in the order of the vertices can give us a bad coloring of the graph.

EXAMPLE ME2.5 **Greed Can Sometimes Be Bad**

The graph in Fig. ME2-8(a) is the same graph as the one in Fig. ME2-7(a) except for the way the vertices are labeled. If we try the same approach we used in Example ME2.4 (color the vertices in alphabetical order and use the same priority order for colors—blue first, red second, green third, yellow fourth) we will get a very different result. We start by coloring vertex A with blue. So far, so good. Now B is not adjacent to A, so we will also color it with blue. Ditto for C and D [Fig. ME2-8(b)]. So far we have used only one color, so we are doing well. When we get to vertex E we have to go down to our second color because E is adjacent to D, which has already been colored blue. Thus, E is colored red. Likewise, F is adjacent to a blue vertex, so we color it red. We are now looking at Fig. ME2-8(c). Since G is now adjacent to a blue and a red vertex, we have to pull out a third color (green) for G [Fig. ME2-8(d)]. Now you can clearly see our bad luck. The next vertex (H) is adjacent to a blue, a red, and a green vertex, so we need a fourth color (yellow) to color it [Fig. ME2-8(e)]. The last two vertices (I and J) are adjacent to blue vertices only, so we will color them red. This gives us the final coloring of the graph [Fig. ME2-8(f)].

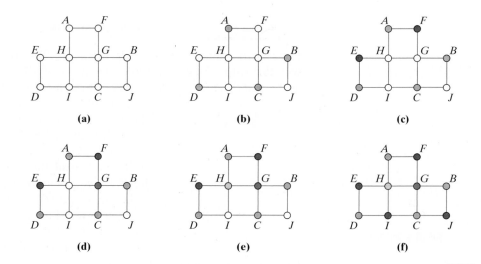

FIGURE ME2-8

See Exercise 21.

Examples ME2.4 and ME2.5 illustrate an important point: The order in which the vertices of the graph are colored can make a big difference in the kind of coloring we get. There is always some way to order the vertices so that we get an optimal coloring, but there is no easy way to figure out what that order is. Moreover, there are $n!$ different ways to order the n vertices, so going through the different orderings to find the best one is out of the question. (Take a quick look at Table 6-4 in Chapter 6 if you need to jog your memory on how quickly factorials grow.)

We will now formalize the ideas introduced in Examples ME2.4 and ME2.5 into what is known as the *greedy algorithm* for graph coloring. To implement the algorithm we assume that the vertices are ordered in some arbitrary order $v_1, v_2, \ldots, v_n$ and that $c_1, c_2, c_3, c_4, \ldots$ represents a priority order for the colors.

GREEDY ALGORITHM FOR GRAPH COLORING

- **Step 1.** Assign the first color (c_1) to the first vertex (v_1).
- **Step 2.** Vertex v_2 is assigned color c_1 if it is not adjacent to v_1; otherwise, it gets assigned color c_2.
- **Steps 3, 4, ..., n.** Vertex v_i is assigned the first possible color in the priority list of colors (i.e., the first color that has not been assigned to one of the already colored neighbors of v_i).

Despite its limitations, the greedy algorithm is a reasonable approach for graph coloring, especially when we use some ingenuity in the way we order the vertices. Since there are more restrictions in coloring vertices of higher degree than there are in coloring vertices of lower degree, one obvious refinement of the greedy algorithm is to order the vertices in *decreasing* order of degrees: color vertices of highest degrees first (if there are ties choose among them at random), color vertices of next highest degree next, and so on. If we were to apply this approach to the graph in Fig. ME2-8, one possible ordering of the vertices would be $G, H, C, I, A, B, D, E, F, J$. If we were to color the vertices in this order using the greedy algorithm, we would get an optimal coloring of the graph in Fig. ME2-8.

See Exercise 20.

The greedy algorithm can also be used to produce upper bounds on the chromatic number of the graph. The best known of these upper bounds is given by the following fact, known as Brook's theorem.

BROOK'S THEOREM

Let $d_1 \geq d_2 \geq \cdots \geq d_n$ be the degrees of the vertices of the graph G listed in decreasing order. Then $\chi(G) \leq d_1 + 1$.

At most d_1 colors

FIGURE ME2-9

Brook's theorem essentially says that *the chromatic number of a graph cannot be more than the largest degree in the graph plus 1*. To see why this is true, think about the worst-case scenario we can run into when we are coloring a vertex using the greedy algorithm: The vertex is adjacent to many other vertices, all of which have been colored with different colors. Since the largest number of adjacent vertices a vertex can have is d_1, the worst thing that could happen is that the first d_1 colors have been used and we would need one more to color our vertex (Fig. ME2-9).

It turns out that for *connected graphs*, the only two cases where we actually have to use the maximum $d_1 + 1$ colors to color the graph are when the graph is the complete graph K_n (all vertices have degree $n - 1$, the chromatic number is n) or the circuit C_n, where n is odd (all vertices have degree 2, and the chromatic number is 3). If we rule these two cases out, we no longer need the +1 in Brook's theorem. We will call this the *strong version* of Brook's theorem.

BROOK'S THEOREM (STRONG VERSION)

Let $d_1 \geq d_2 \geq \cdots \geq d_n$ be the degrees of the vertices of a *connected* graph G listed in decreasing order. If G is not C_n (n odd) or K_n, then $\chi(G) \leq d_1$.

Graph Coloring and Sudoku

The game of Sudoku has become the rage among puzzle and game enthusiasts looking for a more intellectual (and cheaper) challenge than the one provided by an X-Box. Sudoku is addictive, and even ordinary people who are not drawn to video games are hooked on it. These days practically every major newspaper carries a daily Sudoku puzzle.

If you haven't played Sudoku yet, the rules are quite simple: You start with a 9×9 grid of 81 squares called *cells*. The grid is also subdivided into nine 3×3 subgrids called *boxes*. Some of the cells are already filled with the numbers 1 through 9. These are called the *givens*. The challenge of the game is to complete the grid by filling the remaining cells with the numbers 1 through 9. The requirements are that (1) every row and every column of the grid must have the numbers 1 through 9 appear once and (2) each of the nine boxes must have the numbers 1 through 9 appear once.

A typical Sudoku puzzle may have somewhere between 25 and 40 givens, depending on the level of difficulty. Figure ME2-10 is an example of a moderately easy Sudoku puzzle. The labels 1 through 9 on the columns and *a* through *i* on the rows are not part of the puzzle, but they provide a convenient way to refer to the cells. Just for fun, you may want to try this one out before you read on. [*Hint*: Try to figure out what number should go in cell *f*2. Once you have that one figured, go to cell *d*3. That's enough help for now. The solution is given *in* the answer section (see Exercise 23).]

	1	2	3	4	5	6	7	8	9
a			4	8					
b		9		4	6			7	
c		5				6	1	4	
d	2	1		6		5			
e	5	8		7		9		4	1
f		7			8			6	9
g	3	4	5					9	
h		6			3	7		2	
i					4	1			

FIGURE ME2-10

The Sudoku Graph

■ See Exercise 22.

To see the connection between Sudoku and graph coloring, we will first describe the **Sudoku graph**, which for convenience we will refer to as S. The graph S has 81 vertices, with each vertex representing a cell. When two cells *cannot* have the same number (because they are in the same row, in the same column, or in the same box), we put an edge connecting the corresponding vertices of the Sudoku graph S. For example, since cells $a3$ and $a7$ are in the same row, there is an edge joining their corresponding vertices; there is also an edge connecting $a1$ and $b3$ (they are in the same box), and so on. When everything is said and done, each vertex of the Sudoku graph has degree 20, and the graph has a total of 810 edges. S is too large to draw, but we can get a sense of the structure of S by looking at a partial drawing such as the one in Fig. ME2-11. The drawing shows all 81 vertices of S, but only two ($a1$ and $e5$) have their full set of incident edges showing.

The second step in converting a Sudoku puzzle into a graph coloring problem is to assign colors to the numbers 1 through 9. This assignment is arbitrary and is not a priority ordering of the colors as in the greedy algorithm—it's just a simple correspondence between numbers and colors. Figure ME2-12 shows one such assignment.

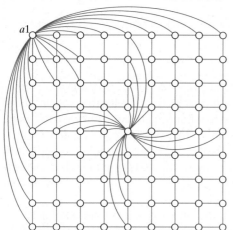

FIGURE ME2-11 A partial drawing of the Sudoku graph.

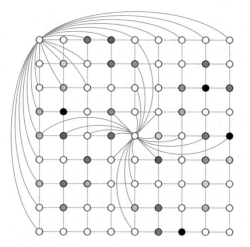

FIGURE ME2-13 A Sudoku puzzle as a graph coloring problem.

Number: 1 2 3 4 5 6 7 8 9

Color: ● ○ ○ ● ○ ● ○ ● ○

FIGURE ME2-12

Once we have the Sudoku graph and an assignment of colors to the numbers 1 through 9, any Sudoku puzzle can be described by a Sudoku graph where some of the vertices are already colored (the ones corresponding to the givens). For example, the Sudoku puzzle shown in Fig. ME2-10 is equivalent to the partial coloring shown in Fig. ME2-13. To solve the Sudoku puzzle all we have to do is color the rest of the vertices using the nine colors in Fig. ME2-12.

Map Coloring

Map drawing and coloring are ancient arts, but the connection between map coloring and mathematics originated in 1852 when a University of London student by the name of Francis Guthrie mentioned to his mathematics professor (well-known mathematician Augustus De Morgan) that he had been coloring many maps of English counties (don't ask why) and noticed that every map he had tried, no matter how complicated, could be colored with four colors

■ A short biographical profile of Hamilton is given in Chapter 6.

(where districts with a common border had to be colored with different colors). He inquired if this was a known mathematical fact, and De Morgan (who didn't know the answer) checked with some of the most famous mathematicians of the time, including his friend William Rowan Hamilton.

Guthrie's notion that any map could be colored with just four colors sounded so simple that everyone assumed it could be easily proved mathematically. After 100 years and many failed attempts at a proof, the *Four-Color Conjecture*, as the problem was famously known, was finally solved in 1976 by Kenneth Appel and Wolfgang Haken of the University of Illinois. Yes, indeed, Guthrie was right: Every map can be colored with four colors or fewer. The complete mathematical proof of this seemingly simple fact took up 500 pages and about 1000 hours of computer time.

Any map coloring problem can be converted into a graph coloring problem by first finding the **dual graph** of the map. In the dual graph, each vertex represents a "state" (remember that this is a metaphor for the political units used in the map), and two vertices are connected by an edge if the corresponding states have a common border. If the common border happens to be just a point (as is the case with Arizona and Colorado), then we do not connect the vertices. The problem of coloring the map using the fewest number of colors now becomes simply the problem of *finding an optimal coloring of the vertices of the dual graph*. We can now take all the tools and techniques we learned for graph coloring and use them for map coloring.

■ For a slightly more meaningful map coloring problem, see Exercise 15.

Our final example is a very small example whose only purpose is to illustrate the idea behind coloring a map by means of its dual graph.

(EXAMPLE ME2.6) **Map Coloring with Dual Graphs**

Figure ME2-14(a) shows a small map with six states, and Fig. ME2-14(b) shows its dual graph. Figure ME2-14(c) shows an optimal 3-coloring of the dual graph. We know that the 3-coloring is optimal because the graph has triangles and thus cannot be colored with 2 colors. The corresponding optimal coloring of the map is shown in Fig. ME2-14(d).

FIGURE ME2-14

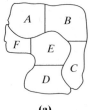
(a)

(b)

(c)

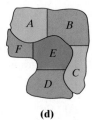
(d)

KEY CONCEPTS

chromatic number, **321**	*k*-coloring, **321**	Sudoku graph, **326**
dual graph, **327**	optimal coloring, **322**	

EXERCISES

A Graph Colorings and Chromatic Numbers

 For Exercises 1 through 6, you can experiment with coloring the graphs using a Java applet available in MyMathLab.* The applet was designed specifically to match these exercises.

1. For the graph G shown in Fig. ME2-15,

   (a) find a 3-coloring of G.

   (b) find a 2-coloring of G.

   (c) find $\chi(G)$. Explain your answer.

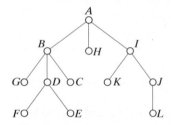

FIGURE ME2-15

2. For the graph G shown in Fig. ME2-16,

   (a) find a 4-coloring of G.

   (b) find a 3-coloring of G.

   (c) find $\chi(G)$. Explain your answer.

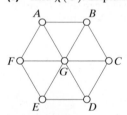

FIGURE ME2-16

3. For the graph G shown in Fig. ME2-17,

   (a) find a 5-coloring of G.

   (b) find a 4-coloring of G.

   (c) find $\chi(G)$. Explain your answer.

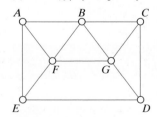

FIGURE ME2-17

*To use MyMathLab, you need a course ID and a student access code. Contact your instructor for more information.

4. For the graph G shown in Fig. ME2-18,

   (a) find a 4-coloring of G.

   (b) find $\chi(G)$. Explain your answer.

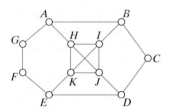

FIGURE ME2-18

5. For the graph G shown in Fig. ME2-19,

   (a) find a 3-coloring of G.

   (b) find $\chi(G)$. Explain your answer.

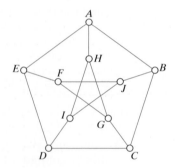

FIGURE ME2-19

6. For the graph G shown in Fig. ME2-20,

   (a) find a 4-coloring of G.

   (b) find $\chi(G)$. Explain your answer.

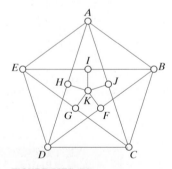

FIGURE ME2-20

7. Let K_n denote the complete graph on n vertices.

   (a) Explain why $\chi(K_n) = n$ (see Example ME2.2).

   (b) Let G be a graph obtained by removing just one edge from K_n. Find $\chi(G)$. Explain your answer.

8. Explain why if $\chi(G) = 1$, then G consists of just isolated vertices.

9. Let C_n denote the circuit with n vertices (see Example ME2.3).

   (a) When n is even, $\chi(C_n) = 2$. Describe a 2-coloring of C_n.

   (b) When n is odd, $\chi(C_n) = 3$. Describe a 3-coloring of C_n. Explain why a 2-coloring is impossible.

10. Let W_n denote the "wheel" with n vertices, as shown in Fig. ME2-21. Find $\chi(W_n)$. Explain your answer. (*Hint: You should do Exercise 9 first.*)

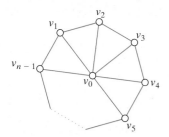

FIGURE ME2-21

11. If G is a tree (i.e., a connected graph with no circuits), find $\chi(G)$. Explain your answer.

12. The graph shown in Fig. ME2-22 is the graph discussed in Example ME2.1. Explain why, except for a change of colors, the 3-coloring shown in the figure is the only possible 3-coloring of the graph.

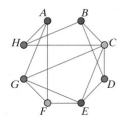

FIGURE ME2-22

B Map Coloring

13. Give an example of a map with 10 states that can be colored with just two colors.

14. Give an example of a map with four states that cannot be colored using fewer than four colors.

15. Find a map of South America and then:

   (a) Find the dual graph for the map.

   (b) Use the greedy algorithm with the vertices ordered by decreasing order of degree to color the dual graph found in (a).

   (c) Find the chromatic number of the dual graph found in (a). What does this imply about the coloring of the original map of South America?

16. Color a map of the contiguous 48 states in the United States using four colors. Explain why it is impossible to color the map using fewer than four colors.

C Miscellaneous

17. A **bipartite graph** is a graph whose vertices can be separated into two sets A and B such that every edge of the graph joins a vertex in A to a vertex in B. A generic bipartite graph is illustrated in Fig. ME2-23.

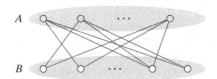

FIGURE ME2-23

   (a) Explain why if G is a bipartite graph, then $\chi(G) = 2$.

   (b) Explain why if $\chi(G) = 2$, then G must be a bipartite graph.

   (c) If G is a bipartite graph, then every circuit in G must have an even number of vertices. True or false? Explain.

18. Suppose G is a graph with n vertices and no loops or multiple edges such that every vertex has degree 3. (These graphs are called 3-*regular* graphs.)

   (a) Explain why n must be even and $n \geq 4$.

   (b) Explain why $\chi(G) \leq 4$.

   (c) Describe the 3-regular graphs for which $\chi(G) = 4$. (*Hint: What types of 3-regular graphs can fail to meet the conditions of the strong version of Brook's theorem?*)

19. Suppose G is a connected graph with n vertices and no loops or multiple edges such that $n - 1$ vertices have degree 3 and one vertex has degree 2.

   (a) Explain why n must be odd and $n \geq 5$.

   (b) Explain why $\chi(G) = 3$. (*Hint: Use the strong version of Brook's theorem together with Exercises 8 and 17.*)

20. This exercise refers to the graph in Fig. ME2-8(a) (see Example ME2.5). Show that if the vertices of the graph are ordered in decreasing order, the greedy algorithm will always give an optimal coloring of the graph.

21. Given a graph G with n vertices show that there is a way to order the vertices in the order $v_1, v_2, \ldots, v_n$ so that when the greedy algorithm is applied to the vertices in that order, the resulting coloring of the graph uses $\chi(G)$ colors. (*Hint: Assume a coloring of the graph with $\chi(G)$ colors and work your way backward to find an ordering of the vertices that produces that particular coloring.*)

22. Let S denote the Sudoku graph discussed in this mini-excursion.

(a) Explain why every vertex of S has degree 20.

(b) Explain why S has 810 edges. (*Hint: See Euler's theorems in Chapter 5.*)

23. If you haven't done so yet, try the Sudoku puzzle given in Fig. ME2-10.

PROJECTS

Scheduling Committee Meetings of the U.S. Senate

This project is a surprising application of the graph coloring techniques we developed in this mini-excursion and involves an important real-life problem: scheduling meetings for the standing committees of the United States Senate.

The U.S. Senate has 16 different standing committees that meet on a regular basis. The business of these committees represents a very important part of the Senate's work, since legislation typically originates in a standing committee and only if it gets approved there moves on to the full Senate. Scheduling meetings of the 16 standing committees is complicated because many of these committees have members in common, and in such cases the committee meetings cannot be scheduled at the same time. An easy solution would be to schedule the meetings of the committees all at different time slots that do not overlap, but this would eat up the lion's share of the Senate's schedule, leaving little time for the many other activities that the Senate has to take on. The optimal solution would be to schedule committees that do not have a member in common for the same time slot and committees that do have a member in common for different time slots. This is where graph coloring comes in.

Imagine a graph where the vertices of the graph represent the 16 committees, and two vertices are adjacent if the corresponding committees have one or more members in common. (For convenience, call this graph the *Senate Committees graph*.) If we think of the possible time slots as colors, a k-coloring of the Senate Committees graph gives a way to schedule the committee meetings using k different time slots. In this project you are to find a meetings schedule for the 16 standing committees of the U.S. Senate using the fewest possible number of time slots.

[*Hints*: (1) Find the membership lists for each of the 16 standing committees of the U.S. Senate. For the most up-to-date information, go to *http://www.senate.gov* and click on the Committees tab. (2) Create the Senate Committees graph. (You can use a spreadsheet instead of drawing the graph—it is quite a large graph—and get all your work done through the spreadsheet.) (3) List the vertices of the graph in decreasing order of degrees and use the greedy algorithm to color the graph.]

REFERENCES AND FURTHER READINGS

1. Delahaye, Jean-Paul, "The Science of Sudoku," *Scientific American*, June 2006, 81–87.

2. *http://www.cut-the-knot.org/Curriculum/Combinatorics/ColorGraph.shtml* (graph coloring applet created by Alex Bogomolny).

3. Herzberg, Agnes M., and M. Ram Murty, "Sudoku Squares and Chromatic Polynomials," *Notices of the American Mathematical Society*, 54 (June–July 2007), 708–717.

4. Porter, M. A., P. J. Mucha, M. E. J. Newman, and C. M. Warmbrand, "A Network Analysis of Committees in the United States House of Representatives," *Proceedings of the National Academy of Sciences*, 102 (2005), 7057–7062.

5. West, Douglas, *Introduction to Graph Theory*. Upper Saddle River, N.J.: Prentice-Hall, 1996.

6. Wilson, Robin, "The Sudoku Epidemic," *MAA Focus*, January 2006, 5–7.

part 3

Shape, Growth, and Form

9 The Mathematics of Spiral Growth

Fibonacci Numbers and the Golden Ratio

In nature's portfolio of architectural works, the shell of the chambered nautilus holds a place of special distinction. The spiral-shaped shell, with its revolving interior stairwell of ever-growing chambers, is more than a splendid piece of natural architecture — it is also a work of remarkable mathematical ingenuity.

Humans have imitated nature's wondrous spiral designs in their own architecture for centuries, all the while trying to understand the magic behind them. What are the physical laws that govern spiral growth in nature? Why do so many different and unusual mathematical concepts come into play? How do the mathematical concepts and physical laws mesh together? What is the source of the intrinsic beauty of nature's spirals? Trying to answer these questions is the goal of this chapter.

More than 2000 years ago, the ancient Greeks got us off to a great head start in our quest to understand nature with two great contributions: Euclidean geometry and irrational numbers. The next important mathematical connection came 800 years ago, with the serendipitous discovery, by a medieval scholar named Leonardo Fibonacci, of an amazing group of numbers now called Fibonacci numbers. (For more on Fibonacci, see the biographical profile at the end of this chapter.) Next came the discovery, by the French philosopher and mathematician René Descartes in 1638, that an equation he had studied for purely theoretical reasons ($r = ae^{\theta}$) is the very same equation that describes the spirals generated by seashells. Since then, other surprising connections between spiral-growing organisms and seemingly abstract mathematical concepts have been discovered, some quite recently. Exactly why and how these concepts play such a crucial role in the development of natural forms is not yet fully understood—not by humans, that is—a humbling reminder that nature still is the oldest and wisest of teachers.

We start the chapter with a discussion in Section 9.1 of a famous problem about the growth of a hypothetical population of rabbits best known for historical reasons as *Fibonacci's rabbits*. Although the rabbits themselves are not directly relevant to the topic of this chapter, the *Fibonacci numbers* that arise in the solution to the rabbit's problem are very relevant, and in Sections 9.2 and 9.3 we discuss in greater detail these numbers and their properties. In Section 9.3 we also introduce the *golden ratio* $\phi = \left(1 + \sqrt{5}\right)/2$—one of the most fascinating and ubiquitous numbers in nature—and discuss the various connections between the golden ratio and Fibonacci numbers. In Section 9.4 we introduce *gnomons* and the important concept of *gnomonic growth*. In Section 9.5 we use the ideas of the preceding sections to give a geometric description of the mechanism that underlies the growth of some of nature's spiral shapes.

9.1   Fibonacci's Rabbits

■ For the details on Fibonacci's life and how *Liber Abaci* affected Western culture, see the biographical profile at the end of this chapter.

In 1202 a young Italian named Leonardo Fibonacci published a book titled *Liber Abaci* (which roughly translated from Latin means "The Book of Calculation"). Although not an immediate success, *Liber Abaci* turned out to be one of the most important books in the history of Western civilization.

Liber Abaci was a remarkable book full of wonderful ideas and problems, but our story in this chapter focuses on just one of those problems—a purely hypothetical question about the growth of a very special family of rabbits. Here is the question, presented in Fibonacci's own (translated) words:

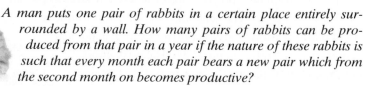

> *A man puts one pair of rabbits in a certain place entirely surrounded by a wall. How many pairs of rabbits can be produced from that pair in a year if the nature of these rabbits is such that every month each pair bears a new pair which from the second month on becomes productive?*

In this section we will give the answer to this question. Why the question is of any interest to us will become clear in later sections. Since the question deals with *pairs* of rabbits (presumably one male and one female), we will adopt the following notation: We will call P_1 the number of *pairs* of rabbits in the first month, P_2 the number of *pairs* of rabbits in the second month, P_3 the number of *pairs* of rabbits in the third month, and so on. With this notation the question asked by Fibonacci (*. . . how many pairs of rabbits can be produced from [the original] pair in a year?*) is answered by the value P_{12} (the number of pairs of rabbits in month 12). For good measure we will add one more value, $P_0 = 1$, representing the original pair of rabbits introduced by "the man" at the start.

Let's see now how the number of pairs of rabbits grows month by month. We start with the original pair, which we will assume is a pair of *young* rabbits. In the first month we still have just the original pair (for convenience, let's call them Pair A), so $P_1 = 1$. By the second month the original pair matures, becomes "productive," and generates a new pair of *young* rabbits. Thus, by the second month we have the original *mature* Pair A plus the new *young* pair we will call Pair B, so $P_2 = 2$. By the third month Pair B is still too young to breed, but Pair A generates another new young pair, Pair C, so $P_3 = 3$. By the fourth month Pair C is still young, but both Pair A and Pair B are mature and generate a new pair each (Pairs D and E). It follows that $P_4 = 5$.

■ For the purposes of our discussion we will use the terms *young* to describe rabbits that are less than two months old and *mature* to describe rabbits that are two months old or older (and can therefore procreate and produce additional rabbits).

We could continue this way, but our analysis can be greatly simplified by the following two observations:

1. In any given month (call it month N) the number of pairs of rabbits equals the total number of pairs in the previous month (i.e., in month $N - 1$) plus the number of mature pairs of rabbits in month N (these are the pairs that produce offspring—one new pair for each mature pair).

2. The number of mature rabbits in month N equals the total number of rabbits in month $N - 2$ (it takes two months for newborn rabbits to become mature).

Observations 1 and 2 can be combined and simplified into a single mathematical formula:

$$P_N = P_{N-1} + P_{N-2}$$

[The above formula reads as follows: The number of pairs of rabbits in any given month (P_N) equals the number of pairs of rabbits the previous month (P_{N-1}) plus the number of pairs of rabbits two months back (P_{N-2}).]

It follows, in order, that

$P_5 = P_4 + P_3 = 5 + 3 = 8$ (we had found earlier that $P_4 = 5$ and $P_3 = 3$),
$P_6 = P_5 + P_4 = 8 + 5 = 13$,
$P_7 = P_6 + P_5 = 13 + 8 = 21$, and continuing in this manner we get
$P_8 = 21 + 13 = 34$, $P_9 = 34 + 21 = 55$, $P_{10} = 55 + 34 = 89$,
$P_{11} = 89 + 55 = 144$, and finally $P_{12} = 144 + 89 = 233$

So there is the answer to Fibonacci's question: In one year the man will have raised 233 *pairs* of rabbits.

This is the end of the story about Fibonacci's rabbits and also the beginning of a much more interesting story about a truly remarkable sequence of numbers called Fibonacci numbers.

9.2 Fibonacci Numbers

THE FIBONACCI SEQUENCE

1, 1, 2, 3, 5, 8, 13, 21, 34, 55, 89, 144, 233, . . .

The sequence of numbers shown above is called the **Fibonacci sequence**, and the individual numbers in the sequence are known as the **Fibonacci numbers**. (If you read Section 9.1, you should recognize these numbers as the number of *pairs* of rabbits in Fibonacci's rabbit problem as we counted them from one month to the next.) The Fibonacci sequence is infinite, and except for the first two 1s, each number in the sequence is the sum of the two numbers before it.

We will denote each Fibonacci number by using the letter F (for Fibonacci) and a *subscript* that indicates the *position* of the number in the sequence. In other words, the first Fibonacci number is $F_1 = 1$, the second Fibonacci number is $F_2 = 1$, the third Fibonacci number is $F_3 = 2$, the tenth Fibonacci number is $F_{10} = 55$. We may not know (yet) the numerical value of the 100th Fibonacci number, but at least we can describe it as F_{100}.

A generic Fibonacci number is usually written as F_N (where N represents a generic position). If we want to describe the Fibonacci number that comes before F_N, we write F_{N-1}; the Fibonacci number two places before F_N is F_{N-2}, and so on. Clearly, this notation allows us to describe relations among the Fibonacci numbers in a clear and concise way that would be hard to match by just using words.

The rule that generates Fibonacci numbers—*a Fibonacci number equals the sum of the two preceding Fibonacci numbers*—is called a **recursive rule** because it defines a number in the sequence using earlier numbers in the sequence. Using subscript notation, the above recursive rule can be expressed by the simple and concise formula $F_N = F_{N-1} + F_{N-2}$ (we already saw a version of this formula in Section 9.1).

There is one thing still missing. The formula $F_N = F_{N-1} + F_{N-2}$ requires two consecutive Fibonacci numbers before it can be used and therefore cannot be applied to generate the first two Fibonacci numbers, F_1 and F_2. For a complete definition we must also explicitly give the values of the first two Fibonacci numbers, namely $F_1 = 1$ and $F_2 = 1$. These first two values serve as "anchors" for the recursive rule and are called the *seeds* of the Fibonacci sequence.

The *recursive rule* $F_N = F_{N-1} + F_{N-2}$ combined with the seeds $F_1 = 1$ and $F_2 = 1$ give us a full definition of the Fibonacci numbers. Since this is not the only definition, we will describe it as the **recursive definition**.

■ This is a good time to try Exercises 1 through 4 (to get better acquainted with the subscript notation) and Exercises 15 through 18 (to understand why the notation is so useful).

> **FIBONACCI NUMBERS (RECURSIVE DEFINITION)**
> - $F_1 = 1$, $F_2 = 1$ (the *seeds*)
> - $F_N = F_{N-1} + F_{N-2}$ (the *recursive rule*)

Using the above recursive definition, we can, in theory, compute any Fibonacci number, but in practice that is easier said than done.

EXAMPLE 9.1   **Cranking Out Large Fibonacci Numbers**

Suppose you were given the following choice: You can have F_{100} pennies or $100 billion. Which one would you choose? Surely, this is a no brainer—how could you pass on the $100 billion? (By the way, that much money would make you richer than Bill Gates.) But before you make a rash decision, let's see if we can figure out the dollar value of the first option. To do so, we need to compute F_{100}.

How could one find the value of F_{100}? With a little patience (and a calculator) we could use the recursive definition as a "crank" that we repeatedly turn to ratchet our way up the sequence: From the seeds F_1 and F_2 we compute F_3, then use F_3 and F_2 to compute F_4, and so on. If all goes well, after many turns of the crank (we will skip the details) you will eventually get to

$$F_{97} = 83,621,143,489,848,422,977$$

and

$$F_{98} = 135,301,852,344,706,746,049$$

One more turn of the crank gives

$$F_{99} = F_{98} + F_{97} = 218,922,995,834,555,169,026$$

and the last turn gives

$$F_{100} = F_{99} + F_{98} = 354,224,848,179,261,915,075$$

Thus, the F_{100} cents can be rounded nicely to $3,542,248,481,792,619,150. How much money is that? If you take $100 billion for yourself and then divide what's left evenly among every man, woman, and child on Earth (about 6.7 billion people), each person would get more than *$500 million!*

The most obvious lesson to be drawn from Example 9.1 is that Fibonacci numbers grow very large very quickly. A more subtle lesson (less obvious because we cheated in Example 9.1 and skipped most of the work) is that computing Fibonacci numbers using the recursive definition takes an enormous amount of effort (each turn of the crank involves just one addition, but as we noted, the numbers being added get very large very quickly).

Is there a more convenient way to compute Fibonacci numbers without the need to repeatedly turn the crank in the recursive definition? There is. In 1736 Leonhard Euler discovered a formula for the Fibonacci numbers that does not rely on previous Fibonacci numbers. The formula was lost and rediscovered 100 years later by French mathematician and astronomer Jacques Binet, who somehow ended up getting all the credit, as the formula is now known as **Binet's formula**.

■ This is the same Euler behind the namesake theorems in Chapter 5.

> **BINET'S FORMULA**
>
> $$F_N = \left(\frac{1}{\sqrt{5}}\right)\left[\left(\frac{1 + \sqrt{5}}{2}\right)^N - \left(\frac{1 - \sqrt{5}}{2}\right)^N\right]$$

Admittedly, Binet's formula looks complicated and intimidating, but it has a clear advantage over the recursive definition in the sense that it is not based on turning a crank—it allows us to calculate a Fibonacci number without needing to calculate any of the preceding Fibonacci numbers. From a practical point of view this is a big advantage. Because of its explicit nature (it defines a Fibonacci number without using previous Fibonacci numbers) we say that Binet's formula is an **explicit formula**.

Binet's formula can be made a lot less intimidating by simply observing that the messy part of the formula consists of constants. If we just let A denote $(1 + \sqrt{5})/2$ and B denote $(1 - \sqrt{5})/2$, the formula takes a somewhat friendlier look: $F_N = (A^N - B^N)/\sqrt{5}$.

Using a programmable or a graphing calculator (you need a calculator that can handle large numbers) you can use the following shortcut of Binet's formula to quickly find the Nth Fibonacci number for large values of N:

■ For details that justify why this shortcut is a variation of Binet's formula, see Exercise 56.

Step 1. Store $A = (1 + \sqrt{5})/2$ in the calculator's memory.

Step 2. Compute A^N.

Step 3. Divide the result in Step 2 by $\sqrt{5}$.

Step 4. Round the result in Step 3 to the nearest whole number. This will give you F_N.

(**EXAMPLE 9.2**) **Computing Large Fibonacci Numbers: Part 2**

In this example we will use the shortcut to Binet's formula described above together with a good programmable calculator to compute F_{100}.

Step 1. Start by computing the number $(1 + \sqrt{5}) \div 2$. The calculator should give you something like 1.6180339887498948482.

Step 2. Using the power key, raise the previous number to the power 100. The calculator should show 792,070,839,848,372,253,127.

Step 3. Divide the previous number by $\sqrt{5}$. The calculator should show 354,224,848, 179,261,915,075.

Step 4. The last step would be to round the number in Step 3 to the nearest whole number. In this case the decimal part is so tiny that the calculator will not show it, so the number already shows up as a whole number and we are done.

We are now at the point in this chapter where we can begin to address a question that may have already crossed your mind: What is so special about Fibonacci numbers?

It will take the rest of this chapter to answer the question, but for starters consider the remarkable affinity that nature has for Fibonacci numbers. Consistently, we find Fibonacci numbers when we count the number of petals in certain varieties of flowers: lilies and irises have 3 petals; buttercups and columbines have 5 petals; cosmos and rue anemones have 8 petals; yellow daisies and marigolds have 13 petals; English daisies and asters have 21 petals; oxeye daisies have 34 petals, and there are other daisies with 55 and 89 petals (Fig. 9-1). Fibonacci numbers also appear

FIGURE 9-1 Daisies, from left to right: white rue anemone, yellow daisy, English daisy, white oxeye daisy, coral Gerber daisy.

consistently in conifers, seeds, and fruits (Fig. 9-2). The bracts in a pinecone, for example, spiral in two different directions in 8 and 13 rows; the scales in a pineapple spiral in three different directions in 8, 13, and 21 rows; the seeds in the center of a sunflower spiral in 55 and 89 rows. Is it all a coincidence? Hardly.

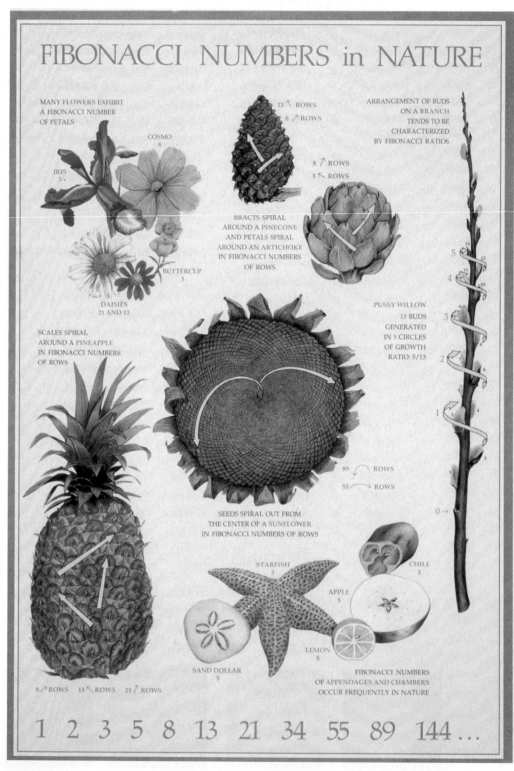

FIGURE 9-2

9.3 The Golden Ratio

In this section we will introduce the second fundamental concept of the chapter, the funky but beautiful number $\frac{1 + \sqrt{5}}{2}$. This number (it's important that you think of it as a single number) is one of the most famous and most studied numbers in all mathematics. The ancient Greeks gave it mystical properties and called it the *divine proportion*, and over the years, the number has taken many different names: the *golden number*, the *golden section*, and in modern times the **golden ratio**, the name that we will use from here on. The customary notation is to use the Greek lowercase letter ϕ (phi) to denote the golden ratio, so from here on $\phi = \frac{1 + \sqrt{5}}{2}$.

The golden ratio is an irrational number—it cannot be simplified into a fraction, and if you want to write it as a decimal, you can only approximate it to so many decimal places. In Example 9.2 we saw that if you enter $(1 + \sqrt{5}) \div 2$, a good calculator might return something like 1.6180339887498948482 (the exact number of decimal places will depend on the calculator). Remember that this is not an exact answer—it is a very good approximation to the golden ratio but still just an approximation. For most practical purposes, a good enough approximation is 1.618. (It is also a lot easier to remember, in case you were planning to memorize it.)

> ┌─ **THE GOLDEN RATIO** ─
> $$\phi = \frac{1 + \sqrt{5}}{2}$$
>
> $$\phi \approx 1.618$$

To better understand the golden ratio we will take a brief detour into the world of quadratic equations.

The Golden Property

Here is a prototypical high school algebra problem—the kind you are likely to find in most algebra books: *Find a positive number such that when you add 1 to it you get the square of the number.*

■ For a quick review of the quadratic formula see Exercises 29 through 32.

To solve this problem we let x be the desired number. The problem then translates into solving the quadratic equation $x^2 = x + 1$. To solve this equation we first rewrite it in the form $x^2 - x - 1 = 0$ and then use the quadratic formula. In this case the quadratic formula gives the solutions

$$\frac{-(-1) \pm \sqrt{(-1)^2 - 4(1)(-1)}}{2} = \frac{1 \pm \sqrt{5}}{2}$$

Of the two solutions, one is negative $[(1 - \sqrt{5})/2 \approx -0.618]$ and the other is the golden ratio $\phi = (1 + \sqrt{5})/2$. It follows that $\phi = (1 + \sqrt{5})/2$ is *the only positive number with the property that when you add one to the number you get the square of the number*, that is, $\phi^2 = \phi + 1$. We will call this property the **golden property**. As we will soon see, the golden property has really important algebraic and geometric implications.

Fibonacci Numbers and the Golden Property

We will use the golden property $\phi^2 = \phi + 1$ to recursively compute higher and higher powers of ϕ. Here is how:

- If we multiply both sides of $\phi^2 = \phi + 1$ by ϕ, we get

$$\phi^3 = \phi^2 + \phi$$

Replacing ϕ^2 by $\phi + 1$ on the right-hand side gives

$$\phi^3 = (\phi + 1) + \phi = 2\phi + 1$$

- If we multiply both sides of $\phi^3 = 2\phi + 1$ by ϕ, we get

$$\phi^4 = 2\phi^2 + \phi$$

Replacing ϕ^2 by $\phi + 1$ on the right-hand side gives

$$\phi^4 = 2(\phi + 1) + \phi = 3\phi + 2$$

- If we multiply both sides of $\phi^4 = 3\phi + 2$ by ϕ, we get

$$\phi^5 = 3\phi^2 + 2\phi$$

Replacing ϕ^2 by $\phi + 1$ on the right-hand side gives

$$\phi^5 = 3(\phi + 1) + 2\phi = 5\phi + 3$$

If we continue this way, we can express every power of ϕ in terms of ϕ:

$$\phi^6 = 8\phi + 5$$
$$\phi^7 = 13\phi + 8$$
$$\phi^8 = 21\phi + 13 \text{ and so on.}$$

■ See Exercises 23 and 24.

Notice that on the right-hand side we always get an expression involving two consecutive Fibonacci numbers. The general formula that expresses higher powers of ϕ in terms of ϕ and Fibonacci numbers is as follows.

┌─ **POWERS OF THE GOLDEN RATIO** ─────────

$$\phi^N = F_N\phi + F_{N-1}$$

Ratios of Consecutive Fibonacci Numbers

We will now explore what is probably the most surprising connection between the Fibonacci numbers and the golden ratio. Let's grab a calculator and take a

TABLE 9-1 Ratios of Consecutive Fibonacci Numbers

N	F_N	F_{N-1}	F_N/F_{N-1}	N	F_N	F_{N-1}	F_N/F_{N-1}
2	1	1	$1/1 = 1$	10	55	34	$55/34 = 1.61764\ldots$
3	2	1	$2/1 = 2$	11	89	55	$89/55 = 1.61818$
4	3	2	$3/2 = 1.5$	12	144	89	$144/89 = 1.61797\ldots$
5	5	3	$5/3 = 1.666\ldots$	13	233	144	$233/144 = 1.61805\ldots$
6	8	5	$8/5 = 1.6$	14	377	233	$377/233 = 1.61802\ldots$
7	13	8	$13/8 = 1.625$	15	610	377	$610/377 = 1.61803\ldots$
8	21	13	$21/13 = 1.61538\ldots$	16	987	610	$987/610 = 1.61803\ldots$
9	34	21	$34/21 = 1.61904\ldots$	17	1597	987	$1597/987 = 1.61803\ldots$

look at what happens when we take the ratio of consecutive Fibonacci numbers. Table 9-1 shows the first 16 values of the ratio F_N/F_{N-1}. The table shows an interesting pattern: As N gets bigger, the ratio of consecutive Fibonacci numbers F_N/F_{N-1} appears to settle down to a fixed value, and that *fixed value turns out to be the golden ratio*!

RATIO OF CONSECUTIVE FIBONACCI NUMBERS

$$F_N/F_{N-1} \approx \phi$$

and the larger the value of N, the better the approximation.

9.4 Gnomons

The most common usage of the word *gnomon* is to describe the pin of a sundial—the part that casts the shadow that shows the time of day. The original Greek meaning of the word *gnomon* is "one who knows," so it's not surprising that the word should find its way into the vocabulary of mathematics.

In this section we will discuss a different meaning for the word *gnomon*. Before we do so, we will take a brief detour to review a fundamental concept of high school geometry—*similarity*.

We know from geometry that two objects are said to be **similar** if one is a scaled version of the other. (When a slide projector takes the image in a slide and blows it up onto a screen, it creates a *similar* but larger image. When a photocopy machine reduces the image on a sheet of paper, it creates a *similar* but smaller image.)

The following important facts about similarity of basic two-dimensional figures will come in handy later in the chapter:

- **Triangles:** Two triangles are similar if and only if the measures of their respective angles are the same. Alternatively, two triangles are similar if and only if their sides are proportional. In other words, if Triangle 1 has sides of length a, b, and c, then Triangle 2 is similar to Triangle 1 if and only if its sides have length ka, kb, and kc for some positive constant k (Fig. 9-3).

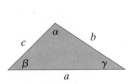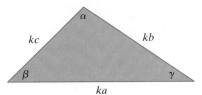

FIGURE 9-3 Similar triangles ($k = 1.6$).

- **Squares:** Two squares are always similar.
- **Rectangles:** Two rectangles are similar if their corresponding sides are proportional (Fig. 9-4).

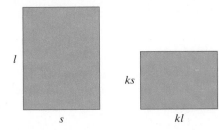

FIGURE 9-4 Similar rectangles ($k = 0.75$).

- **Circles and disks:** Two circles are always similar. Any circular disk (a circle plus all its interior) is similar to any other circular disk.
- **Circular rings:** Two circular rings are similar if and only if their inner and outer radii are proportional (Fig. 9-5).

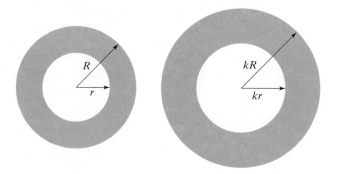

FIGURE 9-5 Similar rings ($k = 4/3$).

We will now return to the main topic of this section—gnomons. In geometry, a **gnomon** G to a figure A is a connected figure that, when suitably *attached* to A, produces a new figure similar to A. By "attached," we mean that the two figures are coupled into one figure without any overlap. Informally, we will describe it this way: G is a gnomon to A if G & A *is similar to* A (Fig. 9-6). Here the symbol "&" should be taken to mean "attached in some suitable way."

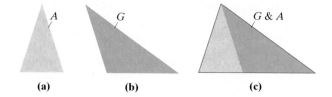

FIGURE 9-6 (a) The original object A. (b) The gnomon G. (c) G & A is similar to A.

(a)   **(b)**   **(c)**

EXAMPLE 9.3   **Gnomons to Squares**

Consider the square S in Fig. 9-7(a). The L-shaped figure G in Fig. 9-7(b) is a gnomon to the square—when G is attached to S as shown in Fig. 9-7(c), we get the square S'.

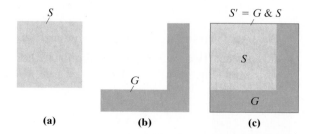

FIGURE 9-7 (a) A square S. (b) The gnomon G. (c) G & S form a larger square.

(a)   **(b)**   **(c)**

Note that the wording is *not* reversible. The square S *is not* a gnomon to the L-shaped figure G, since there is no way to attach the two to form an L-shaped figure similar to G.

EXAMPLE 9.4   **Gnomons to Circular Disks**

Consider the circular disk C with radius r in Fig. 9-8(a). The O-ring G in Fig. 9-8(b) with inner radius r is a gnomon to C. Clearly, G & C form the circular disk C' shown in Fig. 9-8(c). Since all circular disks are similar, C' is similar to C.

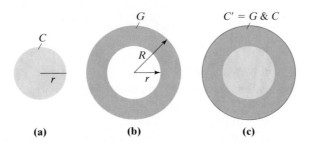

FIGURE 9-8 (a) A circular disk C. (b) The gnomon G. (c) G & C form a larger circular disk.

(a)   (b)   (c)

EXAMPLE 9.5   **Gnomons to Rectangles**

Consider a rectangle R of height h and base b as shown in Fig. 9-9(a). The L-shaped figure G shown in Fig. 9-9(b) can clearly be attached to R to form the larger rectangle R' shown in Fig. 9-9(c). This does not, in and of itself, guarantee that G is a gnomon to R. The rectangle R' [with height $(h + x)$ and base $(b + y)$] is similar to R if and only if their corresponding sides are proportional, which requires that $\dfrac{b}{h} = \dfrac{(b + y)}{(h + x)}$. With a little algebraic manipulation, this can be simplified to $\dfrac{b}{h} = \dfrac{y}{x}$.

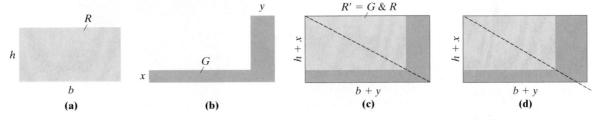

FIGURE 9-9 (a) A rectangle R. (b) A candidate for gnomon G. (c) G & R is similar to R. (d) G & R is not similar to R.

There is a simple geometric way to determine if the L-shaped G is a gnomon to R—just extend the diagonal of R in G & R. If the extended diagonal passes through the outside corner of G, then G is a gnomon [Fig. 9-9(c)]; if it doesn't, then it isn't [Fig. 9-9(d)].

EXAMPLE 9.6   **A Golden Triangle**

In this example, we are going to do things a little bit backward. Let's start with an isosceles triangle T, with vertices B, C, and D whose angles measure $72°$, $72°$, and $36°$, respectively, as shown in Fig. 9-10(a). On side CD we mark the point A so that BA is congruent to BC [Fig. 9-10(b)]. (A is the point of intersection of side CD and the circle of radius BC and center B.) Since T' is an isosceles triangle, angle BAC measures $72°$ and it follows that angle ABC measures $36°$. This

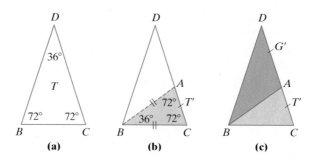

FIGURE 9-10 (a) A 72-72-36 isosceles triangle *T*. (b) *T'* is similar to *T*. (c) *G'* & *T'* = *T*.

implies that triangle *T'* has equal angles as triangle *T* and thus they are similar triangles.

"So what?" you may ask. Where is the gnomon to triangle *T*? We don't have one yet! But we *do* have a gnomon to triangle *T'*—it is triangle *BAD*, labeled *G'* in Fig. 9-10(c). After all, *G'* & *T'* is a triangle similar to *T'*. Note that gnomon *G'* is an isosceles triangle with angles that measure 36°, 36°, and 108°.

We now know how to find a gnomon not only to triangle *T'* but also to any 72-72-36 triangle, including the original triangle *T*: Attach a 36-36-108 triangle to one of the longer sides [Fig. 9-11(a)]. If we repeat this process indefinitely, we get a spiraling series of ever increasing 72-72-36 triangles [Fig. 9-11(b)]. It's not too far-fetched to use a family analogy: Triangles *T* and *G* are the "parents," with *T* having the "dominant genes;" the "offspring" of their union looks just like *T* (but bigger). The offspring then has offspring of its own (looking exactly like its grandparent *T*), and so on ad infinitum.

FIGURE 9-11 The process of adding a 36-36-108 gnomon *G* to a 72-72-36 triangle *T* can be repeated indefinitely, producing a spiraling chain of ever-increasing similar triangles.

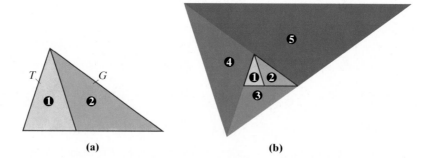

See Exercise 65.

Example 9.4 is of special interest to us for two reasons. First, this is the first time we have an example in which the figure and its gnomon are of the same type (isosceles triangles). Second, the isosceles triangles in this story (72-72-36 and 36-36-108) have a property that makes them unique: In both cases, the ratio of their sides (longer side over shorter side) is the golden ratio. These are the only two isosceles triangles with this property, and for this reason they are called **golden triangles**.

<hr>

(**EXAMPLE 9.7**)   **Square Gnomons to Rectangles**

We saw in Example 9.5 that *any* rectangle can have an L-shaped gnomon. Much more interesting is the case when a rectangle has a square gnomon. Not every rectangle can have a square gnomon, and the ones that do are quite special.

Consider a rectangle *R* with sides of length *l* (long side) and *s* (short side) [Fig. 9-12(a)], and suppose that the square *G* with sides of length *l* shown in

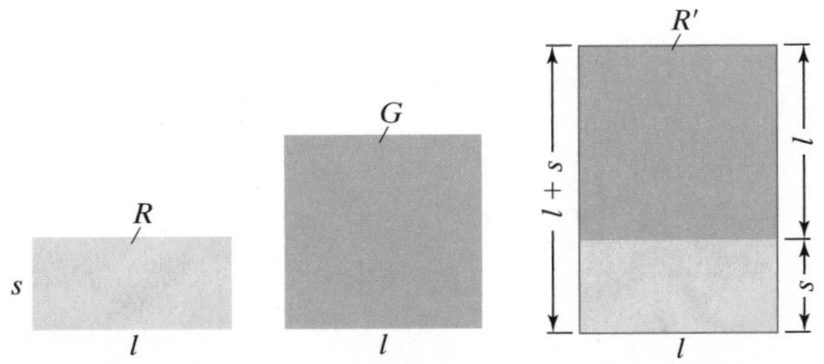

FIGURE 9-12

Fig. 9-12(b) is a gnomon to R. If so, then the rectangle R' shown in Fig. 9-12(c) must be similar to R, which implies that their corresponding sides must be proportional (long side of R'/short side of R' = long side of R/short side of R):

$$\frac{l + s}{l} = \frac{l}{s}$$

After some algebraic manipulation (see Exercise 58), the preceding equation can be rewritten in the form

$$\left(\frac{l}{s}\right)^2 = \frac{l}{s} + 1$$

Since (1) l/s is positive (l and s are the lengths of the sides of a rectangle), (2) this last equation essentially says that l/s satisfies the *golden property*, and (3) the only positive number that satisfies the golden property is ϕ, we can conclude that

$$\frac{l}{s} = \phi$$

We can summarize all the above with the following conclusion: *A rectangle with sides of length l and s (long side and short side, respectively) has a square gnomon if and only if l/s = ϕ.*

Golden and Fibonacci Rectangles

A rectangle whose sides are in the proportion of the golden ratio is called a **golden rectangle**. In other words, a golden rectangle is a rectangle with sides l (long side) and s (short side) satisfying $l/s = \phi$. A close relative to a golden rectangle is a **Fibonacci rectangle**—a rectangle whose sides are consecutive Fibonacci numbers.

(**EXAMPLE 9.8**)   **Golden and Almost Golden Rectangles**

Figure 9-13 shows an assortment of rectangles (Please note that the rectangles are not drawn to the same scale). Some are golden, some are close.

■ The rectangle in Fig. 9-13(a) has $l = 1$ and $s = 1/\phi$. Since $l/s = 1/(1/\phi) = \phi$, this is a golden rectangle.

■ The rectangle in Fig. 9-13(b) has $l = \phi + 1$ and $s = \phi$. Here $l/s = (\phi + 1)/\phi$. Since $\phi + 1 = \phi^2$, this is another golden rectangle.

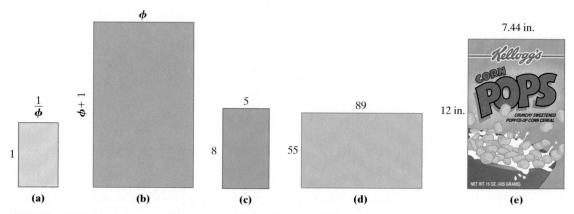

FIGURE 9-13 Golden and "almost golden" rectangles. To the naked eye, they all appear similar.

- The rectangle in Fig. 9-13(c) has $l = 8$ and $s = 5$. This is a Fibonacci rectangle, since 5 and 8 are consecutive Fibonacci numbers. The ratio of the sides is $l/s = 8/5 = 1.6$, so this is not a golden rectangle. On the other hand, the ratio 1.6 is reasonably close to $\phi = 1.618\ldots$, so we will think of this rectangle as "almost golden."

- The rectangle in Fig. 9-13(d) with $l = 89$ and $s = 55$ is another Fibonacci rectangle. Since $89/55 = 1.61818\ldots$, in theory this rectangle is not a golden rectangle. In practice, this rectangle is as good as golden—the ratio of the sides is the same as the golden ratio up to three decimal places.

- The rectangle in Fig. 9-13(e) shows the front of a box of Corn Pops. The dimensions are 12 in. by 7.44 in. This rectangle is neither a golden nor a Fibonacci rectangle. On the other hand, the ratio of the sides ($12/7.44 \approx 1.613$) is very close to the golden ratio. It is safe to say that, sitting on a supermarket shelf, that box of Corn Pops looks temptingly golden.

From a design perspective, golden (and almost golden) rectangles have a special appeal, and they show up in many everyday objects, from posters to cereal boxes. In some sense, golden rectangles strike the perfect middle ground between being too "skinny" and being too "squarish." A prevalent theory, known as the *golden ratio hypothesis*, is that human beings have an innate aesthetic bias in favor of golden rectangles, which, so the theory goes, appeal to our natural sense of beauty and proportion.

■ The golden ratio hypothesis originated with the experiments of German psychologist Gustav Fechner in 1876. For more on this topic, take a look at Project E.

9.5 Spiral Growth in Nature

In nature, where form usually follows function, the perfect balance of a golden rectangle shows up in spiral-growing organisms, often in the form of consecutive Fibonacci numbers. To see how this connection works, consider the following example, which serves as a model for certain natural growth processes.

(**EXAMPLE 9.9**)   **Stacking Squares on Fibonacci Rectangles**

Start with a 1 by 1 square [square ❶ in Fig. 9-14(a)]. Attach to it a 1 by 1 square [square ❷ in Fig. 9-14(b)]. Squares ❶ and ❷ together form a 1 by 2 Fibonacci rec-

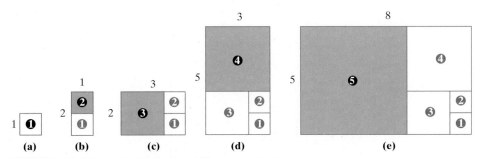

FIGURE 9-14 Fibonacci rectangles beget Fibonacci rectangles.

tangle. We will call this the "second-generation" shape. For the third generation, tack on a 2 by 2 square [square ❸ in Fig. 9-14(c)]. The "third-generation" shape (❶, ❷, and ❸ together) is the 3 by 2 Fibonacci rectangle in Fig. 9-14(c). Next, tack onto it a 3 by 3 square [square ❹ in Fig. 9-14(d)], giving a 3 by 5 Fibonacci rectangle. Then tack on a 5 by 5 square [square ❺ in Fig. 9-14(e)], resulting in an 8 by 5 Fibonacci rectangle. You get the picture—we can keep doing this as long as we want.

We might imagine these growing Fibonacci rectangles as a living organism. At each step, the organism grows by adding a square (a very simple, basic shape). The interesting feature of this growth is that as the Fibonacci rectangles grow larger, they become very close to golden rectangles, and as such, they become essentially similar to one another. This kind of growth—getting bigger while maintaining the same overall shape—is characteristic of the way many natural organisms grow.

The next example is a simple variation of Example 9.9.

(**EXAMPLE 9.10**) **The Growth of a "Chambered" Fibonacci Rectangle**

Let's revisit the growth process of the previous example, except now let's create within each of the squares being added an interior "chamber" in the form of a quarter-circle. We need to be a little more careful about how we attach the chambered square in each successive generation, but other than that, we can repeat the sequence of steps in Example 9.9 to get the sequence of shapes shown in Fig. 9-15. These figures depict the consecutive generations in the evolution of the *chambered Fibonacci rectangle*. The outer spiral formed by the circular arcs is often called a **Fibonacci spiral**, shown in Fig. 9-16.

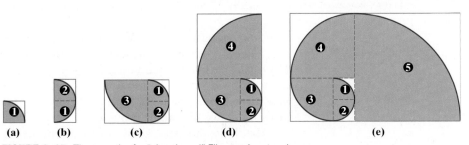

FIGURE 9-15 The growth of a "chambered" Fibonacci rectangle.

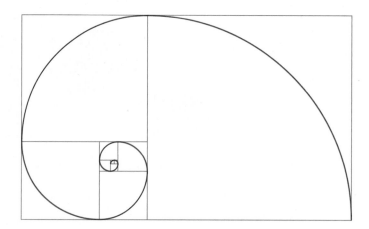

FIGURE 9-16 A Fibonacci spiral after 10 "generations."

Gnomonic Growth

Natural organisms grow in essentially two different ways. Humans, most animals, and many plants grow following what can informally be described as an

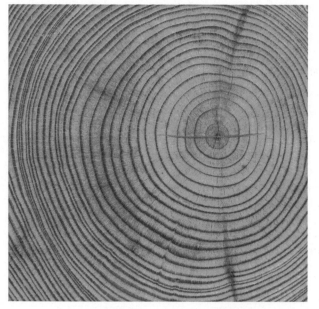

all-around growth rule. In this type of growth, all living parts of the organism grow simultaneously—but not necessarily at the same rate. One characteristic of this type of growth is that there is no obvious way to distinguish between the newer and the older parts of the organism. In fact, the distinction between new and old parts does not make much sense. The historical record (so to speak) of the organism's growth is lost. By the time the child becomes an adult, no identifiable traces of the child (as an organism) remain—that's why we need photographs!

Contrast this with the kind of growth exemplified by the shell of the chambered nautilus, a ram's horn, or the trunk of a redwood tree. These organisms grow following a *one-sided* or *asymmetric growth* rule, meaning that the organism has a part added to it (either by its own or outside forces) in such a way that the old organism together with the added part form the new organism. At any stage of the growth process, we can see not only the present form of the organism but also the organism's entire past. All the previous stages of growth are the building blocks that make up the present structure.

The other important aspect of natural growth is the principle of *self-similarity*: Organisms like to maintain their overall shape as they grow. This is where gnomons come into the picture. For the organism to retain its shape as it grows, the new growth must be a *gnomon* of the entire organism. We will call this kind of growth process **gnomonic growth**.

We have already seen abstract mathematical examples of gnomonic growth (Examples 9.9 and 9.10). Here is a pair of more realistic examples.

FIGURE 9-17 The growth rings in a redwood tree — an example of circular gnomonic growth.

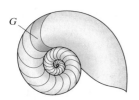

FIGURE 9-18 Gnomonic growth of a chambered nautilus.

FIGURE 9-19

EXAMPLE 9.11   **Circular Gnomonic Growth**

We know from Example 9.4 that the gnomon to a circular disk is an O-ring with an inner radius equal to the radius of the circle. We can thus have circular gnomonic growth (Fig. 9-17) by the regular addition of O-rings. O-rings added one layer at a time to a starting circular structure preserve the circular shape throughout the structure's growth. When carried to three dimensions, this is a good model for the way the trunk of a redwood tree grows. And this is why we can "read" the history of a felled redwood tree by studying its rings.

EXAMPLE 9.12   **Spiral Gnomonic Growth**

Figure 9-18 shows a diagram of a cross section of the chambered nautilus. The chambered nautilus builds its shell in stages, each time adding another chamber to the already existing shell. At every stage of its growth, the shape of the chambered nautilus shell remains the same—the beautiful and distinctive spiral shown in Fig. 9-19. This is a classic example of gnomonic growth—each new chamber added to the shell is a gnomon of the entire shell. The gnomonic growth of the shell proceeds, in essence, as follows: Starting with its initial shell (a tiny spiral similar in all respects to the adult spiral shape), the animal builds a chamber (by producing a special secretion around its body that calcifies and hardens). The resulting, slightly enlarged spiral shell is similar to the original one. The process then repeats itself over many stages, each one a season in the growth of the animal. Each new chamber adds a gnomon to the shell, so the shell grows and yet remains similar to itself. This process is a real-life variation of the mathematical spiral-building process discussed in Example 9.10. The curve generated by the outer edge of a nautilus shell's cross section is called a *logarithmic spiral*. (For more on logarithmic spirals, see Project C.)

More complex examples of gnomonic growth occur in sunflowers, daisies, pineapples, pinecones, and so on. Here, the rules that govern growth are somewhat more involved, but Fibonacci numbers and the golden ratio once again play a prominent role. The interested reader is encouraged to follow up on this topic by way of Project A.

CONCLUSION

Some of the most beautiful shapes in nature arise from a basic principle of design: *Form follows function.* The beauty of natural shapes is a result of their inherent elegance and efficiency, and imitating nature's designs has helped humans design and build beautiful and efficient structures of their own.

In this chapter, we examined a special type of growth—gnomonic growth—where an organism grows by the addition of gnomons, thereby preserving its basic shape even as it grows. Many beautiful spiral-shaped organisms, from seashells to flowers, exhibit this type of growth.

Left: Sunflower. Right: Dome design (Pier Paolo Nervi, architect).

To us, understanding the basic principles behind spiral growth was relevant because it introduced us to some wonderful mathematical concepts that have been known and studied in their own right for centuries—*Fibonacci numbers*, *the golden ratio*, *golden rectangles*, and *gnomons*.

To humans, these abstract mathematical concepts have been, by and large, intellectual curiosities. To nature—the consummate artist and builder—they are the building tools for some of its most beautiful creations. Whatever lesson one draws from this, it should include something about the inherent value of mathematical ideas.

P ROFILE: Leonardo Fibonacci (circa 1175–1250)

Leonardo Fibonacci was born in Pisa, Italy, sometime around 1175 (there are no reliable records of Leonardo's exact date of birth). His actual family name was Bonaccio, so the name Fibonacci, which he preferred and by which he is now known, is a sort of nickname derived from the Latin for *filius Bonacci,* meaning "son of Bonaccio." Pisa, best known nowadays as the home of the famous leaning tower, was at the time an independent and prosperous city-state with important trade routes throughout the Mediterranean. Leonardo's father, Guglielmo Bonaccio, was a civil servant/diplomat who served as the agent and representative for Pisan business interests in North Africa. When Leonardo was a young boy he moved with his father to North Africa, where he spent most of his youth. Thus, rather than receiving a traditional European education, Leonardo was educated in the Moorish tradition, a happenstance that might very well have affected the entire course of Western civilization.

As his father wanted him to be a merchant, a significant part of Leonardo's education was the study of mathematics and accounting. Under the Moors, Leonardo learned arithmetic and algebra using the Hindu-Arabic

decimal system of numeration—the system we all use today but unknown in Europe at the time. Leonardo was not the first European to learn the Hindu-Arabic system of numbers and their arithmetic, but he was the first one to truly grasp their extraordinary potential.

Sometime around the year 1200, when he was about 25, Leonardo returned to Pisa. For the next 25 years, he dedicated himself to bringing the Hindu-Arabic methods of mathematics to Europe. During this period he published at least four books (these are the ones for which manuscripts still exist—it is believed that there were two other books for which the manuscripts are lost). His first book, *Liber Abaci* (*The Book of Calculation*), was published in 1202. (A long-awaited English translation of *Liber Abaci* was published in 2002 [reference 18].)

It is not too far-fetched to claim that the publication of *Liber Abaci* is one of those little-known events that changed the course of history. In *Liber Abaci,* Fibonacci described the workings of the Hindu-Arabic number system and richly illustrated it with marvelous problems and examples. The advantages of this place-value-based decimal system for accounting and bookkeeping were so obvious that in a very short time it became the de facto standard for conducting business, the Roman system was abandoned, and the rest is history.

One of the many examples that Fibonacci introduced in *Liber Abaci* was a simple rabbit-breeding problem: *A certain man put a pair of rabbits in a place surrounded on all sides by a wall. How many pairs of rabbits can be produced from that pair in a year if it is supposed that every month each pair begets a new pair which from the second month on becomes productive?* The solution to this famous problem generates the numbers 1, 1, 2, 3, 5, 8, 13, 21, 34, 55, 89, . . . , which we now call *Fibonacci numbers* (see Section 9.1 for details). To

Fibonacci, the rabbit problem was just another example among the many discussed in his book—little did he know that the numbers generated in the solution would be his ticket to immortality. It is a great irony that the man who brought the Western world its modern system for doing mathematics is best remembered now for the reproductive habits of a bunch of mythical bunnies.

By the time Fibonacci published his last book, *Liber Quadratorum,* he was a famous man, but surprisingly little is known about the latter part of his life or about the date and circumstances of his death, believed to be between 1240 and 1250.

KEY CONCEPTS

EXERCISES

WALKING

A Fibonacci Numbers

1. Compute the value of each of the following.

   (a) F_{10}

   (b) $F_{10} + 2$

   (c) F_{10+2}

   (d) $F_{10}/2$

   (e) $F_{10/2}$

2. Compute the value of each of the following.

   (a) F_{12}

   (b) $F_{12} - 1$

   (c) F_{12-1}

   (d) $F_{12}/4$

   (e) $F_{12/4}$

3. Compute the value of each of the following.

   (a) $F_1 + F_2 + F_3 + F_4 + F_5$

   (b) $F_{1+2+3+4+5}$

   (c) $F_3 \times F_4$

   (d) $F_{3 \times 4}$

   (e) F_{F_4}

4. Compute the value of each of the following.

   (a) $F_1 + F_3 + F_5 + F_7$

   (b) $F_{1+3+5+7}$

   (c) F_{10}/F_5

   (d) F_{10/F_5}

   (e) F_{F_7}

5. Describe in words what each of the expressions represents.

   (a) $3F_N + 1$

   (b) $3F_{N+1}$

   (c) $F_{3N} + 1$

   (d) F_{3N+1}

6. Describe in words what each of the expressions represents.

(a) $F_{2N} - 3$

(b) F_{2N-3}

(c) $2F_N - 3$

(d) $2F_{N-3}$

7. Given that $F_{36} = 14{,}930{,}352$ and $F_{37} = 24{,}157{,}817$,

(a) find F_{38}.

(b) find F_{35}.

8. Given that $F_{31} = 1{,}346{,}269$ and $F_{33} = 3{,}524{,}578$,

(a) find F_{32}.

(b) find F_{34}.

9. Which of the two rules (I or II) is equivalent to the recursive rule $F_N = F_{N-1} + F_{N-2}, N \geq 3$? Explain.

(I) $F_{N+2} = F_{N+1} + F_N, N > 0$

(II) $F_N = F_{N+1} + F_{N+2}, N > 0$

10. Which of the two rules (I or II) is equivalent to the recursive rule $F_N = F_{N-1} + F_{N-2}, N \geq 3$? Explain.

(I) $F_{N-1} - F_N = F_{N+1}, N > 1$

(II) $F_{N+1} - F_N = F_{N-1}, N > 1$

11. Write each of the following integers as the sum of *distinct* Fibonacci numbers.

(a) 47   (b) 48

(c) 207   (d) 210

12. Write each of the following integers as the sum of *distinct* Fibonacci numbers.

(a) 52   (b) 53

(c) 107   (d) 112

13. Consider the following sequence of equations involving Fibonacci numbers.

$$1 + 2 = 3$$
$$1 + 2 + 5 = 8$$
$$1 + 2 + 5 + 13 = 21$$
$$1 + 2 + 5 + 13 + 34 = 55$$
$$\vdots$$

(a) Write down a reasonable choice for the fifth equation in this sequence.

(b) Find the subscript that will make the following equation true.

$$F_1 + F_3 + F_5 + \cdots + F_{21} = F_?$$

(c) Find the subscript that will make the following equation true (assume N is odd).

$$F_1 + F_3 + F_5 + \cdots + F_N = F_?$$

14. Consider the following sequence of equations involving Fibonacci numbers.

$$2(2) - 3 = 1$$
$$2(3) - 5 = 1$$
$$2(5) - 8 = 2$$
$$2(8) - 13 = 3$$
$$\vdots$$

(a) Write down a reasonable choice for the fifth equation in this sequence.

(b) Find the subscript that will make the following equation true.

$$2(F_?) - F_{15} = F_{12}$$

(c) Find the subscript that will make the following equation true.

$$2(F_{N+2}) - F_{N+3} = F_?$$

15. Fact: *If we add any Fibonacci number to the Fibonacci number three positions before it, we get twice the Fibonacci number before it.*

(a) Verify this fact for four different Fibonacci numbers of your choice.

(b) Using F_N as a generic Fibonacci number, write this fact as a mathematical formula.

16. Fact: *If we make a list of any 10 consecutive Fibonacci numbers, the sum of all these numbers divided by 11 is always equal to the seventh number on the list.*

(a) Verify this fact for the list consisting of the first 10 Fibonacci numbers.

(b) Using F_N as the first Fibonacci number on the list, write this fact as a mathematical formula.

17. Fact: *If we make a list of any four consecutive Fibonacci numbers, twice the third one minus the fourth one is always equal to the first one.*

(a) Verify this fact for the list that starts with F_4.

(b) Using F_N as the first Fibonacci number on the list, write this fact as a mathematical formula.

18. Fact: *If we make a list of any four consecutive Fibonacci numbers, the first one times the fourth one is always equal to the third one squared minus the second one squared.*

(a) Verify this fact for the list that starts with F_8.

(b) Using F_N as the first Fibonacci number on the list, write this fact as a mathematical formula.

B The Golden Ratio

For Exercises 19 through 22 you will need a scientific or graphing calculator (or a computer with a mathematics software package).

19. Calculate each of the following to five decimal places.

(a) $21\left(\dfrac{1 + \sqrt{5}}{2}\right) + 13$

(b) $\left(\dfrac{1 + \sqrt{5}}{2}\right)^8$

(c) $\dfrac{\left(\dfrac{1 + \sqrt{5}}{2}\right)^8 - \left(\dfrac{1 - \sqrt{5}}{2}\right)^8}{\sqrt{5}}$

20. Calculate each of the following to five decimal places.

(a) $55\left(\dfrac{1 + \sqrt{5}}{2}\right) + 34$

(b) $\left(\dfrac{1 + \sqrt{5}}{2}\right)^{10}$

(c) $\dfrac{\left(\dfrac{1 + \sqrt{5}}{2}\right)^{10} - \left(\dfrac{1 - \sqrt{5}}{2}\right)^{10}}{\sqrt{5}}$

21. Using $\phi = (1 + \sqrt{5})/2$, calculate each of the following, rounded to the nearest integer.

(a) $\phi^8/\sqrt{5}$

(b) $\phi^9/\sqrt{5}$

(c) Without using a calculator, first try to guess the value of $\phi^7/\sqrt{5}$, rounded to the nearest integer. Verify your guess with a calculator.

22. Using $\phi = (1 + \sqrt{5})/2$, calculate each of the following, rounded to the nearest integer.

(a) $\phi^{10}/\sqrt{5}$

(b) $\phi^{11}/\sqrt{5}$

(c) Without using a calculator, first try to guess the value of $\phi^{12}/\sqrt{5}$, rounded to the nearest integer. Verify your guess with a calculator.

23. (a) Given that $\phi^5 = 5\phi + 3$, show that $\phi^6 = 8\phi + 5$. (Show each step of the derivation.)

(b) Using $\phi^6 = 8\phi + 5$, find the values of the integers a and b in the equation $\phi^6 = a\sqrt{5} + b$.

24. (a) Given that $\phi^8 = 21\phi + 13$, show that $\phi^9 = 34\phi + 21$. (Show each step of the derivation.)

(b) Using $\phi^9 = 34\phi + 21$, find the values of the integers a and b in the equation $\phi^9 = a\sqrt{5} + b$.

25. Given that $F_{499} \approx 8.6168 \times 10^{103}$, find an approximate value for F_{500} in scientific notation. (*Hint*: $F_N/F_{N-1} \approx \phi$.)

26. Given that $F_{1002} \approx 1.138 \times 10^{209}$, find an approximate value for F_{1000} in scientific notation. (*Hint*: $F_N/F_{N-1} \approx \phi$.)

27. The *Fibonacci sequence of order 2* is the sequence of numbers $1, 2, 5, 12, 29, 70, \ldots$. Each term in this sequence (from the third term on) equals two times the term before it plus the term two places before it; in other words, $A_N = 2A_{N-1} + A_{N-2}$ $(N \geq 3)$.

(a) Compute A_7.

(b) Use your calculator to compute to five decimal places the ratio A_N/A_{N-1} for $N = 7$.

(c) Use your calculator to compute to five decimal places the ratio A_N/A_{N-1} for $N = 11$.

(d) Guess the value (to five decimal places) of the ratio A_N/A_{N-1} when $N > 11$.

28. The *Fibonacci sequence of order 3* is the sequence of numbers $1, 3, 10, 33, 109, \ldots$. Each term in this sequence (from the third term on) equals three times the term before it plus the term two places before it; in other words, $A_N = 3A_{N-1} + A_{N-2}$ $(N \geq 3)$.

(a) Compute A_6.

(b) Use your calculator to compute to five decimal places the ratio A_N/A_{N-1} for $N = 6$.

(c) Guess the value (to five decimal places) of the ratio A_N/A_{N-1} when $N > 6$.

C Fibonacci Numbers and Quadratic Equations

Exercises 29 and 30 involve solving quadratic equations using the quadratic formula. Recall that the quadratic formula tells us that the solutions of the standard quadratic equation $ax^2 + bx + c = 0$ are given by $x = \dfrac{-b \pm \sqrt{b^2 - 4ac}}{2a}$.

29. Find the solutions of each quadratic equation. Use a calculator to approximate the solutions to five decimal places.

(a) $x^2 = 2x + 1$

(b) $5x^2 = 2x + 1$

(c) $3x^2 = 8x + 5$

30. Find the solutions of each quadratic equation. Use a calculator to approximate the solutions to five decimal places.

(a) $x^2 = 3x + 1$

(b) $2x^2 = 3x + 1$

(c) $8x^2 = 5x + 2$

31. Consider the quadratic equation $55x^2 = 34x + 21$.

(a) Explain why $x = 1$ is a solution to this equation. (Do not use the quadratic formula.)

(b) Without using the quadratic formula, find the other solution to the equation. (*Hint*: The sum of the two solutions of the quadratic equation $ax^2 + bx + c = 0$ equals $-b/a$.)

32. Consider the quadratic equation $21x^2 = 34x + 55$.

(a) Explain why $x = -1$ is a solution to this equation. (Do not use the quadratic formula.)

(b) Without using the quadratic formula, find the other solution to the equation. (*Hint*: The sum of the two solutions of the quadratic equation $ax^2 + bx + c = 0$ equals $-b/a$.)

33. Consider the quadratic equation $F_N x^2 = F_{N-1}x + F_{N-2}$.

(a) Explain why $x = 1$ is one of the solutions to the equation.

(b) Explain why the other solution is given by $x = (F_{N-1}/F_N) - 1$. [*Hint*: See the hint for Exercise 31(b).]

34. Consider the quadratic equation $F_N x^2 = F_{N+1}x + F_{N+2}$.

(a) Explain why $x = -1$ is one of the solutions to the equation.

(b) Explain why the other solution is given by $x = (F_{N+1}/F_N) + 1$. [*Hint*: See the hint for Exercise 32(b).]

FIGURE 9-21

(a) If the perimeter of P is 10, what is the perimeter of P'?

(b) If the area of P is 30, what is the area of P'?

39. R is a 3 by x rectangle. R' is a 5 by $8 - x$ rectangle. Determine the value(s) of x that make R and R' similar.

40. R is a 2 by x rectangle. R' is a 6 by $x + 1$ rectangle. Determine all value(s) of x that make R and R' similar.

E Gnomons

41. Find the value of c so that the shaded rectangle in Fig. 9-22 is a gnomon to the white 3 by 9 rectangle. (Figure is not drawn to scale.)

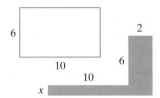

FIGURE 9-22

42. Find the value of x so that the shaded figure in Fig. 9-23 is a gnomon to the white rectangle. (Figure is not drawn to scale.)

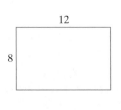

FIGURE 9-23

D Similarity

35. R and R' are similar rectangles. Suppose that the width of R is a and the width of R' is $3a$.

(a) If the perimeter of R is 41.5 in., what is the perimeter of R'?

(b) If the area of R is 105 sq. in., what is the area of R'?

36. O and O' are similar O-rings. The inner radius of O is 5 ft, and the inner radius of O' is 15 ft.

(a) If the circumference of the outer circle of O is 14π ft, what is the circumference of the outer circle of O'?

(b) Suppose that it takes 1.5 gallons of paint to paint the O-ring O. If the paint is used at the same rate, how much paint is needed to paint the O-ring O'?

37. Triangles T and T' shown in Fig. 9-20 are similar triangles. (Note that the triangles are not drawn to scale.)

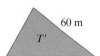

FIGURE 9-20

(a) If the perimeter of T is 13 in., what is the perimeter of T' (in meters)?

(b) If the area of T is 20 sq. in., what is the area of T' (in square meters)?

38. Polygons P and P' shown in Fig. 9-21 are similar polygons.

43. Find the value of x so that the shaded figure in Fig. 9-24 is a gnomon to the white rectangle. (Figure is not drawn to scale.)

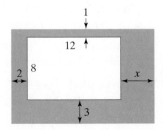

FIGURE 9-24

44. Find the value of x so that the shaded figure in Fig. 9-25 is a gnomon to the white rectangle. (Figure is not drawn to scale.)

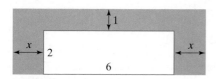

FIGURE 9-25

45. Rectangle A is 10 by 20. Rectangle B is a gnomon to rectangle A. What are the dimensions of rectangle B?

46. Find the value of x so that the shaded "rectangular ring" in Fig. 9-26 is a gnomon to the white rectangle. (Figure is not drawn to scale.)

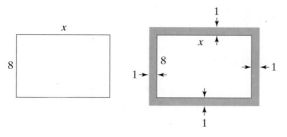

FIGURE 9-26

47. In Fig. 9-27 triangle BCA is a 36-36-108 triangle with sides of length ϕ and 1. Suppose that triangle ACD is a gnomon to triangle BCA.

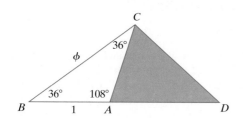

FIGURE 9-27

(a) Find the measure of the angles of triangle ACD.

(b) Find the length of the three sides of triangle ACD.

48. Find the values of x and y so that in Fig. 9-28 the shaded triangle is a gnomon to the white triangle ABC.

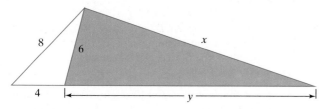

FIGURE 9-28

49. Find the values of x and y so that in Fig. 9-29 the shaded figure is a gnomon to the white triangle.

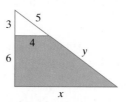

FIGURE 9-29

50. Find the values of x and y so that in Fig. 9-30 the shaded triangle is a gnomon to the white triangle.

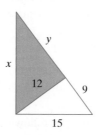

FIGURE 9-30

JOGGING

51. Consider the following sequence of numbers: 5, 5, 10, 15, 25, 40, 65, ... If A_N is the Nth term of this sequence, write A_N in terms of F_N.

52. The *Lucas numbers* are a sequence of numbers defined by the same recursive rule as the Fibonacci numbers but with different seeds. The recursive definition of the Lucas numbers is $L_N = L_{N-1} + L_{N-2}$ ($N \geq 3$) and $L_1 = 1$, $L_2 = 3$. Thus, the first few terms of the Lucas sequence are 1, 3, 4, 7, 11, 18, 29, 47, ...

(a) Show that $L_N = 2F_{N+1} - F_N$. (*Hint*: Show that the expression $2F_{N+1} - F_N$ satisfies the recursive rule and the definition of the seeds of the Lucas sequence.)

(b) Show that an explicit formula for the Lucas numbers is given by $L_N = \left(\dfrac{1 + \sqrt{5}}{2} \right)^N + \left(\dfrac{1 - \sqrt{5}}{2} \right)^N$.
[*Hint*: Use the result of (a) combined with Binet's formula.]

53. Let $T_N = 7F_{N+1} + 4F_N$.

(a) Find the values of T_1 through T_8.

(b) Show that the numbers T_N satisfy the same recursive rule as the Fibonacci numbers $T_N = T_{N-1} + T_{N-2}$.

(c) Give a complete recursive definition (recursive rule and seeds) of the sequence of numbers T_N.

54. (This exercise is a generalization of the previous exercise. You should do Exercise 53 before you try this one.) Let $T_N = aF_{N+1} + bF_N$, where a and b are fixed integers.

(a) Find T_1 through T_5.

(b) Show that the numbers T_N satisfy the same recursive rule as the Fibonacci numbers $T_N = T_{N-1} + T_{N-2}$.

(c) Consider the sequence of numbers 5, 11, 16, 27, 43, ... Find the values of a and b such that these numbers fit the description $T_N = aF_{N+1} + bF_N$.

55. Explain why the irrational numbers $(1 + \sqrt{5})/2$ and $(1 - \sqrt{5})/2$ have exactly the same decimal expansion. (*Hint*: What do you get when you add the two numbers?)

56. (a) Explain what happens to the values of $\left(\dfrac{1 - \sqrt{5}}{2}\right)^N$ as N gets larger. (*Hint*: Get a calculator and experiment with $N = 6, 7, 8, \ldots$ until you get the picture.)

(b) Using your answer in (a), justify the following assertion: $F_N \approx \phi^N / \sqrt{5}$.

(c) Using (b), explain why $F_N / F_{N-1} \approx \phi$.

57. A rectangle R has sides of length $l = 144\phi + 89$ and $s = 89\phi + 55$. Explain why R is a golden rectangle.

58. Explain why the equation $\dfrac{l}{s} = \dfrac{l + s}{l}$ is equivalent to the equation $\left(\dfrac{l}{s}\right)^2 = \dfrac{l}{s} + 1$. (Show each step of the derivation.)

59. A rectangle has dimensions l (long side) by s (short side). If the rectangle is a gnomon to itself, find the value of the ratio l/s.

60. Explain why the shaded figure in Fig. 9-31 cannot have a square gnomon.

FIGURE 9-31

61. Find the values of x, y, and z so that in Fig. 9-32 the shaded figure has an area eight times the area of the white triangle and at the same time is a gnomon to the white triangle. (Figure is not drawn to scale.)

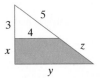

FIGURE 9-32

62. Find the values of x and y so that in Fig. 9-33 the shaded triangle is a gnomon to the white triangle. (Figure is not drawn to scale.)

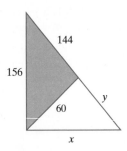

FIGURE 9-33

63. Find the values of x and y so that in Fig. 9-34 the shaded figure has an area of 75 and at the same time is a gnomon to the white rectangle. (Figure is not drawn to scale.)

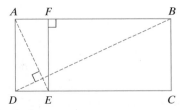

FIGURE 9-34

64. Let $ABCD$ be an arbitrary rectangle as shown in Fig. 9-35. Let AE be perpendicular to the diagonal BD and EF perpendicular to AB as shown. Show that the rectangle $BCEF$ is a gnomon to the rectangle $ADEF$.

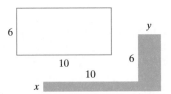

FIGURE 9-35

65. In Fig. 9-36 triangle BCD is a 72-72-36 triangle with base of length 1 and longer side of length x. (Using this choice of values, the ratio of the longer side to the shorter side is $x/1 = x$.)

(a) Show that $x = \phi$. (*Hint*: Triangle ACB is similar to triangle BCD.)

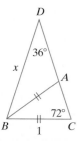

FIGURE 9-36

(b) What are the interior angles of triangle DAB?

(c) Show that in the isosceles triangle DAB, the ratio of the longer to the shorter side is also ϕ.

66. Show that each of the diagonals of the regular pentagon shown in Fig. 9-37 has length ϕ.

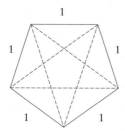

FIGURE 9-37

67. (a) A regular decagon (10 sides) is inscribed in a circle of radius 1. Find the perimeter in terms of ϕ.

(b) Repeat (a) with radius r. Find the perimeter in terms of ϕ and r.

68. (a) Without using a calculator, find the exact value of
$$\sqrt{1 + \sqrt{1 + \sqrt{1 + \phi}}}.$$ (*Hint*: $\phi^2 = \phi + 1$.)

(b) Without using a calculator, find the exact value of
$$1 + \cfrac{1}{1 + \cfrac{1}{1 + (1/\phi)}}.$$

69. Explain why the only even Fibonacci numbers are those having a subscript that is a multiple of 3.

RUNNING

70. You are designing a straight path 2 ft wide using rectangular paving stones with dimensions 1 ft by 2 ft. How many different designs are possible for a path of length

(a) 4 ft.

(b) 8 ft.

(c) N ft. (*Hint*: Give the answer in terms of Fibonacci numbers.)

71. Show that $(F_1)^2 + (F_2)^2 + (F_3)^2 + \cdots + (F_N)^2 = F_N \cdot F_{N+1}$.

72. Show that $F_1 + F_2 + F_3 + \cdots + F_N = F_{N+2} - 1$.

73. Show that $F_1 + F_3 + F_5 + \cdots + F_N = F_{N+1}$. (Note that on the left side of the equation we are adding the Fibonacci numbers with odd subscripts up to N.)

74. Show that every positive integer greater than 2 can be written as the sum of distinct Fibonacci numbers.

75. Show that the sum of any 10 consecutive Fibonacci numbers is a multiple of 11.

76. Suppose that T_N represents a sequence of numbers satisfying the same recursive rule as the Fibonacci numbers ($T_N = T_{N-1} + T_{N-2}$) but with seeds $T_1 = c$ and $T_2 = d$, where c and d are arbitrary positive integers.

(a) Show that there are constants a and b such that $T_N = aF_{N+1} + bF_N$. (*Hint*: See Exercises 53 and 54.)

(b) Show that as N gets larger, and larger, the ratios T_{N+1}/T_N get closer and closer to ϕ.

77. (a) Show that as N gets larger and larger, the ratios F_{N+2}/F_N get closer and closer to ϕ^2.

(b) What number do the ratios F_{N+3}/F_N approximate as N increases? Explain.

78. In Fig. 9-38, $ABCD$ is a square and the three triangles I, II, and III have equal areas. Show that $x/y = \phi$.

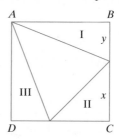

FIGURE 9-38

79. During the time of the Greeks the star pentagram shown in Fig. 9-39 was a symbol of the Brotherhood of Pythagoras. Consider the three segments of lengths x, y, and z shown in the figure.

FIGURE 9-39

(a) Show that $x/y = \phi$, $(x + y)/z = \phi$, and $(x + y + z)/(x + y) = \phi$.

(b) Show that if $y = 1$, then $x = \phi$, $(x + y) = \phi^2$, and $x + y + z = \phi^3$.

PROJECTS AND PAPERS

A Fibonacci Numbers, the Golden Ratio, and Phyllotaxis

In this chapter we mentioned that Fibonacci numbers and the golden ratio often show up in both the plant and animal worlds. In this project you are asked to expand on this topic.

1. Give several detailed examples of the appearance of Fibonacci numbers and the golden ratio in the plant world. Include examples of branch formation in plants, leaf arrangements around stems, and seed arrangements on circular seedheads (such as sunflower heads).

2. Discuss the concept of *phyllotaxis*. What is it, and what are some of the mathematical theories behind it?

(*Notes*: The literature on Fibonacci numbers, the golden ratio, and phyllotaxis is extensive. A search on the Web should provide plenty of information. Two excellent Web sites on this subject are Ron Knott's site at the University of Surrey, England (http://www.mcs.surrey.ac.uk/Personal/R.Knott/Fibonacci), and the site *Phyllotaxis: An Interactive Site for the Mathematical Study of Plant Pattern Formation* (http://www.math.smith.edu/~phyllo/) based at Smith College.

B The Golden Ratio in Art, Architecture, and Music

It is often claimed that from the time of the ancient Greeks through the Renaissance to modern times, artists, architects, and musicians have been fascinated by the golden ratio. Choose one of the three fields (art, architecture, or music) and write a paper discussing the history of the golden ratio in that field. Describe famous works of art, architecture, or music in which the golden ratio is alleged to have been used. How? Who were the artists, architects, and composers?

Be forewarned that there are plenty of conjectures, unsubstantiated historical facts, controversies, claims, and counterclaims surrounding some of the alleged uses of the golden ratio and Fibonacci numbers. Whenever appropriate, you should present both sides to a story.

C Logarithmic Spirals

In this project you are to investigate logarithmic spirals. You should explain what a logarithmic spiral is and how it is related to Fibonacci numbers and the golden ratio, and you should reference several examples of where they occur in the natural world.

D Figurate Numbers

The *triangular* numbers are numbers in the sequence $1, 3, 6, 10, 15, \ldots$. (The Nth **triangular** number T_N is given by the sum $1 + 2 + 3 + \cdots + N$.) In a similar way, *square*, *pentagonal*, and *hexagonal* numbers can be defined. In this project you are to investigate these types of numbers (called *figurate* numbers), give some of their more interesting properties, and discuss the relationship between these numbers and *gnomons*.

E The Golden Ratio Hypothesis

A long-held belief among those who study how humans perceive the outside world (mostly psychologists and psychobiologists) is that the *golden ratio* plays a special and prominent role in the human interpretation of "beauty." Shapes and objects whose proportions are close to the golden ratio are believed to be more pleasing to human sensibilities than those that are not. This theory, generally known as *the golden ratio hypothesis*, originated with the experiments of famous psychologist Gustav Fechner in the late 1800s. In a classic experiment, Fechner showed rectangles of various proportions to hundreds of subjects and asked them to choose. His results showed that the rectangles that were close to the proportions of the golden ratio were overwhelmingly preferred over the rest. Since Fechner's original experiment, there has been a lot of controversy about the golden ratio hypothesis, and many modern experiments have cast serious doubts about its validity.

Write a paper describing the history of the golden ratio hypothesis. Start with a description of Fechner's original experiment. Follow up with other experiments duplicating Fechner's results and some of the more recent experiments that seem to disprove the golden ratio hypothesis. Conclude with your own analysis.

REFERENCES AND FURTHER READINGS

1. Adam, John, *Mathematics in Nature: Modeling Patterns in the Natural World*. Princeton, NJ: Princeton University Press, 2003.

2. Ball, Philip, *The Self-made Tapestry: Pattern Formation in Nature*. New York: Oxford University Press, 2001.

3. Conway, J. H., and R. K. Guy, *The Book of Numbers*. New York: Springer-Verlag, 1996.

4. Douady, S., and Y. Couder, "Phyllotaxis as a Self-Organized Growth Process," in *Growth Patterns in Physical Sciences and Biology*, eds. J. M. Garcia-Ruiz et al. New York: Plenum Press, 1983.

5. Erickson, R. O., "The Geometry of Phyllotaxis," in *The Growth and Functioning of Leaves*, eds. J. E. Dale and F. L. Milthrope. New York: Cambridge University Press, 1983.

6. Gardner, Martin, "About Phi, an Irrational Number That Has Some Remarkable Geometrical Expressions," *Scientific American*, 201 (August 1959), 128–134.

7. Gardner, Martin, "The Multiple Fascinations of the Fibonacci Sequence," *Scientific American*, 220 (March 1969), 116–120.

8. Gazalé, M. J., *Gnomon: From Pharaohs to Fractals*. Princeton, NJ: Princeton University Press, 1999.

9. Gullberg, Jan, *Mathematics: From the Birth of Numbers*. New York: W. W. Norton, 1997.

10. Herz-Fischler, Roger, *A Mathematical History of the Golden Number*. New York: Dover, 1998.

11. Jean, R. V., *Mathematical Approach to Pattern and Form in Plant Growth*. New York: John Wiley & Sons, 1984.

12. Livio, Mario, *The Golden Ratio: The Story of Phi, the World's Most Astonishing Number*. New York: Random House, 2002.

13. Markovsky, George, "Misconceptions About the Golden Ratio," *College Mathematics Journal*, 23 (1992), 2–19.

14. May, Mike, "Did Mozart Use the Golden Section?" *American Scientist*, 84 (March–April 1996), 118.

15. Neill, William, and Pat Murphy, *By Nature's Design*. San Francisco, CA: Chronicle Books, 1993.

16. Olsen, Scott, *The Golden Section: Nature's Greatest Secret*. New York: Wooden Books, 2006.

17. Putz, John F., "The Golden Section and the Piano Sonatas of Mozart," *Mathematics Magazine*, 68 (1995), 275–282.

18. Sigler, Laurence, *Fibonacci's Liber Abaci*. New York: Springer-Verlag, 2002.

19. Stewart, Ian, "Daisy, Daisy, Give Me Your Answer, Do," *Scientific American* (January 1995), 96–99.

20. Stewart, Ian, *Life's Other Secret: The New Mathematics of the Living World*. New York: John Wiley & Sons, 1999, chap. 6.

21. Thompson, D'Arcy, *On Growth and Form*. New York: Dover, 1992, chaps. 11, 13, and 14.

22. Walser, Hans, *The Golden Section*. Stuttgart, Germany: B. G. Teubner, 1996.

10 The Mathematics of Money

Spending It, Saving It, and Growing It

The purpose of this chapter is to make you rich. If that fails, then at the very least I hope this chapter will give you some understanding of the basic principles of money management, the tools to make sound financial decisions, and ideas on how to avoid wasting your hard-earned money. Not surprisingly, there is a good amount of mathematics to learn along the way, but when it comes to financial matters, mathematics alone is not sufficient. As a complement to your understanding of the numbers, to be a good money manager you will need a dose of common sense, a dash of skepticism, and plenty of self-discipline.

A s a consumer, you make decisions about money every day. Some are minor ("Should I get gas at the station on the right or make a U-turn and go to the station across the highway where gas is 5¢ a gallon cheaper?"), but others are much more significant ("If I buy that new red Mustang, should I take the $2000 dealer's rebate or the 0% financing for 60 months option?"). Decisions of the first type usually involve just a little arithmetic and some common sense (on a 20 gallon fill-up you are saving $1 to make that U-turn—is it worth it?); decisions of the second type involve a more sophisticated understanding of the time value of money (is $2000 up front worth more or less than saving the interest on payments over the next five years?). This latter type of question and others similar to it, are the focus of this chapter.

One of the key ideas of this chapter is that money has a *present value* and a *future value*. If you invest X dollars today, you expect a return of more than X dollars at some future time (otherwise you would be better off keeping all your money in a piggy bank). Conversely, if you borrow X dollars today, you expect you will have to repay more than X dollars over time (otherwise your lender would be better off putting his money in a piggy bank). Both scenarios are variations on one mathematical theme: How much are X dollars today worth at some specified time in the future? As we will soon see, the answer to this question depends on many factors.

Throughout this chapter, the computation and manipulation of *percentages* will play a critical role, so we start the chapter with a thorough review of the meaning and use of percentages (Section 10.1). This leads us directly to Section 10.2 and a discussion of *simple interest* and the *simple interest formula*. In Section 10.3 we introduce the concept of *compound interest* (if not the most powerful force in the universe, at the very least one of the most important concepts in the world of finance) and the *compound interest formula*. In Section 10.4 we take a brief detour from money matters and introduce *geometric sequences* and a very important formula we call the *geometric sum formula*. In Sections 10.5 and 10.6 we apply the concepts developed in Sections 10.3 and 10.4 to discuss *annuities*. In Section 10.5 we discuss *deferred annuities* (such as retirement accounts and college trust funds), where savings are made in regular installments so that one can get a lump sum at some specified time in the future. In Section 10.6 we discuss *installment loans* (such as car loans and home loans), where one gets a lump sum in the present and then pays it back by making regular payments over a specified time in the future. In these last two sections we are able to discuss many interesting and sophisticated real-life questions about finances (such as the decision on the $2000 dealer's rebate versus the 60-month 0% interest loan on that red Mustang).

(A concluding bit of advice about this chapter: This chapter has a lot of formulas and many of them look quite complicated, but complicated is not the same as difficult. As we progress through the chapter and face increasingly more complicated formulas, every effort is made to deconstruct these formulas for you. Above all, don't let the formulas intimidate you—you are their master and they are there to serve you.)

10.1   Percentages

A general truism is that people don't like dealing with fractions. There are exceptions, of course, but most people would rather avoid fractions whenever possible. The most likely culprit for "fraction phobia" is the difficulty of dealing with fractions with different denominators. One way to get around this difficulty is to express fractions using a common, standard denominator, and in modern life the commonly used standard is the denominator 100. A "fraction" with denominator 100 can be interpreted as a **percentage**, and the percentage symbol (%) is used to indicate the presence of the hidden denominator 100. Thus,

$$x\% = \frac{x}{100}$$

Percentages are useful for many reasons. They give us a common yardstick to compare different ratios and proportions; they provide a useful way of dealing with fees, taxes, and tips; and they help us better understand how things increase or decrease relative to some given baseline. The next few examples explore these ideas.

❝ **percent:** One part in a hundred. From the Latin *per centum*, or "by the hundred." ❞

—American Heritage Dictionary

(**EXAMPLE 10.1**)   **Comparing Test Scores**

Suppose that in your English Lit class you scored 19 out of 25 on the quiz, 49.2 out of 60 on the midterm, and 124.8 out of 150 on the final exam. Without reading further, can you guess which one was your best score? Not easy, right?

The numbers 19, 49.2, and 124.8 are called *raw scores*. Since each raw score is based on a different total, it is hard to compare them directly, but we can do it easily once we express each score as a percentage of the total number of points possible.

- Quiz score = 19/25: Here we can do the arithmetic in our heads. If we just multiply both numerator and denominator by 4, we get $19/25 = 76/100 = 76\%$.

- Midterm score = 49.2/60: Here the arithmetic is a little harder, so one might want to use a calculator. $49.2 \div 60 = 0.82 = 82\%$. (Since $0.82 = 82/100$, it follows that $0.82 = 82\%$.) This score is a definite improvement over the quiz score.

- Final exam = 124.8/150: Once again, we use a calculator and get $124.8 \div 150 = 0.832 = 83.2\%$. This score is the best one. ⊂⊃

Example 10.1 illustrates the simple but important relation between decimals and percentages: decimals can be converted to percentages through multiplication by 100 (as in $0.76 = 76\%$, $1.325 = 132.5\%$, and $0.005 = 0.5\%$), and conversely, percentages can be converted to decimals through division by 100 (as in $100\% = 1.0$, $83.2\% = 0.832$, $7\frac{1}{2}\% = 0.075$).

A note of caution: Later in the chapter we deal with very small percentages such as 0.035%, $\frac{1}{5}\%$, and even $\frac{1}{100}\%$. In these cases there is a natural tendency to "lose" some zeroes along the way. Please be extra careful when converting very small percentages to decimals—$0.035\% = 0.00035$, $\frac{1}{5}\% = 0.2\% = 0.002$, and $\frac{1}{100}\% = 0.01\% = 0.0001$.

Percentages are most convenient when we want to describe fees, taxes, and tips.

(**EXAMPLE 10.2**)   **Is 3/20th a Reasonable Restaurant Tip?**

Imagine you take an old friend out to dinner at a nice restaurant for her birthday. The final bill comes to $56.80. Your friend suggests that since the service was good, you should tip 3/20th of the bill. What kind of tip is that?

After a moment's thought, you realize that your friend, who can be a bit annoying at times, is simply suggesting you should tip the standard 15%. After all, $3/20 = 15/100 = 15\%$.

Although 3/20 and 15% are mathematically equivalent, the latter is a much more convenient and familiar way to express the amount of the tip. To compute the actual tip, you simply multiply the amount of the bill by 0.15. In this case we get $0.15 \times \$56.80 = \$8.52$. (People use various tricks to do this calculation in their heads. For an exact calculation you can first take 10% of the bill—in this case $5.68—and then add the other 5% by computing half of the 10%—in this case $2.84. If you don't want to worry about nickels and dimes you can just round the $5.68 to $6, add another $3 for the 5%, and make the tip $9.) ⊂⊃

(**EXAMPLE 10.3**)   **Shopping for an iPod**

Imagine you have a little discretionary money saved up and you decide to buy yourself the latest iPod. After a little research you find the following options:

- Option 1: You can buy the iPod at Optimal Buy, a local electronics store. The price is $399. There is an additional 6.75% sales tax. Your total cost out the door is

$$\$399 + (0.0675)\$399 = \$399 + \$26.9325 = \$399 + \$26.94 = \$425.94$$

The above calculation can be shortened by observing that the original price (100%) plus the sales tax (6.75%) can be combined for a total of 106.75% of the original price. Thus, the entire calculation can be carried out by a single multiplication:

$$(1.0675)\$399 = \$425.94 \quad \text{(rounded } up \text{ to the nearest penny)}$$

- Option 2: At Hamiltonian Circuits, another local electronic store, the sales price is $415, but you happen to have a 5% off coupon good for all electronic

■ Note that sales taxes (in this case $26.9325) are always rounded *up* to the nearest penny ($26.94). An extra penny means little to the individual paying it but generates a lot of revenue for the state.

products. Taking the 5% off from the coupon gives the sale price, which is 95% of the original price.

Sale price: $(0.95)\$415 = \$394.25$

We still have to add the 6.75% sales tax on top of that, and as we saw in Option 1, the quick way to do so is to multiply by 1.0675.

Final price including taxes: $(1.0675)\$394.25 = \$420.87$

For efficiency we can combine the two separate calculations (take the discount and add the sales tax) into one:

$$(1.0675)(0.95)\$415 = \$420.87$$

- **Option 3:** You found an online merchant in Portland, Oregon, that will sell you the iPod for $441. This price includes a 5% shipping/processing charge that you wouldn't have to pay if you picked up the iPod at the store in Portland (there is no sales tax in Oregon). The $441 is much higher than the price at either local store, but you are in luck: your best friend from Portland is coming to visit and can pick up the iPod for you and save you the 5% shipping/processing charge. What would your cost be then?

  Unlike option 2, in this situation we do *not* take a 5% discount on the $441. Here the 5% was added to the iPod's base price to come up with the final cost of $441, that is, 105% of the base price equals $441. Using P for the unknown base price, we have

$$(1.05)P = \$441 \quad \text{or} \quad P = \frac{\$441}{1.05} = \$420$$

Although option 3 is your cheapest option, it is hardly worth the few pennies you save to inconvenience your friend. Your best bet is to head to Hamiltonian Circuits with your 5% off coupon.   ⊂⊃

The three options discussed in Example 10.3 can be generalized into the following useful facts about computing percentage increases and decreases.

┌ PERCENTAGE INCREASES AND DECREASES ─────

- If you start with a quantity Q and *increase* that quantity by $x\%$, you end up with the quantity

$$I = \left(1 + \frac{x}{100}\right)Q.$$

- If you start with a quantity Q and *decrease* that quantity by $x\%$, you end up with the quantity

$$D = \left(1 - \frac{x}{100}\right)Q.$$

- If I is the quantity you get when you *increase* an unknown quantity Q by $x\%$, then

$$Q = \frac{I}{1 + (x/100)}.$$

(Notice that this last formula is equivalent to the formula given in the first bullet.)

■ See Exercises 9 and 10.

We can also use the preceding formulas to solve for x (the percentage increase or decrease) in terms of the other two quantities (Q and I or Q and D, depending on whether we have an increase or a decrease). This is exactly what's behind commonly heard statements such as "retail sales were *up* by 3% this month" or "the median price of homes is *down* by 15% from last year," describing, respectively, a 3% monthly increase or a 15% annual decrease.

Percentage increases and decreases are particularly useful when describing or summarizing the ebb and flow of some large quantity over time, as illustrated in the next example.

EXAMPLE 10.4   **The Dow Jones Industrial Average**

The Dow Jones Industrial Average (DJIA) is one of the most commonly used indicators of the overall state of the stock market in the United States. (As of the writing of this material the DJIA hovered around 13,000.) We are going to illustrate the ups and downs of the DJIA with fictitious numbers.

- **Day 1:** On a particular day, the DJIA closed at 12,875.
- **Day 2:** The stock market has a good day and the DJIA closes at 13,029.50. This is an *increase* of 154.50 from the previous day. To express the increase as a percentage, we ask, 154.50 is what percent of 12,875 (the day 1 value that serves as our baseline)? The answer is obtained by simply dividing 154.50 into 12,875 (and then rewriting it as a percentage). Thus, the percentage increase from day 1 to day 2 is

$$\frac{154.50}{12,875} = 0.012 = 1.2\%$$

Here is a little shortcut for the same computation, particularly convenient when you use a calculator (all it takes is one division):

$$13,029.50 \div 12,875 = 1.012$$

All we have to do now is to mentally subtract 1 from the above number. This gives us once again $0.012 = 1.2\%$.]

- **Day 3:** The stock market has a pretty bad day and the DJIA closes at 12,508.32. This represents a change from the *previous day* of $12,508.32 - 13,029.50 = -521.18$, in other words, a *decrease* of 521.18. Once again, this decrease can be expressed as a percentage of the 13,029.50 by dividing:

$$\frac{521.18}{13,029.50} = 0.04 = 4\%$$

Percentage decreases are often used incorrectly, mostly intentionally and in an effort to exaggerate or mislead. The misuse is usually framed by the claim that if an $x\%$ increase changes A to B, then an $x\%$ decrease changes B to A. Not true!

EXAMPLE 10.5   **The Bogus 200% Decrease**

With great fanfare, the police chief of Happyville reports that crime decreased by 200% in one year. He came up with this number based on reported crimes in Happyville going down from 450 one year to 150 the next year. Since an increase from 150 to 450 is a 200% increase (true), a decrease from 450 to 150 must surely be a 200% decrease, right? Wrong.

The critical thing to keep in mind when computing a decrease (or for that matter an increase) between two quantities is that these quantities are not interchangeable. In this particular example the baseline is 450 and not 150, so the correct computation of the decrease in reported crimes is $300/450 = 0.666\ldots \approx 66.67\%$.

There is a moral to Example 10.5: Be wary of any extravagant claims about the percentage decrease of something (be it reported crimes, traffic accidents, pollution, or any other nonnegative quantity). Always keep in mind that a percentage decrease can never exceed 100%, once you reduce something by 100%, you have reduced it to zero.

An important part of being a smart shopper is understanding how markups (profit margins) and markdowns (sales) affect the price of consumer goods. Example 10.6 illustrates the arithmetic of markups and markdowns.

EXAMPLE 10.6   **Combining Markups and Markdowns**

A toy store buys a certain toy from the distributor to sell during the Christmas season. The store marks up the price of the toy by 80% (the intended profit margin). Unfortunately for the toy store, the toy is a bust and doesn't sell well. After Christmas, it goes on sale for 40% off the marked price. After a while, an additional 25% markdown is taken off the sale price and the toy is put on the clearance table. With all the markups and markdowns, what is the percentage profit/loss to the toy store?

The answer to this question is independent of the original cost of the toy to the store. Let's just call this cost C.

- After adding an 80% markup to their cost C, the toy store retails the toy for a price of $(1.8)C$.

- After Christmas, the toy is marked down and put on sale with a "40% off" tag. The sale price is 60% of the retail price. This gives $(0.6)(1.8)C = (1.08)C$, (which represents a net markup of 8% on the original cost to the store).

- Finally, the toy is put on clearance with an "additional 25% off" tag. The clearance price is $(0.75)(1.08)C = 0.81C$. (The clearance price is now 81% of the original cost to the store—a net loss of 19%! That's what happens when toys don't sell.)

10.2   Simple Interest

We now return to an important concept first mentioned in the introduction to this chapter—that money has a *present value* and a *future value*. Unless you are lending money to a friend, if you invest $P today (the **present value**) for a promise of getting $F at some future date (the **future value**), you expect F to be more than P. Otherwise, why do it? The same principle also works in reverse. If you are getting a present value of P today from someone else (either in cash or in goods), you expect to have to pay a future value of F back at some time in the future.

We will spend most of the rest of this chapter exploring the relationship between the present value and the future value of a sum of money. In particular, if we are given the present value P, how do we find the future value F (and vice versa)? The answer depends on several variables, the most important of which is the interest rate. Interest is the return the lender or investor expects as a reward for the use of his or her money, and the standard way to describe an interest rate is as a yearly rate commonly called the **annual percentage rate (APR)**. Thus, we can say, "I am investing my money in an account that pays an APR of 5%," or "I have to pay a 24% APR on the balance on my credit card." (Ouch!)

The APR is the most important variable in computing the return on an investment or the cost of a loan, but several other questions come into play and must be considered. Is the interest *simple* or *compounded*? If compounded, how often is it compounded? Are there additional fees? If so, are they in addition to the interest or are they included in the APR? We will consider these questions in Sections 10.2 and 10.3.

Simple Interest

■ The term *principal* is the standard word used to describe the original sum invested or borrowed. For the purposes of our discussion, we consider the principal to be the same as the present value. Fortunately, both start with the letter *P*.

In **simple interest**, only the original money invested or borrowed (called the **principal**) generates interest over time. This is in contrast to *compound interest*, where the principal generates interest, then the principal plus the interest generate more interest, and so on. (A full discussion of compound interest is coming up in Section 10.3.)

The next series of examples illustrate the way simple interest works.

(**EXAMPLE 10.7**) **Savings Bonds**

■ Savings bonds are the prototypical safe, conservative investment — they pay little interest but carry little risk. The general rule of thumb is that return and risk go hand in hand: Low risk goes with low return, high return requires higher risk.

Imagine that on the day you were born your parents purchased a $1000 savings bond that pays 5% annual simple interest. What is the value of the bond on your 18th birthday? What is the value of the bond on any given birthday?

Here the principal is $P = \$1000$ and the annual percentage rate is 5%. This means that the interest the bond earns in one year is 5% of $1000, or $(0.05)\$1000 = \$50$. Because the bond pays simple interest, the interest earned by the bond is the same every year. Thus,

- Value of the bond on your 1st birthday = $1000 + $50 = $1050.
- Value of the bond on your 2nd birthday = $1000 + (2 × $50) = $1100.
  ⋮
- Value of the bond on your 18th birthday = $1000 + (18 × $50) = $1900.
- Value of the bond when you become t years old = $1000 + ($t$ × $50). ⊂⊃

We can generalize the computation used in Example 10.7 as follows: Change the principal of $1000 to a generic P, change the 5% APR to a generic $R\%$, and let the time period be t years. Then,

- Annual interest = $R\%$ of $P = P\left(\dfrac{R}{100}\right)$.

- Interest for t years = annual interest $\times t = P\left(\dfrac{R}{100}\right)t$.

- Future value of the investment in t years = principal + interest for t years =

$$P + P\left(\frac{R}{100}\right)t = P\left[1 + \left(\frac{R}{100}\right)t\right]$$

Note: The fraction $R/100$ in all the above formulas is just the $R\%$ APR written as a decimal. This fraction will appear regularly throughout the chapter, and as things grow more complicated, the 100 in the denominator starts to add unnecessary clutter to our formulas. For convenience, from here on we will use r as shorthand for the decimal form of $R\%$.

Combining the preceding observation leads us to the following **simple interest formula**:

SIMPLE INTEREST FORMULA

The future value F of P dollars invested under simple interest for t years at an APR of $R\%$ is given by

$$F = P(1 + r \cdot t)$$

(where r denotes the $R\%$ APR written as a decimal).

You should think of the simple interest formula as a formula relating four variables: P (the present value), F (the future value), t (the length of the investment in years), and r (the APR). Given any three of these variables you can find the fourth one using the formula. The next example illustrates how to use the simple interest formula to find a present value P given F, t, and r.

(EXAMPLE 10.8)   Government Bonds: Part 2

Government bonds are often sold based on their future value. Suppose that you want to buy a five-year $1000 U.S. Treasury bond paying 4.28% annual simple interest (so that in five years you can cash in the bond for $1000). Here $1000 is the future value of the bond, and the price you pay for this bond is its present value.

To find the present value of the bond, we let $F = \$1000$, $R = 4.28\%$, and $t = 5$ and use the simple interest formula. This gives

$$\$1000 = P[1 + 5(0.0428)] = P(1.214)$$

Solving the above equation for P gives

$$P = \frac{\$1000}{1.214} = \$823.72 \quad \text{(rounded to the nearest penny).}$$

We end this section with a discussion of interest on credit cards. Generally speaking, credit cards charge exceptionally high interest rates, but you only have to pay interest if you don't pay your monthly balance in full. Thus, a credit card is a two-edged sword: if you make minimum payments or carry a balance from one month to the next, you will be paying a lot of interest; if you pay your balance in full, you pay no interest. In the latter case you got a free, short-term loan from the credit card company. When used wisely, a credit card gives you a rare opportunity — you get to use someone else's money for free. When used unwisely and carelessly, a credit card is a financial trap. Our next example illustrates this.

■ The method for computing interest charges on credit cards varies from one credit card company to another and can be very convoluted (try to read the fine print in your monthly credit card bill), so our discussion will focus on the general idea.

(EXAMPLE 10.9)   Credit Card Use: The Good, the Bad, and the Ugly

Imagine that you recently got a new credit card. Like most people, you did not pay much attention to the terms of use or to the APR, which with this card is a whopping 24%. To make matters worse, you went out and spent a little more than you should have the first month, and when your first statement comes in you are surprised to find out that your *new balance* is $876.

Like with most credit cards, you have a little time from the time you got the statement to the *payment due date* (this *grace period* is usually around 20 days). You can pay a *minimum payment* of $20, the full balance of $876, or any other amount in between. Let's consider these three different scenarios separately.

- Option 1: Pay the full balance of $876 before the payment due date.

  This one is easy. You owe no interest and you got free use of the credit card company's money for a short period of time. When your next monthly bill comes, the only charges will be for your new purchases.

- Option 2: Pay the minimum payment of $20.

  When your next monthly bill comes, you have a new balance of $1165 consisting of:

1. The *previous balance* of $856. (The $876 you previously owed minus your payment of $20.)

2. The charges for your *new purchases*. Let's say, for the sake of argument, that you were a little more careful with your card and your new purchases for this period were $288.

3. The *finance charge* of $21 calculated as follows:

   (i) *Periodic rate* = 0.02

   (ii) *Balance subject to finance charge* = $1050

   (iii) *Finance charge* = (0.02)$1050 = $21

   You might wonder, together with millions of other credit card users, where these numbers come from. Let's take them one at a time.

   (i) The *periodic rate* is obtained by dividing the annual percentage rate (APR) by the number of billing periods. Almost all credit cards use monthly billing periods, so the periodic rate on a credit card is the APR divided by 12. Your credit card has an APR of 24%, thus yielding a periodic rate of 2% = 0.02.

   (ii) The *balance subject to finance charge* (an official credit card term) is obtained by taking the average of the daily balances over the previous billing period. Since this balance includes your new purchases, all of a sudden you are paying interest on all your purchases and lost your grace period! In your case, the balance subject to finance charge came to $1050.

   (iii) The *finance charge* is obtained by multiplying the *periodic rate* times the *balance subject to finance charge*. In this case, (0.02)$1050 = $21.

- Option 3: You make a payment that is more than the minimum payment but less than the full payment.

  Let's say for the sake of argument that you make a payment of $400. When your next monthly bill comes, you have a new balance of $777.64. As in option 2, this new balance consists of:

1. The previous balance, in this case $476 (the $876 you previously owed minus the $400 payment you made)

2. The new purchases of $288

3. The finance charges, obtained once again by multiplying the *periodic rate* (2% = 0.02) times the *balance subject to finance charges*, which in this case came out to $682.

Thus, your finance charges turn out to be (0.02)$682 = $13.64, less than under option 2 but still a pretty hefty finance charge.

Two important general lessons can be drawn from Example 10.9: (1) *Make sure you understand the terms of your credit card agreement.* Know the APR (which can range widely from less than 10% to 24% or even more), know the length of your grace period, and try to understand as much of the fine print as you can. (2) *Make a real effort to pay your credit card balance in full each month.* This practice will help you avoid finance charges and keep you from getting yourself into a financial hole. If you can't make your credit card payments in full each month, you are living beyond your means and you may consider putting your credit card away until your balance is paid.

> " ■ 76% of college undergraduates have at least one credit card; 43% have four or more.
> ■ Only 21% of college students pay their balance in full each month. "
>
> —Nellie Mae

10.3 Compound Interest

Under simple interest the gains on an investment are constant—only the principal generates interest. Under **compound interest**, not only does the original principal generate interest, so does the previously accumulated interest. All other things being equal, money invested under compound interest grows a lot faster than money invested under simple interest, and this difference gets magnified over time. If you are investing for the long haul (a college trust fund, a retirement account, etc.), always look for compound interest.

(**EXAMPLE 10.10**) **Your Trust Fund Found!**

Imagine that you have just discovered the following bit of startling news: On the day you were born, your Uncle Nick deposited $5000 in your name in a trust fund that pays a 6% APR. One of the provisions of the trust fund was that you couldn't touch the money until you turned 18. You are now 18 years, 10 months old and you are wondering, How much money is in the trust fund now? How much money would there be in the trust fund if I waited until my next birthday when I turn 19? How much money would there be in the trust fund if I left the money in for retirement and waited until I turned 60?

Here is an abbreviated timeline of the money in your trust fund, starting with the day you were born:

- Day you were born: Uncle Nick deposits $5000 in trust fund.
- First birthday: 6% interest is added to the account. Balance in account is $(1.06)\$5000$. (We'll leave the expression alone and do the arithmetic later.)
- Second birthday: 6% interest is added to the previous balance (in red). Balance in account is $(1.06)(1.06)\$5000 = (1.06)^2\$5000$.
- Third birthday: 6% interest is added to the previous balance (again in red). Balance in account is $(1.06)(1.06)^2\$5000 = (1.06)^3\$5000$.

At this point you might have noticed that the exponent of (1.06) in the right-hand expression goes up by 1 on each birthday and in fact matches the birthday. Thus,

- Eighteenth birthday: The balance in the account is $(1.06)^{18}\$5000$. It is now finally time to pull out a calculator and do the computation:

$$(1.06)^{18}\$5000 = \$14,271.70 \quad \text{(rounded to the nearest penny)}$$

- Today: Since the bank only credits interest to your account once a year and you haven't turned 19 yet, the balance in the account is still $14,271.70.
- Nineteenth birthday: The future value of the account is

$$(1.06)^{19}\$5000 = \$15,128 \quad \text{(as before, rounded to the nearest penny)}$$

Moving further along into the future, we have

- 60th birthday: The future value of the account is

$$(1.06)^{60}\$5000 = \$164,938.45$$

which is an amazing return for a $5000 investment (if you are willing to wait, of course)!

Figure 10-1(a) plots the growth of the money in the account for the first 18 years. Figure 10-1(b) plots the growth of the money in the account for 60 years.

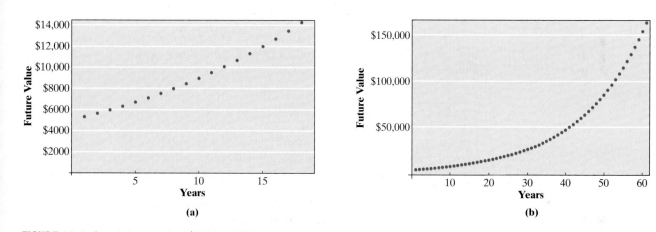

FIGURE 10-1 Cumulative growth of $5000 at 6% interest compounded annually.

We can generalize our computations in Example 10.10 by changing the $5000 principal to a generic P, changing the 6% APR to a generic $R\%$, and letting t denote the whole number of years the principal is invested. (If t is not a whole number we must *round down* to the nearest whole number.) If we do this, we get the following **annual compounding formula**.

ANNUAL COMPOUNDING FORMULA

The future value F of P dollars compounded annually for t years at an APR of $R\%$ is given by

$$F = P(1 + r)^t$$

■ Recall that r is the APR written in decimal form.

(**EXAMPLE 10.11**) **Saving for a Cruise**

Imagine that you have $875 in savings that you want to invest. Your goal is to have $2000 saved in $7\frac{1}{2}$ years. (You want to send your mom on a cruise on her 50th birthday.) Imagine now that the credit union around the corner offers a certificate of deposit (CD) with an APR of $6\frac{3}{4}\%$ compounded annually. What is the future value of your $875 in $7\frac{1}{2}$ years? If you are short of your $2000 target, how much more would you need to invest to meet that target?

To answer the first question, we just apply the annual compounding formula with $P = \$875$, $R = 6.75$ (i.e., $r = 0.0675$), and $t = 7$ (recall that with annual compounding, fractions of a year don't count) and get

$$\$875(1.0675)^7 = \$1382.24 \quad \text{(rounded to the nearest penny)}$$

Unfortunately, this is quite a bit short of the $2000 you want to have saved. To determine how much principal to start with to reach a future value target of $F = \$2000$ in 7 years at 6.75% annual interest, we solve for P in terms of F in the annual compounding formula. In this case substituting $2000 for F gives

$$\$2000 = P(1.0675)^7$$

and solving the above for P gives

$$P = \frac{\$2000}{(1.0675)^7} = \$1266.06$$

This is quite a bit more than the $875 you have right now, so this option is not viable. Don't despair—we'll explore some other options throughout this chapter. ⬭

(**EXAMPLE 10.12**)   **Saving for a Cruise: Part 2**

Let's now return to our story from Example 10.11: You have $875 saved up and a $7\frac{1}{2}$-year window in which to invest your money. As discussed in Example 10.11, the 6.75% APR compounded annually gives a future value of only $1382.24—far short of your goal of $2000.

Now imagine that you find another bank that is advertising a 6.75% APR that is *compounded monthly* (i.e., the interest is computed and added to the principal at the end of each month). It seems reasonable to expect that the monthly compounding could make a difference and make this a better investment. Moreover, unlike the case of annual compounding, you get interest for that extra half a year at the end.

To do the computation we will have to use a variation of the annual compounding formula. The key observation is that since the interest is compounded 12 times a year, the *monthly interest rate* is $6.75\% \div 12 = 0.5625\%$ (0.005625 when written in decimal form). An abbreviated chronology of how the money grows looks something like this:

- Original deposit: $875.
- Month 1: 0.5625% interest is added to the account. The balance in the account is now

$$(1.005625)\$875.$$

- Month 2: 0.5625% interest is added to the previous balance. The balance in the account is now

$$(1.005625)^2\$875.$$

- Month 3: 0.5625% interest is added to the previous balance. The balance in the account is now

$$(1.005625)^3\$875.$$

⋮

- Month 12: At the end of the first year the balance in the account is

$$(1.005625)^{12}\$875 = \$935.92.$$

After $7\frac{1}{2}$ years, or 90 months,

- Month 90: The balance in the account is now

$$(1.005625)^{90}\$875 = \$1449.62.$$

The conclusion is that the future value of $875 in $7\frac{1}{2}$ years at an APR of 6.75% *compounded monthly* is $1449.62. Although this is a slight improvement over the $1382.24 future value under annual compounding, you are still far short of the $2000 target. So you keep looking.

(**EXAMPLE 10.13**) **Saving for a Cruise: Part 3**

The story continues. Imagine you find a bank that pays a 6.75% APR that is *compounded daily*. You are excited! This will surely bring you a lot closer to your $2000 goal. Let's try to compute the future value of $875 in $7\frac{1}{2}$ years.

The analysis is the same as in Example 10.12 (be sure you reread Example 10.12 carefully) except now the interest is compounded 365 times a year (never mind leap years—they don't count in banking), and the numbers are not as nice.

First, we divide the APR of 6.75% by 365. This gives a *daily interest rate* of

■ Make sure you don't lose any zeroes along the way!

$$6.75\% \div 365 \approx 0.01849315\% = 0.0001849315$$

Next, we compute the number of days in the $7\frac{1}{2}$-year life of the investment:

$$365 \times 7.5 = 2737.5$$

Since parts of a day don't count, we round down to 2737.

Taking our cue from Example 10.12, we conclude that the original $875 will be multiplied by the factor (1.0001849315) a total of 2737 times. Thus,

$$F = (1.0001849315)^{2737}\$875 = \$1451.47$$

When we compare the future value obtained in Example 10.13 with the future value obtained in Example 10.12, we are in for somewhat of a surprise. The logical expectation is that the change from monthly compounding to daily compounding should result in a significant boost to the future value of the investment, but all we got was a couple of extra bucks for our troubles.

Let's summarize the results of Examples 10.11, 10.12, and 10.13. Each example represents a scenario in which the present value is $P = \$875$, the APR is 6.75% ($r = 0.0675$), and the length of the investment is $t = 7\frac{1}{2}$ years. The difference is the frequency of compounding during the year.

- Annual compounding (Example 10.11): Future value is $F = \$1382.24$.
- Monthly compounding (Example 10.12): Future value is $F = \$1449.62$.
- Daily compounding (Example 10.13): Future value is $F = \$1451.47$.

A reasonable conclusion from these numbers is that increasing the frequency of compounding (hourly, every minute, every second, every nanosecond) is not going to increase the ending balance by very much. In fact, we leave it as an exercise for the reader to confirm how remarkably meager are the returns of ratcheting up the frequency of compounding:

■ See Exercise 39.
■ See Exercise 40.

- Hourly compounding: Future value is $F = \$1451.67$.
- Compounding by the minute: Future value is $F = \$1451.68$.

The explanation for this surprising law of diminishing returns will be given shortly.

The General Compounding Formula

When an investment is compounded more than once a year, the annual compounding formula needs to be modified accordingly. In Examples 10.12 (monthly compounding) and 10.13 (daily compounding) we got a glimpse of how this might be done. First, we computed the **periodic interest rate** (the interest rate that applies to each compounding period) by dividing the APR by the number of times the money is compounded in a year. (In Example 10.12 we divided the APR by 12; in Example 10.13 we divided the APR by 365.) Second, we calculated the total number of times the money was compounded over the length of the investment. (In Example 10.12 we had 90 months; in Example 10.13 we had 2737.5 days, which we then had to round down to 2737 days.)

The above observations lead us to the following **general compounding formula** for the future value of P dollars invested for t years at an APR of $R\%$ compounded n times a year.

GENERAL COMPOUNDING FORMULA

The future value of P dollars in t years at an APR of $R\%$ compounded n times a year is

$$F = P\left(1 + \frac{r}{n}\right)^{nt}$$

In the general compounding formula, r/n represents the *periodic interest* rate expressed as a decimal, and the exponent $n \cdot t$ represents the total number of compounding periods over the life of the investment. If we use p to denote the periodic interest rate and T to denote the total number of times the principal is compounded over the life of the investment, the general compounding formula takes the following particularly nice form.

GENERAL COMPOUNDING FORMULA (VERSION 2)

The future value F of P dollars compounded a total of T times at a periodic interest rate p is

$$F = P(1 + p)^T$$

One of the remarkable properties of the general compounding formula is that even as n (the frequency of compounding) grows without limit, the future value F approaches a limiting value L. In other words, even if the money were compounded a billion times a second, the future value of the investment could not be more than L. This limiting value represents the future value of an investment under **continuous compounding** (i.e., the compounding occurs over infinitely short time intervals) and is given by the following continuous compounding formula.

CONTINUOUS COMPOUNDING FORMULA

The future value F of P dollars *compounded continuously* for t years at an APR of $R\%$ is

$$F = Pe^{rt}$$

Notes: (1) $e = 2.71828\ldots$ is the base of the natural logarithms. Since e is an irrational number, it has an infinite, nonrepeating decimal expansion, and your calculator can only provide an approximation. (Incidentally, a calculator with an e^x key is essential for financial calculations.) In general, a decent scientific calculator should give an answer that is accurate to the nearest penny, but be alert to the possibility of rounding errors. (2) To understand the derivation of the continuous compounding formula and the mysterious appearance of e requires some calculus, so we will stay away from the derivation of the formula. If you are curious, you can find it in most financial mathematics textbooks (see, for example, reference 1).

(**EXAMPLE 10.14**)   **Saving for a Cruise: Part 4**

You finally found a bank that offers an APR of 6.75% compounded continuously. Using the continuous compounding formula and a calculator, you find that the future value of your $875 in $7\frac{1}{2}$ years is

$$F = \$875(e^{7.5 \times 0.0675}) = \$875(e^{0.50625}) = \$1451.68$$

The most disappointing thing is that when you compare this future value with the future value under daily compounding (Example 10.13), the difference is 21¢. ⬭

The Annual Percentage Yield (APY)

The **annual percentage yield (APY)** of an investment (sometimes called the *effective rate*) is the percentage of profit that the investment generates in a one-year period. For example, if you start with $1000 and after one year you have $1099.60, you have made a profit of $99.60. The $99.60 expressed as a percentage of the $1000 principal is 9.96%—this is your APY. This computation was easy because the principal was $1000, but how do we handle more realistic numbers, say a principal of $835.25 that in one year grows to $932.80?

(**EXAMPLE 10.15**)   **Computing an APY**

Suppose that you invest $835.25. At the end of a year your money grows to $932.80. (The details of how your money grew to $932.80 are irrelevant for the purposes of our computation.) Here is how you compute the APY:

$$APY = \frac{\$932.80 - \$835.25}{\$835.25} \approx 0.1168 = 11.68\%$$ ⬭

In general, if you start with S dollars at the beginning of the year and your investment grows to E dollars by the end of the year, the APY is the ratio $(E - S)/S$. You may recognize this ratio from Section 10.1—it is the *annual percentage increase* of your investment.

The APY is a particularly useful yardstick with which to compare investments with different APRs and different compounding frequencies. For example, which is better, an APR of 6.8% *compounded quarterly* or an APR of 6.7% *compounded continuously*? To answer this question we compare the APYs of the two investments. The simplest way to find the APY in these cases is to compute the future value of $1 in one year. Example 10.16 illustrates the details.

(**EXAMPLE 10.16**)   **Comparing Investments Through APYs**

Which of the following three investments is better: (a) 6.7% APR *compounded continuously*, (b) 6.75% APR *compounded monthly*, or (c) 6.8% APR *compounded quarterly*? Notice that the question is independent of the principal P and the length of the investment t. To compare these investments we will compute their APYs.

(a) The future value of $1 in 1 year at 6.7% interest compounded *continuously* is given by $e^{0.067} \approx 1.06930$. (Here we used the continuous compounding formula).

   The APY in this case is 6.93%. (The beauty of using $1 as the principal is that this last computation is trivial.)

(b) The future value of $1 in 1 year at 6.75% interest compounded *monthly* is $(1 + 0.0675/12)^{12} = 1.005625^{12} \approx 1.06963$. (Here we used the general compounding formula).

   The APY in this case is 6.963%.

(c) The future value of $1 in 1 year at 6.8% interest compounded *quarterly* is $(1 + 0.068/4)^4 = 1.017^4 \approx 1.06975$.

   The APY in this case is 6.975%.

Although they are all quite close, we can now see that (c) is the best choice, (b) is the second-best choice, and (a) is the worst choice. Although the differences between the three investments may appear insignificant when we look at the effect over one year, these differences become quite significant when we invest over longer periods.   ⊂⊃

10.4   Geometric Sequences

In this section we will take a brief break from matters of finance and discuss *geometric sequences*. (We introduced sequences and sequence notation in Chapter 9, and if a quick review is needed, the reader is encouraged to reread Sections 9.1 and 9.2.) An understanding of geometric sequences will give us the mathematical tools needed for our discussions on annuities and installment loans coming up in Sections 10.5 and 10.6.

A **geometric sequence** starts with an *initial term P* and from then on every term in the sequence is obtained by multiplying the preceding term by the same constant c: The second term equals the first term times c, the third term equals the second term times c, and so on. The number c is called the **common ratio** of the geometric sequence.

(**EXAMPLE 10.17**)   **Some Simple Geometric Sequences**

■ 5, 10, 20, 40, 80, . . .

   The above is a geometric sequence with initial term 5 and common ratio $c = 2$. Notice that since the initial term and the common ratio are both positive, every term of the sequence will be positive. Also notice that the sequence is an *increasing* sequence: Every term is bigger than the preceding term. This will happen every time the common ratio c is bigger than 1.

- $27, 9, 3, 1, \frac{1}{3}, \frac{1}{9}, \ldots$

  The above is a geometric sequence with initial term 27 and common ratio $c = \frac{1}{3}$. Notice that this is a *decreasing* sequence, a consequence of the common ratio being between 0 and 1.

- $27, -9, 3, -1, \frac{1}{3}, -\frac{1}{9}, \ldots$

  The above is a geometric sequence with initial term 27 and common ratio $c = -\frac{1}{3}$. Notice that this sequence alternates between positive and negative terms, a consequence of the common ratio being a negative number.

A generic geometric sequence with initial term P and common ratio c can be written in the form

$$P, cP, c^2P, c^3P, c^4P, \ldots$$

As we did with the Fibonacci sequence in Chapter 9, we will use a common letter—in this case, G for geometric—to label the terms of a generic geometric sequence, with subscripts conveniently chosen to start at 0. In other words,

$$G_0 = P, \quad G_1 = cP, \quad G_2 = c^2P, \quad G_3 = c^3P, \quad G_4 = c^4P, \ldots$$

■ One advantage of starting the count at G_0 rather than at G_1 is that now the subscript of G matches the power of c, as in $G_{17} = c^{17}P$, making things a lot easier to remember.

Like the Fibonacci sequence, a geometric sequence can be defined by a recursive and an explicit formula.

> **GEOMETRIC SEQUENCE**
>
> - $G_N = cG_{N-1}; G_0 = P$ (recursive formula)
> - $G_N = c^N P$ (explicit formula)

The recursive formula is just a restatement of the definition of a geometric sequence—an arbitrary term in the sequence (G_N) is the common ratio c times the preceding term in the sequence (G_{N-1}). (This statement cannot be applied to the initial term, which has no preceding term. Thus, we have to add the additional fact that the initial term is P.) The explicit formula says that each term of the sequence is the initial term P times the corresponding power of the common ratio c. By far, the explicit description is the more useful of the two.

Our next example illustrates the connection between geometric sequences and compound interest calculations.

(**EXAMPLE 10.18**)   **A Familiar Geometric Sequence**

Consider the geometric sequence with initial term $P = 5000$ and common ratio $c = 1.06$. The first few terms of this sequence are

$$G_0 = 5000,$$
$$G_1 = (1.06)5000 = 5300,$$
$$G_2 = (1.06)^2 5000 = 5618,$$
$$G_3 = (1.06)^3 5000 = 5955.08,$$
$$\vdots$$

If we put dollar signs in front of these numbers, we get the principal and the balances over the first three years on an investment with a principal of $5000 and with an APR of 6% compounded annually. These numbers might look familiar to you—they come from Uncle Nick's trust fund example (Example 10.10). In fact, the Nth term of the above geometric sequence (rounded to two decimal places) will give the balance in the trust fund on your Nth birthday.

Example 10.18 illustrates the important role that geometric sequences play in the world of finance. If you look at the chronology of a compound interest account started with a principal of P and a periodic interest rate p, the balances in the account at the end of each compounding period are the terms of a geometric sequence with initial term P and common ratio $(1 + p)$:

$$P, \quad P(1 + p), \quad P(1 + p)^2, \quad P(1 + p)^3, \dots$$

Geometric sequences have important applications beyond financial mathematics. The main purpose of our next example is to illustrate this point.

(**EXAMPLE 10.19**)   **Eradicating the Gamma Virus**

Thanks to improved vaccines and good public health policy, the number of reported cases of the gamma virus has been dropping by 70% a year since 2008, when there were 1 million reported cases of the virus. If the present rate continues, how many reported cases of the virus can we predict by the year 2014? How long will it take to eradicate the virus?

Because the number of reported cases of the gamma virus *decreases* by 70% each year, we can model this number by a geometric sequence with common ratio $c = 0.3$ (a 70% decrease means that the number of reported cases is 30% of what it was the preceding year).

We will start the count in 2008 with the initial term $G_0 = P = 1,000,000$ reported cases. In 2009 the numbers will drop to $G_1 = 300,000$ reported cases, in 2010 the numbers will drop further to $G_2 = 90,000$ reported cases, and so on.

By the year 2014 we will be in the sixth iteration of this process, and thus the number of reported cases of the gamma virus will be $G_6 = (0.3)^6 \times 1,000,000 = 729$. By 2015 this number will drop to about 219 cases $(0.3 \times 729 = 218.7)$, by 2016 to about 66 cases $(0.3 \times 219 = 65.7)$, by 2017 to about 20 cases, and by 2018 to about 6 cases.

The Geometric Sum Formula

We will now discuss a very important and useful formula—the **geometric sum formula**—that allows us to add a large number of terms in a geometric sequence without having to add the terms one by one.

THE GEOMETRIC SUM FORMULA

$$P + cP + c^2P + \cdots + c^{N-1}P = P\left(\frac{c^N - 1}{c - 1}\right)$$

Notes: (1) The geometric sum formula works for all values of the common ratio c except $c = 1$. For $c = 1$, the formula breaks down because we get a zero denominator on the right-hand side. (2) There is a total of N terms being added on the left-hand side of the formula. Thus, you can think of the left-hand side of the formula as "the sum of the first N terms of a geometric sequence." (3) You can think of the exponent of c on the right-hand side as being "one more than the highest exponent of c on the left-hand side." (4) You can verify that the geometric sum formula is true by multiplying both sides of the formula by $(c - 1)$ and simplifying the left-hand side.

■ For details, see Exercise 88.

In the next section we will see how the geometric sum formula plays a key role in financial applications. For now, we will conclude this section with an application of the geometric sum formula to public health.

(**EXAMPLE 10.20**) **Tracking the Spread of a Virus**

At the emerging stages, the spread of many infectious diseases—such as HIV and the West Nile virus—often follows a geometric sequence. Let's consider the case of an imaginary infectious disease called the X-virus, for which no vaccine is known. The first appearance of the X-virus occurred in 2008 (year 0), when 5000 cases of the disease were recorded in the United States. Epidemiologists estimate that until a vaccine is developed, the virus will spread at a 40% *annual rate of growth*, and it is expected that it will take at least 10 years until an effective vaccine becomes available. Under these assumptions, how many estimated cases of the X-virus will occur in the United States over the 10-year period from 2008 to 2017?

We can track the spread of the virus by looking at the number of new cases of the virus reported each year. These numbers are given by a geometric sequence with $P = 5000$ and common ratio $c = 1.4$ (40% annual growth):

5000 cases in 2008

$(1.4) \times 5000 = 7000$ new cases in 2009

$(1.4)^2 \times 5000 = 9800$ new cases in 2010

$\vdots$

$(1.4)^9 \times 5000$ new cases in 2017

It follows that the total number of cases over the 10-year period is given by the sum

$$5000 + (1.4) \times 5000 + (1.4)^2 \times 5000 + \cdots + (1.4)^9 \times 5000$$

Using the geometric sum formula, this sum (rounded to the nearest whole number) equals

$$5000 \times \frac{(1.4)^{10} - 1}{1.4 - 1} \approx 349{,}068$$

Our computation shows that about 350,000 people will contract the X-virus over the 10-year period. What would happen if, due to budgetary or technical problems, it takes 15 years to develop a vaccine? All we have to do is change N to 15 in the geometric sum formula:

$$5000 \times \frac{(1.4)^{15} - 1}{1.4 - 1} \approx 1{,}932{,}101$$

These are sobering numbers: The geometric sum formula predicts that if the development of the vaccine is delayed for an extra five years, the number of cases of X-virus cases would grow from 350,000 to almost 2 million!

10.5 Deferred Annuities: Planned Savings for the Future

A **fixed annuity** is a sequence of *equal* payments made or received over regular (monthly, quarterly, annually) time intervals. Annuities (often disguised under different names) are so common in today's financial world that there is a good chance you may be currently involved in one or more annuities and not even realize it. You may be *making* regular deposits to save for a vacation, a wedding, or college, or you may be making regular payments on a car loan or a home mortgage. You could also be at the receiving end of an annuity, *getting* regular payments from an inheritance, a college trust fund set up on your behalf, or a lottery jackpot.

When payments are made so as to produce a lump-sum payout at a later date (e.g., making regular payments into a college trust fund), we call the annuity a *deferred annuity*; when a lump sum is paid to generate a series of regular payments later (e.g., a car loan), we call the annuity an *installment loan*. Simply stated, in a deferred annuity the pain (in the form of payments) comes first and the reward (a lump-sum payout) comes in the future, whereas in an installment loan the reward (car, boat, house) comes in the present and the pain (payments again) is stretched out into the future.

In this section we will discuss deferred annuities. In the next section we will take a look at installment loans. We will start with an example that illustrates the key ideas behind a deferred annuity.

(**EXAMPLE 10.21**)   **Setting Up a College Trust Fund**

Given the cost of college, parents often set up college trust funds for their children by setting aside a little money each month over the years. A college trust fund is a form of forced savings toward a specific goal, and it is generally agreed to be a very good use of a parent's money—it spreads out the pain of college costs over time, generates significant interest income, and has valuable tax benefits.

Let's imagine a mother decides to set up a college trust fund for her newborn child. Her plan is to have $100 withdrawn from her paycheck each month for the next 18 years and deposited in a savings account that pays 6% annual interest *compounded monthly*. What is the future value of this trust fund in 18 years?

What makes this example different from Uncle Nick's trust fund example (Example 10.10) is that money is being added to the account in regular installments of $100 per month. Each $100 monthly installment has a different "lifespan": The first $100 compounds for 216 months (12 times a year for 18 years), the second $100 compounds for only 215 months, the third $100 compounds for only 214 months, and so on. Thus, the future value of each $100 installment is different. To compute the future value of the trust fund we will have to compute the future value of each of the 216 installments separately and add. Sounds like a tall order, but the geometric sum formula will help us out.

Critical to our calculations are that each installment is for a fixed amount ($100) and that the periodic interest rate p is always the same (6% ÷ 12 = 0.5% = 0.005). Thus, when we use the general compounding formula, each future value looks the same except for the compounding exponent:

- Future value of the *first* installment ($100 compounded for 216 months):
$$(1.005)^{216}\$100.$$

   (*Note:* As usual, we will wait until the end to do the arithmetic.)

- Future value of the *second* installment ($100 compounded for 215 months):
$$(1.005)^{215}\$100.$$

- Future value of the *third* installment ($100 compounded for 214 months):
$$(1.005)^{214}\$100.$$

   $\vdots$

- Future value of the last installment ($100 compounded for one month):
$$(1.005)\$100 = \$100.50.$$

The future value F of this trust fund at the end of 18 years is the sum of all the above future values. If we write the sum in reverse chronological order (starting with the last installment and ending with the first), we get

$$F = (1.005)\$100 + (1.005)^2\$100 + \cdots + (1.005)^{215}\$100 + (1.005)^{216}\$100$$

A more convenient way to deal with the above sum is to first observe that the last installment of $(1.005)\$100 = \$100.50$ is a common factor of every term in the sum; therefore,

$$F = \$100.50[1 + 1.005 + (1.005)^2 + \cdots + (1.005)^{214} + (1.005)^{215}]$$

You might now notice that the sum inside the brackets is a geometric sum with common ratio $c = 1.005$ and a total of $N = 216$ terms. Applying the geometric sum formula to this sum gives

$$F = \$100.50\left[\frac{(1.005)^{216} - 1}{1.005 - 1}\right] = \$38,929$$

We can generalize the ideas of Example 10.21 by deconstructing the expression

$$F = \$100.50\left[\frac{(1.005)^{216} - 1}{1.005 - 1}\right]$$

The $100.50 represents the *future value of the last payment*. (Since the last payment is compounded for one month, we get $(1.005)\$100 = \$100.50$.) Inside the brackets we have a $(1.005)^{216} - 1$ in the numerator and a $1.005 - 1$ in the denominator. The 1.005 in both cases represents the usual $(1 + p)$ where p is the periodic interest rate. The exponent 216 represents the *total number of installment* payments made over the life of the investment (in this case, 12 monthly installments made over a period of 18 years). The -1's are inherited from the geometric sum formula.

Putting all the above observations together, we get the following **fixed deferred annuity formula**.

FIXED DEFERRED ANNUITY FORMULA

The future value F of a fixed deferred annuity consisting of T payments of $\$P$ having a periodic interest of p (written in decimal form) is

$$F = L\left[\frac{(1 + p)^T - 1}{p}\right]$$

where L denotes the *future value of the last payment*.

Notes: (1) The lonely p in the denominator comes from simplifying $(1 + p) - 1$. (2) The value of L depends on whether the payments of P are made at the start or at the end of the period and consequently whether the last payment generates interest or not. If the payments are made at the start of the period, then $L = (1 + p)P$; on the other hand, if the payments are made at the end of each period (i.e., the last payment generates no interest), then $L = P$. (3) By far the most common form of annuity is based on monthly installments, with the annual interest rate also compounded monthly. In this case, $p = r/12$, where r is the annual APR written in decimal form, and $T = 12t$, where t is the number of years.

■ See Exercises 63 through 66 for a comparison of the two situations.

$\boxed{\text{EXAMPLE 10.22}}$ **Setting Up a College Trust Fund: Part 2**

In Example 10.21 we saw that an 18-year annuity of $100 monthly payments at an APR of 6% compounded monthly is $38,929. For the same APR and the same number of years, how much should the monthly payments be if our goal is an annuity with a future value of $50,000?

If we use the fixed deferred annuity formula with $F = \$50,000$, we get

$$\$50,000 = L\left[\frac{(1.005)^{216} - 1}{1.005 - 1}\right] = L\left[\frac{(1.005)^{216} - 1}{0.005}\right]$$

■ If this last part is not clear, you may want to take another look at Example 10.21.

Solving for L gives

$$L = \$50,000\left[\frac{0.005}{(1.005)^{216} - 1}\right] = \$129.08$$

Recall now that L is the future value of the last payment, and since the payments are made at the beginning of each month, $L = (1.005)P$. Thus,

$$P = \frac{\$129.08}{1.005} = \$128.44$$

The main point of Example 10.22 is to illustrate that the *fixed deferred annuity formula* establishes a relationship between the future value F of the annuity and the fixed payment P required to achieve that future value. If we know one, we can solve for the other (assuming, of course, a specified number of payments T and a specified periodic interest rate p). The one minor complication (it doesn't look minor, but it really is) is that the fixed annuity formula is given in terms of L rather than P. It is important to remember that either $P = L$ (when the payments are made at the end of each period and thus the last payment generates no interest) or $P = [L/(1 + p)]$ (when the payments are made at the start of each period and thus the last payment does generate interest). Thus, once we solve the fixed deferred annuity formula for L, we either have P or are one step removed from finding P.

$\boxed{\text{EXAMPLE 10.23}}$ **Saving for a Cruise: Part 5**

In Section 10.3 we discussed a series of examples based on the same storyline: You have $875 saved up and a $7\frac{1}{2}$-year window in which to come up with $2000 to send your mom on a cruise for her 50th birthday. (Promise—this is the last time we will revisit this story!)

We saw (Example 10.12) that if you invest the $875 at a 6.75% APR compounded monthly, the future value of your investment is $1449.62—for simplicity, let's call it $1450. This is $550 short of the $2000 you will need. Imagine you want to come up with the additional $550 by making regular monthly deposits into the savings account, essentially creating a small annuity. How much would you have to deposit each month to generate the $550 that you will need? Basically, the cost of one frapuccino.

Using the fixed deferred annuity formula with $F = \$550$, $T = 90$ (12 installments a year for $7\frac{1}{2}$ years), and a periodic rate of $p = 0.005625$ (obtained by taking the 6.75% APR and dividing by 12), we have

■ We are assuming that the last installment will generate no interest; thus, $L = P$. In reality, whatever assumption you make about L makes only a few pennies' difference in this case.

$$\$550 = P\left[\frac{(1.005625)^{90} - 1}{0.005625}\right]$$

Solving the above for P gives

$$P = \$550\left[\frac{0.005625}{(1.005625)^{90} - 1}\right] \approx \$4.72$$

In conclusion, all you have to do to come up with the $2000 that you will need to send Mom on a cruise in $7\frac{1}{2}$ years is the following: (1) Invest your $875 savings in a safe investment such as a CD offered by a bank or a credit union and (2) save about $5 a month and put the money into a fixed deferred annuity.

10.6   Installment Loans: The Cost of Financing the Present

An **installment loan** (also know as a *fixed immediate annuity*) is a series of equal payments made at equal time intervals for the purpose of paying off a lump sum of money received up front. Typical installment loans are the purchase of a car on credit or a mortgage on a home. The most important distinction between an installment loan and a fixed deferred annuity is that an installment loan has a *present value* that we compute by adding the present value of each payment, whereas a fixed deferred annuity has a *future* value that we compute by adding the future value of each payment.

Example 10.24 illustrates how we can compute the monthly payments on an installment loan using the mathematics we have developed in this chapter.

(**EXAMPLE 10.24**)   **Financing That Red Mustang**

You've just landed a really good job and decide to buy the car of your dreams, a brand-new red Mustang. You negotiate a good price ($23,995 including taxes and license fees). You have $5000 saved for a down payment, and you can get a car loan from the dealer for 60 months at 6.48% annual interest. If you take out the loan from the dealer for the balance of $18,995, what would the monthly payments be? Can you afford them? Typically, buyers blindly trust the finance

department at the dealership to provide this information accurately, but as an educated consumer, wouldn't you feel more comfortable doing the calculation yourself? Now you can, and here is how it goes.

Every time you make a future payment on an installment loan, that payment has a present value, and the sum of the present values of all the payments equals the present value of the loan—in this case, the $18,995 that you are financing. Although each monthly loan payment of F has a different present value, each of these present values can be computed using the general compounding formula: *The present value P of a payment of F paid T months in the future is $P = F/(1 + p)^T$, where p denotes the monthly interest rate.*

■ This formula is just a variation of version 2 of the general compounding formula, $F = P(1 + p)^T$.

In this example, the monthly interest rate is $p = 0.0648/12 = 0.0054$. Thus,

- Present value of the *first* payment of F: $F/1.0054$

- Present value of the *second* payment of F: $F/(1.0054)^2$

- Present value of the *third* payment of F: $F/(1.0054)^3$

  $\vdots$

- Present value of the *last* payment of F: $F/(1.0054)^{60}$.

Adding all the above present values gives

$$\frac{F}{1.0054} + \frac{F}{(1.0054)^2} + \frac{F}{(1.0054)^3} + \cdots + \frac{F}{(1.0054)^{60}} = \$18{,}995$$

Notice that the left-hand side of the above equation is a geometric sum of $T = 60$ terms with initial term $(F/1.0054)$ and common ratio $(1/1.0054)$. Using the geometric sum formula, the equation can be rewritten as

$$\left(\frac{F}{1.0054}\right)\left[\frac{\left(\frac{1}{1.0054}\right)^{60} - 1}{\left(\frac{1}{1.0054}\right) - 1}\right] = \$18{,}995$$

From here on there is a bit of messy arithmetic to take care of, but we can solve for F and get $F = \$371.48$. (For good measure, you should double check the computation on your own.)

Although you may have negotiated a good price for the car, when you throw in the financing costs this may not be such a great deal after all (60 payments of $371.48 equals $22,288.80, and even when adjusting for inflation that is a lot more than the $18,995 present value of the loan). At the end, you decide to look around for a better deal. (Don't worry, there is a part 2 coming.)

Let's generalize the results of Example 10.24. Suppose that you get a loan for P dollars (the present value of the loan), and you want to pay off the loan by making T monthly payments of F dollars, with a periodic (monthly in this case) interest rate of p. The relationship between P and F is given by the formula

$$P = \frac{F}{(1 + p)} \left[\frac{\left(\dfrac{1}{1 + p}\right)^T - 1}{\left(\dfrac{1}{1 + p}\right) - 1} \right]$$

The above formula is a direct generalization of the last formula in Example 10.24. This formula looks a lot more complicated than it is, or needs to be. We can make life much easier once we notice the repeated occurrence in the formula of the expression $1/(1 + p)$, which for simplicity we will denote by q. With this change of notation we get the following formula, called the **amortization formula**.

AMORTIZATION FORMULA

If an installment loan of P dollars is paid off in T payments of F dollars at a periodic interest of p (written in decimal form), then

$$P = Fq\left[\frac{q^T - 1}{q - 1}\right]$$

where $q = 1/(1 + p)$.

Notes: (1) The amortization formula has several other equivalent versions—see Exercise 90 for some of these alternatives. (2) There are many amortization calculators available online that allow you to compute the monthly payments on an installment loan without you having to do anything other than entering the amount you want to finance, the number of monthly payments you want to make, and the APR. The calculator does the rest. Several good amortization calculators are listed in the references at the end of the chapter. (3) The Fq in front of the brackets represents the present value of the first installment. In almost all installment loans, payments are due at the end rather than the beginning of each period. Thus, the first payment is due one month after the loan is taken out and has a present value of $F/(1 + p) = Fq$. In some situations, however, payments are made at the beginning of each period. This happens when the first payment of F is made up front and thus has a present value of F. In these situations we must use a modified version of the amortization formula with the Fq replaced by just F.

(**EXAMPLE 10.25**)   **Financing That Red Mustang: Part 2**

During certain times of the year automobile dealers offer incentives in the form of cash rebates or reduced financing costs (including 0% APR), and often the buyer can choose between those two options. Given a choice between a cash rebate or cheap financing, a savvy buyer should be able to figure out which of the two is better, and we are now in a position to do that.

We will consider the same situation we discussed in Example 10.24. You have negotiated a price of $23,995 (including taxes and license fees) for a brand-new red Mustang, and you have $5000 for a down payment. The big break for you is that this dealer is offering two great incentives: a choice between a cash rebate of $2000 or a 0% APR for 60 months. If you choose the cash rebate, you will have a balance of $16,995 that you will have to finance at the dealer's standard interest rate of 6.48%. If you choose the free financing, you will have a 0% APR for 60 months on a balance of $18,995. Which is a better deal?

To answer this question we will compare the monthly payments under both options.

- Option 1: Take the $2000 rebate. Here the present value is $P = \$16,995$, amortized over 60 months at 6.48% APR. The periodic rate is $p = 0.0054$. Applying the amortization formula (see Example 10.24 for the details), we get

$$\left(\frac{F}{1.0054}\right)\left[\frac{\left(\frac{1}{1.0054}\right)^{60} - 1}{\left(\frac{1}{1.0054}\right) - 1}\right] = \$16,995$$

Solving the above equation for F (you should try it yourself) gives the monthly payment under the rebate option:

$$F = \$332.37 \quad \text{(rounded to the nearest penny)}$$

- Option 2: Take the 0% APR. Here the present value is $P = \$18,995$, amortized over 60 months at 0% APR. There is no need for any formulas here: With no financing costs, your monthly payment amortized over 60 months is

$$F = \frac{\$18,995}{60} = \$316.59 \quad \text{(rounded to the nearest penny)}$$

Clearly, in this particular situation the 0% APR option is a lot better than the $2000 rebate option (you are saving approximately $16 a month, which over 60 months is a decent piece of change—see Exercise 74). Of course, each situation is different, and it can also be true that the rebate offer is better than the cheap financing option (see Exercise 93).   ⬭

An interesting application of the amortization formula occurs when someone wins a major lottery prize. Chances are that it will never happen to you or anyone else you know, but just in case, let's look at it. (Remember the promise in the introduction—this chapter aims to make you rich, and if that fails, at least to make you financially smart. There is no reason you can't be both.)

If you were to win a major lottery jackpot, you would immediately face a critical decision: Take a single lump-sum payment up front for about 50% to 60% of the jackpot (the *lump-sum option*), or take the full jackpot but paid out in 25 equal annual payments (the *annuity* option). Most people opt for the "instant gratification" approach and choose the lump-sum option without giving any consideration to the financial value of the annuity option. But is the lump-sum option always the right choice? Many factors that come into play here (life expectancy, family and friends, current financial situation, etc.), but at least from a purely mathematical point of view, we can answer this question by comparing the lump-sum amount to the present value of the annuity.

Our next example represents an imaginary situation, but the numbers involved are taken from a real lottery.

■ In some lotteries the annual payments are graduated so that each year you get a little more than the previous year, which makes the analysis a bit more complicated.

(**EXAMPLE 10.26**)   **The Lottery Winner's Dilemma**

Imagine the unimaginable—you own the only winning ticket to a $9 million lottery jackpot. After taxes are taken out, your winnings are $6.8 million, paid out in 25 annual installments of $272,000 a year (the *annuity option*). You can also choose to take your winnings as a single, tax-free lump-sum payment of $3.75 million (the *lump sum option*). From a purely financial point of view, which of these options is better?

You can answer this question by thinking of the annuity option as an installment loan in which you are the lender getting monthly payments for the money you gave up. You can then compute the *present value* of this installment loan using the amortization formula and compare this present value with the present value of the lump-sum option, which is $3.75 million.

Before we continue, two important observations are in order. First, since payments are made annually, the annual interest rate equals the periodic interest rate. Second, when a lottery prize is paid off under the annuity option, the first installment check is handed out immediately, and all future installments are made at the beginning of the year. So, the version of the amortization formula that applies in this case is

$$P = F\left[\frac{q^T - 1}{q - 1}\right]$$

In this particular example, $F = \$272{,}000$ and $T = 25$. All we need now is the annual interest rate.

To determine a reasonable annual interest rate to use in this calculation, we need to make some basic assumptions. A reasonable assumption is that for a safe investment made over 25 years one should expect an annual rate of return in the range between 4% and 6%, with 4% representing a very conservative approach and 6% representing the less conservative end of the spectrum. To get a good picture of the situation, let's consider five separate cases using annual rates of return of 4%, 4.5%, 5%, 5.5%, and 6%, respectively.

Case 1. APR = 4% ($p = 0.04$). In this case the present value of the annuity is

$$P = \$272{,}000\left[\frac{\left(\frac{1}{1.04}\right)^{25} - 1}{\left(\frac{1}{1.04}\right) - 1}\right] = \$4{,}419{,}170$$

Case 2. APR = 4.5% ($p = 0.045$). In this case the present value of the annuity is

$$P = \$272{,}000\left[\frac{\left(\frac{1}{1.045}\right)^{25} - 1}{\left(\frac{1}{1.045}\right) - 1}\right] = \$4{,}214{,}770$$

Case 3. APR = 5% ($p = 0.05$). In this case the present value of the annuity is

$$P = \$272{,}000\left[\frac{\left(\frac{1}{1.05}\right)^{25} - 1}{\left(\frac{1}{1.05}\right) - 1}\right] = \$4{,}025{,}230$$

Case 4. APR = 5.5% ($p = 0.055$). In this case the present value of the annuity is

$$P = \$272{,}000\left[\frac{\left(\frac{1}{1.055}\right)^{25} - 1}{\left(\frac{1}{1.055}\right) - 1}\right] = \$3{,}849{,}260$$

Case 5. APR $= 6\%$ ($p = 0.06$). In this case the present value of the annuity is

$$P = \$272{,}000 \left[\frac{\left(\dfrac{1}{1.06}\right)^{25} - 1}{\left(\dfrac{1}{1.06}\right) - 1} \right] = \$3{,}685{,}700$$

The bottom line is that under the more conservative assumptions (cases 1, 2, and 3), the annuity option appears to be significantly better than the lump-sum option. For the less conservative assumptions (cases 4 and 5), the lump-sum option is about the same or slightly better than the annuity option. This analysis gives us a good picture of the mathematical part of the story. There are, however, many nonmathematical reasons people tend to choose the lump-sum option over the annuity—control of all the money, the ability to spend as much as we want whenever we want, and the realistic observation that we may not be around long enough to collect on the annuity.

Our final example for this chapter is about refinancing a home mortgage loan, a decision almost all homeowners face at one time or another: Should I take that "great" deal being offered to refinance my home at a lower APR? Is it worth it if I have to pay refinancing costs? If you own a home, you probably have received some of these refinancing offers. If you don't, you no doubt know someone who has. In any case, once you face one of these situations, you are trapped—no matter what you do, you will be making a decision and a big one at that! If you ignore the offer and take no action, you could be passing up an opportunity to save a lot of money over the life of the loan; if you take the offer and it is a bad offer, you could be on your way to financial ruin. These are depressing thoughts, except now we have all the mathematical tools to allow us to make the right decision.

(**EXAMPLE 10.27**)   **To Refinance or Not to Refinance: A Homeowner's Dilemma**

Imagine that you are a homeowner and have just received an offer in the mail to refinance your home loan. The offer is for a 6% APR on a 30-year mortgage. (As with all mortgage loans, the interest is compounded monthly.) In addition, there are loan origination costs: $1500 closing costs plus 1 *point* (1% of the amount of the loan). You are trying to decide if this offer is worth pursuing by comparing it with your current mortgage—a 30-year mortgage for $180,000 with a 6.75% APR. Obviously, a 6% APR is a lot better than a 6.75% APR, but do the savings justify your up-front expenses for taking out the new loan? Besides, you have made 30 monthly payments on your current loan already (you have lived in the house for $2\frac{1}{2}$ years). Will all these payments be wasted?

For a fair comparison between the two options (take out a new loan or keep the current loan), we will compare the monthly payments on your current loan with the monthly payments you would be making if you took out the new loan *for the balance of what you owe on your current loan.* (This way we are truly comparing apples with apples.) We can then determine if the monthly savings on your payments justify the up-front loan origination costs of the new loan. The computation will involve several steps, but fortunately for us, each step is based on an application of the amortization formula.

- **Step 1.** Compute the monthly payment F on the current 30-year mortgage with the 6.75% APR. Here we use the amortization formula with $P = \$180{,}000$, $T = 360$, and $p = 0.0675/12 = 0.005625$.

$$\$180{,}000 = \frac{F}{1.005625}\left[\frac{\left(\dfrac{1}{1.005625}\right)^{360} - 1}{\left(\dfrac{1}{1.005625}\right) - 1}\right]$$

Solving for F (you should verify the computation) gives

$$F = \$1167.48 \quad \text{(rounded to the nearest penny)}$$

- **Step 2.** Compute the balance on your current mortgage after having made the first 30 monthly payments. (This is the present value P of 330 monthly payments of $1167.48.)

$$P = \frac{\$1167.48}{1.005625}\left[\frac{\left(\dfrac{1}{1.005625}\right)^{330} - 1}{\left(\dfrac{1}{1.005625}\right) - 1}\right] = \$174{,}951 \quad \text{(rounded to the nearest dollar)}$$

- **Step 3.** Compute the monthly payment F^* on a new mortgage for $174,951 with a 6% APR. Here $P = \$174{,}951$, $T = 330$ (the number of payments left on the current loan), and $p = 0.06/12 = 0.005$.

$$\$174{,}951 = \frac{F^*}{1.005}\left[\frac{\left(\dfrac{1}{1.005}\right)^{330} - 1}{\left(\dfrac{1}{1.005}\right) - 1}\right]$$

Solving for F^* gives

$$\$F^* = \$1083.75 \quad \text{(rounded to the nearest penny)}$$

At this point, we know that if you refinance and take out a comparable loan (i.e., the new loan picks up exactly where you were on the original loan), you would be saving $1167.48 - \$1083.75 = \$83.73$ a month. How much are these monthly savings worth over time? This is an immediate annuity question, essentially the same as asking for the present value of a series of monthly payments of $83.73. But how many payments are we talking about? And at what APR? The answer to these two questions will determine the present value of your monthly savings.

First, let's assume that you will be keeping the new loan for the full 330 months over which it is amortized ($T = 330$), and let's further assume an APR of 6%. Under these assumptions, the computation of the present value of your savings is given in Step 4a.

■ The APR for this calculation does not have to match the APR on the loan. It represents the return that you might get if you were investing the money, and 6% happens to be a reasonable assumption.

- **Step 4a.** Compute the present value P^* of 330 monthly payments of $F = \$83.73$ at an APR of 6% ($p = 0.005$).

$$P^* = \frac{\$83.73}{1.005}\left[\frac{\left(\dfrac{1}{1.005}\right)^{330} - 1}{\left(\dfrac{1}{1.005}\right) - 1}\right] = \$13{,}516.70$$

Now the up-front loan origination costs are about $3250: $1500 for closing costs plus 1 point (1% of $174,951 is $1749.51, but for simplicity we'll call it $1750). Thus, the loan origination costs of $3250 are more than offset by the long-term savings of refinancing. If you think you would keep the new loan for the long haul, by all means you should refinance! The benefit ($13,516.70) by far outweighs the cost ($3250).

If you are getting a good offer now, though, you might be getting an even better offer in one, two, or three years from now. If you think that you might be refinancing again in the near future, the present value of your monthly savings (were you to refinance now) is considerably less. For example, let's imagine that you refinance again in 24 months and assume as before an APR of 6%. Under these assumptions, the present value of refinancing now is given in Step 4b.

- **Step 4b.** Present value P^{**} of 24 monthly payments of $F = \$83.73$ at an APR of 6% ($p = 0.005$).

$$P^{**} = \frac{\$83.73}{1.005}\left[\frac{\left(\frac{1}{1.005}\right)^{24} - 1}{\left(\frac{1}{1.005}\right) - 1}\right] = \$1889.19$$

Here is a case in which your refinancing savings are much less than your up-front loan origination costs. In this case, refinancing is a bad idea.

So, what should you do? Refinance or not refinance? As we have seen from the preceding discussion, for a definitive answer you would need a crystal ball. If you think that an even better refinancing offer might be coming your way within the next few years, you should not refinance. On the other hand, if you think that once you refinance you are likely to keep the new loan for a long time, you should definitely do so.

▢

◼ Using the amortization formula and logarithms, one can show that it would take 44 months before the benefit of refinancing exceeds the $3250 in loan origination costs. See Exercise 94.

CONCLUSION

> ❝ Waste neither time nor money, but make the best use of both. ❞
>
> —Ben Franklin

Whether it's finding a good deal on an iPod, saving for a vacation, or figuring out the financing of a new car, some of the most confounding decisions we make as independent adults are about money: how to make it first, how to spend it next, and, as we get more mature, how to manage it. Good money management involves many intangible things—luck, timing, common sense, self-discipline, and so forth—and one very tangible thing—understanding the basics of financial mathematics.

This chapter started with a discussion of percentages. In everyday life percentages offer a convenient way to understand ratios and proportions (as in "I got a 76% on the quiz") as well as increases and decreases (as in "We are offering a 30% discount on all Rolling Stone CDs"). In financial mathematics percentages are the standard way to describe profits and losses as well as *interest rates* paid on investments or charged on loans.

Given the introductory nature of this chapter, we focused on the most basic types of financial transactions: investments and loans (which are essentially flip sides of the same coin). Whether an investment or a loan, the basic principle behind the transaction is the same: one party (the investor/lender) gives *money* to another party (the investee/borrower), with the hope that over *time* the investment/loan turns a *profit*. The three key words in the previous sentence (money, time, and profit) represent the three key ingredients in all the formulas we discussed in this chapter: (1) the sum of money invested, (2) the length of the investment, and (3) the interest expected by the lender or investor as a reward for the willingness to temporarily part with money. Because of interest, money has a *present value* and a *future value*, and understanding the mathematical relation between these values was the underlying theme of the chapter.

With some investments (particularly some government bonds), the interest paid on the investment is *simple interest*. We saw that under simple interest only the original investment (called the *principal*) generates interest. This means that the money in an account that pays simple interest grows by the same amount (the interest generated by the principal) every year. The relation between the present and future values of an investment under simple interest was given by the *simple interest formula*.

A better (for the investor) type of interest is *compound interest*, in which not only the principal but also the interest generates additional interest. Compound interest calculations depend not only on the annual interest rate (APR) but also on the frequency of compounding (the number of times a year the interest is compounded). These two variables are combined into a single variable called the *periodic interest rate*, and using the periodic interest rate we derived the *general compounding formula*. For the special case that interest is compounded continuously, we introduced (without giving the details, which require some knowledge of calculus) the *continuous compounding formula*. We also combined the APR and the frequency of compounding to get the *annual percentage yield* (APY), which represents the *effective* annual interest rate of an investment.

In the last part of this chapter we discussed *fixed annuities*—investments in which equal payments are made or received over regular time intervals. With annuities, two mathematical formulas come into play—the *general compounding formula* and the *geometric sum formula*. (For the purposes of understanding the geometric sum formula, we took a brief detour and introduced *geometric sequences*.) The geometric sum formula was the main tool we then used to derive formulas for the *future value of a deferred annuity* (when regular payments are made to receive a lump-sum payoff at some future date) as well as for the *present value of an installment loan* (when a lump sum is paid up front and payments are made over time to amortize it). This latter formula, called the *amortization formula*, has many important applications, and it is the formula that allows us to calculate payments on a car loan or a home mortgage.

The material covered in this chapter represents a good start toward helping you become a good manager of your money and make educated financial decisions. Although there is always going to be an element of uncertainty in most financial decisions, the mathematics you learned here should allow you to understand the nature of this uncertainty so that you can make the most informed possible choice. This, and a little luck, might indeed make you rich.

KEY CONCEPTS

EXERCISES

WALKING

A Percentages

1. Write each of the following percentages as a decimal.
   (a) 2.25%
   (b) 0.75%

2. Express each of the following percentages as a decimal.
   (a) 8.75%
   (b) 0.8%

3. Express each of the following percentages as a simplified fraction.
   (a) 2.25%
   (b) 0.75%

4. Express each of the following percentages as a simplified fraction.
   (a) 8.75%
   (b) 0.8%

5. Your score in the midterm exam was 60 out of 75. Express your score as a percentage.

6. Out of a group of 320 students who took a Calculus Readiness Test, 248 passed. What percentage of the students passed the test?

7. A 250-piece puzzle is missing 14% of its pieces from its box. How many pieces are in the box?

8. Jefferson Elementary School has 750 students. The Friday before spring break 22% of the student body was absent. How many students were in school that day?

9. At the Happyville Mall, you buy a pair of earrings that are marked $6.95. After sales tax, the bill was $7.61. What is the tax rate in Happyville (to the nearest tenth of a percent)?

10. Arvin's tuition bill for last semester was $5760. If he paid $6048 in tuition this semester, what was the percentage increase on his tuition?

11. Total health care spending in the United States during 2005 was $2 trillion, representing approximately 16% of the gross domestic product. Estimate the gross domestic product of the United States in 2005 to the nearest billion dollars.

12. There are 25% more girls than boys in Ms. Jones's class. The girls represent what fraction of the class?

13. For three consecutive years the tuition at Tasmania State University increased by 10%, 15%, and 10%, respectively. What was the overall percentage increase of tuition during the three-year period?

14. For three consecutive years the cost of gasoline increased by 8%, 15%, and 20%, respectively. What was the overall percentage increase of the cost of gasoline during the three-year period?

15. A shoe store marks up the price of its shoes at 120% over cost. A pair of shoes goes on sale for 20% off and then goes on the clearance rack for an additional 30% off. A customer walks in with a 10% off coupon good on all clearance items and buys the shoes. Express the store's profit on these shoes as a percentage of the original cost.

16. Home values in Middletown have declined 7% per year for each of the past four years. What was the total percentage decrease in home values during the four-year period? Round your answer to the nearest tenth of a percentage point.

17. Over a period of one week, the Dow Jones Industrial Average (DJIA) did the following: On Monday the DJIA went up by 2.5%, on Tuesday it went up by 12.1%, on Wednesday it went down by 4.7%, on Thursday it went up by 0.8%, and on Friday it went down by 5.4%. What was the percentage increase/decrease of the DJIA over the week? Round your answer to the nearest tenth of a percentage point.

18. Over a period of one week, the Dow Jones Industrial Average (DJIA) did the following: On Monday the DJIA went down by 3.2%, on Tuesday it went up by 11.2%, on Wednesday it went up by 0.7%, on Thursday it went down by 4.8%, and on Friday it went down by 9.4%. What was the percentage increase/decrease of the DJIA over the week? Round your answer to the nearest tenth of a percentage point.

19. Warren Buffet, who paid an estimated $8,142,000 in federal taxes in 2006, recently commented that he pays 17.7% of his income in federal income tax, whereas his employees pay an average of 32.9% of their wages in federal income taxes.

   (a) Determine Buffet's 2006 income.

   (b) In 2006, the social security tax rate (paid only on the first $94,200 of income) was 6.2%. What percentage of Buffet's income was paid in Social Security taxes in 2006?

20. Willy's commission on selling new cars is a percentage of his sales total. His commission on used cars is 50% more than that on new cars. During the month of January, Willy sold new cars worth a total of $80,000 and used cars worth a total of $50,000. His commission check for the month was $3100.

(a) What percentage commission does Willy earn on used cars?

(b) What percentage commission does Willy earn on new cars?

B Simple Interest

21. Suppose that you buy an $875 savings bond that pays 4.28% annual simple interest. Determine the future value of the bond after

(a) four years.

(b) five years.

(c) ten years.

22. Suppose that you buy a $1250 savings bond that pays 5.1% annual simple interest. Determine the future value of the bond after

(a) three years.

(b) six years.

(c) twelve years.

23. Suppose that you buy a $10,000 bond that pays 5% annual simple interest. When you cash in the bond, you have to pay 15% federal taxes on the interest you earned. How much money would you net if you cash in the bond after five years? Round your answer to the nearest penny.

24. Suppose that you buy a $3000 savings bond that pays 4.5% annual simple interest. When you cash in the bond, you have to pay 18% federal taxes on the interest you earned. How much money would you net if you cash in the bond after four years? Round your answer to the nearest penny.

25. In four years you plan to use $6000 for a down payment on a cherry red Mustang. To accomplish this, you purchase a government bond that pays 5.75% annual simple interest each year. What is the present value of this bond (having a future value of $6000 in four years)?

26. Grandpa Jones purchased a bond paying 4% annual simple interest. It is now $45\frac{1}{2}$ years later, and the investment has grown to a value of $23,800. How much did Grandpa Jones pay for the bond? Round your answer to the nearest dollar.

27. A loan of $5400 collects simple interest each year for eight years. At the end of that time, a total of $8316 is paid back. Determine the APR for this loan.

28. A loan of $6000 collects simple interest each year for six years. At the end of that time, a total of $7620 is paid back. Determine the APR for this loan.

29. You would like to buy a bond earning simple annual interest that will double its value in 12 years. Determine the APR (to the nearest hundredth of a percent) that would be needed.

30. You would like to buy a bond earning simple annual interest that will triple its value in 20 years. Determine the APR (to the nearest hundredth of a percent) that would be needed.

C Compound Interest

31. Suppose that you deposit $3250 in a savings account that pays 9% annual interest, with interest credited to the account at the end of each year. Assuming that no withdrawals are made, find the balance in the account after

(a) 4 years.

(b) 5 years and 7 months.

32. Suppose that you deposit $1237.50 in a savings account that pays 8.25% annual interest, with interest credited to the account at the end of each year. Assuming that no withdrawals are made, find the balance in the account after

(a) 3 years.

(b) $4\frac{1}{2}$ years.

33. Suppose that you deposited $3420 on January 1, 2008, in an account paying $6\frac{5}{8}$% annual interest, with interest credited to the account on December 31 of each year. On January 1, 2011, the interest rate drops to $5\frac{3}{4}$%. What will be the balance in your account on January 1, 2015?

34. Suppose that you deposited $2500 on January 1, 2008, in an account paying $5\frac{3}{8}$% annual interest, with interest credited to the account on December 31 of each year. On January 1, 2012, the interest rate drops to $4\frac{3}{4}$%. What will be the balance in your account on January 1, 2015?

35. Suppose that you deposited $3420 on January 1, 2005, in a savings account paying $6\frac{5}{8}$% annual interest, with interest credited to the account on December 31 of each year. On January 1, 2007, you withdrew $1500, and on January 1, 2008, you withdrew $1000. If you make no other withdrawals, what will be the balance in your account on January 1, 2011?

36. Suppose that you deposited $2500 on January 1, 2005, in a savings account paying $5\frac{3}{8}$% annual interest, with interest credited to the account on December 31 of each year. On January 1, 2008, you withdrew $850. If you make no other withdrawals, what will be the balance in your account on January 1, 2013?

37. Suppose that $5000 is invested in an account with an APR of 12% compounded monthly.

  (a) Find the future value of the account in 5 years.

  (b) Find the APY for this investment.

38. Suppose that $874.83 is invested in an account with an APR of 7.75% compounded daily.

  (a) Find the future value of the account in 2 years.

  (b) Find the APY for this investment.

39. Suppose that $875 is invested in an account with an APR of 6.75%.

  (a) Find the future value of the account in $7\frac{1}{2}$ years if the interest is compounded hourly.

  (b) Find the APY for this investment.

40. Suppose that $875 is invested in an account with an APR of 6.75%.

  (a) Find the future value of the account in $7\frac{1}{2}$ years if the interest is compounded every minute.

  (b) Find the APY for this investment.

41. Which of the following options would you choose to invest your money in: a 6% APR compounded yearly, a 5.75% APR compounded monthly, or a 5.5% APR compounded continuously? Explain.

42. Which of the following options would you choose to invest your money in: a 9% APR compounded yearly, an 8.75% APR compounded monthly, or an 8.6% APR compounded continuously? Explain.

43. Fill in the entries in the last column of the table.

APR	Compounding	APY
12%	Yearly	
12%	Semiannually	
12%	Quarterly	
12%	Monthly	
12%	Daily	
12%	Hourly	
12%	Continuously	

44. Fill in the entries in the last column of the table.

APR	Compounding	APY
8%	Yearly	
8%	Semiannually	
8%	Quarterly	
8%	Monthly	
8%	Daily	
8%	Hourly	
8%	Continuously	

45. Your bank is offering a special promotion for its preferred customers. If you buy a $750 certificate of deposit, at the end of the year you can cash in the CD for $810. What is the APY of this investment?

46. Your bank is offering a special promotion for its preferred customers. If you buy an $800 certificate of deposit, at the end of the year you can cash in the CD for $855. What is the APY of this investment?

47. In an investment that has a 6% APY, how many years will it take for you to at least double your money

  (a) if you invest $1000?

  (b) if you invest P?

48. In an investment that has a 6.75% APY, how many years will it take for you to at least double your money

  (a) if you invest $540?

  (b) if you invest P?

49. A sum of money is invested for three years in a CD paying an APY of 4.75%. At the end of the three years you get a check for $4060.17. How much was the original principal?

50. A sum of money is invested for three years in a CD paying an APY of 10%. At the end of the three years you get a check for $732.05. How much was the original principal?

D **Geometric Sequences and the Geometric Sum Formula**

51. A geometric sequence has initial term $G_0 = 11$ and common ratio $c = 1.25$.

  (a) Find G_1.

  (b) Find G_6.

  (c) Give an explicit formula for G_N.

52. A geometric sequence has initial term $G_0 = 1000$ and common ratio $c = 0.8$.

   (a) Find G_1.

   (b) Find G_5.

   (c) Give an explicit formula for G_N.

53. The first two terms of a geometric sequence are $G_0 = 3072$ and $G_1 = 2304$.

   (a) Find the common ratio c.

   (b) Find G_4.

   (c) Give an explicit formula for G_N.

54. The first two terms of a geometric sequence are $G_0 = 8$ and $G_1 = 12$.

   (a) Find the common ratio c.

   (b) Find G_5.

   (c) Give an explicit formula for G_N.

55. Crime in Happyville is on the rise. Each year the number of crimes committed increases by 50%. Assume that there were 256 crimes committed in 2000, and let G_N denote the number of crimes committed in the year $2000 + N$.

   (a) How many crimes were committed in 2005?

   (b) Give an explicit formula for G_N.

   (c) If the trend continues, approximately how many crimes will be committed in Happyville in 2010?

56. Since 2000, when 100,000 cases were reported, the number of reported new cases each year of the *ZZZ*-virus has decreased by 20%. Let G_N denote the number of reported new cases of the *ZZZ*-virus in the year $2000 + N$.

   (a) How many new cases of the *ZZZ*-virus occurred in 2005?

   (b) Give an explicit formula for G_N.

   (c) If the trend continues, approximately how many new cases of the *ZZZ*-virus will be reported in 2015?

57. Consider the geometric sequence with initial term $G_0 = 3$ and common ratio $c = 2$.

   (a) Find G_{100}.

   (b) Give an explicit formula for G_N.

   (c) Use the geometric sum formula to compute $G_0 + G_1 + \cdots + G_{100}$. (*Hint:* There are 101 terms in this sum.)

   (d) Use the geometric sum formula to compute $G_{50} + G_{51} + \cdots + G_{100}$. (*Hint:* This is a sum of terms in a geometric sequence with initial term G_{50}.)

58. Consider the geometric sequence with first two terms $G_0 = 2.5$ and $G_1 = 10$.

   (a) Find G_{20}.

   (b) Give an explicit formula for G_N.

   (c) Use the geometric sum formula to compute $G_0 + G_1 + \cdots + G_{20}$. (*Hint:* There are 21 terms in this sum.)

   (d) Use the geometric sum formula to compute $G_8 + G_9 + \cdots + G_{20}$. (*Hint:* This is a sum of terms in a geometric sequence with initial term G_8.)

59. Use the geometric sum formula to compute $\$10(1.05) + \$10(1.05)^2 + \$10(1.05)^3 + \cdots + \$10(1.05)^{36}$.

60. Use the geometric sum formula to compute $\$500(1.075) + \$500(1.075)^2 + \$500(1.075)^3 + \cdots + \$500(1.075)^{60}$.

61. Use the geometric sum formula to compute $\dfrac{\$10}{(1.05)} + \dfrac{\$10}{(1.05)^2} + \dfrac{\$10}{(1.05)^3} + \cdots + \dfrac{\$10}{(1.05)^{36}}$.

62. Use the geometric sum formula to compute $\dfrac{\$500}{(1.075)} + \dfrac{\$500}{(1.075)^2} + \dfrac{\$500}{(1.075)^3} + \cdots + \dfrac{\$500}{(1.075)^{60}}$.

E Deferred Annuities

63. Starting at age 25, Markus invests $2000 at the *beginning* of each year in an IRA (individual retirement account) with an APR of 7.5% compounded annually. How much money will there be in Markus's retirement account when he retires at the age of 65? (Assume that his 40th and last deposit generates interest for one year.)

64. Suppose that you save $10 a month and deposit the money at the *beginning* of each month in an IRA with an APR of 8% compounded monthly. How much money will there be in your retirement account in 30 years? (Assume the last deposit generates interest for one month.)

65. Starting at age 25, Markus invests $2000 at the *end* of each year in an IRA with an APR of 7.5% compounded annually. How much money is in Markus's retirement account when he retires at the age of 65? (Assume that his 40th and last deposit generates no interest.)

66. Suppose that you save $10 a month and put the money at the *end* of each month in an IRA with an APR of 8% compounded monthly. How much money will there be in your retirement account in 30 years? (Assume that the last deposit generates no interest.)

67. You decide to open a Christmas Club account at a bank that pays 6% annual interest compounded monthly.

You deposit $100 on the first of January and on the first of each succeeding month through November. How much will you have in your account on the first of December?

68. You decide to save money to buy a car by opening a special account at a bank that offers an 8% APR compounded monthly. You deposit $300 on the first of each month for 36 months. How much will you have in your account at the end of the 36th month?

69. Celine deposits $400 at the end of each month into an account that returns 4.5% annual interest (compounded monthly). At the end of three years she wants to take the money in the account and use it for a 20% down payment on a new home. What is the maximum price of a home that Celine will be able to buy?

70. Donald would like to retire with a $1 million nest egg. He plans to put money at the end of each month into an account earning 6% annual interest compounded monthly. If Donald plans to retire in 35 years, how much does he need to put away each month?

71. Layla plans to send her daughter to Tasmania State University in 12 years. Her goal is to create a college trust fund worth $150,000 in 12 years. If she plans to make weekly deposits into a trust fund earning 4.68% annual interest compounded weekly, how much should her weekly deposits to the trust fund be? (Assume that the deposits are made at the start of each week.)

72. Freddy just remembered that he had deposited money in a savings account earning 7% annual interest at the Middletown bank 15 years ago, but he forgot exactly how much. Today he closed the account and the bank gave him a check for $1172.59. How much was Freddy's initial deposit?

F ▮ **Installment Loans**

73. Find the present value of an installment loan consisting of 25 annual payments of $3000 assuming

(a) an APR of 6% compounded annually. (Assume that the payments are made at the end of each year.)

(b) an APR of 7% compounded annually. (Assume that the payments are made at the end of each year.)

74. Find the present value of a 60-month installment loan consisting of 60 monthly payments of $16 assuming

(a) an APR of 6% compounded monthly. (Assume that the payments are made at the end of each month.)

(b) an annual interest rate of 4% compounded monthly. (Assume that the payments are made at the end of each month.)

75. Ned plans to donate $50 per week to his church for the next 60 years. Assuming an annual interest rate of 5.2% compounded weekly, find the present value of this annuity to Ned's church.

76. Michael Dell, founder of Dell computers, was reputed to have a net worth of $16 billion in 2005. If he were to take the $16 billion and set up a 40-year annuity for himself, what would his yearly payment from the annuity be? Assume a 3% annual rate of return on his money.

77. The Simpsons are planning to purchase a new home. To do so, they will need to take out a 30-year home mortgage loan of $160,000 through Middletown Bank. Annual interest rates for 30-year mortgages at Middletown Bank are 5.75% compounded monthly.

(a) Compute the Simpsons' monthly mortgage payment under this loan.

(b) How much interest will the Simpsons pay over the life of the loan?

78. The Smiths are refinancing their home mortgage to a 15-year loan at 5.25% annual interest compounded monthly. Their outstanding balance on the loan is $95,000.

(a) Under their current loan, the Smith's monthly mortgage payment is $1104. How much will the Smiths be saving in their monthly mortgage payments by refinancing? (Round your answer to the nearest dollar.)

(b) How much interest will the Smiths pay over the life of the new loan?

79. Ken just bought a house. He made a $25,000 down payment and financed the balance with a 20-year home mortgage loan with an interest rate of 5.5% compounded monthly. His monthly mortgage payment is $950. What was the selling price of the house?

80. Cari just bought a house. She made a $35,000 down payment and financed the balance with a 30-year home mortgage loan with an interest rate of 5.75% compounded monthly. Her monthly mortgage payment is $877. What was the selling price of the house?

JOGGING

81. How much should a retailer mark up her goods so that when she has a 25% off sale, the resulting prices will still reflect a 50% markup (on her cost)?

82. Joe, a math major, calculates that in the last three years tuition at Tasmania State University has increased a total of exactly 12.4864%. He also knows that tuition increased by the same percentage each year. Determine this percentage.

83. You have a coupon worth $x\%$ off any item (including sale items) in a store. The particular item you want is on sale at $y\%$ off the marked price of $\$P$. (Assume that both x and y are positive integers smaller than 100).

(a) Give an expression for the price of the item assuming that you first got the $y\%$ off sale price and then had the additional $x\%$ taken off using your coupon.

(b) Give an expression for the price of the item assuming that you first got the $x\%$ off the original price using your coupon and then had the $y\%$ taken off from the sale.

(c) Explain why it makes no difference in which order you have the discounts taken.

84. Zero coupon bonds. A zero coupon bond is a bond that is sold now at a discount and will pay its face value at some time in the future when it matures. Suppose that a zero coupon bond matures to a value of $\$10,000$ in seven years. If the bond earns 3.5% annual interest, what is the purchase price of the bond?

85. On September 19, 1997, CNN reported the following: "CNN founder and Time Warner vice chairman Ted Turner announced Thursday night that he will donate $\$1$ billion over the next decade to United Nations programs. The donation will be made in 10 annual installments of $\$100$ million in Time Warner stock, he said. 'Present-day value that's about $\$600,000$,' he joked." Find the present value of this immediate annuity to the United Nations assuming an annual interest rate of 10.5%.

86. You buy a $\$500$ certificate of deposit (CD), and at the end of two years you cash it for $\$561.80$. What is the annual yield of this investment?

87. Explain why the present value of an immediate annuity goes down as the interest rate goes up. (*Hint*: Try Exercise 73 first.)

88. The geometric sum formula. Verify that $P + cP + c^2P + \cdots + c^{N-1}P = P\left(\dfrac{c^N - 1}{c - 1}\right)$.

[*Hint*: Multiply both sides by $(c - 1)$ and simplify the expression you get on the left-hand side. There will be a lot of cancellations, and it will simplify nicely.]

89. Justify each of the following statements.

(a) $P(1 + p) + P(1 + p)^2 + \cdots + P(1 + p)^T =$

$$P(1 + p)\left[\frac{(1 + p)^T - 1}{p}\right]$$

(b) $P(1 + p)\left[\dfrac{(1 + p)^T - 1}{p}\right] =$

$$P\left(1 + \frac{1}{p}\right)[(1 + p)^T - 1]$$

90. Justify each of the following statements.

(a) $\dfrac{F}{(1 + p)} + \dfrac{F}{(1 + p)^2} + \cdots + \dfrac{F}{(1 + p)^T} =$

$$\frac{F}{(1 + p)}\left[\frac{\left(\dfrac{1}{1 + p}\right)^T - 1}{\left(\dfrac{1}{1 + p}\right) - 1}\right]$$

(b) $\dfrac{F}{(1 + p)} + \dfrac{F}{(1 + p)^2} + \cdots + \dfrac{F}{(1 + p)^T} =$

$$\frac{F}{(1 + p)^T}\left[\frac{(1 + p)^T - 1}{p}\right]$$

(c) $\dfrac{F}{(1 + p)^T}\left[\dfrac{(1 + p)^T - 1}{p}\right] = F\dfrac{\left[1 - \left(\dfrac{1}{1 + p}\right)^T\right]}{p}$

RUNNING

91. You are purchasing a home for $\$120,000$ and are shopping for a loan. You have a total of $\$31,000$ to put down, including the closing costs of $\$1000$ and any loan fee that might be charged. Bank A offers a 10% APR amortized over 30 years with 360 equal monthly payments. There is no loan fee. Bank B offers a 9.5% APR amortized over 30 years with 360 equal monthly payments. There is a 3% loan fee (i.e., a one-time up-front charge of 3% of the loan). Which loan is better?

92. A friend of yours sells his car to a college student and takes a personal note (cosigned by the student's rich uncle) for $\$1200$ with no interest, payable at $\$100$ per month for 12 months. Your friend immediately approaches you and offers to sell you this note. How much should you pay for the note if you want an annual yield of 12% on your investment?

93. You want to purchase a new car. The price of the car is $\$24,035$. The dealer is currently offering a special promotion: You can choose a $\$1500$ rebate up front or 0% financing for the first 36 months and 6% financing for the remaining 24 months of your loan. Which is the better deal, the $\$1500$ rebate or the 0% financing for the first three years followed by 6% financing for the last two years of the loan? Justify your answer by computing your monthly payments over 60 months under each of the two options.

94. What is the smallest positive integer T for which the inequality is true? (*Hint*: Solve for T using logarithms.)

$$\frac{\$83.73}{1.005}\left[\frac{\left(\dfrac{1}{1.005}\right)^T - 1}{\left(\dfrac{1}{1.005}\right) - 1}\right] \geq 3250$$

95. According to a *Philadelphia Inquirer* finance column by Jeff Brown on November 1, 2005, "Borrow $100,000 with a 6% fixed-rate mortgage and you'll pay nearly $116,000 in interest over 30 years. Put an extra $100 a month into principal payments and you'd pay just $76,000—and be done with mortgage payments nine years earlier."

(a) Verify that the increase in the monthly payment that is needed to pay off the mortgage in 21 years is indeed close to $100 and that roughly $40,000 will be saved in interest.

(b) How much should the monthly payment be increased so as to pay off the mortgage in 15 years? How much interest is saved in doing so?

(c) How much should the monthly payment be increased so as to pay off the mortgage in *t* years (*t* < 30)? Express the answer in terms of *t*.

96. Sam started his new job as the mathematical consultant for the XYZ Corporation on July 1, 2005. The company retirement plan works as follows: On July 1, 2006, the company deposits $1000 in Sam's retirement account, and each year thereafter on July 1 the company deposits the amount deposited the previous year plus an additional 6%. The last deposit is to be made on July 1, 2035. In addition, the retirement account earns an annual interest rate of 6% compounded monthly. On July 1, 2035, all deposits and interest paid to the account will stop. What is the future value of Sam's retirement account?

97. Perpetuities. A perpetuity is a constant stream of identical annual payments with no end. The present value of a perpetuity of $C, given an annual interest rate *p* (expressed as a decimal), is given by the formula

$$P = C \cdot \frac{1}{(1 + p)} + C \cdot \frac{1}{(1 + p)^2} + C \cdot \frac{1}{(1 + p)^3} + \cdots$$

(a) Use the geometric sum formula to simplify the sum

$$C \cdot \frac{1}{(1 + p)} + C \cdot \frac{1}{(1 + p)^2} + C \cdot \frac{1}{(1 + p)^3} + \cdots$$
$$+ C \cdot \frac{1}{(1 + p)^T}$$

(b) Explain why as *T* gets larger, the value obtained in (a) gets closer and closer to C/p. (Assume that $0 < p < 1$.)

PROJECTS AND PAPERS

A   The Many Faces of e

The irrational number *e* has many remarkable mathematical properties. In this project you are asked to present and discuss five of the more interesting mathematical properties of *e*. For each property, give a historical background (if possible), a simple mathematical explanation (avoid technical details if you can), and a real-life application (see, for example, reference 3).

B   Growing Annuities

Over time, the fixed payment from an annuity (such as a retirement account) will get you fewer and fewer goods and services. If prices rise 3% a year, items that cost $1000 today will cost more than $1300 in 10 years and more than $1800 in 20 years. To combat this phenomenon, *growing* annuities—in which the payments rise by a fixed percentage, say 3%, each year—have been developed. In this project you are to discuss the mathematics behind growing annuities. (See references 2 and 5.)

C   A Future Home Buyer's Must-Do Project

In this project, you will investigate the process of buying a house. For the purposes of this project, we will assume that you are going to graduate at the end of this semester, get a job, and begin saving for your first house. You will plan to buy your house five years after graduation. At that time, you will use the money you've saved as a down payment and take out a loan for the rest of the cost of the house.

As part of the project, you should address each of the following:

1. *Your projected salary for the first five years after you graduate.* Describe what kind of a job you expect to have and what starting monthly salary is realistic for that type of job. Project the salary over five years, including raises and cost-of-living increases. (Support all your salary assumptions with data and cite your sources.)

2. *Your savings and investments over the five-year period.* For the purposes of this project, assume that 10% of the monthly salary you projected in Step 1 will be invested in a deferred annuity. Research the types of deferred annuities available to you and describe the terms, fees, and APR. Use the amount of money you will be able to save each month, the interest rate you have found, and your knowledge of annuities to compute how much money you will have after five years. Don't forget to deduct any state and federal taxes you will be paying on your income from the annuities. Estimate your taxes on

income from the annuities at 20% federal and 7% state. (If you expect to have a job in a no-income-tax state, disregard the 7% state tax.)

3. *How much you can afford to spend on a home and what your monthly mortgage expenses will be.* To determine how much you can afford to spend on a home, use your estimate of how much you have saved for a down payment and assume that your down payment is 15% of the cost of the home. Research the types of home loans currently available

to someone with your credit rating and income (assume the income projected at the end of the five-year period). Describe the terms of a loan you might be able to take out, including the interest rate, type and length of loan, and loan origination fees. Once you found a loan, project your monthly mortgage payments (be sure to include property taxes and insurance). Once again, cite all the sources for your data and show all the formulas you are using for your calculations.

REFERENCES AND FURTHER READINGS

1. Capinski, Marek, and Tomasz Zastawniak, *Mathematics for Finance: An Introduction to Financial Engineering.* London, England: Springer-Verlag, 2003.

2. Davis, Morton, *The Math of Money.* New York: Springer-Verlag, 2001.

3. Fabozzi, Frank, *Fixed Income Mathematics.* New York: McGraw-Hill, 2006.

4. Hershey, Robert, *All the Math You Need to Get Rich: Thinking with Numbers for Financial Success.* Peru, IL: Open Court Publishing, 2002.

5. Kaminsky, Kenneth, *Financial Literacy: Introduction to the Mathematics of Interest, Annuities, and Insurance,* Lankham, MD: University Press of America, 2003.

6. Lovelock, David, Marilou Mendel, and A. Larry Wright, *An Introduction to the Mathematics of Money: Saving and Investing.* New York: Springer Science & Business Media, 2007.

7. Maor, Eli, *e: The Story of a Number.* Princeton, NJ: Princeton University Press, 1998.

8. Shapiro, David, and Thomas Streiff, *Annuities.* Chicago, IL: Dearborn Financial Publishing, 2001.

9. Taylor, Richard W., "Future Value of a Growing Annuity: A Note," *Journal of Financial Education,* Fall (1986), 17–21.

Mortgage Calculators on the Web

- http://www.amortization-calc.com/
- http://calculator-loan.info/
- http://www.mortgageloan.com/calculator/amortization-calculator
- http://www.vertex42.com/ExcelTemplates/excel-amortization-spreadsheet.html
- https://www.federalreserve.gov/apps/mortcalc/

11 The Mathematics of Symmetry

Beyond Reflection

It is said that Eskimos have dozens of different words for ice. Ice is, after all, a universal theme in the Eskimos' world. By the same token, one would expect our own vocabulary to have dozens of different words to describe the notion of symmetry — symmetry is as pervasive to our world as ice is to the Eskimos'. Surprisingly, just the opposite is the case. In everyday conversation we use a single word — *symmetry* — to cover an incredibly diverse set of situations and ideas.

E xactly what is symmetry? The answer depends very much on the context of the question. In everyday language, *symmetry* is often taken to mean *mirror* or *bilateral symmetry*, the left-right symmetry exhibited (at least on the outside) by the human body. But in everyday language the word *symmetry* also has an aesthetic connotation—a snowflake is often used as the shining example of symmetry because it is so well proportioned and beautiful. This is often how the word is used in art and architecture. Along similar lines, the word *symmetry* is used in music, poetry, and literature to describe particular elements of style. In all these contexts, symmetry is perceived as something good—even when we can't put our finger on exactly what it is.

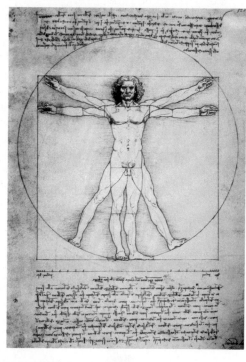

At its very heart, however, symmetry is a mathematical notion, and to grasp the full meaning and relevance of symmetry in our everyday world it is helpful to have some understanding of the mathematics behind it. In this chapter we will take a look at symmetry from a mathematical perspective. For technical reasons (i.e., to keep the theory reasonably simple) we will focus on symmetry in the context of *two-dimensional* objects and shapes. (Since we live in a three-dimensional world, the objects we discuss will be only idealized versions of reality.)

The chapter starts by introducing the concept of a *rigid motion* (Section 11.1), a fundamental idea behind the mathematical definition of symmetry. There are four basic types of rigid motions of two-dimensional objects living in two-dimensional space, and these are discussed in the next four sections — *reflections* in Section 11.2, *rotations* in Section 11.3, *translations* in Section 11.4, and *glide reflections* in Section 11.5. In Section 11.6 we use the rigid motions to formalize the mathematical interpretation of symmetry and to develop the concept of the *symmetry type* of a shape or an object. In the last section (Section 11.7) we introduce *border* and *wallpaper patterns* and discuss their classification in terms of symmetry types.

11.1   Rigid Motions

As are many other core concepts, symmetry is rather hard to define, and we will not even attempt a proper definition until Section 11.6. We will start our discussion with just an informal stab at the mathematical (or geometric if you prefer) interpretation of symmetry.

(EXAMPLE 11.1)   Symmetries of a Triangle

Figure 11-1 shows three triangles: (a) a scalene triangle (all three sides are different), (b) an isosceles triangle, and (c) an equilateral triangle. In terms of symmetry, how do these triangles differ? Which one is the most symmetric? Least symmetric? (What do you think?)

Even without a formal understanding of what symmetry is, most people would answer that the equilateral triangle in (c) is the most symmetric and the scalene triangle in (a) is the least symmetric. This is in fact correct, but why? Think of an imaginary observer—say a tiny (but very observant) ant—standing at the vertices of each of the triangles, looking toward the opposite side. In the case of the scalene triangle (a), the view from each vertex is different. In the case of the isosceles triangle (b), the view from vertices B and C is the same, but the view from vertex A is different. In the case of the equilateral triangle (c), the view is the same from each of the three vertices.

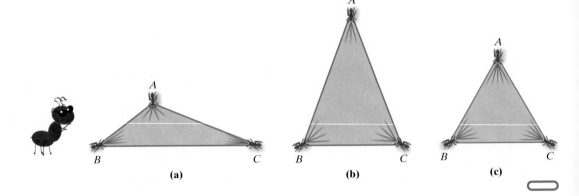

FIGURE 11-1   **(a)**   **(b)**   **(c)**

Let's say, for starters, that *symmetry* is a property of an object that looks the same to an observer standing at different vantage points. This is still pretty vague but a start nonetheless. Now instead of talking about an observer moving around to different vantage points think of the object itself moving—forget the observer. For example, saying that the isosceles triangle in Fig. 11-1(b) looks the same to an observer whether he or she stands at vertex B or vertex C is equivalent to saying that the triangle itself can be moved so that vertices B and C swap locations and the triangle as a whole looks exactly as it did before. And the equilateral triangle in Fig. 11-1(c) can be moved in even more ways so that after the move it looks exactly as it did before. Thus, we might think of symmetry as having to do with ways to move an object so that when all the moving is done, the object looks exactly as it did before.

Given the preceding observations, it is not surprising that to understand symmetry we need to understand the various ways in which we can "move" an object. This is our lead-in to what is a key concept in this chapter—the notion of a *rigid motion*.

The act of taking an object and moving it from some starting position to some ending position *without altering its shape or size* is called a **rigid motion** (and sometimes an *isometry*). If, in the process of moving the object, we stretch it, tear it, or generally alter its shape or size, the motion is *not* a rigid motion. Since in a rigid motion the size and shape of an object are not altered, distances between points are preserved: *The distance between any two points X and Y in the starting position is the same as the distance between the same two points in the ending position.* Figure 11-2 illustrates the difference between a motion that is rigid and one that is not. In (a), the motion does not change the shape of the object; only its position in space has changed. In (b), both position *and* shape have changed.

FIGURE 11-2 (a) A rigid motion preserves distances between points. (b) If the shape is altered, the motion is not rigid.

(a)　(b)

In defining rigid motions we are completely result oriented. We are only concerned with the *net effect* of the motion—where the object started and where the object ended. What happens during the "trip" is irrelevant. This implies that a rigid motion is completely defined by the starting and ending positions of the object being moved, and two rigid motions that move an object from the same starting position to the same ending position are **equivalent rigid motions**—never mind the details of how they go about it (Fig. 11-3).

FIGURE 11-3 Equivalent rigid motions (a) and (b) move an object from the same starting position to the same ending position.

(a)　(b)

Because rigid motions are defined strictly in terms of their net effect, there is a surprisingly small number of scenarios. In the case of two-dimensional objects in a plane, there are only four possibilities: A rigid motion is equivalent to (1) a *reflection*, (2) a *rotation*, (3) a *translation*, or (4) a *glide reflection*. We will call these four types of rigid motions the **basic rigid motions of the plane**.

A rigid motion of the plane—let's call it $\mathcal{M}$—moves each point in the plane from its starting position P to an ending position P', also in the plane. (From here on we will use script letters such as $\mathcal{M}$ and $\mathcal{N}$ to denote rigid motions, which should eliminate any possible confusion between the point M and the rigid motion $\mathcal{M}$.) We will call the point P' the **image** of the point P under the rigid motion $\mathcal{M}$ and describe this informally by saying that $\mathcal{M}$ *moves P to P'*. (We will also stick to the convention that the image point has the same label as the original point but with a prime symbol added.) It may happen that a point P is moved back to itself under $\mathcal{M}$, in which case we call P a **fixed point** of the rigid motion $\mathcal{M}$.

We will now discuss each of the four basic rigid motions of the plane in a little more detail.

■ For three-dimensional objects there are six basic rigid motions: *reflection, rotation, translation, glide reflection, rotary reflection* and *screw displacement* — see Project B.

11.2　Reflections

A **reflection** in the plane is a rigid motion that moves an object into a new position that is a mirror image of the starting position. In two dimensions, the "mirror" is a line called the **axis** of reflection.

From a purely geometric point of view a reflection can be defined by showing how it moves a generic point P in the plane. This is shown in Fig. 11-4: The image of any point P is found by drawing a line through P perpendicular to the axis l and finding the point P' on the opposite side of l at the same distance as P from l. Points on the axis itself are *fixed points* of the reflection [Fig. 11-4(c)].

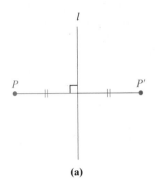

(a)   (b)   (c)

FIGURE 11-4

EXAMPLE 11.2   **Reflections of a Triangle**

Figure 11-5 shows three cases of reflection of a triangle ABC. In all cases the original triangle ABC is shaded in blue and the reflected triangle $A'B'C'$ is shaded in red. In (a), the axis of reflection l does not intersect the triangle ABC. In (b), the axis of reflection l cuts through the triangle ABC—here the points where l intersects the triangle are fixed points of the triangle. In (c), the reflected triangle $A'B'C'$ falls on top of the original triangle ABC. The vertex B is a fixed point of the triangle, but the vertices A and C swap positions under the reflection.

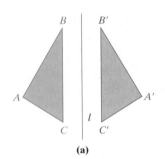
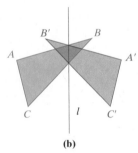
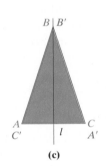

(a)   (b)   (c)

FIGURE 11-5

FIGURE 11-6 *Hand with Reflecting Sphere,* by M. C. Escher, 1935. (© 1935 M. C. Escher/ Cordon Art, Baarn, Holland.) "[A] spherical mirror, resting on a *left* hand. But as print is the reverse of the original stone drawing, it is my *right* hand that you see depicted."   — M. C. Escher

The following are simple but useful properties of a reflection.

Property 1 If we know the axis of reflection, we can find the image of any point P under the reflection (just drop a perpendicular to the axis through P and find the point on the other side of the axis that is at an equal distance). Essentially a reflection is completely determined by its axis l.

Property 2 If we know a point P and its image P' under the reflection (and assuming P' is different from P), we can find the axis l of the reflection (it is the perpendicular bisector of the segment PP'). Once we have the axis l of the reflection, we can find the image of any other point (property 1).

Property 3 The *fixed points* of a reflection are all the points on the axis of reflection l.

Property 4 Reflections are **improper** rigid motions, meaning that they change the *left-right* and *clockwise-counterclockwise* orientations of objects. This property is the reason a left hand reflected in a mirror looks like a right hand (Fig. 11-6) and the hands of a clock reflected in a mirror move counterclockwise.

Property 5 If P' is the image of P under a reflection, then $(P')' = P$ (the image of the image is the original point). Thus, when we apply the same reflection twice, every point ends up in its original position and the rigid motion is equivalent to *not having moved the object at all*.

A rigid motion that is equivalent to not moving the object at all is called the **identity** motion. At first blush it may seem somewhat silly to call the identity motion a motion (after all, nothing moves), but there are very good mathematical reasons to do so, and we will soon see how helpful this convention is for studying and classifying symmetries.

The basic properties of reflections can be summarized as follows.

PROPERTIES OF REFLECTIONS

- A reflection is completely determined by its axis *l*.
- A reflection is completely determined by a single point-image pair *P* and *P'* (as long as *P'* ≠ *P*).
- A reflection has infinitely many fixed points (all points on *l*).
- A reflection is an *improper* rigid motion.
- When the same reflection is applied twice, we get the *identity* motion.

11.3 Rotations

■ We will describe angle measures using degrees, but converting degrees to radians or radians to degrees is easy if you just remember that 180 degrees equals π radians.

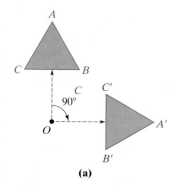

FIGURE 11-7

Informally, a **rotation** in the plane is a rigid motion that pivots or swings an object around a fixed point *O*. A rotation is defined by two pieces of information: (1) the **rotocenter** (the point *O* that acts as the center of the rotation) and (2) the **angle of rotation** (actually the *measure* of an angle indicating the amount of rotation). In addition, it is necessary to specify the direction (clockwise or counterclockwise) associated with the angle of rotation.

Figure 11-7 illustrates geometrically how a *clockwise* rotation with rotocenter *O* and angle of rotation α moves a point *P* to the point *P'*.

(**EXAMPLE 11.3**) **Rotations of a Triangle**

Figure 11-8 illustrates three cases of rotation of a triangle *ABC*. In all cases the original triangle *ABC* is shaded in blue and the reflected triangle *A'B'C'* is shaded in red. In (a), the rotocenter *O* lies outside the triangle *ABC*. The 90° clockwise rotation moved the triangle from the "12 o'clock position" to the "3 o'clock position." (Note that a 90° counterclockwise rotation would have moved the triangle *ABC* to the "9 o'clock position.") In (b), the rotocenter *O* is at the center of the triangle *ABC*. The 180° rotation turns the triangle "upside down." For obvious reasons, a 180° rotation is often called a *half-turn*. (With half-turns the result is the same whether we rotate clockwise or counterclockwise, so it is unnecessary to specify a direction.) In (c), the 360° rotation moves every point back to its original position—from the rigid motion point of view it's as if the triangle had not moved.

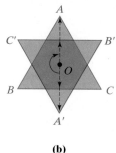

(a)

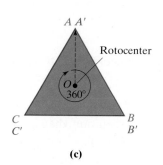

(b)

(c)

FIGURE 11-8

The following are some important properties of a rotation.

Property 6 A 360° rotation is equivalent to a 0° rotation, and a 0° rotation is just the identity motion. (The expression "going around full circle" is the well-known colloquial version of this property.)

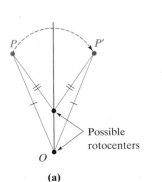

From property 6 we can conclude that all rotations can be described using an angle of rotation between 0° and 360°. For angles larger than 360° we divide the angle by 360° and just use the remainder (for example, a clockwise rotation by 759° is equivalent to a clockwise rotation by 39°; the remaining 720° count as 0°). In addition, we can describe a rotation using clockwise or counterclockwise orientations (for example, a clockwise rotation by 39° is equivalent to a counterclockwise rotation by 321°).

Property 7 When an object is rotated, the left-right and clockwise-counterclockwise orientations are preserved (a rotated left hand remains a left hand, and the hands of a rotated clock still move in the clockwise direction). We will describe this fact by saying that a rotation is a *proper* rigid motion. (Any motion that preserves the left-right and clockwise-counterclockwise orientations of objects is called a **proper** rigid motion.)

A common misconception is to confuse a 180° rotation with a reflection, but we can see that they are very different from just observing that the reflection is an improper rigid motion, whereas the 180° rotation is a proper rigid motion.

Property 8 In every rotation, the rotocenter is a fixed point, and except for the case of the identity (where all points are fixed points) it is the *only* fixed point.

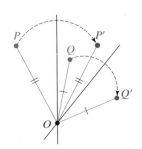

Property 9 Unlike a reflection, a rotation *cannot* be determined by a single point-image pair P and P'; it takes a second point-image pair Q and Q' to nail down the rotation. The reason is that infinitely many rotations can move P to P': Any point located on the perpendicular bisector of the segment PP' can be a rotocenter for such a rotation, as shown in Fig. 11-9(a). Given a second pair of points Q and Q' we can identify the rotocenter O as the point where the perpendicular bisectors of PP' and QQ' meet, as shown in Fig. 11-9(b). Once we have identified the rotocenter O, the angle of rotation α is given by the measure of angle POP' (or for that matter QOQ' — they are the same). [*Note*: In the special case where PP' and QQ' happen to have the same perpendicular bisector, as in Fig. 11-9(c), the rotocenter O is the intersection of PQ and $P'Q'$.]

The basic properties of rotations can be summarized as follows.

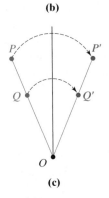

FIGURE 11-9

┌─ **PROPERTIES OF ROTATIONS** ─────────────

- A 360° rotation is equivalent to the *identity* motion.
- A rotation is a *proper* rigid motion.
- A rotation that is not the identity motion has only *one* fixed point, its rotocenter.
- A rotation is completely determined by *two* point-image pairs P, P' and Q, Q'.

11.4 Translations

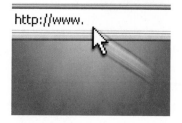

http://www.

FIGURE 11-10

A **translation** consists of essentially dragging an object in a specified *direction* and by a specified amount (the *length* of the translation). The two pieces of information (direction and length of the translation) are combined in the form of a **vector of translation** (usually denoted by v). The vector of translation is represented by an arrow—the arrow points in the direction of translation and the length of the arrow is the length of the translation. A very good illustration of a translation in a two-dimensional plane is the dragging of the cursor on a computer screen (Fig. 11-10). Regardless of what happens in between, the net result when you drag an icon on your screen is a translation in a specific direction and by a specific length.

EXAMPLE 11.4   **Translation of a Triangle**

Figure 11-11 illustrates the translation of a triangle ABC. Two "different" arrows are shown in the figure, but they both have the same length and direction, so they describe the same vector of translation v. As long as the arrow points in the proper direction and has the right length, the placement of the arrow in the picture is immaterial.

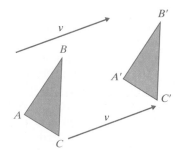

FIGURE 11-11

The following are some important properties of a translation.

Property 10 If we are given a point P and its image P' under a translation, the arrow joining P to P' gives the vector of the translation. Once we know the vector of the translation, we know where the translation moves any other point. Thus, a single point-image pair P and P' is all we need to completely determine the translation.

Property 11 In a translation, every point gets moved some distance and in some direction, so a translation has *no* fixed points.

Property 12 When an object is translated, left-right and clockwise-counterclockwise orientations are preserved: A translated left hand is still a left hand, and the hands of a translated clock still move in the clockwise direction (Fig. 11-12). In other words, translations are proper rigid motions.

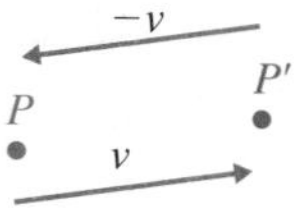

FIGURE 11-12

Property 13 The effect of a translation with vector v can be undone by a translation of the same length but in the opposite direction (Fig. 11-13). The vector for this opposite translation can be conveniently described as $-v$. Thus, a translation with vector v followed with a translation with vector $-v$ is equivalent to the identity motion.

FIGURE 11-13

The basic properties of translations can be summarized as follows.

PROPERTIES OF TRANSLATIONS

- A translation is completely determined by a single point-image pair P and P'.
- A translation has *no* fixed points.
- A translation is a *proper* rigid motion.
- When a translation with vector v is followed with a translation with vector $-v$ we get to the *identity* motion.

11.5 Glide Reflections

A **glide reflection** is a rigid motion obtained by combining a translation (the glide) with a reflection. Moreover, the axis of reflection *must* be parallel to the direction of translation. Thus, a glide reflection is described by two things: the vector of the translation v and the axis of the reflection l, and these two *must* be parallel. The footprints left behind by someone walking on soft sand (Fig. 11-14) are a classic example of a glide reflection: right and left footprints are images of each other under the combined effects of a reflection and a translation.

FIGURE 11-14

> **EXAMPLE 11.5**   **Glide Reflection of a Triangle**

Figure 11-15 illustrates the result of applying the glide reflection with vector v and axis l to the triangle ABC. We can do this in two different ways, but the final result will be the same. In Fig. 11-15(a), the translation is applied first, moving triangle ABC to the intermediate position $A^*B^*C^*$. The reflection is then applied to $A^*B^*C^*$, giving the final position $A'B'C'$. If we apply the reflection first, the triangle ABC gets moved to the intermediate position $A^*B^*C^*$ [Fig. 11-15(b)] and then translated to the final position $A'B'C'$.

Notice that any point and its image under the glide reflection [for example, A and A' in Fig. 11-15(a)] are on opposite sides but equidistant from the axis l. This implies that the midpoint of the segment joining a point and its image under a glide reflection *must* fall on the axis l [point M in Fig. 11-15(a)].

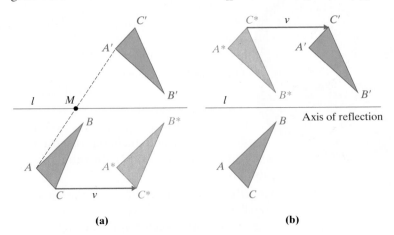

FIGURE 11-15   **(a)**   **(b)**

The following are a few basic properties of a glide reflection.

Property 14 A fixed point of a glide reflection would have to be a point that ends up exactly where it started after it is first translated and then reflected.

This cannot happen because the translation moves every point and the reflection cannot undo the action of the translation. It follows that a glide reflection has *no* fixed points.

Property 15 A glide reflection is a combination of a proper rigid motion (the translation) and an improper rigid motion (the reflection). Since the translation preserves left-right and clockwise-counterclockwise orientations but the reflection reverses them, the net result under a glide reflection is that orientations are reversed. Thus, a glide reflection is an *improper* rigid motion.

Property 16 A glide reflection is completely determined by *two* point-image pairs P, P' and Q, Q'. Given a point-image pair P and P' under a glide reflection, we do not have enough information to determine the glide reflection, but we do know that the axis l must pass through the midpoint of the line segment PP'. Given a second point-image pair Q and Q', we can determine the axis of the reflection: It is the line passing through the points M (midpoint of the line segment PP') and N (midpoint of the line segment QQ'), as shown in Fig. 11-16(a). Once we find the axis of reflection l, we can find the image of one of the points—say P'—under the reflection. This gives the intermediate point P^*, and the vector that moves P to P^* is the vector of translation v, as shown in Fig. 11-16(b). [In the event that the midpoints of PP' and QQ' are the same point M, as shown in Fig. 11-16(c), we can still find the axis l by drawing a line perpendicular to the line PQ passing through the common midpoint M.]

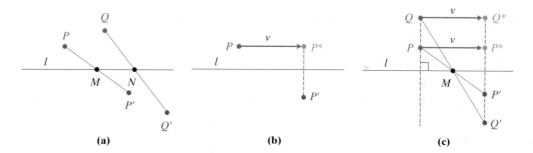

FIGURE 11-16   (a)   (b)   (c)

Property 17 To undo the effects of a glide reflection, we need a second glide reflection in the opposite direction. To be more precise, if we move an object under a glide reflection with vector of translation v and axis of reflection l and then follow it with another glide reflection with vector of translation $-v$ and axis of reflection still l, we get the identity motion. It is as if the object was not moved at all.

The basic properties of glide reflections can be summarized as follows.

> **PROPERTIES OF GLIDE REFLECTIONS**
>
> ■ A glide reflection has *no* fixed points.
> ■ A glide reflection is an *improper* rigid motion.
> ■ A glide reflection is completely determined by *two* point-image pairs P, P' and Q, Q'.
> ■ When a glide reflection with vector v and axis of reflection l is followed with a glide reflection with vector $-v$ and the same axis of reflection l we get the *identity* motion.

11.6   Symmetry as a Rigid Motion

With an understanding of the four basic rigid motions and their properties, we can now look at the concept of symmetry in a much more precise way. Here, finally, is a good definition of **symmetry**, one that probably would not have made much sense at the start of this chapter:

> ┌─ **SYMMETRY** ───
>
> A *symmetry* of an object (or shape) is any *rigid motion* that moves the object back onto itself.

One useful way to think of a symmetry is this: You observe the position of an object, and then, while you are not looking, the object is moved. If you can't tell that the object was moved, the rigid motion is a symmetry. It is important to note that this does not necessarily force the rigid motion to be the identity motion. Individual points may be moved to different positions, even though the whole object is moved back into itself. And, of course, the identity motion is itself a symmetry, one possessed by every object and that from now on we will call simply the *identity*.

Since there are only four basic kinds of rigid motions of two-dimensional objects in two-dimensional space, there are also only four possible types of symmetries: *reflection symmetries, rotation symmetries, translation symmetries*, and *glide reflection symmetries*.

(**EXAMPLE 11.6**)   **The Symmetries of a Square**

What are the possible rigid motions that move the square in Fig. 11-17(a) back onto itself? First, there are *reflection symmetries*. For example, if we use the line l_1 in Fig. 11-17(b) as the axis of reflection, the square falls back into itself with points A and B interchanging places and C and D interchanging places. It is not hard to think of three other reflection symmetries, with axes l_2, l_3, and l_4 as shown in Fig. 11-17(b). Are there any other types of symmetries? Yes—the square has *rotation symmetries* as well. Using the center of the square O as the rotocenter, we can rotate the square by an angle of 90°. This moves the upper-left corner A to the upper-right corner B, B to the lower-right corner C, C to the lower-left corner D, and D to the upper-right corner A. Likewise, rotations with rotocenter O and angles of 180°, 270°, and 360°, respectively, are also symmetries of the square. Notice that the 360° rotation is just the identity.

FIGURE 11-17 (a) The original square, (b) reflection symmetries (axes l_1, l_2, l_3, and l_4), and (c) rotation symmetries (rotocenter O and angles of 90°, 180°, 270°, and 360°, respectively).

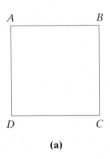

(a)

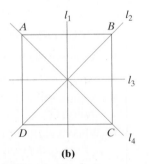

(b)

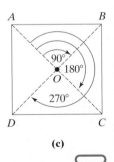

(c)

All in all, we have easily found eight symmetries for the square in Fig. 11-17(a): Four of them are reflections, and the other four are rotations. Could there be more? What if we combined one of the reflections with one of the rotations? A symmetry combined with another symmetry, after all, has to be itself a symmetry. It turns out that the eight symmetries we listed are all there are—no matter how we combine them we always end up with one of the eight.

■ See Exercise 74.

EXAMPLE 11.7 **The Symmetries of a Propeller**

Let's now consider the symmetries of the shape shown in Fig. 11-18(a)—a two-dimensional version of a boat propeller (or a ceiling fan if you prefer) with four blades. Once again, we have a shape with four reflection symmetries [the axes of reflection are l_1, l_2, l_3, and l_4 as shown in Fig. 11-18(b)] and four rotation symmetries [with rotocenter O and angles of $90°$, $180°$, $270°$, and $360°$, respectively, as shown in Fig. 11-18(c)]. And, just as with the square, there are no other possible symmetries.

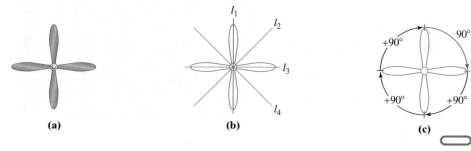

FIGURE 11-18   (a)   (b)   (c)

An important lesson lurks behind Examples 11.6 and 11.7: *Two different-looking objects can have exactly the same set of symmetries.* A good way to think about this is that the square and the propeller, while certainly different objects, are members of the same "symmetry family" and carry exactly the same symmetry genes.

Formally, we will say that two objects or shapes are of the same **symmetry type** if they have *exactly* the same set of symmetries. The symmetry type for the square, the propeller, and each of the objects shown in Fig. 11-19 is called D_4 (shorthand for four reflections plus four rotations).

FIGURE 11-19 Objects with symmetry type D_4.

(a)   (b)   (c)

EXAMPLE 11.8 **The Symmetry Type Z_4**

Let's consider now the propeller shown in Fig. 11-20(a). This object is only slightly different from the one in Example 11.7, but from the symmetry point of

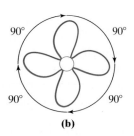
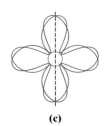

FIGURE 11-20 A propeller with symmetry type Z_4 (four rotation symmetries, no reflection symmetries).

(a)   (b)   (c)

view the difference is significant—here we still have the four rotation symmetries (90°, 180°, 270°, and 360°), but there are no reflection symmetries! [This makes sense because the individual blades of the propeller have no reflection symmetry. As can be seen in Fig. 11-20(c), a vertical reflection is not a symmetry, and neither are any of the other reflections.] This object belongs to a new symmetry family called Z_4 (shorthand for the symmetry type of objects having four rotations only).

EXAMPLE 11.9   **The Symmetry Type Z_2**

Here is one last propeller example. Every once in a while a propeller looks like the one in Fig. 11-21(a), which is kind of a cross between Figs. 11-20(a) and 11-19(a)—only opposite blades are the same. This figure has no reflection symmetries, and a 90° rotation won't work either [Fig. 11-20(b)]. The only symmetries of this shape are a 180° rotation (turn it upside down and it looks the same!) and the 360° rotation (the identity).

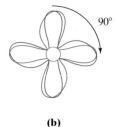
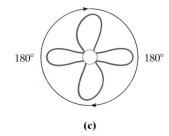

FIGURE 11-21 A propeller with symmetry type Z_2 (two rotation symmetries, no reflection symmetries).

(a)   (b)   (c)

An object having only two rotation symmetries (the identity and a 180° rotation symmetry) is said to be of symmetry type Z_2. Figure 11-22 shows a few additional examples of shapes and objects with symmetry type Z_2.

FIGURE 11-22 Objects with symmetry type Z_2. (a) The letter Z, (b) the letter S (in some fonts but not in others), and (c) the queen of spades (and many other cards in the deck).

(a)   (b)   (c)

(**EXAMPLE 11.10**) **The Symmetry Type D_1**

One of the most common symmetry types occurring in nature is that of objects having a single reflection symmetry plus a single rotation symmetry (the identity). This symmetry type is called D_1. Figure 11-23 shows several examples of shapes and objects having symmetry type D_1. Notice that it doesn't matter if the axis of reflection is vertical, horizontal, or slanted.

(a) (b) (c)

FIGURE 11-23 Objects with symmetry type D_1 (one reflection symmetry plus the identity symmetry).

(**EXAMPLE 11.11**) **The Symmetry Type Z_1**

Many objects and shapes are informally considered to have no symmetry at all, but this is a little misleading, since *every object has at least the identity symmetry*. Objects whose only symmetry is the identity are said to have symmetry type Z_1. Figure 11-24 shows a few examples of objects of symmetry type Z_1—there are plenty of such objects around.

(a) (b) (c)

FIGURE 11-24 Objects with symmetry type Z_1 (only symmetry is the identity symmetry). Why doesn't the six of clubs have a half-turn symmetry? (The answer is given by the two middle clubs.)

(**EXAMPLE 11.12**) **Objects with Lots of Symmetry**

In everyday language, certain objects and shapes are said to be "highly symmetric" when they have lots of rotation and reflection symmetries. Figures 11-25(a) and (b) show two very different looking snowflakes, but from the symmetry point of view they are the same: All snowflakes have six reflection symmetries and six rotation symmetries. Their symmetry type is D_6. (Try to find the six axes of reflection

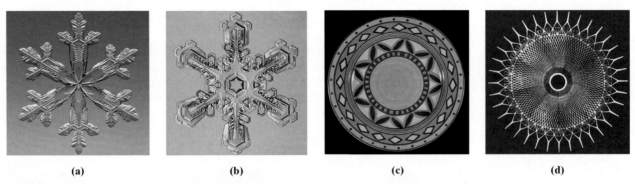

FIGURE 11-25 (a) and (b): Snowflakes (symmetry type D_6). (c): Decorative ceramic plate (symmetry type D_9). (d): Dome design (symmetry type D_{36}).

symmetry and the six angles of rotation symmetry.) Figure 11-25(c) shows a decorative ceramic plate. It has nine reflection symmetries and nine rotation symmetries, and, as you may have guessed, its symmetry type is called D_9. Finally, in Fig. 11-25(d) we have an architectural blueprint of the dome of the Sports Palace in Rome, Italy. The design has 36 reflection and 36 rotation symmetries (symmetry type D_{36}).

In each of the objects in Example 11.12, the number of reflections matches the number of rotations. This was also true in Examples 11.6, 11.7, and 11.10. Coincidence? Not at all. When a finite object or shape has *both* reflection and rotation symmetries, the number of rotation symmetries (which includes the identity) has to match the number of reflection symmetries! Any finite object or shape with exactly N reflection symmetries and N rotation symmetries is said to have symmetry type D_N.

(EXAMPLE 11.13) **The Symmetry Type D_∞**

Are there two-dimensional objects with infinitely many symmetries? The answer is yes—circles. A circle has infinitely many reflection symmetries (any line passing through the center of the circle can serve as an axis) as well as infinitely many rotation symmetries (use the center of the circle as a rotocenter and any angle of rotation will work). We call the symmetry type of the circle D_∞ (the ∞ is the mathematical symbol for "infinity").

(EXAMPLE 11.14) **Shapes with Rotations, but No Reflections**

■ The D comes from *dihedral symmetry* — the technical mathematical term used to describe this group of symmetry types.

We now know that if a finite two-dimensional shape has rotations *and* reflections, it must have exactly the same number of each. In this case, the shape belongs to the D family of symmetries, specifically, it has symmetry type D_N. However, we also saw in Examples 11.8, 11.9, and 11.11 shapes that have rotations, *but no* reflections. In this case, we used the notation Z_N to describe the symmetry type, with the subscript N indicating the actual number of rotations. Figure 11-26(a) shows a flower with five petals that has symmetry type Z_5, Fig. 11-26(b) shows an airplane turbine with 24 blades that has symmetry type Z_{24}. Notice that the absence of reflections is the result of some twist or bump on the petals or blades.

 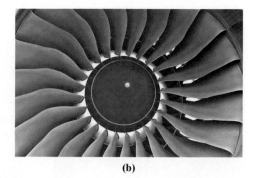

(a)   (b)

FIGURE 11-26 (a) Hibiscus (symmetry type Z_5) and (b) turbine (symmetry type Z_{24}).

We are now in a position to classify the possible symmetries of any *finite* two-dimensional shape or object. (The word *finite* is in there for a reason, which will become clear in the next section.) The possibilities boil down to a surprisingly short list of symmetry types:

- **D_N** This is the symmetry type of shapes with both rotation and reflection symmetries. The subscript N ($N = 1, 2, 3$, etc.) denotes the number of reflection symmetries, which is always equal to the number of rotation symmetries. (The rotations are an automatic consequence of the reflections—an object can't have reflection symmetries without having an equal number of rotation symmetries.)

- **Z_N** This is the symmetry type of shapes with rotation symmetries only. The subscript N ($N = 1, 2, 3$, etc.) denotes the number of rotation symmetries.

- **D_∞** This is the symmetry type of a circle and of circular objects such as rings and washers, the only possible two-dimensional shapes or objects with an infinite number of rotations and reflections.

11.7   Patterns

Well, we've come a long way, but we have yet to see examples of shapes having translation or glide reflection symmetry. If we think of objects and shapes as being finite, translation symmetry is impossible: A slide will always move the object to a new position different from its original position! But if we broaden our horizons and consider infinite "shapes," translation and glide reflection symmetries are indeed possible.

We will formally define a **pattern** as an infinite "shape" consisting of an infinitely repeating basic design called the **motif** of the pattern. The reason we have "shape" in quotation marks is that a pattern is really an abstraction—in the real world there are no infinite objects as such, although the idea of an infinitely repeating motif is familiar to us from such everyday objects as pottery, tile designs, and textiles (Fig. 11-27).

Just like finite shapes, *patterns* can be classified by their symmetries. The classification of patterns according to their symmetry type is of fundamental importance in the study of molecular and crystal organization in chemistry, so it is not surprising that some of the first people to seriously investigate the symmetry types

FIGURE 11-27

of patterns were crystallographers. Archaeologists and anthropologists have also found that analyzing the symmetry types used by a particular culture in their textiles and pottery helps them gain a better understanding of that culture.

We will briefly discuss the symmetry types of *border* and *wallpaper* patterns. A comprehensive study of patterns is beyond the scope of this book, so we will not go into as much detail as we did with finite shapes.

Border Patterns

Border patterns (also called *linear* patterns) are patterns in which a basic *motif* repeats itself indefinitely in a single direction, as in an architectural frieze, a ribbon, or the border design of a ceramic pot (Fig. 11-28).

FIGURE 11-28

The most common direction in a border pattern (what we will call the *direction of the pattern*) is horizontal, but in general a border pattern can be in any direction (vertical, slanted 45°, etc.). (For typesetting in a book, it is much more convenient to display a border pattern horizontally, so you will only see examples of horizontal border patterns.)

We will now discuss what kinds of symmetries are possible in a border pattern. Fortunately, the number of possibilities is quite small.

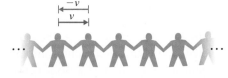

FIGURE 11-29

- **Translations.** A border pattern always has *translation symmetries* — they come with the territory. There is a *basic* translation symmetry v (v moves each copy of the motif one unit to the right), the opposite translation $-v$, and any multiple of v or $-v$ (Fig. 11-29).

- **Reflections.** A border pattern can have (a) no reflection symmetry [Fig. 11-30(a)], (b) *horizontal* reflection symmetry only [Fig. 11-30(b)], (c) *vertical* reflection symmetries [Fig. 11-30(c)] only, or (d) both *horizontal* and *vertical* reflection symmetries [Fig. 11-30(d)]. In this last case the border pattern automatically picks up a half-turn symmetry as well. In terms of reflection symmetries, Fig. 11-30 illustrates the only four possibilities in a border pattern.

■ See Exercise 69.

··· J J J J J J ···   ···B·B·B·B·B·B···   ···A A A A A A···   ···X X X X X X···

     (a)   **(b)**   **(c)**   **(d)**

FIGURE 11-30

- **Rotations.** Like with any other object, the identity (i.e., a 360° rotation) is a rotation symmetry of every border pattern, so every border pattern has at least one rotation symmetry. The only other possible rotation symmetry of a border pattern is a half-turn (180° rotation). Clearly, no other angle of rotation can take a horizontal pattern and move it back onto itself. Thus, in terms of rotation symmetry there are two kinds of border patterns: those whose only rotation symmetry is the identity [Fig. 11-31(a)] and those having half-turn symmetry in addition to the identity [Fig. 11-31(b)].

··· Y Y Y Y Y Y ···   ···Z Z Z·Z Z···
                                           Rotocenter

FIGURE 11-31   **(a)**   **(b)**

- **Glide reflections.** A border pattern can have a *glide reflection symmetry*, but there is only one way this can happen: The axis of reflection *has* to be a line along the center of the pattern, and the *reflection part of the glide reflection is not by itself a symmetry of the pattern*. This means that a border pattern having horizontal reflection symmetry such as the one shown in Fig. 11-32(a) is not considered to have glide reflection symmetry. On the other hand, the border pattern shown in Fig. 11-32(b) does not have horizontal reflection symmetry (the footprints do not fall back onto other footprints), but a glide by the vector w combined with a reflection along the axis l result in an honest-to-goodness glide reflection symmetry. (An important property of the glide reflection symmetry is that the vector w is always half the length of the basic translation symmetry v. This implies that two consecutive glide reflection symmetries are equivalent to the basic translation symmetry.) The border pattern in Fig. 11-32(c) has a vertical reflection symmetry as well as a glide reflection symmetry. In these cases a half-turn symmetry (rotocenter O) comes free in the bargain.

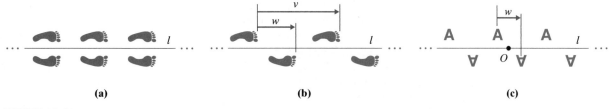

      (a)   **(b)**   **(c)**

FIGURE 11-32

Combining the preceding observations, we get the following list of possible symmetries of a border pattern. (For simplicity, we assume that the border pattern is in a horizontal direction.)

1. The *identity:* All border patterns have it.

2. *Translations:* All border patterns have them. There are a basic translation v, the opposite translation $-v$, and any multiples of these (see Fig. 11-29).

3. *Horizontal reflection:* Some patterns have it, some don't. There is only one possible horizontal axis of reflection, and it must run through the middle of the pattern [see Fig. 11-30(b)].

4. *Vertical reflections:* Some patterns have them, some don't. Vertical axes of reflection (i.e., axes perpendicular to the direction of the pattern) can run through the middle of a motif or between two motifs [see Fig. 11-30(c)].

5. *Half-turns:* Some patterns have them, some don't. Rotocenters must be located at the center of a motif or between two motifs [see Fig. 11-31(b)].

6. *Glide reflections:* Some patterns have them, some don't. Neither the reflection nor the glide can be symmetries on their own. The length of the glide w is half that of the basic translation v. The axis of the reflection runs through the middle of the pattern [see Fig. 11-32(b)].

Based on the various possible combinations of these symmetries, border patterns can be classified into just *seven different symmetry families*, which we list next. (Since all border patterns have identity symmetry and translation symmetry, we will only make reference to the other possible symmetries.)

■ The oddball two-symbol codes are a classification scheme originally developed by crystallographers called the *standard crystallographic notation*. The first symbol is either an *m* (indicating that the pattern has vertical reflections) or a 1 (indicating no vertical reflections). The second symbol specifies the existence of a horizontal reflection (*m*), a glide reflection (*g*), a half-turn (2), or none of these (1).

■ **11.** This symmetry type represents border patterns that have no symmetries other than the identity and translation symmetry. The pattern in Fig. 11-30(a) is an example of this symmetry type.

■ **1m.** This symmetry type represents border patterns with just a horizontal reflection symmetry. The pattern in Fig. 11-30(b) is an example of this symmetry type.

■ **m1.** This symmetry type represents border patterns with just a vertical reflection symmetry. The pattern in Fig. 11-30(c) is an example of this symmetry type.

■ **mm.** This symmetry type represents border patterns with both a horizontal and a vertical reflection symmetry. When both of these symmetries are present, there is also half-turn symmetry. The pattern in Fig. 11-30(d) is an example of this symmetry type.

■ **12.** This symmetry type represents border patterns with only a half-turn symmetry. The pattern in Fig. 11-31(b) is an example of this symmetry type.

■ **1g.** This symmetry type represents border patterns with only a glide reflection symmetry. The pattern in Fig. 11-32(b) is an example of this symmetry type.

■ **mg.** This symmetry type represents border patterns with a vertical reflection and a glide reflection symmetry. When both of these symmetries are present, there is also a half-turn symmetry. The pattern in Fig. 11-32(c) is an example of this symmetry type.

Table 11-1 shows the symmetries (besides the identity) for each of the seven symmetry types.

TABLE 11-1

Symmetry type	Translation	Horizontal reflection	Vertical reflection	Half-turn	Glide reflection	Example
11	Yes	No	No	No	No	Fig. 11-30(a)
1m	Yes	Yes	No	No	No	Fig. 11-30(b)
m1	Yes	No	Yes	No	No	Fig. 11-30(c)
mm	Yes	Yes	Yes	Yes	No	Fig. 11-30(d)
12	Yes	No	No	Yes	No	Fig. 11-31(b)
1g	Yes	No	No	No	Yes	Fig. 11-32(b)
mg	Yes	No	Yes	Yes	Yes	Fig. 11-32(c)

Wallpaper Patterns

Wallpaper patterns are patterns that fill the plane by repeating a *motif* indefinitely along several (two or more) nonparallel directions. Typical examples of such patterns can be found in wallpaper (of course), carpets, and textiles.

With wallpaper patterns things get a bit more complicated, so we will skip the details. The possible symmetries of a wallpaper pattern are as follows:

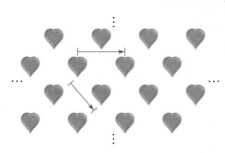

FIGURE 11-33

- **Translations.** Every wallpaper pattern has translation symmetry in at least two different (nonparallel) directions (Fig. 11-33).

- **Reflections.** A wallpaper pattern can have (a) no reflections, (b) reflections in only one direction, (c) reflections in two nonparallel directions, (d) reflections in three nonparallel directions, (e) reflections in four nonparallel directions, and (f) reflections in six nonparallel directions. There are no other possibilities. Note that particularly conspicuous in its absence is the case of reflections in exactly five different directions.

- **Rotations.** In terms of rotation symmetries, a wallpaper pattern can have (a) the identity only, (b) two rotations (identity and 180°), (c) three rotations (identity, 120°, and 240°), (d) four rotations (identity, 90°, 180°, and 270°), and (e) six rotations (identity, 60°, 120°, 180°, 240°, and 300°). There are no other possibilities. Once again, note that a wallpaper pattern cannot have exactly five different rotations.

- **Glide reflections.** A wallpaper pattern can have (a) no glide reflections, (b) glide reflections in only one direction, (c) glide reflections in two nonparallel directions, (d) glide reflections in three nonparallel directions, (e) glide reflections in four nonparallel directions, and (f) glide reflections in six nonparallel directions. There are no other possibilities.

■ The 17 symmetry types of wallpaper patterns are listed and illustrated in Appendix B (p. 625).

In the early 1900s, it was shown mathematically that there are *only 17 possible symmetry types for wallpaper patterns*. This is quite a surprising fact—it means that the hundreds and thousands of wallpaper patterns one can find at a decorating store all fall into just 17 different symmetry families.

CONCLUSION

FIGURE 11-34

There is no doubt that the letter Z and a queen of spades are two very different things (Fig. 11-34)—they look different and they serve completely different purposes. In a very fundamental way, however, they share a very important characteristic: They have exactly the same set of symmetries. Understanding the symmetries of an object and being able to sort and classify the objects in our physical world according to their symmetries is a rare but useful skill, and artists, designers, architects, archaeologists, chemists, and mathematicians often rely on it. In this chapter we explored some of the basic ideas behind the mathematical study of symmetry.

Although we live in a three-dimensional world inhabited by mostly three-dimensional objects and shapes, in this chapter we restricted ourselves to studying the symmetries of two-dimensional objects and shapes for the simple reason that it is a lot easier.

A first important step in understanding symmetries is to understand the different types of *rigid motions*—motions that move an object while preserving its original shape. It is somewhat surprising that despite the infinite freedom we have to move an object, for two-dimensional objects there are just four basic types of rigid motions: *reflections*, *rotations*, *translations*, and *glide reflections*. In other words, no matter how complicated or convoluted the actual "trip" taken by the object might appear to be, in the final analysis the motion is equivalent to a single reflection, a single rotation, a single translation, or a single glide reflection. Those are the only four possibilities.

> **Symmetry is a vast subject, significant in art and nature. Mathematics lies at its root, and it would be hard to find a better one on which to demonstrate the working of the mathematical intellect.**
>
> —Hermann Weyl

Each of the four basic rigid motions is unique in its properties: some are *proper* rigid motions, others are *improper*; some have *no fixed points*, others have *one* or *infinitely many fixed points*. A summary of the key properties of the four basic rigid motions is given in Table 11-2. Notice that no two have exactly the same set of properties, so we can often identify the type of rigid motion by just knowing a couple of its properties. (The rigid motion is improper and has no fixed points, you say? Well, it must be a glide reflection.)

TABLE 11-2

Rigid motion	Specified by	Proper/improper	Fixed points	Point-image pairs needed
Reflection	Axis of reflection l	Improper	All points on l	One
Rotation*	Rotocenter O and angle α	Proper	O only	Two
Translation	Vector of translation v	Proper	None	One
Glide reflection	Vector of translation v and axis of reflection l	Improper	None	Two
***Identity** (0° rotation)		Proper	All points	

When a rigid motion moves an object back onto itself, we call it a *symmetry* of the object, and we say that two different objects have the same *symmetry type* when they have exactly the same set of symmetries. (The letter *Z* and the queen of spades, for example, have the same symmetry type: the *identity* and a *half-turn* are their only symmetries.) When the object is finite, the possible symmetry types for the object fall into two possible categories: The object has *rotation symmetries* only (a *Z-something* kind of object), or it has both rotation and reflection symmetries in equal amounts (a *D-something* kind of object). It is quite remarkable that there are no other possibilities. Nowhere in the universe of finite two-dimensional shapes does there exist, for example, a shape with three reflection symmetries and five rotation symmetries—it is mathematically impossible!

Patterns—infinite shapes consisting of an infinitely repeating *motif*—are even more surprising in terms of their possible symmetry types. *Border patterns*—commonly found in ribbons, baskets, and pottery—are patterns in which the motif is repeated indefinitely in a single direction. With border patterns there are *only seven possible symmetry types*, listed in Table 11-1. *Wallpaper patterns* such as those found in wallpapers and textiles are patterns in which a motif repeats itself in more than one direction and fills the plane. With wallpaper patterns there are only 17 possible symmetry types, listed in Appendix B (p. 625). (If nothing else, this tidbit should help your confidence if you ever find yourself having to choose some wallpaper at the wallpaper store.)

PROFILE: Sir Roger Penrose (1931–   )

Much like the tiling of a bathroom floor, a mathematical *tiling* of the plane is an arrangement of tiles (of one or several shapes) that fully cover the plane without overlaps or gaps. A tiling of the plane is an abstract concept—unlike a bathroom floor, which has a finite area, a plane goes on forever. And yet, there is an easy method for tiling a plane: Find a basic design (a single tile or a combination of several tiles) that can be fitted snugly with other copies of itself and just repeat the same design, mindlessly marching off to infinity in all directions. This kind of tiling, called a *periodic tiling*, is essentially equivalent to a wallpaper pattern. With the right *motif* (the basic design that repeats), incredibly beautiful and exotic periodic tilings are possible (witness the amazing tilings found in some of M. C. Escher's graphic designs). Because tiling a plane is an infinite task, it stands to reason that the only way it can be done is by means of a periodic tiling, and this was the conventional wisdom for centuries. Artists, designers, chemists, and physicists all banked on this assumption. Enter Roger Penrose. In 1974, Penrose, a mathematical physicist already famous for his contributions to cosmology (he proved mathematically that black holes

can exist), found an *aperiodic* tiling of the plane—a method for tiling the plane not based on the periodic repetition of one motif. Using only two basic tile shapes, Penrose was able to show that it is possible to tile a plane with infinitely changing patterns (see Project C for more details).

Roger Penrose was born in Colchester, England. His father, a famous medical geneticist, moved the family to Canada during World War II, and Roger spent his early school years in London, Ontario. After the war, the family returned to the other London, where Penrose began his university studies, earning his B.Sc. degree in mathematics from University College and a Ph.D. in mathematical physics from Cambridge University.

Sir Roger (he was knighted in 1994) has done groundbreaking research in mathematics, mathematical physics, cosmology, and neuroscience and has held appointments at many academic institutions, including Cambridge University, Princeton University, and the University of Texas. He is currently the Rouse Ball Professor of Mathematics at the University of Oxford, England. In addition to his serious professional research in mathematics and physics, Penrose is famous for his contributions to recreational mathematics.

In fact, his original interest in tilings of the plane was strictly recreational (he just loves challenging puzzles). Ironically, his discovery of aperiodic tilings turned out to have amazing and unexpected practical implications. In 1982, following Penrose's lead, chemists working at the National Institute of Standards and Technology discovered the existence of *quasicrystals*, crystal structures with nonrepetitive patterns analogous to Penrose's aperiodic tilings. The discovery of quasicrystals turned the world of crystallography and materials science on its head. The aperiodic structure of quasicrystals defies all the traditional rules of crystal formation in nature and has led to the discovery of heretofore unknown properties of solid matter. As new alloys based on quasicrystalline structure are being developed (the latest is a nonscratch coating for frying pans), it is hard to imagine that we owe it all to one man's singular love of patterns and puzzles.

KEY CONCEPTS

angle of rotation, **405**
axis (of reflection), **403**
basic rigid motions of the plane, **403**
border pattern (one-dimensional pattern), **416**
equivalent rigid motions, **403**
fixed point, **403**
glide reflection, **408**

identity rigid motion, **405**
image, **403**
improper rigid motion, **404**
motif, **415**
pattern, **415**
proper rigid motion, **406**
reflection, **403**
rigid motion, **402**

rotation, **405**
rotocenter, **405**
symmetry, **410**
symmetry type, **411**
translation, **407**
vector of translation, **407**
wallpaper pattern (two-dimensional pattern), **419**

EXERCISES

WALKING

A   Reflections

1. In Fig. 11-35, find the image of P under the indicated reflection.

   (a)   The reflection with axis l_1.

   (b)   The reflection with axis l_2.

   (c)   The reflection with axis l_3.

   (d)   The reflection with axis l_4.

2. In Fig. 11-36, find the image of P under the indicated reflection.

   (a)   The reflection with axis l_1.

   (b)   The reflection with axis l_2.

   (c)   The reflection with axis l_3.

   (d)   The reflection with axis l_4.

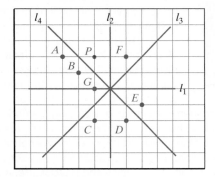

FIGURE 11-35

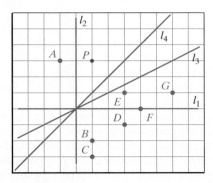

FIGURE 11-36

3. In Fig. 11-37, *l* is the axis of reflection.

  (a) Find the image of *S* under the reflection.

  (b) Find the image of quadrilateral *PQRS* under the reflection.

  (c) Find the fixed point of the reflection that is closest to *Q*.

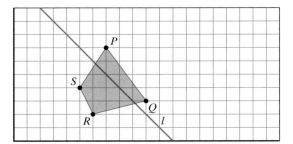

FIGURE 11-37

4. In Fig. 11-38, *l* is the axis of reflection.

  (a) Find the image of *P* under the reflection.

  (b) Find the image of parallelogram *PQRS* under the reflection.

  (c) Find the fixed point of the reflection that is closest to *P*.

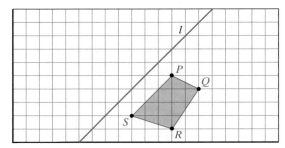

FIGURE 11-38

5. In Fig. 11-39, *P'* is the image of *P* under a reflection.

  (a) Find the axis of the reflection.

  (b) Find the image of *S* under the reflection.

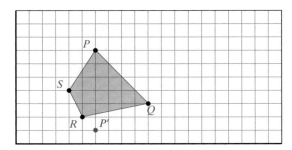

FIGURE 11-39

  (c) Find the image of quadrilateral *PQRS* under the reflection.

  (d) Find a point on the quadrilateral *PQRS* that is a fixed point of the reflection.

6. In Fig. 11-40, *P'* is the image of *P* under a reflection.

  (a) Find the axis of the reflection.

  (b) Find the image of *S* under the reflection.

  (c) Find the image of quadrilateral *PQRS* under the reflection.

  (d) Find a point on the quadrilateral *PQRS* that is a fixed point of the reflection.

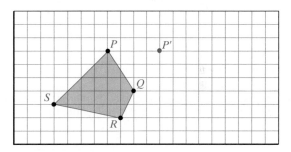

FIGURE 11-40

7. In Fig. 11-41, *P'* is the image of *P* under a reflection.

  (a) Find the axis of the reflection.

  (b) Find the image of the shaded arrow under the reflection.

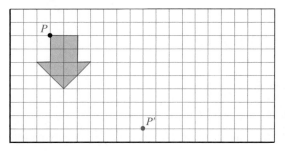

FIGURE 11-41

8. In Fig. 11-42, *R'* is the image of *R* under a reflection.

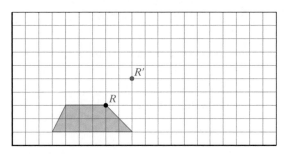

FIGURE 11-42

(a) Find the axis of the reflection.

(b) Find the image of the shaded trapezoid under the reflection.

9. In Fig. 11-43, *A* and *B* are fixed points of a reflection. Find the image of the shaded region under the reflection.

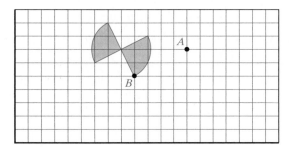

FIGURE 11-43

10. In Fig. 11-44, *A* and *B* are fixed points of a reflection. Find the image of the shaded region under the reflection.

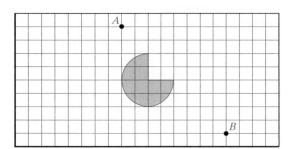

FIGURE 11-44

B Rotations

11. In Fig. 11-45, find

(a) the image of *B* under a 90° clockwise rotation with rotocenter *A*.

(b) the image of *A* under a 90° clockwise rotation with rotocenter *B*.

(c) the image of *D* under a 60° clockwise rotation with rotocenter *B*.

(d) the image of *D* under a 120° clockwise rotation with rotocenter *B*.

(e) the image of *I* under a 3690° clockwise rotation with rotocenter *A*.

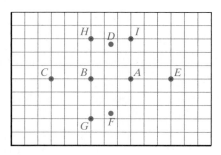

FIGURE 11-45

12. In Fig. 11-45, find

(a) the image of *C* under a 90° counterclockwise rotation with rotocenter *B*.

(b) the image of *F* under a 60° clockwise rotation with rotocenter *A*.

(c) the image of *F* under a 120° clockwise rotation with rotocenter *B*.

(d) the image of *I* under a 90° clockwise rotation with rotocenter *H*.

(e) the image of *G* under a 3870° counterclockwise rotation with rotocenter *B*.

13. In each case, give an answer between 0° and 360°.

(a) A clockwise rotation by an angle of 710° is equivalent to a counterclockwise rotation by an angle of _____.

(b) A clockwise rotation by an angle of 710° is equivalent to a clockwise rotation by an angle of _____.

(c) A counterclockwise rotation by an angle of 7100° is equivalent to a clockwise rotation by an angle of _____.

(d) A clockwise rotation by an angle of 71,000° is equivalent to a clockwise rotation by an angle of _____.

14. In each case, give an answer between 0° and 360°.

(a) A clockwise rotation by an angle of 500° is equivalent to a clockwise rotation by an angle of _____.

(b) A clockwise rotation by an angle of 500° is equivalent to a counterclockwise rotation by an angle of _____.

(c) A clockwise rotation by an angle of 5000° is equivalent to a clockwise rotation by an angle of _____.

(d) A clockwise rotation by an angle of 50,000° is equivalent to a counterclockwise rotation by an angle of _____.

15. In Fig. 11-46, a rotation moves *B* to *B'* and *C* to *C'*.

(a) Find the rotocenter.

(b) Find the image of triangle *ABC* under the rotation.

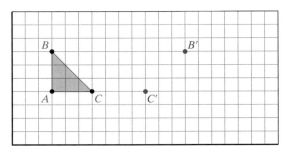

FIGURE 11-46

16. In Fig. 11-47, a rotation moves A to A' and B to B'.

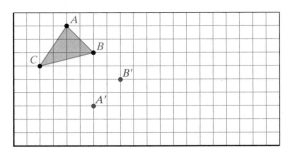

FIGURE 11-47

(a) Find the rotocenter.

(b) Find the image of triangle ABC under the rotation.

17. In Fig. 11-48, a rotation moves P to P' and S to S'.

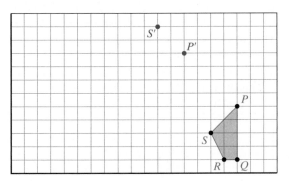

FIGURE 11-48

(a) Find the rotocenter.

(b) Find the angle of rotation.

(c) Find the image of quadrilateral $PQRS$ under the rotation.

18. In Fig. 11-49, a rotation moves Q to Q' and R to R'.

(a) Find the rotocenter.

(b) Find the angle of rotation.

(c) Find the image of quadrilateral $PQRS$ under the rotation.

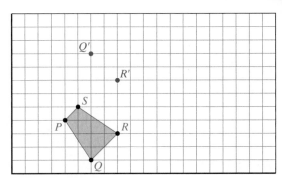

FIGURE 11-49

19. In Fig. 11-50, find the image of triangle ABC under a 60° clockwise rotation with rotocenter O.

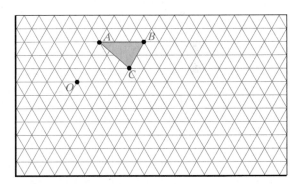

FIGURE 11-50

20. In Fig. 11-51, find the image of $ABCD$ under a 60° counterclockwise rotation with rotocenter O.

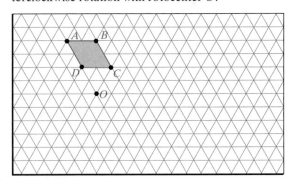

FIGURE 11-51

C Translations

21. In Fig. 11-52, find the image of P under

(a) the translation with vector v_1.

(b) the translation with vector v_2.

(c) the translation with vector v_3.

(d) the translation with vector v_4.

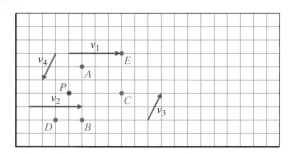

FIGURE 11-52

22. In Fig. 11-53, find the image of P under

(a) the translation with vector v_1.

(b) the translation with vector v_2.

(c) the translation with vector v_3.

(d) the translation with vector v_4.

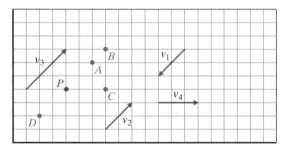

FIGURE 11-53

23. In Fig. 11-54, E' is the image of E under a translation.

(a) Find the image of A under the translation.

(b) Find the image of the shaded figure under the translation.

(c) Draw a vector for the translation.

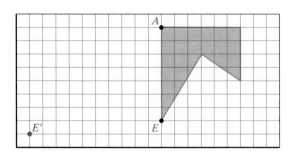

FIGURE 11-54

24. In Fig. 11-55, Q' is the image of Q under a translation.

(a) Find the image of P under the translation.

(b) Find the image of the shaded quadrilateral under the translation.

(c) Draw a vector for the translation.

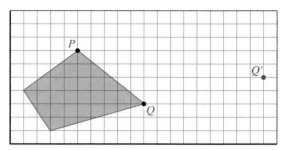

FIGURE 11-55

25. In Fig. 11-56, D' is the image of D under a translation. Find the image of the shaded trapezoid under the translation.

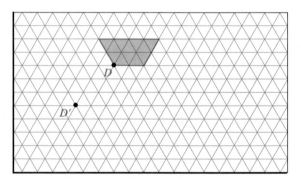

FIGURE 11-56

26. In Fig. 11-57, Q' is the image of Q under a translation. Find the image of the shaded region under the translation.

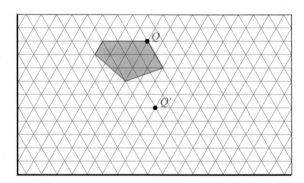

FIGURE 11-57

D Glide Reflections

27. Given a glide reflection with vector v and axis l as shown in Fig. 11-58, find the image of the triangle ABC under the glide reflection.

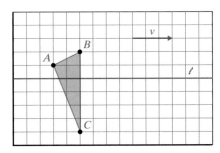

FIGURE 11-58

28. Given a glide reflection with vector v and axis l as shown in Fig. 11-59, find the image of the quadrilateral $ABCD$ under the glide reflection.

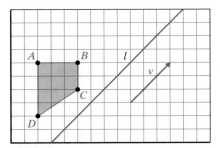

FIGURE 11-59

29. In Fig. 11-60, B' is the image of B and D' is the image of D under a glide reflection.

(a) Find the axis of the glide reflection.

(b) Find the image of A under the glide reflection.

(c) Find the image of the shaded figure under the glide reflection.

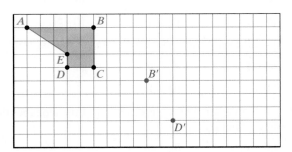

FIGURE 11-60

30. In Fig. 11-61, A' is the image of A and C' is the image of C under a glide reflection.

(a) Find the axis of the glide reflection.

(b) Find the image of B under the glide reflection.

(c) Find the image of the shaded figure under the glide reflection.

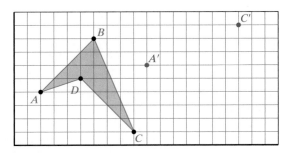

FIGURE 11-61

31. In Fig. 11-62, B' is the image of B and C' is the image of C under a glide reflection. Find the image of the shaded figure under the glide reflection.

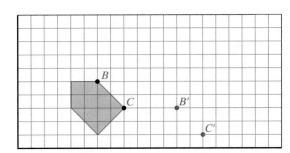

FIGURE 11-62

32. In Fig. 11-63, P' is the image of P and Q' is the image of Q under a glide reflection. Find the image of the shaded figure under the glide reflection.

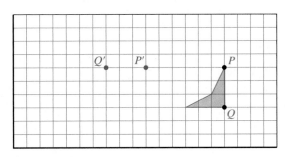

FIGURE 11-63

33. In Fig. 11-64, D' is the image of D and C' is the image of C under a glide reflection. Find the image of the shaded figure under the glide reflection.

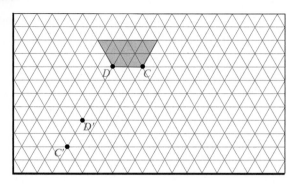

FIGURE 11-64

34. In Fig. 11-65, A' is the image of A and D' is the image of D under a glide reflection. Find the image of the shaded figure under the glide reflection.

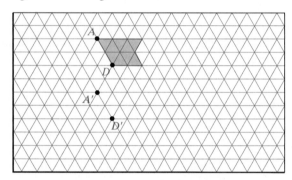

FIGURE 11-65

E Symmetries of Finite Shapes

In Exercises 35 through 38, list all the symmetries of each figure. Describe each symmetry by giving specifics—the axes of reflection, the centers and angles of rotation, etc.

35.

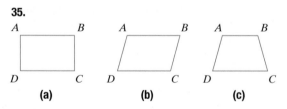

FIGURE 11-66

36.

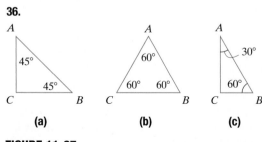

FIGURE 11-67

37.

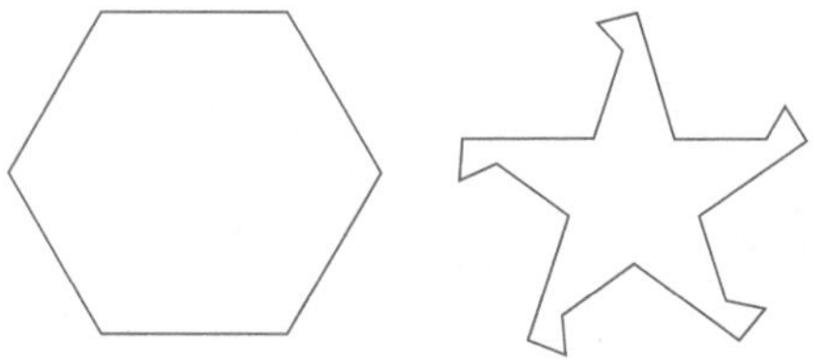

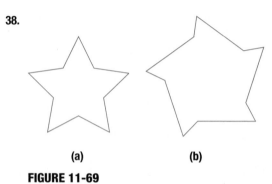

FIGURE 11-68

38.

FIGURE 11-69

39. For each of the figures in Exercise 35, give its symmetry type.

40. For each of the figures in Exercise 36, give its symmetry type.

41. For each of the figures in Exercise 37, give its symmetry type.

42. For each of the figures in Exercise 38, give its symmetry type.

43. Find the symmetry type for each of the following letters.

(a) A   (b) D   (c) L

(d) Z   (e) H   (f) N

44. Find the symmetry type for each of the following symbols.

(a) $ (b) @   (c) %

(d) ×   (e) &

45. Give an example of a capital letter of the alphabet that has symmetry type

(a) Z_1.   (b) D_1.

(c) Z_2.   (d) D_2.

46. Give an example of a one- or two-digit number that has symmetry type

(a) Z_1.   (b) D_1.

(c) Z_2.   (d) D_2.

47. Give an example of

(a) a natural object (plant, animal, mineral) that has symmetry type D_5. Explain your answer.

(b) a human-made object (logo, gadget, consumer product, etc.) that has symmetry type D_5. Explain your answer.

(c) a natural object (plant, animal, mineral) that has symmetry type Z_1. Explain your answer.

(d) a human-made object (logo, gadget, consumer product, etc.) that has symmetry type Z_1. Explain your answer.

48. Give an example of

(a) a natural object (plant, animal, mineral) that has symmetry type D_6. Explain your answer.

(b) a human-made object (logo, gadget, consumer product, etc.) that has symmetry type D_6. Explain your answer.

(c) a natural object (plant, animal, mineral) that has symmetry type Z_2. Explain your answer.

(d) a human-made object (logo, gadget, consumer product, etc.) that has symmetry type Z_2. Explain your answer.

F Symmetries of Border Patterns

In Exercises 49 through 52, classify the border pattern by its symmetry type. Use the standard crystallographic notation (mm, mg, m1, 1m, 1g, 12, or 11).

49. (a) … A A A A A …
(b) … D D D D D …
(c) … Z Z Z Z Z …
(d) … L L L L L …

FIGURE 11-70

50. (a) … J J J J J …
(b) … H H H H H …
(c) … W W W W W …
(d) … N N N N N …

FIGURE 11-71

51. (a) … qpqpqpqp …
(b) … pdpdpdpd …
(c) … pbpbpbpb …
(d) … pqbdpqbd …

FIGURE 11-72

52. (a) … qbqbqbqb …
(b) … qdqdqdqd …
(c) … dbdbdbdb …
(d) … qpdbqpdb …

FIGURE 11-73

53. If a border pattern consists of repeating a motif of symmetry type Z_2, what is the symmetry type of the border pattern?

54. If a border pattern consists of repeating a motif of symmetry type Z_1, what is the symmetry type of the border pattern?

G Miscellaneous

55. In each case, determine whether the rigid motion is a reflection, rotation, translation, or glide reflection or the identity motion.

(a) The rigid motion is proper and has exactly one fixed point.

(b) The rigid motion is proper and has infinitely many fixed points.

(c) The rigid motion is improper and has infinitely many fixed points.

(d) The rigid motion is improper and has no fixed points.

56. In each case, determine whether the rigid motion is a reflection, rotation, translation, or glide reflection or the identity motion.

(a) The rigid motion is proper, and when the same rigid motion is applied twice, we get the identity.

(b) The rigid motion is improper, and when the same rigid motion is applied twice, we get the identity.

(c) The rigid motion is improper, and when the same rigid motion is applied twice, we get a translation.

(d) The rigid motion is proper, and when the same rigid motion is applied twice, we get a translation.

*Exercises 57 through 68 deal with combining rigid motions. Given two rigid motions M and N, we can combine the two rigid motions by first applying M and then applying N to the result. The rigid motion defined by combining M and N (M goes first, N goes second) is called the **product** of M and N.*

57. In Fig. 11-74, l_1, l_2, l_3, and l_4 are axes of reflection. In each case, find the image of P under the indicated product.

(a) The product of the reflection with axis l_1 and the reflection with axis l_2.

(b) The product of the reflection with axis l_2 and the reflection with axis l_1.

(c) The product of the reflection with axis l_2 and the reflection with axis l_3.

(d) The product of the reflection with axis l_3 and the reflection with axis l_2.

(e) The product of the reflection with axis l_1 and the reflection with axis l_4.

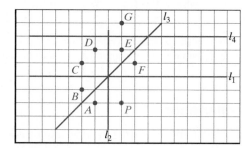

FIGURE 11-74

58. In Fig. 11-75, find the image of P under the indicated product.

(a) The product of the reflection with axis l and the 90° clockwise rotation with rotocenter A.

(b) The product of the 90° clockwise rotation with rotocenter A and the reflection with axis l.

(c) The product of the reflection with axis l and the 180° rotation with rotocenter A.

(d) The product of the 180° rotation with rotocenter A and the reflection with axis l.

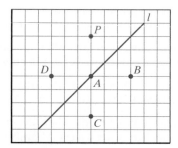

FIGURE 11-75

59. In each case, state whether the rigid motion $\mathcal{M}$ is proper or improper.

(a) $\mathcal{M}$ is the product of a proper rigid motion and an improper rigid motion.

(b) $\mathcal{M}$ is the product of an improper rigid motion and an improper rigid motion.

(c) $\mathcal{M}$ is the product of a reflection and a rotation.

(d) $\mathcal{M}$ is the product of a reflection and a reflection.

60. In each case, state whether the rigid motion $\mathcal{M}$ has (i) no fixed points, (ii) exactly one fixed point, or (iii) infinitely many fixed points.

(a) $\mathcal{M}$ is the product of a reflection with axis l_1 and a reflection with axis l_2. Assume that the lines l_1 and l_2 intersect at a point C.

(b) $\mathcal{M}$ is the product of a reflection with axis l_1 and a reflection with axis l_3. Assume that the lines l_1 and l_3 are parallel.

61. Suppose that a rigid motion $\mathcal{M}$ is the product of a reflection with axis l_1 and a reflection with axis l_2, where l_1 and l_2 intersect at a point C. Explain why $\mathcal{M}$ must be a rotation with center C. [*Hint:* See Exercises 59(d) and 60(a).]

62. Suppose that the rigid motion $\mathcal{M}$ is the product of the reflection with axis l_1 and the reflection with axis l_3, where l_1 and l_3 are parallel. Explain why $\mathcal{M}$ must be a translation. [*Hint:* See Exercises 59(d) and 60(b).]

JOGGING

63. In Fig. 11-76, l_1 and l_2 intersect at C, and the angle between them is α.

(a) Give the rotocenter, angle, and direction of rotation of the product of the reflection with axis l_1 and the reflection with axis l_2.

(b) Give the rotocenter, angle, and direction of rotation of the product of the reflection with axis l_2 and the reflection with axis l_1.

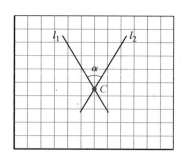

FIGURE 11-76

64. In Fig. 11-77, l_1 and l_3 are parallel and the distance between them is d.

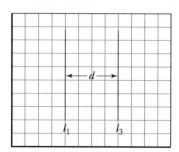

FIGURE 11-77

(a) Give the length and direction of the vector of the product of the reflection with axis l_1 and the reflection with axis l_3.

(b) Give the length and direction of the vector of the product of the reflection with axis l_3 and the reflection with axis l_1.

65. In Fig. 11-78, P' is the image of P under a translation $\mathcal{M}$ and Q' is the image of Q under a translation $\mathcal{N}$.

(a) Find the images of P and Q under the product of $\mathcal{M}$ and $\mathcal{N}$.

(b) Show that the product of $\mathcal{M}$ and $\mathcal{N}$ is a translation. Give a geometric description of the vector of the translation.

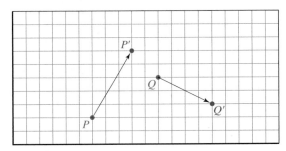

FIGURE 11-78

66. In Fig. 11-79, a glide reflection $\mathcal{M}$ has l as the axis of reflection and v as the vector of translation.

(a) Find the image of the shaded triangle under the product of $\mathcal{M}$ with itself.

(b) Show that the product of a glide reflection with itself is a translation. Describe the direction and amount of the translation in terms of the direction and amount of the original glide.

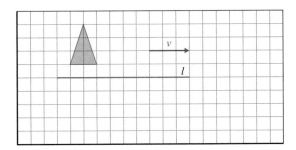

FIGURE 11-79

67. In Fig. 11-80, $\mathcal{M}$ is the translation with vector v and $\mathcal{N}$ is a 90° clockwise rotation with rotocenter O.

(a) Find the image of the shaded triangle under the product of $\mathcal{M}$ and $\mathcal{N}$.

(b) Explain why the product of $\mathcal{M}$ and $\mathcal{N}$ is a rotation and find the rotocenter and angle of the rotation.

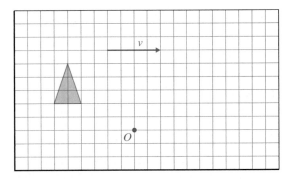

FIGURE 11-80

68. In Fig. 11-81, $\mathcal{M}$ is a 90° clockwise rotation with rotocenter O and $\mathcal{N}$ is the glide reflection with axis l and vector v.

(a) Find the image of the shaded figure under the product of $\mathcal{M}$ and $\mathcal{N}$.

(b) Explain why the product of $\mathcal{M}$ and $\mathcal{N}$ is a reflection and find the axis of the reflection.

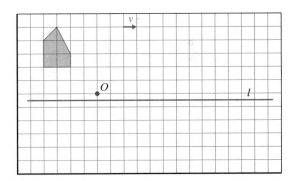

FIGURE 11-81

69. (a) Explain why a border pattern cannot have a reflection symmetry along an axis forming a 45° angle with the direction of the pattern.

(b) Explain why a border pattern can have only horizontal and/or vertical reflection symmetry.

70. Construct border patterns for each of the seven symmetry types using copies of the symbol ♥ (and rotated versions of it).

71. A rigid motion $\mathcal{M}$ moves the triangle PQR into the triangle $P'Q'R'$ as shown in Fig. 11-82. Explain why the rigid motion $\mathcal{M}$ must be a glide reflection.

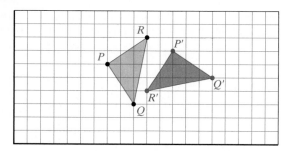

FIGURE 11-82

	r_1	r_2	r_3	R_1	R_2	I
r_1		R_1				
r_2						
r_3						
R_1						
R_2						
I						

72. A *palindrome* is a word that is the same when read forward or backward. MOM is a palindrome, and so is ANNA. (For simplicity, we will assume all letters are capitals.)

(a) Explain why if a word has vertical reflection symmetry, it must be a palindrome.

(b) Give an example of a palindrome (other than ANNA) that doesn't have vertical reflection symmetry.

(c) If a palindrome has vertical reflection symmetry, what can you say about the symmetries of the individual letters in the word?

(d) Find a palindrome with 180° rotational symmetry.

RUNNING

73. Let the six symmetries of the equilateral triangle ABC shown in Fig. 11-83 be denoted as follows: r_1 denotes the reflection with axis l_1; r_2 denotes the reflection with axis l_2; r_3 denotes the reflection with axis l_3; R_1 denotes the 120° clockwise rotation with rotocenter O; R_2 denotes the 240° clockwise rotation with rotocenter O; I denotes the identity symmetry. Complete the symmetry "multiplication table" by entering, in each row and column of the table, the product of the row and the column (i.e., the result of applying first the symmetry in the row and then the symmetry in the column). For example, the entry in row r_1 and column r_2 is R_1 because the product of the reflection r_1 and the reflection r_2 is the rotation R_1.

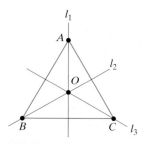

FIGURE 11-83

74. Let the eight symmetries of the square $ABCD$ shown in Fig. 11-84 be denoted as follows: r_1 denotes the reflection with axis l_1; r_2 denotes the reflection with axis l_2; r_3 denotes the reflection with axis l_3; r_4 denotes the reflection with axis l_4; R_1 denotes the 90° clockwise rotation with rotocenter O; R_2 denotes the 180° clockwise rotation with rotocenter O; R_3 denotes the 270° clockwise rotation with rotocenter O; I denotes the identity symmetry.

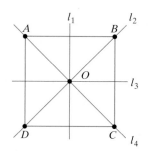

FIGURE 11-84

Complete the symmetry multiplication table below (*Hint*: Try Exercise 73 first.)

	r_1	r_2	r_3	r_4	R_1	R_2	R_3	I
r_1								
r_2								
r_3								
r_4								
R_1								
R_2								
R_3								
I								

75. Find the symmetry type of the wallpaper pattern shown in Fig. 11-85. (The flow chart appears in Appendix B on p. 627.)

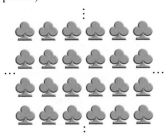

FIGURE 11-85

76. Find the symmetry type of the wallpaper pattern shown in Fig. 11-86. (The flow chart appears in Appendix B on p. 627.)

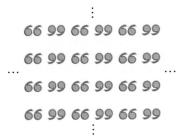

FIGURE 11-86

77. Find the symmetry type of the wallpaper pattern shown in Fig. 11-87.

FIGURE 11-87

78. Find the symmetry type of the wallpaper pattern shown in Fig. 11-88.

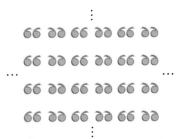

FIGURE 11-88

79. Find the symmetry type of the wallpaper pattern shown in Fig. 11-89.

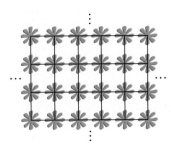

FIGURE 11-89

80. Find the symmetry type of the wallpaper pattern shown in Fig. 11-90.

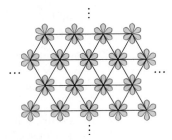

FIGURE 11-90

PROJECTS AND PAPERS

A Patterns Everywhere

Border patterns can be found in many objects from the real world—ribbons, wallpaper borders, and architectural friezes. Even ceramic pots and woven baskets exhibit border patterns (when the pattern goes around in a circle it can be unraveled as if it were going on a straight line). Likewise, wallpaper patterns can be found in wallpapers, textiles, neckties, rugs, and gift-wrapping paper. Patterns are truly everywhere.

This project consists of two separate subprojects.

Part 1. Find examples from the real world of each of the seven possible border-pattern symmetry types. Do not use photographs from a book or designs you have just downloaded from some Web site. This part is not too hard, and it is a warm-up for Part 2, the real challenge.

Part 2. Find examples from the real world of each of the 17 wallpaper-pattern symmetry types. The same rules apply as for Part 1.

Notes: (1) Your best bet is to look at wallpaper patterns and borders at a wallpaper store or gift-wrapping paper and ribbons at a paper store. You will have to do some digging—a few of the wallpaper-pattern symmetry types are hard to find. (2) For ideas, you may want to visit Steve Edwards's excellent Web site *Tiling Plane and Fancy* at *http://www2.spsu.edu/ math/tile/index.htm*.

B Three-Dimensional Rigid Motions

For two-dimensional objects, we have seen that every rigid motion is of one of four basic types. For three-dimensional objects moving in three-dimensional space, there are *six* possible types of rigid motion. Specifically, every rigid motion in three-dimensional space is equivalent to a *reflection*, a *rotation*, a *translation*, a *glide reflection*, a *rotary reflection*, or a *screw displacement*.

Prepare a presentation on the six possible types of rigid motions in three-dimensional space. For each one give a precise definition of the rigid motion, describe its most important properties, and give illustrations as well as real-world examples.

C Penrose Tilings

In the mid-1970s, British mathematician Roger Penrose made a truly remarkable discovery—it is possible to cover the plane using *aperiodic tilings*, something like a wallpaper pattern with a constantly changing motif. (See the biographical profile on page 421.) One of the simplest and most surprising aperiodic tilings discovered by Penrose is based on just two shapes—the two rhombi shown in Figure 11.91.

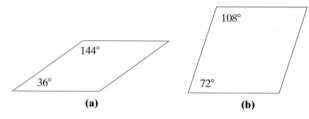

FIGURE 11-91

Prepare a presentation discussing and describing Penrose's tilings based on figures (a) and (b). Include in your presentation the connection between figures (a) and (b) and the golden ratio as well as the connection between Penrose tilings and quasicrystals.

REFERENCES AND FURTHER READINGS

1. Bunch, Bryan, *Reality's Mirror: Exploring the Mathematics of Symmetry*. New York: John Wiley & Sons, 1989.

2. Conway, John H., Heidi Burgiel, and Chaim Goodman-Strauss, *The Symmetry of Things*. New York: A. K. Peters, 2008.

3. Crowe, Donald W., "Symmetry, Rigid Motions, and Patterns," *UMAP Journal*, 8 (1987), 206–236.

4. Du Sautoy, Marcus, *Symmetry: A Journey into the Patterns of Nature*. New York: HarperCollins, 2008.

5. Gardner, Martin, *The New Ambidextrous Universe: Symmetry and Asymmetry from Mirror Reflections to Superstrings*, 3rd ed. New York: W. H. Freeman & Co., 1990.

6. Grünbaum, Branko, and G. C. Shephard, *Tilings and Patterns: An Introduction*. New York: W. H. Freeman & Co., 1989.

7. Hargittai, I., and M. Hargittai, *Symmetry: A Unifying Concept*. Bolinas, CA: Shelter Publications, 1994.

8. Hofstadter, Douglas R., *Gödel, Escher, Bach: An Eternal Golden Braid*. New York: Vintage Books, 1980.

9. Martin, George E., *Transformation Geometry: An Introduction to Symmetry*. New York: Springer-Verlag, 1994.

10. Rose, Bruce, and Robert D. Stafford, "An Elementary Course in Mathematical Symmetry," *American Mathematical Monthly*, 88 (1981), 59–64.

11. Rosen, Joe, *Symmetry Rules: How Science and Nature Are Founded on Symmetry*. Berlin-Heidelberg: Springer-Verlag, 2008.

12. Schattsneider, Doris, *Visions of Symmetry: Notebooks, Periodic Drawings, and Related Work of M. C. Escher*. New York: W. H. Freeman & Co., 1990.

13. Shubnikov, A. V., and V. A. Koptsik, *Symmetry in Science and Art*. New York: Plenum Publishing Corp., 1974.

14. Stewart, Ian, *Why Beauty Is Truth: A History of Symmetry*. New York: Basic Books, 2007.

15. Washburn, Dorothy K., and Donald W. Crowe, *Symmetries of Culture*. Seattle, WA: University of Washington Press, 1988.

16. Washburn, Dorothy K., and Donald W. Crowe, *Symmetry Comes of Age: The Role of Patterns in Culture*. Seattle, WA: University of Washington Press, 2004.

17. Weyl, Hermann, *Symmetry*. Princeton, NJ: Princeton University Press, 1952.

12 The Geometry of Fractal Shapes

Naturally Irregular

The exotic and beautiful images on this chapter opener are meant to grab the reader's attention. (Most of the time it works.) What are these images? What do they have in common? What is their connection to mathematics?

With regard to the first question, a nonspecific answer is that these images come from various and surprisingly different sources. One of them is an abstract painting by the Australian artist Nick Chlebnikowski. A second image is a photograph taken from space by the Landsat Satellite. A third image shows a cross section of lung tissue viewed under a microscope. The fourth image is a piece of computer art generated by a mathematical program. You may want to take a stab at figuring which one is which. (For answers see p. 473.)

The key characteristic these beautiful images share is that they are *fractals*, be it natural fractals or man-made fractals. The purpose of this chapter is to introduce the basic ideas behind fractals and their geometry—a geometry very different from the traditional Euclidean geometry of triangles, squares, and circles.

The chapter is organized around a series of landmark examples of fractal geometry—the *Koch snowflake* in Section 12.1, the *Sierpinski gasket* in Section 12.2, the *chaos game* in Section 12.3, the *twisted Sierpinski gasket* in Section 12.4, and the *Mandelbrot set* in Section 12.5. Each of these examples has something new to add to our fractal excursion. As we explore these examples, we will also learn something about the connection between fractal geometry and the purposeful irregularity of many shapes in nature.

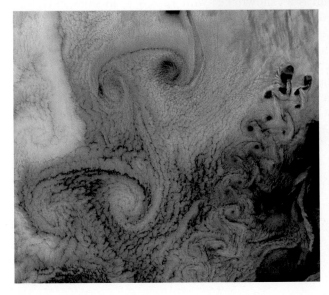

12.1   The Koch Snowflake

Our first example of a *geometric fractal* is a shape known as the **Koch snowflake**, named after the Swedish mathematician Helge von Koch (1870–1954). Like other geometric fractals, the Koch snowflake is constructed by means of a *recursive process*, a process in which the same procedure is applied repeatedly in an infinite feedback loop—the output at one step becomes the input at the next step. (We have seen examples of recursive processes in earlier chapters—recursive ranking methods in Chapter 1 and recursive definitions of sequences in Chapter 9.)

⌐ THE KOCH SNOWFLAKE ─────────────────

- **Start.** Start with a shaded *equilateral* triangle [Fig. 12-1(a)]. We will refer to this starting triangle as the *seed* of the Koch snowflake. The size of the seed triangle is irrelevant, so for simplicity we will assume that the sides are of length 1.

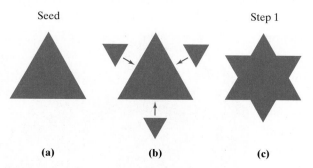

FIGURE 12-1

- **Step 1.** To the middle third of each of the sides of the seed add an equilateral triangle with sides of length 1/3, as shown in Fig. 12-1(b). The result is the 12-sided "snowflake" shown in Fig. 12-1(c).

- **Step 2.** To the middle third of each of the 12 sides of the "snowflake" in Step 1 add an equilateral triangle with sides of length one-third the length of that side. The result is a "snowflake" with $12 \times 4 = 48$ sides, each of length $(1/3)^2 = 1/9$, as shown in Fig. 12-2(a).

    For ease of reference, we will call the procedure of adding an equilateral triangle to the middle third of each side of the figure *procedure KS*. This will make the rest of the construction a lot easier to describe. Notice that *procedure KS* makes each side of the figure "crinkle" into four new sides and that each new side has length one-third the previous side.

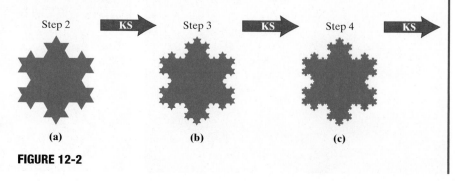

FIGURE 12-2

- **Step 3.** Apply *procedure KS* to the "snowflake" in Step 2. This gives the more elaborate "snowflake" shown in Fig. 12-2(b), with $48 \times 4 = 192$ sides, each of length $(1/3)^3 = 1/27$.
- **Step 4.** Apply *procedure KS* to the "snowflake" in Step 3. This gives the "snowflake" shown in Fig. 12-2(c). (You definitely don't want to do this by hand—there are 192 tiny little equilateral triangles that are being added!)
- **Steps 5, 6, etc.** At each step apply *procedure KS* to the "snowflake" obtained in the previous step.

At every step of this recursive process procedure KS produces a new "snowflake," but after a while it's hard to tell that there are any changes. There is a small difference between the snowflakes in Figs. 12-2(b) and (c)—a few more steps and the triangles being added become tiny specks. Soon enough, the images become *visually stable*: To the naked eye there is no difference between one snowflake and the next. For all practical purposes we are seeing the ultimate destination of this trip: the *Koch snowflake* itself (Fig. 12-3).

FIGURE 12-3

Because the Koch snowflake is constructed in an infinite sequence of steps, an actual rendering of it is impossible—Fig. 12-3 is only an approximation. This is no reason to be concerned or upset—the Koch snowflake is an abstract shape that we can study and explore by looking at imperfect pictures of it (much like we do in high school geometry when we study circles even though it is impossible to draw a perfect circle).

One advantage of recursive processes is that they allow for very simple and efficient definitions, even when the objects being defined are quite complicated. The Koch snowflake, for example, is a fairly complicated shape, but we can define it in two lines using a form of shorthand we will call a **replacement rule**—a rule that specifies how to substitute one piece for another.

THE KOCH SNOWFLAKE (REPLACEMENT RULE)

- **Start:** Start with a shaded equilateral triangle. (This is the seed.)
- **Replacement rule:** Replace each boundary line segment ▬▬ with a "crinkled" version ▬▲▬.

Self-Similarity

If we only consider the boundary of the Koch snowflake and forget about the interior, we get an infinitely jagged curve known as the **Koch curve,** or sometimes the *snowflake curve*. A section of the Koch curve is shown in Fig. 12-4(a). Since this is just a rough rendering of the Koch curve our natural curiosity pushes us to take a closer look. We'll just randomly pick a small part of this section and magnify it [Fig. 12-4(b)]. The surprise (or not!) is that we see nothing new—the small detail looks just like the rough detail. Magnifying further is not going to be much help. Figure 12-4(c) shows a detail of the Koch curve after magnifying it by a factor of almost 100.

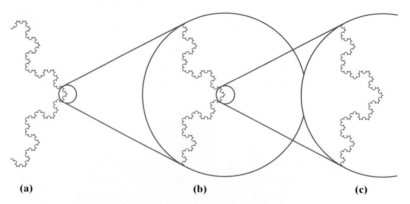

(a)   **(b)**   **(c)**

FIGURE 12-4   (a) A small piece of the Koch curve. (b) The piece in (a) magnified ×9.
(c) The same piece magnified ×81.

This seemingly remarkable characteristic of the Koch curve of looking the same at different scales is called **self-similarity**. As the name suggests, self-similarity implies that the object is similar to a part of itself, which in turn is similar to an even smaller part, and so on. In the case of the Koch curve the self-similarity has three important properties: (1) It is *infinite* (the self-similarity takes place over infinitely many levels of magnification), (2) it is *universal* (the same self-similarity occurs along every part of the Koch curve), and (3) it is **exact** (we see the exact same image at every level of magnification).

Self-similarity is not nearly as rare as it first appears. There are many objects in nature that exhibit some form of self-similarity, even if in nature self-similarity can never be infinite or exact. A good example of natural self-similarity can be found in, of all places, a head of cauliflower. A head of cauliflower is made of individual pieces called florets. Each floret looks much like a smaller head of cauliflower with even smaller florets, and so on (Fig. 12-5). The "so on" here is good only for three or four steps—at some point there are no more florets—so the self-similarity is *finite*. Moreover, the florets look very much like the head of cauliflower but are not exact clones of it. In these cases we describe the self-similarity as **approximate** (as opposed to *exact*) self-similarity.

Although the Koch snowflake is obviously quite different from a head of cauliflower (for one thing you can't make soup with it), the two have a lot in common in terms of structure and form. In fact, one could informally say that the Koch snowflake is a two-dimensional mathematical blueprint for the structure of cauliflower.

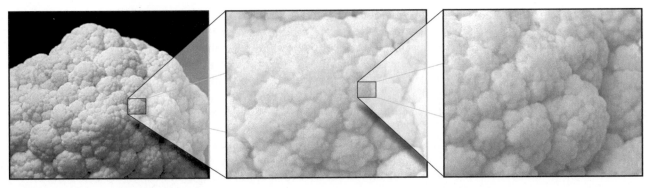

FIGURE 12-5 Approximate self-similarity in a head of cauliflower. At various levels of magnification we see . . . more or less the same thing.

Perimeter and Area of the Koch Snowflake

One of the most surprising facts about the Koch snowflake is that it has a relatively small area but an infinite perimeter and an infinitely long boundary—a notion that seems to defy common sense.

To compute the perimeter of the Koch snowflake, let's look at the perimeter of the "snowflakes" obtained in Steps 1 and 2 of the construction (Fig. 12-6). At each step we replace a side by four sides that are 1/3 as long. Thus, at any given step the perimeter P is 4/3 times the perimeter at the preceding step. This implies that the perimeters keep growing with each step, and growing very fast indeed. After infinitely many steps the perimeter is infinite.

■ See Exercises 1 and 2.

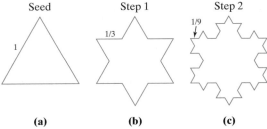

FIGURE 12-6 (a) Three sides of length 1 ($P = 3$). (b) Twelve sides of length 1/3 ($P = 4$). (c) 48 sides of length 1/9 ($P = 48/9 = 5\frac{1}{3}$).

Computing the area of the Koch snowflake is considerably more difficult, but, as we can see from Fig. 12-7, the area of the Koch snowflake is less than the area of the circle that circumscribes the seed triangle and thus, relatively small. Indeed, we can be much more specific: The area of the Koch snowflake is exactly 8/5 (or 1.6) times the area of the seed triangle. We will justify this claim next.

Since the exact computation of the area of the Koch snowflake is a bit technical, we will give a rough sketch here and leave some of the details to the reader. (The reader who wishes to do so can skip the forthcoming calculations without prejudice.)

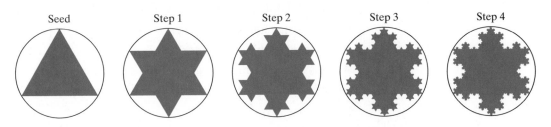

FIGURE 12-7

For simplicity we will call the area of the seed triangle A. Using A as the starting point we can calculate the area of the figures at the early steps of the construction—all we need to know is the number of new triangles being added and the area of each new triangle. A careful look at Fig. 12-8 will show us the pattern: To get from Start to Step 1 we are adding three new triangles, each of area $A/9$. (When the sides of a triangle are scaled by a factor of s, the area is scaled by a factor of s^2. Here $s = 1/3$.) Thus, the total area of the "snowflake" in Step 1 is $A + 3 \cdot (A/9) = A + (A/3)$. To get to Step 2 we add an additional 12 small triangles, each of area $A/81$. The total new area added is $12 \cdot (A/81)$. We will find it convenient to rewrite this in the form $(A/3)(4/9)$. Continuing this way (and after a good look at Fig. 12-8), we can see that in the Nth step of the construction we add $3(4^{N-1})$ new triangles, each having an area of $(1/9)^N A$, for a total added area of $(A/3)(4/9)^{N-1}$.

The total area at the Nth step of the construction is A plus the sum of all the new areas added up to that point:

$$A + \frac{A}{3} + \frac{A}{3}\left(\frac{4}{9}\right) + \frac{A}{3}\left(\frac{4}{9}\right)^2 + \cdots + \frac{A}{3}\left(\frac{4}{9}\right)^{N-1}$$

If we disregard the first A, the remaining terms of the preceding sum are terms in a geometric sequence with starting term $A/3$ and common ratio $r = 4/9$. (We discussed geometric sequences and their sums in Chapter 10). Using the geometric sum formula (see p. 379), the preceding expression becomes

■ For the details, see Exercise 61.

$$A + \frac{3}{5}A\left[1 - \left(\frac{4}{9}\right)^N\right]$$

Now, as N becomes larger and larger, $(4/9)^N$ becomes smaller and smaller (which happens whenever the base is between 0 and 1), and the expression inside the brackets becomes closer and closer to 1. It follows that the entire expression becomes closer and closer to $A + (3/5)A = (8/5)A = (1.6)A$. This implies that after infinitely many steps of adding smaller and smaller areas, we get the surprising result that the area of the Koch snowflake is 1.6 times the area of the seed equilateral triangle.

Having a very large boundary packed within a small area (or volume) is an important characteristic of many self-similar shapes in nature (long boundaries improve functionality, whereas small volumes keep energy costs down). The vascular system of veins and arteries in the human body (see Fig. 12-9) is a perfect example of the tradeoff that nature makes between length and volume: Whereas veins, arteries, and capillaries take up only a small fraction of the body's volume, their reach, by necessity, is enormous. Laid end to end, the veins, arteries, and capillaries of a single human being would extend more than 40,000 miles.

We conclude this section with a summary of the key properties of the Koch snowflake.

PROPERTIES OF THE KOCH SNOWFLAKE

- It has exact and universal self-similarity.
- It has an infinite perimeter.
- Its area is 1.6 times the area of the seed triangle.

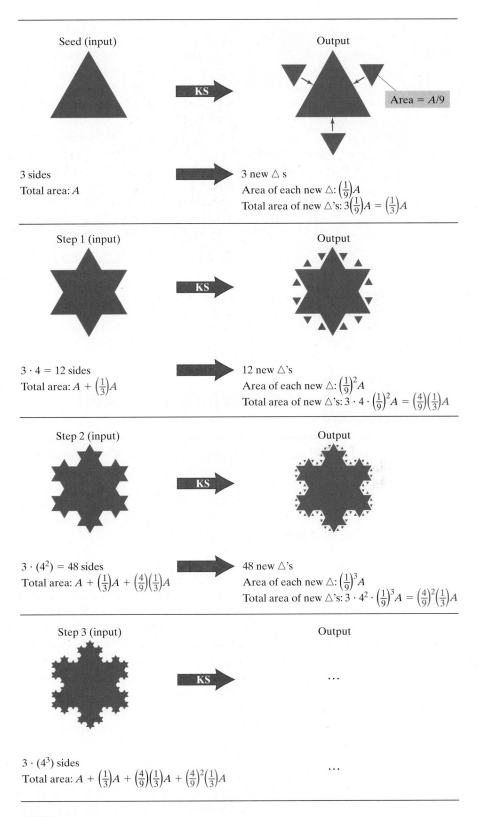

Seed (input)

Output

Area = $A/9$

3 sides
Total area: A

3 new △ s
Area of each new △: $\left(\frac{1}{9}\right)A$
Total area of new △'s: $3\left(\frac{1}{9}\right)A = \left(\frac{1}{3}\right)A$

Step 1 (input)

Output

$3 \cdot 4 = 12$ sides
Total area: $A + \left(\frac{1}{3}\right)A$

12 new △'s
Area of each new △: $\left(\frac{1}{9}\right)^2 A$
Total area of new △'s: $3 \cdot 4 \cdot \left(\frac{1}{9}\right)^2 A = \left(\frac{4}{9}\right)\left(\frac{1}{3}\right)A$

Step 2 (input)

Output

$3 \cdot (4^2) = 48$ sides
Total area: $A + \left(\frac{1}{3}\right)A + \left(\frac{4}{9}\right)\left(\frac{1}{3}\right)A$

48 new △'s
Area of each new △: $\left(\frac{1}{9}\right)^3 A$
Total area of new △'s: $3 \cdot 4^2 \cdot \left(\frac{1}{9}\right)^3 A = \left(\frac{4}{9}\right)^2\left(\frac{1}{3}\right)A$

Step 3 (input)

Output

$\cdots$

$3 \cdot (4^3)$ sides
Total area: $A + \left(\frac{1}{3}\right)A + \left(\frac{4}{9}\right)\left(\frac{1}{3}\right)A + \left(\frac{4}{9}\right)^2\left(\frac{1}{3}\right)A$

$\cdots$

FIGURE 12-8

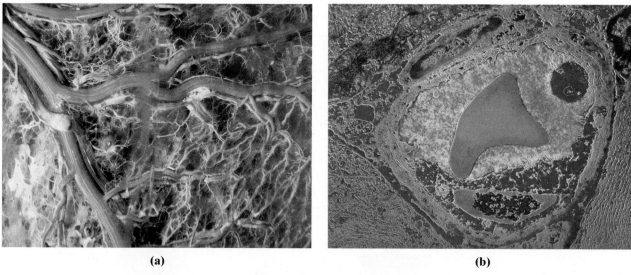

(a)

(b)

FIGURE 12-9 (a) The vascular network of the human body — 40,000 miles of veins, arteries, and capillaries packed inside small quarters. (b) Cross section of a blood capillary, with a single red blood cell in the center (magnification: ×7070).

12.2   The Sierpinski Gasket

With the insight gained by our study of the Koch snowflake, we will now look at another well-known geometric fractal called the **Sierpinski gasket**, named after the Polish mathematician Waclaw Sierpinski (1882 – 1969).

Just like with the Koch snowflake, the construction of the Sierpinski gasket starts with a solid triangle, but this time, instead of repeatedly *adding* smaller and smaller versions of the original triangle, we will *remove* them according to the following procedure:

┌─ **THE SIERPINSKI GASKET** ──────────────

- **Start.** Start with a shaded triangle ABC [Fig. 12-10(a)]. We will call this triangle the *seed* triangle. (Often an equilateral triangle or a right triangle is used as the seed, but here we chose a random triangle to underscore that it can be a triangle of arbitrary shape.)

- **Step 1.** Remove the triangle connecting the midpoints of the sides of the seed triangle. This gives the shape shown in Fig. 12-10(b)—consisting of three shaded triangles, each a half-scale version of the seed and a hole where the middle triangle used to be.

    For convenience, we will call this process of "removing the middle" of a triangle *procedure SG*.

- **Step 2.** To each of the three shaded triangles in Fig. 12-10(b) apply *procedure SG*. The result is the "gasket" shown in Fig. 12-10(c) consisting of $3^2 = 9$ triangles, each at one-fourth the scale of the seed triangle, plus three small holes of the same size and one larger hole in the middle.

- **Step 3.** To each of the nine shaded triangles in Fig. 12-10(c) apply *procedure SG*. The result is the "gasket" shown in Fig. 12-10(d) consisting of $3^3 = 27$

triangles, each at one-eighth the scale of the original triangle, nine small holes of the same size, three medium-sized holes, and one large hole in the middle.

- **Steps 4, 5, etc.** Apply *procedure SG* to each shaded triangle in the "gasket" obtained in the previous step.

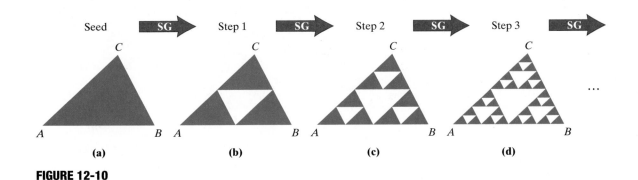

FIGURE 12-10

After a few more steps the figure becomes visually stable — the naked eye cannot tell the difference between the gasket obtained at one step and the gasket obtained at the next step. At this point we have a good rendering of the Sierpinski gasket itself. In your mind's eye you can think of Fig. 12-11 as a picture of the Sierpinski gasket (in reality, it is the gasket obtained at Step 7 of the construction process).

FIGURE 12-11

The Sierpinski gasket is clearly a fairly complicated geometric shape, and yet it can be defined in two lines using the following recursive *replacement rule*.

THE SIERPINSKI GASKET (REPLACEMENT RULE)

- **Start:** Start with a shaded seed triangle.
- **Replacement rule:** Whenever you see a ▲, replace it with a ▲▲.

Looking at Fig. 12-11, we might think that the Sierpinski gasket is made of a huge number of tiny triangles, but this is just an optical illusion — there are no solid triangles in the Sierpinski gasket, just specks of the original triangle surrounded by a sea of white triangular holes. If we were to magnify any one of those small specks, we would see more of the same—specks surrounded by white triangles (Fig. 12-12). This, of course, is another example of *self-similarity*. As with the Koch curve, the self-similarity of the Sierpinski gasket is (1) infinite, (2) universal, and (3) exact.

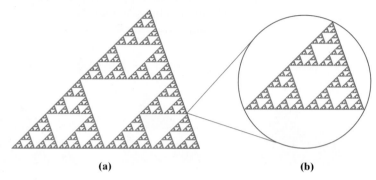

(a)   (b)

FIGURE 12-12 (a) Sierpinski gasket. (b) Sierpinski gasket detail (magnification ×256).

As a geometric object existing in the plane, the Sierpinski gasket should have an area, but it turns out that its area is infinitesimally small, smaller than any positive quantity. Paradoxical as it may first seem, the mathematical formulation of this fact is that the Sierpinski gasket has *zero area*. At the same time, the boundary of the "gaskets" obtained at each step of the construction grows without bound, which implies that the Sierpinski gasket has an infinitely long boundary.

■ See Exercises 21 and 22.

┌─ **PROPERTIES OF THE SIERPINSKI GASKET** ─

- ■ It has exact and universal self-similarity.
- ■ It has infinitely long boundary.
- ■ It has zero area.

Designs that resemble the Sierpinski gasket can be found on the shells of certain families of sea snails and volutes (Fig. 12-13). Exactly why this happens is not fully understood, but the shells sure are beautiful!

FIGURE 12-13 (a) Livonia mamilla. (b) Conus canonicus. (c) Conus ammiralis.

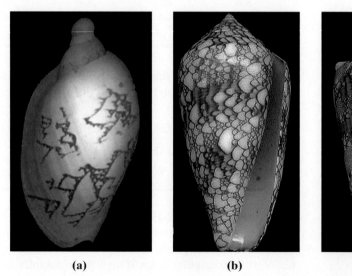

(a)   (b)   (c)

There is a surprising cutting-edge application of the Sierpinski gasket to one of the most vexing problems of modern life: the inconsistent reception on your cell phone. A really efficient cell phone antenna must have a very large boundary (more boundary means stronger reception), a very small area (less area means more energy efficiency), and exact self-similarity (self-similarity means that the antenna works equally well at every frequency of the radio spectrum). What better choice than a design based on a Sierpinski gasket? Fractal antennas based on this concept are now used not only in cell phones but also in wireless modems and GPS receivers (see Project D).

12.3 The Chaos Game

In this section we will introduce a recursive construction rule that involves the element of chance. The construction is known as the **chaos game** and is attributed to Michael Barnsley, a mathematician at the Australian National University.

We start with an arbitrary triangle with vertices A, B, and C and an honest die [Fig. 12-14(a)]. Before we start we assign two of the six possible outcomes of rolling the die to each of the vertices of the triangle. For example, if we roll a 1 or a 2, then A is the chosen vertex; if we roll a 3 or a 4, then B is the chosen vertex; and if we roll a 5 or a 6, then C is the chosen vertex. We are now ready to play the game.

THE CHAOS GAME

- **Start.** Roll the die and mark the chosen vertex. Say we roll a 5. This puts us at vertex C [Fig. 12-14(b)].

- **Step 1.** Roll the die again. Say we roll a 2, so the new chosen vertex is A. We now *move to the point M_1 halfway between the previous position C and the winning vertex A*. Mark the new position M_1 [Fig. 12-14(c)].

- **Step 2.** Roll the die again, and *move to the point halfway between the last position M_1 and the new chosen vertex*. [Say we roll a 3—the move then is to M_2 halfway between M_1 and B as shown in Fig. 12-14(d).] Mark a point at the new position M_2.

- **Steps 3, 4, etc.** Continue rolling the die, each time *moving to a point halfway between the last position and the chosen vertex* and marking that point.

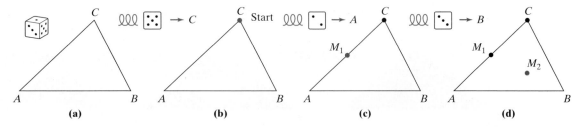

FIGURE 12-14 The chaos game: Always move from the previous position toward the chosen vertex and stop halfway.

What happens after you roll the die many times? Figure 12-15(a) shows the pattern of points after 50 rolls of the die—just a bunch of scattered dots. Figure 12-15(b) shows the pattern of points after 500 rolls, and Fig. 12-15(c) shows the pattern of points after 5000 rolls. The last figure is unmistakable: a Sierpinski gasket! The longer we play the chaos game, the closer we get to a Sierpinski gasket. After 100,000 rolls of the die, it would be impossible to tell the difference between the two.

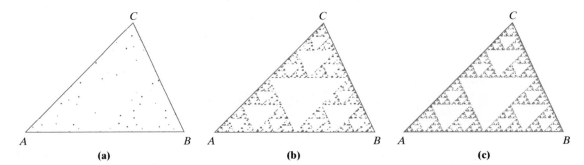

FIGURE 12-15 The "footprint" of the chaos game after (a) 50 rolls of the die, (b) 500 rolls of the die, and (c) 5000 rolls of the die.

This is a truly surprising turn of events. The chaos game is ruled by the laws of chance; thus, we would expect that essentially a random pattern of points would be generated. Instead, we get an approximation to the Sierpinski gasket, and the longer we play the chaos game, the better the approximation gets. An important implication of this is that it is possible to generate geometric fractals using simple rules based on the laws of chance.

Clearly, the best way to see the chaos game in action is with a computer. There are many good Web sites for the chaos game and its variations, and a partial list of them is given in the references at the end of the chapter.

12.4 The Twisted Sierpinski Gasket

Our next construction is a variation of the original Sierpinski gasket. For lack of a better name, we will call it the twisted Sierpinski gasket.

The construction starts out exactly like the one for the regular Sierpinski gasket, with a seed triangle [Fig. 12-16(a)] from which we cut out the middle triangle, whose vertices we will call M, N, and L [Fig. 12-16(b)]. The next move (which we will call the "twist") is new. Each of the points M, N, and L is moved a small amount in a random direction—as if jolted by an earthquake—to new positions M', N', and L'. One possible resulting shape is shown in Fig. 12-16(c).

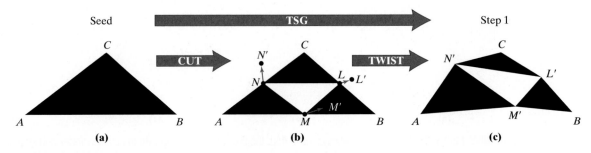

FIGURE 12-16 The two moves in *procedure TSG*: the *cut* and the *twist*.

For convenience, we will use the term *procedure TSG* to describe the combination of the two moves ("cut" and then "twist").

- **Cut.** Remove the "middle" of the triangle [Fig. 12-16(b)].

- **Twist.** Translate each of the midpoints of the sides by a small random amount and in a random direction. [Fig. 12-16(c)].

When we repeat *procedure TSG* in an infinite recursive process, we get the twisted Sierpinski gasket.

THE TWISTED SIERPINSKI GASKET

- **Start.** Start with a shaded seed triangle [Fig. 12-17(a)].

- **Step 1.** Apply *procedure TSG* to the seed triangle. This gives the "twisted gasket" shown in Fig. 12-17(b), with three twisted triangles and a (twisted) hole in the middle.

- **Step 2.** To each of the three shaded triangles in Fig. 12-17(b) apply *procedure TSG*. The result is the "twisted gasket" shown in Fig. 12-17(c), consisting of nine twisted triangles and four holes of various sizes.

- **Steps 3, 4, etc.** Apply *procedure TSG* to each shaded triangle in the "twisted gasket" obtained in the previous step.

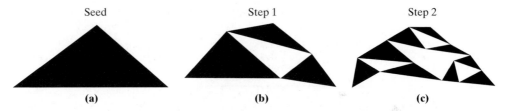

Seed   Step 1   Step 2

(a)   (b)   (c)

FIGURE 12-17 First two steps in the construction of the twisted Sierpinski gasket.

FIGURE 12-18

Figure 12-18 shows Step 7 of the construction of a twisted Sierpinski gasket. Even in this basic form it is remarkable how much the figure resembles a snow-covered mountain. Add a few of the standard tools of computer graphics—color, lighting, and shading—and we can get a very realistic-looking mountain [Fig. 12-19(b)].

(a)   (b)

FIGURE 12-19 (a) Photo of a real mountain range. (b) Computer image created using a variation of the twisted Sierpinski gasket construction.

The construction of the twisted Sierpinski gasket can be also described by a two-line recursive replacement rule.

> ┌─**THE TWISTED SIERPINSKI GASKET (REPLACEMENT RULE)**─────┐
>
> ■ **Start:** Start with a shaded seed triangle.
>
> ■ **Replacement rule:** Whenever you see a shaded triangle, apply *procedure TSG* to it.

The twisted Sierpinski gasket has infinite and universal self-similarity, but due to the randomness of the twisting process the self-similarity is only *approximate*—when we magnify any part of the gasket we see similar but not identical images, as illustrated in Fig. 12-20. It is the approximate self-similarity that gives this shape that natural look that the original Sierpinski gasket lacks, and it illustrates why randomness is an important element in creating artificial imitations of natural-looking shapes (Fig. 12-21).

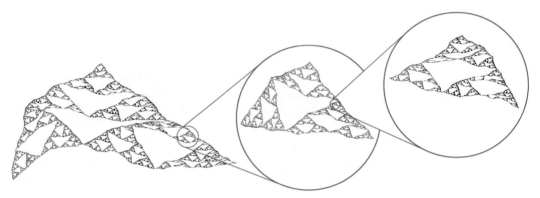

FIGURE 12-20 Approximate self-similarity in the twisted Sierpinski gasket.

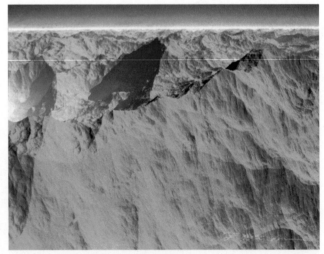

FIGURE 12-21 By changing the shape of the starting triangle, we can change the shape of the mountain, and by changing the rules for how large we allow the random displacements to be, we can change the mountain's texture. Both of these images were created using recursive replacement rules.

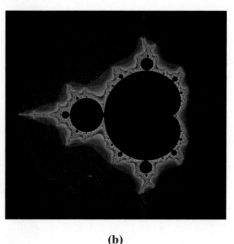 (12.5)

The Mandelbrot Set

■ For more on Mandelbrot, see the biographical profile at the end of this chapter.

In this section we will introduce one of the most interesting and beautiful geometric fractals ever created, an object called the **Mandelbrot set** after Benoit Mandelbrot, a mathematician at Yale University. Some of the mathematics behind the Mandelbrot set goes a bit beyond the level of this book, so we will describe the overall idea and try not to get bogged down in the details.

We will start this section with a brief visual tour. The Mandelbrot set is the shape shown in Fig. 12-22(a), a strange-looking blob of black. Using a strategy we will describe later, the different regions outside the Mandelbrot set can be colored according to their mathematical properties, and when this is done [Fig. 12-22(b)], the Mandelbrot set comes to life like a switched-on neon sign.

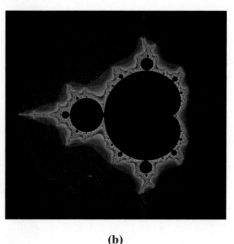

(a)

(b)

FIGURE 12-22 (a) The Mandelbrot set (black) on a white background. (b) The Mandelbrot set comes to life when color is added to the outside.

In the wild imagination of some, the Mandelbrot set looks like some sort of bug—an exotic extraterrestrial flea. The "flea" is made up of a heart-shaped body (called a *cardioid*), a head, and an antenna coming out of the middle of the head. A careful look at Fig. 12-22 shows that the flea has many "smaller fleas that prey on it," but we can only begin to understand the full extent of the infestation when we look at Fig. 12-23(a)—a finely detailed close-up of the boundary of the Mandelbrot set. When we magnify the view around the boundary even further, we can see that these secondary fleas have fleas that "prey" on them [Figs. 12-23(b) and (c)], and further magnification would show this repeats itself ad infinitum. Clearly, Jonathan Swift was onto something!

> " So Natr'alists observe, A Flea Hath smaller Fleas that on him prey and these have smaller Fleas to bite'em and so proceed, ad infinitum. "
>
> — Jonathan Swift

It is clear from Fig. 12-23 that the Mandelbrot set has some strange form of *infinite* and *approximate* self-similarity—at infinitely many levels of magnification we see the same theme—similar but never identical fleas surrounded by similar smaller fleas. At the same time, we can see that there is tremendous variation in the regions surrounding the individual fleas. The images we see are a peek into a psychedelic coral reef—a world of strange "urchins" and "seahorse tails" in Fig. 12-23(b), "anemone" and "starfish" in Fig. 12-23(c). Further magnification shows an even more exotic and beautiful landscape. Figure 12-24(a) is a close-up

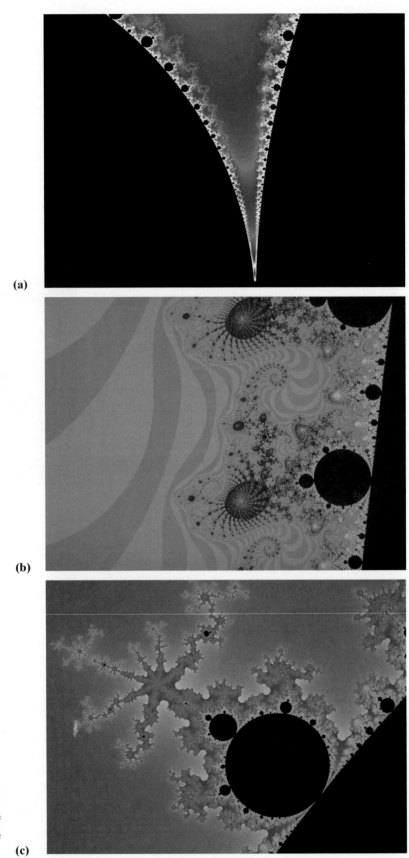

FIGURE 12-23 (a) Detailed close-up of a small region on the boundary. (b) An even tighter close-up of the boundary. (c) A close-up of one of the secondary fleas.

of one of the seahorse tails in Fig. 12-23(b). A further close-up of a section of Fig. 12-24(a) is shown in Fig. 12-24(b), and an even further magnification of it is seen in Fig. 12-24(c), revealing a tiny copy of the Mandelbrot set surrounded by a beautiful arrangement of swirls, spirals, and seahorse tails. [The magnification for Fig. 12-24(c) is approximately 10,000 times the original.]

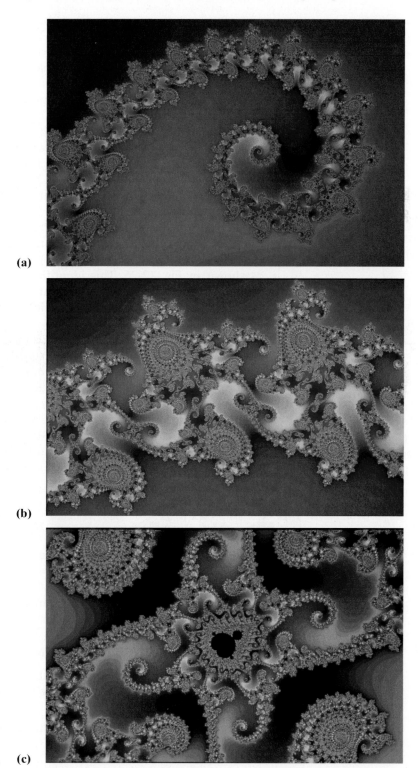

(a)

(b)

(c)

FIGURE 12-24 (a) A close-up of one of the "seahorse tails" in Fig. 12-23(b) (×100 magnification). (b) A further close-up of a section of Fig. 12-24(a) shows the fine detail and more seahorse tails (×200 magnification). (c) At infinitely many levels of magnification, old themes show up in new settings (×10,000 magnification).

Anywhere we choose to look at these pictures, we will find (if we magnify enough) copies of the original Mandelbrot set, always surrounded by an infinitely changing, but always stunning, background. The infinite, approximate self-similarity of the Mandelbrot set manages to blend infinite repetition and infinite variety, creating a landscape as consistently exotic and diverse as nature itself.

Complex Numbers and Mandelbrot Sequences

How does this magnificent mix of beauty and complexity called the Mandelbrot set come about? Incredibly, the Mandelbrot set itself can be described mathematically by a recursive process involving simple computations with *complex numbers*.

You may recall having seen complex numbers in high school algebra. Among other things, complex numbers allow us to take square roots of negative numbers and solve quadratic equations of any kind. The basic building block for complex numbers is the number $i = \sqrt{-1}$. Starting with i we can build all other complex numbers, such as $3 + 2i$, $-0.125 + 0.75i$, and even $1 + 0i = 1$ or $-0.75 + 0i = -0.75$. Just like real numbers, complex numbers can be added, multiplied, divided, and squared. (For a quick review of the basic operations with complex numbers, see Exercises 41 through 44.)

For our purposes, the most important fact about complex numbers is that they have a geometric interpretation: The complex number $(a + bi)$ can be identified with the point (a, b) in a Cartesian coordinate system, as shown in Fig. 12-25. This identification means that every complex number can be thought of as a point in the plane and that operations with complex numbers have geometric interpretations (see Exercises 45 and 46).

The key concept in the construction of the Mandelbrot set is that of a *Mandelbrot sequence*. A **Mandelbrot sequence** is an infinite sequence of complex numbers that starts with an arbitrary complex number s we call the **seed**, and then each successive term in the sequence is obtained recursively by *adding the seed s to the square of the previous term*. Figure 12-26 shows a schematic illustration of how a generic Mandelbrot sequence is generated.

[Notice that the only complex number operations involved are (1) squaring a complex number and (2) adding two complex numbers. If you know how to do these two things, you can compute Mandelbrot sequences.]

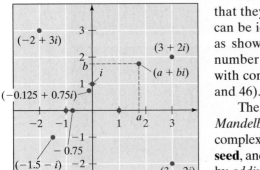

FIGURE 12-25 Every complex number is a point in the Cartesian plane; every point in the Cartesian plane is a complex number.

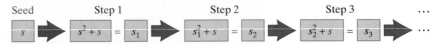

FIGURE 12-26 A Mandelbrot sequence with seed s. The red arrow is shorthand for "square the number and add the seed."

Much like a Koch snowflake and a Sierpinski gasket, a Mandelbrot sequence can be defined by means of a recursive replacement rule:

MANDELBROT SEQUENCE (REPLACEMENT RULE)

■ **Start:** Choose an arbitrary complex number s. We will call s the seed of the Mandelbrot sequence. Set the seed s to be the initial term of the sequence $(s_0 = s)$.

■ **Procedure M:** To find the next term in the sequence, square the preceding term and add the seed [$s_{N+1} = (s_N)^2 + s$].

The following set of examples will illustrate the different patterns of growth exhibited by Mandelbrot sequences as we vary the seeds. These different patterns of growth are going to tell us how to generate the Mandelbrot set itself and the incredible images that we saw in our visual tour. The idea goes like this: Each point in the Cartesian plane is a complex number and thus the seed of some Mandelbrot sequence. The pattern of growth of that Mandelbrot sequence determines whether the seed is inside the Mandelbrot set (black point) or outside (nonblack point). In the case of nonblack points, the color assigned to the point is also determined by the pattern of growth of the corresponding Mandelbrot sequence. Let's try it.

EXAMPLE 12.1   Escaping Mandelbrot Sequences

Figure 12-27 shows the first few terms of the Mandelbrot sequence with seed $s = 1$. (Since integers and decimals are also complex numbers, they make perfectly acceptable seeds.) The pattern of growth of this Mandelbrot sequence is clear—the terms are getting larger and larger, and they are doing so very quickly. Geometrically, it means that the points in the Cartesian plane that represent the numbers in this sequence are getting farther and farther away from the origin. For this reason, we call this sequence an *escaping* Mandelbrot sequence.

Seed $s = 1$ Step 1 $s_1 = 1^2 + 1$ = 2 Step 2 $s_2 = 2^2 + 1$ = 5 Step 3 $s_3 = 5^2 + 1$ = 26 Step 4 $s_4 = 26^2 + 1$ = 677 $\cdots$

FIGURE 12-27 Mandelbrot sequence with seed $s = 1$ (*escaping very quickly*).

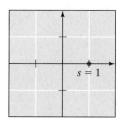

FIGURE 12-28 The seed $s = 1$ is assigned a "cool" color (blue).

In general, when the points that represent the terms of a Mandelbrot sequence move farther and farther away from the origin, we will say that the Mandelbrot sequence is **escaping**. The basic rule that defines the Mandelbrot set is that seeds of escaping Mandelbrot sequences are *not* in the Mandelbrot set and must be assigned some color other than black. While there is no specific rule that tells us what color should be assigned, the overall color palette is based on how fast the sequence is escaping. The typical approach is to use "hot" colors such as reds, yellows, and oranges for seeds that escape slowly and "cool" colors such as blues and purples for seeds that escape quickly. The seed $s = 1$, for example, escapes very quickly, and the corresponding point in the Cartesian plane is painted blue (Fig. 12-28).

EXAMPLE 12.2   Periodic Mandelbrot Sequences

Figure 12-29 shows the first few terms of the Mandelbrot sequence with seed $s = -1$. The pattern that emerges here is also clear—the numbers in the sequence alternate between 0 and -1. In this case, we say that the Mandelbrot sequence is *periodic*.

Seed $s = -1$ Step 1 $s_1 = (-1)^2 + (-1)$ = 0 Step 2 $s_2 = 0^2 + (-1)$ = -1 Step 3 $s_3 = (-1)^2 + (-1)$ = 0 $\cdots$

FIGURE 12-29 Mandelbrot sequence with seed $s = -1$ (*periodic*).

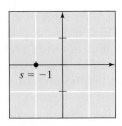

FIGURE 12-30 The seed $s = -1$ is in the Mandelbrot set (black point).

In general, a Mandelbrot sequence is said to be **periodic** if at some point the numbers in the sequence start repeating themselves in a cycle. When the Mandelbrot sequence is periodic, the seed is a point of the Mandelbrot set and thus is assigned the color black (Fig. 12-30).

EXAMPLE 12.3   **Attracted Mandelbrot Sequences**

Figure 12-31 shows the first few terms in the Mandelbrot sequence with seed $s = -0.75$. Here the growth pattern is not obvious, and additional terms of the sequence are needed. Further computation (a calculator will definitely come in handy) shows that as we go farther and farther out in this sequence, the terms get closer and closer to the value -0.5 (see Exercise 62). In this case, we will say that the sequence is *attracted* to the value -0.5.

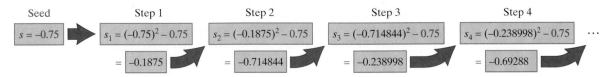

FIGURE 12-31 Mandelbrot sequence with seed $s = -0.75$ (*attracted*).

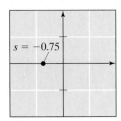

FIGURE 12-32 The seed $s = -0.75$ is in the Mandelbrot set (black point).

In general, when the terms in a Mandelbrot sequence get closer and closer to a fixed complex number a, we say that a is an **attractor** for the sequence or, equivalently, that the sequence is **attracted** to a. Just as with periodic sequences, when a Mandelbrot sequence is attracted, the seed s is in the Mandelbrot set and is colored black (Fig. 12-32).

So far, all our examples have been based on rational number seeds (mostly to keep things simple), but the truly interesting cases occur when the seeds are complex numbers. The next two examples deal with complex number seeds.

EXAMPLE 12.4   **A Periodic Mandelbrot Sequence with Complex Terms**

In this example we will examine the growth of the Mandelbrot sequence with seed $s = i$. Starting with $s = i$ (and using that $i^2 = -1$), we get $s_1 = i^2 + i = -1 + i$. If we now square s_1 and add i, we get $s_2 = (-1 + i)^2 + i = -i$, and repeating the process gives $s_3 = (-i)^2 + i = -1 + i$. At this point we notice that $s_3 = s_1$, which implies $s_4 = s_2, s_5 = s_1$, and so on (Fig. 12-33). This, of course, means that this Mandelbrot sequence is *periodic*, with its terms alternating between the complex numbers $-1 + i$ (odd terms) and $-i$ (even terms).

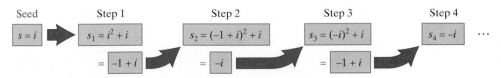

FIGURE 12-33 Mandelbrot sequence with seed $s = i$ (*periodic*).

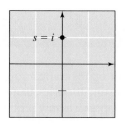

FIGURE 12-34 The seed $s = i$ is in the Mandelbrot set (black point).

The key conclusion from the preceding computations is that the seed i is a black point inside the Mandelbrot set (Fig. 12-34).

The next example illustrates the case of a Mandelbrot sequence with three complex attractors.

EXAMPLE 12.5   **A Mandelbrot Sequence with Three Complex Attractors**

In this example we will examine the growth of the Mandelbrot sequence with seed $s = -0.125 + 0.75i$. The first few terms of the Mandelbrot sequence are shown. We leave it to the enterprising reader to check these calculations. (You will need a scientific calculator that handles complex numbers. If you don't have one, you can download a free virtual calculator for doing complex number arithmetic from http://www.calc3D.com or other similar Web sites.)

$$s_0 = -0.125 + 0.75i \qquad s_1 = -0.671875 + 0.5625i \qquad s_2 = 0.0100098 - 0.00585938i$$

$$s_3 = -0.124934 + 0.749883i \qquad s_4 = -0.671716 + 0.562628i \qquad s_5 = 0.00965136 - 0.00585206i$$

$$s_6 = -0.124941 + 0.749887i \qquad s_7 = -0.67172 + 0.562617i \qquad s_8 = 0.00967074 - 0.00584195i$$

We can see that the terms in this Mandelbrot sequence are complex numbers that essentially cycle around in sets of three and are approaching three different attractors. Since this Mandelbrot sequence is attracted, the seed $s = -0.125 + 0.75i$ represents another point in the Mandelbrot set.

The Mandelbrot Set

Given all the previous examples and discussion, a formal definition of the Mandelbrot set using seeds of Mandelbrot sequences sounds incredibly simple: If the Mandelbrot sequence is *periodic* or *attracted*, the seed is a point of the Mandelbrot set and assigned the color black; if the Mandelbrot sequence is *escaping*, the seed is a point outside the Mandelbrot set and assigned a color that depends on the speed at which the sequence is escaping (hot colors for slowly escaping sequences, cool colors for quickly escaping sequences). There are a few technical details that we omitted, but essentially these are the key ideas behind the amazing pictures that we saw in Figs. 12-23 and 12-24. In addition, of course, a computer is needed to carry out the computations and generate the images.

Because the Mandelbrot set provides a bounty of aesthetic returns for a relatively small mathematical investment, it has become one of the most popular mathematical playthings of our time. Hundreds of software programs that allow one to explore the beautiful landscapes surrounding the Mandelbrot set are available, and many of these programs are freeware. (You can find plenty of these by Googling the term "Mandelbrot set.")

CONCLUSION

The study of fractals and their geometry has become a hot mathematical topic. It is a part of mathematics that combines complex and interesting theories, beautiful graphics, and extreme relevance to the real world. In this chapter we only scratched the surface of this deep and rich topic.

The word **fractal** (from the Latin *fractus*, meaning "broken up or fragmented") was coined by Benoit Mandelbrot in the mid-1970s to describe objects as diverse as the Koch curve, the Sierpinski gasket, the twisted Sierpinski gasket, and the Mandelbrot set, as well as many shapes in nature, such as clouds, mountains, trees, rivers, a head of cauliflower, and the vascular system in the human body.

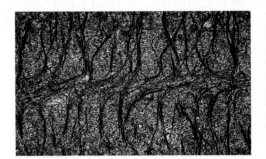  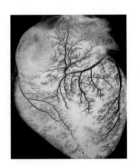

These objects share one key characteristic: They all have some form of self-similarity. (Self-similarity is not the only defining characteristic of a fractal. Others, such as *fractional dimension*, are discussed in Project A.) There is a striking visual difference between the fractal geometry of self-similar shapes and the traditional geometry of lines, circles, spheres, and so on. This visual difference is most apparent when we compare the look and texture of natural objects with those of man-made objects.

Geometry as we have known it in the past was developed by the Greeks about 2000 years ago and passed on to us essentially unchanged. It was (and still is) a great triumph of the human mind, and it has allowed us to develop much of our technology, engineering, architecture, and so on. As a tool and a language for modeling and representing nature, however, Greek geometry has by and large been a failure. The discovery of fractal geometry seems to have given science the right mathematical language to overcome this failure and thus promises to be one of the great achievements of twentieth-century mathematics. Today fractal geometry is used to study the patterns of clouds and how they affect the weather, to diagnose the pattern of contractions of a human heart, to design more efficient antennas, and to create some of the otherworldly computer graphics that animate many of the latest science fiction movies.

PROFILE: Benoit Mandelbrot (1924–)

With the publication in 1983 of his classic book *The Fractal Geometry of Nature*, Benoit Mandelbrot became somewhat of a scientific celebrity. Later, with images of the Mandelbrot set and other exotic fractals becoming part of our popular culture (through screen-savers, T-shirt designs, and television ads), the Mandelbrot name became an icon of the computer age.

Born in Warsaw, Poland, in 1924, Mandelbrot's family moved to France when he was 11 years old. These were difficult times—the beginning of World War II—and Mandelbrot received little formal education during his formative years. One of his uncles was a mathematics professor, and Mandelbrot learned some mathematics from him, but by and large the mathematics that Mandelbrot learned in his youth was self-taught. To Mandelbrot, this turned out to be an asset rather than a liability, as he credits much of his refined mathematical intuition and his ability to think "outside the box" to the unstructured nature of his early education.

After the war, Mandelbrot was able to receive a first-rate university education: an undergraduate degree in mathematics from the École Polytechnique in Paris in 1947, an M.S. degree in aeronautics from the California Institute of Technology in 1948, and a Ph.D. in mathematics from the University of Paris in 1952. After a brief stint doing research and teaching in Europe, Mandelbrot accepted an appointment in the research division of IBM and moved to the United States in 1958. For the next 30 years Mandelbrot worked at IBM's T. J. Watson Research Center in New York State, first as a Research Fellow and eventually as an IBM Fellow, the most prestigious research position within the company. It was during his years at IBM that Mandelbrot developed and refined his theories on fractal geometry.

Upon his retirement from IBM in 1987, Mandelbrot accepted an endowed chair at Yale University, where he is currently the Stirling Professor Emeritus of Mathematical Sciences.

KEY CONCEPTS

approximate self-similarity, **440**
attractor, **456**
attracted (sequence), **456**
chaos game, **446**
escaping (sequence), **455**
exact self-similarity, **440**

fractal, **458**
Koch curve (snowflake curve), **440**
Koch snowflake, **438**
Mandelbrot sequence, **454**
Mandelbrot set, **451**
periodic (sequence), **456**

replacement rule, **439**
seed (Mandelbrot sequence), **454**
self-similarity, **440**
Sierpinski gasket, **444**
twisted Sierpinski gasket, **448**

EXERCISES

WALKING

A The Koch Snowflake and Variations

Exercises 1 through 4 refer to the Koch snowflake discussed in Section 12.1.

1. Assume that the seed triangle of the *Koch snowflake* has sides of length 1 cm. Let M denote the number of sides, l the length of each side, and P the perimeter of the "snowflake" obtained at the indicated step of the construction.

   (a) Complete the missing entries in the table.

	M	l	P
Start	3	1 cm	3 cm
Step 1	12	1/3 cm	4 cm
Step 2			
Step 3			
Step 4			
Step 50			*

   * Give your answer rounded to the nearest kilometer.

   (b) For the "snowflake" obtained at step N of the construction, express M, l, and P in terms of N.

2. Assume that the seed triangle of the *Koch snowflake* has sides of length 6 cm. Let M denote the number of sides, l the length of each side, and P the perimeter of the "snowflake" obtained at the indicated step of the construction.

   (a) Complete the missing entries in the table.

	M	l	P
Start	3	6 cm	18 cm
Step 1	12	2 cm	24 cm
Step 2			
Step 3			
Step 4			
Step 40			*

   * Give your answer rounded to the nearest kilometer.

(b) For the "snowflake" obtained at step N of the construction, express M, l, and P in terms of N.

3. Assume that the seed triangle of the *Koch snowflake* has area $A = 81$ in^2.

   (a) Let R denote the number of triangles added at a particular step, S the area of each added triangle, T the total new area added, and Q the area of the "snowflake" obtained at a particular step of the construction. Complete the missing entries in the table. (*Hint*: See Fig. 12-7.)

	R	S	T	Q
Start	0	0	0	81 in^2
Step 1	3	9 in^2	27 in^2	108 in^2
Step 2	12	1 in^2		
Step 3				
Step 4				
Step 5				

   (b) What is the area of the Koch snowflake?

4. Assume the seed triangle of the *Koch snowflake* has area $A = 729$ in^2.

   (a) Let R denote the number of triangles added at a particular step, S denote the area of each added triangle, T the total new area added, and Q the area of the "snowflake" obtained at a particular step of the construction. Complete the missing entries in the table. (*Hint*: See Fig. 12-7.)

	R	S	T	Q
Start	0	0	0	729 in^2
Step 1	3	81 in^2	243 in^2	972 in^2
Step 2	12	9 in^2		
Step 3				
Step 4				
Step 5				

   (b) What is the area of the Koch snowflake?

*Exercises 5 through 8 refer to a variation of the Koch snowflake called the **quadratic Koch fractal**. The construction of the quadratic Koch fractal is similar to that of the Koch snowflake, but it uses squares instead of equilateral triangles as the shape's building blocks. The following recursive construction rule defines the quadratic Koch fractal.*

QUADRATIC KOCH FRACTAL

- **Start.** *Start with a shaded seed square [Fig. 12-35(a)].*
- **Step 1.** *Attach a smaller square (sides one-third the length of the sides of the seed square) to the middle third of each side [Fig. 12-35(b)].*
- **Step 2.** *Attach a smaller square (sides one-third the length of the sides of the previous side to the middle third of each side [Fig. 12-35(c)]. (Call this procedure QKF.)*

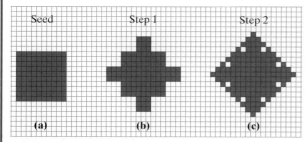

Seed   Step 1   Step 2

(a)   (b)   (c)

FIGURE 12-35

- **Steps 3, 4, etc.** *At each step, apply procedure QKF to the figure obtained in the preceding step.*

---

5. Assume that the seed square of the *quadratic Koch fractal* has sides of length 9 cm. Let M denote the number of sides, l the length of each side, and P the perimeter of the shape obtained at the indicated step of the construction.

(a) Complete the missing entries in the table.

	M	l	P
Start	4	9 cm	36 cm
Step 1	20	3 cm	60 cm
Step 2			
Step 3			
Step 30			*

* Give your answer rounded to the nearest kilometer.

(b) For the shape obtained at Step N of the construction, express M, l, and P in terms of N.

6. Assume that the seed square of the *quadratic Koch fractal* has sides of length a. Let M denote the number of sides, l the length of each side, and P the perimeter of the shape obtained at the indicated step of the construction.

(a) Complete the missing entries in the table.

	M	l	P
Start	4	a	$4a$
Step 1	20	$a/3$	$(20/3)a$
Step 2			
Step 3			
Step 4			

(b) For the shape obtained at Step N of the construction, express M, l, and P in terms of N and a.

7. Assume that the seed square of the *quadratic Koch fractal* has sides of length 9. Let R denote the number of squares added at a particular step, S the area of each added square, T the total new area added, and Q the area of the shape obtained at a particular step of the construction. Complete the missing entries in the table.

	R	S	T	Q
Start	0	0	0	81
Step 1	4	9	36	117
Step 2	20			
Step 3				
Step 4				

8. Assume that the seed square of the *quadratic Koch fractal* has area $A = 243$. Let R denote the number of squares added at a particular step, S the area of each added square, T the total new area added, and Q the area of the shape obtained at a particular step of the construction. Complete the missing entries in the table.

	R	S	T	Q
Start	0	0	0	243
Step 1	4	27	108	351
Step 2	20			
Step 3				
Step 4				

*Exercises 9 through 12 refer to a variation of the Koch snowflake called the **Koch antisnowflake**. The Koch anti-snowflake is much like the Koch snowflake, but it is based on a recursive rule that removes equilateral triangles. The recursive replacement rule for the Koch antisnowflake is as follows.*

KOCH ANTISNOWFLAKE

■ **Start:** *Start with a shaded seed equilateral triangle [Fig. 12-36(a)].*

■ **Replacement rule:** *Replace each boundary line segment* ▬ *with a* ▬\/▬ *. [Figures 12-36(b) and (c) show the shapes obtained at Steps 1 and 2, respectively.]*

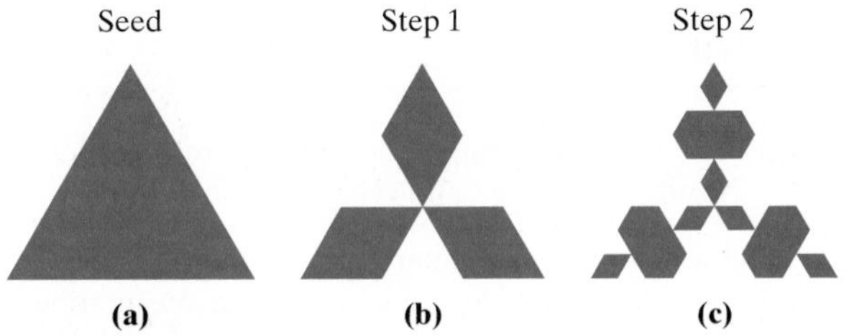

Seed　　　　　Step 1　　　　Step 2

(a)　　　　　　(b)　　　　　　(c)

FIGURE 12-36

---

9. Assume that the seed triangle of the *Koch anti-snowflake* has sides of length 6 cm. Let M denote the number of sides, l the length of each side, and P the perimeter of the shape obtained at the indicated step of the construction.

   (a) Complete the missing entries in the table.

	M	l	P
Start	3	6 cm	18 cm
Step 1	12	2 cm	24 cm
Step 2			
Step 3			
Step 4			
Step 30			*

   * Give your answer rounded to the nearest kilometer.

   (b) For the shape obtained at Step N of the construction, express M, l, and P in terms of N.

10. Assume that the seed triangle of the *Koch anti-snowflake* has sides of length 1 cm. Let M denote the number of sides, l the length of each side, and P the perimeter of the shape obtained at the indicated step of the construction.

    (a) Complete the missing entries in the table.

	M	l	P
Start	3	1 cm	3 cm
Step 1	12	1/3 cm	4 cm
Step 2			
Step 3			
Step 4			
Step 50			*

    * Give your answer rounded to the nearest kilometer.

    (b) For the shape obtained at Step N of the construction, express M, l, and P in terms of N.

11. Assume that the seed triangle of the *Koch anti-snowflake* has area $A = 81$ in^2.

    (a) Let R denote the number of triangles subtracted at a particular step, S the area of each subtracted triangle, T the total area subtracted, and Q the area of the shape obtained at a particular step of the construction. Complete the missing entries in the table.

	R	S	T	Q
Start	0	0	0	81 in^2
Step 1	3	9 in^2	27 in^2	54 in^2
Step 2	12	1 in^2		
Step 3				
Step 4				
Step 5				

    (b) What is the area of the Koch antisnowflake? [*Hint*: Do Exercise 12.1 first and compare your answers to (a) in Exercise 12.1 with your answers to (a) in this exercise.]

12. Assume that the seed triangle of the *Koch antisnowflake* has area $A = 729$ in^2.

    (a) Let R denote the number of triangles subtracted at a particular step, S the area of each subtracted triangle, T the total area subtracted, and Q the area of the shape obtained at a particular step of the construction. Complete the missing entries in the table.

	R	S	T	Q
Start	0	0	0	729 in²
Step 1	3	81 in²	243 in²	486 in²
Step 2	12	9 in²		
Step 3				
Step 4				
Step 5				

(b) What is the area of the Koch antisnowflake? [*Hint*: Do Exercise 12.2 first and compare your answers to (a) in Exercise 12.2 with your answers to (a) in this exercise.]

Exercises 13 through 16 refer to the construction of the **quadratic Koch island**. *The quadratic Koch island is defined by the following recursive replacement rule.*

QUADRATIC KOCH ISLAND

- **Start:** *Start with a seed square [Fig. 12-37(a)]. (Notice that here we are only dealing with the boundary of the square.)*
- **Replacement rule:** *Replace each horizontal boundary segment with the "sawtooth" version shown in Fig. 12-37(b) and each vertical line segment with the "sawtooth" version shown in Fig. 12-37(c).*

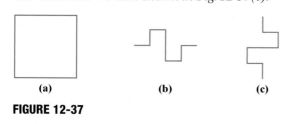

(a)   **(b)**   **(c)**

FIGURE 12-37

---

13. Assume that the seed square of the *quadratic Koch island* has sides of length 16.

(a) Carefully draw the figures obtained in Steps 1 and 2 of the construction. (*Hint:* Use graph paper and make the seed square a 16 by 16 square.)

(b) Find the perimeter of the figure obtained in Step 1 of the construction.

(c) Find the area of the figure obtained in Step 1 of the construction.

(d) Find the perimeter of the figure obtained in Step 2 of the construction.

(e) Find the area of the figure obtained in Step 2 of the construction.

14. Assume that the seed square of the *quadratic Koch island* has sides of length a.

(a) Carefully draw the figures obtained in Steps 1 and 2 of the construction. (*Hint*: Use graph paper and make the seed square a 16 by 16 square.)

(b) Find the perimeter of the figure obtained in Step 1 of the construction.

(c) Find the area of the figure obtained in Step 1 of the construction.

(d) Find the perimeter of the figure obtained in Step 2 of the construction.

(e) Find the area of the figure obtained in Step 2 of the construction.

15. Explain why the area of the quadratic Koch island is the same as the area of the seed square. (*Hint*: Try Exercises 13 and 14 first.)

16. Explain why the quadratic Koch island has infinite perimeter. (*Hint*: Try Exercises 13 and 14 first.)

B The Sierpinski Gasket and Variations

Exercises 17 through 22 refer to the Sierpinski gasket discussed in Section 12.2.

17. Assume that the seed triangle of the *Sierpinski gasket* has area $A = 1$. Let R denote the number of triangles removed at a particular step, S the area of each removed triangle, T the total area removed, and Q the area of the "gasket" obtained at a particular step of the construction. Complete the missing entries in the table.

	R	S	T	Q
Start	0	0	0	1
Step 1	1	1/4	1/4	3/4
Step 2	3	1/16	3/16	9/16
Step 3	9			
Step 4				
Step 5				

18. Assume that the seed triangle of the *Sierpinski gasket* has area A. Let R denote the number of triangles removed at a particular step, S the area of each removed triangle, T the total area removed, and Q the area of the "gasket" obtained at a particular step of the construction. Complete the missing entries in the table.

	R	S	T	Q
Start	0	0	0	A
Step 1	1	A/4	A/4	3A/4
Step 2	3	A/16	3A/16	9A/16
Step 3	9			
Step 4				
Step 5				

19. Assume that the seed triangle of the *Sierpinski gasket* has perimeter of length $P = 8$ cm. Let U denote the number of shaded triangles at a particular step, V the perimeter of each shaded triangle, and W the length of the boundary of the "gasket" obtained at a particular step of the construction. Complete the missing entries in the table.

	U	V	W
Start	1	8 cm	8 cm
Step 1	3	4 cm	12 cm
Step 2	9		
Step 3			
Step 4			
Step 5			

20. Assume that the seed triangle of the *Sierpinski gasket* has perimeter P. Let U denote the number of shaded triangles at a particular step, V the perimeter of each shaded triangle, and W the length of the boundary of the "gasket" obtained at a particular step of the construction. Complete the missing entries in the table.

	U	V	W
Start	1	P	P
Step 1	3	P/2	3P/2
Step 2	9		
Step 3			
Step 4			
Step 5			

21. Let A denote the area of the seed triangle of the *Sierpinski gasket*.

   (a) Find the area of the gasket at step N of the construction expressed in terms of A and N. (*Hint:* Try Exercises 17 and 18 first.)

   (b) Explain why the area of the Sierpinski gasket is infinitesimally small (i.e., smaller than any positive quantity.

22. Let P denote the perimeter of the seed triangle of the *Sierpinski gasket*.

   (a) Find the perimeter of the gasket at step N of the construction expressed in terms of P and N. (*Hint:* Try Exercises 19 and 20 first.)

   (b) Explain why the Sierpinski gasket has an infinitely long perimeter.

*Exercises 23 through 26 refer to a square version of the Sierpinski gasket called the **Sierpinski carpet**. The Sierpinski carpet is defined by the following recursive construction rule.*

SIERPINSKI CARPET

- **Start.** *Start with a shaded seed square [Fig. 12-38(a)].*
- **Step 1.** *Subdivide the seed square into nine equal sub-squares and remove the central subsquare [Fig. 12-38(b)].*
- **Step 2.** *Subdivide each of the remaining shaded squares into nine subsquares and remove the central subsquare [Fig. 12-38(c)]. (Call the procedure of subdividing a shaded square into nine subsquares and removing the middle square procedure SC.)*
- **Steps 3, 4, etc.** *Apply procedure SC to each shaded square of the "carpet" obtained in the previous step.*

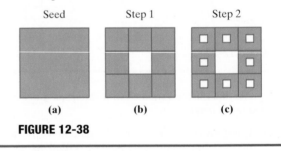

Seed   Step 1   Step 2

(a)   (b)   (c)

FIGURE 12-38

---

23. Assume that the seed square of the *Sierpinski carpet* has area $A = 1$.

   (a) Carefully draw the "carpet" obtained in Step 3 of the construction. [*Hint:* Use a piece of graph paper with a small grid and make the seed square a 27 by 27 square. Use Fig. 12-38(c) as your starting point.]

   (b) Find the area of the "carpet" obtained in Step 1 of the construction.

(c) Find the area of the "carpet" obtained in Step 2 of the construction. (*Hint*: Your best bet is to think of the carpet in Step 2 as made of eight identical one-third scale versions of the carpet in Step 1.)

(d) Find the area of the "carpet" obtained in Step 3 of the construction.

(e) Find the area of the "carpet" obtained in Step N of the construction expressed in terms of N.

24. Assume that the seed square of the *Sierpinski carpet* has area A.

(a) Carefully draw the "carpet" obtained in Step 3 of the construction. [*Hint*: Use a piece of graph paper with a small grid and make the seed square a 27 by 27 square. Use Fig. 12-38(c) as your starting point.]

(b) Find the area of the "carpet" obtained in Step 1 of the construction.

(c) Find the area of the "carpet" obtained in Step 2 of the construction. (*Hint*: Your best bet is to think of the gasket in Step 2 as made of eight identical one-third scale versions of the gasket in Step 1.)

(d) Find the area of the "carpet" obtained in Step 3 of the construction.

(e) Find the area of the "carpet" obtained in Step N of the construction expressed in terms of A and N.

25. Assume that the seed square of the *Sierpinski carpet* has sides of length 1.

(a) Find the length of the boundary of the "carpet" obtained in Step 1 of the construction.

(b) Find the length of the boundary of the "carpet" obtained in Step 2 of the construction. (*Hint*: It is the length of the boundary of the carpet in the previous step plus the perimeter of the "small" white holes introduced in this step.)

(c) Find the length of the boundary of the "carpet" obtained in Step 3 of the construction.

26. Assume that the seed square of the *Sierpinski carpet* has sides of length l.

(a) Find the length of the boundary of the "carpet" obtained in Step 1 of the construction.

(b) Find the length of the boundary of the "carpet" obtained in Step 2 of the construction. (*Hint*: It is the length of the boundary of the carpet in the previous step plus the perimeter of the "small" white holes introduced in this step.)

(c) Find the length of the boundary of the "carpet" obtained in Step 3 of the construction.

*Exercises 27 through 30 refer to the **Sierpinski ternary gasket**, a variation of the Sierpinski gasket defined by the following recursive replacement rule.*

SIERPINSKI TERNARY GASKET

■ **Start:** *Start with a shaded seed equilateral triangle [Fig. 12-39(a)].*

■ **Replacement rule:** *Replace every shaded triangle* ▲ *with a* ▲▲. *[Figures 12-39(b) and (c) show Steps 1 and 2, respectively.]*

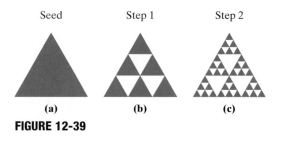

FIGURE 12-39

---

27. Assume that the seed triangle of the *Sierpinski ternary gasket* has area $A = 1$. Let R denote the number of triangles removed at a particular step, S the area of each removed triangle, T the total area removed, and Q the area of the "ternary gasket" obtained at a particular step of the construction. Complete the missing entries in the table.

	R	S	T	Q
Start	0	0	0	1
Step 1	3	1/9	1/3	2/3
Step 2	18	1/81		
Step 3				
Step 4				
Step N				

28. Assume that the seed triangle of the *Sierpinski ternary gasket* has area A. Let R denote the number of triangles removed at a particular step, S the area of each removed triangle, T the total area removed, and Q the area of the "gasket" obtained at a particular step of the construction. Complete the missing entries in the table.

	R	S	T	Q
Start	0	0	0	A
Step 1	3	$A/9$	$A/3$	$2A/3$
Step 2	18	$A/81$		
Step 3				
Step 4				
Step N				

29. Assume that the seed triangle of the *Sierpinski ternary gasket* has perimeter of length $P = 9$ cm. Let U denote the number of shaded triangles at a particular step, V the perimeter of each shaded triangle, and W the length of the boundary of the "ternary gasket" obtained at a particular step of the construction. Complete the missing entries in the table.

	U	V	W
Start	1	9 cm	9 cm
Step 1	6	3 cm	18 cm
Step 2	36		
Step 3			
Step 4			
Step *N*			

30. Assume that the seed triangle of the *Sierpinski ternary gasket* has perimeter P. Let U denote the number of shaded triangles at a particular step, V the perimeter of each shaded triangle, and W the length of the boundary of the "gasket" obtained at a particular step of the construction. Complete the missing entries in the table.

	U	V	W
Start	1	P	P
Step 1	6	$P/3$	$2P$
Step 2	36		
Step 3			
Step 4			
Step *N*			

Exercises 31 and 32 refer to a geometric fractal we will call the **Sierpinski L**. *The Sierpinski L is defined by the following recursive construction rule.*

SIERPINSKI L

- **Start.** *Start with a shaded seed square [Fig. 12-40(a)].*
- **Step 1.** *Subdivide the seed square into four equal subsquares and remove the subsquare on the upper right. This gives an L consisting of the three remaining shaded subsquares [Fig. 12-40(b)].*

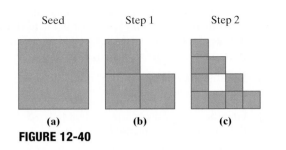

Seed	Step 1	Step 2
(a)	(b)	(c)

FIGURE 12-40

- **Step 2.** *Subdivide each of the subsquares into four subsquares and remove the subsquare on the upper right. [Fig. 12-40(c)]. (Call the procedure of subdividing a square into four subsquares and removing the upper-right subsquare procedure SL.)*
- **Steps 3, 4, etc.** *Apply procedure SL to each square of the shape obtained in the previous step.*

31. Assume that the seed square of the *Sierpinski L* has area A. Let Q denote the area of the shape obtained at a particular step of the construction. Complete the missing entries in the table.

	Q
Start	A
Step 1	$3A/4$
Step 2	
Step 3	
Step 4	
Step *N*	

32. Assume that the seed square of the *Sierpinski L* has perimeter P. Let W denote the length of the boundary of the shape obtained at a particular step of the construction. Complete the missing entries in the table.

	W
Start	P
Step 1	P
Step 2	
Step 3	
Step 4	
Step *N*	

C The Chaos Game and Variations

Exercises 33 through 36 refer to the chaos game as described in Section 12.3. You should use graph paper for these exercises. Start with an isosceles right triangle ABC with AB = AC = 32, as shown in Fig. 12-41. Assume that vertex A corresponds to numbers 1 and 2, vertex B to numbers 3 and 4, and vertex C to numbers 5 and 6.

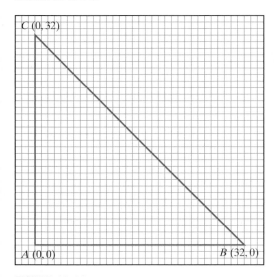

FIGURE 12-41

33. Suppose that an honest die is rolled six times and that the outcomes are 3, 1, 6, 4, 5, and 5. Carefully draw the points P_1 through P_6 corresponding to these outcomes. (*Note*: Each of the points P_1 through P_6 falls on a grid point of the graph. You should be able to identify the location of each point without using a ruler.)

34. Suppose that an honest die is rolled six times and that the outcomes are 2, 6, 1, 4, 3, and 6. Carefully draw the points P_1 through P_6 corresponding to these outcomes. (*Note:* Each of the points P_1 through P_6 falls on a grid point of the graph. You should be able to identify the location of each point without using a ruler.)

35. Using a rectangular coordinate system with A at $(0,0)$, B at $(32,0)$, and C at $(0,32)$, complete the following table.

Number rolled	Point	Coordinates
3	P_1	$(32,0)$
1	P_2	$(16,0)$
2	P_3	
3	P_4	
5	P_5	
5	P_6	

36. Using a rectangular coordinate system with A at $(0,0)$, B at $(32, 0)$, and C at $(0, 32)$, complete the following table.

Number rolled	Point	Coordinates
2	P_1	$(0,0)$
6	P_2	$(0,16)$
5	P_3	
1	P_4	
3	P_5	
6	P_6	

Exercises 37 through 40 refer to a variation of the chaos game discussed in Section 12.3. In this game you start with a square ABCD with sides of length 27 as shown in Fig, 12-42 and a fair die that you will roll many times. When you roll a 1, choose vertex A; when you roll a 2, choose vertex B; when you roll a 3, choose vertex C; and when you roll a 4 choose vertex D. (When you roll a 5 or a 6, disregard the roll and roll again.) A sequence of rolls will generate a sequence of points $P_1, P_2, P_3, \ldots$ inside or on the boundary of the square according to the following rules.

- **Start.** *Roll the die. Mark the chosen vertex and call it P_1.*

- **Step 1.** *Roll the die again. From P_1 move two-thirds of the way toward the new chosen vertex. Mark this point and call it P_2.*

- **Steps 2, 3, etc.** *Each time you roll the die, mark the point two-thirds of the way between the previous point and the chosen vertex.*

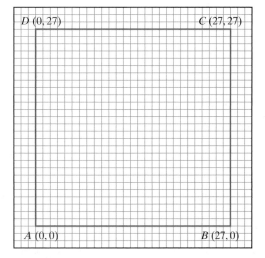

FIGURE 12-42

37. Using graph paper, find the points P_1, P_2, P_3, and P_4 corresponding to

   (a) the sequence of rolls $4, 2, 1, 2$.

   (b) the sequence of rolls $3, 2, 1, 2$.

   (c) the sequence of rolls $3, 3, 1, 1$.

38. Using graph paper, find the points P_1, P_2, P_3, and P_4 corresponding to

   (a) the sequence of rolls $2, 2, 4, 4$.

   (b) the sequence of rolls $2, 3, 4, 1$.

   (c) the sequence of rolls $1, 3, 4, 1$.

39. Using a rectangular coordinate system with A at $(0, 0)$, B at $(27, 0)$, C at $(27, 27)$, and D at $(0, 27)$, find the sequence of rolls that would produce the given sequence of marked points.

   (a) $P_1: (0, 27), P_2: (18, 9), P_3: (6, 3), P_4: (20, 1)$

   (b) $P_1: (27, 27), P_2: (9, 9), P_3: (3, 3), P_4: (19, 19)$

   (c) $P_1: (0, 0), P_2: (18, 18), P_3: (6, 24), P_4: (20, 8)$

40. Using a rectangular coordinate system with A at $(0, 0)$, B at $(27, 0)$, C at $(27, 27)$, and D at $(0, 27)$, find the sequence of rolls that would produce the given sequence of marked points.

   (a) $P_1: (27, 0), P_2: (27, 18), P_3: (9, 24), P_4: (3, 8)$

   (b) $P_1: (0, 27), P_2: (18, 9), P_3: (24, 3), P_4: (8, 19)$

   (c) $P_1: (27, 27), P_2: (9, 9), P_3: (21, 3), P_4: (7, 19)$

D Operations with Complex Numbers

Exercises 41 through 46 are a review of complex number arithmetic. Recall that (1) to add two complex numbers you simply add the real parts and the imaginary parts: e.g., $(2 + 3i) + (5 + 2i) = 7 + 5i$; (2) to multiply two complex numbers you multiply them as if they were polynomials and use the fact that $i^2 = -1$: e.g., $(2 + 3i)(5 + 2i) = 10 + 4i + 15i + 6i^2 = 4 + 19i$. Finally, if you know how to multiply two complex numbers then you also know how to square them, since $(a + bi)^2 = (a + bi)(a + bi)$.

41. Simplify each expression.

   (a) $(-i)^2 + (-i)$

   (b) $(-1 - i)^2 + (-i)$

   (c) $i^2 + (-i)$

42. Simplify each expression.

   (a) $(1 + i)^2 + (1 + i)$

   (b) $(1 + 3i)^2 + (1 + i)$

   (c) $(-7 + 7i)^2 + (1 + i)$

43. Simplify each expression. (Give your answers rounded to three significant digits.)

   (a) $(-0.25 + 0.25i)^2 + (-0.25 + 0.25i)$

   (b) $(-0.25 - 0.25i)^2 + (-0.25 - 0.25i)$

44. Simplify each expression. (Give your answers rounded to three significant digits.)

   (a) $(-0.25 + 0.125i)^2 + (-0.25 + 0.125i)$

   (b) $(-0.2 + 0.8i)^2 + (-0.2 + 0.8i)$

45. (a) Plot the points corresponding to the complex numbers $(1 + i), i(1 + i), i^2(1 + i)$, and $i^3(1 + i)$.

   (b) Plot the points corresponding to the complex numbers $(3 - 2i), i(3 - 2i), i^2(3 - 2i)$, and $i^3(3 - 2i)$.

   (c) What geometric effect does multiplication by i have on a complex number?

46. (a) Plot the points corresponding to the complex numbers $(1 + i), -i(1 + i), (-i)^2(1 + i)$, and $(-i)^3(1 + i)$.

   (b) Plot the points corresponding to the complex numbers $(0.8 + 1.2i), -i(0.8 + 1.2i), (-i)^2(0.8 + 1.2i)$, and $(-i)^3(0.8 + 1.2i)$.

   (c) What geometric effect does multiplication by $-i$ have on a complex number?

E Mandelbrot Sequences

Exercises 47 through 54 refer to Mandelbrot sequences as discussed in the chapter.

47. Consider the Mandelbrot sequence with seed $s = -2$.

   (a) Find s_1, s_2, s_3, and s_4.

   (b) Find s_{100}.

   (c) Is this Mandelbrot sequence *escaping*, *periodic*, or *attracted*? Explain.

48. Consider the Mandelbrot sequence with seed $s = 2$.

   (a) Find s_1, s_2, s_3, and s_4.

   (b) Is this Mandelbrot sequence *escaping*, *periodic*, or *attracted*? Explain.

49. Consider the Mandelbrot sequence with seed $s = -0.5$.

   (a) Using a calculator find s_1 through s_5, rounded to four decimal places.

   (b) Suppose you are given $s_N = -0.366$. Using a calculator find s_{N+1}, rounded to four decimal places.

   (c) Is this Mandelbrot sequence *escaping*, *periodic*, or *attracted*? Explain.

50. Consider the Mandelbrot sequence with seed $s = -0.25$.

   (a) Using a calculator find s_1 through s_{10}, rounded to six decimal places.

(b) Suppose you are given $s_N = -0.207107$. Using a calculator find s_{N+1}, rounded to six decimal places.

(c) Is this Mandelbrot sequence *escaping*, *periodic*, or *attracted*? Explain.

51. Consider the Mandelbrot sequence with seed $s = -i$.

(a) Find s_1 through s_5. (*Hint*: Try Exercise 41 first.)

(b) Is this Mandelbrot sequence *escaping*, *periodic*, or *attracted*? Explain.

52. Consider the Mandelbrot sequence with seed $s = 1 + i$. Find s_1, s_2 and s_3. (*Hint*: Try Exercise 42 first.)

53. Suppose that $s_N = 6$ and $s_{N+1} = 38$ are two consecutive terms of a Mandelbrot sequence.

(a) Find the seed s.

(b) If $s_N = 6$, what is the value of N? [*Hint:* Use the seed you found in (a).]

54. Suppose that $s_N = -15/16$ and $s_{N+1} = -159/256$ are two consecutive terms of a Mandelbrot sequence.

(a) Find the seed s.

(b) If $s_N = -15/16$, what is the value of N? [*Hint:* Use the seed you found in (a).]

JOGGING

55. Assume that the seed triangle of the *Koch snowflake* has sides of length a. Let R denote the number of triangles added at a particular step, S the area of each added triangle, T the total new area added, and Q the area of the "snowflake" obtained at a particular step of the construction. Complete the missing entries in the table. [*Hint*: An equilateral triangle with sides of length a has area $A = a^2\left(\dfrac{\sqrt{3}}{4}\right)$.]

	R	S	T	Q
Start	0	0	0	$a^2\left(\dfrac{\sqrt{3}}{4}\right)$
Step 1				
Step 2				
Step 3				
Step 4				
Step N				

56. Let H denote the total number of "holes" (i.e., white triangles) in the "gasket" obtained at a particular step of the construction of the Sierpinski gasket.

(a) Complete the entries in the following table.

	H
Start	0
Step 1	1
Step 2	$1 + 3$
Step 3	
Step 4	
Step 5	

(b) At Step N of the construction, the value of H is given by $(3^N - 1)/2$. Explain this formula. (*Hint*: You will need to use the geometric sum formula introduced in Chapter 10, p. 378.)

*Exercises 57 and 58 refer to the Menger sponge, a three-dimensional cousin of the Sierpinski carpet. (See Exercises 23–26.) The **Menger sponge** is defined by the following recursive construction rule.*

MENGER SPONGE

- **Start.** *Start with a solid seed cube [Fig. 12-43(a)].*
- **Step 1.** *Subdivide the seed cube into 27 equal subcubes and remove the central cube and the six cubes in the centers of each face. This leaves a "sponge" consisting of 20 solid subcubes, as shown in Fig. 12-43(b).*
- **Step 2.** *Subdivide each solid subcube into 27 subcubes and remove the central cube and the six cubes in the centers of each face. This gives the "sponge" shown in Fig. 12-43(c). (Call the procedure of removing the central cube and the cubes in the center of each face procedure MS.)*

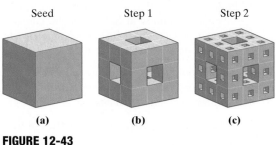

Seed   Step 1   Step 2

(a)   (b)   (c)

FIGURE 12-43

- **Steps 3, 4, etc.** *Apply procedure MS to each cube of the "sponge" obtained in the previous step.*
---

57. Assume that the seed cube of the Menger sponge has volume 1.

    (a) Let C denote the total number of cubes removed at a particular step of the construction, U the volume of each removed cube, and V the volume of the sponge at that particular step of the construction. Complete the entries in the following table.

	C	U	V
Start	0	0	1
Step 1	7	1/27	20/27
Step 2			
Step 3			
Step 4			

    (b) Explain why the Menger sponge has infinitesimally small volume.

58. Let H denote total number of cubic holes in the "sponge" obtained at a particular step of the construction of the Menger sponge.

    (a) Complete the entries in the following table.

	H
Start	0
Step 1	7
Step 2	$7 + 20 \times 7$
Step 3	
Step 4	
Step 5	

    (b) Find a formula that gives the value of H for the "sponge" obtained at Step N of the construction. (*Hint:* You will need to use the geometric sum formula introduced in Chapter 10, p. 378.)

Exercises 59 and 60 refer to reflection and rotation symmetries as discussed in Chapter 11 and thus require a good understanding of the material in that chapter.

59. (a) Describe all the reflection symmetries of the Koch snowflake.

    (b) Describe all the rotation symmetries of the Koch snowflake.

    (c) What is the symmetry type of the Koch snowflake?

60. This exercise refers to the Sierpinski carpet discussed in Exercises 25 through 28.

    (a) Describe all the reflection symmetries of the Sierpinski carpet.

    (b) Describe all the rotation symmetries of the Sierpinski carpet.

    (c) What is the symmetry type of the Sierpinski carpet?

61. Explain why each of the following statements is true. (You will need to use the geometric sum formula from Chapter 10.)

    (a) $1 + \left(\dfrac{4}{9}\right) + \left(\dfrac{4}{9}\right)^2 + \cdots + \left(\dfrac{4}{9}\right)^{N-1} = \dfrac{9}{5}\left[1 - \left(\dfrac{4}{9}\right)^N\right]$

    (b) $\dfrac{A}{3} + \dfrac{A}{3}\cdot\left(\dfrac{4}{9}\right) + \dfrac{A}{3}\cdot\left(\dfrac{4}{9}\right)^2 + \cdots + \dfrac{A}{3}\cdot\left(\dfrac{4}{9}\right)^{N-1} = \dfrac{3}{5}A\left[1 - \left(\dfrac{4}{9}\right)^N\right]$

62. Consider the Mandelbrot sequence with seed $s = -0.75$. Show that this Mandelbrot sequence is attracted to the value -0.5. (*Hint:* Consider the quadratic equation $x^2 - 0.75 = x$ and consider why solving this equation helps.)

63. Consider the Mandelbrot sequence with seed $s = 0.25$. Is this Mandelbrot sequence *escaping*, *periodic*, or *attracted*? If attracted, to what number? (*Hint:* Consider the quadratic equation $x^2 + 0.25 = x$ and consider why solving this equation helps.)

64. Consider the Mandelbrot sequence with seed $s = -1.25$. Is this Mandelbrot sequence *escaping*, *periodic*, or *attracted*? If attracted, to what number?

65. Consider the Mandelbrot sequence with seed $s = \sqrt{2}$. Is this Mandelbrot sequence *escaping*, *periodic*, or *attracted*? If attracted, to what number?

RUNNING

66. Find the area of a quadratic Koch fractal (see Exercises 5 through 8) having a seed square with sides of length 1.

67. Suppose that we play the chaos game using triangle ABC and that M_1, M_2, and M_3 are the midpoints of the three sides of the triangle. Explain why it is impossible at any time during the game to land inside triangle $M_1M_2M_3$.

68. Consider the following variation of the chaos game discussed in Section 12.3. The game is played just like with the ordinary chaos game but with the following change in rules: If you roll a 1, 2, or 3, move halfway toward vertex A; if you roll a 4, move halfway toward vertex B; and if you roll a 5 or 6, move halfway toward vertex C. What familiar geometric fractal is approximated by repeated rolls in this game? Explain.

69. (a) Show that the complex number $s = -0.25 + 0.25i$ is in the Mandelbrot set.

(b) Show that the complex number $s = -0.25 - 0.25i$ is in the Mandelbrot set. [*Hint:* Your work for (a) can help you here.]

70. Show that the Mandelbrot set has a reflection symmetry. (*Hint:* Compare the Mandelbrot sequences with seeds $a + bi$ and $a - bi$.)

Exercises 71 through 73 refer to the concept of **fractal dimension**. *The fractal dimension of a geometric fractal consisting of N self-similar copies of itself each reduced by a scaling factor of S is D = log N/log S. (The fractal dimension is described in a little more detail in Project A below.)*

71. Compute the fractal dimension of the *Koch curve*.

72. Compute the fractal dimension of the *Sierpinski carpet*. (The Sierpinski carpet is discussed in Exercises 23 through 26.)

73. Compute the fractal dimension of the *Menger sponge*. (The Menger sponge is discussed in Exercises 57 and 58.)

PROJECTS AND PAPERS

A Fractal Dimension

The dimensions of a line segment, a square, and a cube are, as we all learned in school, 1, 2, and 3, respectively. But what is the dimension of the Sierpinski gasket?

The line segment of size 4 shown in Fig. 12-44(a) is made of four smaller copies of itself each scaled down by a factor of four; the square shown in Fig. 12-44(b) is made of $16 = 4^2$ smaller copies of itself each scaled down by a factor of four; and the cube shown in Fig. 12-44(c) is made of $64 = 4^3$ smaller copies of itself each scaled down by a factor of four. In all these cases, if N is the number of smaller copies of the object reduced by a scaling factor S, the dimension D is the exponent to which we need to raise S to get N (i.e., $N = S^D$). If we apply the same argument to the Sierpinski gasket shown in Fig. 12-44(d), we see that the Sierpinski gasket is made of $N = 3$ smaller copies of itself and that each copy has been reduced by a scaling factor $S = 2$. If we want to be consistent, the dimension of the Sierpinski gasket should be the exponent D in the equation $3 = 2^D$. To solve for D you have to use *logarithms*. When you do, you get $D = \log 3/\log 2$. Crazy but true: The dimension of the Sierpinski gasket is not a whole number, not even a rational number. It is the irrational number log 3/log 2 (about 1.585)!

For a geometric fractal with exact self-similarity, we will define its dimension as $D = \log N/\log S$, where N is the number of self-similar pieces that the parent fractal is built out of and S is the scale by which the pieces are reduced (if the pieces are one-half the size of the parent fractal $S = 2$, if the pieces are one-third the size of the parent fractal $S = 3$, and so on).

In this project you should discuss the meaning and importance of the concept of dimension as it applies to geometric fractals having exact self-similarity.

B Fractals and Music

The hallmark of a fractal shape is the property of *self-similarity*—there are themes that repeat themselves (either exactly or approximately) at many different scales. This type of repetition also works in music, and the application of fractal concepts to musical composition has produced many intriguing results.

Write a paper discussing the connections between fractals and music.

C Book Review: *The Fractal Murders*

If you enjoy mystery novels, this project is for you.

The Fractal Murders by Mark Cohen (Muddy Gap Press, 2002) is a *whodunit* with a mathematical backdrop. In addition to the standard elements of a classic murder mystery (including a brilliant but eccentric detective), this novel has a fractal twist: The victims are all mathematicians doing research in the field of fractal geometry.

Read the novel and write a review of it. Include in your review a critique of both the literary and the mathematical merits of the book. To get some ideas as to how to write a good book review, you should check out the *New York Times*

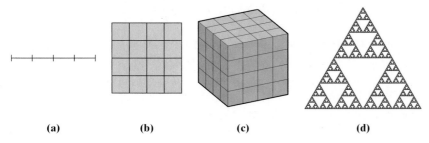

(a)   (b)   (c)   (d)

FIGURE 12-44

Book Review section, which appears every Sunday in the *New York Times* (*www.nytimes.com*).

D   Fractal Antennas

One of the truly innovative practical uses of fractals is in the design of small but powerful antennas that go inside wireless communication devices such as cell phones, wireless modems,

and GPS receivers. The application of fractal geometry to antenna design follows from the discovery in 1999 by radio astronomers Nathan Cohen and Robert Hohlfeld of Boston University that an antenna that has a self-similar shape has the ability to work equally well at many different frequencies of the radio spectrum.

Write a paper discussing the application of the concepts of fractal geometry to the design of antennas.

REFERENCES AND FURTHER READINGS

1. Berkowitz, Jeff, *Fractal Cosmos: The Art of Mathematical Design*. Oakland, CA: Amber Lotus, 1998.

2. Briggs, John, *Fractals: The Patterns of Chaos*. New York: Touchstone Books, 1992.

3. Cohen, Mark, *The Fractal Murders*. Boulder, CO: Muddy Gap Press, 2002.

4. Dewdney, A. K., "Computer Recreations: A Computer Microscope Zooms in for a Look at the Most Complex Object in Mathematics," *Scientific American*, 253 (August 1985), 16–24.

5. Dewdney, A. K., "Computer Recreations: A Tour of the Mandelbrot Set Aboard the Mandelbus," *Scientific American*, 260 (February 1989), 108–111.

6. Dewdney, A. K., "Computer Recreations: Beauty and Profundity. The Mandelbrot Set and a Flock of Its Cousins Called Julia," *Scientific American*, 257 (November 1987), 140–145.

7. Flake, Gary W., *The Computational Beauty of Nature: Computer Explorations of Fractals, Chaos, Complex Systems, and Adaptation*. Cambridge, MA: MIT Press, 2000.

8. Gleick, James, *Chaos: Making a New Science*. New York: Viking Penguin, 1987, Chap. 4.

9. Hastings, Harold, and G. Sugihara, *Fractals: A User's Guide for the Natural Sciences*. New York: Oxford University Press, 1995.

10. Jurgens, H., H. O. Peitgen, and D. Saupe, "The Language of Fractals," *Scientific American*, 263 (August 1990), 60–67.

11. Mandelbrot, Benoit, *The Fractal Geometry of Nature*. New York: W. H. Freeman, 1983.

12. Musser, George, "Practical Fractals," *Scientific American*, 281 (July 1999), 38.

13. Peitgen, H. O., H. Jurgens, and D. Saupe, *Chaos and Fractals: New Frontiers of Science*. New York: Springer-Verlag, 1992.

14. Peitgen, H. O., H. Jurgens, and D. Saupe, *Fractals for the Classroom*. New York: Springer-Verlag, 1992.

15. Peitgen, H. O., and P. H. Richter, *The Beauty of Fractals*. New York: Springer-Verlag, 1986.

16. Peterson, Ivars, *The Mathematical Tourist*. New York: W. H. Freeman, 1988, Chap. 5.

17. Schechter, Bruce, "A New Geometry of Nature," *Discover*, 3 (June 1982), 66–68.

18. Schroeder, Manfred, *Fractals, Chaos, Power Laws: Minutes from an Infinite Paradise*. New York: W. H. Freeman, 1991.

19. Wahl, Bernt, *Exploring Fractals on the Macintosh*. Reading, MA: Addison-Wesley, 1994.

Web Sites and Applets

Koch Snowflake

http://www.shodor.org/interactivate/activities/KochSnowflake/

http://www.3rd-imperium.com/Java/Fractals/KF.html

Sierpinski Gasket

http://www.shodor.org/interactivate/activities/SierpinskiTriangle/

http://www.youtube.com/watch?v=6AYddwLyJq8

Chaos Game

http://www.shodor.org/interactivate/activities/TheChaosGame/

http://www.geoastro.de/ChaosSpiel/ChaosEnglish.html

http://www.cut-the-knot.org/Curriculum/Geometry/SierpinskiChaosGame
.shtml

Mandelbrot set

http://www.youtube.com/watch?v=G_GBwuYuOOs&NR=1

http://www.youtube.com/watch?v=gEw8xpb1aRA

http://www.youtube.com/watch?v=WAJE35wX1nQ

http://www.youtube.com/watch?v=y2oa5DkJcLA

Chapter-Opener Photos

- p. 436. *Monkey's Tail.* This is a computer-generated image of a very small region near the boundary of the Mandelbrot set (p. 451). The image is magnified approximately 50,000 times. (If this image had been rendered at the same scale as the image of the full Mandelbrot set on p. 451 it would be about 4 miles wide.) Image courtesy of Rollo Silver.
- p. 437 (top). *Lung Tissue.* Photograph of a small section of human lung tissue taken through a microscope. The magnification here is 50 times. The white regions are the lung's air passages, the red regions are the *alveoli* (small air spaces), and the blue regions are the blood capillaries. Image courtesy of Astrid and Hanss Frieder-Michler.
- p. 437 (bottom right). *DNA Sequence*, by Nick Chlebnikowski. Oil on board. 30 cm × 30 cm. This painting is part of a series of 500 separate panels, each representing a nucleotide of DNA. To see more of Nick Chlebnikowski's fractal paintings go to *www.fractalpainting.com.* Image courtesy of Nick Chlebnikowski.
- p. 437 (bottom left). *Karman Vortices.* This photograph shows special air vortices called von Karman vortices created high in the atmosphere by winds sweeping over Alaska's Aleutian Islands. The photograph was taken by the Landsat 7 satellite orbiting above Earth. Image courtesy of NASA Landsat Project Science Office and U.S. Geological Survey.

mini-excursion 3

The Mathematics of Population Growth

There Is Strength in Numbers

The connection between mathematics and the study of populations goes back to the very beginnings of civilization. One of the reasons that humans invented the first numbering systems was their need to handle the rudiments of counting populations—how many sheep in the flock, how many people in the tribe, and so on. By biblical times, simple models of population growth were being used to measure crop production and even to estimate the yields of future crops.

Today, mathematical models of population growth are a fundamental tool in our efforts to understand the rise and fall of endangered wildlife populations, fishery stocks, agricultural pests, infectious diseases, radioactive waste, and so on. Entire modern disciplines, such as *mathematical ecology*, *population biology*, and *epidemiology*, are built around the mathematics of population growth. In this mini-excursion we will discuss three basic but important models of population growth—linear, exponential, and logistic—and examine a few of their many real-life applications.

The Dynamics of Population Growth

In its modern usage, the term *population growth* has a very broad meaning, due primarily to the loose way that both words—*population* and *growth*—are interpreted. The Latin root of *population* is *populus* (which means "people"), so in its original interpretation the word refers to human populations. Over time, this scope has been expanded to apply to any collection of objects, be it animal populations, material objects, natural resources, and even waste and pollution.

Second, we normally think of the word *growth* as being applied to things that get bigger, but in its more general meaning, growth can mean *positive growth* (i.e., a population getting bigger) as well as *negative growth* (which applies to populations whose numbers are getting smaller). This is convenient because in many models populations fluctuate and often we don't know ahead of time how a population is going to change: Is it going to have positive growth or negative growth? By allowing *growth* to mean either, we need not concern ourselves with making the distinction.

The growth of a population is a *dynamical process*, meaning that it represents a situation that changes over time. Mathematicians distinguish between two kinds of population growth: *continuous* and *discrete*. In **continuous growth** the dynamics of change are in effect all the time—every hour, every minute, every second, there is change. The classic example of this kind of growth is represented by money left in an account under *continuous compounding*. We will not discuss continuous growth here because the mathematics involved (calculus) is beyond the scope of this book.

■ See Chapter 10, Section 10.3.

The second type of growth, **discrete growth**, is the most common and natural way by which populations change. We can think of it as a *stop-and-go* type of situation. For a while nothing happens, then there is a sudden change in the population. We call such a change a **transition**. Then for a while nothing happens again, then another transition takes place, and so on. This discrete, stop-and-go model of population growth can be used to simulate many real-life situations: animal populations that follow consistent breeding patterns (*breeding seasons*), money in an account that draws interest at regular intervals, the volume of garbage in a landfill, and so on. In all of these examples, the critical feature is the existence of transitions that occur at regular intervals, and the length of time between transitions will not make much of a difference. Thus, the same discrete model can be used when the period between transitions is measured in years, hours, or seconds.

The basic problem of population growth is to predict the future: What will happen to a given population over time? What are the long-term patterns of growth (if any)? In the case of discrete growth models we deal with these questions by finding the rules that govern the transitions. We will call these the **transition rules**. After all, if we have a way to figure out how the population changes each time there is a transition, then (with a little help from mathematics) we can usually figure out how the population changes after many transitions.

The ebb and flow of a particular population over time can be conveniently thought of as a list of numbers called the **population sequence**. Every population sequence starts with an initial population P_0 (called the *seed* of the population sequence) and continues with P_1, P_2, and so on, where P_N is the size of the population in the Nth generation. The reason a population sequence starts with P_0 rather than P_1 is convenience—with this notation the subscripts match the generations: P_1 denotes the population in the first generation, P_2 the population in the second generation, and so on. Figure ME3-1 is a schematic illustration of how a population sequence is generated.

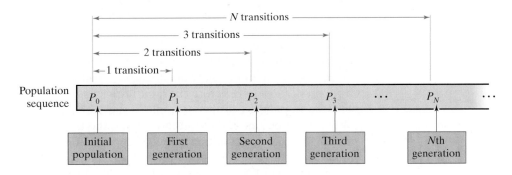

FIGURE ME3-1 A generic population sequence. P_N is the population in the Nth generation.

A very convenient way to describe the population sequence is by means of a *time-series graph*. In a time-series graph the horizontal axis usually represents time (with the tick marks generally corresponding to the transitions), and the vertical axis usually represents the size of the population. A time-series graph can consist

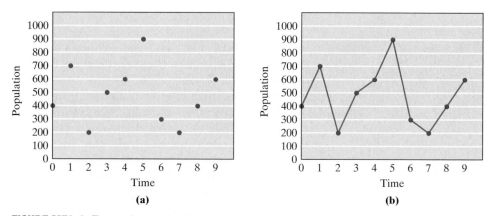

FIGURE ME3-2 Time-series graphs: (a) scatter plot, (b) line graph.

of just marks (such as dots) indicating the population size at each generation or of dots joined by lines, which sometimes helps the visual effect. The former is called a **scatter plot**, the latter a **line graph**. Figure ME3-2 shows a scatter plot and a line graph for the same population sequence: $P_0 = 400$, $P_1 = 700$, $P_2 = 200$, etc.

The Linear Growth Model

The linear growth model is the simplest of all models of population growth. In this model, in each generation the population increases (or decreases) by a fixed amount called the *common difference*. The easiest way to see how the model works is with an example.

EXAMPLE ME3.1   **How Much Garbage Can We Take?**

The city of Cleansburg is considering a new law that would restrict the monthly amount of garbage allowed to be dumped in the local landfill to a maximum of 120 tons a month. There is concern among local officials that, unless this restriction on dumping is imposed, the landfill will reach its maximum capacity of 20,000 tons in a few years. Currently, there are 8000 tons of garbage already in the landfill. Assuming that the law is passed and the landfill collects exactly 120 tons of garbage each month, how much garbage will there be in the landfill five years from now? How long before the landfill reaches its 20,000-ton capacity?

The population in this example is the garbage in the landfill, and since we only care about monthly totals, we define the transitions as taking place once a month. The key assumption about this population growth problem is that the monthly garbage at the landfill grows by 120 tons a month.

From a starting population of $P_0 = 8000$ tons, the next few terms of the population sequence are $P_1 = 8000 + 120 = 8120$, $P_2 = 8120 + 120 = 8240$, $P_3 = 8240 + 120 = 8360$, and so on. After five years of this garbage we will have had 60 transitions, each representing an increase of 120 tons, giving a population of $P_{60} = 8000 + 60(120) = 15,200$ tons.

To determine the number of transitions it would take for the landfill to reach its 20,000-ton maximum, we set up the equation $8000 + 120N = 20,000$, which has as a solution $N = 100$. This means that it will take 100 monthly transitions (eight years and four months) for the landfill to reach its maximum capacity.

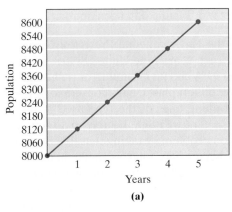
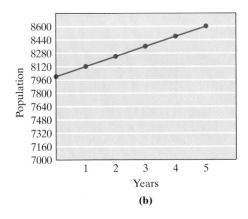

(a)   (b)

FIGURE ME3-3

Based on this information, local officials should start making plans soon for a new landfill.

The line graph in Fig. ME3-3(a) illustrates the growth of the garbage in the dump over the five years in question, and, by the way, so does the line graph in Fig. ME3-3(b). These two line graphs illustrate why linear growth is called linear growth—no matter how we scale it, stretch it or slide it, those "red dots" will always line up.

Example ME3.1 is a typical example of the general **linear growth model**, whose basic characteristic is that in each transition a constant amount—call it d— is added (or subtracted) to the previous population. Thus, when a population grows according to a linear growth model, the population sequence takes the form

$$P_0, \; P_1 = P_0 + d, \; P_2 = P_1 + d, \ldots, P_N = P_{N-1} + d, \ldots$$

A sequence of numbers that takes the above form is called an **arithmetic sequence**, and the number d is called the **common difference** of the arithmetic sequence. (Technically speaking, the arithmetic sequence is just the numerical description of a population growing according to a linear growth model— informally, linear growth and arithmetic sequences can be considered synonymous.)

An alternative definition of an arithmetic sequence with initial term P_0 and common difference d is given by the *explicit formula* $P_N = P_0 + Nd$, as illustrated in Fig. ME3-4.

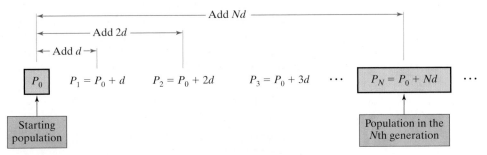

FIGURE ME3-4 Population growth based on a linear growth model.

> **EXAMPLE ME3.2** **Plastic Bag Consumption**

According to the Environmental Protection Agency, about 380 billion plastic bags are consumed each year in the United States. Of these, only about 1% are recycled. The remaining 99% end up in landfills or polluting our waterways, rivers, and oceans.

Of the 380 billion plastic bags consumed, it is estimated that about 100 billion are plastic shopping bags of the type used at the grocery store for bagging groceries. As consumers, we might be able to do something about this part of the plastic bag pollution problem by taking reusable cloth bags with us when we go grocery shopping, but what effect can this small change in grocery shopping habits have?

It is estimated that the typical American consumes about six new plastic grocery bags a week (say two trips to the grocery store and three plastic bags per trip). Roughly, that equals 300 new plastic bags per person per year. Thus, for each person who can be persuaded to switch to reusable cloth bags, we can subtract 300 from the 100 billion annual consumption figures. To put it in slightly more mathematical terms, if we use P to denote the population of plastic grocery bags consumed in the United States in one year and N the number of people who switch to reusable cloth bags, then we get the linear "growth" model

$$P_N = 100,000,000,000 - 300N$$

Notice that the above formula corresponds to an arithmetic sequence with negative common difference $d = -300$. (In any linear growth model where the common difference d is a negative number the population gets smaller with each generation.)

Unfortunately, the 300 is a drop in the bucket when measured against 100 billion, so let's look at a slightly different model based on a more global approach. One percent of the U.S. population equals roughly 3 million people. Thus, for each 1% of the population that can be persuaded to switch to the use of reusable cloth bags at the grocery store, we can subtract 900 million (3 million times 300) = 0.9 billion plastic grocery bags from the 100 billion current consumption figures. We can represent this situation by another linear "growth" model:

$$P_M = 100 - 0.9M$$

where P_M denotes the number of annual plastic grocery bags consumed (in billions) and M is the percent of the population using reusable cloth bags.

Under this model we can at least get a global perspective of how we can make a dent on the plastic bag problem. If just 25% of the U.S. population (one in four people) could be persuaded to habitually grocery shop with reusable cloth bags, we would have 22.5 billion fewer plastic bags filling out our landfills and polluting the environment. ⬭

■ You can find a "population clock" giving the current estimate of the U.S. population at www.census.gov/main/www/popclock.html.

The Arithmetic Sum Formula

Given a population that grows according to a linear growth model, we often need to know what is the sum of the first so many terms of the population sequence—say the sum of the first 10 terms, or the first 100 terms, or the first 500 terms. Of course, in the case of the sum of the first 10 terms we can simply write the terms down and add them the long way, but this is not a reasonable approach when dealing with the addition of hundreds of terms. The next example illustrates how we can find such sums in an efficient way.

> **EXAMPLE ME3.3** **The Cost of Building Up Inventory**

The E-car is a new electric car about to be introduced into the market. To build up the inventory of E-cars and meet the anticipated demand, the manufacturer is planning a 72-week production schedule. The plan is to manufacture 30 E-cars each week for the next 72 weeks. After they are manufactured, the E-cars will be stored in one of several warehouses for the remainder of the time between production and when the cars are introduced into the market. If the storage costs are $10 per E-car per week, what is the total storage cost to the company over the 72-week period?

In this problem the weekly storage costs for weeks 1 through 72 are given by the sequence

$$P_0 = \$0 \text{ (week 0)}$$
$$P_1 = 30 \times \$10 = \$300$$
$$P_2 = 60 \times \$10 = \$600$$
$$\vdots$$
$$P_{71} = 2130 \times \$10 = \$21{,}300$$
$$P_{72} = 2160 \times \$10 = \$21{,}600$$

The weekly storage costs form an arithmetic sequence with initial term $P_0 = 0$ and common difference $d = 300$. (Starting the count at $0 is a little contrived here, but it is consistent with the way we look at population growth.) The new wrinkle in this problem is that to find the total storage cost over the 72-week period we will have to compute the rather long sum $S = 300 + 600 + 900 + \cdots + 21{,}300 + 21{,}600$. We definitely don't want to add these 72 numbers the old-fashioned way—even with a calculator this is a horrible thought!

Luckily, there is a wonderful arithmetic trick that will do the job. The key idea is to write out the sum twice, once forward and once backward.

$$S = \quad 300 + \quad 600 + \quad 900 + \cdots + 21{,}600 \quad \text{(forward)}$$
$$S = 21{,}600 + 21{,}300 + 21{,}000 + \cdots + \quad 300 \quad \text{(backward)}$$

Once they are lined up this way, we notice that each of the 72 columns on the right-hand side of the equal signs adds up to 21,900. Thus, the grand total of both lines is $72 \times 21{,}900$. This is twice the sum S we are trying to compute, so dividing by 2 gives S:

$$S = \frac{72 \times 21{,}900}{2} = \$788{,}400$$

The approach used in Example ME3.3 works with *any* arithmetic sequence $A_0, A_1, A_2, \ldots$. We can add up any number of consecutive terms easily with the following formula, which we will informally refer to as the **arithmetic sum formula**.

ARITHMETIC SUM FORMULA

$$A_0 + A_1 + \cdots + A_{N-1} = \frac{(A_0 + A_{N-1})N}{2}$$

Before we go on, a couple of comments about the arithmetic sum formula are in order. First, it is an extremely useful and important formula. It's not known who first discovered it, but it goes back a long way, and the first known reference to it can be found in Book IX of Euclid's *Elements*, written around 300 B.C. At a minimum the formula has been known for 2300 years, possibly much longer. Second, why did we write the last term of the sum as A_{N-1} rather than A_N? (Surely, A_N seems a lot more user friendly.) The reason is that the sum starts with

A_0, which makes the number of terms and the subscript for the last term offset by 1. In short, the Nth term is A_{N-1}, and if you want A_N you need to go to the $(N + 1)$st term.

You may find the following informal version of the arithmetic sum formula more to your liking: *To find the sum of terms of an arithmetic sequence, add the first term and the last term, multiply the result by the number of terms, and then divide by 2.* (Please remember that this rule applies *only* when you are adding *consecutive* terms of an *arithmetic sequence*.)

The Exponential Growth Model

■ Note the difference between linear and exponential growth: In linear growth we *add* a fixed constant, but in exponential growth we *multiply* by a fixed constant.

The **exponential growth model** is one of the most commonly used (some would say overused) models of population growth. Exponential growth is based on the idea of a *constant rate of growth*—in each transition the population is multiplied by a fixed *factor r* called the **common ratio**.

When a population grows under an exponential growth model with seed P_0 and common ratio r, it generates a population sequence of the form

$$P_0, \ P_1 = rP_0, \ P_2 = rP_1, \ P_3 = rP_2, \ldots, \ P_N = rP_{N-1}, \ldots$$

(i.e., in each generation *the population equals the population in the previous generation multiplied by the common ratio r*).

An alternative, more convenient description of the same population sequence is given by

$$P_0, \ P_1 = rP_0, \ P_2 = r^2P_0, \ P_3 = r^3P_0, \ldots, \ P_N = r^NP_0, \ldots$$

(here the formula $P_N = r^N \cdot P_0$ gives us an *explicit* description for the population sequence: We can compute the population in any generation from just the seed P_0 and the common ratio r.)

If all the above looks familiar, it is because we already studied in detail sequences of this type (known as **geometric sequences**) in Chapter 10, where we used geometric sequences and the geometric sum formula to model the growth of an investment under compound interest. Because most of the basic mathematics behind exponential growth and geometric sequences has been covered in Sections 10.3 and 10.4, our discussion here will be very brief. The key point to be made is that the exponential growth model is important and useful in areas beyond financial mathematics.

(**EXAMPLE ME3.4**)   **The Spread of an Epidemic**

In their early stages, many infectious diseases spread among the population following an exponential growth model—each infected individual is able to pass on the disease to an approximately constant number of individuals over a specified period of time.

Every epidemic starts with an original group of infected individuals (sometimes just one) called "population zero." Let's consider an epidemic in which "population zero" consisted of just one infected individual ($P_0 = 1$). Let's assume, moreover, that on the average each infected individual transmits the infection to one other individual each month. So, every month the number of infected individuals doubles—that is, N months after the epidemic starts the number of infected individuals can be estimated to be $P_N = 1 \cdot (2)^N = 2^N$.

Essentially, the population of infected individuals is a geometric sequence with $P_0 = 1$ and $r = 2$.

The scatter plot in Fig. ME3-5 shows the growth of the epidemic during its first year. By the end of the first year we have $P_{12} = 2^{12} = 4096$ infected individuals. This doesn't seem too alarming until we check what happens in the second year and beyond.

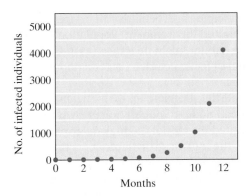

FIGURE ME3-5

If the epidemic continues growing at the same rate, by the end of the second year the number of infected individuals is $2^{24} = 16,777,216$. Even more surprising is that less than nine months after that, every one of the roughly seven billion people on the planet would be infected, since $2^{33} \approx 8.5$ billion. It is clear that as time goes by, our assumption of continued exponential growth is not very realistic. Something has to give.

Example ME3.4 illustrates a basic fact: in biological situations an exponential growth model cannot continue indefinitely. In the case of an epidemic there is always some point in time when the rate of growth of the epidemic starts to taper off, or even reverse itself; otherwise, the human race would have been wiped out many times over. Exactly how this switch from exponential to nonexponential growth happens depends on many variables, and there are many complicated population growth models that deal with this. In the next section we will discuss one of the simplest of such models.

The Logistic Growth Model

One of the key tenets of population biology is the idea that there is an inverse relation between the rate of growth and the density of a population. Small populations have plenty of room to spread out and grow, and thus their growth rates tend to be high. As the population density increases there is less room to grow and there is more competition for resources—the growth rate tends to taper off. Sometimes the population density is so high that resources become scarce or depleted, leading to a decline in the population or even to extinction.

The effects of population density on growth rates were studied in the 1950s by behavioral psychologist John B. Calhoun. Calhoun's now classic studies showed that when rats were placed in a closed environment, their behavior and growth rates were normal as long as the rats were not too crowded. When their environment became too crowded, the rats started to exhibit abnormal behaviors, such as infertility and cannibalism, which effectively put a brake on the population growth rate. In extreme cases, the entire rat population became extinct.

Calhoun's experiments with rats are but one classic illustration of the general principle that a *population's growth rate is negatively impacted by the population's density.*

This principle is particularly important in cases in which the population is confined to a limited environment. Population biologists call such an environment the **habitat**. The habitat might be a cage (as in Calhoun's rat experiments), a lake (as for a population of fish), a garden (as for a population of snails), and, of course, Earth itself (everyone's habitat).

In 1838, the Belgian mathematician Pierre François Verhulst proposed a mathematical model of population growth for species living within a fixed habitat. Verhulst called his model the **logistic growth model**. To put it very informally, the key idea in the logistic growth model is that the rate of growth of the population is directly proportional to the amount of "elbow room" available in the population's habitat. Thus, lots of elbow room means a high growth rate; little elbow room means a low growth rate (possibly less than 1, which means that the population is actually decreasing); and if there is no elbow room at all, the population becomes extinct.

Population biologists use the term **carrying capacity** to describe the total saturation point of a habitat. The carrying capacity (denoted by C) is a ceiling for a population living in a given habitat—by definition the population size will never exceed C ($P_N \leq C$). For a population of size P_N, we will think of the difference $C - P_N$ as the absolute amount of elbow room remaining for this population. An even more convenient way to think of the elbow room is in relative terms: The ratio $(C - P_N)/C$ represents the amount of elbow room available expressed as a fraction (percentage) of the carrying capacity.

When the relative elbow room for a population is very high (close to 100%), the population can breed without constraints, and the classic exponential growth formula $P_{N+1} = rP_N$ does a good job of describing the growth of the population. (Here the growth rate r is a constant called the **growth parameter** that depends only on the type of population we are dealing with.) The key idea behind the logistic growth model is that this growth rate decreases in direct proportion to the amount of elbow room—if the elbow room is 40%, then the growth rate is $(0.4)r$, and so on. Thus,

$$\text{growth rate for period } N = r \cdot \left(\frac{C - P_N}{C} \right)$$

and from this, we get the following transition rule for the logistic growth model:

$$P_{N+1} = r \cdot \left(\frac{C - P_N}{C} \right) \cdot P_N$$

This transition rule can be simplified considerably by switching the notation and using relative population figures $p_N = P_N/C$ to replace the absolute population figures P_N. The lowercase population figures p_N represent the fraction (or percentage) of the carrying capacity taken up by the population. We will call these values the **p-values** of the population sequence. Note that these p-values will always fall between 0 (zero population) and 1 (complete saturation of the habitat).

To express the preceding transition rule in term of p-values, we divide both sides by C and get

$$\frac{P_{N+1}}{C} = r \cdot \left(\frac{C - P_N}{C}\right) \cdot \frac{P_N}{C} = r \cdot \left(1 - \frac{P_N}{C}\right) \cdot \frac{P_N}{C}$$

Substituting p_{N+1} and p_N for P_{N+1}/C and P_N/C, respectively, gives the following equation, called the **logistic equation**.

■ The values of r must be restricted to be between 0 and 4 because for r bigger than 4 the p-values can fall outside the 0 to 1 range.

> **THE LOGISTIC EQUATION**
>
> $$p_{N+1} = r \cdot (1 - p_N) \cdot p_N$$

In the examples that follow, we will look at the growth pattern of an imaginary population using the logistic equation. In each case, all we need to get started is the seed p_0 and the growth parameter r. The logistic equation and a good calculator—or better yet, a spreadsheet—will do the rest. A note of warning: The calculations shown in the examples that follow were done with a computer and carried to 16 decimal places before being rounded off to 3 or 4 decimal places. Thus, they may not match exactly the numbers you get out of a calculator.

(EXAMPLE ME3.5) A Stable Equilibrium

Fish farming is big business these days, so you decide to give it a try. You have access to a large, natural pond in which you plan to set up a rainbow trout hatchery. The carrying capacity of the pond is $C = 10,000$ fish, and the growth parameter of this type of rainbow trout is $r = 2.5$. We will use the logistic equation to model the growth of the fish population in your pond.

You start by seeding the pond with an initial population of 2000 rainbow trout (i.e., 20% of the pond's carrying capacity, or $p_0 = 0.2$). After the first year (trout have an annual hatching season) the population is given by

$$p_1 = 2.5 \times (1 - 0.2) \times (0.2) = 0.4$$

The population of the pond has doubled, and things are looking good! Unfortunately, most of the fish are small fry and not ready to be sent to market. After the second year the population of the pond is given by

$$p_2 = 2.5 \times (1 - 0.4) \times (0.4) = 0.6$$

The population is no longer doubling, but the hatchery is still doing well. You are looking forward to even better yields after the third year. But on the third year you get a big surprise:

$$p_3 = 2.5 \times (1 - 0.6) \times (0.6) = 0.6$$

Stubbornly, you wait for better luck the next year, but

$$p_4 = 2.5 \times (1 - 0.6) \times (0.6) = 0.6$$

From the second year on, the hatchery is stuck at 60% of the pond capacity—nothing is going to change unless external forces come into play. We describe this situation as one in which the population is at a *stable equilibrium*. Figure ME3-6 shows a line graph of the pond's fish population for the first four years.

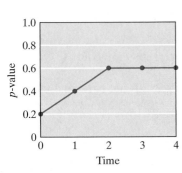

FIGURE ME3-6 $r = 2.5$, $p_0 = 0.2$.

EXAMPLE ME3.6   **An Attracting Point**

Consider the same setting as in Example 10.18 (same pond and the same variety of rainbow trout with $r = 2.5$), but suppose you initially seed the pond with 3000 rainbow trout (30% of the pond's carrying capacity). How will the fish population grow if we start with $p_0 = 0.3$?

The first six years of population growth are as follows:

$$p_1 = 2.5 \times (1 - 0.3) \times (0.3) = 0.525$$
$$p_2 = 2.5 \times (1 - 0.525) \times (0.525) \approx 0.6234$$
$$p_3 = 2.5 \times (1 - 0.6234) \times (0.6234) \approx 0.5869$$
$$p_4 = 2.5 \times (1 - 0.5869) \times (0.5869) \approx 0.6061$$
$$p_5 = 2.5 \times (1 - 0.6061) \times (0.6061) \approx 0.5968$$
$$p_6 = 2.5 \times (1 - 0.5968) \times (0.5968) \approx 0.6016$$

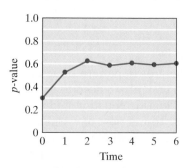

FIGURE ME3-7 $r = 2.5$, $p_0 = 0.3$.

Clearly, something different is happening here. The trout population appears to be fluctuating—up, down, up again, back down—but always hovering near the value of 0.6. We leave it to the reader to verify that as one continues with the population sequence, the p-values inch closer and closer to 0.6 in an oscillating (up, down, up, down, ...) manner. The value 0.6 is called an *attracting point* of the population sequence. Figure ME3-7 shows a line graph of the pond's fish population for the first six years.

EXAMPLE ME3.7   **Complementary Seeds**

In Example ME3.6 we seeded the pond at 30% of its carrying capacity ($p_0 = 0.3$). If we seed the pond with the complementary seed $p_0 = 1 - 0.3 = 0.7$, we end up with the same populations:

$$p_1 = 2.5 \times 0.3 \times 0.7 = 2.5 \times 0.7 \times 0.3 = 0.525$$
$$p_2 = 2.5 \times (1 - 0.525) \times (0.525) \approx 0.6234$$

and so on.

Example ME3.7 points to a simple but useful general rule about logistic growth—the seeds p_0 and $(1 - p_0)$ always produce the same population sequence. This follows because in the expression $p_1 = r \cdot (1 - p_0) \cdot p_0$, p_0 and $(1 - p_0)$ play interchangeable roles—if you change p_0 to $(1 - p_0)$, then you are also changing $(1 - p_0)$ to p_0. Nothing gained, nothing lost! Once the values of the p_1's are the same, the rest of the p-values follow suit. The moral of this observation is that you should never seed your pond at higher than 50% of its carrying capacity.

EXAMPLE ME3.8   **The Two-Cycle Pattern**

You decided that farming rainbow trout is too difficult. You are moving on to raising something easier—goldfish. The particular variety of goldfish you will grow has growth parameter $r = 3.1$.

Suppose you start by seeding a tank at 20% of its carrying capacity ($p_0 = 0.2$). Following the logistic growth model, the first 16 p-values of the goldfish population are

■ For the sake of brevity, the details are left to the reader.

$p_0 = 0.2$,	$p_1 = 0.496$,	$p_2 \approx 0.775$,	$p_3 \approx 0.541$,
$p_4 \approx 0.770$,	$p_5 \approx 0.549$,	$p_6 \approx 0.767$,	$p_7 \approx 0.553$,
$p_8 \approx 0.766$,	$p_9 \approx 0.555$,	$p_{10} \approx 0.766$,	$p_{11} \approx 0.556$,
$p_{12} \approx 0.765$,	$p_{13} \approx 0.557$,	$p_{14} \approx 0.765$,	$p_{15} \approx 0.557$, ...

An interesting pattern emerges here. After a few breeding seasons, the population settles into a two-cycle pattern, alternating between a high-population period at 0.765 and a low-population period at 0.557. Figure ME3-8 convincingly illustrates the oscillating nature of the population sequence.

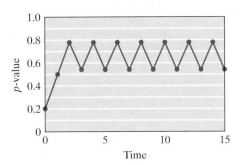

FIGURE ME3-8
$r = 3.1$, $p_0 = 0.2$.

Example ME3.8 describes a situation not unusual in population biology—animal populations that alternate cyclically between two different levels of population density. Even more complex cyclical patterns are possible when we increase the growth parameter just a little.

(**EXAMPLE ME3.9**)   **A Four-Cycle Pattern**

You are now out of the fish-farming business and have acquired an interest in entomology—the study of insects. Let's apply the logistic growth model to study the population growth of a type of flour beetle with growth parameter $r = 3.5$. The seed will be $p_0 = 0.44$. (There is no particular significance to the choice of the seed—you can change the seed and you will still get an interesting population sequence.)

Following are a few specially selected p-values. We leave it to the reader to verify these numbers and fill in the missing details.

$$
\begin{array}{llll}
p_0 = 0.440, & p_1 \approx 0.862, & p_2 \approx 0.415, & p_3 \approx 0.850, \\
p_4 \approx 0.446, & p_5 \approx 0.865, & \ldots & p_{20} \approx 0.497, \\
p_{21} \approx 0.875, & p_{22} \approx 0.383, & p_{23} \approx 0.827, & p_{24} \approx 0.501, \\
p_{25} \approx 0.875, & \ldots
\end{array}
$$

It took a while, but we can now see a pattern: Since $p_{25} = p_{21}$, the population will repeat itself in a four-period cycle ($p_{26} = p_{22}$, $p_{27} = p_{23}$, $p_{28} = p_{24}$, $p_{29} = p_{25} = p_{21}$, etc.), an interesting and surprising turn of events. Figure ME3-9 shows the line graph of the first 26 p-values.

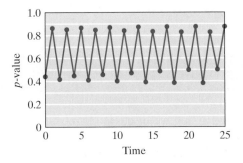

FIGURE ME3-9
$r = 3.5$, $p_0 = 0.44$.

The cyclical behavior exhibited in Example ME3.9 is not unusual, and many insect populations follow cyclical patterns of various lengths—7-year locusts, 17-year cicadas, and so on.

In the logistic growth model, the highest allowed value of the growth parameter r is $r = 4$. Example ME3.10 illustrates what happens in this case.

EXAMPLE ME3.10   **A Glimpse of Chaos**

Let's start with $p_0 = 0.2$ and look at the p-values generated by the logistic equation when $r = 4$. Following are the first 20 p-values.

$$p_0 = 0.2000, \quad p_1 = 0.6400, \quad p_2 \approx 0.9216, \quad p_3 \approx 0.2890,$$
$$p_4 \approx 0.8219, \quad p_5 \approx 0.5854, \quad p_6 \approx 0.9708, \quad p_7 \approx 0.1133,$$
$$p_8 \approx 0.4020, \quad p_9 \approx 0.9616, \quad p_{10} \approx 0.1478, \quad p_{11} \approx 0.5039,$$
$$p_{12} \approx 0.9999, \quad p_{13} \approx 0.0004, \quad p_{14} \approx 0.0010, \quad p_{15} \approx 0.0039,$$
$$p_{16} \approx 0.0157, \quad p_{17} \approx 0.0617, \quad p_{18} \approx 0.2317, \quad p_{19} \approx 0.7121$$

Figure ME3-10 is a line graph showing these first 20 p-values.

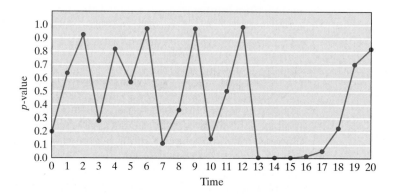

FIGURE ME3-10
$r = 4.0$, $p_0 = 0.2$.

The surprise here is the absence of any apparent pattern. In fact, no matter how much further we continue computing p-values, we will find no predictable pattern—to an outside observer the p-values for this population sequence appear to be quite erratic and seemingly random. Of course, we know better—they are all coming from the logistic equation.

The logistic growth model exhibits many interesting surprises. In addition to Exercises 15 through 24 at the end of this mini-excursion, you are encouraged to experiment on your own much like we did in the preceding examples: Choose a p_0 between 0 and 1, choose an r between 3 and 4, and fire up your calculator!

■ An excellent nontechnical account of the surprising patterns produced by the logistic growth model can be found in references 1 and 2. More technical accounts of the logistic equation can be found in references 3 and 8.

CONCLUSION

In this mini-excursion we discussed three simple but important models of population growth.

In the *linear model*, the population is described by an *arithmetic sequence* of the form $P_0, P_0 + d, P_0 + 2d, P_0 + 3d, \ldots$. In each transition period the population grows by the addition of a fixed amount d called the *common difference.* Linear growth is most common in situations in which there is no "breeding" such as populations of inanimate objects—commodities, resources, garbage, and so on.

In the *exponential model,* the population is described by a *geometric sequence* of the form $P_0, P_0r, P_0r^2, P_0r^3, \ldots$. In each transition period the population grows by multiplication by a positive constant r called the *common ratio.* Exponential growth is typical of situations in which there is some type of "breeding" in the population and the amount of breeding is directly proportional to the size of the population—money drawing interest in a bank account, the early stages of spread of an epidemic, the decay of radioactive materials, and so on.

The *logistic model* of population growth is described by the logistic equation $p_{N+1} = r(1 - p_N)p_N$. This model is used to describe the growth of biological populations whose growth rate is in direct proportion to the amount of space available in the population's habitat. When confined to a single-species habitat, many animal populations (including human populations) are governed by the logistic model or simple variations of it.

Most serious studies of population growth involve models with much more complicated mathematical descriptions, but to us, that is neither here nor there. Ultimately, the details are not as important as the overall picture: a realization that mathematics can be useful even in its most simplistic forms to describe and predict the rise and fall of populations in many fields—from the human realms of industry, finance, and public health to the natural world of population biology and ecology.

KEY CONCEPTS

arithmetic sequence, **477**
arithmetic sum formula, **479**
carrying capacity, **482**
common difference, **477**
common ratio, **480**
continuous growth, **474**
discrete growth, **475**

exponential growth model, **480**
geometric sequence, **480**
growth parameter, **482**
habitat, **482**
linear growth model, **477**
line graph, **476**
logistic equation, **483**

logistic growth model, **482**
population sequence, **475**
p-values, **482**
scatter plot, **476**
transition, **475**
transition rule, **475**

EXERCISES

A Linear Growth and Arithmetic Sequences

1. Consider a population that grows according to a linear growth model. The initial population is $P_0 = 75$, and the common difference is $d = 5$.

(a) Find P_{30}.

(b) How many generations will it take for the population to reach 1000?

(c) How many generations will it take for the population to reach 1002?

2. Consider a population that decays according to a linear model. The initial population is $P_0 = 520$, and the common difference is $d = -20$.

(a) Find P_{24}.

(b) How many generations will it take for the population to reach 10?

(c) How many generations will it take for the population to become extinct?

3. Consider a population that grows according to a linear growth model. The initial population is $P_0 = 8$, and the population in the 10th generation is $P_{10} = 38$.

(a) Find the common difference d.

(b) Find P_{50}.

(c) Give an explicit description of the population sequence.

4. Consider a population that grows according to a linear growth model. The population in the fifth generation is $P_5 = 37$, and the population in the seventh generation is $P_7 = 47$.

(a) Find the common difference d.

(b) Find the initial population P_0.

(c) Give an explicit description of the population sequence.

5. (a) Find $\underbrace{2 + 5 + 5 + \cdots + 5}_{100\ \text{terms}}$.

(b) Find $\underbrace{2 + 7 + 12 + \cdots}_{100\ \text{terms}}$.

6. (a) Find $\underbrace{21 + 7 + 7 + \cdots + 7}_{57\ \text{terms}}$.

(b) Find $\underbrace{21 + 28 + 35 + \cdots}_{57\ \text{terms}}$.

7. (a) The first two terms of an arithmetic sequence are 12 and 15. The number 309 is which term of the arithmetic sequence?

(b) Find $12 + 15 + 18 + \cdots + 309$.

8. (a) An arithmetic sequence has first term 1 and common difference 9. The number 2701 is which term of the arithmetic sequence?

(b) Find $1 + 10 + 19 + \cdots + 2701$.

9. The city of Lightsville currently has 137 streetlights. As part of an urban renewal program, the city council has decided to install and have operational 2 additional streetlights at the end of each week for the next 52 weeks. Each streetlight costs $1 to operate for 1 week.

(a) How many streetlights will the city have at the end of 38 weeks?

(b) How many streetlights will the city have at the end of N weeks? (Assume that $N \leq 52$.)

(c) What is the cost of operating the original 137 lights for 52 weeks?

(d) What is the additional cost for operating the newly installed lights for the 52-week period during which they are being installed?

10. A manufacturer currently has on hand 387 widgets. During the next 2 years, the manufacturer will be increasing his inventory by 37 widgets per week. (Assume that there are exactly 52 weeks in one year.) Each widget costs 10 cents a week to store.

(a) How many widgets will the manufacturer have on hand after 20 weeks?

(b) How many widgets will the manufacturer have on hand after N weeks? (Assume that $N \leq 104$.)

(c) What is the cost of storing the original 387 widgets for 2 years (104 weeks)?

(d) What is the additional cost of storing the increased inventory of widgets for the next 2 years?

B Exponential Growth and Geometric Sequences

11. A population grows according to an exponential growth model. The initial population is $P_0 = 11$ and the common ratio is $r = 1.25$.

(a) Find P_1.

(b) Find P_9.

(c) Give an explicit formula for P_N.

12. A population grows according to an exponential growth model, with $P_0 = 8$ and $P_1 = 12$.

(a) Find the common ratio r.

(b) Find P_9.

(c) Give an explicit formula for P_N.

13. Crime in Happyville is on the rise. Each year the number of crimes committed increases by 50%. Assume that there were 200 crimes committed in 2009, and let P_N denote the number of crimes committed in the year $2009 + N$.

(a) Give a recursive description of P_N.

(b) Give an explicit description of P_N.

(c) If the trend continues, approximately how many crimes will be committed in Happyville in the year 2019?

14. Since 2000, when 100,000 cases were reported, each year the number of new cases of equine flu has decreased by 20%. Let P_N denote the number of new cases of equine flu in the year $2000 + N$.

   (a) Give a recursive description of P_N.

   (b) Give an explicit description of P_N.

   (c) If the trend continues, approximately how many new cases of equine flu will be reported in the year 2015?

C Logistic Growth Model

15. A population grows according to the logistic growth model, with growth parameter $r = 0.8$. Starting with an initial population given by $p_0 = 0.3$,

   (a) find p_1.

   (b) find p_2.

   (c) determine what percent of the habitat's carrying capacity is taken up by the third generation.

16. A population grows according to the logistic growth model, with growth parameter $r = 0.6$. Starting with an initial population given by $p_0 = 0.7$,

   (a) find p_1.

   (b) find p_2.

   (c) determine what percent of the habitat's carrying capacity is taken up by the third generation.

17. For the population discussed in Exercise 15 ($r = 0.8$, $p_0 = 0.3$),

   (a) find the values of p_1 through p_{10}.

   (b) what does the logistic growth model predict in the long term for this population?

18. For the population discussed in Exercise 16 ($r = 0.6$, $p_0 = 0.7$),

   (a) find the values of p_1 through p_{10}.

   (b) what does the logistic growth model predict in the long term for this population?

19. A population grows according to the logistic growth model, with growth parameter $r = 1.8$. Starting with an initial population given by $p_0 = 0.4$,

   (a) find the values of p_1 through p_{10}.

   (b) what does the logistic growth model predict in the long term for this population?

20. A population grows according to the logistic growth model, with growth parameter $r = 1.5$. Starting with an initial population given by $p_0 = 0.8$,

   (a) find the values of p_1 through p_{10}.

   (b) what does the logistic growth model predict in the long term for this population?

21. A population grows according to the logistic growth model, with growth parameter $r = 2.8$. Starting with an initial population given by $p_0 = 0.15$,

   (a) find the values of p_1 through p_{10}.

   (b) what does the logistic growth model predict in the long term for this population?

22. A population grows according to the logistic growth model, with growth parameter $r = 2.5$. Starting with an initial population given by $p_0 = 0.2$,

   (a) find the values of p_1 through p_{10}.

   (b) what does the logistic growth model predict in the long term for this population?

23. A population grows according to the logistic growth model, with growth parameter $r = 3.25$. Starting with an initial population given by $p_0 = 0.2$,

   (a) find the values of p_1 through p_{10}.

   (b) what does the logistic growth model predict in the long term for this population?

24. A population grows according to the logistic growth model, with growth parameter $r = 3.51$. Starting with an initial population given by $p_0 = 0.4$,

   (a) find the values of p_1 through p_{10}.

   (b) what does the logistic growth model predict in the long term for this population?

D Miscellaneous

25. Each of the following sequences follows a linear, an exponential, or a logistic growth model. For each sequence, determine which model applies (if more than one applies, then indicate all the ones that apply).

   (a) $2, 4, 8, 16, 32, \ldots$

   (b) $2, 4, 6, 8, 10, \ldots$

   (c) $0.8, 0.4, 0.6, 0.6, 0.6, \ldots$

   (d) $0.81, 0.27, 0.09, 0.03, 0.01, \ldots$

   (e) $0.49512, 0.81242, 0.49528, 0.81243, 0.49528, \ldots$

   (f) $0.9, 0.75, 0.6, 0.45, 0.3, \ldots$

   (g) $0.7, 0.7, 0.7, 0.7, 0.7, \ldots$

26. Each of the line graphs shown in Figs ME3-11 through ME3-16 describes a population that grows according to a linear, an exponential, or a logistic model. For each line graph, determine which model applies.

(a)

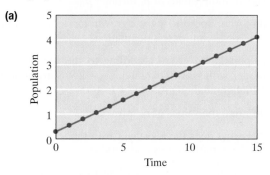

FIGURE ME3-11

(b)

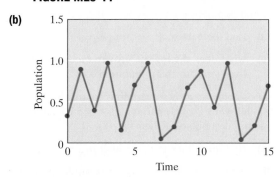

FIGURE ME3-12

(c)

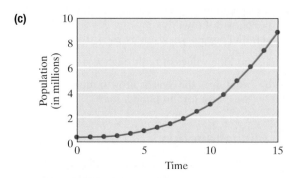

FIGURE ME3-13

(d)

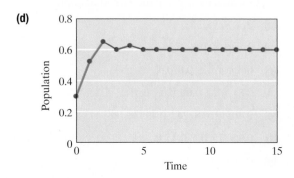

FIGURE ME3-14

(e)

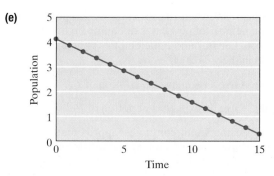

FIGURE ME3-15

(f)

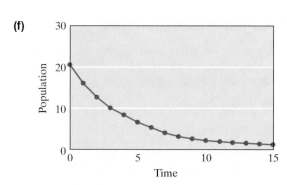

FIGURE ME3-16

27. Show that the sum of the first N terms of an arithmetic sequence with first term c and common difference d is
$$\frac{N}{2}[2c + (N - 1)d].$$

28. Consider a population that grows according to the logistic growth model with initial population given by $p_0 = 0.7$. What growth parameter r would keep the population constant?

29. Suppose that you are in charge of stocking a lake with a certain type of alligator with a growth parameter $r = 0.8$. Assuming that the population of alligators grows according to the logistic growth model, is it possible for you to stock the lake so that the alligator population is constant? Explain.

30. Consider a population that grows according to the logistic growth model with growth parameter $r(r > 1)$. Find p_0 in terms of r so that the population is constant.

31. The purpose of this exercise is to understand why we assume that, under the logistic growth model, the growth parameter r is between 0 and 4.

(a) What does the logistic equation give for p_{N+1} if $p_N = 0.5$ and $r > 4$? Is this a problem?

(b) What does the logistic equation predict for future generations if $p_N = 0.5$ and $r = 4$?

(c) If $0 \le p \le 1$, what is the largest possible value of $(1 - p)p$?

(d) Explain why, if $0 < p_0 < 1$ and $0 < r < 4$, then $0 < p_N < 1$, for every positive integer N.

32. Suppose that $r > 3$. Using the logistic growth model, find a population p_0 such that $p_0 = p_2 = p_4 \ldots$, but $p_0 \neq p_1$.

33. Show that if $P_0, P_1, P_2, \ldots$ is an arithmetic sequence, then $2^{P_0}, 2^{P_1}, 2^{P_2}, \ldots$ must be a geometric sequence.

PROJECTS AND PAPERS

A The Malthusian Doctrine

In 1798, Thomas Malthus wrote his famous *Essay on the Principle of Population*. In this essay, Malthus put forth the principle that population grows according to an exponential growth model, whereas food and resources grow according to a linear growth model. Based on this doctrine, Malthus predicted that humankind was doomed to a future where the supply of food and other resources would be unable to keep pace with the needs of the world's population.

Write an analysis paper detailing some of the consequences of Malthus's doctrine. Does the doctrine apply in a modern technological world? Can the doctrine be the explanation for the famines in sub-Saharan Africa? Discuss the many possible criticisms that can be leveled against Malthus's doctrine. To what extent do you agree with Malthus's doctrine?

B The Logistic Equation and the United States Population

The logistic growth model, first discovered by Verhulst, was rediscovered in 1920 by the American population ecologists Raymond Pearl and Lowell Reed. Pearl and Reed compared the population data for the United States between 1790 and 1920 with what would be predicted using a logistic equation and found that the numbers produced by the equation and the real data matched quite well.

In this project, you are to discuss and analyze Pearl and Reed's 1920s paper (reference 10). Here are some suggested questions you may want to discuss: Is the logistic model a good model to use with human populations? What might be a reasonable estimate for the carrying capacity of the United States? What happens with Pearl and Reed's model when you expand the census population data all the way to the latest population figures available? Note: Current and historical U.S. population data can be found at *www.census.gov*.

REFERENCES AND FURTHER READINGS

1. Cipra, Barry, "Beetlemania: Chaos in Ecology," in *What's Happening in the Mathematical Sciences 1998–1999*. Providence, RI: American Mathematical Society, 1999.

2. Gleick, James, *Chaos: Making a New Science*. New York: Viking Penguin, 1987, chap. 3.

3. Gordon, W. B., "Period Three Trajectories of the Logistic Map," *Mathematics Magazine*, 69 (1996), 118–120.

4. Hoppensteadt, Frank, *Mathematical Methods of Population Biology*. Cambridge: Cambridge University Press, 1982.

5. Hoppensteadt, Frank, *Mathematical Theories of Populations: Demographics, Genetics and Epidemics*. Philadelphia: Society for Industrial and Applied Mathematics, 1975.

6. Hoppensteadt, Frank, and Charles Peskin, *Mathematics in Medicine and the Life Sciences*. New York: Springer-Verlag, 1992.

7. Kingsland, Sharon E., *Modeling Nature: Episodes in the History of Population Ecology*. Chicago: University of Chicago Press, 1985.

8. May, Robert M., "Biological Populations with Nonoverlapping Generations: Stable Points, Stable Cycles and Chaos," *Science*, 186 (1974), 645–647.

9. May, Robert M., "Simple Mathematical Models with Very Complicated Dynamics," *Nature*, 261 (1976), 459–467.

10. Pearl, Raymond, and Lowell J. Reed, "On the Rate of Growth of the Population of the United States since 1790 and Its Mathematical Representation," *Proceedings of the National Academy of Sciences USA*, 6 (June 1920), 275–288.

11. Smith, J. Maynard, *Mathematical Ideas in Biology*. Cambridge: Cambridge University Press, 1968.

part 4

Statistics

13 Collecting Statistical Data

Censuses, Surveys, and Clinical Studies

Information has become the primary currency of the twenty-first century, and, by and large, wherever there is information, statistics are not far behind. Open today's paper and look at the business section — you'll find plenty of statistics there. Not interested in the stock market? Check the health section or the sports section. All of them are spiked with statistics.

W

hat is *statistics*? At its most basic level, statistics is the blending of two fundamental skills we learn separately in school: communicating and manipulating numbers. When we use numbers as a tool to transmit information, we are doing something statistical. If you prefer a more formal description, here it is: *Statistics is the science of dealing with data.* And what is *data*? Data is any type of information packaged in numerical form. [In modern usage, the word *data* is used for the singular form (a single piece of numerical information) as well as the plural form (many pieces of numerical information).]

Behind every statistical statement there is a story, and, like any story, it has a beginning, a middle, an end, and a moral. In this chapter we will discuss the beginning of the story, which, in statistics, typically means the process of gathering or collecting data. Collecting data seems deceptively simple, but history has repeatedly shown that doing so in an accurate, efficient, and timely manner can be the most difficult part of the statistical story. One of the special features of this chapter is the use of case studies that will help us learn a few of the do's and don'ts of data collection from the past experience of others.

The first step in proper data collection is to identify the population to which the data apply, and the concepts of *population*, *N-value*, and *census* are discussed in Section 13.1. In Sections 13.2, 13.3, and 13.4 the basic concepts of sampling are introduced (*sampling frame*, *selection bias*, *nonresponse bias*, *sampling error*, *sampling variability*, *chance error*) as well as various sampling methods (*quota sampling*, *simple random sampling*, *stratified sampling*). A method for estimating the size of a population using sampling (the *capture-recapture method*) is discussed in Section 13.5. When the goal of data collection is to study the relationship between a cause and an effect (e.g., does taking a math class improve your chances of getting a high-paying job?), a special type of data collection process called a *clinical study* is required. Some of the key concepts behind clinical studies (*control group*, *treatment group*, *randomization*, *placebo effect*, *blind and double-blind studies*) are discussed in Section 13.6.

13.1 The Population

Every statistical statement refers, directly or indirectly, to some group of individuals or objects. In statistical terminology, this collection of individuals or objects is called the **population**. The first question we should ask ourselves when trying to make sense of a statistical statement is, "What is the population to which the statement applies?"

In an ideal world, the specific population to which a statistical statement applies is clearly identified within the story itself. In the real world, this rarely happens because the details are skipped (mostly to keep the story moving along but sometimes with an intent to confuse or deceive) or, alternatively, because two (or more) related populations are involved in the story.

> **EXAMPLE 13.1**   The Return of the Bald Eagle

Because of its iconic importance—the bald eagle is the national bird and the emblem of the United States—much has been said and written about bald eagle populations in North America, from their near extinction to their miraculous recovery (the details are explained briefly in the excerpt on the left, taken from a 2002 *National Geographic* article).

Bald Eagles Come Back from the Brink

BY JOHN L. ELIOT

They ruled the skies on seven-foot (two-meter) wingspans when 17th-century Europeans arrived in North America. Throughout the continent, half a million bald eagles may have soared. But settlers blamed them for killing livestock, so shooting began—and the proud birds' numbers began to plunge....

The Bald Eagle Protection Act of 1940 prohibited shooting or otherwise harming the birds in the [lower 48 states] but didn't cover the pesticides that with-

in a decade began to destroy eagles' eggs. By the 1960s only about 400 breeding pairs of bald eagles remained in the lower 48 [states].... The banning of DDT in 1972 and other measures launched an amazing comeback by the eagles, whose status changed from endangered to threatened in 1995. Today, with more than 6,000 breeding pairs, bald eagles may soon be taken off the endangered species list entirely, their survival as an icon secured—for now.

National Geographic Magazine, July 2002

In the context of populations, this is, in spirit, a story about the bald eagle population in the United States—more specifically the contiguous 48 states. (The bald eagle population of Alaska is not really a part of the story, as bald eagles never reached endangered levels there, and by the way, there are no bald eagles in Hawaii.) Unlike many other animals and birds, bald eagles stay within the confines of a reasonably small geographical area, so it is possible to discuss bald eagle populations in a regional context such as the contiguous 48 states, or even populations within a specific state.

Interestingly enough, the case about the comeback of the bald eagle population is often made through the use of a *proxy*—the number of bald eagle *breeding pairs* (400 in the 1960s; over 6000 in 2000). Thus, upon closer scrutiny the story ties together two different but closely related populations: the overall bald eagle population within the contiguous 48 states (including chicks, adolescent birds, etc.) and the population of breeding pairs. The former is the population of interest, but the latter is the population of convenience because breeding pairs are much easier to identify, track, and count.

The *N*-Value

Given a specific population, an obviously relevant question is, "How many individuals or objects are there in that population?" This number is called the **N-value** of the population. (It is common practice in statistics to use capital *N* to denote population sizes.) It is important to keep in mind the distinction between the *N*-value—a number specifying the size of the population—and the population itself.

As we learned in Chapter 10, populations change, and thus *N*-values must often be discussed within the context of time. Example 13.1 illustrates this point.

EXAMPLE 13.2   **The Return of the Bald Eagle: Part 2**

Over a period of many years, the United States Fish and Wildlife Service was able to keep a remarkably accurate tally of the number of bald eagle breeding pairs in the contiguous 48 states. (As we discussed in Example 13.1, breeding pairs are used as a useful proxy for the health of the overall population.) A tremendous amount of effort has gone into collecting and verifying these *N*-values, which, for a wildlife population, are of remarkable accuracy. Figure 13-1 summarizes the population numbers over the period 1963–2000. (No tallies were conducted in 1964–1973, 1975–1980, 1983, and 1985.) Since 2000 the bald eagle population has grown to the point that the U.S. Fish and Wildlife Service has discontinued the annual tallies.

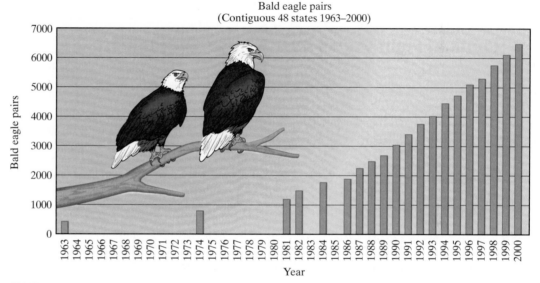

Bald eagle pairs
(Contiguous 48 states 1963–2000)

FIGURE 13-1
Source: United States Fish and Wildlife Service.

Our next example illustrates a basic truism: We cannot determine an *N*-value if we do not first identify the population.

EXAMPLE 13.3   ***N* Is in the Eye of the Beholder**

Andy has a coin jar full of quarters. He is hoping that there is enough money in the jar to pay for a new baseball glove. Dad says to go count them, and if there isn't enough, he will lend Andy the difference. Andy dumps the quarters out of the jar, makes a careful tally, and comes up with a count of 116 quarters.

What is the *N*-value here? The answer depends on how we define the population. Are we counting coins or money? To Dad, who will end up stuck with all the quarters, the total number of coins might be the most relevant issue. Thus, to Dad, $N = 116$. Andy, on the other hand, is concerned with how much money is in the jar. If he were to articulate his point of view in statistical language, he would say that $N = 29$ (dollars).

The word **data** is the plural of the Latin word **datum**, meaning "something given," and in ordinary usage has a somewhat broader meaning than the one we will give it in this chapter. For our purposes we will use the word *data* as any type of information packaged in numerical form, and we will adhere to the standard convention that as a noun it can be used both in singular ("the data *is* . . ."), and plural ("the data *are* . . .") forms. The process of collecting data by going through every member of the population is called a **census**. (When Andy dumped all the quarters out of the jar and counted them, he was essentially conducting a small census.) The idea behind a census is simple enough, but in practice a census requires a great deal of "cooperation" from the population. (The quarters in Andy's jar were nice enough to stay put and let themselves be counted.) For larger, more dynamic populations (wildlife, humans, etc.), accurate tallies are inherently difficult if not impossible, and in these cases the best we can hope for is a good estimate of the N-value.

> **EXAMPLE 13.4**   **2000 Census Undercounts**

The most notoriously difficult N-value question around is, "What is the N-value of the national population of the United States?" This is a question the United States Census tries to answer every 10 years—with very little success.

The 2000 U.S. Census was the largest single peacetime undertaking of the federal government—it employed over 850,000 people and cost about $6.5 billion—and yet it missed counting between 3 and 4 million people (the excerpt on the left from a 2001 *New York Times* article gives some of the details).

Given the critical importance of the U.S. Census and given the tremendous resources put behind the effort by the federal government, why is the head count so far off? How can the best intentions and tremendous resources of our government fail so miserably in an

Political Fight Brews Over Census Correction

By Haya El Nasser

The 2000 Census did a better job counting people than the last Census, especially minorities and children, but it still missed about 2.7 million to 4 million people. . . .

Preliminary estimates show that the net number of people missed falls between 0.96% and 1.4%. In 1990, the undercount was 1.6%, or 4 million people. There was a significant drop in the undercount of blacks, Hispanics, American Indians and children, population groups that were disproportionately missed in 1990 . . . . The estimates are bound to heat up political infighting. The Census

Bureau must decide whether the numbers should be adjusted to compensate for the undercount. . . .

Census numbers are used to redraw political districts. An adjusted count would include more minorities, which could reshape key political districts. Republicans worry an adjusted count would help Democrats . . . . The Census Bureau estimates the number of people missed through the same method that would be used to adjust the numbers. It surveys 314,000 sample households and checks to see whether those households filled out Census forms.

Source: *New York Times, February 15, 2001*

activity that on a smaller scale can be carried out by a child trying to buy a baseball glove?

Our first case study gives a brief overview of the ins and outs of the U.S. Census and illustrates the difficulties faced by censuses in general.

> **CASE STUDY 1**   THE U.S. CENSUS

Article 1, Section 2, of the Constitution of the United States mandates that a national census be conducted every 10 years. The original intent of the census was to "count heads" for a twofold purpose: taxes and political representation. Like everything else in the Constitution, Article 1, Section 2, was a compromise of many competing interests: The count was to exclude "Indians not taxed" and to count slaves as "three-fifths of a free Person." Since then, the scope and purpose of the U.S. Census have been modified and expanded by the 14th Amendment and the courts in many ways:

■ Besides counting heads, the U.S. Census Bureau now collects additional information about the population: sex, age, race, ethnicity, marital status, housing, income, and employment data. Some of this information is updated on a regular basis, not just every 10 years.

- Census data are now used for many important purposes beyond the original ones of *taxation* and *representation*: the allocation of billions of federal dollars to states, counties, cities, and municipalities; the collection of other important government statistics such as the Consumer Price Index and the Current Population Survey; the redrawing of legislative districts within each state; and the strategic planning of production and services by business and industry.

- For the purposes of the Census, the United States population is defined as consisting of "all persons *physically present* and *permanently residing* in the United States." Citizens, legal resident aliens, and even illegal aliens are meant to be included.

Nowadays, the notion that if we put enough money and effort into it, all individuals living in the United States can be counted like coins in a jar is unrealistic. In 1790, when the first U.S. Census was carried out, the population was smaller and relatively homogeneous, as people tended to stay in one place, and, by and large, they felt comfortable in their dealings with the government. Under these conditions it might have been possible for census takers to count heads accurately. Today's conditions are completely different. People are constantly on the move. Many distrust the government. In large urban areas many people are homeless or don't want to be counted. And then there is the apathy of many people who think of a census form as another piece of junk mail.

If the Census undercount were consistent among all segments of the population, the undercount problem could be solved easily. Unfortunately, the modern U.S. Census is plagued by what is known as a *differential undercount*. Ethnic minorities, migrant workers, and the urban poor populations have significantly larger undercount rates than the population at large, and the undercount rates vary significantly within these groups. Using modern statistical techniques, it is possible to make adjustments to the raw Census figures that correct some of the inaccuracy caused by the differential undercount, but in 1999 the Supreme Court ruled in *Department of Commerce et al. v. United States House of Representatives et al.* that only the raw numbers, and not statistically adjusted numbers, can be used for the purposes of apportionment of Congressional seats among the states.

13.2 Sampling

The practical alternative to a census is to collect data only from some members of the population and use that data to draw conclusions and make inferences about the entire population. Statisticians call this approach a **survey** (or a **poll** when the data collection is done by *asking questions*). The subgroup chosen to provide the data is called the **sample**, and the act of selecting a sample is called **sampling**.

Ideally, every member of the population should have an opportunity to be chosen as part of the sample, but this is possible only if we have a mechanism to identify each and every member of the population. In many situations this is impossible. Say we want to conduct a public opinion poll before an election. The population for the poll consists of all voters in the upcoming election, but how can we identify who is and is not going to vote ahead of time? We know who the registered voters are, but among this group there are still many nonvoters.

The first important step in a survey is to distinguish the population for which the survey applies (the **target population**) and the actual subset of the population from which the sample will be drawn, called the **sampling frame**. The ideal scenario is when the sampling frame is the same as the target population—that would mean that every member of the target population is a candidate for the sample. When this is impossible (or impractical), an appropriate sampling frame must be chosen.

Polls are most famously (and infamously) used to predict the outcome of political elections. Among the many issues that make pre-election polls particularly delicate is the problem of identifying the members of the target population—namely, the people who are going to end up voting. The conventional approach is to use registered voters as the sampling frame, but the use of registered voters can lead to some bad data. Our next example illustrates this point.

(**EXAMPLE 13.5**)   **Sampling Frames Can Make a Difference**

A CNN/USA Today/Gallup poll conducted right before the November 2, 2004, national election asked the following question: "If the election for Congress were being held today, which party's candidate would you vote for in your congressional district, the Democratic Party's candidate or the Republican Party's candidate?"

When the question was asked of 1866 *registered* voters nationwide, the results of the poll were 49% for the Democratic Party candidate, 47% for the Republican Party candidate, 4% undecided.

When exactly the same question was asked of 1573 *likely* voters nationwide, the results of the poll were 50% for the Republican Party candidate, 47% for the Democratic Party candidate, 3% undecided.

Clearly, one of the two polls had to be wrong, because in the first poll the Democrats beat out the Republicans, whereas in the second poll it was the other way around. The only significant difference between the two polls was the choice of the sampling frame—in the first poll the sampling frame used was *all registered voters*, and in the second poll the sampling frame used was *all likely voters*. Although neither one faithfully represents the target population of actual voters, using likely voters instead of registered voters for the sampling frame gives much more reliable data. (The second poll predicted very closely the average results of the 2004 congressional races across the nation.)

So, why don't all pre-election polls use likely voters as a sampling frame instead of registered voters? The answer is economics. Registered voters are relatively easy to identify—every county registrar can produce an accurate list of registered voters. Not every registered voter votes, though, and it is much harder to identify those who are "likely" to vote. (What is the definition of *likely* anyway?) Typically, one has to look at demographic factors (age, ethnicity, etc.) as well as past voting behavior to figure out who is likely to vote and who isn't. Doing that takes a lot more effort, time, and money.   ⊂⊃

The basic philosophy behind sampling is simple and well understood—if we have a sample that is "representative" of the entire population, then whatever we want to know about a population can be found out by getting the information from the sample. In practice, however, things are not always as simple. If we are to draw reliable data from a sample, we must (a) find a sample that is representative of the population, and (b) determine how big the sample should be. These two issues go hand in hand, and we will discuss them next.

Sometimes a very small sample can be used to get reliable information about a population, no matter how large the population is. This is the case when the population is highly homogeneous. For example, a person's blood is essentially the same everywhere in the body, which explains why a small blood sample drawn from an arm can be used to draw reliable data about the patient's blood sugar levels, cholesterol levels, and so on. In the extreme case of a population of identical individuals, all it takes is one individual to make up a representative sample of the population.

The more heterogeneous a population gets, the more difficult it is to find a representative sample. The difficulties can be well illustrated by taking a look at the history of *public opinion polls*.

Public Opinion Polls

You are probably familiar with public opinion polls, such as the Gallup poll, the Harris poll, and many others. A public opinion poll is a survey in which the members of the sample provide information by answering specific questions from an "interviewer." The question-and-answer exchange can be done through a questionnaire, a personal telephone interview, or a direct face-to-face interview.

Nowadays, public opinion polls are used regularly to measure "the pulse of the nation." They give us statistical information ranging from voters' preferences before an election to opinions on issues such as the environment, abortion, and the economy.

Given their widespread use and the influence they exert, it is important to ask how much we can trust the information that we get from public opinion polls. This is a complex question that goes to the very heart of mathematical statistics. We'll start our exploration of it with some historical examples.

CASE STUDY 2   THE 1936 *LITERARY DIGEST* POLL

The U.S. presidential election of 1936 pitted Alfred Landon, the Republican governor of Kansas, against the incumbent Democratic President, Franklin D. Roosevelt. At the time of the election, the nation had not yet emerged from the Great Depression, and economic issues such as unemployment and government spending were the dominant themes of the campaign.

The *Literary Digest*, one of the most respected magazines of the time, conducted a poll a couple of weeks before the election. The magazine had used polls to accurately predict the results of every presidential election since 1916, and their 1936 poll was the largest and most ambitious poll ever. The *sampling frame* for the *Literary Digest* poll consisted of an enormous list of names that included (1) every person listed in a telephone directory anywhere in the United States, (2) every person on a magazine subscription list, and (3) every person listed on the roster of a club or professional association. From this sampling frame a list of about 10 million names was created, and every name on this list was mailed a mock ballot and asked to mark it and return it to the magazine.

One cannot help but be impressed by the sheer scope and ambition of the 1936 *Literary Digest* poll, as well as the magazine's unbounded confidence in its accuracy. In its issue of August 22, 1936, the *Literary Digest* crowed:

> Once again, [we are] asking more than ten million voters—one out of four, representing every county in the United States—to settle November's election in October.
>
> Next week, the first answers from these ten million will begin the incoming tide of marked ballots, to be triple-checked, verified, five-times cross-classified and totaled. When the last figure has been totted and checked, if past experience is a criterion, the country will know to within a fraction of 1 percent the actual popular vote of forty million [voters].

Based on the poll results, the *Literary Digest* predicted a landslide victory for Landon with 57% of the vote, against Roosevelt's 43%. Amazingly, the election turned out to be a landslide victory for Roosevelt with 62% of the vote, against 38%

for Landon. The difference between the poll's prediction and the actual election results was a whopping 19%, the largest error ever in a major public opinion poll. The results damaged the credibility of the magazine so much so that soon after the election its sales dried up and it went out of business—the victim of a major statistical blunder.

For the same election, a young pollster named George Gallup (for more on Gallup, see the biographical profile at the end of the chapter) was able to predict accurately a victory for Roosevelt using a sample of "only" 50,000 people. In fact, Gallup *also* publicly predicted, to within 1%, the incorrect results that the *Literary Digest* would get using a sample of just 3000 people taken from the same sampling frame the magazine was using. What went wrong with the *Literary Digest* poll and why was Gallup able to do so much better?

The first thing seriously wrong with the *Literary Digest* poll was the sampling frame, consisting of names taken from telephone directories, lists of magazine subscribers, rosters of club members, and so on. Telephones in 1936 were something of a luxury, and magazine subscriptions and club memberships even more so, at a time when 9 million people were unemployed. When it came to economic status the *Literary Digest* sample was far from being a representative cross section of the voters. This was a critical problem, because voters often vote on economic issues, and given the economic conditions of the time, this was especially true in 1936.

When the choice of the sample has a built-in tendency (whether intentional or not) to exclude a particular group or characteristic within the population, we say that a survey suffers from **selection bias**. It is obvious that selection bias must be avoided, but it is not always easy to detect it ahead of time. Even the most scrupulous attempts to eliminate selection bias can fall short (as will become apparent in our next case study).

The second serious problem with the *Literary Digest* poll was the issue of *nonresponse bias*. In a typical survey it is understood that not every individual is willing to respond to the survey request (and in a democracy we cannot force them to do so). Those individuals who do not respond to the survey request are called *nonrespondents*, and those who do are called *respondents*. The percentage of respondents out of the total sample is called the **response rate**. For the *Literary Digest* poll, out of a sample of 10 million people who were mailed a mock ballot only about 2.4 million mailed a ballot back, resulting in a 24% response rate. When the response rate to a survey is low, the survey is said to suffer from **nonresponse bias**. (Exactly at what point the response rate is to be considered low depends on the circumstances and nature of the survey, but a response rate of 24% is generally considered very low.)

Nonresponse bias can be viewed as a special type of selection bias—it excludes from the sample reluctant and uninterested people. Since reluctant and uninterested people can represent a significant slice of the population, we don't want them excluded from the sample. But getting reluctant, uninterested, and apathetic slugs to participate in a survey is a conundrum—in a free country we cannot force people to participate, and bribing them with money or chocolate chip cookies is not always a practical solution.

One of the significant problems with the *Literary Digest* poll was that the poll was conducted by mail. This approach is the most likely to magnify nonresponse bias, because people often consider a mailed questionnaire just another form of junk mail. Of course, given the size of their sample, the *Literary Digest* hardly had a choice. This illustrates another important point: Bigger is not better, and a big sample can be more of a liability than an asset.

The *Literary Digest* story has two morals: (1) *You'll do better with a well-chosen small sample than with a badly chosen large one*, and (2) *watch out for selection bias and nonresponse bias*.

Convenience Sampling

There is always a cost (effort, time, money) associated with collecting data, and it is a truism that this cost is proportional to the quality of the data collected—the better the data, the more effort required to collect it. It follows that the temptation to take shortcuts when collecting data is always there and that data collected "on the cheap" should always be scrutinized carefully. One commonly used shortcut in sampling is known as **convenience sampling**. In convenience sampling the selection of which individuals are in the sample is dictated by what is easiest or cheapest for the data collector, never mind trying to get a representative sample.

A classic example of convenience sampling is when interviewers set up at a fixed location such as a mall or outside a supermarket and ask passersby to be part of a public opinion poll. A different type of convenience sampling occurs when the sample is based on *self-selection*—the sample consists of those individuals who volunteer to be in it. Self-selection is the reason why many Area Code 800 polls ("Call 1-800-YOU-NUTS to express your opinion on the new tax proposal...") are not to be trusted. Even worse are the Area Code 900 polls, for which an individual has to actually pay (sometimes as much as $2) to be in the sample. A sample consisting entirely of individuals who paid to be in the sample is not likely to be a representative sample of general public opinion.

Convenience sampling is not always bad—at times there is no other choice or the alternatives are so expensive that they have to be ruled out. We should keep in mind, however, that data collected through convenience sampling are naturally tainted and should always be scrutinized (that's why we always want to get to the details of *how* the data were collected). More often than not, convenience sampling gives us data that are too unreliable to be of any scientific value. With data, as with so many other things, *you get what you pay for*.

Quota Sampling

Quota sampling is a systematic effort to force the sample to be representative of a given population through the use of quotas—the sample should have so many women, so many men, so many blacks, so many whites, so many people living in urban areas, so many people living in rural areas, and so on. The proportions in each category in the sample should be the same as those in the population. If we can assume that every important characteristic of the population is taken into account when the quotas are set up, it is reasonable to expect that the sample will be representative of the population and produce reliable data.

Our next case study illustrates some of the difficulties with the assumptions behind quota sampling.

CASE STUDY 3 | THE 1948 PRESIDENTIAL ELECTION

George Gallup had introduced quota sampling as early as 1935 and had used it successfully to predict the winner of the 1936, 1940, and 1944 presidential elections. Quota sampling thus acquired the reputation of being a "scientifically reliable" sampling method, and by the 1948 presidential election all three major national polls—the Gallup poll, the Roper poll, and the Crossley poll—used quota sampling to make their predictions.

For the 1948 election between Thomas Dewey and Harry Truman, Gallup conducted a poll with a sample of approximately 3250 people. Each individual in the sample was interviewed in person by a professional interviewer to minimize nonresponse bias, and each interviewer was given a very detailed set of quotas to meet—for example, 7 white males under 40 living in a rural area, 5 black males over 40 living in a rural area, 6 white females under 40 living in an urban area, and so on. By the time all the interviewers met their quotas, the entire sample was expected to accurately represent the entire population in every respect: gender, race, age, and so on.

Based on his sample, Gallup predicted that Dewey, the Republican candidate, would win the election with 49.5% of the vote to Truman's 44.5% (with third-party candidates Strom Thurmond and Henry Wallace accounting for the remaining 6%). The Roper and Crossley polls also predicted an easy victory for Dewey. (In fact, after an early September poll showed Truman trailing Dewey by 13 percentage points, Roper announced that he would discontinue polling since the outcome was already so obvious.) The actual results of the election turned out to be almost the exact reverse of Gallup's prediction: Truman got 49.9% and Dewey 44.5% of the national vote.

Truman's victory was a great surprise to the nation as a whole. So convinced was the *Chicago Daily Tribune* of Dewey's victory that it went to press on its early edition for November 4, 1948, with the headline "Dewey defeats Truman." The picture of Truman holding aloft a copy of the *Tribune* and his famous retort "Ain't the way I heard it" have become part of our national folklore.

To pollsters and statisticians, the erroneous predictions of the 1948 election had two lessons: (1) *Poll until election day*, and (2) *quota sampling is intrinsically flawed*.

What's wrong with quota sampling? After all, the basic idea behind it appears to be a good one: Force the sample to be a representative cross section of the population by having each important characteristic of the population proportionally represented in the sample. Since income is an important factor in determining how people vote, the sample should have all income groups represented in the same proportion as the population at large. The same should be true for gender, race, age, and so on. Right away, we can see a potential problem: Where do we stop? No matter how careful we might be, we might miss some criterion that would affect the way people vote, and the sample could be deficient in this regard.

"Ain't the way I heard it," Truman gloats while holding an early edition of the *Chicago Daily Tribune* in which the headline erroneously claimed a Dewey victory based on the predictions of all the polls.

An even more serious flaw in quota sampling is that, other than meeting the quotas, the interviewers are free to choose whom they interview. This opens the door to selection bias. Looking back over the history of quota sampling, we can see a clear tendency to overestimate the Republican vote. In 1936, using quota sampling, Gallup predicted that the Republican candidate would get 44% of the vote, but the actual number was 38%. In 1940, the prediction was 48%, and the actual vote was 45%; in 1944, the prediction was 48%, and the actual vote was 46%. Gallup was able to predict the winner correctly in each of these elections, mostly because the spread between the candidates was large enough to cover the error. In 1948, Gallup (and all the other pollsters) simply ran out of luck. It was time to ditch quota sampling.

The failure of quota sampling as a method for getting representative samples has a simple moral: *Even with the most carefully laid plans, human intervention in choosing the sample can result in selection bias.*

13.3 Random Sampling

The best alternative to human selection is to let the *laws of chance* determine the selection of a sample. Sampling methods that use randomness as part of their design are known as **random sampling** methods, and any sample obtained through random sampling is called a **random sample** (or a *probability sample*).

The idea behind random sampling is that the decision as to which individuals should or should not be in the sample is best left to chance because the laws of chance are better than human design in coming up with a representative sample. At first, this idea seems somewhat counterintuitive. How can a process based on random selection guarantee an unbiased sample? Isn't it possible to get by sheer bad luck a sample that is very biased (e.g., mostly male or all college students)? In theory, such an outcome is possible, but in practice, when the sample is large enough, the odds of it happening are so low that we can pretty much rule it out.

Most present-day methods of quality control in industry, corporate audits in business, and public opinion polling are based on random sampling. The reliability of data collected by random sampling methods is supported by both practical experience and mathematical theory. (We will discuss some of the details of this theory in Chapter 16.)

Simple Random Sampling

The most basic form of random sampling is called **simple random sampling**. It is based on the same principle a lottery is. Any set of numbers of a given size has an equal chance of being chosen as any other set of numbers of that size. Thus, if a lottery ticket consists of six winning numbers, a fair lottery is one in which any combination of six numbers has the same chance of winning as any other combination of six numbers. In sampling, this means that any group of members of the population should have the same chance of being the sample as any other group of the same size.

In theory, simple random sampling is easy to implement. We put the name of each individual in the population in "a hat," mix the names well, and then draw as many names as we need for our sample. Of course "a hat" is just a metaphor. If our population is 100 million voters and we want to choose a simple random sample of 2000, we will not be putting all 100 million names in a real hat and then drawing 2000 names one by one. These days, the "hat" is a computer database containing a list of members of the population. A computer program then randomly selects the names. This is a fine idea for small, compact populations, but a hopeless one when it comes to national surveys and public opinion polls.

Implementing simple random sampling in national public opinion polls raises problems of expediency and cost. Interviewing hundreds of individuals chosen by simple random sampling means chasing people all over the country, a task that requires an inordinate amount of time and money. For most public opinion polls—especially those done on a regular basis—the time and money needed to do this are simply not available.

Stratified Sampling

The alternative to simple random sampling used nowadays for national surveys and public opinion polls is a sampling method known as **stratified sampling**. The basic idea of stratified sampling is to break the sampling frame into categories, called **strata**, and then (unlike quota sampling) *randomly* choose a sample from these strata. The chosen strata are then further divided into categories, called substrata, and a random sample is taken from these substrata. The selected substrata are further subdivided, a random sample is taken from them, and so on. The process goes on for a predetermined number of steps (usually four or five).

Our next case study illustrates how stratified sampling works in the case of a national public opinion poll. Basic variations of the same idea can be used at the state, city, or local level. The specific details, of course, will be different.

> **CASE STUDY 4** NATIONAL PUBLIC OPINION POLLS

In national public opinion polls the *strata* and *substrata* are defined by a combination of geographic and demographic criteria. For example, the nation is first divided into "size of community" *strata* (big cities, medium cities, small cities, villages, rural areas, etc.). The strata are then subdivided by geographical region (New England, Middle Atlantic, East Central, etc.). This is the first layer of substrata. Within each geographical region and within each size of community stratum some communities (called *sampling locations*) are selected by simple random sampling. The selected sampling locations are the only places where interviews will be conducted. Next, each of the selected sampling locations is further subdivided into geographical units called *wards*. This is the

The Gallup Poll

THE GALLUP REPORT. PRINCETON, NJ: AMERICAN INSTITUTE OF PUBLIC OPINION

The design of the sample used by the Gallup Poll for its standard surveys of public opinion is that of a replicated area probability *stratified sample* ... After stratifying the nation geographically and by size of community in order to insure conformity of the sample with the 2000 Census distribution of the population, over 360 different sampling locations or areas are selected on a mathematically random basis from within cities, towns, and countries which have in turn, been selected on a mathematically random basis. The interviewers have no choice whatsoever concerning the part of the city, town, or country in which they conduct their interviews.

Approximately five interviews are conducted in each randomly selected sampling point. Interviewers are given maps of the area to which they are assigned and are required to follow a specific travel pattern on contacting households. At each occupied dwelling unit, interviewers are instructed to select respondents by following a prescribed systematic method. This procedure is followed until the assigned number of interviews with male and female adults have been completed. . . .

second layer of substrata. Within each sampling location some of the wards are selected using simple random sampling. The selected wards are then divided into smaller units, called *precincts* (third layer), and within each ward some of its precincts are selected by simple random sampling. At the last stage, *households* (fourth layer) are selected from within each precinct by simple random sampling. The interviewers are then given specific instructions as to which households in their assigned area they must conduct interviews in and the order that they must follow.

The efficiency of stratified sampling compared with simple random sampling in terms of cost and time is clear. The members of the sample are clustered in well-defined and easily manageable areas, significantly reducing the cost of conducting interviews as well as the response time needed to collect the data. For a large, heterogeneous nation like the United States, stratified sampling has generally proved to be a reliable way to collect national data.

What about the size of the sample? Surprisingly, it does not have to be very large. Typically, a Gallup poll is based on samples consisting of approximately 1500 individuals, and roughly the same size sample can be used to poll the popu-

■ Whether you poll the United States or New York State or Baton Rouge (Louisiana) . . . you need . . . the same number of interviews or samples. It's no mystery really — if a cook has two pots of soup on the stove, one far larger than the other, and thoroughly stirs them both, he doesn't have to take more spoonfuls from one than the other to sample the taste accurately.

—George Gallup

lation of a small city as the population of the United States. *The size of the sample does not have to be proportional to the size of the population.*

13.4 Sampling: Terminology and Key Concepts

As we now know, except for a *census*, the common way to collect statistical information about a population is by means of a *survey*. (When the survey consists of asking people their opinion on some issue, we refer to it as a *public opinion poll*.) In a survey, we use a subset of the population, called a *sample*, as the source of our information, and from this sample, we try to generalize and draw conclusions about the entire population. Statisticians use the term **statistic** to describe any kind of numerical information drawn from a sample. A statistic is always an estimate for some unknown measure, called a **parameter**, of the population. Let's put it this way: A *parameter* is the numerical information we would like to have—the pot of gold at the end of the statistical rainbow, so to speak. Calculating a parameter is difficult and often impossible, since the only way to get the exact value for a parameter is to use a census. If we use a sample, then we can get only an estimate for the parameter, and this estimate is called a *statistic*.

We will use the term **sampling error** to describe the difference between a parameter and a statistic used to estimate that parameter. In other words, the sampling error measures how much the data from a survey differs from the data that would have been obtained if a census had been used. Of course, the very point of sampling is to avoid using a census, so sampling errors can only be estimated. Usually the estimates are given in terms of a margin of error, as in "The margin of sampling error for the poll was plus or minus 3%." We will discuss in greater detail the exact meaning of such statements in Chapter 16.

Sampling error can be attributed to two factors: *chance error* and *sampling bias*. **Chance error** is the result of the basic fact that a sample, being just a sample, can only give us approximate information about the population. In fact, different samples are likely to produce different statistics for the same population, even when the samples are chosen in exactly the same way—a phenomenon known as **sampling variability**. While sampling variability, and thus chance error, are unavoidable, with careful selection of the sample and the right choice of sample size they can be kept to a minimum.

Sample bias is the result of choosing a bad sample and is a much more serious problem than chance error. Even with the best intentions, getting a sample that is representative of the entire population can be very difficult and can be affected by many subtle factors. Sample bias is the result. As opposed to chance error, sample bias can be eliminated by using proper methods of sample selection.

Last, we shall make a few comments about the size of the sample, typically denoted by the letter n (to contrast with N, the size of the population). The ratio n/N is called the **sampling proportion**. A sampling proportion of x% tells us that the size of the sample is intended to be x% of the population. Some sampling methods are conducive for choosing the sample so that a given sampling proportion is obtained, but in many sampling situations it is very difficult to predetermine what the exact sampling proportion is going to be (we would have to know the exact values of both N and n). In any case, it is not the sampling proportion that matters but rather the absolute sample size and quality. Typically, modern public opinion polls use samples of n between 1000 and 1500 to get statistics that have a margin of error of less than 5%, be it for the population of a city, a region, or the entire country. (We will see more details about this idea in Chapter 16.)

13.5 The Capture-Recapture Method

We have already observed that finding the exact N-value of a large and elusive population can be extremely difficult and sometimes impossible. In many cases, a good estimate is all we really need, and such estimates are possible through sampling methods. The simplest sampling method for estimating the N-value of a population is called the **capture-recapture method**. (This method is commonly used by biologists to estimate the size of wildlife populations, which explains the unusual choice of the phrase *capture-recapture* in the name.).

> ┌─ **THE CAPTURE-RECAPTURE METHOD** ────────
>
> - **Step 1. Capture (sample):** Capture (choose) a sample of size n_1, *tag* (mark, identify) the animals (objects, people), and release them back into the general population.
>
> - **Step 2. Recapture (resample):** After a certain period of time, capture a new sample of size n_2 and take an exact head count of the *tagged* individuals (i.e., those that were also in the first sample). Call this number k.
>
> - **Step 3. Estimate:** The N-value of the population can be estimated to be approximately $(n_1 \cdot n_2)/k$.

The capture-recapture method is based on the assumption that both the captured and recaptured samples are representative of the entire population. Under these assumptions, the proportion of tagged individuals in the recaptured sample is approximately equal to the proportion of the tagged individuals in the population. In other words, the ratio k/n_2 is approximately equal to the ratio n_1/N. From this we can solve for N and get $N \approx (n_1 \cdot n_2)/k$.

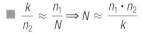

■ $\dfrac{k}{n_2} \approx \dfrac{n_1}{N} \Rightarrow N \approx \dfrac{n_1 \cdot n_2}{k}$

(**EXAMPLE 13.6**)   **Small Fish in a Big Pond**

A large pond is stocked with catfish. As part of a research project we need to estimate the number of catfish in the pond. An actual head count is out of the question (short of draining the pond), so our best bet is the capture-recapture method.

Step 1. For our first sample we capture a predetermined number n_1 of catfish, say $n_1 = 200$. The fish are tagged and released unharmed back in the pond.

Step 2. After giving enough time for the released fish to mingle and disperse throughout the pond, we capture a second sample of n_2 catfish. While n_2 does not have to equal n_1, it is a good idea for the two samples to be of approximately the same order of magnitude. Let's say that $n_2 = 150$. Of the 150 catfish in the second sample, 21 have tags (were part of the original sample).

Assuming the second sample is representative of the catfish population in the pond, the ratio of tagged fish in the second sample $(21/150)$ is approximately the same as the ratio of tagged fish in the pond $(200/N)$. This gives the approximate proportion

$$21/150 \approx 200/N$$

which in turn gives

$$N \approx 200 \times 150/21 \approx 1428.57$$

Obviously, the value $N = 1428.57$ cannot be taken literally, since N must be a whole number. Besides, even in the best of cases, the computation is only an estimate. A sensible conclusion is that there are approximately $N = 1400$ catfish in the pond.

13.6 Clinical Studies

A survey typically deals with issues and questions that have direct and measurable answers. *If the election were held today, would you vote for candidate X or candidate Y? How many people live in your household? How many catfish have tags?* In these situations data collection involves some combination of *observing*, *measuring*, and *recording* but no active involvement or interference with the phenomenon being observed.

A different type of data collection process is needed when we are trying to establish connections between a cause and an effect. *Does taking a math class increase your chances of getting a good paying job? Does repeated exposure to secondhand smoke significantly increase your risk for developing lung cancer? Does a daily dose of aspirin reduce your chances of a heart attack? Do the benefits of hormone replacement therapy for women over 50 outweigh the risks?* These kinds of cause-and-effect questions cannot be answered by means of an immediate measurement and require observation over an extended period of time. Moreover, in these situations the data collection process requires the active involvement of the experimenter—in addition to *observation*, *measurement*, and *recording*, there is also *treatment*.

When we want to know if a certain cause X produces a certain effect Y, we set up a *study* in which cause X is produced and its effects are observed. If the effect Y is observed, then it is possible that X was indeed the cause of Y. We have established an *association* between the cause X and the effect Y. The problem, however, is the nagging possibility that some other cause Z different from X produced the effect Y and that X had nothing to do with it. Just because we established an association, we have not established a cause-and-effect relation between the variables. Statisticians like to explain this by a simple saying: *Association is not causation*.

Let's illustrate with a fictitious example. Suppose we want to find out if eating too much candy increases the chances of developing diabetes in adults. Here the cause X is eating candy, and the effect Y is developing diabetes. We set up an experiment in which 1000 adult laboratory rats are given, in addition to their normal diet, 8 ounces of candy a day for a period of six months. At the end of the six-month period, 150 of the 1000 rats have developed diabetes. Since in the general rat population only 3% of rats are diabetic, we have definitely established an association between the cause (candy) and the effect (increased incidence of diabetes). It is very tempting to conclude that the diabetes in the rats was indeed caused by the candy, but if we did that, we would be doing really bad science. How can we be so sure that the candy was really the cause of the increased diabetes? Could there be another hidden cause that we haven't noticed? What about their regular food? Lack of exercise? Cage conditions?

If you think that our fictitious candy story is too far-fetched to be realistic, consider the next case study.

CASE STUDY 5   THE ALAR SCARE

Alar is a chemical used by apple growers to regulate the rate at which apples ripen. Until 1989, practically all apples sold in grocery stores were sprayed with Alar. But in 1989 Alar became bad news, denounced in newspapers and on TV as a potent cancer-causing agent and a primary cause of cancer in children. As a result of these reports, people stopped buying apples, schools all over the country removed apple juice from their lunch menus, and the Washington State apple industry lost an estimated $375 million.

The case against Alar was based on a single 1973 study in which laboratory mice were exposed to the active chemicals in Alar. The dosage used in the study was eight times greater than the maximum tolerated dosage—a concentration at which even harmless substances can produce tissue damage. In fact, a child would have to eat about 200,000 apples a day to be exposed to an equivalent dosage of the chemical. Subsequent studies conducted by the National Cancer Institute and the Environmental Protection Agency failed to show any cause-and-effect relationship between Alar and cancer in children.

While it is generally accepted now that Alar does not cause cancer, because of potential legal liability, it is no longer used. The Alar scare turned out to be a false alarm based on a poor understanding of the statistical evidence. Unfortunately, it left in its wake a long list of casualties, among them the apple industry, the product's manufacturer, the media, and the public's confidence in the system.

For most cause-and-effect situations, especially those complicated by the involvement of human beings, a single effect can have many possible and actual causes. What causes cancer? Unfortunately, there is no single cause—diet, lifestyle, the environment, stress, and heredity are all known to be contributory causes. The extent to which each of these causes contributes individually and the extent to which they interact with each other are extremely difficult questions that can be answered only by means of carefully designed statistical studies.

For the remainder of this chapter we will illustrate an important type of study called a **clinical study** or **clinical trial**. Generally, clinical studies are concerned with determining whether a single variable or treatment (usually a vaccine, a drug, therapy, etc.) can cause a certain effect (a disease, a symptom, a cure, etc.). The importance of such clinical studies is self-evident: Every new vaccine, drug, or treatment must prove itself by means of a clinical study before it is officially approved for public use. Likewise, almost everything that is bad for us (cigarettes, caffeine, trans fats, etc.) gets its official certification of badness by means of a clinical study.

Properly designing a clinical study can be both difficult and controversial, and as a result we are often bombarded with conflicting information produced by different studies examining the same cause-and-effect question. The basic principles guiding a clinical study, however, are pretty much established by statistical practice.

The first and most important issue in any clinical study is to isolate the cause (treatment, drug, vaccine, therapy, etc.) that is under investigation from all other possible contributing causes (called **confounding variables**) that could produce the same effect. Generally, this is best done by *controlling* the study.

In a **controlled study**, the subjects are divided into two different groups: the *treatment group* and the *control group*. The **treatment group** consists of those subjects receiving the actual treatment; the **control group** consists of subjects that are not receiving any treatment—they are there for comparison purposes only (that's why the control group is sometimes also called the *comparison* group). If a real cause-and-effect relationship exists between the treatment and the effect being studied, then the treatment group should show the effects of the treatment and the control group should not.

To eliminate the many potential confounding variables that can bias its results, a well-designed controlled study should have control and treatment groups that are similar in every characteristic other than the fact that one group is being treated and the other one is not. (It would be a very bad idea, for example, to have a treatment group that is all female and a control group that is all male.) The most reliable way to get equally representative treatment and control groups is to use a *randomized controlled study*. In a **randomized controlled study**, the subjects are assigned to the treatment group or the control group randomly.

When the randomization part of a randomized controlled study is properly done, treatment and control groups can be assumed to be statistically similar. But there is still one major difference between the two groups that can significantly affect the validity of the study—a critical confounding variable known as the *placebo effect*. The **placebo effect** follows from the generally accepted principle that *just the idea that one is getting a treatment can produce positive results*. Thus, when subjects in a study are getting a pill or a vaccine or some other kind of treatment, how can the researchers separate positive results that are consequences of the treatment itself from those that might be caused by the placebo effect? When possible, the standard way to handle this problem is to give the control group a *placebo*. A **placebo** is a *make-believe* form of treatment—a harmless pill, an injection of saline solution, or any other fake type of treatment intended to look like the real treatment. A controlled study in which the subjects in the control group are given a placebo is called a **controlled placebo study**.

By giving all subjects a seemingly equal treatment (the treatment group gets the real treatment and the control group gets a placebo that looks like the real treatment), we do not eliminate the placebo effect but rather control it—whatever its effect might be, it affects all subjects equally. It goes without saying that the use of placebos is pointless if the subject knows he or she is getting a placebo. Thus, a second key element of a good controlled placebo study is that all subjects be kept in the dark as to whether they are being treated with a real treatment or a placebo. A study in which neither the members of the treatment group nor the members of the control group know to which of the two groups they belong is called a **blind study**.

Blindness is a key requirement of a controlled placebo study but not the only one. To keep the interpretation of the results (which can often be ambiguous) totally objective, it is important that the scientists conducting the study and collecting the data also be in the dark when it comes to who got the treatment and who got the placebo. A controlled placebo study in which neither the subjects nor the scientists conducting the experiment know which subjects are in the treatment group and which are in the control group is called a **double-blind study**.

Our next case study illustrates one of the most famous and important double-blind studies in the annals of clinical research.

CASE STUDY 6   THE 1954 SALK POLIO VACCINE FIELD TRIALS

Polio (infantile paralysis) has been practically eradicated in the Western world. In the first half of the twentieth century, however, it was a major public health problem. Over one-half million cases of polio were reported between 1930 and 1950, and the actual number may have been considerably higher.

Because polio attacks mostly children and because its effects can be so serious (paralysis or death), eradication of the disease became a top public health priority in the United States. By the late 1940s, it was known that polio is a virus and, as such, can best be treated by a vaccine that is itself made up of a virus. The vaccine virus can be a closely related virus that does not have the same harmful effects, or it can be the actual virus that produces the disease but that has been killed by a special treatment. The former is known as a *live-virus vaccine*, the latter as a *killed-virus vaccine*. In response to either vaccine, the body is known to produce *antibodies* that remain in the system and give the individual immunity against an attack by the real virus.

Both the live-virus and the killed-virus approaches have their advantages and disadvantages. The live-virus approach produces a stronger reaction and better immunity, but at the same time, it is also more likely to cause a harmful reaction

and, in some cases, even to produce the very disease it is supposed to prevent. The killed-virus approach is safer in terms of the likelihood of producing a harmful reaction, but it is also less effective in providing the desired level of immunity.

These facts are important because they help us understand the extraordinary amount of caution that went into the design of the study that tested the effectiveness of the polio vaccine. By 1953, several potential vaccines had been developed, one of the more promising of which was a killed-virus vaccine developed by Jonas Salk at the University of Pittsburgh. The killed-virus approach was chosen because there was a great potential risk in testing a live-virus vaccine in a large-scale study. (A large-scale study was needed to collect enough information on polio, which, in the 1950s, had a rate of incidence among children of about 1 in 2000.)

The testing of any new vaccine or drug creates many ethical dilemmas that have to be taken into account in the design of the study. With a killed-virus vaccine the risk of harmful consequences produced by the vaccine itself is small, so one possible approach would have been to distribute the vaccine widely among the population and then follow up on whether there was a decline in the national incidence of polio in subsequent years. This approach, which was not possible at the time because supplies were limited, is called the *vital statistics* approach and is the simplest way to test a vaccine. This is essentially the way the smallpox vaccine was determined to be effective. The problem with such an approach for polio is that polio is an epidemic type of disease, which means that there is a great variation in the incidence of the disease from one year to the next. In 1952, there were close to 60,000 reported cases of polio in the United States, but in 1953, the number of reported cases had dropped to almost half that (about 35,000). Since no vaccine or treatment was used, the cause of the drop was the natural variability typical of epidemic diseases. But if an ineffective polio vaccine had been tested in 1952 without a control group, the observed effect of a large drop in the incidence of polio in 1953 could have been incorrectly interpreted as statistical evidence that the vaccine worked.

The final decision on how best to test the effectiveness of the Salk vaccine was left to an advisory committee of doctors, public officials, and statisticians convened by the National Foundation for Infantile Paralysis and the Public Health Service. It was a highly controversial decision, but at the end, a large-scale, randomized, double-blind, controlled placebo study was chosen. Approximately 750,000 children were randomly selected to participate in the study. Of these, about 340,000 declined to participate, and another 8500 dropped out in the middle of the experiment. The remaining children were randomly divided into two groups—a treatment group and a control group—with approximately 200,000 children in each group. Neither the families of the children nor the researchers collecting the data knew if a particular child was getting the actual vaccine or a shot of harmless solution. The latter was critical because polio is not an easy disease to diagnose—it comes in many different forms and degrees. Sometimes it can be a borderline call, and if the doctor collecting the data had prior knowledge of whether the subject had received the real vaccine or the placebo, the diagnosis could have been subjectively tipped one way or the other.

A summary of the results of the Salk vaccine field trials is shown in Table 13-1. These data were taken as conclusive evidence that the Salk vaccine was an effective treatment for polio, and on the basis of this study, a massive inoculation campaign was put into effect. Today, all children are routinely inoculated against polio, and the disease has essentially been eradicated in the United States. Statistics played a key role in this important public health breakthrough.

TABLE 13-1	Results of the Salk Vaccine Field Trials			
	Number of children	Number of reported cases of polio	Number of paralytic cases of polio	Number of fatal cases of polio
Treatment group	200,745	82	33	0
Control group	201,229	162	115	4
Declined to participate in the study	338,778	182*	121*	0*
Dropped out in the middle	8,484	2*	1*	0*
Total	749,236	428	270	4

*These figures are not a reliable indicator of the actual number of cases—they are only self-reported cases.
Source: Adapted from Thomas Francis, Jr., et al., "An Evaluation of the 1954 Poliomyelitis Vaccine Trials—Summary Report," *American Journal of Public Health*, 45 (1955), 25.

CONCLUSION

In this chapter we have discussed different methods for collecting data. In principle, the most accurate method is a *census*, a method that relies on collecting data from each member of the population. In most cases, because of considerations of cost and time, a census is an unrealistic strategy. When data are collected from only a subset of the population (called a *sample*), the data collection method is called a *survey*. The most important rule in designing good surveys is to eliminate or minimize *sample bias*. Today, almost all strategies for collecting data are based on surveys in which the laws of chance are used to determine how the sample is selected, and these methods for collecting data are called *random sampling* methods. Random sampling is the best way known to minimize or eliminate sample bias. Two of the most common random sampling methods are *simple random sampling* and *stratified sampling*. In some special situations, other more complicated types of random sampling can be used.

Sometimes identifying the sample is not enough. In cases in which cause-and-effect questions are involved, the data may come to the surface only after an extensive study has been carried out. In these cases, isolating the cause variable under consideration from other possible causes (called *confounding variables*) is an essential prerequisite for getting reliable data. The standard strategy for doing this is a *controlled study* in which the sample is broken up into a *treatment group* and a *control group*. Controlled studies are now used (and sometimes abused) to settle issues affecting every aspect of our lives. We can thank this area of statistics for many breakthroughs in social science, medicine, and public health, as well as for the constant and dire warnings about our health, our diet, and practically anything that is fun.

> " Statistics show that of those who contract the habit of eating, very few survive. "
>
> —Wallace Irwin

PROFILE: George Gallup (1901–1984)

George Gallup was a man of many talents—journalist, sociologist, political analyst, businessman, and statistician. It was the combination of all these talents, together with a strong entrepreneurial spirit and an unlimited amount of self-confidence, that made him a unique fixture of twentieth-century American life—the man trusted by generations to best measure the nation's pulse. Gallup did not invent the modern-day public opinion poll, but he certainly set the standard for how to do it right. Under his tenure, the poll that bears his name became the most widely read and credible public opinion poll in the world.

George Horace Gallup Jr. was born in Jefferson, Iowa, a small farm town in America's heartland. He studied journalism at the University of Iowa, paying his way through school partly by working as the editor of the student newspaper. He earned a bachelor's degree in journalism in 1923, a master's degree in psychology in 1925, and a Ph.D. in journalism in 1928. In his doctoral thesis, entitled *About Unbiased Methodology of Exploring Readers' Interest in the Content of Newspapers,* Gallup showed that public opinion could be scientifically collected from a very small sample. His doctoral work was a precursor of the sampling methods he would later develop to conduct public opinion polls.

After working as a professor of journalism at Drake University and at Northwestern University, Gallup was recruited by New York advertising agency Young and Rubicam to become the head of its newly developed market-research department. But Gallup was a journalist at heart, and by 1935 he had had enough of the advertising business, notwithstanding the fact that he had made millions in it. He moved to Princeton, New Jersey, where he founded the American Institute of Public Opinion and started a weekly syndicated newspaper column immodestly entitled "America Speaks," which featured the results of public opinion polls he designed and conducted. From these humble beginnings, the Gallup poll was born. The organization he founded (now called the Gallup Organization and run by his two sons, George III and Alec) has grown to become the largest and most respected private polling organization in the world.

Today, public opinion polling is used extensively throughout the world, but the increased use of polls in modern life has generated many criticisms about their impact and influence on the political process. As the father of polling, Gallup spent his later years trying to defend polls as an important and useful instrument of democracy. "When a president, or any other leader, pays attention to poll results, he is, in effect, paying attention to the views of the people," he once said.

George Gallup died in 1984 in Switzerland, where he had spent a good part of his last years, partly for its beauty and partly because, as he put it, the country was "virtually run by polls." Once, as a young editor of the University of Iowa student newspaper, Gallup wrote, "Doubt everything, Question everything. Be a radical!" Even as he became an American institution, George Gallup remained always true to this credo. ◼

KEY CONCEPTS

EXERCISES

WALKING

A Surveys, Polls, and Sampling

Exercises 1 through 4 refer to the following story: As part of a sixth-grade statistics project, the teacher brings to class a candy jar full of gumballs of two different colors, red and green. The assignment is to estimate the proportion of red gumballs in the jar. Suppose that after the jar is shaken well, one of the students draws 25 gumballs from the jar: 8 are red, and 17 are green.

1. (a) Describe the population for this survey.

    (b) Describe the sample for this survey.

    (c) Give the sample statistic for the percentage of red gumballs in the jar.

    (d) Name the sampling method used for this survey.

2. Suppose that the jar contains a total of 200 gumballs.

    (a) Give the sampling proportion for this survey.

    (b) Give the sample statistic for the number of red gumballs in the jar.

3. Suppose that the jar contains 50 red gumballs and 150 green gumballs.

    (a) Give the parameter for the percentage of red gumballs in the jar.

    (b) Give the sampling error, expressed as a percent.

    (c) Is the sampling error found in (b) a result of sampling variability or of sampling bias? Explain.

4. (a) Compare and contrast the target population and sampling frame for this survey.

    (b) What data collection method should be used to find the parameter for the percentage of red gumballs in the jar?

Exercises 5 through 8 refer to the following story: The city of Cleansburg has 8325 registered voters. There is an election for mayor of Cleansburg, and there are three candidates for the position: Smith, Jones, and Brown. The day before the election, a telephone poll of 680 randomly chosen registered voters produced the following results: 306 people surveyed indicated that they would vote for Smith, 272 indicated that they would vote for Jones, and 102 indicated that they would vote for Brown.

5. (a) Give the sampling proportion for this survey.

    (b) Give the sample statistic estimating the percentage of the vote going to Smith.

6. (a) Describe the population for this survey.

    (b) Describe the sample for this survey.

    (c) Name the sampling method used for this survey.

7. Given that in the actual election Smith received 42% of the vote, Jones 43% of the vote, and Brown 15% of the vote, find the sampling errors in the survey expressed as percentages.

8. Do you think that the sampling error in this example is due primarily to chance error or to sample bias? Explain your answer.

Exercises 9 through 12 refer to the following story: The 1250 students at Eureka High School are having an election for Homecoming King. The candidates are Tomlinson (captain of the football team), Garcia (class president), and Marsalis (member of the marching band). At the football game a week before the election, a pre-election poll was taken of students as they entered the stadium gates. Of the students who attended the game, 203 planned to vote for Tomlinson, 42 planned to vote for Garcia, and 105 planned to vote for Marsalis.

9. (a) Describe the sample for this survey.

    (b) Give the sampling proportion for this survey.

10. Name the sampling method used for this survey.

11. (a) Compare and contrast the population and the sampling frame for this survey.

    (b) Is the sampling error a result of sampling variability or of sample bias? Explain.

12. (a) Give the sample statistics estimating the percentage of the vote going to each candidate.

    (b) A week after this survey, Garcia was elected Homecoming King with 51% of the vote, Marsalis got 30% of the vote, and Tomlinson came in last with 19% of the vote. Find the sampling errors in the survey expressed as percentages.

Exercises 13 through 16 refer to a poll about extramarital affairs conducted in 1988 by the columnist Abigail Van Buren, better known as "Dear Abby." Dear Abby asked her readers to let her know whether they had been faithful or unfaithful in their marriages. The readers' responses are summarized in the table.

Status	Women	Men
Faithful	127,318	44,807
Unfaithful	22,468	15,743
Total	149,786	60,550

Based on the results of this survey, Dear Abby concluded that extramarital affairs among married couples are much less common than people believe. ("The results were astonishing," she wrote. "There are far more faithfully wed couples than I had surmised.")

13. (a) Describe as specifically as you can the sampling frame for this survey.

(b) Compare and contrast the target population and the sampling frame for this survey.

(c) How was the sample chosen?

(d) Eighty-five percent of the women who responded to this survey claimed to be faithful. Is 85% a parameter? A statistic? Neither? Explain your answer.

14. (a) Explain why this survey was subject to selection bias.

(b) Explain why this survey was subject to nonresponse bias.

15. (a) Based on the Dear Abby data, estimate the percentage of married men who are faithful to their spouses.

(b) Based on the Dear Abby data, estimate the percentage of married people who are faithful to their spouses.

(c) How accurate do you think the estimates in (a) and (b) are? Explain.

16. If money were no object, could you devise a survey that might give more reliable results than the Dear Abby survey? Describe briefly what you would do.

Exercises 17 through 20 refer to the following story: The Cleansburg Planning Department is trying to determine what percent of the people in the city want to spend public funds to revitalize the downtown mall. To do so, the department decides to conduct a survey. Five professional interviewers are hired. Each interviewer is asked to pick a street corner of his or her choice within the city limits, and every day between 4:00 P.M. and 6:00 P.M. the interviewers are supposed to ask each passerby if he or she wishes to respond to a survey sponsored by Cleansburg City Hall. If the response is yes, the follow-up question is asked: Are you in favor of spending public funds to revitalize the downtown mall? The interviewers are asked to return to the same street corner as many days as are necessary until each has conducted a total of 100 interviews. The results of the survey are shown in the table.

Interviewer	Yes[a]	No[b]	Nonrespondents[c]
A	35	65	321
B	21	79	208
C	58	42	103
D	78	22	87
E[d]	12	63	594

[a]In favor of spending public funds to revitalize the downtown mall.
[b]Opposed to spending public funds to revitalize the downtown mall.
[c]Declined to be interviewed or had no opinion.
[d]Got frustrated and quit.

17. (a) Describe as specifically as you can the target population for this survey.

(b) Compare and contrast the target population and the sampling frame for this survey.

18. (a) What is the size of the sample?

(b) Calculate the response rate in this survey. Was this survey subject to nonresponse bias?

19. (a) Can you explain the big difference in the data from interviewer to interviewer?

(b) One of the interviewers conducted the interviews at a street corner downtown. Which interviewer? Explain.

(c) Do you think the survey was subject to selection bias? Explain.

(d) Was the sampling method used in this survey the same as quota sampling? Explain.

20. Do you think this was a good survey? If you were a consultant to the Cleansburg Planning Department, could you suggest some improvements? Be specific.

*Exercises 21 through 24 refer to the following story: The dean of students at Tasmania State University wants to determine how many undergraduates at TSU are familiar with a new financial aid program offered by the university. There are 15,000 undergraduates at TSU, so it is too expensive to conduct a census. The following sampling method is used to choose a representative sample of undergraduates to poll. Start with the registrar's alphabetical listing containing the names of all undergraduates. Randomly pick a number between 1 and 100, and count that far down the list. Take that name and every 100th name after it. For example, if the random number chosen is 73, then pick the 73rd, 173rd, 273rd, and so forth, names on the list. (The sampling method illustrated in this survey is known as **systematic sampling**.)*

21. (a) Compare and contrast the sampling frame and the target population for this survey.

(b) Give the exact N-value of the population.

22. (a) Find the sampling proportion.

(b) Suppose that the survey had a response rate of 90%. Find the size n of the sample.

23. (a) Explain why the method used for choosing the sample is not simple random sampling.

(b) If 100% of those responding claimed that they were not familiar with the new financial aid program offered by the university, is this result more likely due to sampling variability or to sample bias? Explain.

24. (a) Suppose that the survey had a response rate of 90% and that 108 students responded that they were not familiar with the new financial aid program. Give a statistic for the total number of students at the university who were not familiar with the new financial aid program.

(b) Do you think the results of this survey will be reliable? Explain.

Exercises 25 and 26 refer to the following story: An orange grower wishes to compute the average yield from his orchard. The orchard contains three varieties of trees: 50% of his trees are of variety A, 25% of variety B, and 25% of variety C.

25. (a) Suppose that the grower samples randomly from 300 trees of variety A, 150 trees of variety B, and 150 trees of variety C. What type of sampling is being used?

(b) Suppose that the grower selects for his sample a 10 by 30 rectangular block of 300 trees of variety A, a 10 by 15 rectangular block of 150 trees of variety B, and a 10 by 15 rectangular block of 150 trees of variety C. What type of sampling is being used?

26. (a) Suppose that in his survey, the grower found that each tree of variety A averages 100 oranges, each tree of variety B averages 50 oranges, and each tree of variety C averages 70 oranges. Estimate the average yield per tree of his orchard.

(b) Is the yield you found in (a) a parameter or a statistic? Explain.

27. Name the sampling method that best describes each situation. Choose your answer from the following list: simple random sampling, convenience sampling, quota sampling, stratified sampling, census.

(a) George wants to know how the rest of the class did on the last quiz. He peeks at the scores of a few students sitting right next to him. Based on what he sees, he concludes that nobody did very well.

(b) Eureka High School has 400 freshmen, 300 sophomores, 300 juniors, and 200 seniors. The student newspaper conducts a poll asking students if the football coach should be fired. The student newspaper randomly selects 20 freshmen, 15 sophomores, 15 juniors, and 10 seniors for the poll.

(c) For the last football game of the season, the coach chooses the three captains by putting the names of all the players in a hat and drawing three names. (Maybe that's why they are trying to fire him!)

(d) For the last football game of the season, the coach chooses the three captains by putting the names of all the *seniors* in a hat and drawing three names.

28. Dr. Blackbeard's Statistics 101 class has 120 students (40 freshmen, 30 sophomores, 30 juniors, and 20 seniors). Toward the end of the semester, Dr. Blackbeard is expected to have students in the class fill out a teacher rating questionnaire. The purpose of the questionnaire is to collect data on how the students rate the instructor. For each situation described below, name the sampling method that best describes the method used by Dr. Blackbeard to collect the data. Choose your answer from the following list: simple random sampling, convenience sampling, quota sampling, stratified sampling, census.

(a) Dr. Blackbeard has every student in class (on a day when everyone is present) fill out the questionnaire.

(b) Dr. Blackbeard randomly selects 8 freshmen, 6 sophomores, 6 juniors, and 4 seniors and has them fill out the questionnaire.

(c) Dr. Blackbeard selects 8 freshmen, 6 sophomores, 6 juniors, and 4 seniors in alphabetical order from the class roster and has them fill out the questionnaire.

(d) Dr. Blackbeard has the first 24 students walking into the classroom one day fill out the questionnaire.

29. You are a fruit wholesaler. You have just received 250 crates of pineapples: 75 crates came from supplier A, 75 crates from supplier B, and 100 crates from supplier C. You wish to determine if the pineapples are good enough to ship to your best customers by inspecting a sample of $n = 20$ crates. Describe how you might implement each of the following sampling methods.

(a) Simple random sampling

(b) Convenience sampling

(c) Stratified sampling

(d) Quota sampling

30. For each of the following situations, determine if the sample is representative of the population. Explain your answer.

(a) To determine if the bowl of soup has enough salt, you stir the bowl well and then sample a spoonful.

(b) To determine if the statistics test you are taking is easy or difficult, you look at the first three questions on the test and then make up your mind.

(c) To determine if you have a viral infection, the doctor draws 10 ml of blood from your arm and tests the blood.

(d) To determine how well liked a person is, you look up the number of friends the person has on Facebook.com.

B The Capture-Recapture Method

31. You want to estimate how many fish there are in a small pond. Let's suppose that you first capture $n_1 = 500$ fish, tag them, and throw them back into the pond. After a couple of days you go back to the pond and capture $n_2 = 120$ fish, of which $k = 30$ are tagged. Give an estimate of the N-value of the fish population in the pond.

32. To estimate the population in a rookery, 4965 fur seal pups were captured and tagged in early August. In late August, 900 fur seal pups were captured. Of these, 218 had been tagged. Based on these figures, estimate the population of fur seal pups in the rookery to the nearest hundred.

Source: Chapman and Johnson, "Estimation of Fur Seal Pup Populations by Randomized Sampling," *Transactions of the American Fisheries Society*, 97 (July 1968), 264–270.

Exercises 33 through 36 refer to the following story: You have an extremely large coin jar full of nickels, dimes, and quarters. You want to have an approximate idea of how much money you have, but you don't want to go through the trouble of counting all the coins, so you decide to use the capture-recapture method. For the first sample, you shake the jar well and randomly draw 150 coins. You get 36 quarters, 45 nickels, and 69 dimes. Using a black marker, you mark the 150 coins with a black dot and put them back in the jar. For the second sample, you shake the jar well and randomly draw another set of 100 coins. You get 28 quarters, 4 of which have black dots; 29 nickels, 5 of which have black dots; and 43 dimes, 8 of which have black dots.

33. Estimate the total number of quarters in the jar. (*Hint*: Disregard the nickels and dimes. They are irrelevant.)

34. Estimate the total number of nickels in the jar. (*Hint*: Disregard the quarters and dimes. They are irrelevant.)

35. Estimate the total amount of money in the jar. (*Hint*: Do Exercises 33 and 34 first.)

36. Do you think the capture-recapture method is a reliable way to estimate the amount of money in the jar? Explain your answer. Discuss some of the potential pitfalls and issues one should be concerned about.

37. Starting in 2004, a study to determine the number of lake sturgeon on Rainy River and Lake of the Woods on the United States–Canada border was conducted by the Canadian Ministry of Natural Resources, the Minnesota Department of Natural Resources, and the Rainy River First Nations. Using the capture-recapture method, the study concluded that the size of the population of lake sturgeon on Rainy River and Lake of the Woods was approximately 160,000. In the capture phase of the study, 1700 lake sturgeon were caught, tagged, and released. Of these tagged sturgeon, seven were recaptured during the recapture phase of the study. Based on these figures, estimate the number of sturgeon caught in the recapture phase of the study.

Source: Dan Gauthier, "Lake of the Woods Sturgeon Population Recovering," *Daily Miner and News* (Kenora, Ont.), June 11, 2005, p. 31.

38. A 2004 study conducted at Utah Lake using the capture-recapture method estimated the carp population in the lake to be about 1.1 million. Over a period of 15 days, workers captured, tagged, and released 24,000 carp. In the recapture phase of the study 10,300 carp were recaptured. Estimate how many of the carp that were recaptured had tags.

Source: Brett Prettyman, "With Carp Cooking Utah Lake, It's Time to Eat," *Salt Lake Tribune*, July 15, 2004, p. D3.

39. One implicit assumption when using the capture-recapture method to estimate the size of a population is that the capture process is truly random, with all individuals having the same likelihood of being captured. Sometimes that is not true, and some individuals are more prone to capture than others (more likely to take the bait, less cagey, slower, dumber, etc.). If that were the case, would the capture-recapture method be likely to *underestimate* or *overestimate* the size of the population? Explain your answer.

40. One implicit assumption when using the capture-recapture method to estimate the size of a population is that when individuals are tagged in the capture stage, these individuals are not affected in any harmful way by the tags. Sometimes, though, tagged individuals become affected, with the tags often making them more likely prey to predators (imagine, for example, tagging fish with bright yellow tags that make them stand out or tagging a bird on a wing in such a way that it affects its ability to fly). If that were the case, would the capture-recapture method be likely to *underestimate* or *overestimate* the size of the population? Explain your answer.

C Clinical Studies

Exercises 41 through 44 refer to the following story: The manufacturer of a new vitamin (vitamin X) decides to sponsor a study to determine the vitamin's effectiveness in curing the common cold. Five hundred of the 50,000 college students in the San Diego area were paid to participate as subjects in this study. All the participating subjects were suffering from a cold at the start of the study. The subjects were each given two tablets of vitamin X a day. Based on information provided by the subjects themselves, 457 of the 500 subjects were cured of their colds within 3 days. (The average number of days a cold lasts is 4.87 days.) As a result of this study, the manufacturer launched an advertising campaign based on the claim that "vitamin X is more than 90% effective in curing the common cold."

41. (a) Describe as specifically as you can the target population for the study.

(b) Compare and contrast the target population, the sampling frame, and the sample for the study.

(c) Is selection bias present in the sample?

42. (a) Was the study a controlled study? Explain.

(b) List three possible causes other than the effectiveness of vitamin X itself that could have confounded the results of the study.

43. List four different problems with the study that indicate poor design.

44. Make some suggestions for improving the study.

Exercises 45 through 48 refer to a clinical study conducted at the Houston Veterans Administration Medical Center on the effectiveness of knee surgery to cure degenerative arthritis (osteoarthritis) of the knee. Of the 324 individuals who met the inclusion criteria for the study, 144 declined to participate. The researchers randomly divided the remaining 180 subjects into three groups: One group received a type of arthroscopic knee surgery called debridement; a second group received a type of arthroscopic knee surgery called lavage; and a third group received skin incisions to make it look like they had had arthroscopic knee surgery, but no actual surgery was performed. The patients in the study did not know which group they were in and in particular did not know if they were receiving the real surgery or simulated surgery. All the patients who participated in the study were evaluated for two years after the procedure. In the two-year followup, all three groups said that they had slightly less pain and better knee movement, but the "fake" surgery group often reported the best results.

Source: New England Journal of Medicine, 347, no. 2 (July 11, 2002): 81–88.

45. Describe as specifically as you can the target population for this study.

46. (a) Describe the sample.

(b) What was the size of the sample?

(c) Was the sample chosen by random sampling? Explain.

47. (a) Was this study a controlled placebo experiment? Explain.

(b) Describe the treatment group(s) in this study.

(c) Could this study be considered a randomized controlled experiment? Explain.

(d) Was this experiment blind, double blind, or neither?

48. As a result of this study, the Department of Veterans Affairs issued an advisory to its doctors recommending that they stop using arthroscopic knee surgery for patients suffering from osteoarthritis. Do you agree or disagree with the advisory? Explain your answer.

Exercises 49 through 52 refer to a clinical trial named APPROVe designed to determine whether Vioxx, a medication used for arthritis and acute pain, was effective in preventing the recurrence of colorectal polyps in patients with a history of colorectal adenomas. APPROVe was conducted between 2002 and 2003 and involved 2586 participants, all of whom had a history of colorectal adenomas. The participants were randomly divided into two groups: 1287 were given 25 milligrams of Vioxx daily for the duration of the clinical trial (originally intended to last three years), and 1299 patients were given a placebo. Neither the participants nor the doctors involved in the clinical trial knew who was in which group. During the trial, 72 of the participants had cardiovascular events (mostly heart attacks or strokes). Later it was found

that 46 of these people were from the group taking the Vioxx and only 26 were from the group taking the placebo. Based on these results, the clinical trial was stopped in 2003 and Vioxx was taken off the market in 2004.

49. Describe as specifically as you can the target population for APPROVe.

50. (a) Describe the sample for APPROVe.

(b) Give the size n of the sample.

51. (a) Describe the control and treatment groups in APPROVe.

(b) APPROVe can be described as a *double-blind, randomized controlled placebo experiment*. Explain why each of these terms applies.

52. What conclusions would you draw from APPROVe?

Exercises 53 through 56 refer to a study on the effectiveness of an HPV (human papilloma virus) vaccine conducted between October 1998 and November 1999. HPV is the most common sexually transmitted infection—more than 20 million Americans are infected with HPV—but most HPV infections are benign, and in most cases infected individuals are not even aware they are infected. (On the other hand, some HPV infections can lead to cervical cancer in women.) The researchers recruited 2392 women from 16 different centers across the United States to participate in the study through advertisements on college campuses and in the surrounding communities. To be eligible to participate in the study, the subjects had to meet the following criteria: (1) be a female between 16 and 23 years of age, (2) not be pregnant, (3) have no prior abnormal Pap smears, and (4) report to have had sexual relations with no more than five men. At each center, half of the participants were randomly selected to receive the HPV vaccine, and the other half received a placebo injection. After 17.4 months, the incidence of HPV infection was 3.8 per 100 woman-years at risk in the placebo group and 0 per 100 woman-years at risk in the vaccine group. In addition, all nine cases of HPV-related cervical precancerous growths occurred among the placebo recipients.

Source: New England Journal of Medicine, 347, no. 21 (November 21, 2002): 1645–1651.

53. (a) Describe as specifically as you can the target population for the study.

(b) Compare and contrast the sampling frame and target population for the study.

54. (a) Describe the sample for the study.

(b) Was the sample chosen using random sampling? Explain.

55. (a) Describe the treatment group in the study.

(b) Could this study be considered a double-blind, randomized controlled placebo study? Explain.

56. Carefully state what a legitimate conclusion from this study might be.

Exercises 57 through 60 refer to a landmark study conducted in 1896 in Denmark by Dr. Johannes Fibiger, who went on to receive the Nobel Prize in Medicine in 1926. The purpose of the study was to determine the effectiveness of a new serum for treating diphtheria, a common and often deadly respiratory disease in those days. Fibiger conducted his study over a one-year period in one particular Copenhagen hospital. New diphtheria patients admitted to the hospital received different treatments based on the day of admission. In one set of days (call them "even" days for convenience), the patients were treated with the serum daily and also received the standard treatment. Patients admitted on alternate days (the "odd" days) received just the standard treatment. Over the one-year period of the study, eight of the 239 patients admitted on the "even" days and treated with the serum died, whereas 30 of the 245 patients admitted on the "odd" days died.

57. **(a)** Describe as specifically as you can the target population for Fibiger's study.

(b) Compare and contrast the sampling frame and target population for the study.

58. **(a)** Describe the sample for Fibiger's study.

(b) Is selection bias a possible problem in this study? Explain.

59. **(a)** Describe the treatment and control groups in Fibiger's study.

(b) What conclusions would you draw from Fibiger's study? Explain.

60. In a study on the effectiveness of the diphtheria serum conducted before Fibiger's study, patients in one Copenhagen hospital were chosen to be in the treatment group and were given the serum, whereas patients in a different Copenhagen hospital were chosen to be in the control group and were given the standard treatment. Fibiger did not believe that the results of this earlier study could be trusted. What are some possible confounding variables that may have affected the results of this earlier study?

D Miscellaneous

61. In 2003, Morgan Spurlock set out to study the effect McDonald's has on the average American, a study he documented in the award-winning film *Super Size Me*. Spurlock ate three meals at McDonald's every day for 30 days, taking the "supersize" option whenever it was offered. He also curtailed his physical activity to better match the exercise habits of the average American. In the end, his health declined dramatically: he gained 25 pounds, suffered severe liver dysfunction, and developed symptoms of depression.

(a) Was Spurlock's study a survey or a clinical trial? Explain.

(b) Describe as specifically as you can the target population for the study.

(c) Describe the sample for the study.

(d) List three problems with this study that indicate poor design.

62. Thinking that Spurlock's documentary *Super Size Me* (see Exercise 61) unfairly targeted McDonald's, Merab Morgan set out to do a fast-food study of her own. Morgan ate only at McDonald's for 90 days, sticking to meal plans of no more than 1400 calories a day and choosing to eat mostly burgers and salads (no french fries!). By the end of her study, Morgan had dropped 37 pounds and felt well.

(a) Describe the treatment group in Morgan's study.

(b) List some of the possible confounding variables in Morgan's study.

(c) Carefully state what a legitimate conclusion from this study might be.

63. A study by the *Center for Academic Transformation* showed that students using electronic learning tools have seen marked progress in their test scores. At the University of Alabama, the passing rate for an intermediate algebra class doubled from 40 percent to 80 percent when the class was redesigned to rely heavily on supplemental materials, online practice exercises, and interactive tutorials.

(a) Was this study a survey or a clinical trial? Explain.

(b) List some of the possible confounding variables in this study.

64. There are 50 students—30 females and 20 males—in the Math 101 class at Tasmania State University. The professor chooses a sample of 10 students from the class as follows: 6 students are randomly chosen from among the females, and 4 students are randomly chosen from among the males.

(a) Does every student in the class have an equal likelihood of being selected for the sample? Explain.

(b) Does every set of 10 students in the class have an equal likelihood of being selected as the sample? Explain.

65. Determine in each case if the data is a population parameter or a sample statistic.

(a) In 2007, 25% of students taking the SAT math test scored above 590.

(b) In crash testing of a new automobile model, 20% of the crashes would have caused severe neck injury.

(c) Mr. Johnson's blood tested positive for type II diabetes.

(d) A recent Gallup poll shows that 6 in 10 Americans have attempted to lose weight.

66. Choose the most appropriate concept for each statement from the following list: association is not causation, sampling variability, selection bias, placebo effect.

(a) "In the early part of the twentieth century it was discovered that when viewed over time, the number of crimes increased with membership in the Church of England."

(b) "Half of the subjects in the study were given sugar pills. These subjects responded with fewer days ill than those receiving the treatment."

(c) "Jack chose a simple random sample of 100 widgets, and 11 of the widgets in his sample were defective. Jill chose a simple random sample of 100 widgets, and 19 of the widgets in her sample were defective."

(d) "Jay didn't understand how Jones could have been elected mayor when almost every person he knows voted for Brown."

JOGGING

67. Informal surveys. In everyday life we are constantly involved in activities that can be described as *informal surveys*, often without even realizing it. Here are some examples.

(i) Al gets up in the morning and wants to know what kind of day it is going to be, so he peeks out the window. He doesn't see any dark clouds, so he figures it's not going to rain.

(ii) Betty takes a sip from a cup of coffee and burns her lips. She concludes that the coffee is too hot and decides to add a tad of cold water to it.

(iii) Carla got her first Math 101 exam back with a C grade on it. The students sitting on each side of her also received C grades. She concludes that the entire Math 101 class received a C on the first exam.

For each of the preceding examples,

(a) describe the population.

(b) discuss whether the sample is random or not.

(c) discuss the validity of the conclusions drawn. (There is no right or wrong answer to this question, but you should be able to make a reasonable case for your position.)

68. Read the examples of informal surveys given in Exercise 67. Give three new examples of your own. Make them as different as possible from the ones given in Exercise 67 [changing coffee to soup in (ii) is not a new example].

69. Leading-question bias. The way the questions in many surveys are phrased can itself be a source of bias. When a question is worded in such a way as to predispose the respondent to provide a particular response, the results of the survey are tainted by a special type of bias called *leading-question bias*. The following is an extreme hypothetical situation intended to drive the point home.

In an effort to find out how the American taxpayer feels about a tax increase, the Institute for Tax Reform conducts a "scientific" one-question poll.

Are you in favor of paying higher taxes to bail the federal government out of its disastrous economic policies and its mismanagement of the federal budget? Yes ___. No ___.

Ninety-five percent of the respondents answered no.

(a) Explain why the results of this survey might be invalid.

(b) Rephrase the question in a neutral way. Pay particular attention to highly charged words.

(c) Make up your own (more subtle) example of leading-question bias. Analyze the critical words that are the cause of bias.

70. Question order bias. In July 1999, a Gallup poll of 1061 people asked the following two questions:

■ *As you may know, former Major League Baseball player Pete Rose is ineligible for baseball's Hall of Fame because of charges that he gambled on baseball games. Do you think he should or should not be eligible for admission to the Hall of Fame?*

■ *As you may know, former Major League Baseball player Shoeless Joe Jackson is ineligible for baseball's Hall of Fame because of charges that he took money from gamblers in exchange for fixing the 1919 World Series. Do you think he should or should not be eligible for admission to the Hall of Fame?*

The order in which the questions were asked was random: approximately half of the people polled were asked about Rose first and Jackson second; the other half were asked about Jackson first and Rose second. When the order of the questions was Rose first and Jackson second, 64% of the respondents said that Rose should be eligible for admission to the Hall of Fame and 33% said that Jackson should be eligible for admission to the Hall of Fame. When the order of the questions was Jackson first and Rose second, 52% said that Rose should be eligible for admission to the Hall of Fame and 45% said that Jackson should be eligible for admission to the Hall of Fame. Explain why you think each player's support for eligibility was less (by 12% in each case) when the player was second in the order of the questions.

71. Today, most consumer marketing surveys are conducted by telephone. In selecting a sample of households that are representative of all the households in a given geographical area, the two basic techniques used are (1) randomly selecting telephone numbers to call from the local telephone directory or directories and (2) using a computer to randomly generate seven-digit numbers to try that are compatible with the local phone numbers.

(a) Briefly discuss the advantages and disadvantages of each technique. In your opinion, which of the two will produce the more reliable data? Explain.

(b) Suppose that you are trying to market burglar alarms in New York City. Which of the two techniques for selecting the sample would you use? Explain your reasons.

72. The following two surveys were conducted in January 1991 to assess how the American public viewed media coverage of the Persian Gulf war. Survey 1 was an Area Code 900 telephone poll survey conducted by *ABC News*. Viewers were asked to call a certain 900 number if they believed that the media were doing a good job of covering the war and a different 900 number if they believed that the media were not doing a good job in covering the war. Each call cost 50 cents. Of the 60,000 respondents, 83% believed that the media were not doing a good job. Survey 2 was a telephone poll of 1500 randomly selected households across the United States conducted by the *Times-Mirror* survey organization. In this poll, 80% of the respondents indicated that they approved of the press coverage of the war.

(a) Briefly discuss survey 1, indicating any possible types of bias.

(b) Briefly discuss survey 2, indicating any possible types of bias.

(c) Can you explain the discrepancy between the results of the two surveys?

(d) In your opinion, which of the two surveys gives the more reliable data?

73. An article in the *Providence Journal* about automobile accident fatalities includes the following observation:

"Forty-two percent of all fatalities occurred on Friday, Saturday, and Sunday, apparently because of increased drinking on the weekends."

(a) Give a possible argument as to why the conclusion drawn may not be justified by the data.

(b) Give a different possible argument as to why the conclusion drawn may be justified by the data after all.

74. (a) For the capture-recapture method to give a reasonable estimate of *N*, what assumptions about the two samples must be true?

(b) Give reasons why the assumptions in (a) may not hold true in many situations.

75. Consider the following hypothetical survey designed to find out what percentage of people cheat on their income taxes.

Fifteen hundred taxpayers are randomly selected from the Internal Revenue Service (IRS) rolls. These individuals are then interviewed in person by representatives of the IRS and read the following statement.

"This survey is for information purposes only. Your answer will be held in strict confidence. Have you ever cheated on your income taxes? Yes____. No____."

Twelve percent of the respondents answered yes.

(a) Explain why the above figure might be unreliable.

(b) Can you think of ways in which a survey of this type might be designed so that more reliable information could be obtained? In particular, discuss who should be sponsoring the survey and how the interviews should be carried out.

PROJECTS AND PAPERS

A What's the Latest?

In this project you are to do an in-depth report on a recent study. Find a recent article from a newspaper or news magazine reporting the results of a major study and write an analysis of the study. Discuss the extent to which the article gives the reader enough information to assess the validity of the study's conclusions. If in your opinion there is information missing in the article, generate a list of questions that would help you further assess the validity of the study's conclusions. Pay particular attention to the ideas and concepts discussed in this chapter, including the target population, sample size, sampling bias, randomness, controls, and so on.

Note: The best reporting on clinical studies can be found in major newspapers such as the *New York Times*, *Washington Post*, and *Los Angeles Times* or weekly news magazines such as *Time* or *Newsweek*. All of these have Web sites where their most recent articles can be downloaded for free.

B The U.S. Census: A Continuing Political Battle

In January 1999, the United States Supreme Court ruled 5 to 4 (*Department of Commerce et al. v. U.S. House of Representatives et al.*) that the state population figures used for the apportionment of seats in the House of Representatives cannot be determined by means of statistical sampling methods and that an actual census of the population is required by the Census Act. The ruling was based on arguments presented in the earlier case *Glavin v. Clinton*. In an effort to gain a seat in the House of Representatives, the state of Utah has recently unsuccessfully raised similar arguments with the Court (*Utah et al. v. Evans, Secretary of Commerce, No. 01-283, 2002*, and *Utah et al. v. Evans, Secretary of Commerce, No. 01-714, 2002*).

In this project you are to use the Court decisions cited as a guide in writing a summary of arguments both for and against the Census Bureau's current methodologies. Such arguments may mirror those made by Republicans and Democrats in this continuing political battle.

C The Placebo Effect: Myth or Reality?

There is no consensus among researchers conducting clinical studies as to the true impact of the placebo effect. According to some researchers, as many as 30% of patients in a clinical trial can be affected by it; according to others, the placebo effect is a myth.

Write a paper on the *placebo effect*. Discuss the history of this idea, the most recent controversies regarding whether it truly exists or not and why is it important to determine its true impact.

D Ethical Issues in Clinical Studies

Until the world was exposed to the atrocities committed by Nazi physicians during World War II, there was little consensus regarding the ethics of medical experiments involving human subjects. Following the trial of the Nazi physicians in 1946, the Nuremberg Code of Ethics was developed to deal with the ethics of clinical studies on human subjects. Sadly, unethical experimentation continued. In the Tuskegee syphilis study starting in 1932, 600 low-income African American males (400 infected with syphilis) were monitored. Even though a proven cure (penicillin) became available in the 1950s, the study continued until 1972 with participants denied treatment. As many as 100 subjects died from the disease.

In this project you are to write a paper summarizing current ethical standards for the treatment of humans in experimental studies. Sources for your summary may include the Belmont Report (*http://ohsr.od.nih.gov/guidelines/belmont.html*), the Declaration of Helsinki (*www.cirp.org/library/ethics/helsinki*), and the Nuremberg Code (*http://ohsr.od.nih.gov/guidelines/nuremberg.html*).

REFERENCES AND FURTHER READINGS

1. Anderson, M. J., and S. E., Fienberg, *Who Counts? The Politics of Census-Taking in Contemporary America*. New York: Russell Sage Foundation, 1999.

2. Day, Simon, *Dictionary for Clinical Trials*. New York: John Wiley and Sons, 1999.

3. Francis, Thomas, Jr., et al., "An Evaluation of the 1954 Poliomyelitis Vaccine Trials—Summary Report, *American Journal of Public Health*, 45 (1955), 1–63.

4. Freedman, D., R. Pisani, R. Purves, and A. Adhikari, *Statistics*, 2nd ed. New York: W. W. Norton, 1991, chaps. 19 and 20.

5. Gallup, George, *The Sophisticated Poll Watcher's Guide*. Princeton, NJ: Princeton Public Opinion Press, 1972.

6. Gleick, James, "The Census: Why We Can't Count," *New York Times Magazine* (July 15, 1990), 22–26, 54.

7. Kalton, Graham, *Introduction to Survey Sampling*. Newbury Park, CA: Sage Publications, 1983.

8. Kish, Leslie, *Survey Sampling*. New York: Wiley Interscience, 1995.

9. Kolata, Gina, "Searching For Clarity: A Primer on Studies," *The New York Times* (Sept. 30, 2008), D1, D6.

10. Meier, Paul, "The Biggest Public Health Experiment Ever: The 1954 Field Trial of the Salk Poliomyelitis Vaccine," in *Statistics: A Guide to the Unknown*, 3rd ed., eds. Judith M. Tanur et al. Belmont, CA: Wadsworth, 1989, 3–14.

11. Mosteller, F., et al., *The Pre-election Polls of 1948*. New York: Social Science Research Council, 1949.

12. Paul, John, *A History of Poliomyelitis*. New Haven, CT: Yale University Press, 1971.

13. Scheaffer, R. L., W. Mendenhall, and L. Ott, *Elementary Survey Sampling*. Boston: PWS-Kent, 1990.

14. Utts, Jessica M., *Seeing Through Statistics*. Belmont, CA: Wadsworth, 1996.

15. Yates, Frank, *Sampling Methods for Censuses and Surveys*. New York: Macmillan Publishing Co., 1981.

16. Zivin, Justin A., "Understanding Clinical Trials," *Scientific American*, 282 (April 2000), 69–75.

14 Descriptive Statistics

Graphing and Summarizing Data

The primary purpose of collecting data is to give meaning to a statistical story, to uncover some new fact about our world, and — last but certainly not least — to make a point, no matter how outlandish. But what do we do when we have too much data? One important purpose of statistics is to describe large amounts of data in a way that is understandable, useful, and, if need be, convincing. This is called *descriptive statistics* and is the subject of this chapter.

Imagine that our data consist of the test scores of a group of students in a standardized exam. If we are dealing with a small group of students — say a class — then it is reasonable to look at the collection of test scores of the group and get the "big picture" (how the group performed compared with other groups, how many are at grade level, etc.). On the other hand, if we are dealing with a large group (hundreds, thousands, or even millions), trying to get to the big picture by looking at the individual scores of the students is hopeless. The amount of data to deal with becomes overwhelming — a huge babble of numbers.

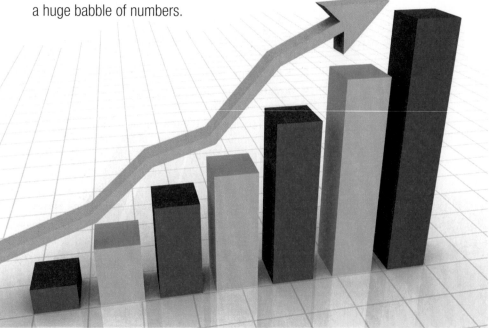

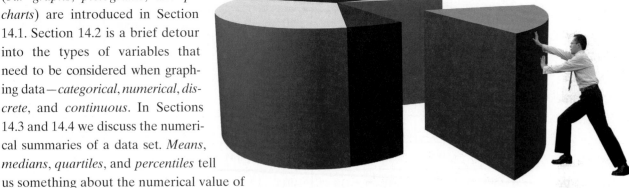

> 66 Numbers are like people. Torture them enough
> and they'll tell you anything. 99
>
> —Anonymous

here are two strategies when trying to make some sense out of a large
set of numbers. One is to present the data in the form of pictures or
graphs; the other is to use numerical summaries that serve as "snapshots" of the
data set. Graphical descriptions of data
(*bar graphs*, *pictograms*, and *pie
charts*) are introduced in Section
14.1. Section 14.2 is a brief detour
into the types of variables that
need to be considered when graphing data—*categorical*, *numerical*, *discrete*, and *continuous*. In Sections
14.3 and 14.4 we discuss the numerical summaries of a data set. *Means*,
medians, *quartiles*, and *percentiles* tell
us something about the numerical value of
the data (they are called *measures of location*) and are discussed in Section 14.3.
Ranges, *interquartile ranges*, and *standard deviations* provide information about
the spread within the data (they are known as *measures of spread*) and are discussed in Section 14.4.

14.1 Graphical Descriptions of Data

Data Sets

A **data set** is a collection of data values. Statisticians often refer to the individual
data values in a data set as **data points**. For the sake of simplicity, we will work
with data sets in which each data point consists of a single number, but in more
complicated settings, a single data point can consist of many numbers.

As usual, we will use the letter N to represent the size of the data set. In real-
life applications, data sets can range in size from reasonably small (a dozen or so
data points) to very large (hundreds of millions of data points), and the larger the
data set is, the more we need a good way to describe and summarize it.

To illustrate many of the ideas of this chapter we will need a reasonable
data set—big enough to be realistic, but not so big that it will bog us down.
Example 14.1, which we will revisit several times in the chapter, provides such

a data set. This is a fictitious data set from a hypothetical statistics class, but except for the details, it describes a situation that is familiar to every college student.

(**EXAMPLE 14.1**)   **Stat 101 Test Scores**

As usual, the day after the midterm exam in his Stat 101 class, Dr. Blackbeard has posted the results online (Table 14-1). The data set consists of $N = 75$ data points (the number of students who took the test). Each data point (listed in the second column) is a score between 0 and 25 (Dr. Blackbeard gives no partial credit). Notice that the numbers listed in the first column are not data points—they are numerical IDs used as substitutes for names to protect the students' rights of privacy.

TABLE 14-1   Stat 101/Midterm Scores

ID	Score	ID	Score	ID	Score	ID	Score	ID	Score
1257	12	2651	10	4355	8	6336	11	8007	13
1297	16	2658	11	4396	7	6510	13	8041	9
1348	11	2794	9	4445	11	6622	11	8129	11
1379	24	2795	13	4787	11	6754	8	8366	13
1450	9	2833	10	4855	14	6798	9	8493	8
1506	10	2905	10	4944	6	6873	9	8522	8
1731	14	3269	13	5298	11	6931	12	8664	10
1753	8	3284	15	5434	13	7041	13	8767	7
1818	12	3310	11	5604	10	7196	13	9128	10
2030	12	3596	9	5644	9	7292	12	9380	9
2058	11	3906	14	5689	11	7362	10	9424	10
2462	10	4042	10	5736	10	7503	10	9541	8
2489	11	4124	12	5852	9	7616	14	9928	15
2542	10	4204	12	5877	9	7629	14	9953	11
2619	1	4224	10	5906	12	7961	12	9973	10

Like students everywhere, the students in the Stat 101 class have one question foremost on their mind when they look at Table 14-1: How did I do? Each student can answer this question directly from the table. It's the next question that is statistically much more interesting. How did the class as a whole do? To answer this last question, we will have to find a way to package the information in Table 14-1 into a compact, organized, and intelligible whole.

Bar Graphs and Variations Thereof

(EXAMPLE 14.2)   **Stat 101 Test Scores: Part 2**

The first step in summarizing the information in Table 14-1 is to organize the scores in a **frequency table** such as Table 14-2. In this table, the number below each score gives the **frequency** of the score—that is, the number of students getting that particular score. We can readily see from Table 14-2 that there was one student with a score of 1, one with a score of 6, two with a score of 7, six with a score of 8, and so on. Notice that the scores with a frequency of zero are not listed in the table.

TABLE 14-2	Frequency Table for the Stat 101 Data Set												
Score	1	6	7	8	9	10	11	12	13	14	15	16	24
Frequency	1	1	2	6	10	16	13	9	8	5	2	1	1

While Table 14-2 is a considerable improvement over Table 14-1, we can do even better. Figure 14-1 shows the same information in a much more visual way called a **bar graph**, with the test scores listed in increasing order on a horizontal axis and the frequency of each test score displayed by the *height* of the column above that test score. Notice that in the bar graph, even the test scores with a frequency of zero show up—there simply is no column above these scores.

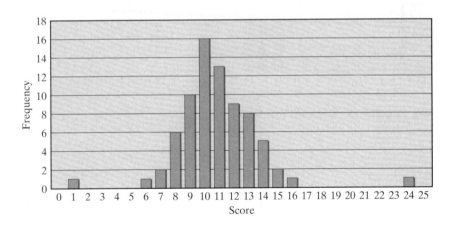

FIGURE 14-1 Bar graph for the Stat 101 data set.

Bar graphs are easy to read, and they are a nice way to present a good general picture of the data. With a bar graph, for example, it is easy to detect **outliers**—extreme data points that do not fit into the overall pattern of the data. In this example there are two obvious outliers—the score of 24 (head and shoulders above the rest of the class) and the score of 1 (lagging way behind the pack).

Sometimes it is more convenient to express the bar graph in terms of *relative frequencies*—that is, the frequencies given in terms of percentages of the total population. Figure 14-2 shows a *relative frequency bar graph* for the Stat 101 data set. Notice that we indicated on the graph that we are dealing with percentages rather than total counts and that the size of the data set is

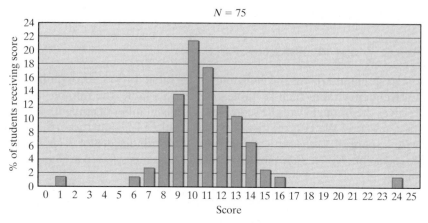

FIGURE 14-2 Relative frequency bar graph for the Stat 101 data set.

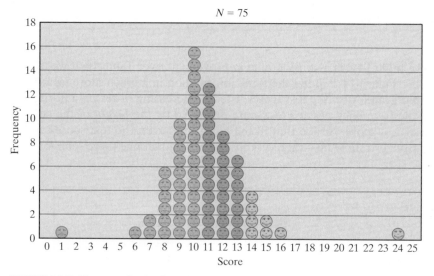

FIGURE 14-3 Pictogram for the Stat 101 data set.

$N = 75$. This allows anyone who wishes to do so to compute the actual frequencies. For example, Fig. 14-2 indicates that 12% of the 75 students scored a 12 on the exam, so the actual frequency is given by $75 \times 0.12 = 9$ students. The change from actual frequencies to percentages (or vice versa) does not change the shape of the graph—it is basically a change of scale.

While the term *bar graph* is most commonly used for graphs like the ones in Figs. 14-1 and 14-2, devices other than bars can be used to add a little extra flair or to subtly influence the content of the information given by the raw data. Dr. Blackbeard, for example, might have chosen to display the midterm data using a graph like the one shown in Fig. 14-3, which conveys all the information of the original bar graph (Fig. 14-1) and also includes a subtle individual message to each student.

Frequency charts that use icons or pictures instead of bars to show the frequencies are commonly referred to as **pictograms**. The point of a pictogram is that a graph is often used not only to inform but also to impress and persuade, and, in such cases, a well-chosen icon or picture can be a more effective tool than just a bar.

EXAMPLE 14.3   Selling the XYZ Corporation

Figure 14-4(a) is a pictogram showing the growth in yearly sales of the XYZ Corporation between 2001 and 2006. It's a good picture to show at a shareholders meeting, but the picture is actually quite misleading. Figure 14-4(b) shows a pictogram for exactly the same data with a much more accurate and sobering picture of how well the XYZ Corporation had been doing.

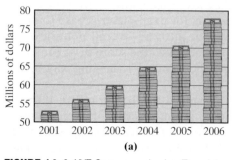

(a)

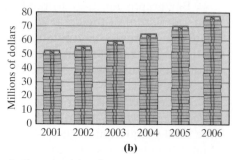

(b)

FIGURE 14-4 XYZ Corp. annual sales. Two pictograms for the same data set.

The difference between the two pictograms can be attributed to a couple of standard tricks of the trade: (1) stretching the scale of the vertical axis and (2) "cheating" on the choice of starting value on the vertical axis. As an educated consumer, you should always be on the lookout for these tricks. In graphical descriptions of data, a fine line separates objectivity from propaganda. (For more on this topic, see Project A.)

14.2 Variables

Before we continue with our discussion of graphs, we need to discuss briefly the concept of a **variable**. In statistical usage, a variable is any characteristic that varies with the members of a population. The students in Dr. Blackbeard's Stat 101 course (the population) did not all perform equally on the exam. Thus, the *test score* is a variable, which in this particular case is a whole number between 0 and 25. In some instances, such as when the instructor gives partial credit or when there is subjective grading, a test score may take on a fractional value, such as 18.5 or 18.25. Even in these cases, however, the possible increments for the values of the variable are given by some minimum amount—a quarter-point, a half-point, whatever. In contrast to this situation, consider a different variable: the *amount of time* each student studied for the exam. In this case the variable can take on values that differ by any amount: an hour, a minute, a second, a tenth of a second, and so on.

A variable that represents a measurable quantity is called a **numerical** (or **quantitative**) **variable**. When the difference between the values of a numerical variable can be arbitrarily small, we call the variable **continuous**; when possible values of the numerical variable change by minimum increments, the variable is called **discrete**. Examples of *discrete* variables are a person's IQ, an SAT score, a person's shoe size, and the number of points scored in a basketball game. Examples of *continuous* variables are a person's height, weight, foot size (as opposed to shoe size), and the time it takes him or her to run a mile.

Sometimes in the real world the distinction between continuous and discrete variables is blurred. Height, weight, and age are all continuous variables in theory, but in practice they are frequently rounded to the nearest inch, ounce, and year (or month in the case of babies), at which point they become discrete variables. On the other hand, money, which is in theory a discrete variable (because the difference between two values cannot be less than a penny), is almost always thought of as continuous, because in most real-life situations a penny can be thought of as an infinitesimally small amount of money.

Variables can also describe characteristics that cannot be measured numerically: nationality, gender, hair color, and so on. Variables of this type are called **categorical** (or **qualitative**) **variables**.

In some ways, categorical variables must be treated differently from numerical variables—they cannot, for example, be added, multiplied, or averaged. In other ways, categorical variables can be treated much like discrete numerical variables, particularly when it comes to graphical descriptions, such as bar graphs and pictograms.

EXAMPLE 14.4 **Enrollments at Tasmania State University**

Table 14-3 shows undergraduate enrollments in each of the five schools at Tasmania State University. A sixth category ("Other") includes undeclared students, interdisciplinary majors, and so on.

TABLE 14-3

School	Enrollment
Agriculture	2400
Business	1250
Education	2840
Humanities	3350
Science	4870
Other	290

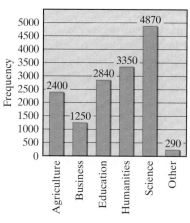

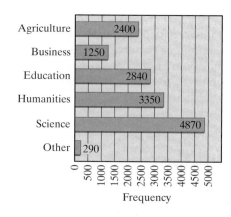

FIGURE 14-5   **(a)**   **(b)**

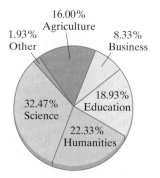

FIGURE 14-6

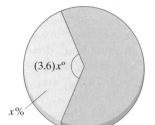

FIGURE 14-7

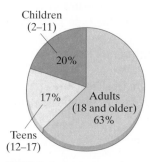

FIGURE 14-8 Audience composition for prime-time TV viewership by age group.

Figure 14-5 shows two equivalent bar graphs describing the information in Table 14-3. The only difference between the two graphs is that in Fig. 14-5(b) the values of the categorical variable are on the vertical axis. There is no prohibition against doing this, and, in fact, many people prefer this approach when dealing with categorical variables.

When the number of categories is small, as is the case here, another common way to describe the relative frequencies of the categories is by using a **pie chart**. In a pie chart the "pie" represents the entire population (100%), and the "slices" represent the **categories** (or **classes**), with the size (angle) of each slice being proportional to the relative frequency of the corresponding category.

Some relative frequencies, such as 50% and 25%, are very easy to sketch, but how do we accurately draw the slice corresponding to a more complicated frequency, say, 32.47%? Here, a little elementary geometry comes in handy. Since 100% equals 360°, 1% corresponds to an angle of 360°/100 = 3.6°. It follows that the frequency 32.47% is given by 32.47 × 3.6° = 117° (rounded to the nearest degree, which is generally good enough for most practical purposes). Figure 14-6 shows an accurate pie chart for the school-enrollment data given in Table 14-3.

PIE CHARTS

The general rule in drawing pie charts is that a slice representing x% is given by an angle of $(3.6)x$ degrees (see Fig. 14-7).

Bar graphs and pie charts are excellent ways to graphically display categorical data, but graphs and charts can be deceiving, and we should be very careful about the conclusions we draw when the data are used to compare different populations. Our next example illustrates this point.

EXAMPLE 14.5   **Who's Watching the Boob Tube Tonight?**

According to Nielsen Media Research data, the percentages of the TV audience watching TV during prime time (8 P.M. to 11 P.M.), broken up by age group, are as follows: adults (18 years and older), 63%; teenagers (12–17 years), 17%; children (2–11 years), 20%. (The exact figures vary from year to year. These figures are averaged over several years.)

The pie chart in Fig. 14-8 shows this breakdown of audience composition by age group. A pie chart such as this one might be used to make the point that children and teenagers really do not watch as much TV as it is generally believed. The problem with this conclusion is that children make up only 15% of the population at large and teens only 8%. In relative terms, a higher percentage of teenagers (taken out of the total teenage population) watch prime-time TV than any other group, with children second and adults last.

The moral of Example 14.5 is that using absolute percentages, as we did in Fig. 14-8, can be quite misleading. When comparing characteristics of a population that is broken up into categories, it is essential to take into account the relative sizes of the various categories.

Class Intervals

While the distinction between qualitative and quantitative data is important in many aspects of statistics, when it comes to deciding how best to display graphically the frequencies of a population, a critical issue is the number of categories into which the data can fall. When the number of categories is too big (say, in the dozens), a bar graph or pictogram can become muddled and ineffective. This happens more often than not with numerical data—numerical variables can take on infinitely many values, and even when they don't, the number of values can be too large for any reasonable graph. Our next example illustrates how to deal with this situation.

TABLE 14-4	2007 SAT Math Scores
Score range	Frequency
200–249	13,159
250–299	23,083
300–349	59,672
350–399	124,616
400–449	202,883
450–499	243,569
500–549	247,435
550–599	213,880
600–649	156,057
650–699	120,933
700–749	54,108
750–800	35,136
Total	1,494,531

EXAMPLE 14.6   2007 SAT Math Scores

The college dreams and aspirations of millions of high school seniors often ride on their SAT scores. The SAT consists of three sections: a math section, a writing section, and a critical reading section, with the scores for each section ranging from a minimum of 200 to a maximum of 800 and going up in increments of 10 points.

In 2007, there were 1,494,531 college-bound seniors who took the SAT. How do we describe the math section results for this group of students? In one way, this is the same problem as the one illustrated in Example 14.1 (Stat 101 midterm scores)—just a different test and a much larger number of test-takers. We could set up a frequency table (or a bar graph) with the number of students scoring each of the possible scores—200, 210, 220, . . . , 790, 800. The problem is that there are 61 different possible scores between 200 and 800, and this number is too large for an effective bar graph.

In situations such as this one it is customary to present a more compact picture of the data by grouping together, or *aggregating*, sets of scores into categories called **class intervals**. The decision as to how the class intervals are defined and how many there are will depend on how much or how little detail is desired, but as a general rule of thumb, the number of class intervals should be somewhere between 5 and 20.

SAT scores are usually aggregated into 12 class intervals of essentially the same size: 200–249, 250–299, 300–349, . . . , 700–749, 750–800. Using these class intervals, the distribution of scores in the math section of the 2007 SAT is given in Table 14-4, and the associated bar graph is shown in Fig. 14-9 (*Source*: The College Board).

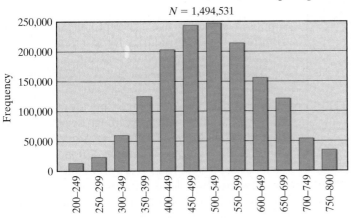

FIGURE 14-9 2007 SAT Math: Distribution of scores.

Our next example deals with a topic every college student can relate to—grades!

EXAMPLE 14.7 **Stat 101 Test Scores: Part 3**

The process of converting test scores (a *numerical* variable) into grades (a *categorical* variable) requires setting up *class intervals* for the various letter grades. Typically, the professor has the latitude to decide how to do this. One standard approach is to use an *absolute* grading scale, usually with class intervals of (almost) equal length for all grades except F (e.g., A = 90–100%, B = 80–89%, C = 70–79%, D = 60–69%, F = 0–59%). Another frequently used approach is to use a *relative* grading scale. Here the professor fits the class intervals for the grades to the performance of the class in the test, often using class intervals of varying lengths. Some people call this "grading on the curve," although this terminology is somewhat misused.

To illustrate relative grading in action, let's revisit the Stat 101 midterm scores discussed in Example 14.1. After looking at the overall class performance (see Example 14.2), Dr. Blackbeard chooses to "curve" the test scores using class intervals of his own creation. The class intervals and corresponding results are shown in Table 14-5.

The grade distribution in the Stat 101 midterm can now be best seen by means of the bar graph shown in Fig. 14-10. The picture speaks for itself—this was a very tough exam!

TABLE 14-5	Stat 101 Grades				
Grade	A	B	C	D	F
Score	18–25	14–17	11–13	9–10	8 or less
Frequency	1	8	30	26	10
Percent	1.33%	10.67%	40%	34.67%	13.33%

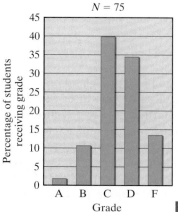

FIGURE 14-10 Stat 101 grades.

Histograms

When a numerical variable is continuous, its possible values can vary by infinitesimally small increments. As a consequence, there are no gaps between the class intervals, and our old way of doing things (using separated columns or stacks) will no longer work. In this case we use a variation of a bar graph called a **histogram**. We illustrate the concept of a histogram in the next example.

TABLE 14-6	Starting Salaries of First-Year TSU Graduates	
Salary	Number of students	Percentage
40,000⁺–45,000	228	7%
45,000⁺–50,000	456	14%
50,000⁺–55,000	1043	32%
55,000⁺–60,000	912	28%
60,000⁺–65,000	391	12%
65,000⁺–70,000	163	5%
70,000⁺–75,000	65	2%
Total	3258	100%

EXAMPLE 14.8 **Starting Salaries of TSU Graduates**

Suppose we want to use a graph to display the distribution of starting salaries for last year's graduating class at Tasmania State University.

The starting salaries of the $N = 3258$ graduates range from a low of \$40,350 to a high of \$74,800. Based on this range and the amount of detail we want to show, we must decide on the length of the class intervals. A reasonable choice would be to use class intervals defined in increments of \$5000. Table 14-6 is a frequency table for the data based on these class intervals. We chose a starting value of \$40,000 for convenience. (The third column in the table shows the data as a percentage of the population.)

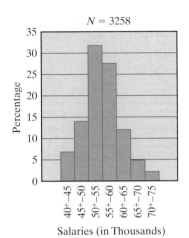

$N = 3258$

Salaries (in Thousands)

FIGURE 14-11 Histogram for start-ing salaries of first-year graduates of TSU (with class intervals of $5000).

■ For more details on histograms with class intervals of unequal lengths, see Exercises 73 and 74.

The histogram showing the relative frequency of each class interval is shown in Fig. 14-11. As we can see, a histogram is very similar to a bar graph. Several important distinctions must be made, however. To begin with, because a histogram is used for continuous variables, there can be no gaps between the class intervals, and it follows, therefore, that the columns of a histogram must touch each other. Among other things, this forces us to make an arbitrary decision as to what happens to a value that falls exactly on the boundary between two class intervals. Should it always belong to the class interval to the left or to the one to the right? This is called the *endpoint convention*. The su-perscript "plus" marks in Table 14-6 indicate how we chose to deal with the endpoint convention in Fig. 14-11. A starting salary of exactly $50,000, for example, would be listed under the $45,000^+$–$50,000$ class interval rather than the $50,000^+$–$55,000$ class interval.

When creating histograms, we should try, as much as possible, to define class intervals of equal length. When the class intervals are of unequal length, the rules for creating a histogram are considerably more complicated, since it is no longer appropriate to use the heights of the columns to indicate the frequencies of the class intervals.

14.3   Numerical Summaries of Data

As we have seen, a picture can be an excellent tool for summarizing large data sets. Unfortunately, circumstances do not always lend themselves equally well to the use of pictures, and bar graphs and pie charts cannot be readily used in everyday conversation. A different and very important approach is to use a few well-chosen numbers to summarize an entire data set.

In the next couple of sections we will discuss two types of *numerical summaries* of a data set: **measures of location** and **measures of spread**. Measures of location such as the *mean* (or *average*), the *median*, and the *quartiles*, are numbers that provide information about the values of the data. Measures of spread such as the *range*, the *interquartile range*, and the *standard deviation* are numbers that pro-vide information about the spread within the data set. In this section we will focus on measures of location. In Section 14.4 we will discuss measures of spread.

The Mean (Average)

The best known of all numerical summaries of data is the *average*, also called the *mean*. (There is no universal agreement as to which of these names is a better choice—in some settings *mean* is a better choice than *average*, in other settings it's the other way around. In this chapter we will use whichever seems the better choice at the moment.)

The **average** (or **mean**) of a set of N numbers is found by adding the numbers and dividing the total by N. In other words, the average of the numbers $d_1, d_2, d_3, \ldots, d_N$ is $A = (d_1 + d_2 + \cdots + d_N)/N$.

⬭ **EXAMPLE 14.9**   **Stat 101 Test Scores: Part 4**

In this example we will find the average test score in the Stat 101 exam first intro-duced in Example 14.1. To find this average we need to add all the test scores and divide by 75. The addition of the 75 test scores can be simplified considerably if we use a frequency table. (Table 14-7 is the same as Table 14-2, shown again for the reader's convenience.)

TABLE 14-7	Frequency Table for the Stat 101 Data Set											
Score	1	6	7	8	9	10	11	12	13	14	15	16	24
Frequency	1	1	2	6	10	16	13	9	8	5	2	1	1

From the frequency table we can find the sum S of all the test scores as follows: Multiply each test score by its corresponding frequency and then add these products. Thus, the sum of all the test scores is

$$S = (1 \times 1) + (6 \times 1) + (7 \times 2) + (8 \times 6) + \cdots + (16 \times 1) + (24 \times 1) = 814$$

If we divide this sum by $N = 75$, we get the average test score A (rounded to two decimal places)

$$A = 814/75 \approx 10.85 \text{ points}$$

TABLE 14-8	
Value	Frequency
d_1	f_1
d_2	f_2
$\vdots$	$\vdots$
d_k	f_k

In general, to find the average A of a data set given by a frequency table such as Table 14-8 we do the following:

■ **Step 1.** $S = d_1 \cdot f_1 + d_2 \cdot f_2 + \cdots + d_k \cdot f_k$.
■ **Step 2.** $N = f_1 + f_2 + \cdots + f_k$.
■ **Step 3.** $A = S/N$.

When dealing with data sets that have outliers, averages can be quite misleading. As our next example illustrates, even a single outlier can have a big effect on the average.

(**EXAMPLE 14.10**)   **Starting Salaries of Philosophy Majors**

Imagine that you just read in the paper the following remarkable tidbit: *The average starting salary of philosophy majors who recently graduated from Tasmania State University is $76,400 a year!* This is quite an impressive number, but before we all rush out to change majors, let's point out that one of the graduating philosophy majors happens to be basketball star "Hoops" Tallman, who is doing his thing in the NBA for a starting salary of $3.5 million a year.

If we were to take this one outlier out of the population of 75 philosophy majors, we would have a more realistic picture of what philosophy majors are making. Here is how we can do it:

■ The total of all 75 salaries is 75 times the average salary:

$$75 \times \$76,400 = \$5,730,000$$

■ The total of the other 74 salaries (excluding Hoops's cool 3.5 mill) is

$$\$5,730,000 - \$3,500,000 = \$2,230,000$$

■ The average of the remaining 74 salaries is

$$\$2,230,000/74 \approx \$30,135$$

So far, all our examples have involved data values that are positive, but negative data values are also possible, and when both negative and positive data values are averaged, the results can be misleading.

TABLE 14-9

Month	Balance
Jan.	−732 ← Christmas bills come in!
Feb.	−158
Mar.	−71
Apr.	−238
May	1839 ← $2000 lottery winnings
Jun.	−103
Jul.	−148
Aug.	−162
Sep.	−85
Oct.	−147
Nov.	−183
Dec.	500 ← Christmas present from mom

(EXAMPLE 14.11) **Living Beyond Your Means**

Table 14-9 shows the monthly balance (monthly income minus monthly spending) in Billy's budget over the past year. A negative amount indicates that Billy spent more than what he had coming in (adding to his credit card debt).

In spite of his consistent overspending, Billy's average monthly balance for the year is $26 (check it out!). This average hides the true picture of what is going on. Billy is living well beyond his means but was bailed out by a lucky break and a generous mom.

Percentiles

While a single numerical summary—such as the average—can be useful, it is rarely sufficient to give a meaningful description of a data set. A better picture of the data set can be presented by using a well-organized cadre of numerical summaries. The most common way to do this is by means of *percentiles*.

The **pth percentile** of a data set is a value such that p percent of the numbers fall *at or below* this value and the rest fall *at or above* it. It essentially splits a data set into two parts: the lower $p\%$ of the data values and the upper $(100 - p)\%$ of the data values.

Many college students are familiar with percentiles, if for no other reason than the way they pop up in SAT reports. In all SAT reports, a given score—say a score of 620 in the math section—is identified with a percentile—say the 81st percentile. This can be interpreted to mean that 81% of those taking the test scored 620 or less, or, looking up instead of down, that 19% of those taking the test scored 620 or more.

There are several different ways to compute percentiles that will satisfy the definition, and different statistics books describe different methods. We will illustrate one such method below.

The first step in finding the pth *percentile* of a data set of N numbers is to *sort the numbers from smallest to largest*. Let's denote the sorted data values by $d_1, d_2, d_3, \ldots, d_N$, where d_1 represents the smallest number in the data set, d_2 the second smallest number, and so on. Sometimes we will also need to talk about the average of two consecutive numbers in the sorted list, so we will use more unusual subscripts such as $d_{3.5}$ to represent the average of the data values d_3 and d_4, $d_{7.5}$ to represent the average of the data values d_7 and d_8, and so on.

The next, and most important, step is to identify which d represents the pth percentile of the data set. To do this, we compute the pth *percent of N*, which we will call the **locator** and denote by the letter L. [In other words, $L = (p/100) \cdot N$.] If L happens to be a whole number, then the pth *percentile* will be $d_{L.5}$ (the average of d_L and d_{L+1}). If L is not a whole number, then the pth *percentile* will be d_{L^+}, where L^+ represents the value of L rounded up.

The procedure for finding the *p*th *percentile* of a data set is summarized as follows:

FINDING THE *p*TH PERCENTILE OF A DATA SET

- **Step 0.** Sort the data set from smallest to largest. Let $d_1, d_2, d_3, \ldots, d_N$ represent the sorted data.
- **Step 1.** Find the locator $L = (p/100) \cdot N$.
- **Step 2.** Depending on whether L is a whole number or not, the *p*th *percentile* is given by
  - $d_{L.5}$ if L is a whole number.
  - d_{L^+} if L is not a whole number (L^+ is L rounded up).

The following example illustrates the procedure for finding percentiles of a data set.

(**EXAMPLE 14.12**)   **Scholarships by Percentile**

To reward good academic performance from its athletes, Tasmania State University has a program in which athletes with GPAs in the top 20th percentile of their team's GPAs get a $5000 scholarship and athletes with GPAs in the top forty-fifth percentile of their team's GPAs who did not get the $5000 scholarship get a $2000 scholarship.

The women's soccer team has $N = 15$ players. A list of their GPAs is as follows:

3.42, 3.91, 3.33, 3.65, 3.57, 3.45, 4.0, 3.71, 3.35, 3.82, 3.67, 3.88, 3.76, 3.41, 3.62

When we sort these GPAs we get the list

3.33, 3.35, 3.41, 3.42, 3.45, 3.57, 3.62, 3.65, 3.67, 3.71, 3.76, 3.82, 3.88, 3.91, 4.0

Since this list goes from lowest to highest GPA, we are looking for the 80th percentile and above (top 20th percentile) for the $5000 scholarships and the 55th percentile and above (top 45th percentile) for the $2000 scholarships.

$5000 scholarships: The locator for the 80th percentile is $(0.8) \times 15 = 12$. Here the locator is a whole number, so the 80th percentile is given by $d_{12.5} = 3.85$ (the average between $d_{12} = 3.82$ and $d_{13} = 3.88$). Thus, three students (the ones with GPAs of 3.88, 3.91 and 4.0) get $5000 scholarships.

$2000 scholarships: The locator for the 55th percentile is $(0.55) \times 15 = 8.25$. This locator is not a whole number, so we round it up to 9, and the 55th percentile is given by $d_9 = 3.67$. Thus, the students with GPAs of 3.67, 3.71, 3.76, and 3.82 (all students with GPAs of 3.67 or higher except the ones that already received $5000 scholarships) get $2000 scholarships.

The Median and the Quartiles

The 50th percentile of a data set is known as the **median** and denoted by M. The median splits a data set into two halves—half of the data is at or below the median and half of the data is at or above the median.

We can find the median by simply applying the definition of percentile with $p = 50$, but the bottom line comes down to this: (1) when N is *odd*, the median is the data value in position $(N + 1)/2$ of the sorted data set; (2) when N is *even*, the median is the average of the data values in position $N/2$ and $(N/2) + 1$ of the sorted data set. [All of the preceding follows from the fact that the locator for the median is $L = N/2$. When N is even, L is an integer; when N is odd, L is not an integer.]

> **FINDING THE MEDIAN OF A DATA SET**
>
> - Sort the data set from smallest to largest. Let $d_1, d_2, d_3, \ldots, d_N$ represent the sorted data.
> - If N is odd, the median is $d_{\frac{N+1}{2}}$.
> - If N is even, the median is the average of $d_{\frac{N}{2}}$ and $d_{\frac{N}{2}+1}$.

After the median, the next most commonly used set of percentiles are the first and third **quartiles**. The *first quartile* (denoted by Q_1) is the 25th percentile, and the *third quartile* (denoted by Q_3) is the 75th percentile.

(**EXAMPLE 14.13**)   **Home Prices in Green Hills**

During the last year, 11 homes sold in the Green Hills subdivision. The selling prices, in chronological order, were $267,000, $252,000, $228,000, $234,000, $292,000, $263,000, $221,000, $245,000, $270,000, $238,000, and $255,000. We are going to find the *median* and the *quartiles* of the $N = 11$ home prices.

Sorting the home prices from smallest to largest (and dropping the 000's) gives the sorted list

$$221, 228, 234, 238, 245, 252, 255, 263, 267, 270, 292$$

The locator for the median is $(0.5) \times 11 = 5.5$, the locator for the first quartile is $(0.25) \times 11 = 2.75$, and the locator for the third quartile is $(0.75) \times 11 = 8.25$. Since these locators are not whole numbers, they must be rounded up: 5.5 to 6, 2.75 to 3, and 8.25 to 9. Thus, the median home price is given by $d_6 = 252$ (i.e., $M = \$252,000$), the first quartile is given by $d_3 = 234$ (i.e., $Q_1 = \$234,000$), and the third quartile is given by $d_9 = 267$ (i.e., $Q_3 = \$267,000$).

(**EXAMPLE 14.13 (CONTINUED)**)   **Another Home Sells in Green Hills**

Oops! Just this morning a home sold in Green Hills for $264,000. We need to re-calculate the median and quartiles for what are now $N = 12$ home prices.

We can use the sorted data set that we already had—all we have to do is insert the new home price (264) in the right spot (remember, we drop the 000's!). This gives

$$221, 228, 234, 238, 245, 252, 255, 263, \mathbf{264}, 267, 270, 292$$

Now $N = 12$ and in this case the median is the average of $d_6 = 252$ and $d_7 = 255$. It follows that the median home price is $M = \$253,500$. The locator for the first quartile is $0.25 \times 12 = 3$. Since the locator is a whole number, the first quartile is the average of $d_3 = 234$ and $d_4 = 238$ (i.e., $Q_1 = \$236,000$). Similarly, the third quartile is $Q_3 = 265,500$ (the average of $d_9 = 264$ and $d_{10} = 267$). ⊂⊃

(**EXAMPLE 14.14**)   **Stat 101 Test Scores: Part 5**

We will now find the median and quartile scores for the Stat 101 data set (shown again in Table 14-10).

TABLE 14-10	Frequency Table for the Stat 101 Data Set													
Score	1	6	7	8	9	10	11	12	13	14	15	16	24	
Frequency	1	1	2	6	10	16	13	9	8	5	2	1	1	

Having the frequency table available eliminates the need for sorting the scores—the frequency table has, in fact, done this for us. Here $N = 75$ (odd), so the median is the thirty-eighth score (counting from the left) in the frequency table. To find the thirty-eighth number in Table 14-10, we tally frequencies as we move from left to right: $1 + 1 = 2$; $1 + 1 + 2 = 4$; $1 + 1 + 2 + 6 = 10$; $1 + 1 + 2 + 6 + 10 = 20$; $1 + 1 + 2 + 6 + 10 + 16 = 36$. At this point, we know that the 36th test score on the list is a 10 (the last of the 10's) and the next 13 scores are all 11's. We can conclude that the 38th test score is 11. Thus, $M = 11$.

The locator for the first quartile is $L = (0.25) \times 75 = 18.75$. Thus, $Q_1 = d_{19}$. To find the nineteenth score in the frequency table, we tally frequencies from left to right: $1 + 1 = 2$; $1 + 1 + 2 = 4$; $1 + 1 + 2 + 6 = 10$; $1 + 1 + 2 + 6 + 10 = 20$. At this point we realize that $d_{10} = 8$ (the last of the 8's) and that d_{11} through d_{20} all equal 9. Hence, the first quartile of the Stat 101 midterm scores is $Q_1 = d_{19} = 9$.

Since the first and third quartiles are at an equal "distance" from the two ends of the sorted data set, a quick way to locate the third quartile now is to look for the nineteenth score in the frequency table when we count frequencies *from right to left*. We leave it to the reader to verify that the third quartile of the Stat 101 data set is $Q_3 = 12$.

(**EXAMPLE 14.15**)　**2007 SAT Math Scores: Part 2**

In this example we continue the discussion of the 2007 SAT math scores introduced in Example 14.6. Recall that the number of college-bound high school seniors taking the test was $N = 1{,}494{,}531$. As reported by the College Board, the median score in the test was $M = 510$, the first quartile score was $Q_1 = 430$, and the third quartile was $Q_3 = 590$. What can we make of this information?

Let's start with the median. From $N = 1{,}494{,}531$ (an odd number), we can conclude that the median (510 points) is the 747,266th score in the sorted list of test scores. This means that there were *at least* 747,266 students who scored 510 or less in the math section of the 2007 SAT. Why did we use "at least" in the preceding sentence? Could there have been more than that number who scored 510 or less? Yes, almost surely. Since the number of students who scored 510 is in the thousands, it is very unlikely that the 747,266th score is the last of the 510s.

■ The details are left to the reader — see Exercise 39.

In a similar vein, we can conclude that there were at least 373,633 scores of $Q_1 = 430$ or less [the locator for the first quartile is $(0.25) \times 1{,}494{,}531 = 373{,}632.75$] and at least 1,120,899 scores of $Q_3 = 590$ or less.

---

*A **note of warning:*** Medians, quartiles, and general percentiles are often computed using statistical calculators or statistical software packages, which is all well and fine since the whole process can be a bit tedious. The problem is that there is no universally agreed upon procedure for computing percentiles, so different types of calculators and different statistical packages may give different answers from each other and from those given in this book for quartiles and other percentiles (everyone agrees on the median). *Keep this in mind when doing the exercises*—the answer given by your calculator may be slightly different from the one you would get from the procedure we use in the book.

The Five-Number Summary

A common way to summarize a large data set is by means of its *five-number summary*. The **five-number summary** is given by (1) the smallest value in the data set (called the *Min*), (2) the *first quartile* Q_1, (3) the *median M*, (4) the *third* *quartile* Q_3 and (5) the largest value in the data set (called the *Max*). These five numbers together often tell us a great deal about the data.

(**EXAMPLE 14.16**)   **Stat 101 Test Scores: Part 6**

For the Stat 101 data set, the five-number summary is $Min = 1, Q_1 = 9, M = 11,$ $Q_3 = 12, Max = 24$ (see Example 14.14). What useful information can we get out of this?

Right away we can see that the $N = 75$ test scores were not evenly spread out over the range of possible scores. For example, from $M = 11$ and $Q_3 = 12$ we can conclude that at least 25% of the class (that means at least 19 students) scored either 11 or 12 on the test. At the same time, from $Q_3 = 12$ and $Max = 24$ we can conclude that less than one-fourth of the class (i.e., at most 18 students) had scores in the 13–24 point range. Using similar arguments, we can conclude that at least 19 students had scores between $Q_1 = 9$ and $M = 11$ points and no more than 18 students scored in the 1–8 point range.

The "big picture" we get from the five-number summary of the Stat 101 test scores is that there was a lot of bunching up in a narrow band of scores (at least half of the students in the class scored in the range 9–12 points), and the rest of the class was all over the place. In general, this type of "lumpy" distribution of test scores is indicative of a test with an uneven level of difficulty—a bunch of easy questions and a bunch of really hard questions with little in between. (Having seen the data, we know that the *Min* and *Max* scores were both outliers and that if we disregard these two outliers, the test results don't look quite so bad. Of course, there is no way to pick this up from just the five-number summary.)

Box Plots

Invented in 1977 by statistician John Tukey, a *box plot* (also known as a *box-and-whisker* plot) is a picture of the five-number summary of a data set. The **box plot** consists of a rectangular box that sits above a scale and extends from the first quartile Q_1 to the third quartile Q_3 on that scale. A vertical line crosses the box, indicating the position of the median M. On both sides of the box are "whiskers" extending to the smallest value, *Min*, and largest value, *Max*, of the data. Figure 14-12 shows a generic box plot for a data set.

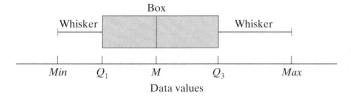

FIGURE 14-12

Figure 14-13(a) shows a box plot for the Stat 101 data set (see Example 14.14). The long whiskers in this box plot are largely due to the outliers 1 and 24. Figure 14-13(b) shows a variation of the same box plot, but with the two outliers, marked with two crosses, segregated from the rest of the data. (When there are outliers it is useful to segregate them from the rest of the data set—we think of outliers as "anomalies" within the data set.)

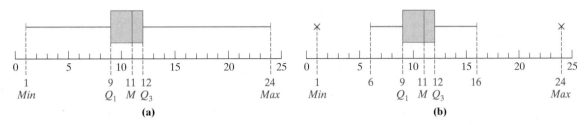

FIGURE 14-13 (a) Box plot for the Stat 101 data set. (b) Same box plot with the outliers separated from the rest of the data.

Box plots are particularly useful when comparing similar data for two or more populations. This is illustrated in the next example.

> **EXAMPLE 14.17**   **Comparing Agriculture and Engineering Salaries**

Figure 14-14 shows box plots for the starting salaries of two different populations: first-year agriculture and engineering graduates of Tasmania State University.

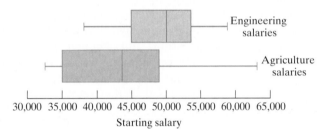

FIGURE 14-14 Comparison of starting salaries of first-year graduates in agriculture and engineering.

Superimposing the two box plots on the same scale allows us to make some useful comparisons. It is clear, for instance, that engineering graduates are doing better overall than agriculture graduates, even though at the very top levels agriculture graduates are better paid. Another interesting point is that the median salary of agriculture graduates ($43,000) is less than the first quartile of the salaries of engineering graduates ($45,000). The very short whisker on the left side of the agriculture box plot tells us that the bottom 25% of agriculture salaries are concentrated in a very narrow salary range ($32,500–$35,000). We can also see that agriculture salaries are much more spread out than engineering salaries, even though most of the spread occurs at the higher end of the salary scale.

14.4   Measures of Spread

There are several different ways to describe the spread of a data set; in this section we will describe the three most commonly used ones.

The Range

An obvious approach to describing the spread of a data set is to take the difference between the highest and lowest values of the data. This difference is called the **range** of the data set and usually denoted by R. Thus, $R = Max - Min$.

The range of a data set is a useful piece of information when there are no outliers in the data. In the presence of outliers the range tells a distorted story. For example, the range of the test scores in the Stat 101 exam is $24 - 1 = 23$ points, an indication of a big spread within the scores (i.e., a very heterogeneous group of students). True enough, but if we discount the two outliers, the remaining 73 test scores would have a much smaller range of $16 - 6 = 10$ points.

The Interquartile Range

To eliminate the possible distortion caused by outliers, a common practice when measuring the spread of a data set is to use the **interquartile range**, denoted by the acronym IQR. The interquartile range is the difference between the third quartile and the first quartile ($IQR = Q_3 - Q_1$), and it tells us how spread out the middle 50% of the data values are. For many types of real-world data, the interquartile range is a useful measure of spread.

> **EXAMPLE 14.18** **2007 SAT Math Scores: Part 3**

The five-number summary for the 2007 SAT math scores (see Example 14.15) was $Min = 200$ (yes, there were a few jokers who missed every question!), $Q_1 = 430$, $M = 510$, $Q_3 = 590$, $Max = 800$ (there are still a few geniuses around!). It follows that the 2007 SAT math scores had a range of 600 points $800 - 200 = 600$) and an interquartile range of 160 points ($IQR = 590 - 430 = 160$).

The Standard Deviation

The most important and most commonly used measure of spread for a data set is the *standard deviation*. The key concept for understanding the standard deviation is the concept of *deviation from the mean*. If A is the average of the data set and x is an arbitrary data value, the difference $x - A$ is x's **deviation from the mean**. The deviations from the mean tell us how "far" the data values are from the average value of the data. The idea is to use this information to figure out how spread out the data is. There are, unfortunately, several steps before we can get there.

The deviations from the mean are themselves a data set, which we would like to summarize. One way would be to average them, but if we do that, the negative deviations and the positive deviations will always cancel each other out so that we end up with an average of 0. This, of course, makes the average useless in this case. The cancellation of positive and negative deviations can be avoided by squaring each of the deviations. The squared deviations are never negative, and if we average them out, we get an important measure of spread called the **variance**, denoted by V. Finally, we take the square root of the variance and get the **standard deviation**, denoted by the Greek letter σ (and sometimes by the acronym SD).

The following is an outline of the definition of the standard deviation of a data set.

◼ In some definitions of the variance you divide the sum of the squared deviations by $N - 1$ (instead of by N, as one would in an ordinary average). There are reasons why this definition is appropriate in some circumstances, but a full explanation would take us beyond the purpose and scope of this chapter. In any case, except for small values of N, the difference between the two definitions tends to be very small.

┌─ **THE STANDARD DEVIATION OF A DATA SET** ─────────

■ Let A denote the mean of the data set. For each number x in the data set, compute its *deviation from the mean* $(x - A)$ and *square* each of these numbers. These numbers are called the *squared deviations*.

■ Find the average of the squared deviations. This number is called the *variance V*.

■ The *standard deviation* is the square root of the variance $\left(\sigma = \sqrt{V}\right)$.

■ See Exercises 55–58.

Standard deviations of large data sets are not fun to calculate by hand, and they are rarely found that way. The standard procedure for calculating standard deviations is to use a computer or a good scientific or business calculator, often preprogrammed to do all the steps automatically. Be that as it may, it is still important to understand what's behind the computation of a standard deviation, even when the actual grunt work is going to be performed by a machine.

TABLE 14-11

x	$(x - 90)$	$(x - 90)^2$
85	−5	25
86	−4	16
87	−3	9
88	−2	4
89	−1	1
91	1	1
92	2	4
93	3	9
94	4	16
95	5	25

(**EXAMPLE 14.19**)   **Calculation of a SD**

Over the course of the semester, Angela turned in all of her homework assignments. Her grades in the 10 assignments (sorted from lowest to highest) were 85, 86, 87, 88, 89, 91, 92, 93, 94, and 95. Our goal in this example is to calculate the standard deviation of this data set the old-fashioned way (i.e., doing our own grunt work).

The first step is to find the mean A of the data set. It's not hard to see that $A = 90$. We are lucky—this is a nice round number! The second step is to calculate the *deviations from the mean* and then the *squared deviations*. The details are shown in the second and third columns of Table 14-11. When we average the squared deviations, we get $(25 + 16 + 9 + 4 + 1 + 1 + 4 + 9 + 16 + 25)/10 = 11$. This means that the variance is $V = 11$ and thus the standard deviation (rounded to one decimal place) is $\sigma = \sqrt{11} \approx 3.3$ points.

Standard deviations are measured in the same units as the original data, so in Example 14.19 the standard deviation of Angela's homework scores was roughly 3.3 points. What should we make of this fact? It is clear from just a casual look at Angela's homework scores that she was pretty consistent in her homework, never straying too much above or below her average score of 90 points. The standard deviation is, in effect, a way to measure this degree of consistency (or lack thereof). A small standard deviation tells us that the data are consistent and the spread of the data is small, as is the case with Angela's homework scores.

The ultimate in consistency within a data set is when all the data values are the same (like Angela's friend Chloe, who got a 20 in every homework assignment). When this happens the standard deviation is 0. On the other hand, when there is a lot of inconsistency within the data set, we are going to get a large standard deviation. This is illustrated by Angela's other friend, Tiki, whose homework scores were 5, 15, 25, 35, 45, 55, 65, 75, 85, and 95. We would expect the standard deviation of this data set to be quite large—in fact, it is almost 29 points.

The standard deviation is arguably the most important and frequently used measure of data spread. Yet it is not a particularly intuitive concept. Here are a few basic guidelines that recap our preceding discussion:

■ The standard deviation of a data set is measured in the same units as the original data. For example, if the data are points on a test, then the standard deviation is also given in points. Conversely, if the standard deviation is given in dollars, then we can conclude that the original data must have been money—home prices, salaries, or something like that. For sure, the data couldn't have been test scores on an exam.

■ It is pointless to compare standard deviations of data sets that are given in different units. Even for data sets that are given in the same units—say, for example, test scores—the underlying scale should be the same. We should not try to compare standard deviations for SAT scores measured on a scale of 200–800 points with standard deviations of a set of homework assignments measured on a scale of 0–100 points.

■ For data sets that are based on the same underlying scale, a comparison of standard deviations can tell us something about the spread of the data. If the standard deviation is small, we can conclude that the data points are all bunched together—there is very little spread. As the standard deviation increases, we can conclude that the data points are beginning to spread out. The more spread out they are, the larger the standard deviation becomes. A standard deviation of 0, means that all data values are the same.

As a measure of spread, the standard deviation is particularly useful for analyzing real-life data. We will come to appreciate its importance in this context in Chapter 16.

CONCLUSION

Whether we like it or not, as we navigate through life in the information age, we are awash in a sea of data. Today, data are the common currency of scientific, social, and economic discourse. Powerful satellites constantly scan our planet, collecting prodigious amounts of weather, geological, and geographical data. Government agencies, such as the Bureau of the Census and the Bureau of Labor Statistics, collect millions of numbers a year about our living, working, spending, and dying habits. Even in our less serious pursuits, such as sports, we are flooded with data, not all of it great.

Faced with the common problem of data overload, statisticians and scientists have devised many ingenious ways to organize, display, and summarize large amounts of data. In this chapter we discussed some of the basic concepts in this area of statistics.

Graphical summaries of data can be produced by bar graphs, pictograms, pie charts, histograms, and so on. (There are many other types of graphical descriptions that we did not discuss in the chapter.) The kind of graph that is the most appropriate for a situation depends on many factors, and creating a good "picture" of a data set is as much an art as a science.

> Statistical reasoning will one day be as necessary for efficient citizenship as the ability to read and write.
>
> — H. G. Wells

Numerical summaries of data, when properly used, help us understand the overall pattern of a data set without getting bogged down in the details. They fall into two categories: (1) measures of location, such as the *average*, the *median*, and the *quartiles*, and (2) measures of spread, such as the *range*, the *interquartile range*, and the *standard deviation*. Sometimes we even combine numerical summaries and graphical displays, as in the case of the *box plot*. We touched upon all of these in this chapter, but the subject is a big one, and by necessity we only scratched the surface.

In this day and age, we are all consumers of data, and at one time or another, we are likely to be providers of data as well. Thus, understanding the basics of how data are organized and summarized has become an essential requirement for personal success and good citizenship.

PROFILE: W. Edwards Deming (1900–1993)

W. E. Deming was a pioneer in the application of statistics to industry. He is best known for his theories of quality control in manufacturing, which have been widely adopted, first by Japanese and later by American companies such as Xerox and Ford. Like all great ideas, Deming's ideas about quality control were built on a simple observation: All industrial processes are subject to some level of statistical variation, and this variation negatively impacts quality. From this Deming developed the principle that to improve manufacturing quality one has to reduce the causes of statistical variation.

William Edwards Deming was born in Sioux City, Iowa, in 1900, into a family of very modest means. As a young man, much of his life revolved around study and work. As an undergraduate at the University of Wyoming, he supported himself by doing all sorts of odd jobs, from janitor's aide to cleaning boilers at an oil refinery. He earned a bachelor's of science in electrical engineering from Wyoming in 1921, a master's degree in mathematics and physics from the University of Colorado in 1925, and a Ph.D. in mathematical physics from Yale University in 1928. Deming became famous as a statistician and management guru, but he was trained in classical mathematics and physics and was an accomplished musician who played several instruments and composed religious music.

With a doctorate from Yale in hand and a family to support, Deming joined the Department of Agriculture in Wash-

ington, D.C., where he performed laboratory research on fertilizers and developed statistical methods to boost productivity. In 1939, Deming joined the Census Bureau, where, as head mathematician, he helped develop many currently used statistical sampling methods. In 1946 he retired from the Census Bureau and became a statistical consultant and professor at New York University.

After World War II, the methods of *statistical process control* developed by Deming were quickly implemented by Japanese industry. Deming's methods focused on reducing variation in assembly line production, thereby minimizing defects and increasing productivity. Though a folk hero in Japan, Deming was virtually ignored by the American business community until 1980, when NBC aired the documentary "If Japan Can, Why Can't We?" detailing the superior quality and growing popularity of Japanese products. Though Deming was already 80 years old at the time, many American industrial giants belatedly adopted his techniques.

Deming remained active until his death at the age of 93, and his lectures and management seminars were attended by thousands. He was the author of many successful books and over 170 papers and was the recipient of many honors and awards. The Deming Prize, established in his honor by the Japanese Union of Scientists and Engineers, is today the highest award a company can achieve for its excellence in management.

KEY CONCEPTS

EXERCISES

WALKING

A Tables, Bar Graphs, Pie Charts, and Histograms

Exercises 1 through 4 refer to the data set shown in Table 14-12. The table shows the scores on a Chem 103 test consisting of 10 questions worth 10 points each.

TABLE 14-12	Chem 103 Test Scores		
Student ID	Score	Student ID	Score
1362	50	4315	70
1486	70	4719	70
1721	80	4951	60
1932	60	5321	60
2489	70	5872	100
2766	10	6433	50
2877	80	6921	50
2964	60	8317	70
3217	70	8854	100
3588	80	8964	80
3780	80	9158	60
3921	60	9347	60

1. Make a frequency table for the Chem 103 test scores.

2. Make a bar graph showing the actual frequencies of the scores on the test.

3. Suppose that the grading scale for the test is A: 80–100; B: 70–79; C: 60–69; D: 50–59; and F: 0–49.

   (a) Make a frequency table for the distribution of the test grades.

   (b) Make a bar graph showing the distribution of the test grades.

4. Suppose that the grading scale for the test is A: 80–100; B: 70–79; C: 60–69; D: 50–59; and F: 0–49.

   (a) What percentage of the students who took the test got a grade of D?

   (b) In a pie chart showing the distribution of the test grades, what is the size of the central angle (in degrees) of the "wedge" representing the grade of D?

Exercises 5 and 6 refer to the following situation. Every year, the first-grade students at Cleansburg Elementary are given a musical aptitude test. Based on the results of the test, the children are scored from 0 (no musical aptitude) to 5 (extremely talented). This year's results are given in Table 14-13.

TABLE 14-13	Musical Aptitude Scores of First-Grade Students					
Aptitude score	0	1	2	3	4	5
Frequency	24	16	20	12	5	3

5. Make a pie chart showing the results of the musical aptitude test.

6. Make a relative frequency bar graph showing the results of the musical aptitude test.

Exercises 7 through 10 refer to Table 14-14, which gives the home-to-school distance d (measured to the closest one-half mile) for each kindergarten student at Cleansburg Elementary School.

TABLE 14-14	Home-to-School Distance		
Student ID	d	Student ID	d
1362	1.5	3921	5.0
1486	2.0	4355	1.0
1587	1.0	4454	1.5
1877	0.0	4561	1.5
1932	1.5	5482	2.5
1946	0.0	5533	1.0
2103	2.5	5717	8.5
2877	1.0	6307	1.5
2964	0.5	6573	0.5
3491	0.0	8436	3.0
3588	0.5	8592	0.0
3711	1.5	8964	2.0
3780	2.0	9205	0.5
		9658	6.0

7. Make a bar graph for the home-to-school distances for the kindergarteners at Cleansburg Elementary School using the following class intervals:

> *Very close*: Less than 1 mile
>
> *Close*: 1 mile up to and including 1.5 miles
>
> *Nearby*: 2 miles up to and including 2.5 miles
>
> *Not too far*: 3 miles up to and including 4.5 miles
>
> *Far*: 5 miles or more

8. Make a bar graph for the home-to-school distances for the kindergarteners at Cleansburg Elementary School using the following class intervals:

> *Zone A*: 1.5 miles or less
>
> *Zone B*: more than 1.5 miles up to and including 2.5 miles
>
> *Zone C*: more than 2.5 miles up to and including 3.5 miles
>
> *Zone D*: more than 3.5 miles

9. Using the class intervals given in Exercise 7, make a pie chart for the home-to-school distances for the kindergarteners at Cleansburg Elementary School.

10. Using the class intervals given in Exercise 8, make a pie chart for the home-to-school distances for the kindergarteners at Cleansburg Elementary School.

Exercises 11 and 12 refer to the bar graph shown in Fig. 14-15 describing the scores of a group of students on a 10-point math quiz.

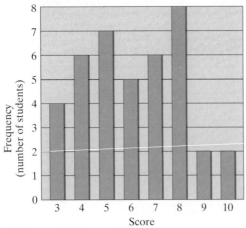

FIGURE 14-15

11. (a) How many students took the math quiz?

(b) What percentage of the students scored 2 points?

(c) If a grade of 6 or more was needed to pass the quiz, what percentage of the students passed? (Round your answer to the nearest percent.)

12. Make a relative frequency bar graph showing the results of the quiz.

Exercises 13 and 14 refer to the pie chart in Fig. 14-16.

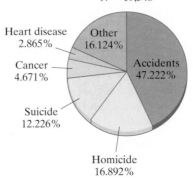

Cause of Death in U.S. Among 18- to 22-Year-Olds (2005)
$N = 19{,}548$

FIGURE 14-16 (*Source: Centers for Disease Control and Prevention, www.cdc.gov.*)

13. (a) Is cause of death a quantitative or a qualitative variable?

(b) Based on the data provided in the pie chart, estimate the number of 18- to 22-year-olds who died in the United States in 2005 due to an accident.

(c) Estimate the size of the central angle (to the nearest degree) of the slice representing the category "Other."

14. (a) Based on the data provided in the pie chart, estimate the number of 18- to 22-year-olds who died in the United States in 2005 due to cancer.

(b) Calculate the size of the central angle (to the nearest degree) for each of the slices in Fig. 14-16.

Exercises 15 and 16 refer to the pie chart in Fig. 14-17 showing the breakdown of the $2.9 trillion 2007 federal budget. Note: In the United States, 1 trillion = 1000 billion.

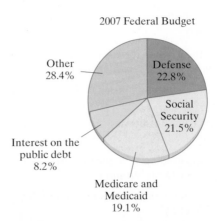

2007 Federal Budget

FIGURE 14-17 (*Source: Office of Management and Budget, www.gpo.gov/usbudget/index.html.*)

15. (a) Estimate how much money the federal government spent on defense in 2007. (Round your answer to the nearest billion.)

(b) Calculate the size of the central angle (to the nearest degree) for each of the slices in Fig. 14-17.

16. (a) Estimate how much money the federal government spent on Medicare and Medicaid in 2007. (Round your answer to the nearest billion.)

(b) Suppose that you need to draw a bar graph showing the same data shown in Fig. 14-17 using the scale $100 billion = 1 cm. Give the height, to the nearest centimeter, of the bar that represents the category "Other."

17. Table 14-15 shows the percentage of U.S. working married couples in which the wife's income is higher than the husband's (1999–2005). Using the ideas of Example 14.2, make two different-looking pictograms for the data in Table 14-15. In the first pictogram you are trying to convince your audience that things are looking great for women in the workplace and that women's salaries are catching up to men's very quickly. In the second pictogram you are trying to give a much more conservative view and convince your audience that women's salaries are catching up with men's very slowly.

TABLE 14-15	Percentage of Wives Making More Than Husbands, 1999–2005						
Year	1999	2000	2001	2002	2003	2004	2005
Percent	28.9	29.9	30.7	31.9	32.4	32.6	33.0

(*Source: Bureau of Labor Statistics, www.bls.gov.*)

18. Table 14-16 shows the percentage of U.S. workers who are members of unions (1997–2006). Using the ideas of Example 14.2, make two different-looking pictograms for the data in Table 14-16. In the first pictogram you are trying to convince your audience that unions are holding their own and that the percentage of union members in the workforce is steady. In the second bar graph you are trying to convince your audience that there is a steep decline in union membership in the U.S. workforce.

TABLE 14-16	Percentage of Unionized U.S. Workers, 1997–2006									
Year	1997	1998	1999	2000	2001	2002	2003	2004	2005	2006
Percent	14.1	13.9	13.9	13.4	13.3	13.3	12.9	12.5	12.5	12.0

(*Source: Bureau of Labor Statistics, www.bls.gov.*)

Exercises 19 and 20 refer to Table 14-17, which shows the birth weights (in ounces) of the 625 babies born in Cleansburg hospitals in 2008.

TABLE 14-17	Cleansburg Birth Weights ($N = 625$), 2008 (in ounces)				
More than	Less than or equal to	Frequencies	More than	Less than or equal to	Frequencies
48	60	15	108	120	184
60	72	24	120	132	142
72	84	41	132	144	26
84	96	67	144	156	5
96	108	119	156	168	2

19. (a) Give the length of each class interval (in ounces).

(b) Suppose that a baby weighs exactly 5 pounds 4 ounces. To what class interval does she belong? Describe the endpoint convention.

(c) Draw the histogram describing the 2008 birth weights in Cleansburg using the class intervals given in Table 14-17.

20. (a) Write a new frequency table for the birth weights in Cleansburg using class intervals of length equal to 24 ounces. Use the same endpoint convention as the one used in Table 14-17.

(b) Draw the histogram corresponding to the frequency table found in (a).

Exercises 21 and 22 refer to the two histograms shown in Fig. 14-18 summarizing the team payrolls in Major League Baseball (2008). The two histograms are based on the same data set but use slightly different class intervals. (You can assume that no team had a payroll that was exactly equal to a whole number of millions of dollars.)

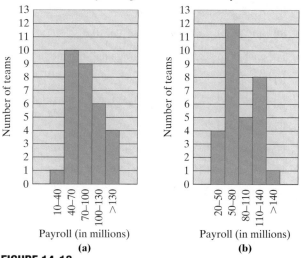

2008 Major League Baseball Team Payrolls

FIGURE 14-18

(*Source: sports.espn.go.com/mlb/teams/salaries?team=min*).

21. (a) How many teams in Major League Baseball had a payroll of more than $110 million in 2008?

(b) How many teams had a payroll of between $130 million and $140 million?

22. (a) How many teams in Major League Baseball had a payroll of less than $100 million?

(b) Given that the team with the lowest payroll (the Florida Marlins) had a payroll of $22,650,000, determine the number of teams with a payroll of between $40 million and $50 million?

B Means and Medians

23. Consider the data set $\{3, -5, 7, 4, 8, 2, 8, -3, -6\}$.

(a) Find the average.

(b) Find the median.

(c) Consider the data set $\{3, -5, 7, 4, 8, 2, 8, -3, -6, 2\}$ obtained by adding one more data point to the original data set. Find the average and median of this data set.

24. Consider the data set $\{-3.8, -7.3, -4.5, 8.3, 8.3, -9.1, -3.8, 13.2\}$.

(a) Find the average.

(b) Find the median.

(c) Consider the data set $\{-3.8, -7.3, -4.5, 8.3, 8.3, -9.1, -3.8\}$ having one less data point than the original set. Find the average and the median of this data set.

25. For each data set, find the average and the median.

(a) $\{0, 1, 2, 3, 4, 5, 6, 7, 8, 9\}$

(b) $\{1, 2, 3, 4, 5, 6, 7, 8, 9\}$

(c) $\{1, 2, 3, 4, 5, 6, 7, 8, 9, 10\}$

(d) $\{a, 2a, 3a, 4a, 5a, 6a, 7a, 8a, 9a, 10a\}$

26. For each data set, find the average and the median.

(a) $\{1, 2, 1, 2, 1, 2, 1, 2, 1, 2\}$

(b) $\{1, 2, 3, 4, 1, 2, 3, 4, 1, 2, 3, 4, 1, 2, 3, 4\}$

(c) $\{1, 2, 3, 4, 5, 5, 4, 3, 2, 1\}$

(d) $\{a, b, c, d, d, c, b, a, a, b, c, d\}$, where $a < b < c < d$.

27. For the data set $\{1, 2, 3, 4, 5, \ldots, 98, 99\}$, find

(a) the average.

(b) the median.

28. For the data set $\{1, 2, 3, 4, 5, \ldots, 997, 998, 999, 1000\}$, find

(a) the average.

(b) the median.

29. This exercise refers to the musical aptitude test given to Cleansburg Elementary first-graders first discussed in Exercises 5 and 6. The results of the test are shown once again in Table 14-18.

TABLE 14-18	Musical Aptitude Scores of First-Grade Students					
Aptitude score	0	1	2	3	4	5
Frequency	24	16	20	12	5	3

(a) Find the average aptitude score.

(b) Find the median aptitude score.

30. Table 14-19 shows the ages of the firefighters in the Cleansburg Fire Department.

TABLE 14-19	Age Distribution in Cleansburg Fire Department									
Age	25	27	28	29	30	31	32	33	37	39
Frequency	2	7	6	9	15	12	9	9	6	4

(a) Find the average age of the Cleansburg firefighters rounded to two decimal places.

(b) Find the median age of the Cleansburg firefighters.

31. Table 14-20 shows the relative frequencies of the scores of a group of students on a philosophy quiz.

TABLE 14-20	Relative Distribution of Philosophy Quiz Scores				
Score	4	5	6	7	8
Relative frequency	7%	11%	19%	24%	39%

(a) Find the average quiz score.

(b) Find the median quiz score.

32. Table 14-21 shows the relative frequencies of the scores of a group of students on a 10-point math quiz.

TABLE 14-21	Relative Distribution of Math Quiz Scores						
Score	3	4	5	6	7	8	9
Relative frequency	8%	12%	16%	20%	18%	14%	12%

(a) Find the average quiz score rounded to two decimal places.

(b) Find the median quiz score.

C Percentiles and Quartiles

33. Consider the data set $\{3, -5, 7, 4, 8, 2, 8, -3, -6\}$.

    (a) Find the first quartile.

    (b) Find the third quartile.

    (c) Consider the data set $\{3, -5, 7, 4, 8, 2, 8, -3, -6, 2\}$ obtained by adding one more data point to the original data set. Find the first and third quartiles of this data set.

34. Consider the data set $\{-3.8, -7.3, -4.5, 8.3, 8.3, -9.1, -3.8, 13.2\}$.

    (a) Find the first quartile.

    (b) Find the third quartile.

    (c) Consider the data set $\{-3.8, -7.3, -4.5, 8.3, 8.3, -9.1, -3.8\}$ obtained by deleting one data point from the original data set. Find the first and third quartiles of this data set.

35. For each data set, find the 75th and the 90th percentiles.

    (a) $\{1, 2, 3, 4, \ldots, 98, 99, 100\}$

    (b) $\{0, 1, 2, 3, 4, \ldots, 98, 99, 100\}$

    (c) $\{1, 2, 3, 4, \ldots, 98, 99\}$

    (d) $\{1, 2, 3, 4, \ldots, 98\}$

36. For each data set, find the 10th and the 25th percentiles.

    (a) $\{1, 2, 3, \ldots, 49, 50, 50, 49, \ldots, 3, 2, 1\}$

    (b) $\{1, 2, 3, \ldots, 49, 50, 49, \ldots, 3, 2, 1\}$

    (c) $\{1, 2, 3, \ldots, 49, 49, \ldots, 3, 2, 1\}$

37. This exercise refers to the age distribution in the Cleansburg Fire Department shown in Table 14-19 (Exercise 30).

    (a) Find the first quartile of the data set.

    (b) Find the third quartile of the data set.

    (c) Find the 90th percentile of the data set.

38. This exercise refers to the math quiz scores shown in Table 14-21 (Exercise 32).

    (a) Find the first quartile of the data set.

    (b) Find the third quartile of the data set.

    (c) Find the 70th percentile of the data set.

39. In 2007, a total of $N = 1,494,531$ college-bound seniors took the SAT test. Assume that the test scores are sorted from lowest to highest and that the sorted data set is $\{d_1, d_2, \ldots, d_{1,494,531}\}$.

    (a) Determine the position of the median.

    (b) Determine the position of the first quartile.

    (c) Determine the position of the third quartile.

40. In 2006, a total of $N = 1,465,744$ college-bound seniors took the SAT test. Assume that the test scores are sorted from lowest to highest and that the sorted data set is $\{d_1, d_2, \ldots, d_{1,465,744}\}$.

    (a) Determine the position of the median.

    (b) Determine the position of the first quartile.

    (c) Determine the position of the third quartile.

D Box Plots and Five-Number Summaries

41. For the data set $\{3, -5, 7, 4, 8, 2, 8, -3, -6\}$,

    (a) find the five-number summary. [Use the results of Exercises 23(b) and 33.]

    (b) draw a box plot.

42. For the data set $\{-3.8, -7.3, -4.5, 8.3, 8.3, -9.1, -3.8, 13.2\}$,

    (a) find the five-number summary. [See Exercises 24(b) and 34.]

    (b) draw a box plot.

43. This exercise refers to the distribution of ages in the Cleansburg Fire Department discussed in Exercises 30 and 37.

    (a) Find the five-number summary of the data set.

    (b) Draw a box plot.

44. This exercise refers to the distribution of math quiz scores discussed in Exercises 32 and 38.

    (a) Find the five-number summary of the data set.

    (b) Draw a box plot.

Exercises 45 and 46 refer to the two box plots in Fig. 14-19 showing the starting salaries of Tasmania State University first-year graduates in agriculture and engineering. (These are the two box plots discussed in Example 14.18.)

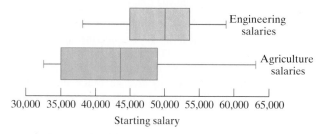

FIGURE 14-19

45. (a) Approximately what is the median salary for agriculture majors?

    (b) Approximately what is the median salary for engineering majors?

(c) Explain how we can tell that the median salary for engineering majors is more than the third quartile of the salaries for agriculture majors.

46. (a) Fill in the blank: Of the 612 engineering graduates, at most ___ had a starting salary greater than $35,000.

(b) Fill in the blank: If there were 240 agriculture graduates with starting salaries of $25,000 or less, the total number of agriculture graduates is approximately ___.

E Ranges and Interquartile Ranges

47. For the data set $\{3, -5, 7, 4, 8, 2, 8, -3, -6\}$, find

(a) the range.

(b) the interquartile range (see Exercise 33).

48. For the data set $\{-3.8, -7.3, -4.5, 8.3, 8.3, -9.1, -3.8, 13.2\}$, find

(a) the range.

(b) the interquartile range (see Exercise 34).

49. A realty company has sold $N = 341$ homes in the last year. The five-number summary for the sale prices is $Min = \$97{,}000$, $Q_1 = \$115{,}000$, $M = \$143{,}000$, $Q_3 = \$156{,}000$, and $Max = \$249{,}000$.

(a) Find the interquartile range of the home sale prices.

(b) How many homes sold for a price between $115,000 and $156,000 (inclusive)? (*Note*: If you don't believe that you have enough information to give an exact answer, you should give the answer in the form of "at least ___" or "at most ___.")

50. This exercise refers to the starting salaries of Tasmania State University first-year graduates in agriculture and engineering discussed in Exercises 45 and 46.

(a) Estimate the range for the starting salaries of agriculture majors.

(b) Estimate the interquartile range for the starting salaries of engineering majors.

(c) There were 612 engineering majors. Determine how many starting salaries of engineering majors were between $Q_1 = \$35{,}000$ and $Q_3 = \$43{,}500$ (inclusive). (*Note*: If you don't believe that you have enough information to give an exact answer, you should give the answer in the form of "at least ___" or "at most ___.")

*For Exercises 51 through 54, you should use the following definition of an outlier: An **outlier** is any data value that is above the third quartile by more than 1.5 times the IQR [Outlier $> Q_3 + 1.5(IQR)$] or below the first quartile by more than 1.5 times the IQR [Outlier $< Q_1 - 1.5(IQR)$]. (Note: There is no one universally agreed upon definition of an outlier; this is but one of several definitions used by statisticians.)*

51. Suppose that the preceding definition of outlier is applied to the Stat 101 data set discussed in Example 14.16.

(a) Fill in the blank: Any score bigger than or equal to ___ is an outlier.

(b) Fill in the blank: Any score smaller than or equal to ___ is an outlier.

(c) Find the outliers (if there are any) in the Stat 101 data set.

52. Using the preceding definition, find the outliers (if there are any) in the City of Cleansburg Fire Department data set discussed in Exercises 30 and 37. (*Hint*: Do Exercise 37 first.)

53. The distribution of the heights (in inches) of 18-year-old U.S. males has first quartile $Q_1 = 67$ in. and third quartile $Q_3 = 71$ in. Using the preceding definition, determine which heights correspond to outliers.

54. The distribution of the heights (in inches) of 18-year-old U.S. females has first quartile $Q_1 = 62.5$ in. and third quartile $Q_3 = 66$ in. Using the preceding definition, determine which heights correspond to outliers.

F Standard Deviations

The purpose of Exercises 55 through 58 is to practice computing standard deviations by using the definition. Granted, computing standard deviations this way is not the way it is generally done in practice; a good calculator (or a computer package) will do it much faster and more accurately. The point is that computing a few standard deviations the old-fashioned way should help you understand the concept a little better. If you use a calculator or a computer to answer these exercises, you are defeating their purpose.

55. Find the standard deviation of each of the following data sets.

(a) $\{5, 5, 5, 5\}$

(b) $\{0, 5, 5, 10\}$

(c) $\{0, 10, 10, 20\}$

56. Find the standard deviation of each of the following data sets.

(a) $\{3, 3, 3, 3\}$

(b) $\{0, 6, 6, 8\}$

(c) $\{-6, 0, 0, 18\}$

57. Find the standard deviation of each of the following data sets.

(a) $\{0, 1, 2, 3, 4, 5, 6, 7, 8, 9\}$

(b) $\{1, 2, 3, 4, 5, 6, 7, 8, 9, 10\}$

(c) $\{6, 7, 8, 9, 10, 11, 12, 13, 14, 15\}$

58. Find the standard deviation of each of the following data sets.

(a) $\{-3, -2, -1, 0, 1, 2, 3\}$

(b) $\{-2, -1, 0, 1, 2, 3, 4\}$

(c) $\{0, 1, 2, 3, 4, 5, 6, 7\}$

G Miscellaneous

The Mode. *The mode of a data set is the data point that occurs with the highest frequency. When there are several data points (or categories) tied for the most frequent, each of them is a mode, but if all data points have the same frequency, rather than say that every data point is a mode, it is customary to say that there is no mode.*

In Exercises 59 through 64, you should find the mode or modes of the given data sets. If there is no mode, your answer should indicate so.

59. Find the mode of the data set in Table 14-13 (Exercises 5 and 6).

60. Find the mode of the data set in Table 14-19 (Exercise 30).

61. Find the mode of the data set shown in Fig. 14-20. If there is no mode, your answer should indicate so.

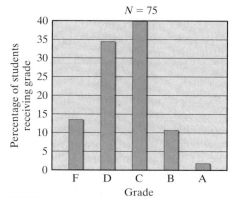

FIGURE 14-20

62. Find the mode of the data set shown in Fig. 14-21. If there is no mode, your answer should indicate so.

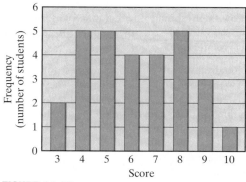

FIGURE 14-21

63. Find the mode of the data set shown in Fig. 14-22. If there is no mode, your answer should indicate so.

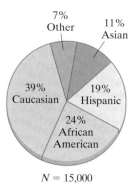

FIGURE 14-22

64. Find the mode of the data set shown in Fig. 14-23. If there is no mode, your answer should indicate so.

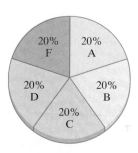

FIGURE 14-23

JOGGING

65. Mike's average on the first five exams in Econ 1A is 88. What must he earn on the next exam to raise his overall average to 90?

66. Sarah's overall average in Physics 101 was 93%. Her average was based on four exams each worth 100 points and a final worth 200 points. What is the lowest possible score she could have made on the first exam?

67. In 2006, the median SAT score was the average of $d_{732,872}$ and $d_{732,873}$, where $\{d_1, d_2, \ldots, d_N\}$ denotes the data set of all SAT scores ordered from lowest to highest. Determine the number of students N who took the SAT in 2006.

68. In 2004, the third quartile of the SAT scores was $d_{1,064,256}$, where $\{d_1, d_2, \ldots, d_N\}$ denotes the data set of all SAT scores ordered from lowest to highest. Determine the number of students N who took the SAT in 2004.

69. (a) Give an example of 10 numbers with an average less than the median.

(b) Give an example of 10 numbers with a median less than the average.

(c) Give an example of 10 numbers with an average less than the first quartile.

(d) Give an example of 10 numbers with an average more than the third quartile.

70. Suppose that the average of 10 numbers is 7.5 and that the smallest of them is $Min = 3$.

(a) What is the smallest possible value of Max?

(b) What is the largest possible value of Max?

71. This exercise refers to the 2008 payrolls of major league baseball teams summarized by the histograms in Fig. 14-24. (This is Fig. 14-18 in Exercises 21 and 22.) Using the information shown in the figure, it can be determined that the median payroll of 2008 baseball teams falls somewhere between $70 million and $80 million. Explain how.

2008 Major League Baseball Team Payrolls

FIGURE 14-24

72. A data set is called **constant** if every value in the data set is the same. Explain why any data set with standard deviation 0 must be a constant data set.

Exercises 73 and 74 refer to histograms with unequal class intervals. When sketching such histograms, the columns must be drawn so that the frequencies or percentages are proportional to the area of the column. Figure 14-25 illustrates this. If the column over class interval 1 represents 10% of the population, then the column over class interval 2, also representing 10% of the population, must be one-

third as high, because the class interval is three times as large (Fig. 14-25).

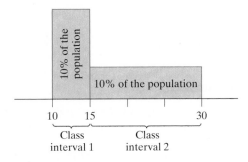

FIGURE 14-25

73. If the height of the column over the class interval 20–30 is one unit and the column represents 25% of the population, then

(a) how high should the column over the interval 30–35 be if 50% of the population falls into this class interval?

(b) how high should the column over the interval 35–45 be if 10% of the population falls into this class interval?

(c) how high should the column over the interval 45–60 be if 15% of the population falls into this class interval?

74. Two hundred senior citizens are tested for fitness and rated on their times on a one-mile walk. These ratings and associated frequencies are given in Table 14-22. Draw a histogram for these data based on the categories defined by the ratings in the table.

TABLE 14-22 One-mile walk times

Time	Rating	Frequency
6^+ to 10 minutes	Fast	10
10^+ to 16 minutes	Fit	90
16^+ to 24 minutes	Average	80
24^+ to 40 minutes	Slow	20

75. News media have accused Tasmania State University of discriminating against women in the admission policies in its schools of architecture and engineering. The *Tasmania Gazette* states that "68% of all male applicants to the schools of architecture or engineering are admitted, while only 51% of the female applicants to these same schools are admitted." The actual data are given in Table 14-23.

TABLE 14-23 Applications/admissions by gender

	School of Architecture		School of Engineering	
	Applied	Admitted	Applied	Admitted
Male	200	20	1000	800
Female	500	100	400	360

(a) What percent of the male applicants to the School of Architecture were admitted? What percent of the female applicants to this same school were admitted?

(b) What percent of the male applicants to the School of Engineering were admitted? What percent of the female applicants to this same school were admitted?

(c) How did the *Tasmania Gazette* come up with its figures?

(d) Explain how it is possible for the results in (a) and (b) and the *Tasmania Gazette* statement to all be true.

76. What happens to the five-number summary of the Stat 101 data set (see Example 14.16) if

(a) two points are added to each score?

(b) 10% is added to each score?

77. Let A denote the mean of the data set $\{x_1, x_2, x_3, \ldots, x_N\}$.

(a) Find the mean of the data set $\{x_1 + c, x_2 + c, x_3 + c, \ldots, x_N + c\}$.

(b) Use the results of (a) to explain why the mean of $\{x_1 - A, x_2 - A, x_3 - A, \ldots, x_N - A\}$ is 0. (*Note*: the data set $\{x_1 - A, x_2 - A, x_3 - A, \ldots, x_N - A\}$ represents the set of *deviations from the mean*.)

78. Let M denote the median of the data set $\{x_1, x_2, x_3, \ldots, x_N\}$. Find the median of the data set $\{x_1 + c, x_2 + c, x_3 + c, \ldots, x_N + c\}$. Explain your answer.

79. Explain why the data sets $\{x_1, x_2, x_3, \ldots, x_N\}$ and $\{x_1 + c, x_2 + c, x_3 + c, \ldots, x_N + c\}$ have

(a) the same range.

(b) the same standard deviation. (*Hint*: Try Exercise 57 or 58 first.)

RUNNING

80. Consider a data set of 10 numbers with $Min = 2$, $Max = 12$, and mean $A = 7$. Let σ denote the standard deviation of this data set.

(a) What is the smallest possible value of σ?

(b) What is the largest possible value of σ?

81. Show that the standard deviation of any set of numbers is always less than or equal to the range of the set of numbers.

82. (a) Show that if $\{x_1, x_2, x_3, \ldots, x_N\}$ is a data set with mean A and standard deviation σ, then $\sigma\sqrt{N} \geq |x_i - A|$ for every data value x_i.

(b) Use (a) to show that every data value is bigger than or equal to $A - \sigma\sqrt{N}$ and smaller than or equal to $A - \sigma\sqrt{N}$ (i.e., for every data value x, $A - \sigma\sqrt{N} \leq x \leq A + \sigma\sqrt{N}$).

83. (a) Given two numbers A and $\sigma > 0$, find two other numbers (expressed in terms of A and σ) whose mean is A and whose standard deviation is σ.

(b) Find three equally spaced numbers whose mean is A and whose standard deviation is σ.

(c) Generalize the preceding by finding N equally spaced numbers whose mean is A and whose standard deviation is σ. (*Hint*: Consider N even and N odd separately.)

84. Show that if A is the mean and M is the median of the data set $\{1, 2, 3, \ldots, N\}$, then for all values of N, $A = M$.

85. Show that if A is the mean and M is the median of a data set consisting of the first N terms of an arithmetic sequence, then $A = M$. (*Note*: Arithmetic sequences are discussed in Mini-Excursion 3.)

86. Suppose that the mean of the numbers $x_1, x_2, x_3, \ldots, x_N$ is A and that the variance of these same numbers is V. Suppose also that the mean of the numbers $x_1^2, x_2^2, x_3^2, \ldots, x_N^2$ is B. Show that $V = B - A^2$. (In other words, for any data set, if we take the mean of the squared data values and subtract the square of the mean of the data values, we get the variance.)

87. Suppose that the standard deviation of the data set $\{x_1, x_2, x_3, \ldots, x_N\}$ is σ. Explain why the standard deviation of the data set $\{a \cdot x_1, a \cdot x_2, a \cdot x_3, \ldots, a \cdot x_N\}$ (where a is a positive number) is $a \cdot \sigma$.

88. Using the formula $1^2 + 2^2 + 3^2 + \cdots + N^2 = N(N+1)(2N+1)/6$,

(a) find the standard deviation of the data set $\{1, 2, 3 \ldots, 98, 99\}$. (*Hint*: Use Exercise 85.)

(b) find the standard deviation of the data set $\{1, 2, 3 \ldots, N\}$.

89. (a) Find the standard deviation of the data set $\{315, 316, \ldots, 412, 413\}$. [*Hint*: Use Exercises 79(b) and 88.]

(b) Find the standard deviation of the data set $\{k + 1, k + 2, \ldots, k + N\}$.

90. Chebyshev's theorem. The Russian mathematician P. L. Chebyshev (1821–1894) showed that for any data set and any constant k greater than 1, at least $1 - (1/k^2)$ of the data must lie within k standard deviations on either side of the mean A. For example, when $k = 2$, this says that $1 - \dfrac{1}{4} = \dfrac{3}{4}$ (i.e., 75%) of the data must lie within two standard deviations of A (i.e., somewhere between $A - 2 \cdot \sigma$ and $A + 2 \cdot \sigma$).

(a) Using Chebyshev's theorem, what percentage of a data set must lie within three standard deviations of the mean?

(b) How many standard deviations on each side of the mean must we take to be assured of including 99% of the data?

(c) Suppose that the average of a data set is A. Explain why there is no number k of standard deviations for which we can be certain that 100% of the data lies within k standard deviations on either side of the mean A.

PROJECTS AND PAPERS

A Lies, Damn Lies, and Statistics

Statistics are often used to exaggerate, distort, and misinform, and this is most commonly done by the misuse of graphs and charts. In this project you are to discuss the different graphical "tricks" that can be used to mislead or slant the information presented in a picture. Attempt to include items from recent newspapers, magazines, and other media.

Note: *References 2, 9, and 14 are good sources for this project.*

B Data in Your Daily Life

Which month is the best one to invest in the stock market? During which day of the week are you most likely to get into an automobile accident? Which airline is the safest to travel with? Who is the best place-kicker in the National Football League?

In this project, you are to formulate a question from everyday life (similar to one of the aforementioned questions) that is amenable to a statistical analysis. Then you will need to find relevant data that attempt to answer this question and summarize these data using the methods discussed in this chapter. Present your final conclusions and defend them using appropriate charts and graphs.

C Book Review: *Curve Ball*

Baseball is the ultimate statistical sport, and statistics are as much a part of baseball as peanuts and Cracker Jack. The book *Curve Ball: Baseball, Statistics, and the Role of Chance in the Game* (reference 1), written by statisticians and baseball lovers Jim Albert and Jay Bennett, is a compilation of all sorts of fascinating statistical baseball issues, from how to better measure a player's offensive performance to what is the true value of home field advantage.

In this project you are to pick one of the many topics discussed in *Curve Ball* and write an analysis paper on the topic.

REFERENCES AND FURTHER READINGS

1. Albert, Jim, and Jay Bennett, *Curve Ball: Baseball, Statistics, and the Role of Chance in the Game*. New York: Springer-Verlag, 2001, chap. 2.
2. Cleveland, W. S., *The Elements of Graphing Data*, rev. ed. New York: Van Nostrand Reinhold Co., 1994.
3. Cleveland, W. S., *Visualizing Data*. Summit, NJ: Hobart Press, 1993.
4. Gabor, Andrea, *The Man Who Discovered Quality*. New York: Penguin Books, 1992.
5. Harris, Robert, *Information Graphics: A Comprehensive Illustrated Reference*. New York: Oxford University Press, 2000.
6. Mosteller, F., and W. Kruskal, et al., *Statistics by Example: Exploring Data*. Reading, MA: Addison-Wesley, 1973.
7. Tanner, Martin, *Investigations for a Course in Statistics*. New York: Macmillan Publishing Co., 1990.
8. Tufte, Edward, *Envisioning Information*. Cheshire, CT: Graphics Press, 1990.

9. Tufte, Edward, *The Visual Display of Quantitative Information*. Cheshire, CT: Graphics Press, 1983.

10. Tufte, Edward, *Visual Explanations: Images and Quantities, Evidence and Narrative*. Cheshire, CT: Graphics Press, 1997.

11. Tukey, John W., *Exploratory Data Analysis*. Reading, MA: Addison-Wesley, 1977.

12. Utts, Jessica, *Seeing Through Statistics*. Belmont, CA: Wadsworth Publishing Co., 1996.

13. Wainer, H., "How to Display Data Badly," *The American Statistician*, 38 (1984), 137–147.

14. Wainer, H., *Visual Revelations*. New York: Springer-Verlag, 1997.

15. Wildbur, Peter, *Information Graphics*. New York: Van Nostrand Reinhold Co., 1989.

15 Chances, Probabilities, and Odds

Measuring Uncertainty

With the possible exception of death and taxes, pretty much everything else that happens to us in our lives is layered with some degree of *uncertainty*. That's why we pay attention to the weather report, buy insurance, and constantly wish our friends "good luck."

Although we are all familiar with uncertainty, we don't always have a good grasp on how to measure it. Some situations involve only a small degree of uncertainty ("I'm pretty sure I aced the midterm"), some situations involve a large degree of uncertainty ("I have no clue how I did in that midterm"), and some situations fall in between ("I think I did well in the midterm, but . . ."). To quantify and measure more precisely the amount of uncertainty in the many uncertain events that affect our lives we use the concept of *probability*.

<blockquote>
❝ Nothing in life is certain except death and taxes. ❞
—Ben Franklin
</blockquote>

*C*hance, *probability, odds*—we use these words carelessly in casual conversation, and most of the time we can get away with it. Technically speaking, however, each of these words represents a slightly different way to measure the *likelihood* of an event. When we speak of the *chance* of some event happening we express it in terms of percentages, such as when the weatherperson reports that "the chance of rain tomorrow is 40%." When we speak of the *probability* of some event happening we express it in terms of a ratio ("the probability is 2 out of 5") or, equivalently as a fraction $\left(\frac{2}{5}\right)$ or a decimal (0.4). Finally, when we speak of the *odds in favor* of an event, we use a pair of numbers, as in "the odds of the Red Sox winning the World Series are 2 to 3."

In this chapter we will learn how to interpret and work with probabilities, chances, and odds in a formal mathematical context. This will be our very brief introduction to the mathematical theory of probability, a relatively young branch of mathematics that has become critically important to many aspects of modern life. Insurance, public health, science, sports, gambling, the stock market—wherever there is uncertainty to be tamed—the mathematical theory of probability plays a significant role.

Our discussion in this chapter is divided into two parts. In the first part we introduce the basic theoretical framework needed for a meaningful discussion of probabilities: the concepts of *random experiment* and *sample space* (Section 15.1), the basic rules of *counting* (Section 15.2), and the dual concepts of *permutation* and *combination* (Section 15.3). In the second part of the chapter we discuss general *probability spaces* (Section 15.4) and probabilities in spaces in which all outcomes are equally likely (Section 15.5). The chapter concludes with a brief discussion of *odds* and their relationship to probabilities (Section 15.6).

15.1 Random Experiments and Sample Spaces

In broad terms, probability is the *quantification of uncertainty*. To understand what that means, we may start by formalizing the notion of uncertainty.

We will use the term **random experiment** to describe an activity or a process *whose outcome cannot be predicted ahead of time*. Typical examples of random experiments are tossing a coin, rolling a pair of dice, drawing cards out of a deck of cards, predicting the result of a football game, and forecasting the path of a hurricane.

Associated with every random experiment is the *set* of all of its possible outcomes, called the **sample space** of the experiment. For the sake of simplicity, we will concentrate on experiments for which there is only a finite set of outcomes, although experiments with infinitely many outcomes are both possible and important.

We illustrate the importance of the sample space by means of several examples. Since the sample space of any experiment is a set of outcomes, we will use set notation to describe it. We will consistently use the letter S to denote a sample space and the letter N to denote the *size* of the sample space S (i.e., the number of outcomes in S).

EXAMPLE 15.1   **Tossing a Coin**

One simple random experiment is to *toss a quarter* and *observe whether it lands heads or tails*. The sample space can be described by $S = \{H, T\}$, where H stands for *Heads* and T for *Tails*. Here $N = 2$.

A couple of comments about coins are in order here. First, the fact that the coin in Example 15.1 was a quarter is essentially irrelevant. Practically all coins have an obvious "heads" side (and thus a "tails" side), and even when they don't—as in a "buffalo nickel"—we can agree ahead of time which side is which. Second, there are fake coins out there on which both sides are "heads." Tossing such a coin does not fit our definition of a random experiment, so from now on, we will assume that all coins used in our experiments have two different sides, which we will call H and T.

EXAMPLE 15.2   **More Coin Tossing**

Suppose we toss a coin *twice* and *record* the outcome of each toss (H or T) in the order it happens. The sample space now is $S = \{HH, HT, TH, TT\}$, where HT means that the first toss came up H and the second toss came up T, which is a different outcome from TH (first toss T and second toss H). In this sample space $N = 4$.

Suppose now we *toss two distinguishable coins* (say, a nickel and a quarter) *at the same time* (tricky but definitely possible). This random experiment appears different from the one where we toss one coin twice, but the sample space is still $S = \{HH, HT, TH, TT\}$. (Here we must agree what the order of the symbols is—for example, the first symbol describes the quarter and the second the nickel.)

Since they have the same sample space, we will consider the two random experiments just described as the same random experiment.

Now let's consider a different random experiment. We are still tossing a coin twice, but we only care now about the *number of heads* that come up. Here there are only three possible outcomes (no heads, one head, or both heads), and symbolically we might describe this sample space as $S = \{0, 1, 2\}$.　　⬭

The important point made in Example 15.2 is that a random experiment is defined by two things: the action (such as tossing coins) and what it is that we are interested in observing from the action.

(**EXAMPLE 15.3**)　**Shooting Free Throws**

Here is a familiar scenario: Your favorite basketball team is down by 1, clock running out, and one of your players is fouled and goes to the line to shoot free throws, with the game riding on the outcome. It's not a good time to think of sample spaces, but let's do it anyway.

Clearly the shooting of free throws is a random experiment, but what is the sample space? As in Example 15.2, the answer depends on a few subtleties.

In one scenario (the *penalty situation*) your player is going to shoot two free throws no matter what. In this case one could argue that what really matters is how many free throws he or she makes (make both and win the game, miss one and tie and go to overtime, miss both and lose the game). When we look at it this way the sample space is $S = \{0, 1, 2\}$.

A somewhat more stressful scenario is when your player is shooting a *one-and-one*. This means that the player gets to shoot the second free throw only if he or she makes the first one. In this case there are also three possible outcomes, but the circumstances are different because the order of events is relevant (miss the first free throw and lose the game, make the first free throw but miss the second one and tie the game, make both and win the game). We can describe this sample space as $S = \{f, sf, ss\}$, where we use f to indicate *failure* (missed the free throw) and s to indicate *success*.　　⬭

We will now discuss a couple of examples of random experiments involving dice. A die is a cube, usually made of plastic, whose six faces are marked with dots (from 1 to 6) called "pips." Random experiments using dice have a long-standing tradition in our culture and are a part of both gambling and recreational games such as Monopoly or Yahtzee.

(**EXAMPLE 15.4**)　**Rolling a Pair of Dice**

The most common scenario when rolling a pair of dice is to only consider the *total* of the two numbers rolled. In this situation we don't really care how a particular total comes about. We can "roll a seven" in various paired combinations—a 3 and a 4, a 2 and a 5, a 1 and a 6. No matter how the individual dice come up, the only thing that matters is the total rolled.

The possible outcomes in this scenario range from "rolling a two" to "rolling a twelve," and the sample space can be described by $S = \{2, 3, 4, 5, 6, 7, 8, 9, 10, 11, 12\}$.　　⬭

(**EXAMPLE 15.5**)　**More Dice Rolling**

A more general scenario when rolling a pair of dice is when we do care what number each individual die turns up (in certain bets in craps, for example, two

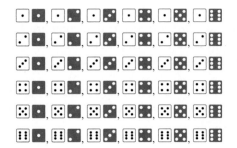

FIGURE 15-1

"fours" is a winner but a "five" and a "three" is not). Here we have a sample space with 36 different outcomes, shown in Fig. 15-1. Notice that in Fig. 15-1 the dice are colored white and red, a symbolic way to emphasize the fact that we are treating the dice as *distinguishable* objects. This is why the rolls ⚀⚄ and ⚄⚀ are considered different outcomes.

> **EXAMPLE 15.6**   **Ranking the Candidates in an Election**

Five candidates (A, B, C, D, and E) are running in an election. The top three vote getters are chosen President, Vice President, and Secretary, in that order. Such an election can be considered a random experiment with sample space $S = \{ABC, ACB, BAC, BCA, CAB, CBA, ABD, ADB, \ldots, CDE\}$. We will assume that outcome ABC denotes that candidate A is elected President, B is elected Vice President, and C is elected Secretary. This is important because the outcomes ABC and BAC are different outcomes and must appear separately in the sample space.

Because this sample space is a bit too large to write out in full (we will soon learn that its size is $N = 60$), we use the "..." notation. It is a way of saying "and so on—you get the point."

Example 15.6 illustrates the point that sample spaces can have a lot of outcomes and that we are often justified in not writing each and every one of them down. This is where the "..." comes in handy. The key thing is to understand what the sample space looks like without necessarily writing all the outcomes down. Our real goal is to find N, the size of the sample space. If we can do it without having to list all the outcomes, then so much the better. We will discuss how this is done in the next two sections.

15.2   Counting Outcomes in Sample Spaces

> **EXAMPLE 15.7**   **Tossing More Coins**

If we *toss a coin three times* and separately record the outcome of each toss, the sample space is given by $S = \{HHH, HHT, HTH, HTT, THH, THT, TTH, TTT\}$. Here we can just count the outcomes and get $N = 8$.

Now let's look at an expanded version of the same idea and *toss a coin 10 times* (and record the outcome of each toss). In this case the sample space S is too big to write down, but we can "count" the number of outcomes in S without having to tally them one by one. Here is how we do it.

Each outcome in S can be described by a string of 10 consecutive letters, where each letter is either H or T. (This is a natural extension of the ideas introduced in Example 15.2.) For example, $THHTHTHHTT$ represents a *single* outcome in our sample space—the one in which the first toss came up T, the second toss came up H, the third toss came up H, and so on. To count all the outcomes, we will argue as follows: There are two choices for the first letter (H or T), two choices for the second letter, two choices for the third letter, and so on down to the tenth letter. The total number of possible strings is found by *multiplying* the number of choices for each letter. Thus $N = 2^{10} = 1024$.

The rule we used in Example 15.6 is an old friend from Chapter 2—the **multiplication rule**, and for a formal definition of the multiplication rule the reader is referred to Section 2.4. Informally, the multiplication rule simply says that *when something is done in stages, the number of ways it can be done is found by multiplying the number of ways each of the stages can be done.* The easiest way to understand this simple but powerful idea is by looking at some examples.

(**EXAMPLE 15.8**)   **The Making of a Wardrobe**

Dolores is a young saleswoman planning her next business trip. She is thinking about packing three different pairs of shoes, four skirts, six blouses, and two jackets. If all the items are color coordinated, how many different *outfits* will she be able to create by combining these items?

To answer this question, we must first define what we mean by an "outfit." Let's assume that an outfit consists of one pair of shoes, one skirt, one blouse, and one jacket. Then to make an outfit Dolores must choose a pair of shoes (three choices), a skirt (four choices), a blouse (six choices), and a jacket (two choices). By the multiplication rule the total number of possible outfits is $3 \times 4 \times 6 \times 2 = 144$. (Think about it—Dolores can be on the road for over four months and never have to wear the same outfit twice! And it all fits in a small suitcase.)

(**EXAMPLE 15.9**)   **The Making of a Wardrobe: Part 2**

Once again, Dolores is packing for a business trip. This time, she packs three pairs of shoes, four skirts, three pairs of slacks, six blouses, three turtlenecks, and two jackets. As before, we can assume that she coordinates the colors so that everything goes with everything else. This time, we will define an outfit as consisting of a pair of shoes, a choice of "lower wear" (either a skirt *or* a pair of slacks), and a choice of "upper wear" (it could be a blouse *or* a turtleneck *or both*), and, finally, she may or may not choose to wear a jacket. How many different such outfits are possible?

This is a more sophisticated variation of Example 15.8. Our strategy will be to think of an outfit as being put together in stages and to draw a box for each of the stages. We then separately count the number of choices at each stage and enter that number in the corresponding box. (Some of these calculations can themselves be mini-counting problems.) The last step is to multiply the numbers in each box. The details are illustrated in Fig. 15-2. The final count for the number of different outfits is $N = 3 \times 7 \times 27 \times 3 = 1701$.

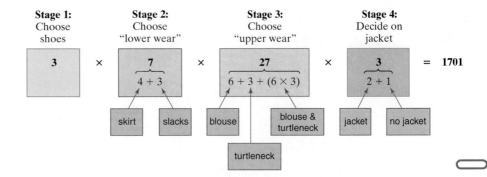

FIGURE 15-2

The method of drawing boxes representing the successive stages in a process, and putting the number of choices for each stage inside the box is a convenient strategy that often helps clarify one's thinking. Silly as it may seem, we strongly recommend it. For ease of reference, we will call it the *box model* for counting.

(**EXAMPLE 15.10**)   **Ranking the Candidates in an Election: Part 2**

This example is a follow-up to Example 15.6. Five candidates are running in an election, with the top three vote getters elected (in order) as President, Vice President, and Secretary. We want to know how big the sample space is. Using a box model, we see that this becomes a reasonably easy counting problem, as illustrated in Fig. 15-3.

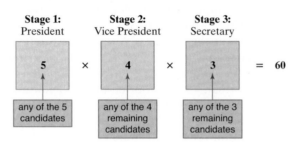

FIGURE 15-3

15.3 Permutations and Combinations

Many counting problems can be reduced to a question of counting the number of ways in which we can choose groups of objects from a larger group of objects. Often these problems require somewhat more sophisticated counting methods than the plain vanilla multiplication rule. In this section we will discuss the dual concepts of *permutation* (a group of objects in which the ordering of the objects within the group *makes a difference*) and *combination* (a group of objects in which the ordering of the objects is *irrelevant*).

(**EXAMPLE 15.11**)   **The Pleasures of Ice Cream**

Baskin-Robbins offers 31 different flavors of ice cream. A "true double" is the name we will use for two scoops of ice cream of two *different* flavors. Say you want a true double in a bowl—how many different choices do you have?

   The natural impulse is to count the number of choices using the multiplication rule (and a box model) as shown in Fig. 15-4. This would give an answer of $31 \times 30 = 930$ true doubles. Unfortunately, this answer is *double counting* each of the true doubles. Why?

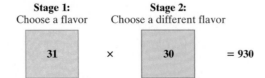

FIGURE 15-4

When we use the multiplication rule, *there is a well-defined order to things*, and a scoop of strawberry followed by a scoop of chocolate is counted separately from a scoop of chocolate followed by a scoop of strawberry. But by all reasonable standards, a bowl of chocolate-strawberry is the same as a bowl of strawberry-chocolate. (This is not necessarily true when the ice cream is served in a cone. Fussy people can be very picky about the order of the flavors when the scoops are stacked up.)

The good news is that now that we understand why the count of 930 is wrong we can fix it. All we have to do to get the correct answer is to divide the original count by 2. It follows that the number of true double choices at Baskin-Robbins is $(31 \times 30)/2 = 465$.

Example 15.11 is an important one. It warns us that we have to be careful about how we use the multiplication rule and box models in counting problems where the order in which we choose the objects (ice cream flavors) does not affect the answer. Let's take this idea to the next level.

EXAMPLE 15.12 **The Pleasures of Ice Cream: Part 2**

Some days you have a real craving for ice cream, and on such days you like to go to Baskin-Robbins and order a *true triple* (in a bowl). How many different choices do you have? (As you might have guessed, a "true triple" consists of three scoops of ice cream each of a different flavor.)

Starting with the multiplication rule, we have 31 choices for the "first" flavor, 30 choices for the "second" flavor, and 29 choices for the "third" flavor, for an apparent grand total of $31 \times 30 \times 29 = 26{,}970$. But once again this answer counts each true triple more than once; in fact, it does so *six times*! (More on that shortly.) If we accept this, the correct answer must be $26{,}970/6 = 4495$.

Why is it that the count of 26,970 counts each true triple *six* times? The answer comes down to this: Any three flavors (call them X, Y, and Z) can be listed in $3 \times 2 \times 1 = 6$ different ways (XYZ, XZY, YXZ, YZX, ZXY, and ZYX). The multiplication rule counts each of these separately, but when you think in terms of ice cream scoops in a bowl, they are the same true triple regardless of the order.

The bottom line is that there are 4495 different possibilities for a true triple at Baskin-Robbins. We can better understand where this number comes from by looking at it in its raw, uncalculated form $[(31 \times 30 \times 29)/(3 \times 2 \times 1) = 4495]$. The numerator ($31 \times 30 \times 29$) comes from counting *ordered* triples using the multiplication rule; the denominator ($3 \times 2 \times 1$) comes from counting the number of ways in which three things (in this case, the three flavors in a triple) can be rearranged. The denominator $3 \times 2 \times 1$ is already familiar to us—it is the *factorial* of 3. (We discussed the factorial in Chapters 2 and 6, so we won't dwell on it here.)

Our next example will deal with the game of poker. Despite the great deal of exposure poker gets on television these days, there are a lot of misconceptions about the mathematics behind poker hands. For readers not familiar with the game, poker is a betting game played with a standard deck of cards [a standard deck of cards has 52 cards divided into 4 *suits* (clubs, diamonds, hearts, and spades) and with 13 *values* in each suit (2, 3, ..., 10, J, Q, K, A)]. There are many variations of poker depending on how many cards are dealt and whether all the cards are *down* cards (only the player receiving the card can see it) or some of the cards are *up* cards (dealt face up so that they can be seen by all the players).

(EXAMPLE 15.13)   **Five-Card Poker Hands**

In this example we will compare two types of games: five-card *stud poker* and five-card *draw poker*. In both of these games a player ends up with five cards, but there is an important difference when analyzing the mathematics behind the games: In five-card *draw* the order in which the cards come up is irrelevant; in five-card *stud* the order in which the cards come up is extremely relevant. The reason for this is that in five-card draw all cards are dealt down, but in five-card stud only the first card is dealt down—the remaining four cards are dealt up, one at a time. This means that players can assess the relative strengths of the other players' hands as the game progresses and play their hands accordingly.

Left: Five-card *draw* poker hand: The order in which the cards are dealt is irrelevant. *Center and right*: Five-card *stud* poker hands. The order of the up cards is important. The hand on the right is better than the hand in the center.

Counting the number of five-card *stud* poker hands is a direct application of the multiplication rule: 52 possibilities for the first card, 51 for the second card, 50 for the third card, 49 for the fourth card, and 48 for the fifth card, for an awesome total of $52 \times 51 \times 50 \times 49 \times 48 = 311{,}875{,}200$ possible hands.

Counting the number of five-card *draw* poker hands requires a little more finesse. Here a player gets five down cards and the hand is the same regardless of the order in which the cards are dealt. There are $5! = 120$ different ways in which the same set of five cards can be ordered, so that one draw hand corresponds to 120 different stud hands. Thus, the stud hands count is exactly 120 times bigger than the draw hands count. Great! All we have to do then is divide the 311,875,200 (number of stud hands) by 120 and get our answer: There are 2,598,960 possible five-card draw hands. [As before, it's more telling to look at this answer in the uncalculated form $(52 \times 51 \times 50 \times 49 \times 48)/5! = 2{,}598{,}960$.]   ⬭

We are now ready to generalize the ideas developed in Examples 15.12 and 15.13. Suppose that we have a set of n distinct objects and we want to select r different objects from this set. The number of ways that this can be done depends on whether the selections are *ordered* or *unordered*. Ordered selections are the generalization of stud poker hands—selecting the same objects in different order gets you something different. Unordered selections are the generalization of draw poker hands—selecting the same objects in different order gets you nothing new. To distinguish between these two scenarios, we use the terms **permutation** to describe an ordered selection and **combination** to describe an unordered selection. (One way to remember which is which is to remember that there are many more permutations than there are combinations.)

■ You should take a look at your calculator and figure out the correct sequence of keystrokes to compute these numbers, and then try Exercises 31 and 32.

For a given number of objects n and a given selection size r (where $0 \leq r \leq n$), we can talk about the "number of permutations of n objects taken r at a time" and the "number of combinations of n objects taken r at a time," and these two extremely important families of numbers are denoted $_nP_r$ and $_nC_r$, respectively. (Some calculators use variations of this notation, such as $P_{n,r}$ and $C_{n,r}$, respectively.)

A summary of the essential facts about the numbers $_nP_r$ and $_nC_r$ is given in Table 15-1.

TABLE 15-1	Permutations and Combinations	
Notation	$_nP_r$	$_nC_r$
Formula 1	$_nP_r = n(n-1)(n-2)\cdots(n-r+1)$	$_nC_r = \dfrac{n(n-1)(n-2)\cdots(n-r+1)}{r!}$
Formula 2	$_nP_r = \dfrac{n!}{(n-r)!}$	$_nC_r = \dfrac{n!}{(n-r)!r!}$
Applications	Stud poker hands, rankings, committees with assignments	Draw poker hands, lottery tickets, coalitions, subsets

EXAMPLE 15.14 The Florida Lotto

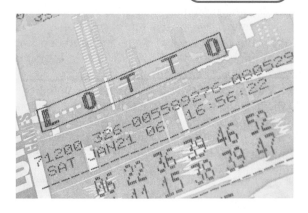

Like many other state lotteries, the Florida Lotto is a game in which for a small investment of just one dollar a player has a chance of winning tens of millions of dollars. Enormous upside, hardly any downside—that's why people love playing the lottery and, like they say, "everybody has to have a dream." But, in general, lotteries are a very bad investment, even if it's only a dollar, and the dreams can turn to nightmares. Why so?

In a Florida Lotto ticket, one gets to select *six* numbers from 1 through 53. To win the jackpot (there are other lesser prizes we won't discuss here), those six numbers have to match the winning numbers drawn by the lottery in any order. Since a lottery draw is just an unordered selection of six objects (the winning numbers) out of 53 objects (the numbers 1 through 53), the number of possible draws is $_{53}C_6$ (see Table 15-1). Doesn't sound too bad until we do the computation (or use a calculator) and realize that

$$_{53}C_6 = \frac{53 \times 52 \times 51 \times 50 \times 49 \times 48}{6!} = 22,957,480.$$

15.4 Probability Spaces

What Is a Probability?

If we toss a coin in the air, *what is the probability that it will land heads up*? This one is not a very profound question, and almost everybody agrees on the answer, although not necessarily for the same reason. The standard answer given is 1 out of 2, or 1/2. But why is the answer 1/2, and what does such an answer mean?

One common explanation given for the answer of 1/2 is that when we toss a coin, there are two possible outcomes (H and T), and since H represents one of the two possibilities, the probability of the outcome H must be 1 out of 2 or 1/2. This logic, while correct in the case of an honest coin, has a lot of holes in it.

Consider how the same argument would sound in a different scenario.

(**EXAMPLE 15.15**)   **Who Is Shooting Those Free Throws?**

Imagine an NBA player in the act of shooting a free throw. Just like with a coin toss, there are two possible outcomes to the free-throw shot (success or failure), but it would be absurd to conclude that the probability of making the free throw is therefore 1 out of 2, or 1/2. Here the two outcomes are not both equally likely, and their probabilities should reflect that.

The probability of a basketball player making a free throw very much depends on the abilities of the player doing the shooting—it makes a difference if it's Steve Nash or Shaquille O'Neal. Nash is one of the best free-throw shooters in the history of the NBA, with a career average of 90%, while Shaq is a notoriously poor free-throw shooter (52% career average). These percentages, over the long term, represent an approximation of the true probability each of them has of making a free throw—about 0.90 and 0.52, respectively. In either case, the probability is not 0.5.

Example 15.15 leads us to what is known as the *empirical* interpretation of the concept of probability. Under this interpretation when tossing an honest coin the probability of *Heads* is 1/2 not because *Heads* is one out of two possible outcomes but because, if we were to toss the coin over and over—hundreds, possibly thousands, of times—in the long run about half of the tosses will turn out to be heads, a fact that has been confirmed by experiment many times.

The argument as to exactly how to interpret the statement "the probability of X is such and such" goes back to the late 1600s, and it wasn't until the 1930s that a formal theory for dealing with probabilities was developed by the Russian mathematician A. N. Kolmogorov (1903–1987). This theory has made probability one of the most useful and important concepts in modern mathematics. In the remainder of this chapter we will discuss some of the basic concepts of *probability theory*.

Events

An **event** is any subset of the sample space. That is, an event is any set of individual outcomes. (This definition includes the possibility of an "event" that has no outcomes as well as events consisting of a single outcome.) By definition, events are sets (subsets of the sample space), and we will deal with events using set notation as well as the basic rules of set theory.

A convenient way to think of an event is as a package in which outcomes with some common characteristic are bundled together. Say you are rolling a pair of dice and hoping that on the next roll "you roll a 7"—that would make you temporarily rich. The best way to describe what really matters (to you) is by packaging together all the different ways to "roll a 7" as a single event E:

$$E = \{ \boxdot\boxed{::}, \boxed{\because}\boxed{\because}, \boxed{\therefore}\boxed{\vdots}, \boxed{::}\boxed{\therefore}, \boxed{\ddots}\boxed{\cdot\cdot}, \boxed{::}\boxed{\cdot} \}$$

Sometimes an event consists of just one outcome. We will call such an event a **simple event**. In some sense simple events are the building blocks for all other events (more on that later). There is also the special case of the empty set { }, corresponding to an event with no outcomes. Such an event can never happen, and thus we call it the **impossible event**.

EXAMPLE 15.16 **Coin-Tossing Events**

Let's revisit the experiment of tossing a coin three times and recording the result of each toss (Example 15.7). The sample space for this experiment is $S = \{HHH, HHT, HTH, HTT, THH, THT, TTH, TTT\}$. The set S has hundreds of subsets (256 to be exact), each of which represents a different event. Table 15-2 shows just a few of these events.

TABLE 15-2

Event (in words)	Event (as a set)
Toss 2 or more heads.	$\{HHT, HTH, THH, HHH\}$
Toss more than 2 heads.	$\{HHH\}$
Toss 2 heads or fewer.	$\{TTT, TTH, THT, HTT, THH, HTH, HHT\}$
Toss no tails.	$\{HHH\}$
Toss exactly 1 tail.	$\{HHT, HTH, THH\}$
Toss exactly 1 head.	$\{HTT, THT, TTH\}$
First toss is heads.	$\{HHH, HHT, HTH, HTT\}$
Toss same number of heads as tails.	$\{\ \}$ (*Note:* This is the *impossible event.*)
Toss 3 heads or fewer.	S (*Note:* This event is called the *certain event.*)

Probability Assignments

Let's return to free-throw shooting, as it is a useful metaphor for many probability questions.

EXAMPLE 15.17 **The Unknown Free-Throw Shooter**

A player is going to shoot a free throw. We know nothing about his or her abilities—for all we know, the player could be Steve Nash, or Shaquille O'Neal, or Joe Schmoe, or you, or me.

How can we describe the probability that he or she will make that free throw? It seems that there is no way to answer this question, since we know nothing about the ability of the shooter. We could argue that the probability could be just about any number between 0 and 1. No problem—we make our unknown probability a variable, say p.

What can we say about the probability that our shooter misses the free throw? A lot. Since there are only two possible outcomes in the sample space $S = \{s, f\}$, the probability of success (s) and the probability of failure (f) must complement each other—in other words, must add up to 1. This means that the probability of missing the free throw must be $1 - p$.

Table 15-3 is a summary of the line on a generic free-throw shooter. Humble as it may seem, Table 15-3 gives a complete model of free-throw shooting. It works when the free-throw shooter is Steve Nash (make it $p = 0.90$), Shaquille O'Neal (make it $p = 0.52$), or the author of this book (make it $p = 0.30$). Each one of the choices results in a different assignment of numbers to the outcomes in the sample space.

TABLE 15-3

Event	Probability
$\{\ \}$	0
$\{s\}$	p
$\{f\}$	$1 - p$
$\{s, f\}$	1

Example 15.17 illustrates the concept of a *probability assignment*. A **probability assignment** is a function that assigns to each event E a number between

0 and 1, which represents the probability of the event E and which we denote by $\Pr(E)$. A probability assignment always assigns probability 0 to the *impossible event* $[\Pr(\{\ \}) = 0]$ and probability 1 to the whole sample space $[\Pr(S) = 1]$.

■ In the case of a simple event $\{a\}$ we cheat a little and for the sake of simplicity we speak of "the probability of the outcome a" when we really should say "the probability of the event $\{a\}$." Accordingly, we will write $\Pr(a)$ in lieu of the technically correct but awkward $\Pr(\{a\})$.

With finite sample spaces a probability assignment is defined by assigning probabilities to just the simple events in the sample space. Once we do this, we can find the probability of any event by simply *adding the probabilities of the individual outcomes that make up that event.* There are only two requirements for a valid probability assignment: (1) *All probabilities are numbers between 0 and 1,* and (2) *the sum of the probabilities of the simple events equals 1.*

When bookmakers or professional odds-makers handicap a sporting event, they do so by essentially defining a probability assignment for the sample space of all possible outcomes of that event. The next example is a simple illustration of how this might be done.

(**EXAMPLE 15.18**) **Handicapping a Tennis Tournament**

There are six players playing in a tennis tournament: A (Russian, female), B (Croatian, male), C (Australian, male), D (Swiss, male), E (American, female), and F (American, female).

To handicap the winner of the tournament we need a probability assignment on the sample space $S = \{A, B, C, D, E, F\}$. With sporting events the probability assignment is subjective (it reflects an opinion), but professional odds-makers are usually very good at getting close to the right probabilities. For example, imagine that a professional odds-maker comes up with the following probability assignment: $\Pr(A) = 0.08$, $\Pr(B) = 0.16$, $\Pr(C) = 0.20$, $\Pr(D) = 0.25$, $\Pr(E) = 0.16$. [We are missing $\Pr(F)$ from the list, but since the probabilities of the simple events must add up to 1, we can do the arithmetic: $\Pr(F) = 0.15$.]

Once we have the probabilities of the simple events, the probabilities of all other events follow by addition. For example, the probability that an American will win the tournament is given by $\Pr(E) + \Pr(F) = 0.16 + 0.15 = 0.31$. Likewise, the probability that a male will win the tournament is given by $\Pr(B) + \Pr(C) + \Pr(D) = 0.16 + 0.20 + 0.25 = 0.61$. The probability that an American male will win the tournament is $\Pr(\{\ \}) = 0$, since this one is an impossible event—there are no American males in the tournament! ⊂⊃

The probability assignment discussed in Example 15.18 reflects the opinion of one specific observer. A different odds-maker might have a slightly different perspective and come up with a different probability assignment. This underscores the fact that sometimes there is no one single "correct" probability assignment on a sample space.

■ In this chapter the term *probability space* will always refer to a *finite* probability space. Infinite probability spaces are important and interesting but require a much higher level of mathematics and we will not discuss them here.

Once a specific probability assignment is made on a sample space, the combination of the sample space and the probability assignment is called a **probability space**. The following is a summary of the key facts related to probability spaces.

┌─ **ELEMENTS OF A PROBABILITY SPACE** ─────────────────

- **Sample space:** $S = \{o_1, o_2, \ldots, o_N\}$.
- **Probability assignment:** $\Pr(o_1), \Pr(o_2), \ldots, \Pr(o_N)$.
  [Each of these is a number between 0 and 1 satisfying $\Pr(o_1) + \Pr(o_2) + \cdots + \Pr(o_N) = 1$.]
- **Events:** These are all the subsets of S, including $\{\ \}$ and S itself. The probability of an event is given by the sum of the probabilities of the individual outcomes that make up the event. [In particular, $\Pr(\{\ \}) = 0$ and $\Pr(S) = 1$.]

15.5 Equiprobable Spaces

One of the most common uses of randomness in the real world is as a mechanism to guarantee fairness. That's why it is universally accepted that tossing a coin, rolling a die, or drawing cards is a fair way of choosing among equally deserving choices. This is true as long as the coin, die, or deck of cards is "honest."

What does *honesty* mean when applied to coins, dice, or decks of cards? It essentially means that all individual outcomes in the sample space are equally probable. Thus, an *honest* coin is one in which H and T have the same probability of coming up, and an *honest* die is one in which each of the numbers 1 through 6 is equally likely to be rolled. A probability space in which each simple event has an equal probability is called an **equiprobable space**. (Informally, you can think of an equiprobable space as an "equal opportunity" probability space.)

In equiprobable spaces, calculating probabilities of events becomes simply a matter of counting. First, we need to find N, the size of the sample space. Each individual outcome in the sample space will have probability equal to $1/N$ (they all have the same probability, and the sum of the probabilities must equal 1). To find the probability of the event E we then find k, the number of outcomes in E. Since each of these outcomes has probability $1/N$, the probability of E is then k/N.

> **PROBABILITIES IN EQUIPROBABLE SPACES**
>
> If k denotes the size of an event E and N denotes the size of the sample space S, then in an equiprobable space
> $$\Pr(E) = \frac{k}{N}$$

EXAMPLE 15.19 Honest Coin Tossing

This example is a follow-up of Example 15.16. Suppose that a coin is tossed three times, and we have been assured that the coin is an honest coin. If this is true, then each of the eight possible outcomes in the sample space has probability 1/8. (Recall that there are $N = 8$ outcomes in the sample space.)

The probability of any event E is given by the number of outcomes in E divided by 8. Table 15-4 shows each of the events in Table 15-2 with their respective probabilities.

TABLE 15-4

Event (in words)	Event (as a set)	Probability
Toss 2 or more heads.	$\{HHT, HTH, THH, HHH\}$	$4/8 = 1/2$
Toss more than 2 heads.	$\{HHH\}$	$1/8$
Toss 2 heads or fewer.	$\{TTT, TTH, THT, HTT, THH, HTH, HHT\}$	$7/8$
Toss no tails.	$\{HHH\}$	$1/8$
Toss exactly 1 tail.	$\{HHT, HTH, THH\}$	$3/8$
Toss exactly 1 head.	$\{HTT, THT, TTH\}$	$3/8$
First toss is heads.	$\{HHH, HHT, HTH, HTT\}$	$4/8 = 1/2$
Toss same number of heads as tails.	$\{\ \}$ (*Note:* This is the *impossible event.*)	0
Toss 3 heads or fewer.	S (*Note:* This event is called the *certain event.*)	1

(**EXAMPLE 15.20**)   **Rolling a Pair of Honest Dice**

This example is a follow-up of Examples 15.4 and 15.5. Imagine that you are play-ing some game that involves rolling a pair of honest dice, and the only thing that matters is the total of the two numbers rolled. As we saw in Example 15.4 the sample space in this situation is $S = \{2, 3, 4, 5, 6, 7, 8, 9, 10, 11, 12\}$, where the outcomes are the possible totals that could be rolled. This sample space has $N = 11$ possible outcomes, but the outcomes are not equally likely, so it would be wrong to assign to each outcome the probability 1/11. In fact, most people with some experience rolling a pair of dice know that the likelihood of rolling a 7 is much higher than that of rolling a 12.

How can we find the exact probabilities for the various totals that one might roll? We can answer this question by considering the sample space discussed in Example 15.5, where every one of the $N = 36$ possible rolls is listed as a separate outcome (see Fig. 15-1 in Example 15.5). Because the dice are honest, each of these 36 possible outcomes is equally likely to occur, so the probability of each is 1/36. Now we have something!

Table 15-5 shows the probability of rolling a 2, 3, 4, . . . , 12. In each case the numerator represents the number of ways that particular total can be rolled. For example, the event "roll a 7" consists of six distinct possible rolls: $\{ \boxdot\boxplus, \boxdot\boxtimes, \boxdot\boxtimes, \boxcolon\boxtimes, \boxtimes\boxtimes, \boxplus\boxdot \}$. Thus, the probability of "rolling a 7" is 6/36 = 1/6. The other probabilities are computed in a similar manner.

TABLE 15-5	
Event	Probability
"Roll a 2": $\{\boxdot\ \boxdot\}$	1/36
"Roll a 3": $\{\boxdot\ \boxtimes, \boxdot\ \boxdot\}$	2/36
"Roll a 4": $\{\boxdot\ \boxtimes, \boxdot\ \boxtimes, \boxtimes\ \boxdot\}$	3/36
"Roll a 5": $\{\boxdot\ \boxtimes, \boxdot\ \boxtimes, \boxtimes\ \boxtimes, \boxcolon\ \boxdot\}$	4/36
"Roll a 6": $\{\boxdot\ \boxtimes, \boxdot\ \boxtimes, \boxtimes\ \boxtimes, \boxcolon\ \boxtimes, \boxtimes\ \boxdot\}$	5/36
"Roll a 7": $\{\boxdot\ \boxplus, \boxdot\ \boxtimes, \boxtimes\ \boxtimes, \boxcolon\ \boxtimes, \boxtimes\ \boxtimes, \boxplus\ \boxdot\}$	6/36
"Roll an 8": $\{\boxdot\ \boxplus, \boxtimes\ \boxtimes, \boxcolon\ \boxtimes, \boxtimes\ \boxtimes, \boxplus\ \boxtimes\}$	5/36
"Roll a 9": $\{\boxtimes\ \boxplus, \boxcolon\ \boxtimes, \boxtimes\ \boxtimes, \boxplus\ \boxtimes\}$	4/36
"Roll a 10": $\{\boxcolon\ \boxplus, \boxtimes\ \boxtimes, \boxplus\ \boxtimes\}$	3/36
"Roll an 11": $\{\boxtimes\ \boxplus, \boxplus\ \boxtimes\}$	2/36
"Roll a 12": $\{\boxplus\ \boxplus\}$	1/36

(**EXAMPLE 15.21**)   **Rolling a Pair of Honest Dice: Part 2**

Once again, we are rolling a pair of honest dice. We are now going to find the probability of the event E: "at least one of the dice comes up an ace." (In dice jar-gon, a $\boxdot$ is called an ace.)

This is a slightly more sophisticated counting question than the ones in Example 15.20. We will show three different ways to answer the question.

- **Tallying.** We can just write down all the individual outcomes in the event E and tally their number. This approach gives

$$E = \{\ \boxdot\ \blacksquare,\ \boxdot\ \blacksquare,\ \boxdot\ \blacksquare,\ \boxdot\ \blacksquare,\ \boxdot\ \blacksquare,\ \boxdot\ \blacksquare,\ \boxdot\ \blacksquare,\ \boxdot\ \blacksquare,\ \boxdot\ \blacksquare,\ \boxdot\ \blacksquare,\ \boxdot\ \blacksquare\ \}$$

and $\Pr(E) = 11/36$. There is not much to it, but in a larger sample space it will take a lot of time and effort to list every individual outcome in the event.

- **Complementary Event.** Behind this approach is the germ of a really important idea. Imagine that you are playing a game, and you win if at least one of the two numbers comes up an ace (that's event E). Otherwise you lose (call that event F). The two events E and F are called **complementary events**, and E is called the **complement** of F (and vice versa). The key idea is that *the probabilities of complementary events add up to 1*. Thus, when you want $\Pr(E)$ but it's actually easier to find $\Pr(F)$, then you first do that. Once you have $\Pr(F)$, you are in business: $\Pr(E) = 1 - \Pr(F)$.

  Here we can find $\Pr(F)$ by a direct application of the multiplication principle. There are 5 possibilities for the first die (it can't be an ace but it can be anything else) and likewise there are five possibilities for the second die. This means that there are 25 different outcomes in the event F, and thus $\Pr(F) = 25/36$. It follows that $\Pr(E) = 1 - (25/36) = 11/36$.

- **Independent Events.** In this approach, we look at each die separately and, as usual, pretend that one die is white and the other one is red. We will let F_1 denote the event "the white die does *not* come up an ace" and F_2 denote the event "the red die does *not* come up an ace." Clearly, $\Pr(F_1) = 5/6$ and $\Pr(F_2) = 5/6$. Now comes a critical observation: The probability that both events F_1 and F_2 happen can be found by *multiplying* their respective probabilities. This means that $\Pr(F) = (5/6) \times (5/6) = 25/36$. We can now find $\Pr(E)$ exactly as before: $\Pr(E) = 1 - \Pr(F) = 11/36$. ⊂⊃

Of the three approaches used in Example 15.20, the last approach appears to be the most convoluted, but, in fact, it is the one with the most promise. It is based on the concept of *independence* of events. Two events are said to be **independent events** if the occurrence of one event does not affect the probability of the occurrence of the other. When events E and F are independent, the probability that both occur is the product of their respective probabilities; in other words, $\Pr(E \text{ and } F) = \Pr(E) \cdot \Pr(F)$. This is called the **multiplication principle for independent events**.

The multiplication principle for independent events in an important and useful rule, but be forewarned—it works *only* with independent events! For events that are *not* independent, multiplying their respective probabilities gives us a bunch of nonsense.

Our next example illustrates the real power of the multiplication principle for independent events. As we just mentioned, this principle can only be applied when the events in question are independent, and in many circumstances this is not the case. Fortunately, in the examples we are going to consider the independence of the events in question is intuitively obvious. For example, if we roll an honest die several times, what happens on each roll has no impact whatsoever on what is going to happen on subsequent rolls (i.e., the rolls are independent events).

EXAMPLE 15.22   **Rolling a Pair of Honest Dice: Part 3**

Imagine a game in which you roll an honest die four times. If at least one of your rolls comes up an ace, you are a winner. Let E denote the event "you win" and F denote the event "you lose." We will find $\Pr(E)$ by first finding $\Pr(F)$, using the same ideas we developed in Example 15.21.

We will let F_1, F_2, F_3, and F_4 denote the events "first roll is not an ace," "second roll is not an ace," "third roll is not an ace," and "fourth roll is not an ace," respectively. Then

$$\Pr(F_1) = 5/6, \quad \Pr(F_2) = 5/6, \quad \Pr(F_3) = 5/6, \quad \Pr(F_4) = 5/6$$

Now we use the multiplication principle for independent events:

$$\Pr(F) = (5/6) \times (5/6) \times (5/6) \times (5/6) = (5/6)^4 \approx 0.482$$

Finally, we find $\Pr(E)$:

$$\Pr(E) = 1 - \Pr(F) \approx 0.518$$

EXAMPLE 15.23   **Five-Card Poker Hands: Part 2**

One of the best hands in five-card draw poker is a *four-of-a-kind* (four cards of the same value plus a fifth card called a *kicker*). Among all four-of-a-kind hands, the very best is the one consisting of four aces plus a kicker. We will let F denote the event of drawing a hand with four aces plus a kicker in five-card draw poker. (The F stands for "fabulously lucky.") Our goal in this example is to find $\Pr(F)$.

The size of our sample space is $N = {}_{52}C_5 = 2{,}598{,}960$ (see Example 15.13). Of these roughly 2.6 million possible hands, there are only 48 hands in F: four of the five cards are the aces; the kicker can be any one of the other 48 cards in the deck. Thus, $\Pr(F) = 48/2{,}598{,}960 \approx 0.0000185$, roughly 1 in 50,000.

EXAMPLE 15.24   **The Sucker's Bet**

Imagine that a friend offers you the following bet: Toss an honest coin 10 times. If the tosses come out in an even split (5 *Heads* and 5 *Tails*), you win and your friend buys the pizza; otherwise you lose (and you buy the pizza). Sounds like a reasonable offer—let's check it out.

We are going to let E denote the event "5 *Heads* and 5 *Tails* are tossed," and we will compute $\Pr(E)$. We already found (Example 15.7) that there are $N = 1024$ equally likely outcomes when a coin is tossed 10 times and that each of these outcomes can be described by a string of 10 *H*s and *T*s.

We now need to figure out how many of the 1024 strings of *H*s and *T*s have 5 *H*s and 5 *T*s. There are of course, quite a few: *HHHHHTTTTT*, *HTHTHTHTHT*, *TTHHHTTTHH*, and so on. Trying to make a list of all of them is not a very practical idea. The trick is to think about these strings as 10 slots each taken up by an *H* or a *T*, but once you determine which slots have *H*s you automatically know which slots have *T*s. Thus, each string of 5 *H*s and 5 *T*s is determined by the choice of the 5 slots for the *H*s. These are unordered choices of 5 out of 10 slots, and the number of such choices is ${}_{10}C_5 = 252$.

Now we are in business: $\Pr(E) = 252/1024 \approx 0.246$.

The moral of this example is that your friend is a shark. When you toss an honest coin 10 times, about 25% of the time you are going to end up with an even split, and the other 75% of the time the split is uneven.

■ You may want to think about this probability and take a guess before you read on — you might be surprised when you see the answer.

15.6 Odds

Dealing with probabilities as numbers that are always between 0 and 1 is the mathematician's way of having a consistent terminology. To the everyday user, consistency is not that much of a concern, and we know that people often express the likelihood of an event in terms of *odds*. Probabilities and odds are not identical concepts but are closely related.

EXAMPLE 15.25 **Odds of Making Free Throws**

In Example 15.15 we discussed the fact that Steve Nash (one of the most accurate free-throw shooters in NBA history) shoots free throws with a probability of $p = 0.90$. We can interpret this to mean that on the average, out of every 100 free throws, Nash is going to make 90 and miss about 10, for a hit/miss ratio of 9 to 1. This ratio of hits to misses gives what is known as the *odds of* the event (in this case Nash making the free throw). We can also express the odds in a negative context and say that the *odds against* Nash making the free throw are 1 to 9.

In contrast, the odds of Shaquille O'Neal making a free throw can be described as being *roughly* 1 to 1. To be more precise, O'Neal shoots free throws with a probability of $p = 0.52$ (see Example 15.15), which represents a hit-to-miss ratio of 13 to 12 (52 to 48 simplified). Thus, the exact odds of Shaq making a free throw are 13 to 12. ⬭

■ When F and U have a common factor, it is customary (and desirable) to simplify the numbers before giving the odds. One rarely hears odds of 15 to 5 — they are much better expressed as odds of 3 to 1.

ODDS

Let E be an arbitrary event. If F denotes the number of ways that event E can occur (the *favorable* outcomes or *hits*) and U denotes the number of ways that event E does not occur (the *unfavorable* outcomes, or *misses*), then the **odds of** (also called the **odds in favor of**) the event E are given by the ratio F to U, and the **odds against** the event E are given by the ratio U to F.

EXAMPLE 15.26 **Odds of Rolling a "Natural"**

Suppose that you are playing a game in which you roll a pair of dice, presumably honest. In this game, when you roll a "natural" (i.e., roll a 7 or an 11) you automatically win.

If we let E denote the event "roll a natural," we can check that out of 36 possible outcomes 8 are favorable (6 ways to "roll a 7" and two ways to "roll an 11" — see Table 15.5) and the other 28 are unfavorable. It follows that the odds of rolling a "natural" are 2 to 7 (originally 8 to 28 and then simplified to 2 to 7). ⬭

It is easy to convert odds into probabilities: If the odds of the event E are F to U, then $\Pr(E) = F/(F + U)$.

Converting probabilities into odds is also easy when the probability is given in the form of a fraction: If $\Pr(E) = A/B$, then the odds of E are A to $B - A$. (When the probability is given in decimal form, the best thing to do is to first convert the decimal form into fractional form.)

EXAMPLE 15.27 **Handicapping a Tennis Tournament: Part 2**

This example is a follow-up to Example 15.18. Recall that the probability assignment for the tennis tournament was as follows: $\Pr(A) = 0.08$, $\Pr(B) = 0.16$,

$\Pr(C) = 0.20, \Pr(D) = 0.25, \Pr(E) = 0.16,$ and $\Pr(F) = 0.15$. We will now express each of these probabilities as odds. (Notice that to do so we first convert the decimals into fractions in reduced form.)

- $\Pr(A) = 0.08 = 8/100 = 2/25$. Thus, the odds of A winning the tournament are 2 to 23.
- $\Pr(B) = 0.16 = 16/100 = 4/25$. The odds of B winning the tournament are 4 to 21.
- $\Pr(C) = 0.20 = 20/100 = 1/5$. The odds of C winning the tournament are 1 to 4.
- $\Pr(D) = 0.25 = 25/100 = 1/4$. The odds of D winning the tournament are 1 to 3.
- $\Pr(E) = \Pr(B) = 4/25$. The odds of E winning the tournament are the same as B's (4 to 21).
- $\Pr(F) = 0.15 = 15/100 = 3/20$. The odds of F winning the tournament are 3 to 17.

A final word of caution: There is a difference between odds as discussed in this section and the *payoff odds* posted by casinos or bookmakers in sports gambling situations. Suppose we read in the newspaper, for example, that the Las Vegas bookmakers have established that "the odds that the Boston Celtics will win the NBA championship are 5 to 2." What this means is that if you want to bet in favor of the Celtics, for every $2 that you bet, you can win $5 if the Celtics win. This ratio may be taken as some indication of the actual odds in favor of the Celtics winning, but several other factors affect payoff odds, and the connection between payoff odds and actual odds is tenuous at best.

CONCLUSION

While the average citizen thinks of probabilities, chances, and odds as vague, informal concepts that are useful primarily when discussing the weather or playing the lottery, scientists and mathematicians think of probability as a formal framework within which the laws that govern chance events can be understood. The basic elements of this framework are a *sample space* (which represents a precise mathematical description of all the possible outcomes of a *random experiment*), *events* (collections of these outcomes), and a *probability assignment* (which associates a numerical value to each of these events).

Of the many ways in which probabilities can be assigned, a particularly important case is the one in which all individual outcomes have the same probability (*equiprobable spaces*). When this happens, the critical steps in calculating probabilities revolve around two basic (but not necessarily easy) questions: (1) What is the size of the sample space, and (2) what is the size of the event in question? To answer these kinds of questions, knowing the basic principles of "counting" is critical.

> 66 His Sacred Majesty, Chance, decides everything. 99
>
> —Voltaire

When we stop to think how much of our lives is ruled by fate and chance, the importance of probability theory in almost every walk of life is hardly surprising. Understanding the basic mathematical principles behind this theory can help us better judge when taking a chance is a smart move and when it is not. In the long run, this will make us not just better card players, but also better and more successful citizens.

PROFILE: Persi Diaconis (1945–   )

Persi Diaconis picks up an ordinary deck of cards, fresh from the box, and writes a word in Magic Marker on one side: RANDOM. He shuffles the deck once. The letters have re-formed themselves into six bizarre runes that still look vaguely like the letters R, A, and so on. Diaconis shuffles again, and the markings on the side become indecipherable. After two more shuffles, you can't even tell that there used to be six letters. The side of the pack looks just like the static on a television set. It didn't look random before, but it sure looks random now.

Keep watching. After two more shuffles, the word RANDOM miraculously reappears on the side of the deck—only it is written twice, in letters half the original size. After one more shuffle, the original letters materialize at the original size. Diaconis turns the cards over and spreads them out with a magician's flourish, and there they are in their exact original sequence, from the ace of spades to the king of diamonds.

Diaconis has just performed eight perfect shuffles in a row. There's no hocus-pocus, just skill perfected in his youth: Diaconis ran away from home at 14 to become a magician's assistant and later became a professional magician and blackjack player. Even now at 57, he is one of a couple of dozen people on the planet who can do eight perfect shuffles in less than a minute.

Diaconis's work these days involves much more than nimbleness of hand. He is a professor of mathematics and statistics at Stanford University. But he is also the world's leading expert on shuffling. He knows that what seems to be random often isn't, and he has devoted much of his career to exploring the difference. His work has applications to filing systems for computers and the reshuffling of the genome during evolution. And it has led him back to Las Vegas, where, instead of trying to beat the casinos, he now works for them.

A card counter in blackjack memorizes the cards that have already been played to get better odds by making bets based on his knowledge of what has yet to turn up. If the deck has a lot of face cards and 10's left in it, for instance, and he needs a 10 for a good hand, he will bet more because he's more likely to get it. A good card counter, Diaconis estimates, has a 1 to 2 percent advantage over the casino. On a bad day, a good card counter can still lose $10,000 in a hurry. And on a good day, he may get a tap on the shoulder by a

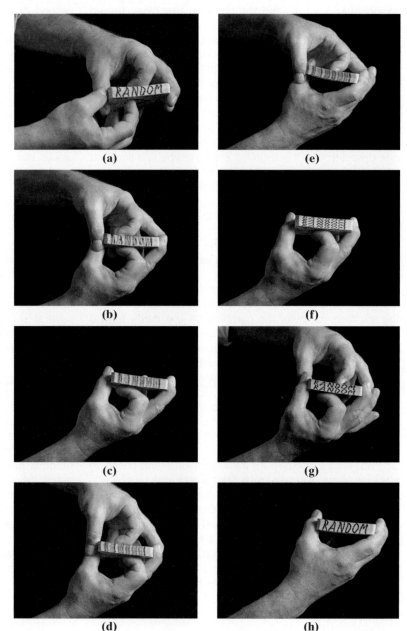

(a)   (e)

(b)   (f)

(c)   (g)

(d)   (h)

large person who will say, "You can call it a day now." By his mid-twenties, Diaconis had figured out that doing mathematics was an easier way to make a living.

Two years ago, Diaconis himself got a tap on the shoulder. A letter arrived from a manufacturer of casino equipment, asking him to figure out whether its card-shuffling machines produced random shuffles. To Diaconis's surprise, the company gave him and his Stanford colleague, Susan Holmes, carte blanche to study the inner workings of the machine. It was like taking a Russian spy on a tour of the CIA and asking him to find the leaks.

When shuffling machines first came out, Diaconis says, they were transparent, so gamblers could actually see the cutting and riffling inside. But gamblers stopped caring after a while, and the shuffling machines turned into closed boxes. They also stopped shuffling cards the way humans do. In the machine that Diaconis and Holmes looked at, each card gets randomly directed, one at a time, to one of 10 shelves. The shuffling machine can put each new card either on the top of the cards already on the shelf or on the bottom, but not between them.

"Already I could see there was something wrong," says Holmes. If you start out with all the red cards at the top of the deck and all the black cards at the bottom, after one pass through the shuffling machine you will find that each shelf contains a red-black sandwich. The red cards, which got placed on the shelves first, form the middle of each sandwich. The black cards, which came later, form the outside. Since there are only 10 shelves, there are at most 20 places where a red card is followed by a black one or vice versa—fewer than the average number of color changes (26) that one would expect from a random shuffle.

The nonrandomness can be seen more vividly if the cards are numbered 1 to 52. After they have passed through the shuffling machine, the numbers on the cards form a zigzag pattern. The top card on the top shelf is usually a high number. Then the numbers decrease until they hit the middle of the first red-black sandwich; then they increase and decrease again, and so on, at most 10 times.

Diaconis and Holmes figured out the exact probability that any given card would end up in any given location after one pass through the machine. But that didn't indicate whether a gambler could use this information to beat the house.

So Holmes worked out a demonstration. It was based on a simple game. You take cards from a deck one by one and each time try to predict what you've selected before you look at it. If you keep track of all the cards, you'll always get the last one right. You'll guess the second-to-last card right half the time, the third-to-last a third of the time, and so on. On average, you will guess about 4.5 cards correctly out of 52.

By exploiting the zigzag pattern in the cards that pass through the shuffling machine, Holmes found a way to double the success rate. She started by predicting that the highest possible card (52) would be on top. If it turned out to be 49, then she predicted 48—the next highest number—for the second card. She kept going this way until her prediction was too low—predicting, say, 15 when the card was actually 18. That meant the shuffling machine had reached the bottom of a zigzag and the numbers would start climbing again. So she would predict 19 for the next card. Over the long run, Holmes (or, more precisely, her computer) could guess nine out of every 52 cards correctly.

To a gambler, the implications are staggering. Imagine playing blackjack and knowing one-sixth of the cards before they are turned over! In reality, a blackjack player would not have such a big advantage, because some cards are hidden and six full decks are used. Still, Diaconis says, "I'm sure it would double or triple the advantage of the ordinary card counter."

Diaconis and Holmes offered the equipment manufacturer some advice: Feed the cards through the machine twice. The alternative would be more expensive: Build a 52-shelf machine.

A small victory for shuffling theory, one might say. But randomization applies to more than just cards. Evolution randomizes the order of genes on a chromosome in several ways. One of the most common mutations is called a "chromosome inversion," in which the arm of a chromosome gets cut in two random places, flipped over end-to-end, and reattached, with the genes in reverse order. In fruit flies, inversions happen at a rate of roughly one per every million years. This is very similar to a shuffling method called transposition that Diaconis studied 20 years ago. Using his methods, mathematical biologists have estimated how many inversions it takes to get from one species of fruit fly to another, or to a completely random genome. That, Diaconis suggests, is the real magic he ran away from home to find. "I find it amazing," he says, "that mathematics developed for purely aesthetic reasons would mesh perfectly with what engineers or chromosomes do when they want to make a mess."

Source: Reprinted by permission from Dana Mackenzie, "The Mathematics of Shuffling," *Discover* (October 2002), pp. 22–23.

KEY CONCEPTS

EXERCISES

WALKING

A | Random Experiments and Sample Spaces

1. Write out the sample space for each of the following random experiments.

   (a) A coin is tossed three times in a row. The observation is how the coin lands (H or T) on each toss.

   (b) A basketball player shoots three consecutive free throws. The observation is the result of each free throw (s for success, f for failure).

   (c) A coin is tossed three times in a row. The observation is the number of times the coin comes up tails.

   (d) A basketball player shoots three consecutive free throws. The observation is the number of successes.

2. Write out the sample space for each of the following random experiments.

   (a) A coin is tossed four times in a row. The observation is how the coin lands (H or T) on each toss.

   (b) A student randomly guesses the answers to a four-question true-or-false quiz. The observation is the student's answer (T or F) for each question.

   (c) A coin is tossed four times in a row. The observation is the percentage of tosses that are heads.

   (d) A student randomly guesses the answers to a four-question true-or-false quiz. The observation is the percentage of correct answers in the test.

3. Four runners ($A, B, C,$ and D) are running in a race. The observation is the order in which the four runners cross the finish line (assume that there are no ties). Write out the sample space for this random experiment.

4. A gumball machine has gumballs of four different flavors: apple (A), blackberry (B), cherry (C), and Doublemint (D). The gumballs are well mixed, and when you drop a quarter in the machine you get two random gumballs. The observation is the flavors of the two gumballs. Write out the sample space for this random experiment. (*Hint:* The size of the sample space is $N = 10$.)

In Exercises 5 through 8, the sample spaces are too big to write down in full. In these exercises, you should describe the sample space either by describing a generic outcome or by listing some outcomes and then using the . . . notation. In the latter case, you should write down enough outcomes to make the description reasonably clear.

5. A coin is tossed 10 times in a row. The observation is how the coin lands (H or T) on each toss. Describe the sample space.

6. A student randomly guesses the answers to a 10-question true-or-false quiz. The observation is the student's answer (T or F) for each question. Describe the sample space.

7. A die is rolled four times in a row. The observation is the number that comes up on each roll. Describe the sample space.

8. A student randomly guesses the answers to a multiple-choice quiz consisting of 10 questions. The observation is the student's answer ($A, B, C, D,$ or E) for each question. Describe the sample space.

B | The Multiplication Rule

9. A California license plate starts with a digit other than 0 followed by three capital letters followed by three more digits (0 through 9).

   (a) How many different California license plates are possible?

   (b) How many different California license plates start with a 5 and end with a 9?

   (c) How many different California license plates have no repeated symbols (all the digits are different and all the letters are different)?

10. A computer password consists of four letters (A through Z) followed by a single digit (0 through 9). Assume that the passwords are not case sensitive (i.e., that an uppercase letter is the same as a lowercase letter).

   (a) How many different passwords are possible?

   (b) How many different passwords end in 1?

   (c) How many different passwords do not start with Z?

   (d) How many different passwords have no Z's in them?

11. Dolores packs two pairs of high-heel shoes, two pairs of tennis shoes, four formal dresses, three pairs of jeans, five T-shirts, and four silk blouses to go on a vacation.

   (a) How many different "outfits" can Dolores make with these items? (Assume the following fashion standards: tennis shoes with a formal dress are not OK, tennis shoes with a silk blouse are not OK, high-heels with jeans and a T-shirt are not OK, high-heels with jeans and a silk blouse are fine.)

(b) Suppose that at the last minute Dolores adds a sweater that she could wear on top of a T-shirt. How many "outfits" could she make then?

12. A French restaurant offers a menu consisting of three different appetizers, two different soups, four different salads, nine different main courses, and five different desserts.

(a) A fixed-price lunch meal consists of a choice of appetizer, salad, and main course. How many different fixed-price lunch meals are possible?

(b) A fixed-price dinner meal consists of a choice of appetizer, a choice of soup or salad, a main course, and a dessert. How many different fixed-price dinner meals are possible?

(c) A dinner special consists of a choice of soup, salad, or both, plus a main course. How many dinner specials are possible?

13. A set of reference books consists of eight volumes numbered 1 through 8.

(a) In how many ways can the eight books be arranged on a shelf?

(b) In how many ways can the eight books be arranged on a shelf so that at least one book is out of order?

14. Eight people (four men and four women) line up at a checkout stand in a grocery store.

(a) In how many ways can they line up?

(b) In how many ways can they line up if the first person in line must be a woman?

(c) In how many ways can they line up if they must alternate woman, man, woman, man, and so on and if a woman must always be first in line?

15. The ski club at Tasmania State University has 35 members (15 females and 20 males). A committee of three members—a President, a Vice President, and a Treasurer—must be chosen.

(a) How many different three-member committees can be chosen?

(b) How many different three-member committees can be chosen in which the committee members are all females?

(c) How many different three-member committees can be chosen in which the committee members are all the same gender?

(d) How many different three-member committees can be chosen in which the committee members are *not* all the same gender?

16. The ski club at Tasmania State University has 35 members (15 females and 20 males). A committee of four

members—a President, a Vice President, a Treasurer, and a Secretary—must be chosen.

(a) How many different four-member committees can be chosen?

(b) How many different four-member committees can be chosen in which the President and Treasurer are both females?

(c) How many different four-member committees can be chosen if the President and Treasurer are both females and the Vice President and Secretary are both males?

17. How many seven-digit numbers (i.e., numbers from 1,000,000 through 9,999,999)

(a) are even?

(b) are divisible by 5?

(c) are divisible by 25?

18. How many nine-digit numbers (i.e., numbers between 100,000,000 and 999,999,999)

(a) have no repeated digits?

(b) are number palindromes? (A *number palindrome* is a number such as 374323473 that reads the same whether read from left to right or from right to left.)

C Permutations and Combinations

19. Compute each of the following without the use of a calculator.

(a) $_{10}P_2$

(b) $_{10}C_2$

(c) $_{10}P_3$

(d) $_{10}C_3$

20. Compute each of the following without the use of a calculator.

(a) $_{11}P_2$

(b) $_{11}C_2$

(c) $_{20}C_2$

(d) $_{20}P_3$

21. Compute each of the following without the use of a calculator.

(a) $_{10}C_9$

(b) $_{10}C_8$

(c) $_{200}C_{199}$

(d) $_{200}C_{198}$

22. Compute each of the following without the use of a calculator.

(a) $_{21}C_{20}$

(b) $_{21}C_{19}$

(c) $_{201}C_{200}$

(d) $_{201}C_{199}$

23. Find the value of each of the following.

(a) $_{11}C_4$

(b) $_{11}C_7$

(c) $_{50}C_3$

(d) $_{50}C_{47}$

24. Find the value of each of the following.

(a) $_{13}C_4$

(b) $_{13}C_9$

(c) $_{14}C_5$

(d) $_{14}C_9$

25. Find the value of each of the following.

(a) $_{11}C_3 + _{11}C_4$

(b) $_{12}C_4$

(c) $_{12}C_4 + _{12}C_5$

(d) $_{13}C_5$

26. Find the value of each of the following.

(a) $_{10}C_3 + _{10}C_4$

(b) $_{11}C_4$

(c) $_{13}C_8 + _{13}C_9$

(d) $_{14}C_9$

27. Given that $_{150}P_{50} \approx 6.12 \times 10^{104}$, find an approximate value for $_{150}P_{51}$.

28. Given that $_{1500}P_{500} \approx 1.2 \times 10^{1547}$, find an approximate value for $_{1500}P_{501}$.

29. Given that $_{150}C_{50} \approx 2.01 \times 10^{40}$, find an approximate value for $_{150}C_{51}$.

30. Given that $_{1500}C_{500} \approx 9.8 \times 10^{412}$, find an approximate value for $_{1500}C_{501}$.

31. Find the value of each of the following. (You will probably want to use a calculator with $_nP_r$ and $_nC_r$ keys.)

(a) $_{20}P_9$

(b) $_{20}P_9/9!$

(c) $_{20}C_9$

32. Find the value of each of the following. (You will probably want to use a calculator with $_nP_r$ and $_nC_r$ keys.)

(a) $_{19}P_{11}$

(b) $_{19}P_{11}/11!$

(c) $_{19}C_{11}$

In Exercises 33 through 36, you are asked to give your answer using the notation $_nP_r$ or $_nC_r$. You do not have to give an actual numerical answer, so you won't need a calculator.

33. The board of directors of the XYZ Corporation has 15 members.

(a) How many different slates of four members (a President, a Vice President, a Treasurer, and a Secretary) can be chosen?

(b) How many delegations of four members (all members are equal) can be chosen?

34. There are 10 athletes entered in an Olympic event.

(a) In how many ways can one pick the winners of the gold, silver, and bronze medals?

(b) In how many ways can one pick the seven athletes who will not earn any medals?

35. The Brute Force Bandits is a punk rock band planning their next concert tour. The band has a total of 30 new songs in their repertoire.

(a) How many different set lists of 18 songs can the band select to play on the tour? (*Hint*: Assume that the order in which the songs are listed is irrelevant.)

(b) How many different ways are there for the band to record a CD consisting of 18 songs chosen from the 30 new songs? (*Hint*: In recording a CD, the order in which the songs appear on the CD is relevant.)

36. There are 119 Division 1A college football teams.

(a) How many top 25 rankings are possible? (Assume that every team has a chance to be a top 25 team.)

(b) How many ways are there to choose eight teams and seed them (1 through 8) for a playoff?

(c) How many ways are there to choose eight teams (unseeded) for a playoff?

D **General Probability Spaces**

37. Consider the sample space $S = \{o_1, o_2, o_3, o_4, o_5\}$. Suppose that $\Pr(o_1) = 0.22$ and $\Pr(o_2) = 0.24$.

(a) Find the probability assignment for the probability space when o_3, o_4, and o_5 all have the same probability.

(b) Find the probability assignment for the probability space when $\Pr(o_5) = 0.1$ and o_3 has the same probability as o_4 and o_5 combined.

38. Consider the sample space $S = \{o_1, o_2, o_3, o_4\}$. Suppose that $\Pr(o_1) + \Pr(o_2) = \Pr(o_3) + \Pr(o_4)$ and that $\Pr(o_1) = 0.15$.

(a) Find the probability assignment for the probability space when o_2 and o_3 have the same probability.

(b) Find the probability assignment for the probability space when $\Pr(o_3) = 0.22$.

39. Four candidates are running for mayor of Happyville. According to the polls candidate A has a "one in five" probability of winning [i.e., $\Pr(A) = 1/5$]. Of the other three candidates, all we know is that candidate C is twice as likely to win as candidate B and that candidate D is three times as likely to win as candidate B. Find the probability assignment for this probability space.

40. Seven horses (A, B, C, D, E, F, and G) are running in the Boonsville Sweepstakes. According to the oddsmakers, A has a "one in four" probability of winning [i.e., $\Pr(A) = 1/4$], B has a "three in ten" probability of winning, and C has a "one in twenty" probability of winning. The remaining four horses all have the same probability of winning. Find the probability assignment for the probability space.

41. The circular spinner shown in Fig. 15-5 has a red sector with a central angle of $108°$, a blue sector and a white sector each with central angles of $72°$, and a green sector and a yellow sector having the same central angle. Assume that the needle is spun so that it randomly stops at one of these sectors (if it stops exactly on the line between sectors, then the needle is spun again). Find the probability assignment for the probability space.

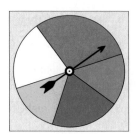

FIGURE 15-5

42. The circular spinner shown in Fig. 15-6 is divided into four sectors. The white sector has a central angle of $36°$, the yellow sector has a central angle of $54°$, and the blue sector has a central angle of $144°$. Assume that the needle is spun so that it randomly stops at one of these sectors (if it stops exactly on the line between sectors then

the needle is spun again). Find the probability assignment for the probability space.

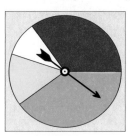

FIGURE 15-6

E Events

43. A coin is tossed three times in a row. The observation is how the coin lands (heads or tails) on each toss [see Exercise 1(a)]. Write out the event described by each of the following statements as a set.

(a) E_1: "the coin comes up heads twice."

(b) E_2: "all three tosses come up the same."

(c) E_3: "exactly half of the tosses come up heads."

(d) E_4: "the first two tosses come up tails."

44. A student randomly guesses the answers to a four-question true-or-false quiz. The observation is the student's answer (T or F) for each question [see Exercise 2(b)]. Write out the event described by each of the following statements as a set.

(a) E_1: "the student answers *True* to two out of the four questions."

(b) E_2: "the student answers *True* to *at least* two out of the four questions."

(c) E_3: "the student answers *True* to *at most* two out of the four questions."

(d) E_4: "the student answers *True* to the first two questions."

45. A pair of dice is rolled. The observation is the number that comes up on each die (see Example 15.5). Write out the event described by each of the following statements as a set.

(a) E_1: "*pairs* are rolled." (A "pair" is a roll in which both dice come up the same number.)

(b) E_2: "*craps* are rolled." ("Craps" is a roll in which the sum of the two numbers rolled is 2, 3, or 12.)

(c) E_3: "a *natural* is rolled." (A "natural" is a roll in which the sum of the two numbers rolled is 7 or 11.)

46. A card is drawn out of a standard deck of 52 cards. Each card can be described by giving its "value" ($A, 2, 3, 4, \ldots, 10, J, Q, K$) and its "suit" ($H$ for hearts, C for clubs,

D for diamonds, and S for spades). For example, $2D$ denotes the two of diamonds and JH denotes the jack of hearts. Write out the event described by each of the following statements as a set.

(a) E_1: "draw a queen."

(b) E_2: "draw a heart."

(c) E_3: "draw a *face* card." (A "face" card is a jack, queen, or king.)

47. A coin is tossed 10 times in a row. The observation is how the coin lands (H or T) on each toss (see Exercise 5). Write out the event described by each of the following statements as a set.

(a) E_1: "none of the tosses comes up tails."

(b) E_2: "one of the 10 tosses comes up tails."

(c) E_3: "nine or more of the tosses come up tails."

48. Five candidates (A, B, C, D, and E) have a chance to be selected to be on *American Idol*. Any subset of them (including none of them or all of them) can be selected. The observation is which subset of individuals is selected. Write out the event described by each of the following statements as a set.

(a) E_1: "two candidates get selected."

(b) E_2: "three candidates get selected."

(c) E_3: "three candidates get selected, and A is not one of them."

F Equiprobable Spaces

49. An honest coin is tossed three times in a row. Find the probability of each of the following events. (*Hint:* Do Exercise 43 first.)

(a) E_1: "the coin comes up heads twice."

(b) E_2: "all three tosses come up the same."

(c) E_3: "exactly half of the tosses come up heads."

(d) E_4: "the first two tosses come up tails."

50. A student randomly guesses the answers to a four-question true-or-false quiz. Find the probability of each of the following events. (*Hint:* Do Exercise 44 first.)

(a) E_1: "the student answers *True* to two out of the four questions."

(b) E_2: "the student answers *True* to *at least* two out of the four questions."

(c) E_3: "the student answers *True* to *at most* two out of the four questions."

(d) E_4: "the student answers *True* to the first two questions."

51. A pair of honest dice is rolled. Find the probability of each of the following events. (*Hint:* Do Exercise 45 first.)

(a) E_1: "*pairs* are rolled." (A "pair" is a roll in which both dice come up the same number.)

(b) E_2: "*craps* are rolled." ("Craps" is a roll in which the sum of the two numbers rolled is 2, 3, or 12.)

(c) E_3: "a *natural* is rolled." (A "natural" is a roll in which the sum of the two numbers rolled is 7 or 11.)

52. A card is drawn at random out of a well-shuffled deck of 52 cards. Find the probability of each of the following events. (*Hint:* Do Exercise 46 first.)

(a) E_1: "draw a queen."

(b) E_2: "draw a heart."

(c) E_3: "draw a *face* card." (A "face" card is a jack, queen, or king.)

53. An honest coin is tossed 10 times in a row. Find the probability of each of the following events. (*Hint:* Do Exercise 47 first.)

(a) E_1: "none of the tosses comes up tails."

(b) E_2: "one of the 10 tosses comes up tails."

(c) E_3: "nine or more of the tosses come up tails."

54. Five candidates (A, B, C, D, and E) have a chance to be selected to be on *American Idol*. Any subset of them (including none of them or all of them) can be selected, and assume that the selection process is completely random (the subsets of candidates are all equally likely). Find the probability of each of the following events. (*Hint:* Do Exercise 48 first.)

(a) E_1: "two candidates get selected."

(b) E_2: "three candidates get selected."

(c) E_3: "three candidates get selected, and A is not one of them."

55. A student takes a 10-question true-or-false quiz and randomly guesses the answer to each question. Suppose that a correct answer is worth 1 point and an incorrect answer is worth -0.5 points. Find the probability that the student

(a) gets 10 points.

(b) gets –5 points.

(c) gets 8.5 points.

(d) gets 8 or more points.

(e) gets 5 points.

(f) gets 7 or more points.

56. Suppose that the probability of giving birth to a boy and the probability of giving birth to a girl are both 0.5. Find the probability that in a family of four children,

   (a) all four children are girls.

   (b) there are two girls and two boys.

   (c) the youngest child is a girl.

   (d) the oldest child is a boy.

57. The Tasmania State University glee club has 15 members. A *quartet* of four members must be chosen to sing at the university president's reception. Assume that the quartet is chosen randomly by drawing the names out of a hat. Find the probability that

   (a) Alice (one of the members of the glee club) is chosen to be in the quartet.

   (b) Alice is not chosen to be in the quartet.

   (c) the four members chosen for the quartet are Alice, Bert, Cathy, and Dale.

58. Ten professional basketball teams are participating in a draft lottery. (A draft lottery is a lottery to determine the order in which teams get to draft players.) Ten balls, each containing the name of one team (call them A, B, C, D, E, F, G, H, I, and J for short), are placed in an urn and thoroughly mixed. Four balls are drawn, one at a time, from the urn. The four teams chosen get to draft first and in the order they are chosen. The remaining six teams have to draft in reverse order of season records. Find the probability that

   (a) A is the first team chosen.

   (b) A is one of the four teams chosen.

   (c) A is not one of the four teams chosen.

G   Odds

59. Find the odds of each of the following events.

   (a) an event E with $\Pr(E) = 4/7$

   (b) an event E with $\Pr(E) = 0.6$

60. Find the odds of each of the following events.

   (a) an event E with $\Pr(E) = 3/11$

   (b) an event E with $\Pr(E) = 0.375$

61. In each case, find the probability of an event E having the given odds.

   (a) The odds in favor of E are 3 to 5.

   (b) The odds against E are 8 to 15.

62. In each case, find the probability of an event E having the given odds.

   (a) The odds in favor of E are 4 to 3.

   (b) The odds against E are 12 to 5.

   (c) The odds in favor of E are the same as the odds against E.

JOGGING

63. The 35-member ski club at Tasmania State University needs to select a planning committee for an upcoming ski trip. The committee will have a chairperson, a secretary, and three at-large members. How many different planning committees can be chosen?

64. Andy and Roger are playing in a tennis match. (A tennis match is a best-of-five contest: The first player to win three games wins the match, and there are no ties.) We can describe the outcome of the tennis match by a string of letters (A or R) that indicate the winner of each game. For example, the string $RARR$ represents an outcome where Roger wins games 1, 3, and 4, at which point the match is over (game 5 is not played).

   (a) Describe the event "Roger wins the match in game 5."

   (b) Describe the event "Roger wins the match."

   (c) Describe the event "the match goes five games."

65. Two teams (call them X and Y) play against each other in the World Series. (The World Series is a best-of-seven series: The first team to win four games wins the series, and games cannot end in a tie.) We can describe an outcome for the World Series by writing a string of letters that indicate (in order) the winner of each game. For example, the string $XYXXYX$ represents the outcome: X wins game 1, Y wins game 2, X wins game 3, and so on.

   (a) Describe the event "X wins in five games."

   (b) Describe the event "the series is over in game 5."

   (c) Describe the event "the series is over in game 6."

   (d) Find the size of the sample space.

66. A pizza parlor offers six toppings—pepperoni, Canadian bacon, sausage, mushroom, anchovies, and olives— that can be put on a basic cheese pizza. How many different pizzas can be made? (A pizza can have anywhere from no toppings to all six toppings.)

67. (a) In how many different ways can 10 people form a line?

   (b) In how many different ways can 10 people be seated around a circular table? [*Hint:* The answer to (b) is much less than the answer to (a). There are many different ways in which the same circle of 10 people can be broken up to form a line. How many?]

   (c) In how many different ways can five boys and five girls get in line so that boys and girls alternate (boy, girl, boy, ..., or girl, boy, girl, ...)?

(d) In how many different ways can five boys and five girls sit around a circular table so that boys and girls alternate?

68. Eight points are taken on a circle.

(a) How many chords can be drawn by joining all possible pairs of the points?

(b) How many triangles can be made using these points as vertices?

69. A study group of 15 students is to be split into three groups of five students each. In how many ways can this be done?

70. An urn contains seven red balls and three blue balls.

(a) If three balls are selected all at once, what is the probability that two are blue and one is red?

(b) If three balls are selected by pulling out a ball, noting its color, and putting it back in the urn before the next selection, what is the probability that only the first and third balls drawn are blue?

(c) If three balls are selected one at a time without putting them back in the urn, what is the probability that only the first and third balls drawn are blue?

71. If we toss an honest coin 20 times, what is the probability of

(a) getting 10 heads and 10 tails?

(b) getting 3 heads and 17 tails?

(c) getting 3 or more heads?

Exercises 72 through 76 refer to five-card draw poker hands. [See Examples 15.13 and 15.23.]

72. What is the probability of getting "four of a kind" (four cards of the same value)?

73. What is the probability of getting all five cards of the same color?

74. What is the probability of getting a "flush" (all five cards of the same suit)?

75. What is the probability of getting an "ace-high straight" (10, J, Q, K, A of any suit but not all the same suit)?

76. What is the probability of getting a "full house" (three cards of equal value and two other cards of equal value)?

77. Which makes the better bet, bet 1 or bet 2? Bet 1: In four consecutive rolls of an honest die, a 6 will come up at least once. Bet 2: In 24 consecutive rolls of a pair of dice, two 6's (a roll called *boxcars*) will come up at least once. [*Historical note*: A French gambler by the name of Chevalier de Méré (1607–1684) wanted to know the answer to this question because he had large sums of money riding on these bets. He asked his friend, the famous mathematician Blaise Pascal, for the answer. Pascal's work on this question is considered by many historians to have laid the groundwork for the formal mathematical theory of probability.]

78. A factory assembles car stereos. From random testing at the factory, it is known that, on the average, 1 out of every 50 car stereos will be defective (which means that the probability that a car stereo randomly chosen from the assembly line will be defective is 0.02). After manufacture, car stereos are packaged in boxes of 12 for delivery to the stores.

(a) What is the probability that in a box of 12 car stereos, there are no defective ones? What assumptions are you making?

(b) What is the probability that in a box of 12 car stereos, there is at most 1 defective one?

79. There are 500 tickets sold in a raffle. If you have three of these tickets and five prizes are to be given, what is the probability that you will win at least one prize? (Give your answer in symbolic notation.)

RUNNING

80. If an honest coin is tossed N times, what is the probability of getting the same number of heads as tails? (*Hint*: Consider two cases: N even and N odd.)

81. How many different "words" (they don't have to mean anything) can be formed using all the letters in

(a) the word PARSLEY.

(b) the word PEPPER. [*Hint*: This question is very different (and harder) than the one in (a). What's the difference?]

82. **Shooting craps.** In the game of craps, the player's first roll of a pair of dice is very important. If the first roll is 7 or 11, the player wins. If the first roll is 2, 3, or 12, the player loses. If the first roll is any other number (4, 5, 6, 8, 9, 10), this number is called the player's "point." The player then continues to roll until either the point reappears, in which case the player wins, or until a 7 shows up before the point, in which case the player loses. What is the probability that the player will win? (Assume that the dice are honest.)

83. **The birthday problem.** There are 30 people in a room. What is the probability that at least two of these people have the same birthday—that is, have their birthdays on the same day and month?

84. **Yahtzee.** Yahtzee is a dice game in which five standard dice are rolled at one time.

(a) What is the probability of scoring "Yahtzee" with one roll of the dice? (You score Yahtzee when all five dice match.)

(b) What is the probability of a *four of a kind* with one roll of the dice? (A *four of a kind* is rolled when four of the five dice match.)

(c) What is the probability of rolling a *large straight* in one roll of the dice? (A *large straight* consists of five numbers in succession on the dice.)

(d) What is the probability of rolling *trips* with one roll of the dice? (*Trips* are rolled when three of the five dice match and the other two dice do not match the trips or each other.)

85. Heartless poker. Heartless poker is played with a deck of cards in which all the hearts have been removed.

Note that the deck now has 39 cards. Find the probability of each of the following five-card draw heartless poker hands.

(a) a *four of a kind*

(b) a *flush* (i.e., five cards of the same suit)

(c) *a full house* (i.e., three cards of equal value and two other cards of equal value)

(d) *trips* (i.e., three cards of equal value and two other cards that are not a pair)

PROJECTS AND PAPERS

A A History of Gambling

Games of chance, be they for religious purposes or for profit, can be traced all the way back to some of the earliest civilizations—the Babylonians, Assyrians, and ancient Egyptians were all known to play primitive dice games as well as board games of various kinds.

In this project you are to write a paper discussing the history of gambling and the historical role of games of chance in the development of probability theory. Good references for this project are references 1, 2, 4, and 5.

B The Letters Between Pascal and Fermat

In the summer of 1654, Pierre de Fermat and Blaise Pascal—two of the most famous mathematicians of their time—exchanged a series of famous letters discussing certain mathematical problems related to gambling. The problems were originally proposed to Pascal by his friend, Chevalier de Méré—a nobleman and notorious gambler with a flair for raising interesting mathematical questions (see Exercise 77). The exchange of letters between Pascal and Fermat is of particular interest because it laid the foundation for the future development of the mathematical theory of probability. While not all of the correspondence between Fermat and

Pascal has survived, enough of the letters exist to see how both mathematicians approached, each in his own way, de Méré's questions.

In this project, you are to write a paper discussing the famous exchange of letters between Pascal and Fermat. Describe the questions that were discussed and the approach that each mathematician took to solve them. Good references for this project are references 4, 5, and 15.

C Book Review: *Conned Again, Watson! Cautionary Tales of Logic, Math, and Probability*

As its title suggests, *Conned Again, Watson! Cautionary Tales of Logic, Math, and Probability* by Colin Bruce (see reference 3) is a collection of mathematical tales set in the form of Sherlock Holmes adventures. Each of the stories has a mathematical connection, and several touch on issues related to the theme of this chapter (particularly relevant are "The Case of the Gambling Nobleman" and "The Case of the Surprise Heir," but several others deal with probability topics as well).

Choose one of the stories in the book with a probabilistic connection and write a review of the story. Pay special attention to the mathematical issues raised in the story.

REFERENCES AND FURTHER READINGS

1. Bennett, Deborah, *Randomness*. Cambridge, MA: Harvard University Press, 1998.

2. Bernstein, Peter, *Against the Gods: The Remarkable Story of Risk*. New York: John Wiley & Sons, 1996.

3. Bruce, Colin, *Conned Again, Watson! Cautionary Tales of Logic, Math, and Probability*. Cambridge, MA: Perseus Publishing, 2002.

4. David, Florence N., *Games, Gods, and Gambling*. New York: Dover Publications, 1998.

5. Devlin, Keith, *The Unfinished Game: Pasal, Fermat, and the Seventeenth-Century Letter that Made the World Modern*. New York: Basic Books, 2008.

6. Everitt, Brian S., *Chance Rules: An Informal Guide to Probability, Risk, and Statistics*. New York: Springer-Verlag, 1999.

7. Gigerenzer, Gerd, *Calculated Risks*. New York: Simon & Schuster, 2002.

8. Haigh, John, *Taking Chances*. New York: Oxford University Press, 1999.

9. Kaplan, Michael, and Ellen Kaplan, *Chances Are: Adventures in Probability*. New York. Penguin Books, 2007.

10. Krantz, Les, *What the Odds Are*. New York: HarperCollins Publishers, 1992.

11. Levinson, Horace, *Chance, Luck and Statistics: The Science of Chance*. New York: Dover Publications, 1963.

12. McGervey, John D., *Probabilities in Everyday Life*. New York: Ivy Books, 1992.

13. Mlodinow, Leonard, *The Drunkard's Walk: How Randomness Rules Our Lives*. New York: Pantheon Books, 2008.

14. Packel, Edward, *The Mathematics of Games and Gambling*. Washington, DC: Mathematical Association of America, 1981.

15. Renyi, Alfred, *Letters on Probability*. Detroit, MI: Wayne State University Press, 1972.

16. Rosenthal, Jeffrey, *Struck by Lightning: The Curious World of Probabilities*. Washington, DC: Joseph Henry Press, 2006.

17. Weaver, Warren, *Lady Luck: The Theory of Probability*. New York: Dover Publications, 1963.

16 The Mathematics of Normal Distributions

The Call of the Bell

More than 60 years ago, during World War II, the South African mathematician John Kerrich was captured by the Germans and spent several years as a prisoner of war. To pass the time, Kerrich tossed a coin — over and over, . . . , and over, keeping meticulous records of the results. By the time he was done he had tossed the same coin 10,000 times. Kerrich broke up his 10,000 tosses into 100 sets of 100 tosses and recorded the number of heads in each set of 100. The results of Kerrich's coin-tossing experiments are shown in Fig. 16-1 on the opposite page. (*Source:* John Kerrich, *An Experimental Introduction to the Theory of Probability*, 1964.) The height of each bar in this bar graph represents the frequency of the corresponding "head-count." Out of the 100 sets of 100 coin tosses, 1 set yielded 34 heads, 2 sets yielded 37 heads, 1 set yielded 40 heads, 3 sets yielded 42 heads, and so on. The largest yield of heads was 68 heads in 1 set.

A dmittedly, the bar graph in Fig. 16-1 does not look particularly interesting, so why does Kerrich's coin-tossing experiments deserve mention? Essentially, the only problem with Kerrich's data is that he didn't toss the coin enough times! What would have happened with the data if Kerrich had continued the coin-tossing experiments for a while? The much more interesting graph shown in Fig. 16-2 shows the results for 10,000 sets of 100 tosses each (that's *one million* coin tosses!). These results were obtained by the author using a computer simulation rather than an actual coin toss marathon but are just as valid as if they were done for real. In fact, had Kerrich tossed a coin one million times, his data would have produced a bell-shaped bar graph looking very much like the one in Fig. 16-2. We can say this with full confidence even though it never happened. Like many other phenomena, the long-term behavior of a tossed coin is guaranteed (when tossed enough times) to produce a bell-shaped bar graph—it is a fundamental law of mathematics.

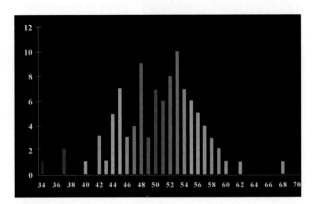

FIGURE 16-1 Number of heads in 100 coin tosses (100 trials).

The bell-shaped bar graph in Fig. 16-2 is but one example of real-world data that approximates what is known as a *normal (bell-shaped) distribution*. More than any other type of regular pattern, normal distributions rule the statistical world, and the purpose of this chapter is to gain an understanding of the meaning of "normal" in the context of statistical data. Section 16.1 is an introductory section illustrating the ubiquity of *approximately normal* distributions among real-world data sets. Sections 16.2 through 16.4 discuss the basic mathematical properties of *normal (bell-shaped) curves*. Section 16.5 revisits the connection between *normal curves* and *real-world data* sets introduced in Section 16.1. Section 16.6 deals with the connection between *random events* and *normal distributions*. The chapter concludes with a brief introduction to the important topic of *statistical inference* (Section 16.7).

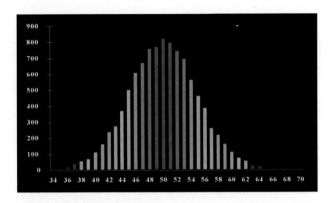

FIGURE 16-2 Number of heads in 100 coin tosses (10,000 trials).

16.1 Approximately Normal Distributions of Data

(**EXAMPLE 16.1**) **Distribution of Heights of NBA Players**

■ The data would be even more bell-shaped if it weren't for the abundance of 6'7" to 7' players in the NBA. This is not a quirk of nature but rather a reflection of the way NBA teams draft players.

Table 16-1 is a frequency table giving the heights of 430 NBA players listed on team rosters at the start of the 2008–2009 season. The bar graph for this data set is shown in Fig. 16-3. We can see that the bar graph fits roughly the pattern of a somewhat skewed (off-center) bell-shaped curve (the orange curve). An idealized bell-shaped curve for this data (the red curve) is shown for comparison purposes.

The long and the short of Table 16-1.

TABLE 16-1		Heights of NBA Players (2008–2009 Season); $N = 430$					
Ht.	Freq.	Ht.	Freq.	Ht.	Freq.	Ht.	Freq.
5'5"	1	6'2"	17	6'8"	45	7'2"	3
5'9"	1	6'3"	24	6'9"	52	7'3"	1
5'10"	3	6'4"	25	6'10"	44	7'6"	1
5'11"	3	6'5"	30	6'11"	41		
6'0"	8	6'6"	36	7'0"	30		
6'1"	15	6'7"	42	7'1"	8		

Source: National Basketball Association, *www.nba.com*

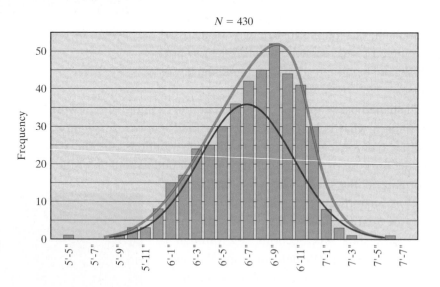

FIGURE 16-3 Height distribution of NBA players (2008–2009 season).

(**EXAMPLE 16.2**) **2007 SAT Math Scores**

Table 16-2 shows the scores of $N = 1,494,531$ college-bound seniors on the mathematics section of the 2007 SAT. (Scores range from 200 to 800 and are grouped in class intervals of 50 points.) The table shows the score distribution and the percentage of test takers in each class interval. A bar graph for the data is shown in Fig. 16-4. The orange bell-shaped curve traces the pattern of the data in the bar

TABLE 16-2	2007 Mathematics SAT Scores ($N = 1,494,531$)	
Score ranges	Frequency	Percentage
750–800	35,136	2.35%
700–740	54,108	3.62%
650–690	120,933	8.09%
600–640	156,057	10.44%
550–590	213,880	14.31%
500–540	247,435	16.56%
450–490	243,569	16.30%
400–440	202,883	13.57%
350–390	124,616	8.34%
300–340	59,672	3.99%
250–290	23,083	1.54%
200–240	13,159	0.88%

Source: *The College Board National Report*, 2007

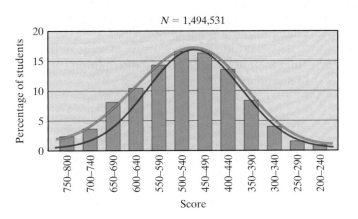

FIGURE 16-4

graph. If the data followed a perfect bell curve, it would follow the red curve shown in the figure. Unlike the curves in Fig. 16-3, here the orange and red curves are very close.

The two very different data sets discussed in Examples 16.1 and 16.2 have one thing in common—both can be described as having bar graphs that *roughly* fit a bell-shaped pattern. In Example 16.1, the fit is crude; in Example 16.2, it is very good. In either case, we say that the data set has an **approximately normal distribution**. The word *normal* in this context is to be interpreted as meaning that the data fits into a special type of bell-shaped curve; the word *approximately* is a reflection of the fact that with real-world data we should not expect an absolutely perfect fit. A distribution of data that has a perfect bell shape is called a **normal distribution**.

Perfect bell-shaped curves are called **normal curves**. Every approximately normal data set can be idealized mathematically by a corresponding normal curve (the red curves in Examples 16.1 and 16.2). This is important because we can then use the mathematical properties of the normal curve to analyze and draw conclusions about the data. The tighter the fit between the approximately normal distribution and the normal curve, the better our analysis and conclusions are going to be. Thus, to understand real-world data sets that have an approximately normal distribution, we first need to understand some of the mathematical properties of normal curves.

16.2 Normal Curves and Normal Distributions

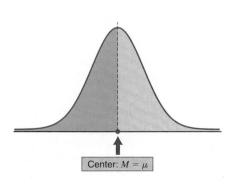

Carl Friedrich Gauss (1777–1855). A biographical profile of Gauss can be found at the end of this chapter.

The study of normal curves can be traced back to the work of the great German mathematician Carl Friedrich Gauss, and for this reason, normal curves are sometimes known as *Gaussian curves*. Normal curves all share the same basic shape—that of a bell—but otherwise they can differ widely in their appearance. Some bells are short and squat, others are tall and skinny, and others fall somewhere in between (Fig. 16-5). Mathematically speaking, however, they all have the same underlying structure. In fact, whether a normal curve is skinny and tall or short and squat depends on the choice of units on the axes, and any two normal curves can be made to look the same by just fiddling with the scales of the axes.

What follows is a summary of some of the essential facts about normal curves and their associated normal distributions. These facts are going to help us greatly later on in the chapter.

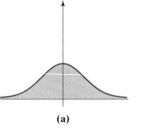   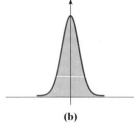   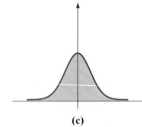

(a)   (b)   (c)

FIGURE 16-5 Normal curves may look different, but they are all cut out of the same cloth.

Center: $M = \mu$

FIGURE 16-6

- **Symmetry.** Every normal curve has a vertical axis of symmetry, splitting the bell-shaped region outlined by the curve into two identical halves. This is the only line of symmetry of a normal curve, so we can refer to it without ambiguity as *the line of symmetry*.

- **Median/mean.** We will call the point of intersection of the horizontal axis and the line of symmetry of the curve the **center** of the distribution (Fig. 16-6). The center represents both the *median M* and the *mean* (average) μ of the data. Thus, in a normal distribution, $M = \mu$. The fact that in a normal distribution the median equals the mean implies that 50% of the data are less than or equal to the mean and 50% of the data are greater than or equal to the mean.

▨ In an approximately normal distribution of real-world data, the median and the mean need not be equal, but one would expect them to be very close.

> ┌─ **MEDIAN AND MEAN OF A NORMAL DISTRIBUTION** ─────
>
> In a normal distribution, $M = \mu$. (If the distribution is approximately normal, then $M \approx \mu$.)

■ **Standard deviation.** The **standard deviation**—traditionally denoted by the Greek letter σ (sigma)—is an important measure of spread, and it is particularly useful when dealing with normal (or approximately normal) distributions, as we will see shortly. The easiest way to describe the standard deviation of a normal distribution is to look at the normal curve. If you were to bend a piece of wire into a bell-shaped normal curve, at the very top you would be bending the wire downward [Fig. 16-7(a)], but at the bottom you would be bending the wire upward [Fig. 16-7(b)]. As you move your hands down the wire, the curvature gradually changes, and there is one point on each side of the curve where the transition from being bent downward to being bent upward takes place. Such a point [P in Fig. 16-7(c)] is called a **point of inflection** of the curve. The standard deviation σ of a normal distribution is the horizontal distance between the line of symmetry of the curve and one of the two points of inflection (P or P' in Fig. 16-8).

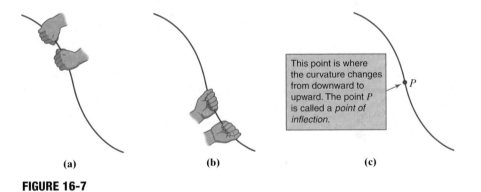

This point is where the curvature changes from downward to upward. The point P is called a *point of inflection.*

    (a)   **(b)**   **(c)**

FIGURE 16-7

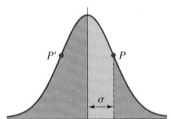

FIGURE 16-8

> ┌─ **STANDARD DEVIATION OF A NORMAL DISTRIBUTION** ─────
>
> In a *normal distribution*, the standard deviation σ equals the distance between a point of inflection and the line of symmetry of the curve.

■ **Quartiles.** We learned in Chapter 14 how to find the quartiles of a data set. When the data set has a normal distribution, the first and third quartiles can be approximated using the mean μ and the standard deviation σ. The magic number to memorize is 0.675. Multiplying the standard deviation by 0.675 tells us how far to go to the right or left of the mean to locate the quartiles.

> ┌─ **QUARTILES OF A NORMAL DISTRIBUTION** ─────
>
> In a *normal distribution*, $Q_3 \approx \mu + (0.675)\sigma$ and $Q_1 \approx \mu - (0.675)\sigma$.

> **EXAMPLE 16.3** **A Mystery Normal Distribution**

Imagine you are told that a data set of $N = 1{,}494{,}531$ numbers has a normal distribution with mean $\mu = 515$ and standard deviation $\sigma = 114$. For now, let's not worry about the source of this data—we'll discuss this soon.

Just knowing the mean and standard deviation of this normal distribution will allow us to draw a few useful conclusions about this data set.

- In a normal distribution, the median equals the mean, so the median value is $M = 515$. This implies that of the 1,494,531 numbers, there are 747,266 that are smaller than or equal to 515 and 747,266 that are greater than or equal to 515.

- The first quartile is given by $Q_1 \approx 515 - 0.675 \times 114 \approx 438$. This implies that 25% of the data set (373,633 numbers) are smaller than or equal to 438.

- The third quartile is given by $Q_3 \approx 515 + 0.675 \times 114 \approx 592$. This implies that 25% of the data set (373,633 numbers) are bigger than or equal to 592.

16.3 Standardizing Normal Data

We have seen that normal curves don't all look alike, but this is only a matter of perception. In fact, all normal distributions tell the same underlying story but use slightly different dialects to do it. One way to understand the story of any given normal distribution is to rephrase it in a simple common language—a language that uses the mean μ and the standard deviation σ as its only vocabulary. This process is called **standardizing** the data.

To standardize a data value x, *we measure how far x has strayed from the mean μ using the standard deviation σ as the unit of measurement*. A standardized data value is often referred to as a **z-value**.

The best way to illustrate the process of standardizing normal data is by means of a few examples.

> **EXAMPLE 16.4** **Standardizing Normal Data**

Let's consider a normally distributed data set with mean $\mu = 45$ ft and standard deviation $\sigma = 10$ ft. We will standardize several data values, starting with a couple of easy cases.

- $x_1 = 55$ ft is a data point located 10 ft *above* the mean $\mu = 45$ ft. Coincidentally, 10 ft happens to be exactly *one* standard deviation. The fact that $x_1 = 55$ ft is located one standard deviation above the mean (A in Fig. 16-9) can be rephrased by saying that the *standardized value* of $x_1 = 55$ is $z_1 = 1$.

- $x_2 = 35$ ft is a data point that is 10 ft (i.e., one standard deviation) *below* the mean (B in Fig. 16-9). This means that the standardized value of $x_2 = 35$ is $z_2 = -1$.

- $x_3 = 50$ ft is a data point that is 5 ft (i.e., half a standard deviation) above the mean (C in Fig. 16-9). This means that the standardized value of $x_3 = 50$ ft is $z_3 = 0.5$. (We could have also said that the standardized value is $z_3 = 1/2$, but it is customary to use decimals to describe standardized values.)

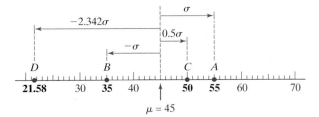

FIGURE 16-9

- $x_4 = 21.58$ is ... uh, this is a slightly more complicated case. How do we handle this one? First, we find the signed distance between the data value and the mean by taking their difference $(x_4 - \mu)$. In this case we get 21.58 ft − 45 ft $= -23.42$ ft. (Notice that for data values smaller than the mean this difference will be negative.) If we now divide this difference by $\sigma = 10$ ft, we get the standardized value $z_4 = -2.342$. This tells us that the data point x_4 is -2.342 standard deviations from the mean $\mu = 45$ ft (D in Fig. 16-9). ⬯

In Example 16.4 we were somewhat fortunate in that the standard deviation was $\sigma = 10$, an especially easy number to work with. It helped us get our feet wet. What do we do in more realistic situations, when the mean and standard deviation may not be such nice round numbers? Other than the fact that we may need a calculator to do the arithmetic, the basic idea we used in Example 16.4 remains the same.

> **STANDARDIZING RULE**
>
> In a normal distribution with mean μ and standard deviation σ, the standardized value of a data point x is $z = (x - \mu)/\sigma$.

(EXAMPLE 16.5)   **Standardizing Normal Data: Part 2**

This time we will consider a normally distributed data set with mean $\mu = 63.18$ lb and standard deviation $\sigma = 13.27$ lb. What is the standardized value of $x = 91.54$ lb?

This looks nasty, but with a calculator, it's a piece of cake:

$$z = (x - \mu)/\sigma = (91.54 - 63.18)/13.27 = 28.36/13.27 \approx 2.14$$

One important point to note is that while the original data is given in pounds, there are no units given for the z-value. (The units for the z-value are standard deviations, and this is implicit in the very fact that it is a z-value.) ⬯

The process of standardizing data can also be reversed, and given a z-value we can go back and find the corresponding x-value. All we have to do is take the formula $z = (x - \mu)/\sigma$ and solve for x in terms of z. When we do this we get the equivalent formula $x = \mu + \sigma \cdot z$. Given μ, σ, and a value for z, this formula allows us to "unstandardize" z and find the original data value x.

(EXAMPLE 16.6)   **"Unstandardizing" a z-Value**

Consider a normal distribution with mean $\mu = 235.7$ m and standard deviation $\sigma = 41.58$ m. What is the data value x that corresponds to the standardized z-value $z = -3.45$?

We first compute the value of -3.45 standard deviations: $-3.45\sigma = -3.45 \times 41.58$ m $= -143.451$ m. The negative sign indicates that the data point is to be located below the mean. Thus, $x = 235.7$ m -143.451 m $= 92.249$ m.

16.4   The 68-95-99.7 Rule

When we look at a typical bell-shaped distribution, we can see that most of the data are concentrated near the center. As we move away from the center the heights of the columns drop rather fast, and if we move far enough away from the center, there are essentially no data to be found. These are all rather informal observations, but there is a more formal way to phrase this, called the **68-95-99.7 rule**. This useful rule is obtained by using one, two, and three standard deviations above and below the mean as special landmarks. In effect, the 68-95-99.7 rule is three separate rules in one.

> **THE 68-95-99.7 RULE**
>
> 1. In every normal distribution, about 68% of all the data values fall within one standard deviation above and below the mean. In other words, 68% of all the data have standardized values between $z = -1$ and $z = 1$. The remaining 32% of the data are divided equally between data with standardized values $z \le -1$ and data with standardized values $z \ge 1$ [Fig. 16-10(a)].
>
> 2. In every normal distribution, about 95% of all the data values fall within two standard deviations above and below the mean. In other words, 95% of all the data have standardized values between $z = -2$ and $z = 2$. The remaining 5% of the data are divided equally between data with standardized values $z \le -2$ and data with standardized values $z \ge 2$ [Fig. 16-10(b)].
>
> 3. In every normal distribution, about 99.7% (i.e., practically 100%) of all the data values fall within three standard deviations above and below the mean. In other words, 99.7% of all the data have standardized values between $z = -3$ and $z = 3$. There is a minuscule amount of data with standardized values outside this range [Fig. 16-10(c)].

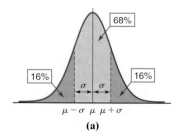
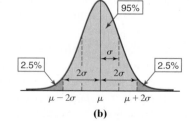
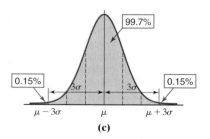

FIGURE 16-10 The 68-95-99.7 rule.

For approximately normal distributions, it is often convenient to round the 99.7% to 100% and work under the assumption that essentially all of the data fall within three standard deviations above and below the mean. This means that if there are no outliers in the data, we can figure that there are approximately six standard deviations separating the smallest (*Min*) and the largest (*Max*) values of the data. In Chapter 14 we defined the range R of a data set ($R = Max - Min$), and, in the case of an approximately normal distribution, we can conclude that the range is about six standard deviations. Remember that this is true as long as we can assume that there are no outliers.

16.5 Normal Curves as Models of Real-Life Data Sets

The reason we like to idealize a real-life, approximately normal data set by means of a normal distribution is that we can use many of the properties we just learned about normal distributions to draw useful conclusions about our data. For example, the 68-95-99.7 rule for normal curves can be reinterpreted in the context of an approximately normal data set as follows.

THE 68-95-99.7 RULE FOR APPROXIMATELY NORMAL DATA

1. In an approximately normal data set, about 68% of the data values fall within (plus or minus) one standard deviation of the mean.

2. In an approximately normal data set, about 95% of the data values fall within (plus or minus) two standard deviations of the mean.

3. In an approximately normal data set, about 99.7%, or practically 100%, of the data values fall within (plus or minus) three standard deviations of the mean.

(EXAMPLE 16.7)   2007 SAT Math Scores: Part 2

This example is a follow-up of Example 16.2. We are now going to use what we learned so far to analyze the 2007 SAT mathematics scores. As you may recall, there were $N = 1,494,531$ scores, distributed in a nice, approximately normal distribution shown in Table 16-2 and Fig. 16-4. The two new pieces of information that we are going to use now are that the mean score was $\mu = 515$ points and the standard deviation was $\sigma = 114$ points. (*Source*: The College Board, National Report, 2007.)

Just knowing the mean and the standard deviation (and that the distribution of test scores is approximately normal) allows us to draw a lot of useful conclusions:

- **Median.** In an approximately normal distribution, the mean and the median should be about the same. Given that the mean score was $\mu = 515$, we can expect the median score to be close to 515. Moreover, the median has to be an actual test score when N is odd (which it is in this example), and SAT scores come in multiples of 10, so a reasonable guess for the median would be either 510 or 520 points.

- **First quartile.** Recall that the first quartile is located 0.675 standard deviation below the mean. This means that in this example the first quartile should be close to $515 - 0.675 \times 114 \approx 438$ points. But again, the first quartile has to be an actual test score (the only time this is not the case is when N is divisible by 4), so a reasonable guess is that the first quartile of the test scores is either 430 or 440 points.

- **Third quartile.** We know that the third quartile is located 0.675 standard deviation above the mean. In this case this gives $515 + 0.675 \times 114 \approx 592$. The most reasonable guess is that the third quartile of the test scores is either 590 or 600 points.

In all three cases our guesses were very good—as reported by the College Board, the 2007 SAT math scores had median $M = 510$, first quartile $Q_1 = 430$, and third quartile $Q_3 = 590$.

We are now going to go analyze the 2007 SAT scores in a little more depth— using the 68-95-99.7 rule.

- **The middle 68.** Approximately 68% of the scores should have fallen within plus or minus *one* standard deviation from the mean. In this case, this range of scores goes from $515 - 114 = 401$ to $515 + 114 = 629$ points. Since SAT scores can only come in multiples of 10, we can estimate that a little over two-thirds of students had scores between 400 and 630 points. The remaining third were equally divided between those scoring 630 points or more (about 16% of test takers) and those scoring 400 points or less (the other 16%).

- **The middle 95.** Approximately 95% of the scores should have fallen within plus or minus *two* standard deviations from the mean, that is, between $515 - 228 = 287$ and $515 + 228 = 743$ points. In practice this means SAT scores between 290 and 740 points. The remaining 5% of the scores were 740 points or above (about 2.5%) and 290 points or below (the other 2.5%).

- **Everyone.** The 99.7 part of the 68-95-99.7 rule is not much help in this example. Essentially, it says that all test scores fell between $515 - 342 = 173$ points and $515 + 342 = 857$ points. Duh! SAT mathematics scores are always between 200 and 800 points, so what else is new? ⊂⊃

16.6 Distributions of Random Events

We are now ready to take up another important aspect of normal curves—their connection with random events and, through that, their critical role in margins of error of public opinion polls. Our starting point is the following important example.

(**EXAMPLE 16.8**)   **Coin-Tossing Experiments**

In the opening of this chapter, we discussed the coin-tossing experiment performed by John Kerrich while he was a prisoner of war during World War II. Kerrich tossed a coin 10,000 times and kept records of the number of heads in groups of 100 tosses.

With modern technology, we can repeat Kerrich's experiment and take it much further. Practically any computer can imitate the tossing of a coin by means of a random-number generator. If we use this technique, we can "toss coins" in mind-boggling numbers—millions of times if we so choose.

We will start modestly. We will toss our make-believe coin 100 times and count the number of heads, which we will denote by X. Before we do that, let's say a few words about X. Since we cannot predict ahead of time its exact value—we are tempted to think that it should be 50, but, in principle, it could be anything from 0 to 100—we call X a **random variable**. The possible values of the random variable X are governed by the laws of probability: Some values of X are extremely unlikely ($X = 0$, $X = 100$) and others are much more likely ($X = 50$), although the likelihood of $X = 50$ is not as great as one would think. It also seems reasonable (assuming that the coin is fair and heads and tails are equally likely) that the likelihood of $X = 49$ should be the same as the likelihood $X = 51$, the likelihood of $X = 48$, should be the same as the likelihood of $X = 52$, and so on.

While all of the preceding statements are true, we still don't have a clue as to what is going to happen when we toss the coin 100 times. One way to get a sense of the probabilities of the different values of X is to repeat the experiment many times and check the frequencies of the various outcomes. Finally, we are ready to do some experimenting!

Our first trial results in 46 heads out of 100 tosses ($X = 46$). The first 10 trials give, in order, $X = 46, 49, 51, 53, 49, 52, 47, 46, 53, 49$. Figure 16-11(a) shows a bar graph for these data.

Continuing this way, we collect data for the values of X in 100, 500, 1000, 5000, and 10,000 trials. The bar graphs are shown in Figs. 16-11(b) through (f), respectively.

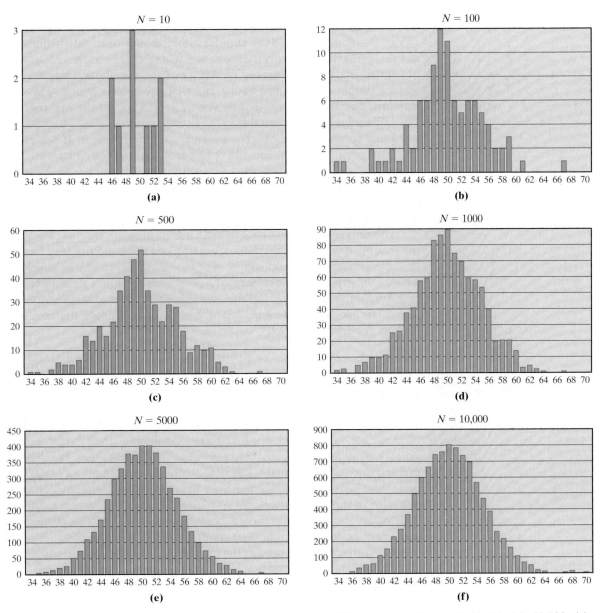

FIGURE 16-11 Distribution of random variable X (number of heads in 100 coin tosses) (a) 10 trials, (b) 100 trials, (c) 500 trials, (d) 1000 trials, (e) 5000 trials, and (f) 10,000 trials.

Figure 16-11 paints a pretty clear picture of what happens: As the number of trials increases, the distribution of the data becomes more and more bell shaped. At the end, we have data from 10,000 trials, and the bar graph gives an almost perfect normal distribution!

What would happen if someone else decided to repeat what we did—toss an honest coin (be it by hand or by computer) 100 times, count the number of heads, and repeat this experiment a few times? The first 10 trials are likely to produce results very different from ours, but as the number of trials increases, their results and our results will begin to look more and more alike. After 10,000 trials, their bar graph will be almost identical to the bar graph shown in Fig. 16-11(f). In a sense, this says that doing the experiments a second time is a total waste of time—in fact, it was even a waste the first time! The outline of the final distribution could have been predicted without ever tossing a coin!

Knowing that the random variable X has an approximately normal distribution is, as we have seen, quite useful. The clincher would be to find out the values of the mean μ and the standard deviation σ of this distribution. Looking at Fig. 16-11(f), we can pretty much see where the mean is—right at 50. This is not surprising, since the axis of symmetry of the distribution has to pass through 50 as a simple consequence of the fact that the coin is honest. The value of the standard deviation is less obvious. For now, let's accept the fact that it is $\sigma = 5$. We will explain how we got this value shortly.

Let's summarize what we now know. An honest coin is tossed 100 times. The number of heads in the 100 tosses is a random variable, which we call X. If we repeat this experiment a large number of times (say N), the random variable X will have an approximately normal distribution with mean $\mu = 50$ and standard deviation $\sigma = 5$, and the larger the value of N is, the better this approximation will be.

The real significance of these facts is that they are true not because we took the trouble to toss a coin a million times. Even if we did not toss a coin at all, all of these statements would still be true. *For a sufficiently large number of repetitions of the experiment of tossing an honest coin 100 times, the number of heads X is a random variable that has an approximately normal distribution with center $\mu = 50$ heads and standard deviation $\sigma = 5$ heads.* This is a mathematical, rather than an experimental, fact.

16.7   Statistical Inference

Next, we are going to take our first tentative leap into statistical inference. Suppose that we have an honest coin and intend to toss it 100 times. We are going to do this just once, and we will let X denote the resulting number of heads. Been there, done that! What's new now is that we a have a solid understanding of the statistical behavior of the random variable X—it has an approximately normal distribution with mean $\mu = 50$ and standard deviation $\sigma = 5$—and this allows us to make some very reasonable predictions about the possible values of X.

For starters, we can predict the chance that X will fall somewhere between 45 and 55 (one standard deviation below and above the mean)—it is 68%. Likewise, we know that the chance that X will fall somewhere between 40 and 60 is 95%, and between 35 and 65 is a whopping 99.7%.

What if, instead of tossing the coin 100 times, we were to toss it n times? (After all, there is nothing unique about tossing 100 times.) Not surprisingly, the bell-shaped distribution we saw in Example 16.8 would still be there—only the values of μ and σ would change. Specifically, for n sufficiently large (typically $n \geq 30$), the number of heads in n tosses would be a random variable with an approximately normal distribution with mean $\mu = n/2$ heads and standard deviation $\sigma = (\sqrt{n})/2$ heads. This is an important fact for which we have coined the name the **honest-coin principle**.

THE HONEST-COIN PRINCIPLE

Let X denote the number of heads in n tosses of an honest coin (assume $n \geq 30$). Then, X has an *approximately normal* distribution with mean $\mu = n/2$ and standard deviation $\sigma = \sqrt{n}/2$.

Note that when we apply the honest-coin principle to $n = 100$ tosses, the distribution of the number of heads has mean $\mu = 100/2 = 50$ heads and standard deviation $\sigma = \sqrt{100}/2 = 10/2 = 5$ heads, confirming what we already knew.

(EXAMPLE 16.9)   **Coin-Tossing Experiments: Part 2**

An honest coin is going to be tossed 256 times. Before this is done, we have the opportunity to make some bets. Let's say that we can make a bet (with even odds) that if the number of heads tossed falls somewhere between 120 and 136, we will win; otherwise, we will lose. Should we make such a bet?

Let X denote the number of heads in 256 tosses of an honest coin. By the honest-coin principle, X is a random variable having a distribution that is approximately normal with mean $\mu = 256/2 = 128$ heads and standard deviation $\sigma = \sqrt{256}/2 = 8$ heads. The values 120 to 136 are exactly one standard deviation below and above the mean of 128, which means that *there is a 68% chance that the number of heads will fall somewhere between 120 and 136*. We should indeed make this bet! A similar calculation tells us that *there is a 95% chance that the number of heads will fall somewhere between 112 and 144*, and *the chance that the number of heads will fall somewhere between 104 and 152 is 99.7%*.   ⊂⊃

What happens when the coin being tossed is not an honest coin? Surprisingly, the distribution of the number of heads X in n tosses of such a coin is still approximately normal, as long as the number n is not too small (a good rule of thumb is $n \geq 30$). All we need now is a **dishonest-coin principle** to tell us how to find the mean and the standard deviation.

THE DISHONEST-COIN PRINCIPLE

Let X denote the number of heads in n tosses of a coin (assume $n \geq 30$). Let p denote the probability of heads on each toss of the coin. Then, X has an approximately normal distribution with mean $\mu = n \cdot P$ and standard deviation $\sigma = \sqrt{n \cdot p \cdot (1 - p)}$.

(EXAMPLE 16.10)   **Coin-Tossing Experiments: Part 3**

A coin is rigged so that it comes up heads only 20% of the time (i.e., $p = 0.20$). The coin is tossed 100 times ($n = 100$), and X is the number of heads in the 100 tosses. What can we say about X?

According to the dishonest-coin principle, the distribution of the random variable X is approximately normal with mean $\mu = 100 \times 0.20 = 20$ and standard deviation $\sigma = \sqrt{100 \times 0.20 \times 0.80} = 4$.

Applying the 68-95-99.7 rule with $\mu = 20$ and $\sigma = 4$ gives the following facts:

- There is about a 68% chance that X will be somewhere between 16 and 24 $(\mu - \sigma \le X \le \mu + \sigma)$.

- There is about a 95% chance that that X will be somewhere between 12 and 28 $(\mu - 2\sigma \le X \le \mu + 2\sigma)$.

- The number of heads is almost guaranteed (about 99.7% chance) to fall somewhere between 8 and 32 $(\mu - 3\sigma \le X \le \mu + 3\sigma)$.

Note that in this example, heads and tails are no longer interchangeable concepts—heads is an outcome with probability $p = 0.2$, while tails is an outcome with much higher probability (0.8). We can, however, apply the principle equally well to describe the distribution of the number of tails in 100 coin tosses of the same dishonest coin: The distribution for the number of tails is approximately normal with mean $\mu = 100 \times 0.80 = 80$ and standard deviation $\sigma = \sqrt{100 \times 0.80 \times 0.20} = 4$. Note that σ is still the same.

■ See Exercise 76.

The dishonest-coin principle can be applied to any coin, even one that is fair (i.e., with $p = 1/2$). In the case $p = 1/2$, the honest- and dishonest-coin principles say the same thing.

The dishonest-coin principle is a special version of one of the most important laws in statistics, a law generally known as the *central limit theorem*. We will now briefly illustrate why the importance of the dishonest-coin principle goes beyond the tossing of coins.

(**EXAMPLE 16.11**)   **Sampling for Defective Light Bulbs**

An assembly line produces 100,000 light bulbs a day, 20% of which generally turn out to be defective. Suppose that we draw a random sample of $n = 100$ light bulbs. Let X represent the *number of defective light bulbs* in the sample. What can we say about X?

A moment's reflection will show that, in a sense, this example is completely parallel to Example 16.10—think of selecting defective light bulbs as analogous to tossing heads with a dishonest coin. We can use the dishonest-coin principle to infer that the number of defective light bulbs in the sample is a random variable having an approximately normal distribution with a mean of 20 light bulbs and standard deviation of 4 light bulbs. Using these facts, we can draw the following conclusions:

- There is a 68% chance that the number of defective light bulbs in the sample will fall somewhere between 16 and 24.

- There is a 95% chance that the number of defective light bulbs in the sample will fall somewhere between 12 and 28.

- The number of defective light bulbs in the sample is practically guaranteed (a 99.7% chance) to fall somewhere between 8 and 32.

Probably the most important point here is that each of the preceding facts can be rephrased in terms of sampling errors, a concept we first discussed in Chapter 13. For example, say we had 24 defective light bulbs in the sample; in other words, 24% of the sample (24 out of 100) are defective light bulbs. If we use this statistic to estimate the percentage of defective light bulbs overall, then the sampling error would be 4% (because the estimate is 24% and the value of the parameter is 20%). By the same token, if we had 16 defective light bulbs in the sample, the sampling error would be −4%. Coincidentally, the standard deviation is $\sigma = 4$ light bulbs, or 4% of the sample. (We computed it in Example 16.10.) Thus, we can rephrase our previous assertions about sampling errors as follows:

- When estimating the proportion of defective light bulbs coming out of the assembly line by using a sample of 100 light bulbs, there is a 68% chance that the sampling error will fall somewhere between −4% and 4%.

- When estimating the proportion of defective light bulbs coming out of the assembly line by using a sample of 100 light bulbs, there is a 95% chance that the sampling error will fall somewhere between −8% and 8%.

- When estimating the proportion of defective light bulbs coming out of the assembly line by using a sample of 100 light bulbs, there is a 99.7% chance that the sampling error will fall somewhere between −12% and 12%.   ⬭

(**EXAMPLE 16.12**)   **Sampling with Larger Samples**

Suppose that we have the same assembly line as in Example 16.11, but this time we are going to take a really big sample of $n = 1600$ light bulbs. Before we even count the number of defective light bulbs in the sample, let's see how much mileage we can get out of the dishonest-coin principle. The standard deviation for the distribution of defective light bulbs in the sample is $\sqrt{1600 \times 0.2 \times 0.8} = 16$, which just happens to be exactly 1% of the sample ($16/1600 = 1\%$). This means that when we estimate the proportion of defective light bulbs coming out of the assembly line using this sample, we can have some sort of a handle on the sampling error.

- We can say with some confidence (68%) that the sampling error will fall somewhere between −1% and 1%.

- We can say with a lot of confidence (95%) that the sampling error will fall somewhere between −2% and 2%.

- We can say with tremendous confidence (99.7%) that the sampling error will fall somewhere between −3% and 3%.   ⬭

The next and last example shows how the dishonest-coin principle can be used to estimate the margin of error in a public opinion poll, an issue of considerable importance in modern statistics.

(**EXAMPLE 16.13**)   **Measuring the Margin of Error of a Poll**

In California, school bond measures require a 66.67% vote for approval. Suppose that an important school bond measure is on the ballot in the upcoming election. In the most recent poll of 1200 randomly chosen voters, 744 of the 1200 voters sampled, or 62%, indicated that they would vote for the school bond measure. Let's assume that the poll was properly conducted and that the 1200 voters sampled represent an unbiased sample of the entire population. What are the chances that the 62% statistic is the result of sampling variability and that the actual vote for the bond measure will be 66.67% or more?

Here, we will use a variation of the dishonest-coin principle, with each vote being likened to a coin toss: A vote for the bond measure is equivalent to flipping heads, a vote against the bond measure is equivalent to flipping tails. In this analogy, the probability (p) of "heads" represents the proportion of voters in the population that support the bond measure: If p turns out to be 0.6667 or more, the bond measure will pass. Our problem is that we don't know p, so how can we use the dishonest-coin principle to estimate the mean and standard deviation of the sampling distribution?

We start by letting the 62% (0.62) statistic from the sample serve as an estimate for the actual value of p in the formula for the standard deviation given by the dishonest-coin principle. (Even though we know that this is only a rough estimate for p, it turns out to give us a good estimate for the standard deviation σ.) Using $p \approx 0.62$ and the dishonest-coin principle, we get

$$\sigma = \sqrt{n \cdot p \cdot (1 - p)} \approx \sqrt{1200 \times 0.62 \times 0.38} \approx 16.8 \text{ votes.}$$

This number represents the approximate standard deviation for the number of "heads" (i.e., voters who will vote for the school bond measure) in the sample. If we express this number as a percentage of the sample size, we can say that the standard deviation represents approximately 1.4% of the sample ($16.8/1200 = 0.014$).

The standard deviation for the sampling distribution of the proportion of voters in favor of the measure expressed as a percentage of the entire sample is called the **standard error**. (For our example, we have found above that the standard error is approximately 1.4%.) In sampling and public opinion polls, it is customary to express the information about the population in terms of **confidence intervals**, which are themselves based on standard errors: A 95% confidence interval is given by two standard errors below and above the statistic obtained from the sample, and a 99.7% confidence interval is given by going three standard errors below and above the sample statistic.

For the school bond measure, a 95% confidence interval is 62% plus or minus $2 \times (1.4\%) = 2.8\%$. This means that we can say with 95% confidence (we would be right approximately 95 out of 100 times) that the actual vote for the bond measure will fall somewhere between 59.2% ($62 - 2.8$) and 64.8% ($62 + 2.8$) and thus that the bond measure will lose. Want even more certainty? Take a 99.7% confidence interval of 62% plus or minus $3 \times (1.4\%) = 4.2\%$—it is almost certain that the actual vote will turn out somewhere in that range. Even in the most optimistic scenario, the vote will not reach the 66.67% needed to pass the bond measure.

CONCLUSION

From coin tosses to test scores, many real-life data sets follow the call of the bell. In this chapter we studied bell-shaped (normal) curves, some of their mathematical properties, and how these properties can be used to analyze real-life bell-shaped data sets. It was a brief introduction to what is undoubtedly one of the most widely used and sophisticated tools of modern mathematical statistics.

In this chapter we also got a brief glimpse of the concept of statistical inference. The process of drawing conclusions and inferences based on limited data is an essential part of statistics. It gives us a way not only to analyze what has already taken place, but also to make reasonably accurate large-scale predictions of what will happen in certain random situations. Casinos know, without any shadow of a doubt, that in the long run they will make a profit—it is a mathematical law! A similar law gives us the confidence to trust the results of surveys and public opinion polls (up to a point!), the quality of the products we buy, and even the statistical data our government uses to make many of its decisions. In all of these cases, bell-shaped distributions of data and the laws of probability come together to give us insight into what was, is, and most likely will be.

PROFILE: Carl Friedrich Gauss (1777–1855)

In spite of the obvious connection between mathematics and money, mathematicians don't usually make the cut when it comes to getting their picture featured in a bank note. The notable exception is Carl Friedrich Gauss, the man often referred to as "the prince of mathematicians." The German 10-mark note shown in the figure celebrates both Gauss and one of his most famous "discoveries"—the normal curve (shown above the 10).

Carl Friedrich Gauss was born in the Duchy of Brunswick (now Germany) into a poor, working-class family of little education. From a very early age, young Gauss showed a prodigious talent for numbers and abstract mathematical thinking. One of the most famous stories about Gauss was his solution to the addition problem $1 + 2 + 3 + \cdots + 99 + 100 = ?$. The problem was assigned one day in class by a teacher hoping to keep the children busy for a while. No sooner had the teacher finished assigning the problem, Gauss had the solution. His insight: The sum consists of 50 pairs $(1 + 100, 2 + 99, \ldots, 50 + 51)$, each of which adds up to 101 and thus equals $50 \times 101 = 5050$. Gauss was seven years old at the time.

Recognizing Gauss's mathematical genius, the Duke of Brunswick agreed to be the young man's sponsor and financially support his education. Gauss first attended the Collegium Carolinum in Brunswick, but eventually transferred to the more prestigious Göttingen University. By the time he was 21, Gauss had produced several major mathematical discoveries, including the complete characterization of all regular polygons that can be constructed with straightedge and compass, a classic problem that could be traced all the way back to the ancient Greeks. In 1799 Gauss received his doctorate in mathematics from the University of Helmstedt for his work on what is now known as the *fundamental theorem of algebra*. Two years later he published his first and greatest book, *Disquisitiones Arithmeticae*, a pioneering treatise on the theory of numbers. With the publication of *Disquisitiones Arithmeticae*, Gauss came to be recognized as one of the greatest mathematicians of his age. He was 24 years old at the time.

Gauss's scientific genius went well beyond theoretical mathematics, and he made major contributions to empirical sciences such as astronomy and surveying, where precise observation and measurement are critical. He greatly added to his growing scientific reputation when he was able to predict the exact orbit of the asteroid Ceres, which had been discovered in 1801 by the Italian astronomer Giuseppe Piazzi. Piazzi had observed and tracked what he thought was a new planet for only a few weeks before it disappeared behind the Sun. Using Piazzi's limited data and a method of his own invention, Gauss predicted where and when Ceres would be seen again. Just about a year after its disappearance Ceres reappeared, almost in the exact position Gauss had predicted. On the strength of this achievement, Gauss was appointed Director of the Göttingen Observatory and Professor of Astronomy at Göttingen University.

As part of his astronomical calculations, Gauss became interested in the study of measurement errors and their probability distributions. This led him to the discovery of the bell-shaped normal distribution, which is also called the *Gaussian distribution* in his honor. Always a practical man, toward the later part of his life Gauss became increasingly interested in the applications of mathematics to physics and made important discoveries in the theory of magnetism, developed Kirchhoff's laws in electricity, and constructed an early version of the telegraph. He also became interested in the analysis and prediction of financial markets and made a tremendous amount of money speculating in stocks.

By the time he died in 1855 at the age of 78, Gauss had achieved a remarkable measure of fame and fortune—he was the preeminent mathematician of his time as well as an extremely wealthy man.

KEY CONCEPTS

EXERCISES

WALKING

A Normal Curves

1. Consider the normal distribution shown in Fig. 16-12, and assume that P is a point of inflection of the curve.

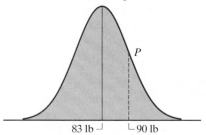

FIGURE 16-12

(a) Find the mean μ of the distribution.

(b) Find the median M of the distribution.

(c) Find the standard deviation σ of the distribution.

2. Consider the normal distribution shown in Fig. 16-13, and assume that P is a point of inflection of the curve.

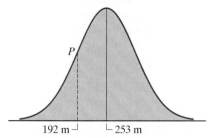

FIGURE 16-13

(a) Find the mean μ of the distribution.

(b) Find the median M of the distribution.

(c) Find the standard deviation σ of the distribution.

3. Consider the normal distribution shown in Fig. 16-14, and assume that P and P' are the two points of inflection of the curve.

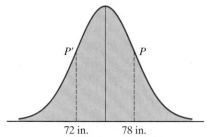

FIGURE 16-14

(a) Find the median M of the distribution.

(b) Find the standard deviation σ of the distribution.

(c) Find the third quartile Q_3 of the distribution rounded to the nearest inch.

(d) Find the first quartile Q_1 of the distribution rounded to the nearest inch.

4. Consider the normal distribution shown in Fig. 16-15, and assume that P and P' are the two points of inflection of the curve.

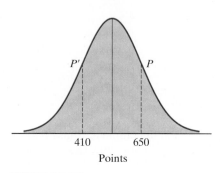

FIGURE 16-15

(a) Find the median M of the distribution.

(b) Find the standard deviation σ of the distribution.

(c) Find the third quartile Q_3 of the distribution rounded to the nearest point.

(d) Find the first quartile Q_1 of the distribution rounded to the nearest point.

5. Consider a normal distribution with mean $\mu = 81.2$ lb and standard deviation $\sigma = 12.4$ lb.

(a) Find the third quartile Q_3 of the distribution rounded to the nearest tenth of a pound.

(b) Find the first quartile Q_1 of the distribution rounded to the nearest tenth of a pound.

6. Consider a normal distribution with mean $\mu = 2354$ points and standard deviation $\sigma = 468$ points.

(a) Find the third quartile Q_3 of the distribution rounded to the nearest point.

(b) Find the first quartile Q_1 of the distribution rounded to the nearest point.

7. Consider a normal distribution with $Q_1 = 432.5$ points and $Q_3 = 567.5$ points.

  (a) Find the mean μ of the distribution.

  (b) Find the standard deviation σ of the distribution rounded to the nearest point.

8. Consider a normal distribution with $Q_1 = 229$ and $Q_3 = 391$.

  (a) Find the mean μ of the distribution.

  (b) Find the standard deviation σ of the distribution rounded to the nearest integer.

9. Estimate the value of the standard deviation σ (rounded to the nearest tenth of an inch) of a normal distribution with $\mu = 81.2$ in. and $Q_3 = 94.7$ in.

10. Estimate the value of the standard deviation σ (rounded to the nearest dollar) of a normal distribution with $\mu = \$18{,}565$ and $Q_1 = \$15{,}514$.

11. Explain why a distribution with median $M = 82$, mean $\mu = 71$, and standard deviation $\sigma = 11$ cannot be a normal distribution.

12. Explain why a distribution with median $M = 453$, mean $\mu = 453$, first quartile $Q_1 = 343$, and third quartile $Q_3 = 553$ cannot be a normal distribution.

13. Explain why a distribution with $\mu = 195$, $Q_1 = 180$, and $Q_3 = 220$ cannot be a normal distribution.

14. Explain why a distribution with $M = \mu = 47$, $Q_1 = 35$, and $\sigma = 10$ cannot be a normal distribution.

B Standardizing Data

15. A normal distribution has $\mu = 30$ kg and $\sigma = 15$ kg. Find the standardized value of

  (a) 45 kg.   **(b)** 0 kg.

  (c) 54 kg.   **(d)** 3 kg.

16. A normal distribution has $\mu = 110$ points and $\sigma = 12$ points. Find the standardized value of

  (a) 98 points.   **(b)** 110 points.

  (c) 128 points.   **(d)** 71 points.

17. A normal distribution has $\mu = 310$ points and $Q_3 = 391$ points. Find the standardized value of

  (a) 490 points.   **(b)** 250 points.

  (c) 220 points.   **(d)** 442 points.

18. A normal distribution has $\mu = 49.5$ lb and $Q_1 = 44.1$ lb. Find the standardized value of

  (a) 41.5 lb.   **(b)** 61.5 lb.

  (c) 35.1 lb.   **(d)** 67.5 lb.

19. Find the standardized value of Q_1 in any normal distribution.

20. Find the standardized value of Q_3 in any normal distribution.

21. In a normal distribution with $\mu = 183.5$ ft and $\sigma = 31.2$ ft, find the data value corresponding to each of the following standardized values.

  (a) -1 **(b)** 0.5

  (c) -2.3 **(d)** 0

22. In a normal distribution with $\mu = 83.2$ and $\sigma = 4.6$, find the data value corresponding to each of the following standardized values.

  (a) 2   **(b)** -1.5

  (c) -0.43 **(d)** 0

23. In a normal distribution with $\mu = 50$ lb, a weight of $x = 84$ lb has the standardized value $z = 2$. Find the standard deviation σ.

24. In a normal distribution with mean $\mu = 30$, the data value $x = -6$ has the standardized value $z = -3$. Find the standard deviation σ.

25. In a normal distribution with standard deviation $\sigma = 15$, the data value $x = 50$ has the standardized value $z = 3$. Find the mean μ.

26. In a normal distribution with standard deviation $\sigma = 20$, the data value $x = 10$ has the standardized value $z = -2$. Find the mean μ.

27. In a normal distribution, the data value $x_1 = 20$ has the standardized value $z_1 = -2$ and the data value $x_2 = 100$ has the standardized value $z_2 = 3$. Find the mean μ and the standard deviation σ.

28. In a normal distribution, the data value $x_1 = -10$ has the standardized value $z_1 = 0$ and the data value $x_2 = 50$ has the standardized value $z_2 = 2$. Find the mean μ and the standard deviation σ.

C The 68-95-99.7 Rule

29. Consider the normal distribution defined by Fig. 16-16. Find the mean μ and the standard deviation σ of the distribution.

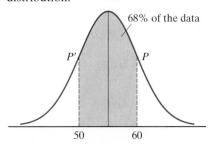

FIGURE 16-16

30. Consider the normal distribution defined by Fig. 16-17. Find the mean μ and the standard deviation σ of the distribution.

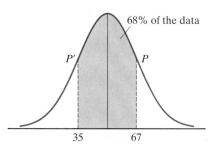

FIGURE 16-17

31. Consider the normal distribution defined by Fig. 16-18.

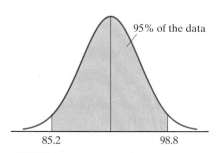

FIGURE 16-18

(a) Find the mean μ and the standard deviation σ of the distribution.

(b) Find the first quartile Q_1 and the third quartile Q_3 of the distribution.

32. Consider the normal distribution defined by Fig. 16-19.

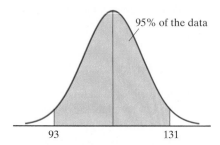

FIGURE 16-19

(a) Find the mean μ and the standard deviation σ of the distribution.

(b) Find the first quartile Q_1 and the third quartile Q_3 of the distribution.

33. Consider the normal distribution defined by Fig. 16-20. Find the mean μ and the standard deviation σ of the distribution.

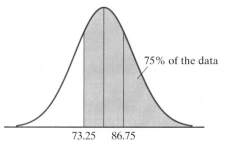

FIGURE 16-20

34. Consider the normal distribution defined by Fig. 16-21. Find the mean μ and the standard deviation σ of the distribution.

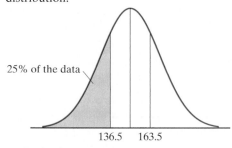

FIGURE 16-21

35. A normal distribution has standard deviation $\sigma = 6.1$ cm, and 84% of the data fall above 50.2 cm. Find the mean μ.

36. A normal distribution has mean $\mu = 56.3$ cm, and 84% of the data fall above 50.2 cm. Find the standard deviation σ.

37. In a normal distribution with mean μ and standard deviation σ, what percent of the data fall

(a) below the value $\mu + 2\sigma$?

(b) between $\mu + \sigma$ and $\mu + 2\sigma$?

38. In a normal distribution, what percent of the data fall

(a) above the value $\mu - 3\sigma$?

(b) between $\mu - 3\sigma$ and $\mu - 2\sigma$?

39. A normal distribution has mean $\mu = 12.6$ and standard deviation $\sigma = 4.0$. Approximately what percent of the data fall between 9.9 and 16.6?

40. A normal distribution has mean $\mu = 500$ and standard deviation $\sigma = 35$. Approximately what percent of the data fall between 465 and 605?

D **Approximately Normal Data Sets**

Exercises 41 through 44 refer to the following: 2500 students take a college entrance exam. The scores on the exam have an approximately normal distribution with mean $\mu = 52$ points and standard deviation $\sigma = 11$ points.

41. (a) Estimate the average score on the exam.

(b) Estimate the percentage of students scoring 52 points or more.

(c) Estimate the percentage of students scoring between 41 and 63 points.

(d) Estimate the percentage of students scoring 63 points or more.

42. (a) Estimate the percentage of students scoring between 30 and 74 points.

(b) Estimate the percentage of students scoring 30 points or less.

(c) Estimate the percentage of students scoring 19 points or less.

43. (a) Estimate the first-quartile score for this exam.

(b) Estimate the third-quartile score for this exam.

44. For each score, estimate the percentile (percentage of students scoring at or below that score).

(a) 52 points   **(b)** 63 points

(c) 60 points   **(d)** 85 points

Exercises 45 through 48 refer to the following: As part of a research project, the blood pressures of 2000 patients in a hospital are recorded. The systolic blood pressures (given in millimeters) have an approximately normal distribution with mean $\mu = 125$ and standard deviation $\sigma = 13$.

45. (a) Estimate the number of patients whose blood pressure was between 99 and 151 millimeters.

(b) Estimate the number of patients whose blood pressure was between 112 and 151 millimeters.

46. (a) Estimate the number of patients whose blood pressure was 99 millimeters or higher.

(b) Estimate the number of patients whose blood pressure was between 99 and 138 millimeters.

47. For each of the following blood pressures, estimate the corresponding percentile.

(a) 112 millimeters

(b) 138 millimeters

(c) 164 millimeters

48. (a) Assuming that there were no outliers, estimate the value of the lowest (*Min*) and the highest (*Max*) blood pressures.

(b) Assuming that there were no outliers, give an estimate of the five-number summary (*Min*, Q_1, μ, Q_3, *Max*) of the distribution of blood pressures.

Exercises 49 through 52 refer to the following: Packaged foods sold at supermarkets are not always the weight indicated on the package. Variability always crops up in the manufacturing and packaging process. Suppose that the exact weight of a "12-ounce" bag of potato chips is a random variable that has an approximately normal distribution with mean $\mu = 12$ ounces and standard deviation $\sigma = 0.5$ ounce.

49. If a "12-ounce" bag of potato chips is chosen at random, what are the chances that

(a) it weighs somewhere between 11 and 13 ounces?

(b) it weighs somewhere between 12 and 13 ounces?

(c) it weighs more than 11 ounces?

50. If a "12-ounce" bag of potato chips is chosen at random, what are the chances that

(a) it weighs somewhere between 11.5 and 12.5 ounces?

(b) it weighs somewhere between 12 and 12.5 ounces?

(c) it weighs more than 12.5 ounces?

51. Suppose that 500 "12-ounce" bags of potato chips are chosen at random. Estimate the number of bags with weight

(a) 11 ounces or less.

(b) 11.5 ounces or less.

(c) 12 ounces or less.

(d) 12.5 ounces or less.

(e) 13 ounces or less.

(f) 13.5 ounces or less.

52. Suppose that 1500 "12-ounce" bags of potato chips are chosen at random. Estimate the number of bags of potato chips with weight

(a) between 11 and 11.5 ounces.

(b) between 11.5 and 12 ounces.

(c) between 12 and 12.5 ounces.

(d) between 12.5 and 13 ounces.

(e) between 13 and 13.5 ounces.

Exercises 53 through 56 refer to the following: The distribution of weights for children of a given age and sex is approximately normal. This fact allows a doctor or nurse to find from a child's weight the weight percentile of the population (all children of the same age and sex) to which the child belongs. Typically, this is done using special charts provided to the doctor or nurse, but these percentiles can also be computed using facts about approximately normal distributions, such as the ones we learned in this chapter. (Note: The numbers in these examples are 1997 figures taken from charts produced by the National Center for Health Statistics, U.S. Department of Health and Human Services.)

53. The distribution of weights for six-month-old baby boys is approximately normal with mean $\mu = 17.25$ pounds and standard deviation $\sigma = 2$ pounds.

(a) Suppose that a six-month-old boy weighs 15.25 pounds. Approximately what weight percentile is he in?

(b) Suppose that a six-month-old boy weighs 21.25 pounds. Approximately what weight percentile is he in?

(c) Suppose that a six-month-old boy is in the 75th percentile in weight. Estimate his weight.

54. The distribution of weights for 12-month-old baby girls is approximately normal with mean $\mu = 21$ pounds and standard deviation $\sigma = 2.2$ pounds.

(a) Suppose that a 12-month-old girl weighs 16.6 pounds. Approximately what weight percentile is she in?

(b) Suppose that a 12-month-old girl weighs 18.8 pounds. Approximately what weight percentile is she in?

(c) Suppose that a 12-month-old girl is in the 75th percentile in weight. Estimate her weight.

55. The distribution of weights for one-month-old baby girls is approximately normal with mean $\mu = 8.75$ pounds and standard deviation $\sigma = 1.1$ pounds.

(a) Suppose that a one-month-old girl weighs 11 pounds. Approximately what weight percentile is she in?

(b) Suppose that a one-month-old girl weighs 12 pounds. Approximately what weight percentile is she in?

(c) Suppose that a one-month-old girl is in the 25th percentile in weight. Estimate her weight.

56. The distribution of weights for 12-month-old baby boys is approximately normal with mean $\mu = 22.5$ pounds and standard deviation $\sigma = 2.2$ pounds.

(a) Suppose that a 12-month-old boy weighs 24 pounds. Approximately what weight percentile is he in?

(b) Suppose that a 12-month-old boy weighs 21 pounds. Approximately what weight percentile is he in?

(c) Suppose that a 12-month-old boy is in the 84th percentile in weight. Estimate his weight.

E The Honest- and Dishonest-Coin Principles

57. An honest coin is tossed $n = 3600$ times. Let the random variable Y denote the number of tails tossed.

(a) Find the mean and the standard deviation of the distribution of the random variable Y.

(b) Estimate the chances that Y will fall somewhere between 1770 and 1830.

(c) Estimate the chances that Y will fall somewhere between 1800 and 1830.

(d) Estimate the chances that Y will fall somewhere between 1830 and 1860.

58. An honest coin is tossed $n = 6400$ times. Let the random variable X denote the number of heads tossed.

(a) Find the mean and the standard deviation of the distribution of the random variable X.

(b) Estimate the chances that X will fall somewhere between 3120 and 3280.

(c) Estimate the chances that X will fall somewhere between 3080 and 3200.

(d) Estimate the chances that X will fall somewhere between 3240 and 3280.

59. Suppose that a random sample of $n = 7056$ adults is to be chosen for a survey. Assume that the gender of each adult in the sample is equally likely to be male as it is female. Estimate the probability that

(a) the number of females in the sample is between 3486 and 3570.

(b) the number of females in the sample is less than 3486.

(c) the percentage of females in the sample is below 50.6%. (*Hint:* Find 50.6% of 7056 first.)

60. An honest die is rolled. If the roll comes out even (2, 4, or 6), you will win \$1; if the roll comes out odd (1, 3, or 5), you will lose \$1. Suppose that in one evening you play this game $n = 2500$ times in a row.

(a) Estimate the probability that by the end of the evening you will not have lost any money.

(b) Estimate the probability that the number of "even rolls" (roll a 2, 4, or 6) will fall between 1250 and 1300.

(c) Estimate the probability that you will win \$100 or more.

61. A dishonest coin with probability of heads $p = 0.4$ is tossed $n = 600$ times. Let the random variable X represent the number of times the coin comes up heads.

(a) Find the mean and standard deviation for the distribution of X.

(b) Find the first and third quartiles for the distribution of X.

(c) Find the probability that the number of heads will fall somewhere between 216 and 264.

62. A dishonest coin with probability of heads $p = 0.75$ is tossed $n = 1200$ times. Let the random variable X represent the number of times the coin comes up heads.

(a) Find the mean and standard deviation for the distribution of X.

(b) Find the first and third quartiles for the distribution of X.

(c) Find the probability that the number of heads will fall somewhere between 900 and 945.

63. Suppose that an honest die is rolled $n = 180$ times. Let the random variable X represent the number of times the number 6 is rolled.

(a) Find the mean and standard deviation for the distribution of X.

(b) Find the probability that a 6 will be rolled more than 40 times.

(c) Find the probability that a 6 will be rolled some-where between 30 and 35 times.

64. Suppose that one out of every ten plasma televisions shipped has a defective speaker. Out of a shipment of $n = 400$ plasma televisions, find the probability that there are

   (a) at most 40 with defective speakers.

   (b) more than 52 with defective speakers.

65. Each day a machine produces 1600 widgets. In 95 out of the last 100 days, the machine has produced somewhere between 117 and 139 defective widgets per day. What are the chances that a randomly selected widget produced by this machine is defective?

66. At Tasmania State University, the probability that an entering freshman will graduate in four years is 0.89. What are the chances that of a class of 2000 freshmen, 1750 or more will graduate in four years?

JOGGING

Percentiles. *The **pth percentile** of a sorted data set is a number x_p such that $p\%$ of the data fall at or below x_p and $(100 - p)\%$ of the data fall at or above x_p. (For details, see Chapter 14, Section 14.3.) For normally distributed data sets, there are detailed statistical tables that give the location of the pth percentile for every possible p between 1 and 99. The following table is an abbreviated version giving the approximate location of some of the more frequently used percentiles in a normal distribution with mean μ and standard deviation σ. For approximately normal distributions, Table 16-3 can be used to estimate these percentiles.*

TABLE 16-3

Percentile	Approximate location	Percentile	Approximate location
99th	$\mu + 2.33\sigma$	1st	$\mu - 2.33\sigma$
95th	$\mu + 1.65\sigma$	5th	$\mu - 1.65\sigma$
90th	$\mu + 1.28\sigma$	10th	$\mu - 1.28\sigma$
80th	$\mu + 0.84\sigma$	20th	$\mu - 0.84\sigma$
75th	$\mu + 0.675\sigma$	25th	$\mu - 0.675\sigma$
70th	$\mu + 0.52\sigma$	30th	$\mu - 0.52\sigma$
60th	$\mu + 0.25\sigma$	40th	$\mu - 0.25\sigma$
50th	μ		

In Exercises 67 through 73, you should use the table to make your estimates.

67. The distribution of weights for six-month-old baby boys is approximately normal with mean $\mu = 17.25$ pounds and standard deviation $\sigma = 2$ pounds.

   (a) Suppose that a six-month-old baby boy weighs in the 95th percentile of his age group. Estimate his weight in pounds approximated to two decimal places.

   (b) Suppose that a six-month-old baby boy weighs in the 40th percentile of his age group. Estimate his weight in pounds approximated to two decimal places.

68. Several thousand students took a college entrance exam. The scores on the exam have an approximately normal distribution with mean $\mu = 55$ points and standard deviation $\sigma = 12$ points.

   (a) For a student who scored in the 99th percentile, estimate the student's score on the exam.

   (b) For a student who scored in the 30th percentile, estimate the student's score on the exam.

69. Consider again the distribution of weights of six-month-old baby boys discussed in Exercise 67.

   (a) Jimmy is a six-month-old-baby who weighs 17.75 lb. Estimate the percentile corresponding to Jimmy's weight.

   (b) David is a six-month-old baby who weighs 16.2 lb. Estimate the percentile corresponding to David's weight.

70. Consider again the college entrance exam discussed in Exercise 68.

   (a) Mary scored 83 points on the exam. Estimate the percentile in which this score places her.

   (b) Adam scored 45 points on the exam. Estimate the percentile in which this score places him.

   (c) If there were 250 students that scored 35 points or less on the exam, estimate the total number of students who took the exam.

71. In 2007, a total of 1,494,531 college-bound seniors took the SAT exam. The distribution of scores in the *Critical Reading* section of the SAT was approximately normal with mean $\mu = 502$ and standard deviation $\sigma = 113$. (Source: *www.collegeboard.org*.)

   (a) Estimate the 75th percentile score on the exam. (*Note*: SAT scores come in multiples of 10.)

   (b) Estimate the 70th percentile score on the exam.

   (c) Estimate the percentile corresponding to a test score of 530.

72. Consider a normal distribution with mean $\mu = 0$ and standard deviation $\sigma = 1$.

   (a) Find the 90th percentile (rounded to two decimal places).

   (b) Find the 10th percentile (rounded to two decimal places).

   (c) Find the 80th percentile (rounded to two decimal places).

(d) Find the 20th percentile (rounded to two decimal places).

(e) Suppose that you are given that the 85th percentile is approximately 1.04. Find the 15th percentile.

73. The grade breakdown in Dr. Blackbeard's Stat 101 class is: 10% A's, 20% B's, 40% C's, 25% D's, and 5% F's. The numerical class scores had an approximately normal distribution with mean $\mu = 65.2$ and standard deviation $\sigma = 10$.

(a) What is the minimum numerical score needed to get an A?

(b) What is the minimum numerical score needed to get a B?

(c) What is the minimum numerical score needed to get a C?

(d) What is the minimum numerical score needed to get a D?

74. An honest coin is tossed n times. Let the random variable X denote the number of heads tossed. Find the value of n so that there is a 95% chance that X will be between $(n/2) - 15$ and $(n/2) + 15$.

75. An honest coin is tossed n times. Let the random variable Y denote the number of tails tossed. Find the value of n so that there is a 16% chance that Y will be at least $(n/2) + 10$.

76. Explain why when the dishonest-coin principle is applied to an honest coin ($p = 1/2$), we get the honest-coin principle.

RUNNING

77. A dishonest coin with probability of heads $p = 0.1$ is tossed n times. Let the random variable X denote the number of heads tossed. Find the value of n so that there is a 95% chance that X will be between $(n/10) - 30$ and $(n/10) + 30$.

78. An honest pair of dice is rolled n times. Let the random variable Y denote the number of times a total of 7 is rolled. Find the value of n so that there is a 95% chance that Y will be between $(n/6) - 20$ and $(n/6) + 20$.

79. In American roulette there are 18 red numbers, 18 black numbers, and 2 white numbers (0 and 00) as illustrated in

Fig. 16-22. The probability of a red number coming up on a single play of roulette is $p = 18/38 \approx 0.47$. Suppose that we go on a binge and bet $1 on red 10,000 times in a row. (A $1 bet on red wins $1 if a red number comes up, if a black or white number comes up, the bettor loses $1.)

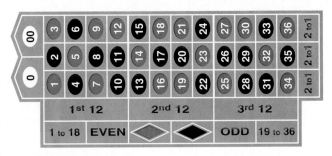

FIGURE 16-22

(a) Let Y represent the number of times we lose (i.e., the number of times that red does not come up). Use the dishonest-coin principle to describe the distribution of the random variable Y.

(b) Approximately what are the chances that we will lose 5300 times or more?

(c) Approximately what are the chances that we will lose somewhere between 5150 and 5450 times?

(d) Explain why the chances that we will break even or win in this situation are essentially zero.

80. After polling a random sample of 800 voters during the most recent gubernatorial race, the *Tasmania Gazette* reports the following:

> As the race for governor of Tasmania heads into its final days, our most recent poll shows Mrs. Butterworth ahead of the incumbent Mrs. Cubbison by 6 percentage points—53% to 47%. The results of the poll indicate with near certainty that if the election had been held at the time the poll was taken, Mrs. Butterworth would be the next governor of Tasmania.

(a) Estimate the standard error for this poll.

(b) Compute a 95% confidence interval for this poll.

(c) Compute a 99.7% confidence interval for this poll.

PROJECTS AND PAPERS

A Confidence Intervals

The concept of a *confidence interval* was introduced in Example 16.13 in this chapter. The two most frequently used levels of confidence intervals are 95% confidence intervals (sometimes described as intervals at a 95% *confidence*

level) and 90% confidence intervals (intervals at a 90% *confidence level*).

In this project, you are to describe the process of constructing confidence intervals in general. Given a target confidence level 95% (or 90%, or x%), how do you construct the corresponding confidence interval? Conversely, given a specified interval, how do you find the confidence level that best fits that interval? Illustrate the relationship between confidence level and the size of the confidence interval using a real-life poll. Conclude with a discussion of some common misconceptions about confidence intervals. When no confidence interval is mentioned, how should we properly interpret the results of the poll?

B Book Review: *The Bell Curve* and Rebuttals

In 1994, the late Richard J. Herrnstein of Harvard and Charles Murray of the Massachusetts Institute of Technology (MIT) wrote a book titled *The Bell Curve: Intelligence and Class Structure in American Life*. The book used social statistics (in particular, bell-shaped distributions of data on IQs and other tests of human intelligence) to justify the conclusion that intelligence levels vary across racial, ethnic, and economic groups. The book became one of the most discussed books of its time and raised a firestorm of controversy and many rebuttals.

In this project you should (1) summarize the main arguments presented by Herrnstein and Murray in their book and the justification given for these arguments and (2) summarize some of the main rebuttal arguments to the premises of the book. Most of the information you will need for this project you will find in references 3, 6, and 8.

REFERENCES AND FURTHER READINGS

1. Clemons, T., and M. Pagano, "Are Babies Normal?" *The American Statistician*, 53:4 (1999), 298–302.

2. Converse, P. E., and M. W. Traugott, "Assessing the Accuracy of Polls and Surveys." *Science*, 234 (1986), 1094–1098.

3. Devlin, B., S. E. Fienberg, D. P. Resnick, and K. Roeder, "Wringing the Bell Curve," *Chance*, 8, (1995), 27–36.

4. Frankel, Max, "Margins of Error," *New York Times Magazine*, December 15, 1996, 34.

5. Freedman, D., R. Pisani, R. Purves, and A. Adhikari, *Statistics*, 2d ed. New York: W. W. Norton, 1991, chaps. 16 and 18.

6. Gould, Stephen Jay, *The Mismeasure of Man*. New York: W. W. Norton, 1996.

7. Herrnstein, R. J., and C. Murray, *The Bell Curve: Intelligence and Class Structure in American Life*. New York: Free Press, 1994.

8. Jacoby, Russell, and Naomi Glauberman (eds.), *The Bell Curve Debate: History, Documents, Opinions*. New York: Times Books, 1995.

9. Kerrich, John, *An Experimental Introduction to the Theory of Probability*. Witwatersrand, South Africa: University of Witwatersrand Press, 1964.

10. Larsen, J., and D. F. Stroup, *Statistics in the Real World*. New York: Macmillan Publishing Co., 1976.

11. Mosteller, F., W. Kruskal, et al., *Statistics by Example: Detecting Patterns*. Reading, MA: Addison-Wesley Publishing Co., 1973.

12. Mosteller, F., R. Rourke, and G. Thomas, *Probability and Statistics*. Reading, MA: Addison-Wesley Publishing Co., 1961.

13. Taleb, Nassim, *The Black Swan: The Impact of the Highly Improbable*. New York: Random House, 2007.

14. Tanner, Martin, *Investigations for a Course in Statistics*. New York: Macmillan Publishing Co., 1990.

15. Tent, M. B. W., *Prince of Mathematics: Carl Friedrich Gauss*. Wellesley, MA: A. K. Peters, 2006.

mini-excursion 4

The Mathematics of Managing Risk

Expected Values

Life is full of risks. Some risks we take because we essentially have no choice—for example, when we fly in an airplane, get in a car, or ride in an elevator to the top of a skyscraper. Some risks we take on willingly and just for thrills—that's why some of us surf big waves, rock climb, or skydive. A third category of risks is that of *calculated risks*, risks that we take because there is a tangible payoff—that's why some of us gamble, invest in the stock market, or try to steal second base.

Implicit in the term *calculated risk* is the idea that there is some type of calculation going on. Often the calculus of risk is informal and fuzzy ("I have a gut feeling my number is coming up," "It's a good time to invest in real estate," etc.), but in many situations both the risk and the associated payoffs can be quantified in precise mathematical terms. In these situations the relationship between the risks we take and the payoffs we expect can be measured using an important mathematical tool called the *expected value*. The notion of expected value gives us the ability to make rational decisions—what risks are worth taking and what risks are just plain foolish.

> *A lot of people approach risk as if it's the enemy, when it's really fortune's accomplice.*
>
> —Sting

In this mini-excursion we will briefly explore the concept of the *expected value* (or *expectation*) of a random variable and illustrate the concept with several real-life applications to gambling, investing, and risk-taking in general. The main prerequisites for this mini-excursion are covered in Chapter 15, in particular the concepts covered in Sections 15.4 and 15.5.

Weighted Averages

(**EXAMPLE ME4.1**) **Computing Class Scores**

Imagine that you are a student in Prof. Goodie's Stat 100 class. The grading for the class is based on two midterms, homework, and a final exam. The breakdown for the scoring is given in the first two rows of Table ME4-1. Your scores are given in the last row. You needed to average 90% or above to get an A in the course. Did you?

One of your friends says you should add all your scores and divide by the total number of possible points to compute your final average. This gives $(82 + 88 + 182 + 190)/600 = 0.9033\cdots \approx 90.33\%$.

TABLE ME4-1

	Midterm 1	Midterm 2	Homework	Final exam
Weight	20%	20%	25%	35%
Possible points	100	100	200	200
Your scores	82	88	182	190

If your friend is right you would end up with an A. A second friend says that you are going to miss an A by just one percentage point, and isn't that too bad! Your second friend's calculation is as follows: You got 82% on the first midterm, 88% on the second midterm, 91% (182/200) on the homework, and 95% (190/200) on the final exam. The average of these four percentages is $(82\% + 88\% + 95\% + 91\%)/4 = 89\%$.

Fortunately, you know more math than either one of your friends. The correct computation, you explain to them patiently, requires that we take into account that (1) the scores are based on different scales (100 points, 200 points) and (2) the weights of the scores (20%, 20%, 25%, 35%) are not all the same. To take care of (1) we express the numerical scores as percentages (82%, 88%, 91%, 95%), and to take care of (2) we multiply these percentages by their respective weights (20% = 0.20, 25% = 0.25, 35% = 0.35). We then add all of these numbers. Thus, your correct average is

$$0.2 \times 82\% + 0.2 \times 88\% + 0.25 \times 91\% + 0.35 \times 95\% = 90\%$$

How sweet it is—you are getting that A after all!   ⊂⊃

(**EXAMPLE ME4.2**)   **Average GPAs at Tasmania State University**

Table ME4-2 shows GPAs at Tasmania State University broken down by class. Our task is to compute the overall GPA of *all undergraduates* at TSU.

TABLE ME4-2   GPAs at Tasmania State University by Class

Class	Freshman	Sophomore	Junior	Senior
Average GPA	2.75	3.08	2.94	3.15

The information given in Table ME4-2 is not enough to compute the overall GPA because the number of students in each class is not the same. After a little digging we find out that the 15,000 undergraduates at Tasmania State are divided by class as follows: 4800 freshmen, 4200 sophomores, 3300 juniors, and 2700 seniors.

To compute the overall school GPA we need to "weigh" the average GPA for each class with the relative size of that class. The relative size of the freshman class is 4800/15,000 = 0.32, the relative size of the sophomore class is 4200/15,000 = 0.28, and so on (0.22 for the junior class and 0.18 for the senior class). When all is said and done, the overall school GPA is

$$0.32 \times 2.75 + 0.28 \times 3.08 + 0.22 \times 2.94 + 0.18 \times 3.15 = 2.9562$$

⊂⊃

Examples ME4.1 and ME4.2 illustrate how to average numbers that have different relative *weights* using the notion of a *weighted average*. In Example ME4.1 the scores (82%, 88%, 91%, and 95%) were multiplied by their corresponding "weights" (0.20, 0.20, 0.25, and 0.35) and then added to compute the weighted average. In Example ME4.2 the class GPAs (2.75, 3.08, 2.94, and 3.15) were multiplied by the corresponding class "weight" (0.32, 0.28, 0.22, 0.18) to compute the *weighted average*. Note that in both examples the **weights** add up to 1, which is not surprising since they represent percentages that must add up to 100%.

WEIGHTED AVERAGE

Let X be a variable that takes the values $v_1, v_2, \ldots, v_N$, and let $w_1, w_2, \ldots, w_N$ denote the respective weights for these values, with $w_1 + w_2 + \cdots + w_N = 1$. The **weighted average** for X is given by

$$w_1 \cdot v_1 + w_2 \cdot v_2 + \cdots + w_N \cdot v_N$$

Expected Values

The idea of a weighted average is particularly useful when the weights represent probabilities.

EXAMPLE ME4.3   **To Guess or Not to Guess**

The SAT is a standardized college entrance exam taken by over a million students each year. In the multiple-choice sections of the SAT each question has five possible answers (A, B, C, D, and E). A correct answer is worth 1 point, and, to discourage indiscriminate guessing, an incorrect answer carries a penalty of $1/4$ point (i.e., it counts as $-1/4$ point).

Imagine you are taking the SAT and are facing a multiple-choice question for which you have no clue as to which of the five choices might be the right answer—they all look equally plausible. You can either play it safe and leave it blank, or you can gamble and take a wild guess. In the latter case, the upside of getting 1 point must be measured against the downside of getting a penalty of $1/4$ point. What should you do?

Table ME4-3 summarizes the possible options and their respective payoffs (for the purposes of illustration assume that the correct answer is B, but of course, you don't know that when you are taking the test).

TABLE ME4-3

Option	Leave blank	A	B	C	D	E
Payoff	0	-0.25	1	-0.25	-0.25	-0.25

We will now need some basic probability concepts introduced in Chapter 15. When you are randomly guessing the answer to this multiple-choice question you are unwittingly conducting a *random experiment* with sample space $S = \{A, B, C, D, E\}$. Since each of the five choices is equally likely to be the correct answer (remember, you are clueless on this one), you assign equal probabilities of $p = 1/5 = 0.2$ to each outcome. This probability assignment, combined with the information in Table ME4-3, gives us Table ME4-4.

■ Also note that standardized tests such as the SAT are designed for "key balance" (each of the possible answers occurs with approximately equal frequency).

TABLE ME4-4

Outcome	Correct answer (B)	Incorrect answer (A, C, D, or E)
Point payoff	1	−0.25
Probability	0.2	0.8

Using Table ME4-4 we can compute something analogous to a weighted average for the point payoffs with the *probabilities as weights*. We will call this the **expected payoff**. Here the expected payoff E is

$$E = 0.2 \times 1 + 0.8 \times (-0.25) = 0.2 - 0.2 = 0 \text{ points}$$

The fact that the expected payoff is 0 points implies that this guessing game is a "fair" game—in the long term (if you were to make these kinds of guesses many times) the penalties that you would accrue for wrong guesses are neutralized by the benefits that you would get for your lucky guesses. This knowledge will not impact what happens with an individual question [the possible outcomes are still a +1 (lucky) or a −0.25 (wrong guess)], but it gives you some strategically useful information: On the average, totally random guessing on the multiple-choice section of the SAT neither helps nor hurts!   ⊂⊃

EXAMPLE ME4.4   **To Guess or Not to Guess: Part 2**

Example ME4.3 illustrated what happens when the multiple-choice question has you completely stumped—each of the five possible choices (A, B, C, D, or E) looks equally likely to be the right answer. To put it bluntly, you are clueless! At a slightly better place on the ignorance scale is a question for which you can definitely rule out one or two of the possible answers. Under these circumstances we must do a different calculation for the risk/benefits of guessing.

Let's consider first a multiple-choice question for which you can safely rule out one of the five choices. For the purposes of illustration let's assume that the correct answer is B and that you can definitely rule out choice E. Among the four possible choices (A, B, C, or D) you have no idea which is most likely to be the correct answer, so you are going to randomly guess. In this scenario we assign the same probability (0.25) to each of the four choices, and the guessing game is described in Table ME4-5.

TABLE ME4-5

Outcome	Correct answer (B)	Incorrect answer (A, C, or D)
Point payoff	1	−0.25
Probability	0.25	0.75

Once again, the *expected payoff* E can be computed as a weighted average:

$$E = 0.25 \times 1 + 0.75 \times (-0.25) = 0.25 - 0.1875 = 0.0625$$

■ $0.0625 = 1/16$.

An expected payoff of 0.0625 points is very small—it takes 16 guesses of this type to generate an expected payoff equivalent to 1 correct answer (1 point). At the same time, the fact that it is a positive expected payoff means that the benefit justifies (barely) the risk.

A much better situation occurs when you can rule out two of the five possible choices (say, D and E). Now the random experiment of guessing the answer is described in Table ME4-6.

TABLE ME4-6

Outcome	Correct answer (B)	Incorrect answer (A or C)
Point payoff	1	$-1/4$
Probability	1/3	2/3

■ The switch from decimals to fractions is to avoid having to round 1/3 and 2/3 in our calculations.

In this situation the expected payoff E is given by

$$E = (1/3) \times 1 + (2/3) \times (-1/4) = 1/3 - 1/6 = 1/6$$

Examples ME4.3 and ME4.4 illustrate the mathematical reasoning behind a commonly used piece of advice given to SAT-takers: *Guess the answer if you can rule out some of the options; otherwise, don't bother.*

■ For a review of *random variables,* see Chapter 16.

The basic idea illustrated in Examples ME4.3 and ME4.4 is that of the *expected value* (or *expectation*) of a random variable.

┌─ **EXPECTED VALUE** ─────────────────────────

Suppose X is a random variable with outcomes $o_1, o_2, o_3, \ldots, o_N$, having probabilities $p_1, p_2, p_3, \ldots, p_N$, respectively. The **expected value** (or **expectation**) of X is given by

$$E = p_1 \cdot o_1 + p_2 \cdot o_2 + p_3 \cdot o_3 + \cdots + p_N \cdot o_N$$

Applications of Expected Value

In many real-life situations, we face decisions that can have many different potential consequences—some good, some bad, some neutral. These kinds of decisions are often quite hard to make because there are so many intangibles, but sometimes the decision comes down to a simple question: Is the reward worth the risk? If we can quantify the risks and the rewards, then we can use the concept of *expected value* to help us make the right decision. The classic illustration of this type of situation is provided by "games" in which there is money riding on the outcome. This includes not only typical gambling situations (dice games, card games, lotteries, etc.) but also investing in real estate or playing the stock market.

Playing the stock market is a legalized form of gambling in which people make investment decisions (buy? sell? when? what?) instead of rolling the dice or spinning a wheel. Some people have a knack for making good investment decisions and in the long run can do very well—others, just the opposite. A lot has to do with what drives the decision-making process—gut feelings, rumors, and can't-miss tips from your cousin Vinny at one end of the spectrum, reliable information combined with sound mathematical principles at the other end. Which approach would you rather trust your money to? (If your answer is the former, then you might as well skip the next example.)

> **EXAMPLE ME4.5**   **To Buy or Not to Buy?**

Fibber Pharmaceutical is a new start-up in the pharmaceutical business. Its stock is currently selling for $10. Fibber's future hinges on a new experimental vaccine it believes has great promise for the treatment of the avian flu virus. Before the vaccine can be approved by the Food and Drug Administration (FDA) for use with the general public it must pass a series of clinical trials known as Phase I, Phase II, and Phase III trials. If the vaccine passes all three trials and is approved by the FDA, shares of Fibber are predicted to jump tenfold to $100 a share. At the other end of the spectrum, the vaccine may turn out to be a complete flop and fail Phase I trials. In this case, Fibber's shares will be worthless. In between these two extremes are two other possibilities: The vaccine will pass Phase I trials but fail Phase II trials or will pass Phase I and Phase II trials but fail Phase III trials. In the former case shares of Fibber are expected to drop to $5 a share; in the latter case shares of Fibber are expected to go up to $15 a share.

Table ME4-7 summarizes the four possible outcomes of the clinical trials. The last row of Table ME4-7 gives the probability of each outcome based on previous experience with similar types of vaccines.

TABLE ME4-7

Outcome of trials	Fail Phase I	Fail Phase II	Fail Phase III	FDA approval
Estimated share price	$0	$5	$15	$100
Probability	0.25	0.45	0.20	0.10

Combining the second and third rows of Table ME4-7, we can compute the expected value E of a future share of Fibber Pharmaceutical:

$$E = 0.25 \times \$0 + 0.45 \times \$5 + 0.20 \times \$15 + 0.10 \times \$100 = \$15.25$$

To better understand the meaning of the $15.25 expected value of a $10 share of Fibber Pharmaceutical, imagine playing the following game: For a cost of $10 you get to roll a die. This is no ordinary die—this die has only four faces (labeled I, II, III, and IV), and the probabilities of each face coming up are different (0.25, 0.45, 0.20, and 0.10, respectively). If you Roll a I, your payoff is $0 (your money is gone); if you roll a II, your payoff is $5 (you are still losing $5); if you roll a III, your payoff is $15 (you made $5); and if you roll a IV, your payoff is $100 (jackpot!). The beauty of this random experiment is that it can be played over and over, hundreds or even thousands of times. If you do this, sometimes you'll roll a I and poof—your money is gone, a few times you'll roll a IV and make out like a bandit, other times you'll roll a III or a II and make or lose a little money. The key observation is that if you play this game long enough, the average payoff is going to be $15.25 per roll of the die, giving you an average profit or gain of $5.25 per roll. In purely mathematical terms, this game is a game well worth playing (but since this book does not condone gambling, this is just a theoretical observation).

Is Fibber Pharmaceutical a good investment or not? At first glance, an expected value of $15.25 per $10 invested looks like a great risk, but we have to balance this with the several years that it might take to collect on the investment (unlike the die game that pays off right away). For example, assuming four years to complete the clinical trials (sometimes it can take even longer), an investment in Fibber shares has a comparable expected payoff as a safe investment at a fixed annual yield of about 11%. This makes for a good, but hardly spectacular, investment.

■ See Exercise 20.

(**EXAMPLE ME4.6**)   **Raffles and Fund-Raisers**

A common event at many fund-raisers is to conduct a raffle—another form of legalized gambling. At this particular fund-raiser, the raffle tickets are going for $2. In this raffle, the organizers will draw one grand-prize winner worth $500, four second-prize winners worth $50 each, and fifteen third-prize winners worth $20 each. Sounds like a pretty good deal for a $2 investment, but is it? The answer, of course, depends on how many tickets are sold in this raffle.

Suppose that this is a big event, and 1000 tickets are sold. Table ME4-8 shows the four possible outcomes for your raffle ticket (first row), their respective net payoffs after subtracting the $2 cost of the ticket (second row), and their respective probabilities (third row). The last column of the table reflects the fact that if your number is not called, your ticket is worthless and you lost $2.

TABLE ME4-8

Outcome	Grand prize	Second prize	Third prize	No prize
Net gain	$498	$48	$18	−$2.00
Probability	1/1000	4/1000	15/1000	980/1000

From Table ME4-8 we can compute the expected value E of the raffle ticket:

$$E = \frac{1}{1000} \times \$498 + \frac{4}{1000} \times \$48 + \frac{15}{1000} \times \$18 + \frac{980}{1000} \times \$(-2)$$

$$= \$0.498 + \$0.192 + \$0.27 - \$1.96 = -\$1.00$$

The negative expected value is an indication that this game favors the people running the raffle. We would expect this—it is, after all, a fund-raiser. But the computation gives us a precise measure of the extent to which the game favors the raffle: on the average we should expect to lose $1 for every $2 raffle ticket purchased. If we purchased, say, 100 raffle tickets, we would in all likelihood have a few winning tickets and plenty of losing tickets, but who cares about the details—at the end we would expect a net loss of about $100.   ⊂⊃

(**EXAMPLE ME4.7**)   **Raffles and Fund-Raisers: Part 2**

Most people buy raffle tickets to support a good cause and not necessarily as a rational investment, but it is worthwhile asking, What should be the price of a raffle ticket if we want the raffle to be a "fair game"? (Let's assume that we are still discussing the raffle in Example ME4.6.) In order to answer this question, we set the price of a raffle ticket to be x. Table ME4-9 is a generalization of Table ME4-8 for a raffle ticket with a cost of x.

TABLE ME4-9

Outcome	Grand prize	Second prize	Third prize	No prize
Net gain	$500 − x$	$50 − x$	$20 − x$	$−x$
Probability	1/1000	4/1000	15/1000	980/1000

If we now set the expected gain to be $0 (that's what will make it a fair game), we get the equation

$$\frac{1}{1000}(500 - x) + \frac{4}{1000}(50 - x) + \frac{15}{1000}(20 - x) + \frac{980}{1000}(-x) = 0$$

Solving the above equation for x gives $x = \$1$.

EXAMPLE ME4.8   Chuck-a-Luck

Chuck-a-luck is an old game, played mostly in carnivals and county fairs. We will discuss it here because it involves a slightly more sophisticated calculation of the probabilities of the various outcomes. To play chuck-a-luck you place a bet, say $1, on one of the numbers 1 through 6. Say that you bet on the number 4. You then roll three dice (presumably honest). If you roll three 4's, you win $3.00; if you roll just two 4's, you win $2; if you roll just one 4, you win $1 (and of course, in all of these cases, you get your original $1 back). If you roll no 4's, you lose your $1. Sounds like a pretty good game, doesn't it?

To compute the expected payoff for chuck-a-luck we will use some of the ideas introduced in Chapter 15, Section 15.5. When we roll three dice, the sample space consists of $6 \times 6 \times 6 = 216$ outcomes. The different events we will consider are shown in the first row of Table ME4-10 (the * indicates any number other than a 4). The second row of Table ME4-10 shows the size (number of outcomes) of that event, and the third row shows the respective probabilities.

TABLE ME4-10

Event	(4, 4, 4)	(4, 4, *)	(4, *, 4)	(*, 4, 4)	(4, *, *)	(*, 4, *)	(*, *, 4)	(*, *, *)
Size	1	5	5	5	5×5	5×5	5×5	$5 \times 5 \times 5$
Probability	1/216	5/216	5/216	5/216	25/216	25/216	25/216	125/216

■ See Exercise 19.

Combining columns in Table ME4-10, we can deduce the probability of rolling three 4's, two 4's, one 4, or no 4's when we roll three dice. This leads to Table ME4-11. The payoffs are based on a $1 bet.

TABLE ME4-11

Roll	Three 4's	Two 4's	One 4	No 4's
Net gain	$3	$2	$1	−$1
Probability	1/216	15/216	75/216	125/216

The expected value of a $1 bet on chuck-a-luck is given by

$$E = (1/216) \times \$3 + (15/216) \times \$2 + (75/216) \times \$1 + (125/216) \times \$(-1)$$
$$= \$(-17/216) \approx \$-0.08$$

Essentially, this means that in the long run, for every $1 bet on chuck-a-luck, the player will lose on the average about 8 cents, or 8%. This, of course, represents the "house" profit and in gambling is commonly referred to as the *house margin* or *house advantage*.

The concept of expected value is used by insurance companies to set the price of premiums (the process is called **expectation based pricing**). Our last example is an oversimplification of how the process works, but it illustrates the key ideas behind the setting of life insurance premiums.

(**EXAMPLE ME4.9**) **Setting Life Insurance Premiums**

A life insurance company is offering a $100,000 one-year term life insurance policy to Janice, a 55-year-old nonsmoking female in moderately good health. What should be a reasonable premium for this policy?

For starters, we will let P denote the *break-even*, or *fair*, premium that the life insurance company should charge Janice if it were not in it to make a profit. This can be done by setting the expected value to be 0. Using mortality tables, the life insurance company can determine that the probability that someone in Janice's demographic group will die within the next year is 1 in 500, or 0.002. The second row of Table ME4-12 gives the value of the two possible outcomes to the life insurance company, and the third row gives the respective probabilities.

TABLE ME4-12

Outcome	Janice dies	Janice doesn't die
Payoff	$(P - 100,000)$	P
Probability	0.002	0.998

■ See Exercise 14.

Setting the expected payoff equal to 0 gives $P = 200$. This is the premium the insurance company should charge to break even, but of course, insurance companies are in business to make a profit. A standard gross profit margin in the insurance industry is 20%, which in this case would tack on $40 to the premium. We can conclude that the premium for Janice's policy should be about $240.   ⊂⊃

KEY CONCEPTS

expectation based pricing, **620**   expected value (expectation), **616**   weighted average, **614**
expected payoff, **615**   weight, **614**

EXERCISES

A Weighted Averages

1. The scoring for a Psych 101 final grade is given in the following table. The last row of the table shows Paul's scores. Find Paul's score in the course, expressed as a percent.

	Test 1	Test 2	Test 3	Quizzes	Paper	Final
Weight (percentage of grade)	15%	15%	15%	10%	25%	20%
Possible points	100	100	100	120	100	180
Paul's scores	77	83	91	90	87	144

2. In his record-setting victory in the 1997 Masters, Tiger Woods had the following distribution of scores:

Score	2	3	4	5	6
Percentage of holes	1.4%	36.1%	50%	11.1%	1.4%

What was Tiger Woods's average score per hole during the 1997 Masters?

3. At Thomas Jefferson High School, the student body is divided by age as follows: 7% of the students are 14, 22% of the students are 15, 24% of the students are 16, 23% of the students are 17, 19% of the students are 18, and the rest of the students are 19. Find the average age of the students at Thomas Jefferson High School.

4. In 2005 the Middletown Zoo averaged 4000 visitors on sunny days, 3000 visitors on cloudy days, 1500 visitors on rainy days, and only 100 visitors on snowy days. The percentage of days of each type in 2005 is given in the following table. Find the average daily attendance at the Middletown Zoo for 2005.

Weather condition	Sunny	Cloudy	Rainy	Snowy
Percentage of days	47%	27%	19%	7%

B Expected Values

5. Find the expected value of the random variable with outcomes and associated probability distribution shown in the following table.

Outcome	5	10	15
Probability	1/5	2/5	2/5

6. Find the expected value of the random variable with outcomes and associated probability distribution shown in the following table.

Outcome	10	20	30	40
Probability	0.2	0.3	0.4	?

7. A box contains twenty $1 bills, ten $5 bills, five $10 bills, four $20 bills, and one $100 bill. You reach into the box

and pull out a bill at random, which you get to keep. Let X represent the value of the bill that you draw.

(a) Find the probability distribution of X.

(b) Find the expected value of X.

(c) How much should you pay for the right to draw from the box if the game is to be a fair game?

8. A basketball player shoots two consecutive free-throws. Each free-throw is worth 1 point and has probability of success $p = 3/4$. Let X denote the number of points scored.

(a) Find the probability distribution of X.

(b) Find the expected value of X.

9. A fair coin is tossed three times. Let X denote the number of *heads* that come up.

(a) Find the probability distribution of X.

(b) Find the expected value of X.

10. Suppose that you roll a pair of honest dice. If you roll a total of 7, you win $18; if you roll a total of 11, you win $54; if you roll any other total, you lose $9. Let X denote the payoff in a single play of this game.

(a) Find the expected value of a play of this game.

(b) How much should you pay for the right to roll the dice if the game is to be a fair game?

11. On an American roulette wheel, there are 18 red numbers, 18 black numbers, plus 2 green numbers (0 and 00). If you bet $N on red, you win $N if a red number comes up (i.e., you get $2N back—your original bet plus your $N profit); if a black or green number comes up, you lose your $N bet.

(a) Find the expected value of a $1 bet on red.

(b) Find the expected value of an $N bet on red.

12. On an American roulette wheel, there are 38 numbers: 00, 0, 1, 2, . . . , 36. If you bet $N on any one number—say, for example, on 10—you win $36N if 10 comes up (i.e., you get $37N back—your original bet plus your $36N profit); if any other number comes up, you lose your $N bet.

(a) Find the expected value of a $1 bet on 10 (or any other number).

(b) Find the expected value of an $N bet on 10.

13. Suppose that you roll a single die. If an odd number (1, 3, or 5) comes up, you win the amount of your roll ($1, $3, or $5, respectively). If an even number (2, 4, or 6) comes up, you have to pay the amount of your roll ($2, $4, or $6, respectively).

(a) Find the expected value of this game.

(b) Find a way to change the rules of this game so that it is a fair game.

14. This exercise refers to Example ME4.9.

(a) Explain how the fair premium of $P = 200$ is obtained.

(b) Find the value of a fair premium for a person with probability of death over the next year estimated to be 3 in 1000.

C Miscellaneous

15. Joe is buying a new plasma TV at Circuit Town. The salesman offers Joe a three-year extended warranty for $80. The salesman tells Joe that 24% of these plasma TVs require repairs within the first three years, and the average cost of a repair is $400. Should Joe buy the extended warranty? Explain your reasoning.

16. Jackie is buying a new MP3 player from Better Buy. The store offers her a two-year extended warranty for $19. Jackie read in a consumer magazine that for this model MP3, 5% require repairs within the first two years at an average cost of $50. Should Jackie buy the extended warranty? Explain your reasoning.

17. The service history of the Prego SUV is as follows: 50% will need no repairs during the first year, 35% will have repair costs of $500 during the first year, 12% will have repair costs of $1500 during the first year, and the remaining SUVs (the real lemons) will have repair costs of $4000 during their first year. Determine the price that the insurance company should charge for a one-year extended warranty on a Prego SUV if it wants to make an average profit of $50 per policy.

18. An insurance company plans to sell a $250,000 one-year term life insurance policy to a 60-year-old male. Of 2.5 million men having similar risk factors, the company estimates that 7500 of them will die in the next year. What is the premium that the insurance company should charge if it would like to make a profit of $50 on each policy?

19. This exercise refers to the game of chuck-a-luck discussed in Example ME4.8. Explain why, when you roll three dice,

(a) the probability of rolling two 4's plus another number (not a 4) is 15/216.

(b) the probability of rolling one 4 plus two other numbers (not 4's) is 75/216.

(c) the probability of rolling no 4's is 125/216.

20. For this exercise you will need to use the general compounding formula introduced in Chapter 10. Explain why the expected value of an investment in Fibber Pharmaceutical (see Example ME4.5) is comparable to an investment with an expected annual yield of about 11%. (Assume that it takes four years to complete the clinical trials.)

21. In the California Super Lotto game you choose 5 numbers between 1 and 47 plus a Mega number between 1 and 27. Suppose that a winning ticket pays $30 million (assume that the prize will not be split among several winners). Find the expected value of a $1 lottery ticket. Round your answer to the nearest penny. (*Hint*: You may want to review Section 15.3 of Chapter 15.)

22. In the California Mega Millions Lottery you choose 5 numbers between 1 and 56 plus a Mega number between 1 and 46. Suppose that a winning ticket pays $64 million (assume that the prize will not be split among several winners). Find the expected value of a $1 lottery ticket. Round your answer to the nearest penny. (*Hint*: You may want to review Section 15.3 of Chapter 15.)

REFERENCES AND FURTHER READINGS

1. Gigerenzer, Gerd, *Calculated Risks*. New York: Simon & Schuster, 2002.

2. Hacking, Ian, *An Introduction to Probability and Inductive Logic.* New York: Cambridge University Press, 2001.

3. Haigh, John, *Taking Chances*. New York: Oxford University Press, 1999.

4. Packel, Edward, *The Mathematics of Games and Gambling*. Washington, DC: Mathematical Association of America, 1981.

5. Weaver, Warren, *Lady Luck: The Theory of Probability*. New York: Dover Publications, 1963.

Power Indexes and the Electoral College (2001–2010)

State	Electoral votes		Power index (%)	
	Number	Percent	Shapley-Shubik	Banzhaf
Alabama	9	1.673	1.639	1.640
Alaska	3	0.558	0.540	0.546
Arizona	10	1.859	1.824	1.823
Arkansas	6	1.115	1.086	1.092
California	55	10.223	11.036	11.402
Colorado	9	1.673	1.639	1.640
Connecticut	7	1.301	1.270	1.274
Delaware	3	0.558	0.540	0.546
District of Columbia	3	0.558	0.540	0.546
Florida	27	5.019	5.087	5.012
Georgia	15	2.788	2.761	2.744
Hawaii	4	0.743	0.722	0.728
Idaho	4	0.743	0.722	0.728
Illinois	21	3.903	3.910	3.865
Indiana	11	2.045	2.010	2.007
Iowa	7	1.301	1.270	1.274
Kansas	6	1.115	1.086	1.092
Kentucky	8	1.487	1.454	1.457
Louisiana	9	1.673	1.639	1.640
Maine	4	0.743	0.722	0.728
Maryland	10	1.859	1.824	1.823
Massachusetts	12	2.230	2.197	2.190
Michigan	17	3.160	3.141	3.116

| State | Electoral votes | | Power index (%) | |
	Number	Percent	Shapley-Shubik	Banzhaf
Minnesota	10	1.859	1.824	1.823
Mississippi	6	1.115	1.086	1.092
Missouri	11	2.045	2.010	2.007
Montana	3	0.558	0.540	0.546
Nebraska	5	0.929	0.904	0.910
Nevada	5	0.929	0.904	0.910
New Hampshire	4	0.743	0.722	0.728
New Jersey	15	2.788	2.761	2.744
New Mexico	5	0.929	0.904	0.910
New York	31	5.762	5.888	5.795
North Carolina	15	2.788	2.761	2.744
North Dakota	3	0.558	0.540	0.546
Ohio	20	3.717	3.717	3.677
Oklahoma	7	1.301	1.270	1.274
Oregon	7	1.301	1.270	1.274
Pennsylvania	21	3.903	3.910	3.865
Rhode Island	4	0.743	0.722	0.728
South Carolina	8	1.487	1.454	1.457
South Dakota	3	0.558	0.540	0.546
Tennessee	11	2.045	2.010	2.007
Texas	34	6.320	6.499	6.393
Utah	5	0.929	0.904	0.910
Vermont	3	0.558	0.540	0.546
Virginia	13	2.416	2.384	2.375
Washington	11	2.045	2.010	2.007
West Virginia	5	0.929	0.904	0.910
Wisconsin	10	1.859	1.824	1.823
Wyoming	3	0.558	0.540	0.546

The 17 Wallpaper Symmetry Types

Symmetry Type	Translation 2 or more directions	Rotations (Given by the smallest angle)					Reflections (Number of directions)					Glide Reflections (Number of directions)					Example
		identity	2-fold	4-fold	3-fold	6-fold	1	2	4	3	6	1	2	4	3	6	
		DEGREES															
		0	180	90	120	60											
p1	✓	✓															
pm	✓	✓					✓										
pg	✓	✓										✓					
cm	✓	✓					✓					✓					
p2	✓	✓	✓														
pmg	✓	✓	✓				✓					✓					
pmm	✓	✓	✓					✓									
pgg	✓	✓	✓										✓				
cmm	✓	✓	✓					✓					✓				

Symmetry Type	Translation (2 or more directions)	Rotations (Given by the smallest angle)					Reflections (Number of directions)					Glide Reflections (Number of directions)					Example
		identity	2-fold	4-fold	3-fold	6-fold											
		D E G R E E S					1	2	4	3	6	1	2	4	3	6	
		0	180	90	120	60											
p4	✓	✓	✓a	✓b													
p4m	✓	✓	✓a	✓b					✓				✓				
p4g	✓	✓	✓a	✓b				✓						✓			
p3	✓	✓			✓												
p3m1	✓	✓			✓c					✓					✓		
p31m	✓	✓			✓d					✓					✓		
p6	✓	✓	✓		✓	✓											
p6m	✓	✓	✓		✓	✓					✓					✓	

a, b : Different rotocenters

c : All rotocenters on axes of reflection

d : Not all rotocenters on axes of reflection

Flowchart for Classifying Wallpaper Patterns

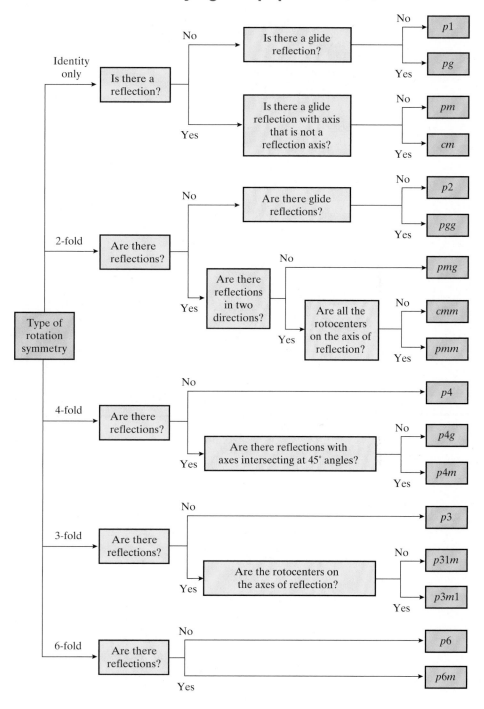

ANSWERS

Chapter 1

WALKING

A Ballots and Preference Schedules

1. (a)

Number of voters	5	3	5	3	2	3
1st choice	A	A	C	D	D	B
2nd choice	B	D	E	C	C	E
3rd choice	C	B	D	B	B	A
4th choice	D	C	A	E	A	C
5th choice	E	E	B	A	E	D

(b) A   **(c)** C

2. (a)

Number of voters	4	5	6	2
1st choice	A	B	C	A
2nd choice	D	C	A	C
3rd choice	B	D	D	D
4th choice	C	A	B	B

(b) None; A and C each have six votes.   **(c)** D

3. (a) 21   **(b)** 11   **(c)** A

4. (a) 1240   **(b)** 621   **(c)** B

5. (a) B and E

(b)

Number of voters	8	3	5	5
1st choice	A	A	C	D
2nd choice	C	D	D	C
3rd choice	D	C	A	A

(c) A

6. (a) A and C

(b)

Number of voters	745	495
1st choice	B	D
2nd choice	D	B

(c) B

7.

Number of voters	255	480	765
1st choice	L	C	M
2nd choice	M	M	L
3rd choice	C	L	C

8.

Number of voters	450	900	225	675
1st choice	A	B	C	C
2nd choice	C	C	B	A
3rd choice	B	A	A	B

9.

Number of voters	47	36	24	13	5
1st choice	B	A	B	E	C
2nd choice	E	B	A	B	E
3rd choice	A	D	D	C	A
4th choice	C	C	E	A	D
5th choice	D	E	C	D	B

10.

Number of voters	47	36	24	13	5
A	1	2	5	2	4
B	3	1	2	4	1
C	2	4	3	1	5
D	5	3	1	5	2
E	4	5	4	3	3

B Plurality Method

11. (a) B and D tie.   **(b)** D   **(c)** B

12. (a) A and B tie.   **(b)** B   **(c)** A

13. (a) 24 votes guarantee A is the outright winner.

(b) 12 votes guarantee C is the outright winner.

14. (a) 28 votes guarantee B is the outright winner.

(b) 20 votes guarantee D is the outright winner.

15. (a) 145   **(b)** 10   **(c)** 7

16. (a) 257   **(b)** 8   **(c)** 5

C Borda Count Method

17. (a) C

(b)

Number of voters	5	3	5	5	3
1st choice	A	A	C	D	B
2nd choice	B	D	D	C	A
3rd choice	C	B	A	B	C
4th choice	D	C	B	A	D

(c) A

(d) independence-of-irrelevant-alternatives criterion (IIA)

18. (a) B

(b)

Number of voters	153	102	55	202	108	20	110	160	175	155
1st choice	A	A	A	B	B	B	C	C	A	B
2nd choice	C	B	C	A	C	C	A	B	C	C
3rd choice	B	C	B	C	A	A	B	A	B	A

(c) A

(d) independence-of-irrelevant-alternatives criterion (IIA)

19. (a) *B*

    (b) *D* has a majority of the first-place votes (13) but does not win the election.

    (c) *D*, having a majority of the first-place votes, is a Condorcet candidate but does not win the election.

20. (a) *D*

    (b) *C* has a majority of the first-place votes but does not win the election.

    (c) *C*, having a majority of the first-place votes, is a Condorcet candidate but does not win the election.

21. *B*

22. *A*

23. (a) 440   **(b)** 110   **(c)** 10   **(d)** 1100   **(e)** 310

24. (a) 100   **(b)** 20   **(c)** 15   **(d)** 300   **(e)** 49

25. *C* with 140 points; the points sum to 500.

26. *E* with 147 points; the points sum to 600.

D   Plurality-with-Elimination Method

27. (a) *A*

    (b)

Number of voters	5	3	5	3	2	3
1st choice	A	A	E	D	D	B
2nd choice	B	D	D	B	B	E
3rd choice	D	B	A	E	A	A
4th choice	E	E	B	A	E	D

    (c) *E*

    (d) independence-of-irrelevant-alternatives criterion (IIA)

28. *B*

29. (a) *D*   **(b)** *D* is the majority winner.

    (c) If there is a choice that has a majority of the first-place votes, then that candidate will be the winner under the plurality-with-elimination method in the first round.

30. (a) *C*   **(b)** *C* is a majority winner.

    (c) If there is a choice that has a majority of the first-place votes, then that candidate will be the winner under the plurality-with-elimination method in the first round.

31. *B*

32. *B*

33. (a) *D*   **(b)** *B*   **(c)** the Condorcet criterion

34. (a) Clinton   **(b)** Buford   **(c)** the monotonicity criterion

E   Pairwise-Comparisons Method

35. (a) *C*

    (b)

Number of voters	8	6	5	5	2
1st choice	C	E	E	D	D
2nd choice	B	D	C	C	E
3rd choice	D	B	D	E	B
4th choice	E	C	B	B	C

    (c) *D*

    (d) independence-of-irrelevant-alternatives criterion (IIA)

36. *C*

37. *A*

38. *D*

39. (a) *A*: 1 point, *B*: 2 points, *C*: 3 points, *D*: $2\frac{1}{2}$ points, *E*: $1\frac{1}{2}$ points

    (b) winner is *C*

40. (a) *A*: 2 points, *B*: 3 points, *C*: 3 points, *D*: $2\frac{1}{2}$ points, *E*: $2\frac{1}{2}$ points, *F*: 2 points

    (b) *B* and *C* tie for first place

F   Ranking Methods

41. (a) *A, B, D, C*   **(b)** *C, B, A, D*   **(c)** *D, A, B, C*

    (d) winner: *B*; second place tie: *A, C, D*

42. (a) *D, B, C, A*   **(b)** *A, B, C, D*

    (c) *C, D, B, A*   **(d)** *B, A, C, D*

43. (a) *A, B, C, D*   **(b)** *B, A, D, C*

    (c) *B, A, C, D*   **(d)** *B, A, D, C*

44. (a) *A, B, C, D*   **(b)** *A, B, D, C*

    (c) *B, A, C, D*   **(d)** *B, A, D, C*

45. *B, C, D, A*

46. *C, B, D, A*

47. *A, C, B, D*

48. *C, B, D, A*

49. first-place: *B*; second place tie: *A, C, D*

50. *D, B, A, C*

G   Miscellaneous

51. 125,250

52. 5,185,810

53. 5,060,560

54. 37,500,535

55. (a) 105   **(b)** 1 hour and 45 minutes

56. (a) 210 matches   **(b)** 3 days

57. 6

58. (a) 24   **(b)** 12

59. (a) A **(b)** B **(c)** A

(d) the Condorcet criterion, majority criterion, and independence-of-irrelevant-alternatives criterion

60. (a) C **(b)** B **(c)** C

(d) From (a) and (b), the Condorcet criterion is violated; from (b) and (c), the independence-of-irrelevant-alternatives criterion is violated.

JOGGING

61. Suppose that A gets a first-place votes and B gets b first-place votes, where $a > b$. It is clear that candidate A wins the election under the plurality method, the plurality-with-elimination method, and the method of pairwise comparisons. Under the Borda count method, A gets $2a + b$ points while B gets $2b + a$ points. Since $a > b$, $2a + b > 2b + a$, and so again A wins the election.

62. One example:

Number of voters	8	4	3	2
1st choice	A	B	B	D
2nd choice	C	D	C	C
3rd choice	B	C	D	B
4th choice	D	A	A	A

63. If X is the winner of an election using the plurality method and, in a reelection, the only changes in the ballots are changes that only favor X, then no candidate other than X can increase his or her first-place votes, and so X is still the winner of the election.

64. If X is the winner of an election using the Borda count method and, in a reelection, the only changes in the ballots are changes that only favor X, then no candidate other than X can increase his or her Borda count, and so X is still the winner of the election.

65. If X is the winner of an election using the method of pairwise comparisons and, in a reelection, the only changes in the ballots are changes that favor X and only favor X, then candidate X will still win every pairwise comparison that he or she won in the original election and possibly even some new ones — while no other candidate will win any new pairwise comparisons. So, X is also the winner of the reelection.

66. (a) In this variation, each candidate gets 1 point less on each ballot. Thus, if there are k voters, each candidate gets a total of k fewer points, i.e., $q = p - k$.

(b) Since each candidate's Borda total is decreased by k, the extended ranking of the candidates is the same.

67. (a) Suppose the candidate is ranked in the xth place in a particular ballot. Under this variation of the Borda count method the candidate gets x points from that ballot; under the Borda count method as originally described in the chapter the candidate gets $(N + 1) - x$ points for that ballot. These two numbers always add up to $(N + 1)$, and this is true for all k ballots. Thus, $p + r = k(N + 1)$.

(b) Suppose that candidates C_1 and C_2 receive p_1 and p_2 points, respectively, using the Borda count as originally described in the chapter and r_1 and r_2 points under the variation described in this exercise. Then, if $p_1 < p_2$, we have $-p_1 > -p_2$ and so $k(N + 1) - p_1 > k(N + 1) - p_2$, which implies, using part (a), that $r_1 > r_2$. Consequently, the relative ranking of the candidates is not changed.

68. The count would be the same as the method described in Exercise 67 except that each candidate's Borda count is divided by a fixed number of ballots. Hence, the ranking remains the same.

69. (a) 65

(b) 34 second-place votes and 31 third-place votes

(c) 31 second-place votes and 34 third-place votes

70. 14 points for each first-place vote, 9 points for each second-place vote, 8 for each third-place vote

71. 5 points for each first-place vote, 3 points for each second-place vote, 1 point for each third-place vote

72. (a) Using plurality with a runoff in the Math Club election results in C winning the election. Recall that D was the winner using the plurality-with-elimination method.

(b) See Example 1.10.

(c)

Number of voters	10	8	6
1st choice	A	C	B
2nd choice	B	B	C
3rd choice	C	A	A

73. (a) C

(b) A is a Condorcet candidate but is eliminated in the first round.

Number of voters	10	6	6	3	3
1st choice	B	A	A	D	C
2nd choice	C	B	C	A	A
3rd choice	D	D	B	C	B
4th choice	A	C	D	B	D

(c) B wins under the Coombs method. However, if 8 voters move B from their third choice to their second choice, then C wins.

Number of voters	10	8	7	4
1st choice	B	C	C	A
2nd choice	A	A	B	B
3rd choice	C	B	A	C

74. (a) B

(b) C is a Condorcet candidate in the Math Club election. However, B wins the election using the Bucklin method.

(c) Suppose that candidate X is the winner of an election under the Bucklin method in round N. If, in a reelection, the only changes in ballots are changes that favor X (and only X), then X will still have a majority of votes in round N. In fact, since the changes favor only X, no other candidate will have more votes in that or any earlier round.

Chapter 2

WALKING

A   Weighted Voting Systems

1. (a) 6   **(b)** 108   **(c)** 28   **(d)** 61%

2. (a) 15   **(b)** 87   **(c)** 5   **(d)** 72%

3. (a) 14   **(b)** 27   **(c)** 18   **(d)** 19

4. (a) 11   **(b)** 20   **(c)** 15   **(d)** 16

5. (a) $[49: 48, 24, 12, 12]$

    (b) $[49: 36, 18, 9, 9]$

    (c) $[49: 32, 16, 8, 8]$

6. (a) $[121: 96, 48, 48, 24, 12, 12]$

    (b) $[121: 72, 36, 36, 18, 9, 9]$

    (c) $[121: 64, 32, 32, 16, 8, 8]$

7. (a) P_1 is a dictator; the rest are dummies.

    (b) P_1 has veto power; P_4 is a dummy.

    (c) P_1 and P_2 have veto power; P_3 and P_4 are dummies.

8. (a) P_1 is a dictator; P_2, P_3, and P_4 are dummies.

    (b) P_1 has veto power.

    (c) P_1 and P_2 have veto power; P_3 and P_4 are dummies.

    (d) All players have veto power.

9. (a) 13   **(b)** 11   **(c)** 11

10. (a) 13   **(b)** 11   **(c)** 12

B   Banzhaf Power

11. (a) 10

    (b) $\{P_1, P_2\}$, $\{P_1, P_3\}$, $\{P_1, P_2, P_3\}$, $\{P_1, P_2, P_4\}$, $\{P_1, P_3, P_4\}$, $\{P_2, P_3, P_4\}$, $\{P_1, P_2, P_3, P_4\}$

    (c) P_1

    (d) $\beta_1 = \frac{5}{12}, \beta_2 = \frac{3}{12}, \beta_3 = \frac{3}{12}, \beta_4 = \frac{1}{12}$

12. (a) 4

    (b) P_1 and P_2

    (c) $P_1, P_3,$ and P_4

    (d) $\{P_1, P_2\}$, $\{P_1, P_2, P_3\}$, $\{P_1, P_2, P_4\}$, $\{P_1, P_3, P_4\}$, $\{P_1, P_2, P_3, P_4\}$

    (e) $\beta_1 = \frac{5}{10}, \beta_2 = \frac{3}{10}, \beta_3 = \frac{1}{10}, \beta_4 = \frac{1}{10}$

13. (a) $\beta_1 = \frac{3}{5}, \beta_2 = \frac{1}{5}, \beta_3 = \frac{1}{5}$ **(b)** $\beta_1 = \frac{3}{5}, \beta_2 = \frac{1}{5}, \beta_3 = \frac{1}{5}$

14. (a) $\beta_1 = \frac{1}{2}, \beta_2 = \frac{1}{2}, \beta_3 = 0$ **(b)** $\beta_1 = \frac{1}{2}, \beta_2 = \frac{1}{2}, \beta_3 = 0$

15. (a) $\beta_1 = \frac{4}{12}, \beta_2 = \frac{3}{12}, \beta_3 = \frac{2}{12}, \beta_4 = \frac{2}{12}, \beta_5 = \frac{1}{12}$

    (b) $\beta_1 = \frac{7}{19}, \beta_2 = \frac{5}{19}, \beta_3 = \frac{3}{19}, \beta_4 = \frac{3}{19}, \beta_5 = \frac{1}{19}$

16. (a) $\beta_1 = \frac{1}{3}, \beta_2 = \frac{1}{3}, \beta_3 = \frac{1}{3}, \beta_4 = 0, \beta_5 = 0$

    (b) $\beta_1 = \frac{4}{13}, \beta_2 = \frac{4}{13}, \beta_3 = \frac{3}{13}, \beta_4 = \frac{1}{13}, \beta_5 = \frac{1}{13}$

17. (a) $\beta_1 = 1, \beta_2 = 0, \beta_3 = 0, \beta_4 = 0$

    (b) $\beta_1 = \frac{7}{10}, \beta_2 = \frac{1}{10}, \beta_3 = \frac{1}{10}, \beta_4 = \frac{1}{10}$

    (c) $\beta_1 = \frac{3}{5}, \beta_2 = \frac{1}{5}, \beta_3 = \frac{1}{5}, \beta_4 = 0$

    (d) $\beta_1 = \frac{1}{2}, \beta_2 = \frac{1}{2}, \beta_3 = 0, \beta_4 = 0$

    (e) $\beta_1 = \frac{1}{3}, \beta_2 = \frac{1}{3}, \beta_3 = \frac{1}{3}, \beta_4 = 0$

18. (a) $\beta_1 = 1, \beta_2 = 0, \beta_3 = 0, \beta_4 = 0$

    (b) $\beta_1 = \frac{7}{10}, \beta_2 = \frac{1}{10}, \beta_3 = \frac{1}{10}, \beta_4 = \frac{1}{10}$

    (c) $\beta_1 = \frac{3}{5}, \beta_2 = \frac{1}{5}, \beta_3 = \frac{1}{5}, \beta_4 = 0$

    (d) $\beta_1 = \frac{1}{2}, \beta_2 = \frac{1}{2}, \beta_3 = 0, \beta_4 = 0$

    (e) $\beta_1 = \frac{1}{3}, \beta_2 = \frac{1}{3}, \beta_3 = \frac{1}{3}, \beta_4 = 0$

19. (a) $\{\underline{P_1}, \underline{P_2}\}, \{\underline{P_1}, \underline{P_3}\}, \{\underline{P_1}, P_2, P_3\}$

    (b) $\beta_1 = \frac{3}{5}, \beta_2 = \frac{1}{5}, \beta_3 = \frac{1}{5}$

20. (a) $\{\underline{P_1}, \underline{P_2}\}, \{\underline{P_1}, \underline{P_2}, P_3\}, \{\underline{P_1}, \underline{P_2}, P_4\}, \{\underline{P_1}, \underline{P_2}, P_3, P_4\}$

    (b) $\beta_1 = \frac{1}{2}, \beta_2 = \frac{1}{2}, \beta_3 = 0$

21. (a) $\{\underline{P_1}, \underline{P_2}\}, \{\underline{P_1}, \underline{P_3}\}, \{\underline{P_2}, \underline{P_3}\}, \{P_1, P_2, P_3\}, \{\underline{P_1}, \underline{P_2}, P_4\}, \{\underline{P_1}, \underline{P_2}, P_5\}, \{\underline{P_1}, \underline{P_2}, P_6\}, \{\underline{P_1}, \underline{P_3}, P_4\}, \{\underline{P_1}, \underline{P_3}, P_5\}, \{\underline{P_1}, \underline{P_3}, P_6\}, \{\underline{P_2}, \underline{P_3}, P_4\}, \{\underline{P_2}, \underline{P_3}, P_5\}, \{\underline{P_2}, \underline{P_3}, P_6\}$

    (b) $\{\underline{P_1}, \underline{P_2}, P_4\}, \{\underline{P_1}, \underline{P_3}, P_4\}, \{\underline{P_2}, \underline{P_3}, P_4\}, \{P_1, P_2, P_3, P_4\}, \{\underline{P_1}, \underline{P_2}, P_4, P_5\}, \{\underline{P_1}, \underline{P_2}, P_4, P_6\}, \{\underline{P_1}, \underline{P_3}, P_4, P_5\}, \{\underline{P_1}, \underline{P_3}, P_4, P_6\}, \{\underline{P_2}, \underline{P_3}, P_4, P_5\}, \{\underline{P_2}, \underline{P_3}, P_4, P_6\}, \{P_1, P_2, P_3, P_4, P_5\}, \{P_1, P_2, P_3, P_4, P_6\}, \{\underline{P_1}, \underline{P_2}, P_4, P_5, P_6\}, \{\underline{P_1}, \underline{P_3}, P_4, P_5, P_6\}, \{\underline{P_2}, \underline{P_3}, P_4, P_5, P_6\}, \{P_1, P_2, P_3, P_4, P_5, P_6\}$

    (c) $\beta_4 = 0$

    (d) $\beta_1 = \frac{1}{3}, \beta_2 = \frac{1}{3}, \beta_3 = \frac{1}{3}, \beta_4 = 0, \beta_5 = 0, \beta_6 = 0$

22. (a) $\{\underline{P_1}, \underline{P_2}, \underline{P_3}\}, \{\underline{P_1}, \underline{P_2}, \underline{P_4}\}, \{\underline{P_1}, \underline{P_2}, \underline{P_5}\}, \{\underline{P_1}, \underline{P_3}, \underline{P_4}\}, \{\underline{P_2}, \underline{P_3}, \underline{P_4}\}.$

    (b) $\{P_1, P_2, P_3, P_4\}, \{\underline{P_1}, \underline{P_2}, \underline{P_3}, P_5\}, \{\underline{P_1}, \underline{P_2}, \underline{P_3}, P_6\}, \{\underline{P_1}, \underline{P_2}, \underline{P_4}, P_5\}, \{\underline{P_1}, \underline{P_2}, \underline{P_4}, P_6\}, \{\underline{P_1}, \underline{P_2}, \underline{P_5}, P_6\}, \{\underline{P_1}, \underline{P_3}, \underline{P_4}, P_5\}, \{\underline{P_1}, \underline{P_3}, \underline{P_4}, P_6\}, \{\underline{P_1}, \underline{P_3}, \underline{P_5}, P_6\}, \{\underline{P_2}, \underline{P_3}, \underline{P_4}, P_5\}, \{\underline{P_2}, \underline{P_3}, \underline{P_4}, P_6\}$

    (c) $\{P_1, P_2, P_3, P_4, P_5\}, \{P_1, P_2, P_3, P_4, P_6\}, \{\underline{P_1}, P_2, P_3, P_5, P_6\}, \{\underline{P_1}, \underline{P_2}, P_4, P_5, P_6\}, \{\underline{P_1}, \underline{P_3}, P_4, P_5, P_6\}, \{\underline{P_2}, \underline{P_3}, \underline{P_4}, P_5, P_6\}$

    (d) $\beta_1 = \frac{15}{52}, \beta_2 = \frac{13}{52}, \beta_3 = \frac{11}{52}, \beta_4 = \frac{9}{52}, \beta_5 = \frac{3}{52}, \beta_6 = \frac{1}{52}$

23. (a) $\{\underline{A}, \underline{B}\}, \{\underline{A}, \underline{C}\}, \{\underline{B}, \underline{C}\}, \{A, B, C\}, \{\underline{A}, \underline{B}, D\}, \{\underline{A}, \underline{C}, D\}, \{\underline{B}, \underline{C}, D\}, \{A, B, C, D\}$

    (b) A, B, and C have Banzhaf power index of $\frac{1}{3}$ each; D is a dummy.

24. (a) $\{\underline{A}, \underline{B}\}, \{\underline{A}, \underline{C}\}, \{\underline{A}, \underline{D}\}, \{\underline{A}, B, C\}, \{\underline{A}, B, D\}, \{\underline{A}, C, D\}, \{\underline{B}, \underline{C}, \underline{D}\}, \{A, B, C, D\}$

    (b) A has Banzhaf power index of $\frac{1}{2}$; B, C, and D have Banzhaf power index of $\frac{1}{6}$ each.

C   Shapley-Shubik Power

25. (a) $\langle P_1, \underline{P_2}, P_3 \rangle, \langle P_1, \underline{P_3}, P_2 \rangle, \langle P_2, \underline{P_1}, P_3 \rangle, \langle P_2, P_3, \underline{P_1} \rangle, \langle P_3, \underline{P_1}, P_2 \rangle, \langle P_3, \overline{P_2}, \underline{P_1} \rangle$

    (b) $\sigma_1 = \frac{4}{6}, \sigma_2 = \frac{1}{6}, \sigma_3 = \frac{1}{6}$

26. (a) $\langle P_1, \underline{P_2}, P_3 \rangle, \langle P_1, \underline{P_3}, P_2 \rangle, \langle P_2, \underline{P_1}, P_3 \rangle, \langle P_2, \underline{P_3}, P_1 \rangle, \langle P_3, \underline{P_1}, P_2 \rangle, \langle P_3, \underline{P_2}, P_1 \rangle$

    (b) $\sigma_1 = \frac{1}{3}, \sigma_2 = \frac{1}{3}, \sigma_3 = \frac{1}{3}$

27. (a) $\sigma_1 = 1, \sigma_2 = 0, \sigma_3 = 0, \sigma_4 = 0$

   (b) $\sigma_1 = \frac{4}{6}, \sigma_2 = \frac{1}{6}, \sigma_3 = \frac{1}{6}, \sigma_4 = 0$

   (c) $\sigma_1 = \frac{1}{2}, \sigma_2 = \frac{1}{2}, \sigma_3 = 0, \sigma_4 = 0$

   (d) $\sigma_1 = \frac{1}{3}, \sigma_2 = \frac{1}{3}, \sigma_3 = \frac{1}{3}, \sigma_4 = 0$

28. (a) $\sigma_1 = 1, \sigma_2 = 0, \sigma_3 = 0, \sigma_4 = 0$

   (b) $\sigma_1 = \frac{9}{12}, \sigma_2 = \frac{1}{12}, \sigma_3 = \frac{1}{12}, \sigma_4 = \frac{1}{12}$

   (c) $\sigma_1 = \frac{1}{2}, \sigma_2 = \frac{1}{2}, \sigma_3 = 0, \sigma_4 = 0$

   (d) $\sigma_1 = \frac{1}{3}, \sigma_2 = \frac{1}{3}, \sigma_3 = \frac{1}{3}, \sigma_4 = 0$

29. (a) $\sigma_1 = \frac{5}{12}, \sigma_2 = \frac{3}{12}, \sigma_3 = \frac{3}{12}, \sigma_4 = \frac{1}{12}$

   (b) same as in (a)

   (c) same as in (a)

30. (a) $\sigma_1 = \frac{9}{12}, \sigma_2 = \frac{1}{12}, \sigma_3 = \frac{1}{12}, \sigma_4 = \frac{1}{12}$

   (b) same as in (a)

   (c) same as in (a)

31. (a) $\langle P_1, \underline{P_2}, P_3 \rangle, \langle P_1, \underline{P_3}, P_2 \rangle, \langle P_2, \underline{P_1}, P_3 \rangle, \langle P_2, P_3, \underline{P_1} \rangle, \langle P_3, \underline{P_1}, P_2 \rangle, \langle P_3, P_2, \underline{P_1} \rangle$

   (b) $\sigma_1 = \frac{4}{6}, \sigma_2 = \frac{1}{6}, \sigma_3 = \frac{1}{6}$

32. (a) $\langle P_1, \underline{P_2}, P_3 \rangle, \langle P_1, P_3, \underline{P_2} \rangle, \langle P_2, \underline{P_1}, P_3 \rangle, \langle P_2, P_3, \underline{P_1} \rangle, \langle P_3, P_1, \underline{P_2} \rangle, \langle P_3, P_2, \underline{P_1} \rangle$

   (b) $\sigma_1 = \frac{1}{2}, \sigma_2 = \frac{1}{2}, \sigma_3 = 0$

33. $\sigma_A = \frac{1}{3}, \sigma_B = \frac{1}{3}, \sigma_C = \frac{1}{3}, \sigma_D = 0$

34. $\sigma_A = \frac{12}{24}, \sigma_B = \frac{4}{24}, \sigma_C = \frac{4}{24}, \sigma_D = \frac{4}{24}$

D Miscellaneous

35. (a) 6,227,020,800

   (b) 6,402,373,705,728,000

   (c) 15,511,210,043,330,985,984,000,000

   (d) 491,857 years

36. (a) 479,001,600   (c) 2,432,902,008,176,640,000

   (b) 1,307,674,368,000   (d) 77 years

37. (a) 362,880 (b) 11 (c) 110 (d) 504 (e) 10,100

38. (a) 121,645,100,408,832,000   (c) 40,200

   (b) 20   (d) 990

39. (a) 11.1   (b) 101.01

40. (a) 21.05   (b) 201.005

41. (a) 63 (b) 31 (c) 31 (d) 15 (e) 16

42. (a) 31 (b) 15 (c) 15 (d) 7 (e) 8

43. (a) 120 (b) 24 (c) 24 (d) 96

44. (a) 720 (b) 120 (c) 120 (d) 600

45. (a) 96

   (b) $\frac{96}{120} = \frac{4}{5}$

   (c) $\sigma_1 = \frac{4}{5}, \sigma_2 = \sigma_3 = \sigma_4 = \sigma_5 = \frac{1}{20}$

46. (a) 600

   (b) $\frac{600}{720} = \frac{5}{6}$

   (c) $\sigma_1 = \frac{5}{6}, \sigma_2 = \sigma_3 = \sigma_4 = \sigma_5 = \sigma_6 = \frac{1}{30}$

JOGGING

47. If x is even, then $q = \frac{15x}{2} + 1 \le 8x$; if x is odd, then $q = \frac{15x+1}{2} \le 8x$.

48. (a) $\beta_1 = \frac{2}{5}, \beta_2 = \beta_3 = \beta_4 = \frac{1}{5}$

   (b) $\sigma_1 = \frac{1}{2}, \sigma_2 = \sigma_3 = \sigma_4 = \frac{1}{6}$

49. $\beta_1 = \frac{3}{5}, \beta_2 = \frac{1}{5}, \beta_3 = \frac{1}{5}$

50. $\beta_1 = \frac{9}{21} = \frac{3}{7}, \beta_2 = \beta_3 = \beta_4 = \beta_5 = \frac{3}{21} = \frac{1}{7}$

51. (a) [4: 2, 1, 1, 1] (Other answers are possible.)

   (b) $\sigma_H = \frac{1}{2}, \sigma_{A_2} = \sigma_{A_3} = \sigma_{A_4} = \frac{1}{6}$

52. (a) P has veto power $\Leftrightarrow$ the coalition formed by all players other than P is a losing coalition $\Leftrightarrow$ P must be a member of any winning coalition.

   (b) P has veto power $\Leftrightarrow$ the coalition formed by all players other than P is a losing coalition $\Leftrightarrow$ P is a critical player in the grand coalition.

53. (a) Suppose that a winning coalition that contains P is not a winning coalition without P. Then P would be a critical player in that coalition, contradicting the fact that P is a dummy.

   (b) P is a dummy $\Leftrightarrow$ P is never critical $\Leftrightarrow$ the numerator of its Banzhaf power index is 0 $\Leftrightarrow$ its Banzhaf power index is 0.

   (c) Suppose that P is not a dummy. Then P is critical in some winning coalition. Let S denote the other players in that winning coalition. The sequential coalition with the players in S first (in any order) followed by P and then followed by the remaining players has P as its pivotal player. Thus, P's Shapley-Shubik power index is not zero. Conversely, if P's Shapley-Shubik power index is not zero, then P is pivotal in some sequential coalition. A coalition consisting of P together with the players preceding P in that sequential coalition is a winning coalition, and P is a critical player in it. Thus, P is not a dummy.

54. (a) P_5 is a dummy. It takes (at least) three of the first four players to pass a motion.

   (b) $\beta_1 = \beta_2 = \beta_3 = \beta_4 = \frac{1}{4}, \beta_5 = 0; \sigma_1 = \sigma_2 = \sigma_3 = \sigma_4 = \frac{1}{4}, \sigma_5 = 0$

   (c) $q = 21, 31,$ or 41

   (d) $w = 1, 2, 3$

55. (a) $7 \le q \le 13$

   (b) For $q = 7$ or $q = 8$, P_1 is a dictator because $\{P_1\}$ is a winning coalition.

   (c) For $q = 9$, only P_1 has veto power because P_2 and P_3 together have only five votes.

   (d) For $10 \le q \le 12$, both P_1 and P_2 have veto power because no motion can pass without both of their votes. For $q = 13$, all three players have veto power.

(e) For $q = 7$ or $q = 8$, both P_2 and P_3 are dummies because P_1 is a dictator. For $10 \leq q \leq 12$, P_3 is a dummy because all winning coalitions contain $\{P_1, P_2\}$, which is itself a winning coalition.

56. (a) $5 \leq w \leq 9$

(b) P_1 is a dictator if $w = 9$.

(c) For $w = 5$, P_1 and P_2 have veto power. If $6 \leq w \leq 8$, then P_1 alone has veto power.

(d) If $w = 5$, then all winning coalitions contain $\{P_1, P_2\}$, which is itself a winning coalition, and so both P_3 and P_4 are dummies. If $w = 9$, then P_2, P_3, and P_4 are all dummies because P_1 is a dictator.

57. (a) Both have $\beta_1 = \frac{2}{5}, \beta_2 = \frac{1}{5}, \beta_3 = \frac{1}{5}$, and $\beta_4 = \frac{1}{5}$.

(b) In the weighted voting system $[q: w_1, w_2, \ldots, w_N]$, if P_k is critical in a coalition, then the sum of the weights of all the players in that coalition (including P_k) is at least q but the sum of the weights of all the players in the coalition except P_k is less than q. Consequently, if the weights of all the players in that coalition are multiplied by $c > 0$ ($c = 0$ would make no sense), then the sum of the weights of all the players in the coalition (including P_k) is at least cq but the sum of the weights of all the players in the coalition except P_k is less than cq. Therefore, P_k is critical in the same coalitions in the weighted voting system $[cq: cw_1, cw_2, \ldots, cw_N]$.

58. (a) Both have $\sigma_1 = \frac{1}{2}, \sigma_2 = \frac{1}{6}, \sigma_3 = \frac{1}{6}$, and $\sigma_4 = \frac{1}{6}$.

(b) In the weighted voting system $[q: w_1, w_2, \ldots, w_N]$, P_k pivotal in the sequential coalition $\langle P_1, P_2, \ldots, P_k, \ldots, P_N \rangle$ means that $w_1 + w_2 + \cdots + w_k \geq q$, but $w_1 + w_2 + \cdots + w_{k-1} < q$. In the weighted voting system $[cq: cw_1, cw_2, \ldots, cw_N]$, P_k pivotal in the sequential coalition $\langle P_1, P_2, \ldots, P_k, \ldots, P_N \rangle$ means that $cw_1 + cw_2 + \cdots + cw_k \geq cq$, but $cw_1 + cw_2 + \cdots + cw_{k-1} < cq$. These two statements are equivalent since $cw_1 + cw_2 + \cdots + cw_k \geq cq \Leftrightarrow w_1 + w_2 + \cdots + w_k \geq q$, and $cw_1 + cw_2 + \cdots + cw_{k-1} < cq \Leftrightarrow w_1 + w_2 + \cdots + w_{k-1} < q$. This same reasoning applies to any sequential coalition, and so the pivotal players are exactly the same.

59. The senior partner has Shapley-Shubik power index $\frac{N}{N+1}$; each of the junior partners has Shapley-Shubik power index $\frac{1}{N(N+1)}$.

60. (a) $\frac{V}{2} < q \leq V - w_1$ (where V denotes the sum of all the weights)

(b) $V - w_N < q \leq V$

(c) $V - w_i < q \leq V - w_{i+1}$

61. You should buy from P_3.

62. You should buy from P_4.

63. (a) You should buy from P_2.

(b) You should buy two votes from P_2.

(c) Buying a single vote from P_2 raises your power from $\frac{1}{25} = 4\%$ to $\frac{3}{25} = 12\%$. Buying a second vote from P_2 raises your power to $\frac{2}{13} \approx 15.4\%$.

64. (a) $\frac{1}{2}$

(b) Before the merger; $\beta_2 = \frac{1}{2}$, $\beta_3 = 0$. After the merger; $\beta^* = \frac{1}{2}$.

(c) Before the merger; $\beta_2 = \frac{1}{3}$, $\beta_3 = \frac{1}{3}$. After the merger; $\beta^* = \frac{1}{2}$.

(d) The Banzhaf power index of the merger of two players can be greater than, equal to, or less than the sum of Banzhaf power indices of the individual players.

65. (a) The losing coalitions are $\{P_1\}, \{P_2\}$, and $\{P_3\}$. The complements of these coalitions are $\{P_2, P_3\}, \{P_1, P_3\}$, and $\{P_1, P_2\}$, respectively, all of which are winning coalitions.

(b) The losing coalitions are $\{P_1\}, \{P_2\}, \{P_3\}, \{P_4\}, \{P_2, P_3\}, \{P_2, P_4\}$, and $\{P_3, P_4\}$. The complements of these coalitions are $\{P_2, P_3, P_4\}, \{P_1, P_3, P_4\}, \{P_1, P_2, P_4\}, \{P_1, P_2, P_3\}, \{P_1, P_4\}, \{P_1, P_3\}$, and $\{P_1, P_2\}$, respectively, all of which are winning coalitions.

(c) If P is a dictator, then the losing coalitions are all the coalitions without P; the winning coalitions are all the coalitions that include P.

(d) 2^{N-1}

66. (a) The winning coalitions are $\{P_1, P_2\}$ and $\{P_1, P_2, P_3\}$.

(b) The winning coalitions are $\{P_1, P_2\}, \{P_1, P_2, P_3\}, \{P_1, P_2, P_4\}, \{P_1, P_3, P_4\}$, and $\{P_1, P_2, P_3, P_4\}$.

(c) The winning coalitions for both weighted voting systems are those consisting of any three of the five players, any four of the five players, and the grand coalition.

(d) If a player is critical in a winning coalition, then that coalition is no longer winning if the player is removed. An equivalent system (with the same winning coalitions) will find that player critical in the same coalitions.

(e) If a player, P, is pivotal in a sequential coalition, then the players to P's left in the sequential coalition do not form a winning coalition, but including P does make it a winning coalition. An equivalent system (with the same winning coalitions) will find the player P pivotal in that same sequential coalition.

67. (a) In a 9-member coalition with 5 permanent members and 4 nonpermanent members, every member is critical. In any winning coalition with 10 or more members only the 5 permanent members are critical.

(b) At least nine members are needed to form a winning coalition. So, there are $210 + 638 = 848$ winning coalitions. Since every member is critical in each nine-member coalition, the nine-member coalitions yield a total of $210 \times 9 = 1890$ critical players. Since only the permanent members are critical in coalitions having 10 or more members, there are $638 \times 5 = 3190$ critical players in these coalitions. Thus, the total number of critical players in all winning coalitions is 5080.

(c) Each permanent member is critical in each of the 848 winning coalitions.

(d) The 5 permanent members together have $5 \times \frac{848}{5080} = \frac{4240}{5080}$ of the power. The remaining $\frac{840}{5080}$ of the power is shared equally among the 10 nonpermanent members.

(e) The total number of votes is 45, so if any one of the permanent members does not vote for a measure, there would be at most $45 - 7 = 38$ votes and the measure would not pass. Thus, all permanent members have veto power. On the other hand, all 5 permanent members' votes only add up to 35, so at least 4 nonpermanent members votes are needed for a measure to pass.

Chapter 3

WALKING

A   Shares, Fair Shares, and Fair Divisions

1. (a) $6   **(b)** $18   **(c)** $2   **(d)** $4.50

2. (a) $19.20   **(b)** $4.80   **(c)** $6.40   **(d)** $1.20

3. (a) $3.60   **(b)** $14.40   **(c)** $1.44   **(d)** $4.32   **(e)** $3

4. (a) $13.50   **(b)** $4.50   **(c)** $5.40   **(d)** $1.35   **(e)** $3

5. (a) s_1: $2.50; s_2: $3.75; s_3: $5.00; s_4: $6.25; s_5: $7.50; s_6: $5.00

  (b) s_3, s_4, s_5, s_6

6. (a) s_1: $9.00; s_2: $5.25; s_3: $1.50; s_4: $3.00; s_5: $4.50; s_6: $6.75

  (b) s_1, s_2, s_6

7. (a) s_2, s_3 **(b)** s_3 **(c)** s_1, s_2, s_3

  (d) Ben must receive s_3. So, Ana must receive s_2 and Cara is left with s_1.

8. (a) s_2, s_3 **(b)** s_1, s_2 **(c)** s_1

  (d) Chia must receive s_1. So, Bella must receive s_2 and Amy is left with s_3.

9. (a) s_2 **(b)** s_2, s_3 **(c)** s_2, s_3

  (d) It is not possible. No player considers s_1 to be a fair share.

10. (a) s_2, s_3 **(b)** s_2, s_3 **(c)** s_2, s_3

  (d) It is not possible. No player considers s_1 a fair share.

11. (a) s_2, s_3 **(b)** s_1, s_2, s_3, s_4 **(c)** s_1, s_2 **(d)** s_1

  (e) Dana must receive s_1. Then, Cory must receive s_2. After that, Abe must receive s_3, leaving Betty with s_4.

12. (a) s_3, s_4 **(b)** s_2, s_3 **(c)** s_1, s_2 **(d)** s_3

  (e) Duncan must receive s_3. Then, Benson must receive s_2 and Adams must receive s_4. This leaves Cagle with s_1.

13. (a) s_1, s_4 **(b)** s_1, s_2 **(c)** s_1, s_3 **(d)** s_4

  (e) Adams: s_1; Benson: s_2; Cagle: s_3; Duncan: s_4

14. (a) s_2, s_3 **(b)** s_3, s_4 **(c)** s_3 **(d)** s_1, s_3

  (e) Abe: s_2; Betty: s_4; Cory: s_3; Dana: s_1

B   The Divider-Chooser Method

15. (a) s_1: $[0, 8]$; s_2: $[8, 12]$ **(b)** s_1; $8.00

16. (a) s_1: $[0, 3]$; s_2: $[3, 12]$ **(b)** s_2; $7.00

17. (a) s_1: $[0, 10]$; s_2: $[10, 28]$ **(b)** s_2; $6.50

18. (a) s_1: $[0, 18]$; s_2: $[18, 28]$ **(b)** s_1; $6.50

19. (a) yes; s_2 (75%)   **(b)** no (David values s_2 more than s_1.)

  (c) yes; s_2 (75%)

20. (a) no (Paula values s_2 more than s_1.)   **(b)** yes; s_2 (60%)

  (c) no (Paula values s_2 more than s_1.)

C   The Lone-Divider Method

21. (a) C_1: s_2, C_2: s_3, D: s_1 or C_1: s_3, C_2: s_2, D: s_1

  (b) C_1: s_2, C_2: s_1, D: s_3 or C_1: s_2, C_2: s_3, D: s_1 or C_1: s_3, C_2: s_1, D: s_2

22. (a) C_1: s_2, C_2: s_1, D: s_3 or C_1: s_2, C_2: s_3, D: s_1

  (b) C_1: s_1, C_2: s_2, D: s_3 or C_1: s_1, C_2: s_3, D: s_2 or C_1: s_2, C_2: s_3, D: s_1

23. (a) C_1: s_2, C_2: s_1, C_3: s_3, D: s_4 (First, C_1 must receive s_2. Then, C_3 must receive s_3. So, C_2 receives s_1 and D receives s_4.)

  (b) C_1: s_2, C_2: s_3, C_3: s_1, D: s_4 or C_1: s_3, C_2: s_1, C_3: s_2, D: s_4

  (c) C_1: s_2, C_2: s_1, C_3: s_4, D: s_3 or C_1: s_2, C_2: s_3, C_3: s_1, D: s_4 or C_1: s_2, C_2: s_3, C_3: s_4, D: s_1

24. (a) C_1: s_2, C_2: s_1, C_3: s_3, D: s_4 (First, C_1 must receive s_2. Then, C_3 must receive s_3. So, C_2 receives s_1 and D receives s_4.)

  (b) C_1: s_2, C_2: s_1, C_3: s_4, D: s_3 or C_1: s_2, C_2: s_3, C_3: s_1, D: s_4 or C_1: s_2, C_2: s_3, C_3: s_4, D: s_1

  (c) C_1: s_2, C_2: s_1, C_3: s_3, D: s_4 or C_1: s_2, C_2: s_1, C_3: s_4, D: s_3 or C_1: s_2, C_2: s_3, C_3: s_4, D: s_1

25. (a) fair   **(b)** not fair ($\{s_2, s_3, s_4\}$ may be worth less than 75% to C_2 and C_3.)   **(c)** fair   **(d)** not fair ($\{s_1, s_4\}$ may be worth less than 50% to C_2 and C_3.)

26. (a) fair   **(b)** not fair ($\{s_1, s_2, s_4\}$ is worth less than 75% to C_3.)   **(c)** fair   **(d)** not fair (C_1 values $\{s_1, s_2\}$ as less than 50%).

27. (a) C_1: s_2, C_2: s_4, C_3: s_5, C_4: s_3, D: s_1; C_1: s_4, C_2: s_2, C_3: s_5, C_4: s_3, D: s_1 (If C_1 is to receive s_2, then C_2 must receive s_4 (and conversely). So, C_4 must receive s_3. It follows that C_3 must receive s_5. This leaves D with s_1.)

  (b) C_1: s_2, C_2: s_4, C_3: s_5, C_4: s_3, D: s_1 (If C_1 is to receive s_2, then C_2 must receive s_4. So, C_4 must receive s_3. It follows that C_3 must receive s_5. This leaves D with s_1.)

28. (a) C_1: s_2, C_2: s_4, C_3: s_1, C_4: s_3, D: s_5 or C_1: s_3, C_2: s_2, C_3: s_1, C_4: s_4, D: s_5 or C_1: s_3, C_2: s_4, C_3: s_2, C_4: s_1, D: s_5. These three divisions are completely determined by the share given to C_4.

  (b) C_1: s_1, C_2: s_4, C_3: s_5, C_4: s_2, D: s_3. If C_4 is to receive s_2, then C_2 must receive s_4. So, C_1 must receive s_1. It follows that C_3 must receive s_5. This leaves D with s_3.

29. (a) Gong (he is the only player who could value each piece equally).

  (b) Egan: $\{s_3, s_4\}$; Fine: $\{s_1, s_3, s_4\}$; Hart: $\{s_3\}$

  (c) Egan: s_4; Fine: s_1; Gong: s_2; Hart: s_3

30. (a) Betty (she is the only player who could value each piece equally).

  (b) Abe: $\{s_1, s_2, s_4\}$; Cory: $\{s_1, s_4\}$; Dana: $\{s_4\}$

  (c) Abe: s_2; Betty: s_3; Cory: s_1; Dana: s_3

D **The Lone-Chooser Method**

33. (a) 40° strawberry; 40° strawberry; 90° vanilla–10° strawberry

  (b) 75° vanilla; 15° vanilla–40° strawberry; 50° strawberry

  (c) Angela: 80° strawberry; Boris: 15° vanilla–90° strawberry; Carlos: 165° vanilla–10° strawberry

  (d) Angela: $12:00; Boris: $10.00; Carlos: $22.67

34. (a) 45° vanilla; 45° vanilla; 90° strawberry

  (b) 90° vanilla–10° vanilla; 40° strawberry; 40° strawberry

  (c) Angela: 80° strawberry; Boris: 90° vanilla–100° strawberry; Carlos: 90° vanilla

  (d) Angela: $12.00; Boris: $16.00; Carlos: $12.00

35. (a) s_2; 90° vanilla; 90° vanilla; 60° strawberry

  (b) 40° strawberry; 40° strawberry; 40° strawberry

  (c) Angela: 80° strawberry; Boris: 90° vanilla–60° strawberry; Carlos: 90° vanilla–40° strawberry

  (d) Angela: $12.00; Boris: $12.00; Carlos: $14.67

36. (a) s_2; 65° strawberry; 65° strawberry; 50° strawberry–45° vanilla

  (b) 45° vanilla; 45° vanilla; 45° vanilla

  (c) Angela: 130° strawberry; Boris: 90° vanilla–50° strawberry; Carlos: 90° vanilla

  (d) Angela: $19.50; Boris: $11.00; Carlos: $12.00

37. (a) s_1 (60° chocolate, 30° chocolate–30° strawberry, 60° strawberry)

  (b) 30° orange; 30° orange; 30° orange–90° vanilla

  (c) Arthur: 60° orange; Brian: 90° strawberry–30° chocolate; Carl: 90° vanilla–30° orange–60° chocolate

  (d) Arthur: $33\frac{1}{3}$%; Brian: $66\frac{2}{3}$%; Carl: $83\frac{1}{3}$%

31. (a) s_1: [0, 4]; s_2: [4, 8]; s_3: [8, 12]   **(b)** s_1, s_2 **(c)** s_2, s_3

  (d) Three possible fair divisions are

    (1) Jared: s_2; Karla; s_1; Lori: s_3

    (2) Jared: s_1; Karla; s_2; Lori: s_3

    (3) Jared: s_3; Karla; s_1; Lori: s_2

32. (a) s_1: [0, 6]; s_2: [6, 9]; s_3: [9, 12]   **(b)** s_1 **(c)** s_1

  (d) s_1 and s_2 are recombined into a single piece and divided fairly by Jared and Karla using the divider-chooser method.

38. (a) s_2 (45° chocolate; 45° orange; 45° orange–45° vanilla)

  (b) 30° chocolate; 15° chocolate–90° strawberry–15° vanilla; 30° vanilla

  (c) Arthur: 90° orange–45° vanilla; Brian: 60° chocolate–90° strawberry–15° vanilla; Carl: 30° chocolate–30° vanilla

  (d) Arthur: 50%; Brian: $83\frac{1}{3}$%; Carl: $33\frac{1}{3}$%

39. (a) either s_1 or s_2 (we assume he picks s_2) 30° chocolate; 30° chocolate; 30° chocolate–90° orange

  (b) 30° strawberry; 30° strawberry; 30° strawberry–90° vanilla

  (c) Arthur: 30° chocolate–90° orange–30° strawberry; Brian: 60° strawberry–90° vanilla; Carl: 60° chocolate

  (d) Arthur: $66\frac{2}{3}$%; Brian: $33\frac{1}{3}$%; Carl: $33\frac{1}{3}$%

40. (a) either s_1 or s_2 (we assume he picks s_1) 30° chocolate; 30° chocolate; 30° chocolate–90° strawberry

  (b) 30° orange; 30° orange; 30° orange–90° vanilla

  (c) Arthur: 60° orange–90° vanilla; Brian: 90° strawberry–30° chocolate–30° orange; Carl: 60° chocolate

  (d) Arthur: $33\frac{1}{3}$%; Brian: $66\frac{2}{3}$%; Carl: $33\frac{1}{3}$%

41. (a) s_1: [0; 3]; s_2: [3, 12]

  (b) Jared picks s_2 (J_1: [3, 6]; J_2: [6, 9]; J_3: [9, 12])

  (c) K_1: [0, 1]; K_2: [1, 2]; K_3: [2, 3]

  (d) Jared: [3, 9]; Karla: [0, 2]; Lori: [2, 3] and [9, 12] Jared: 50%; Karla: $33\frac{1}{3}$%; Lori: $38\frac{8}{9}$%

42. (a) s_1: [0, 7.5]; s_2: [7.5, 12]

  (b) Karla picks s_1 (K_1: [0, 2], K_2: [2, 4], K_3: [4, 7.5])

  (c) L_1: [7.5, 9]; L_2: [9, 10.5]; L_3: [10.5, 12]

  (d) Jared: [4, 7.5] and [10.5, 12]; Karla: [0, 4]; Lori: [7.5, 10.5] Jared: 5/12 = $41\frac{2}{3}$%; Karla: $66\frac{2}{3}$%; Lori: $33\frac{1}{3}$%

E **The Last-Diminisher Method**

43. (a) P_2, P_4 **(b)** P_4; $6.00   **(c)** P_1; $\geq$ $24

44. (a) P_2, P_3 **(b)** P_3; $7.50   **(c)** P_1; $\geq$ $22.50

45. (a) P_3, P_4 **(b)** P_4; $80,000   **(c)** P_1; $\geq$ $225,000

46. (a) P_4, P_5 **(b)** P_5; $70,000   **(c)** P_1; $\geq$ $240,000

47. (a) P_9 **(b)** P_1 **(c)** P_5 **(d)** P_1 **(e)** P_2 **(f)** 11

48. (a) P_6 **(b)** P_1 **(c)** P_1 **(d)** P_5 **(e)** P_2 **(f)** 5

49. (a) Angela; 100° strawberry wedge   **(b)** Carlos

  (c) Carlos; 110° vanilla wedge

  (d) Angela: $\frac{5}{12}$; Boris: $\frac{19}{45}$; Carlos: $\frac{11}{27}$

50. (a) Angela; 80° strawberry wedge   **(b)** Boris

  (c) Carlos; 165° vanilla wedge

  (d) Angela: $\frac{1}{3}$; Boris: $\frac{11}{30}$; Carlos: $\frac{11}{18}$

51. (a) $2\frac{2}{3}$ **(b)** Karla; [0, 2]

  (c) Jared: [5, 12]; Karla: [0, 2]; Lori: [2, 5]

52. (a) 4   **(b)** Karla; [0, $2\frac{2}{3}$]

  (c) Jared: [$7\frac{1}{3}$, 12]; Karla: [0, $2\frac{2}{3}$]; Lori: [$2\frac{2}{3}$, $7\frac{1}{3}$]

F **The Method of Sealed Bids**

53. (a) Ana: $300; Belle: $300; Chloe: $400

(b) In the first settlement, Ana gets the desk and receives $120 in cash, Belle gets the dresser, and Chloe gets the vanity and the tapestry and pays $360.

(c) $240

(d) In the final settlement, Ana gets the desk and receives $200 in cash, Belle gets the dresser and receives $80, and Chloe gets the vanity and the tapestry and pays $280.

54. (a) In the first settlement, Andre gets the farm and pays $220,000, Bea gets the painting and $151,000, and Chad gets the house and $45,000. The surplus is $24,000.

(b) In the final settlement, Andre gets the farm and pays $212,000, Bea gets the painting and $159,000, and Chad gets the house and $53,000.

55. (a) A: $600; B: $582; C: $618; D: $606; E: $600

(b) In the first settlement, A gets items 4 and 5 and pays $765, B gets $582, C gets items 1 and 3 and pays $287, D gets $606, and E gets items 2 and 6 and pays $266.

(c) $130 **(d)** In the final settlement, A gets items 4 and 5 and pays $739, B gets $608, C gets items 1 and 3 and pays $261, D gets $632, and E gets items 2 and 6 and pays $240.

56. (a) Alan: $29,333.33; Bly: $28,333.33; Claire: 35,666.67

(b) In the first settlement, Alan receives $29,333.33, Bly receives items 3 and 5 and pays $13,666.67, and Claire ends up with items 1, 2, and 4 and pays the estate $37,333.33.

(c) $21,666.67 **(d)** In the final settlement, Alan gets $36,555.55, Bly gets items 3 and 5 and pays the estate $6,444.45, and Claire gets items 1, 2, and 4 and pays $30,111.10.

57. $1800

58. $600

59. Anne gets $75,000 and Chia gets $80,000.

60. Al gets $56,000 and Ben gets $54,000.

G **The Method of Markers**

61. (a) 10, 11, 12, 13 **(b)** 1, 2, 3 **(c)** 5, 6, 7 **(d)** 4, 8, 9

62. (a) 4, 5, 6, 7, 8 **(b)** 10, 11, 12, 13 **(c)** 1, 2 **(d)** 3, 9

63. (a) 1, 2 **(b)** 10, 11, 12 **(c)** 4, 5, 6, 7 **(d)** 3, 8, 9

64. (a) 3, 4, 5, 6, 7 **(b)** 10, 11, 12 **(c)** 1 **(d)** 2, 8, 9

65. (a) A: 19, 20; B: 15, 16, 17; C: 1, 2, 3; D: 11, 12, 13; E: 5, 6, 7, 8

(b) 4, 9, 10, 14, 18

66. (a) A: 7; B: 15; C: 1, 2, 3; D: 11

(b) 4, 5, 6, 8, 9, 10, 12, 13, 14

67. (a) W S S G S W W B G G G S G S G S B B

Q_1 R_1 Q_2 R_2 S_2 Q_3 R_3 S_3

S_1 T_2 T_3

T_1

(b) Quintin gets a W, two S's, and a G; Ramon gets two W's and a B; Tim gets three G's and two S's; and Stephone gets an S and two B's. An S and two G's are left over.

68. (a) O O O GD GD GD BB BB BB GD GD GD O O O

S_1 R_1 Q_1 Q_2 R_2 S_2

(b) Sophie gets three Opera and one Grateful Dead CDs; Queenie gets one Beach Boys CD; Roxy gets three Grateful Dead and three Opera CDs. Two Grateful Dead and two Beach Boys CDs are left over.

(c) Queenie would pick a Beach Boys CD, Sophie would pick a Grateful Dead CD, and Roxy would pick either of the remaining Beach Boys or Grateful Dead CDs.

69. (a)

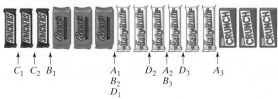

A_1 B_1 B_2 A_2 C_2
       C_1

(b) Ana gets three Snickers, Belle gets one Nestle Crunch, and Chloe gets two Reese's Peanut Butter Cups. Two Nestle Crunch bars and one Reese's Peanut Butter Cup are left over.

(c) Belle would select one of the Nestle Crunch bars, Chloe would then select the Reese's Peanut Butter Cup, and then Ana would be left with the last Nestle Crunch bar.

70. (a) C_3 can be placed anywhere between the last Snickers and the first Nestle Crunch bar.

C_1 C_2 B_1 A_1 D_2 A_2 D_3 A_3
              B_2
              D_1 B_3

(b) Arne gets three Nestle Crunch bars, Bruno gets three Reese's Peanut Butter Cups, Chloe gets one Snickers, and Daphne gets two Baby Ruth bars. Two Snickers and four Baby Ruth bars are left over.

(c) Arne is happy with a Baby Ruth, Bruno is happy with either a Snickers or a Baby Ruth, Chloe is happy with a Snickers, and Daphne is happy with a Baby Ruth.

JOGGING

71. (a) Since P_2 and P_3 value the land uniformly, each thinks that a fair share must have an area of at least 10,000 m². The area of C is only 8000 m².

(b)

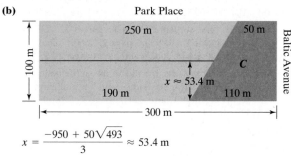

$$x = \frac{-950 + 50\sqrt{493}}{3} \approx 53.4 \text{ m}$$

(c)

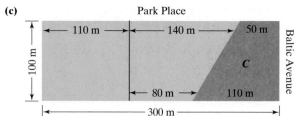

72. (a) Since P_2 and P_3 value the land uniformly, each thinks that a fair share must have an area of at least 10,000 m². The area of C is only 9000 m².

(b)

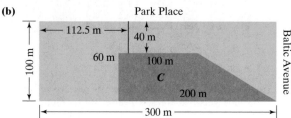

73. B must pay $\dfrac{x+y}{4}$ dollars to A.

74. A will receive $\dfrac{x}{3} + \dfrac{2z-x-y}{9}$ dollars.

B will receive $\dfrac{y}{3} + \dfrac{2z-x-y}{9}$ dollars.

C gets the business and pays

$$\frac{x}{3} + \frac{y}{3} + 2 \cdot \frac{2z-x-y}{9} \text{ dollars.}$$

75. B receives the strawberry part. C will receive $\frac{5}{6}$ of the chocolate part. A receives the vanilla part and $\frac{1}{6}$ of the chocolate part.

76. (a) $y = 180°$ **(b)** $y = 144°$ **(c)** $y = 36°$
   (d) $y = 0°$

77. (a) Dandy: s_4; Burly: s_1; Curly: s_2; Greedy: s_3
   (b) If Greedy only bids on s_1, then the bid lists are as follows: Dandy: $\{s_1, s_2, s_3, s_4\}$; Burly: $\{s_1\}$; Curly: $\{s_1, s_2\}$; Greedy: $\{s_1\}$. Suppose that Curly is given s_2 and Dandy is given s_3. The parcels s_1 and s_4 would then be joined together and split by Burly and Greedy. But a parcel consisting of s_1 and s_4 is only worth \$440,000 to Greedy, and half of that is only worth \$220,000.

78. Anne cleans bathrooms and pays \$11.67 per month; Bess cooks and pays \$11.67 per month; Cindy washes dishes, vacuums, mows the lawn, and receives \$23.33 per month.

79. Arne gets the stereo and the table and pays \$298.75; Brent patches the nail holes, repairs the window, and gets \$6.25 and the couch; Carlos gets \$251.25; Dale cleans the rugs and gets the desk and \$41.25.

80. (a) two, three, or four choosers bidding on the same item; three or four choosers bidding on the same two items; four choosers bidding on the same three items
   (b) 10
   (c) $\dfrac{(N-1)(N-2)}{2}$

Chapter 4

WALKING

A **Standard Divisors and Quotas**

1. (a) 50,000
   (b) Apure: 66.2; Barinas: 53.4; Carabobo: 26.6; Dolores: 13.8
2. (a) 40,000
   (b) Apure: 82.75; Barinas: 66.75; Carabobo: 33.25; Dolores: 17.25
3. (a) 1000   **(b)** the average ridership per bus each day
   (c) A: 45.30; B: 31.07; C: 20.49; D: 14.16; E: 10.26; F: 8.72
4. (a) 12   **(b)** the average number of patients per nurse
   (c) A: 72.583; B: 85.75; C: 50.833; D: 15.833
5. (a) 119   **(b)** 200,000

   (c) A: 8,100,000; B: 5,940,000; C: 4,730,000; D: 2,920,000; E: 2,110,000
6. (a) 250   **(b)** 50
   (c) A: 1646; B: 762; E: 2081; H: 1066; S: 6945
7. 32.32
8. 12.1%
9. (a) 0.5%
   (b) A: 22.74; B: 16.14; C: 77.24; D: 29.96; E: 20.84; F: 33.08
10. (a) 0.8%   **(b)** A: 7.8; B: 32.7; C: 35.6; D: 48.9

B Hamilton's Method

11. $A:66; B:53; C:27; D:14$

12. $A:83; B:67; C:33; D:17$

13. $A:45; B:31; C:21; D:14; E:10; F:9$

14. $A:72; B:86; C:51; D:16$

15. $A:40; B:30; C:24; D:15; E:10$

16. $A:33; B:15; E:42; H:21; S:139$

17. $A:23; B:16; C:77; D:30; E:21; F:33$

18. $A:8; B:33; C:35; D:49$

19. **(a)** Bob: 0; Peter: 3; Ron: 8 **(b)** Bob: 1; Peter: 2; Ron: 8

   (c) The population paradox. For studying 3.70% more, Bob gets a piece of candy. However, Peter, who studies 4.94% more, has to give up a piece.

20. **(a)** Bob: 1; Peter: 2; Ron: 7 **(b)** Bob: 0; Peter: 3; Ron: 8

   (c) The Alabama paradox. When 10 pieces of candy are divided, Bob is to get one piece, but when 11 pieces are divided, Bob ends up with none.

21. **(a)** Bob: 0; Peter: 3; Ron: 8 **(b)** Bob: 1; Peter: 3; Ron: 7; Jim: 6

   (c) The new-states paradox. Ron loses a piece of candy to Bob when Jim enters the discussion and is given his fair share (six pieces) of candy.

22. **(a)** Hamilton's method satisfies the quota rule. **(b)** same as (a)

   (c) Lilly's standard quota must be less than or equal to 8.47. If her standard quota is between 8 and 8.47, then Lilly will only receive 8 pieces since her decimal part is less than Katie's. If Lilly's standard quota is less than 8, then she will not receive 9 pieces because of the quota rule.

C Jefferson's Method

23. $A:67; B:54; C:26; D:13$

24. $A:83; B:67; C:33; D:17$

25. $A:46; B:31; C:21; D:14; E:10; F:8$

26. $A:73; B:86; C:51; D:15$

27. $A:41; B:30; C:24; D:14; E:10$

28. $A:33; B:15; E:41; H:21; S:140$

29. $A:22; B:16; C:78; D:30; E:21; F:33$

30. $A:7; B:33; C:36; D:49$

31. From Exercise 7, the standard quota of Texas was 32.32. But Jefferson's method only allows upper-quota violations.

32. It violates the quota rule.

D Adams's Method

33. $A:66; B:53; C:27; D:14$

34. $A:82; B:67; C:33; D:18$

35. $A:45; B:31; C:20; D:14; E:11; F:9$

36. $A:72; B:86; C:51; D:16$

37. $A:40; B:29; C:24; D:15; E:11$

38. $A:33; B:16; E:42; H:22; S:137$

39. $A:23; B:16; C:77; D:30; E:21; F:33$

40. $A:8; B:33; C:35; D:49$

41. It violates the quota rule.

42. From Exercise 7, the standard quota of Texas was 32.32. But Adams's method only allows lower-quota violations.

E Webster's Method

43. $A:66; B:53; C:27; D:14$

44. $A:83; B:67; C:33; D:17$

45. $A:45; B:31; C:21; D:14; E:10; F:9$

46. $A:72; B:86; C:51; D:16$

47. $A:40; B:30; C:24; D:15; E:10$

48. $A:33; B:15; E:42; H:21; S:139$

49. $A:23; B:16; C:77; D:30; E:21; F:33$

50. $A:8; B:33; C:35; D:49$

JOGGING

51. **(a)** the total number of seats available

   (b) the standard divisor

   (c) the percentage of the total population in state N

52. **(a)** an upper quota violation by state X.

   (b) a lower quota violation by state X.

   (c) The number of seats that state X has received is within 0.5 of its standard quota.

   (d) The number of seats that state X has received satisfies the quota rule, but is not the result of conventional rounding of the standard quota.

53. If the modified divisor is D, then state A has modified quota $p_A/D \geq 3.0$. Also, state B has modified quota $p_B/D < 1.0$. So, $p_A \geq 3D > 3p_B$. That is, more than three-fourths of the population lives in state A.

54. If the modified divisor is D, then state A has modified quota $p_A/D \geq 2.5$. Also, state B has modified quota $p_B/D < 0.5$. So, $p_A \geq 2.5D > 5p_B$. That is, more than five-sixths of the population lives in state A.

55. (a) Since the two fractional parts add up to 1, the surplus to be allocated using Hamilton's method is 1, and it will go to the state with the larger (more than 0.5) fractional part. The apportionment is the same as rounding the quotas conventionally (i.e., Webster's method).

(b) When there are only two states, Hamilton's and Webster's methods agree. Since Webster's method can never suffer from the Alabama or population paradox, neither will Hamilton's in this case.

(c) When there are only two states, Webster's and Hamilton's methods agree. Since Hamilton's method can never violate the quota rule, neither will Webster's method in this case.

56. (a) In Jefferson's method the modified quotas are larger than the standard quotas, so rounding down will give each state at least the integer part of the standard quota for that state.

(b) In Adams's method the modified quotas are smaller than the standard quota, so rounding up will give each state at most one more than the integer part of the standard quota for that state.

(c) If there are only two states, an upper-quota violation for one state results in a lower-quota violation for the other state. Neither Jefferson's nor Adams's method can have both upper- and lower-quota violations. So, when there are only two states, neither method can violate the quota rule.

57. (a) 49,374,462 **(b)** 1,591,833

58. lower bound: 4,528,665; upper bound: 5,822,569

59. (a) E: 5; F: 10; G: 15; H: 19

(b) For $D = 100$, the modified quotas are E: 5, F: 10, G: 15, and H: 20. For a modified divisor $D < 100$, each of the modified quotas will increase, so rounding down will give at least 50

seats. But for $D > 100$, each of the modified quotas will decrease, so rounding down will give at most 46 seats.

(c) There is no divisor such that the sum of the modified lower quotas will be 49. [See (b).]

60. (a) E: 5; F: 10; G: 15; H: 21

(b) For $D = 100$, the modified quotas are E: 5, F: 10, G: 15, and H: 20. For a modified divisor $D < 100$, each of the modified quotas will increase, so rounding up will give at least 54 seats. But for $D > 100$, each of the modified quotas will decrease, so rounding up will give at most 50 seats.

(c) There is no divisor such that the sum of the modified upper quotas will be 51. [See (b).]

61. Of those states for which the standard quota is not an integer, there are K that receive their lower quota. The remaining states (those with the largest fractional parts) receive their upper quota of seats. The states receiving their upper and lower quotas are the same under Hamilton's method or this alternate version of Hamilton's method.

62. (a) A: 33; B: 138; C: 4; D: 41; E: 14; F: 20

(b) State D is apportioned 41 seats under Lowndes's method (all the methods shown in Table 4-17 apportion 42 seats to state D). States C and E do better under Lowndes's method than under Hamilton's method.

63. (a) For example, $q_1 = 3.9$ and $q_2 = 10.1$.

(b) For example, $q_1 = 3.4$ and $q_2 = 10.6$.

(c) Assuming that $f_1 > f_2$, the methods produce different apportionments if $\dfrac{f_2}{q_2 - f_2} > \dfrac{f_1}{q_1 - f_1}$, i.e., if $\dfrac{q_1}{q_2} > \dfrac{f_1}{f_2}$.

Mini-Excursion 1

WALKING

A   The Geometric Mean

1. (a) 100 **(b)** 200 **(c)** $\frac{1}{200}$

2. (a) 4 **(b)** 40 **(c)** $\frac{1}{40}$

3. $2^4 \cdot 3^2 \cdot 7^4 \cdot 11$

4. $11^6 \cdot 13^4 \cdot 17^4 \cdot 19$

5. 18.6%

6. 9.3%

7.

Consecutive integers	Geometric mean (G)	Arithmetic mean (A)	Difference $(A - G)$
15, 16	15.492	15.5	0.008
16, 17	16.492	16.5	0.008
17, 18	17.493	17.5	0.007
18, 19	18.493	18.5	0.007
19, 20	19.494	19.5	0.006
29, 30	29.496	29.5	0.004
39, 40	39.497	39.5	0.003
49, 50	49.497	49.5	0.003

8. (a) $3.75G$ **(b)** $100G$ **(c)** $\frac{1}{G}$ **(d)** G^2

9. (a) $\sqrt{ka \cdot kb} = k\sqrt{a \cdot b} = kG$

(b) $\sqrt{\dfrac{k}{a} \cdot \dfrac{k}{b}} = k\dfrac{1}{\sqrt{a \cdot b}} = \dfrac{k}{G}$

10. (a) $A = \dfrac{a + b}{2} = \dfrac{2a}{2} = a$ and $G = \sqrt{a \cdot b} = \sqrt{a \cdot a} = a$

(b) Assume $a \neq b$. Then,
$$(a - b)^2 > 0 \Rightarrow a^2 - 2ab + b^2 > 0$$
$$\Rightarrow a^2 + b^2 > 2ab$$
$$\Rightarrow \frac{a^2 + b^2}{4} > \frac{ab}{2}$$
$$\Rightarrow \frac{a^2 + 2ab + b^2}{4} > ab$$
$$\Rightarrow \left(\frac{a + b}{2}\right)^2 > ab \Rightarrow A = \frac{a + b}{2} > \sqrt{ab} = G$$

11. (a) $\sqrt{ab}$

(b) Since the length of the legs of a right triangle are always strictly less than the length of the hypotenuse, it follows that $\sqrt{ab} < \dfrac{a + b}{2}$.

B   **The Huntington-Hill Method**

12. (a) 4   **(b)** 5   **(c)** 51   **(d)** 4

13. $A: 25; B: 18; C: 3; D: 37; E: 41$

14. (a) $A: 33; B: 138; C: 3; D: 42; E: 14; F: 20$

    (b) The apportionments are the same.

15. (a) $A: 34; B: 41; C: 22; D: 59; E: 15; F: 29$

    (b) $A: 34; B: 41; C: 22; D: 59; E: 15; F: 29$

    (c) The apportionments are the same.

16. (a) $A: 35; B: 20; C: 52; D: 8; E: 15; F: 70$

    (b) $A: 34; B: 20; C: 52; D: 8; E: 16; F: 70$

    (c) The apportionments are different.

17. (a) Aleta: 88; Bonita: 4; Corona: 5; Doritos: 3

    (b) An upper quota violation occurs.

18. Suppose N is any positive integer. Since the square roots increase at a decreasing rate, we know that
$$\sqrt{N+1} - \sqrt{N} > \sqrt{N+2} - \sqrt{N+1}.$$
Squaring both sides gives
$$(\sqrt{N+1} - \sqrt{N})^2 > (\sqrt{N+2} - \sqrt{N+1})^2$$
$$\Rightarrow N + 1 - 2\sqrt{N(N+1)} + N$$
$$> N + 2 - 2\sqrt{(N+1)(N+2)} + N + 1$$
$$\Rightarrow \frac{(N+1) + N}{2} - \sqrt{N(N+1)}$$
$$> \frac{(N+2) + (N+1)}{2} - \sqrt{(N+1)(N+2)}.$$
That is, the difference between the arithmetic and geometric means of N and $N + 1$ is greater than the difference between the arithmetic and geometric means of $N + 1$ and $N + 2$.

19. If $a < b$, then $\sqrt{a} < \sqrt{b}$. It follows that $-\sqrt{a} > -\sqrt{b}$ and hence $\sqrt{c} - \sqrt{a} > \sqrt{c} - \sqrt{b}$. Squaring both sides yields $(\sqrt{c} - \sqrt{a})^2 > (\sqrt{c} - \sqrt{b})^2$. Multiplying this out gives $c - 2\sqrt{ac} + a > c - 2\sqrt{bc} + b$. Dividing both sides by 2 and rearranging terms produces the inequality we are seeking:
$$\frac{a+c}{2} - \sqrt{ac} > \frac{b+c}{2} - \sqrt{bc}.$$

20. (a) The divisor is too small since two of the standard quotas are exactly at the geometric mean (cutoff) and are rounded up.

State	A	B	C	Total
Standard quota	$6\sqrt{2}$	$\frac{299}{243}\sqrt{2}$	$\sqrt{2}$	
Geometric mean	$6\sqrt{2}$	$\sqrt{2}$	$\sqrt{2}$	
Apportionment	9	2	2	13

    (b) For any divisor $D > \dfrac{1215}{\sqrt{2}}$, two of the modified quotas are certainly less than the geometric mean and are rounded down.

State	A	B	C	Total
Modified Quota	$< 6\sqrt{2}$	$< \frac{1495}{1215}\sqrt{2}$	$< \sqrt{2}$	
Geometric mean	$6\sqrt{2}$	$\sqrt{2}$	$\sqrt{2}$	
Apportionment	8	2	1	11

    (c) Part (a) shows that any modified divisor $D \leq \dfrac{1215}{\sqrt{2}}$ will not work because it is too small. Part (b) shows that any modified divisor $D > \dfrac{1215}{\sqrt{2}}$ will not work because it would be too large.

21. (a)

a	b	Q	a	b	Q
1	2	1.581	9	10	9.513
2	3	2.550	19	20	19.506
3	4	3.536	29	30	29.504
4	5	4.528	99	100	99.501
5	6	5.523	10	20	15.811
6	7	6.519	10	90	64.031
7	8	7.517	10	190	134.536
8	9	8.515	10	990	700.071

    (b) Since $a < b$ and both a and b are positive, it follows that $a^2 < \dfrac{a^2 + b^2}{2} < b^2$. Taking square roots gives
$$a < \sqrt{\frac{a^2 + b^2}{2}} < b.$$

    (c) $Q = \sqrt{\dfrac{a^2 + b^2}{2}} = \sqrt{\left(\dfrac{b-a}{2}\right)^2 + \left(\dfrac{b+a}{2}\right)^2} \geq \sqrt{\left(\dfrac{b+a}{2}\right)^2} = \dfrac{b+a}{2}$

22. (a)

a	b	H	a	b	H
1	2	1.333	9	10	9.474
2	3	2.400	19	20	19.487
3	4	3.429	29	30	29.492
4	5	4.444	99	100	99.497
5	6	5.455	10	20	13.333
6	7	6.462	10	90	18.000
7	8	7.467	10	190	19.000
8	9	8.471	10	990	19.800

    (b) Since $a < b$, we have $2a < a + b < 2b$. Thus, $\dfrac{1}{b} = \dfrac{2a}{2ab} < \dfrac{a+b}{2ab} < \dfrac{2b}{2ab} = \dfrac{1}{a}$. That is, $\dfrac{1}{b} < \dfrac{1}{H} < \dfrac{1}{a}$. It then follows that $b > H > a$.

    (c) First, notice that $H = \dfrac{G^2}{A}$, where H is the harmonic mean, G is the geometric mean, and A is the arithmetic mean. From Exercise 10, $A > G$, so $\dfrac{G}{A} < 1$. Hence, $H = \dfrac{G^2}{A} < G$.

Chapter 5

WALKING

A   Graphs: Basic Concepts

1. (a) $\mathcal{V} = \{A, B, C, X, Y, Z\}$
   (b) $\mathcal{E} = \{AX, AY, AZ, BB, BC, BX, CX, XY\}$
   (c) $\deg(A) = 3, \deg(B) = 4, \deg(C) = 2, \deg(X) = 4,$
   $\deg(Y) = 2, \deg(Z) = 1$

2. (a) $\mathcal{V} = \{A, B, C, D, E\}$
   (b) $\mathcal{E} = \{AB, AC, AD, AE, BC, BE, CC, CE, DE\}$
   (c) $\deg(A) = 4, \deg(B) = 3, \deg(C) = 5, \deg(D) = 2,$
   $\deg(E) = 4$

3. (a) $\mathcal{V} = \{A, B, C, D, X, Y, Z\}$
   (b) $\mathcal{E} = \{AX, AX, AY, BX, BY, DZ, XY\}$
   (c) $\deg(A) = 3, \deg(B) = 2, \deg(C) = 0, \deg(D) = 1,$
   $\deg(X) = 4, \deg(Y) = 3, \deg(Z) = 1$

4. (a) $\mathcal{V} = \{A, B, C, D, E, X, Y, Z\}$
   (b) $\mathcal{E} = \{AB, AD, AE, BE, CE, DE, XY, XY, YZ\}$
   (c) $\deg(A) = 3, \deg(B) = 2, \deg(C) = 1, \deg(D) = 2,$
   $\deg(E) = 4, \deg(X) = 2, \deg(Y) = 3, \deg(Z) = 1$

5.

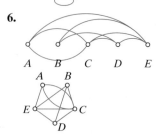

6.

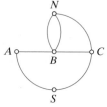

7. (a) A, B, D, E **(b)** AD, BC, DD, DE **(c)** 5   **(d)** 12
8. (a) A, C, Y **(b)** AX, AZ, CY, YY **(c)** 4   **(d)** 16
9. (a)

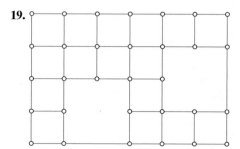

10. (a)

11. (a) C, B, A, H, F **(b)** C, B, D, A, H, F **(c)** C, B, A, H, F
   (d) C, D, B, A, H, G, G, F **(e)** 4   **(f)** 3   **(g)** 12
12. (a) D, A, H, G, F, E **(b)** D, B, A, H, G, G, F, E
   (c) D, A, H, F, E **(d)** $D, C, B, A, H, G, G, F, E$
   (e) 5   **(f)** 3   **(g)** 15
13. (a) G, G **(b)** There are none.
   (c) $A, B, D, A; B, C, D, B; F, G, H, F$
   (d) $A, B, C, D, A; F, G, G, H, F$ **(e)** 6
14. (a) E, E **(b)** F, H, F **(c)** A, B, C, A **(d)** A, B, E, D, A
   (e) $A, C, B, E, D, A; A, B, E, E, D, A$ **(f)** 6
15. (a) AH, EF **(b)** There are none.   **(c)** AB, BC, BE, CD
16. (a) BF, FG **(b)** There are none.   **(c)** AB, BC, CD, DE

B   Graph Models

17.

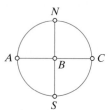

18.

19.

20.

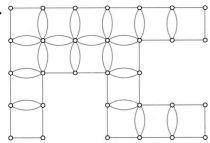

21.

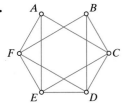

22. (a)

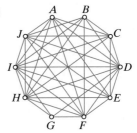

(b)

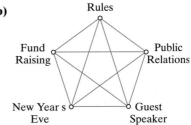

C | **Euler's Theorems**

23. (a) (i); all vertices are even.

  (b) (iii); four vertices are odd.

  (c) (iv); see Exercises 9(a) and 9(b).

24. (a) (ii); exactly two vertices are odd.

  (b) (iii); 10 vertices are odd.

  (c) (v); the graph may be disconnected.

25. (a) (iii); four vertices are odd.

  (b) (i); all vertices are even.

  (c) (iii); the graph is disconnected.

26. (a) (ii); exactly two vertices are odd.

  (b) (i); all vertices are even.

  (c) (iii); the graph is disconnected.

27. (a) (ii); exactly two vertices are odd.

  (b) (i); all vertices are even.

  (c) (iii); more than two vertices are odd.

28. (a) (ii); exactly two vertices are odd.

  (b) (iii); the graph is disconnected.

  (c) (iii); more than two vertices are odd.

D | **Finding Euler Circuits and Euler Paths**

29.

30.

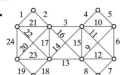

31.

32.

33.

34.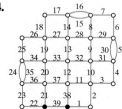

35.

36.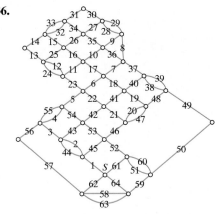

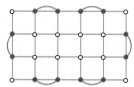

E **Eulerizations and Semi-eulerizations**

37.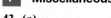

38.

39.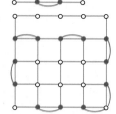

40.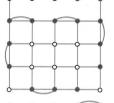

41.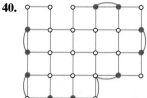

42.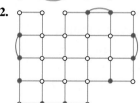

F **Miscellaneous**

43. (a)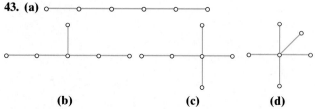

(b)   **(c)**   **(d)**

44. (a)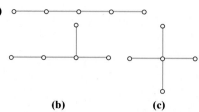

(b)   **(c)**

45.

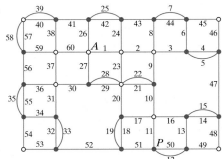

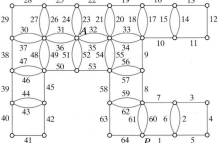

46.

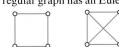

47. 4; there are 10 odd vertices. Two can be used as the starting and ending vertices. The remaining eight can be paired so that each pair forces one lifting of the pencil.

48. 6; there are 14 odd vertices. Two can be used as the starting and ending vertices. The remaining 12 can be paired so that each pair forces one lifting of the pencil.

JOGGING

49. **(a)** Each edge contributes 2 to the sum of the degrees.

(b) If there were an odd number of vertices, then the sum of the degrees of all the vertices would be odd.

50. $(k-2)/2$; see Exercises 47 and 48.

51. None. If a vertex had degree 1, then the edge incident to that vertex would be a bridge.

52. **(a)** If each vertex were of odd degree, then the graph has an odd number of odd vertices. Since this is not possible, every vertex is even and the graph has an Euler circuit.

(b) The first regular graph has an Euler circuit; the second does not.

53. **(a)** Both m and n must be even.

(b) $m = 1$ and $n = 1$, $m = 1$ and $n = 2$, or $m = 2$ and n is odd.

54. **(a)** Keep adding edges as long as you can without creating any circuits.

(b) $N - 1$ dollars

55. **(a)** Make a complete graph with $N - 1$ of the vertices, and leave the other vertex as an isolated vertex.

(b) $(N - 1)(N - 2)/2$ dollars

56. **(a)** $NK, C, B, NK, B, A, SK, C$

(b) $NK, C, B, NK, B, A, SK, C, SK$

(c) $NK, C, B, NK, B, A, SK, C, B$

(d) $NK, C, B, NK, B, A, SK, C, B, A$

57. **(a)** 12

(b)

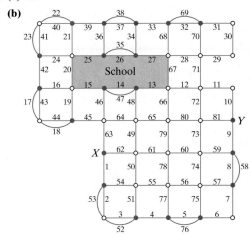

58. **(a)** $R, L, R, A, C, L, C, D, L, D, B, R, B$

(b) $B, R, B, R, L, R, A, C, L, C, D, B, D, L$

59. **(a)** $D, L, R, L, C, A, R, B, R, B, D, C, L, D$

(b) $L, D, L, C, D, B, R, B, R, L, R, A, C, L$

60. Yes; one of the many possible journeys is given by crossing the bridges in the following order: $a, b, c, d, e, f, g, h, i, l, m, n, o, p, k$. (*Note*: Euler did not ask that the journey start and end at the same place.)

Chapter 6

WALKING

A **Hamilton Circuits and Hamilton Paths**

1. (a) 1. A, B, D, C, E, F, G, A;

    2. A, D, C, E, B, G, F, A;

    3. A, D, B, E, C, F, G, A

  (b) A, G, F, E, C, D, B

  (c) D, A, G, B, C, E, F

2. (a) 1. $A, F, E, D, C, H, I, J, G, B, A$;

    2. $A, F, E, I, J, G, H, D, C, B, A$;

    3. $A, F, E, I, H, D, C, B, G, J, A$

  (b) $A, F, E, D, C, H, I, J, G, B$

  (c) $F, J, A, B, G, H, C, D, E, I$

3. (a) A, B, C, D, E, F, A; A, B, E, D, C, F, A and their mirror images

  (b) A, B, C, D, E, F, G, A; A, B, E, D, C, F, G, A and their mirror images

  (c) A, B, C, D, E, F, G, A; A, B, E, D, C, F, G, A; A, F, C, D, E, B, G, A; A, F, E, D, C, B, G, A and their mirror images

4. (a) A, B, C, D, E, A and its mirror image

  (b) A, B, C, D, E, A; A, B, C, E, D, A and their mirror images

5. (a) A, F, B, C, G, D, E **(b)** A, F, B, C, G, D, E, A

  (c) A, F, B, E, D, G, C **(d)** F, A, B, E, D, C, G

6. (a) A, D, G, C, F, B, E **(b)** A, D, G, C, F, B, E, A

  (c) A, D, E, B, F, C, G **(d)** F, B, E, A, D, C, G

7. (a) A, B, C, D, E, F, A; A, B, E, D, C, F, A and their mirror images

  (b) D, E, F, A, B, C, D; D, C, F, A, B, E, D and their mirror images

8. (a) A, B, C, F, E, D, A; A, D, F, E, B, C, A; A, B, E, D, F, C, A and their mirror images

  (b) D, A, B, C, F, E, D; D, F, E, B, C, A, D; D, F, C, A, B, E, D and their mirror images

9. There is no Hamilton circuit since two vertices have degree 1. There is no Hamilton path since any such path must contain edges AB, BE, and BC, which would force vertex B to be visited more than once.

10. There is no Hamilton circuit since two vertices have degree 1. One Hamilton path is F, B, A, E, C, D, G.

11. (a) $A, I, J, H, B, C, F, E, G, D$

  (b) $G, D, E, F, C, B, A, I, J, H$

  (c) If such a path were to start by heading left, it would not contain C, D, E, F, or G, since it would need to pass through B again to do so. On the other hand, if a path were to start by heading right, then it could not contain A, I, H, or J.

  (d) Any circuit would need to cross the bridge BC twice (to return to where it started). But then B and C would be included twice.

12. (a) D, A, G, F, E, C, B **(b)** B, A, G, F, E, C, D

  (c) Any path passing through vertices B, D, and G must contain edges AB, AD, and AG, respectively. Consequently, any path passing through vertices B, D, and G must contain at least three edges meeting at A and hence would pass through vertex A more than once. By symmetry, it is clear that there is no Hamilton path that starts at C.

  (d) Any circuit passing through vertices B, D, and G must contain edges AB, AD, and AG, respectively. Since these three edges all meet at A, they would pass through vertex A more than once.

13. (a) 6   **(b)** B, D, A, E, C, B; weight $= 27$

  (c) The mirror image B, C, E, A, D, B; weight $= 27$

14. (a) 8   **(b)** A, D, C, B, E, A; weight $= 22$

  (c) The mirror image A, E, B, C, D, A; weight $= 22$

15. (a) A, D, F, E, B, C; weight $= 29$

  (b) A, B, E, D, F, C; weight $= 30$

  (c) A, D, F, E, B, C; weight $= 29$

16. (a) B, A, E, C, D; weight $= 19$

  (b) B, A, C, E, D; weight $= 25$

  (c) B, A, E, C, D; weight $= 19$

B **Factorials and Complete Graphs**

17. (a) 362,880   **(b)** 11   **(c)** 110   **(d)** 504   **(e)** 10,100

18. (a) 121,645,100,408,832,000   **(b)** 20   **(c)** 40,200

19. (a) 11.1   **(b)** 101.01

20. (a) 21.05   **(b)** 201.005

21. (a) $2,432,902,008,176,640,000 \approx 2.43 \times 10^{18}$

  (b) $1,124,000,727,777,607,680,000 \approx 1.12 \times 10^{21}$

  (c) $20! = 2,432,902,008,176,640,000 \approx 2.43 \times 10^{18}$

22. (a) $15,511,210,043,330,985,984,000,000 \approx 1.55 \times 10^{25}$

  (b) $10,888,869,450,418,352,160,768,000,000 \approx 1.09 \times 10^{28}$

  (c) $25! = 15,511,210,043,330,985,984,000,000 \approx 1.55 \times 10^{25}$

23. (a) ≈ 77 years   **(b)** ≈ 1620 years

24. (a) $\approx 491,857$ years   **(b)** $\approx 12,788,288$ years

25. (a) 190   **(b)** 210   **(c)** 50

26. (a) 19,900   **(b)** 20,100   **(c)** 500

27. (a) $N = 6$ **(b)** $N = 10$ **(c)** $N = 201$

28. (a) $N = 7$ **(b)** $N = 12$ **(c)** $N = 401$

C Brute-Force and Nearest-Neighbor Algorithms

29. **(a)** A, C, B, D, A; cost = 102
  (b) A, D, C, B, A; cost = 120
  (c) B, D, C, A, B; cost = 114
  (d) C, D, B, A, C; cost = 114

30. **(a)** A, B, C, D, A; cost = 155
  (b) A, B, D, C, A; cost = 190
  (c) C, B, D, A, C; cost = 165
  (d) D, B, C, A, D; cost = 165

31. **(a)** B, C, A, E, D, B; cost = \$722
  (b) C, A, E, D, B, C; cost = \$722
  (c) D, B, C, A, E, D; cost = \$722
  (d) E, C, A, D, B, E; cost = \$741

32. **(a)** A, E, B, C, D, A; total travel time = 113 minutes
  (b) A, E, B, C, D, A; total travel time = 113 minutes
  (c) A, E, B, C, D, A; total travel time = 113 minutes

33. **(a)** A, D, E, C, B, A; cost = \$11,656
  (b) A, D, B, C, E, A; cost = \$9,760
  (c) A, B, C, E, D, A; cost = \$11,656

34. **(a)** E, C, I, G, M, T, E; total travel time = 19.4 years
  (b) E, T, M, I, C, G, E; total travel time = 18.9 years
  (c) E, G, I, M, T, C, E; total travel time = 18.3 years

35. **(a)** Atlanta, Columbus, Kansas City, Tulsa, Minneapolis, Pierre, Atlanta; cost = \$2915.25
  (b) Atlanta, Kansas City, Tulsa, Minneapolis, Pierre, Columbus, Atlanta; cost = \$2804.25

36. **(a)** Nashville, Louisville, St. Louis, Pittsburgh, Boston, Dallas, Houston, Nashville; total length = 4340 miles
  (b) Nashville, Pittsburgh, Boston, Dallas, Houston, St. Louis, Louisville, Nashville; total length = 4315 miles

D Repetitive Nearest-Neighbor Algorithm

37. B, C, D, E, A, B
38. D, A, F, E, B, C, D; cost = \$200
39. A, D, B, C, E, A; cost = \$9760
40. E, T, M, I, C, G, E; total travel time = 18.9 years

41. Atlanta, Columbus, Minneapolis, Pierre, Kansas City, Tulsa, Atlanta; cost = \$2439.00
42. Nashville, Louisville, Pittsburgh, Boston, Houston, Dallas, St. Louis, Nashville; total length = 4093 miles

E Cheapest-Link Algorithm

43. B, E, D, C, A, B
44. B, D, C, A, F, E, B; cost = \$200
45. B, E, C, A, D, B; cost = \$10,000
46. E, C, I, G, M, T, E; total travel time = 19.4 years

47. Atlanta, Columbus, Pierre, Minneapolis, Kansas City, Tulsa, Atlanta; cost = \$2598.75
48. Nashville, Louisville, St. Louis, Dallas, Houston, Boston, Pittsburgh, Nashville; total mileage = 4222 miles

F Miscellaneous

49. 12.5%
50. 6.5%
51. \$1500
52. \$2400
53. **(a)**

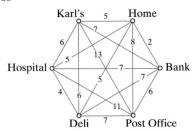

  (b) Home, Bank, Post Office, Deli, Hospital, Karl's, Home

54. **(a)**

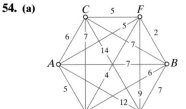

  (b) F, B, E, D, A, C, F

55. **(a)**

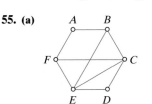

(b) Any Hamilton circuit gives a possible seating arrangement. For example,

(c)

56. (a)

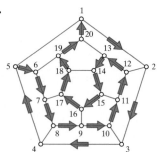

(b) Any Hamilton circuit gives a possible seating arrangement. For example,

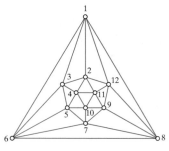

(c)

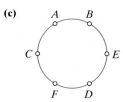

57. The graph describing the friendships among the guests does not have a Hamilton circuit. Thus, it is impossible to seat everyone around the table with friends on both sides.

58. The graph describing the friendships among the guests does not have a Hamilton circuit. Thus, it is impossible to seat everyone around the table with friends on both sides.

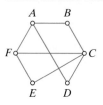

JOGGING

59.

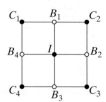

60.

61. (a) 1001   **(b)** 1503

62. (a) $I, B_1, C_2, B_2, C_3, B_3, C_4, B_4, C_1$

   (b) $C_1, B_1, C_2, B_2, I, B_4, C_4, B_3, C_3$

   (c) $C_1, B_4, C_4, B_3, C_3, B_2, C_2, B_1, I$

   (d) Suppose that we color the vertices of the grid graph with two colors [say black (B) and white (W)] with adjacent vertices having different colors (see figure). Since there are five black vertices and four white vertices and in any Hamilton path the vertices must alternate color, the only possible Hamilton paths are of the form $B, W, B, W, B, W, B, W, B$. Since every boundary vertex is white, it is impossible to end a Hamilton path on a boundary vertex.

63. The 2 by 2 grid graph cannot have a Hamilton circuit because each of the four corner vertices as well as the interior vertex I must be preceded and followed by a boundary vertex. But there are only four boundary vertices—not enough to go around.

64. (a) $C_1, B_8, B_7, C_4, B_6, I_3, I_1, I_2, I_4, B_5, C_3, B_4, B_3, C_2, B_2, B_1, C_1$

    (b) $C_1, B_8, B_7, C_4, B_6, I_3, I_1, B_1, B_2, I_2, I_4, B_5, C_3, B_4, B_3, C_2$

    (c) $C_1, B_8, B_7, C_4, B_6, I_3, I_4, B_5, C_3, B_4, B_3, C_2, B_2, B_1, I_1, I_2$

    (d) Suppose that X and Y are any two adjacent vertices. If we pick a Hamilton circuit that contains the edge XY and remove that edge, then we get a Hamilton path that has X and Y as its endpoints. (Finding a Hamilton circuit is relatively easy.)

65. (a)

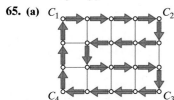

    (b)

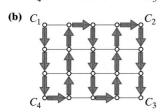

    (c) Think of the vertices of the graph as being colored like a checkerboard, with C_1 being a white vertex. Then each time we move from one vertex to the next, we must move from a white vertex to a black vertex or from a black vertex to a white vertex. Since there are 10 white vertices and 10 black vertices and we are starting with a white vertex, we must end at a black vertex. But C_2 is a white vertex. Therefore, no such Hamilton path is possible.

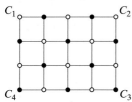

66. Suppose that the cheapest edge in a graph is the edge joining vertices X and Y. Using the nearest-neighbor algorithm, we will eventually visit one of these vertices—suppose that the first one of these vertices we visit is X. Then, since edge XY is the cheapest edge in the graph and since we have not yet visited vertex Y, the nearest-neighbor algorithm will take us to Y.

67.

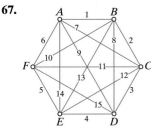

68. (a)

    (b) If a graph has n vertices, then a Hamilton circuit must have n edges. If a graph has m edges, then an Euler circuit must also have m edges. Thus, the only way a graph can have the same circuit be an Euler circuit and a Hamilton circuit is if $n = m$. In a connected graph, this can happen only, when the graph is itself a circuit.

69. (a) Any circuit would need to cross the bridge twice (to return to where it started). But then each vertex at the end of the bridge would be included twice.

    (b)

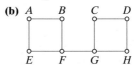

70. $21! = 21 \times 20 \times 19 \times \cdots \times 11 \times 10!$

    $> 10 \times 10 \times 10 \times \cdots \times 10 \times 10!$

    $= 10^{11} \times 10!$

71. Dallas, Houston, Memphis, Louisville, Columbus, Chicago, Kansas City, Denver, Atlanta, Buffalo, Boston, Dallas.

The process starts by finding the smallest number in the Dallas row. This is 243 miles to Houston. We cross out the Dallas column and proceed by finding the smallest number in the Houston row (other than that representing Dallas, which has been crossed out). This is 561 miles to Memphis. We now cross out the Houston column and continue by finding the smallest number in the Memphis row (other than those representing Dallas and Houston, which have been crossed out). The process continues in this fashion until all cities have been reached.

72. Dallas, Houston, Denver, Boston, Buffalo, Columbus, Louisville, Chicago, Atlanta, Memphis, Kansas City, Dallas.

73. (a) Detroit–Flint–Lansing–Grand Rapids–Cheboygan–Sault Ste. Marie–Marquette–Escanaba–Menominee–Sault Ste. Marie–Cheboygan–Detroit. Total cost = $541.86 (1374 miles @ $0.39 + $6.00 for toll fees).

    (b) Detroit–Flint–Lansing–Grand Rapids–Cheboygan–Sault Ste. Marie–Marquette–Escanaba–Menominee. Total cost = $389.61 (789 miles @ $0.49 + $3.00 for the toll fee).

Chapter 7

WALKING

A   Trees

1. (a) tree   **(b)** not a tree (not connected)
(c) not a tree (has a circuit)

2. (a) not a tree (not connected)   **(b)** not a tree (has a circuit)
(c) tree

3. (a) 7   **(b)** 7

4. (a) 10   **(b)** 9

5. (a) (II) Use Tree Property 3.
(b) (III)

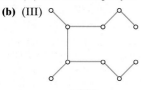

A tree   Not a tree

(c) (I) Use Tree Property 2.   **(d)** (I) Use Tree Property 1.
(e) (I) Use Tree Property 2.

6. (a) (II) Use Tree Property 3.
(b) (III)

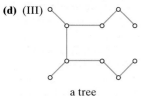

a tree   not a tree

(c) (II) Use Tree Property 1.
(d) (III)

a tree   not a tree

(e) (I) Use Tree Property 1.   **(f)** (I) Use Tree Property 2.

7. (a) (III)

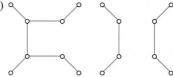

A tree   Not a tree

(b) (II) A tree has no circuits.

(c) (I) A graph with 8 vertices, 7 edges, and no circuits must also be connected and hence must be a tree.

8. (a) (II) A tree has no circuits.

(b) (II) Since the degree of each vertex is 9, the sum of the degrees is 90. However, the sum of degrees for a tree with 10 vertices would be 18.

(c) (III)

a tree   not a tree

(d) (II) Since the degree of each vertex is 2, the sum of the degrees of all 10 vertices must be 20. But a tree with 10 vertices must have 9 edges, and the sum of the degrees of the vertices would be 18.

9. The graph would have an Euler circuit. Any graph with a circuit is not a tree.

10. (a)

(b)

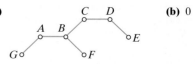

B   Spanning Trees

11. (a)

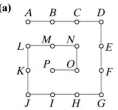

(b) 6

12. (a)

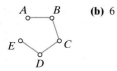

(b) 9

13. (a)

(b) 0

14. (a)

C   D
A   B   E   F

H   G

(b) 0

15. **(a)**

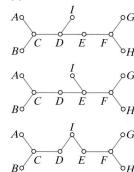

(b)

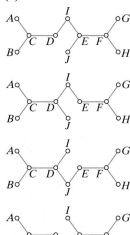

(c) 6

16. **(a)**

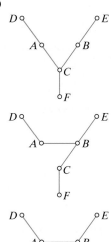

(b)

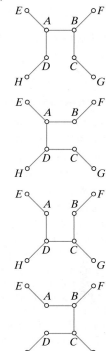

(c) 6

17. **(a)** 12   **(b)** 54

18. **(a)** 16   **(b)** 1024

C **Minimum Spanning Trees and Kruskal's Algorithm**

19. **(a)**

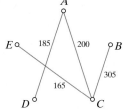

(b) 855

20. **(a)**

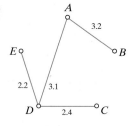

(b) 10.9

21. **(a)**

(b) 9.3

22. **(a)** Several MSTs are possible.   **(b)** 280

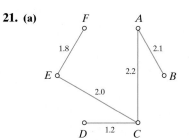

23. Kansas City–Tulsa, Pierre–Minneapolis, Minneapolis–Kansas City, Atlanta–Columbus, Columbus–Kansas City

24. Louisville–Nashville, Dallas–Houston, Louisville–St. Louis, Louisville–Pittsburgh, Pittsburgh–Boston, Dallas–St. Louis

25. $380,000; the MST will have 19 edges at a cost of $20,000 each.

26. $580,000; the MST will have 29 edges at a cost of $20,000 each.

D   **Steiner Points and Shortest Networks**

27. (a) 580 miles   **(b)** 385 miles

28. (a) 642 miles   **(b)** 375 miles

29. (a) 366 km   **(b)** 334 km

30. (a) 430 km   **(b)** 450 km

31. 224 miles

32. 539 miles

33. Z; the sum of the distances from Z to A, B, and C is 232 miles; the sum of the distances from X to A, B, and C is 240 miles; and the sum of the distances from Y to A, B, and C is 243 miles.

34. Y; the sum of the distances from Y to A, B, and C is 1260 miles; the sum of the distances from X to A, B, and C is 1300 miles; and the sum of the distances from Z to A, B, and C is 1510 miles.

35. (a) $CE + ED + EB$ is larger, since CD, DB is the shortest network connecting the cities C, D, and B.

(b) CD, DB form the shortest network connecting the cities C, D, and B, since angle CDB is 120°, causing the shortest network to coincide with the MST.

(c) CE, EB form the shortest network connecting C, E, and B, since angle CEB is more than 120°, causing the shortest network to coincide with the MST.

36. (a) $CA + AB$ is larger than $DC + DA + DB$, since D is a Steiner point for triangle ABC. Therefore, $DC + DA + DB$ is the length of the shortest network connecting A, B, and C.

(b) $EC + EA + EB$ is larger than $DC + DA + DB$. D is a Steiner point for triangle ABC, and so $DC + DA + DB$ is the length of the shortest network connecting A, B, and C.

(c) DC, DA, and DB form the shortest network connecting A, B, and C, since D is a Steiner point.

E   **Miscellaneous**

37. (a) $k = 0, 1, 2, 5$ are all possible.

(b) $0 \le k \le 120$ and $k = 123$ are all possible.

38. (a) $k = 0, 1, 2, 3, 4, 7$ are all possible.

(b) $0 \le k \le 2478$ and $k = 2481$ are all possible.

39. (a) 0   **(b)** 1   **(c)** 118

40. (a) 0   **(b)** 1   **(c)** 2471

41. (a)

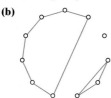

(b)

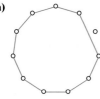

(c)

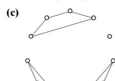

42. (a)

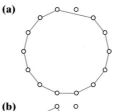

(b)

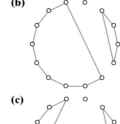

(c)

43. (a) $h = 2s$ **(b)** $l = \sqrt{3}s$ **(c)** $h = 42, l \approx 36.37$

(d) $s = 15.2, l \approx 26.33$

44. (a) $s = \frac{1}{2}h$ **(b)** $l = \frac{\sqrt{3}}{2}h$ **(c)** $h = \frac{2l}{\sqrt{3}}$

(d) $s \approx 100, h \approx 200$

45. 239 miles

46. 440 miles

47. 450 miles

48. 875 miles

49. In a 30-60-90 triangle, the side opposite the 30° angle is one-half the length of the hypotenuse, and the side opposite the 60° angle is $\sqrt{3}/2$ times the length of the hypotenuse [see Exercise 44(b)].

JOGGING

51. (a) 11 **(b)** 29 **(c)** 56

52. (a) $2N - 2$

(b) Let v be the number of vertices in the graph and e the number of edges. Recall that in a tree $v = e + 1$, and in any graph the sum of the degrees of all the vertices is $2e$. Assume that there are exactly k vertices of degree 1. The remaining $v - k$ vertices must have degree at least 2. Therefore, the sum of the degrees of all the vertices must be at least $k + 2(v - k)$. Putting all this together, we have $2e \ge k + 2(v - k) = k + 2(e + 1 - k) = -k + 2e + 2$, so $k \ge 2$.

(c) By part (b), all vertices would need to be of degree 1. But then the graph would be disconnected.

53. (a)

(b)

(c)

(d)

54. (a) They are trees.

(b) If $R = 1$, then $M = N$ (the number of edges is the same as the number of vertices). So, the network is not a tree. From the definition of a tree, there must be at least one circuit. Suppose that the graph had two (or more) circuits. Then, there would be an edge of one of the circuits that is not an edge of another circuit. Removing such an edge would leave us with a connected graph with the number of vertices being one more than the number of edges (i.e., a tree). But this tree would have a circuit. This is impossible, so there cannot be more than one circuit.

(c) The maximum redundancy occurs when the graph is a complete graph and the degree of each vertex is $N - 1$. The redundancy of the complete graph is then, $R = [N(N - 1)/2] - (N - 1) = (N^2 - 3N + 2)/2$.

Therefore, the distance from C to J is $(\sqrt{3}/2) \times 500$ mi ≈ 433 miles, and so the total length of the T-network (rounded to the nearest mile) is 433 mi + 500 mi = 933 mi.

50. In a 30-60-90 triangle, the side opposite the 30° angle is one-half the length of the hypotenuse, and the side opposite the 60° angle is $\sqrt{3}/2$ times the length of the hypotenuse. Therefore, the distance from any one of the three cities to S (the Steiner point) is $(2/\sqrt{3}) \times 250 = 500/\sqrt{3}$ mi, and so the total length of the Y-network (rounded to the nearest mile) is $3 \times (500/\sqrt{3}) = 500\sqrt{3} \approx 866$ mi.

55. If $M \ge N$, then the redundancy of the network is at least 1 and the network has at least one circuit. The shortest length of that circuit is 3. None of the edges in the circuit are bridges, but the remaining $M - 3$ edges could all be bridges.

56. $C_4C_7, C_3C_6, C_1C_8, C_1C_2, C_2C_5, C_2C_6, C_8C_9, C_7C_9$

57. $205 million

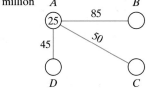

58. (a) The MST is given in Fig. 7-67(a). If $x > 59$, then edge EG will not replace any edge in the existing MST since it has a larger value.

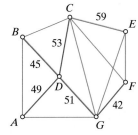

(b) If $x < 59$, then edge EG will replace edge CE in the MST.

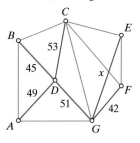

59. (a) and **(b)** $7.5 million

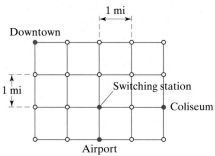

60. (a) and **(b)** $33 million

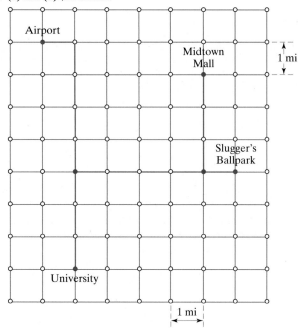

61. (a) $m(\angle BFA) = m(\angle AEC) = m(\angle CGB) = 120°$

(b) In $\triangle ABF$, $m(\angle BFA) = 120°$ and so $m(\angle BAF) + m(\angle ABF) = 60°$. So, $m(\angle BAF) < 60°$ and $m(\angle ABF) < 60°$. Similarly, $m(\angle ACE) < 60°$, $m(\angle CAE) < 60°$, $m(\angle BCG) < 60°$, and $m(\angle CBG) < 60°$. Consequently, $m(\angle A) = m(\angle BAF) + m(\angle CAE) < 60° + 60° = 120°$. Likewise, $m(\angle B) < 120°$ and $m(\angle C) < 120°$. So, triangle ABC has a Steiner point S.

(c) Any point X inside or on $\triangle ABF$ (except vertex F) will have $m(\angle AXB) > 120°$. Any point X inside or on $\triangle ACE$ (except vertex E) will have $m(\angle AXC) > 120°$. Any point X inside or on $\triangle BCG$ (except vertex G) will have $m(\angle BXC) > 120°$. If S is the Steiner point, $m(\angle ASB) = m(\angle ASC) = m(BSC) = 120°$, so S cannot be inside or on $\triangle ABF$, $\triangle ACE$, or $\triangle BCG$. It follows that the Steiner point S must lie inside $\triangle EFG$.

62. (a) 1414 miles

(b) The length of the network is $4x + (500 - x) = 3x + 500$, where $250^2 + (x/2)^2 = x^2$ (see figure). Solving gives

$x = (500\sqrt{3}/3)$, so the length of the network is $3(500\sqrt{3}/3) + 500 = 500\sqrt{3} + 500 \approx 1366$ mi.

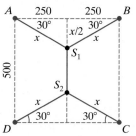

63. (a) The length of the network is $4x + (300 - x) = 3x + 300$ (see figure). The triangle formed by A, S_1, and the midpoint of AB is a 30-60-90 triangle. Therefore, $x = 2/\sqrt{3} \times 200 = (400\sqrt{3}/3)$, so the length of the network is $3(400\sqrt{3}/3) + 300 = 400\sqrt{3} + 300 \approx 993$ mi.

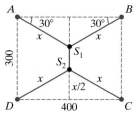

(b) The length of the network is $4x + (400 - x) = 3x + 400$, where $150^2 + (x/2)^2 = x^2$ (see figure). Solving gives $x = (300\sqrt{3}/3)$, so the length of the network is $3(300\sqrt{3}/3) + 400 = 300\sqrt{3} + 400 \approx 919.6$ mi.

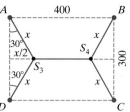

64. The area of $\triangle ABC$ is $\frac{1}{2}\overline{BC} \cdot h$. This area is also the sum of the areas of $\triangle PBC$, $\triangle ABP$, and $\triangle APC$, which are $\frac{1}{2}\overline{BC} \cdot y$, $\frac{1}{2}\overline{AB} \cdot x$, and $\frac{1}{2}\overline{AC} \cdot z$, respectively. So, $\frac{1}{2}\overline{BC} \cdot h = \frac{1}{2}\overline{BC} \cdot y + \frac{1}{2}\overline{AB} \cdot x + \frac{1}{2}\overline{AC} \cdot z = \frac{1}{2}\overline{BC} \cdot (x + y + z)$, since $\triangle ABC$ is equilateral. Thus, $h = x + y + z$.

65. (a)

(b)

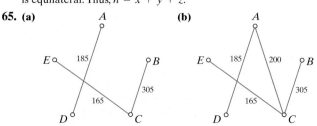

66. (a) According to Cayley's theorem, there are $3^{3-2} = 3$ spanning trees in a complete graph with three vertices.

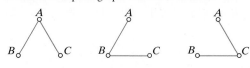

Likewise, for the complete graph with four vertices, Cayley's theorem predicts $4^{4-2} = 16$ spanning trees.

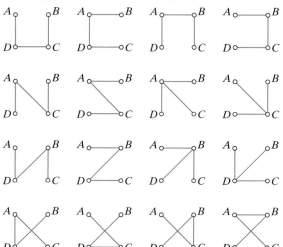

(b) The number of spanning trees is larger than the number of Hamilton circuits. The complete graph with N vertices has $(N - 1)!$ Hamilton circuits and N^{N-2} spanning trees. Since $(N - 1)! = 2 \times 3 \times 4 \times \cdots \times (N - 1)$ and $N^{N-2} = N \times N \times N \times \cdots \times N$. Notice that each factor in $(N - 1)!$ is smaller than the corresponding factor in N^{N-2}. Thus, for $N \geq 3, (N - 1)! < N^{N-2}$.

Chapter 8

WALKING

A Directed Graphs

1. (a) $\mathcal{V} = \{A, B, C, D, E\}$;
   $\mathcal{A} = \{AB, AC, BD, CA, CD, CE, EA, ED\}$

   (b) $\text{indeg}(A) = 2, \text{indeg}(B) = 1, \text{indeg}(C) = 1,$
   $\text{indeg}(D) = 3, \text{indeg}(E) = 1$

   (c) $\text{outdeg}(A) = 2, \text{outdeg}(B) = 1, \text{outdeg}(C) = 3,$
   $\text{outdeg}(D) = 0, \text{outdeg}(E) = 2$

2. (a) $\mathcal{V} = \{A, B, C, D, E\}$;
   $\mathcal{A} = \{AC, AD, AE, BA, BC, BE, EB, ED\}$

   (b) $\text{indeg}(A) = 1, \text{indeg}(B) = 1, \text{indeg}(C) = 2,$
   $\text{indeg}(D) = 2, \text{indeg}(E) = 2$

   (c) $\text{outdeg}(A) = 3, \text{outdeg}(B) = 3, \text{outdeg}(C) = 0,$
   $\text{outdeg}(D) = 0, \text{outdeg}(E) = 2$

3. (a) 8   **(b)** 8   **(c)** 8

4. (a) 8   **(b)** 8   **(c)** 8

5. (a) C, E **(b)** B, C **(c)** B, C, E **(d)** There are none.
   (e) CD, CE, CA **(f)** There are none.

6. (a) B **(b)** C, D, E **(c)** A, E **(d)** There are none.
   (e) AC, AD, AE **(f)** There are none.

7. (a)   **(b)**   **(c)**

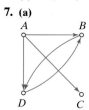 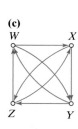 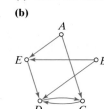

8. (a)   **(b)**   **(c)**

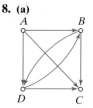 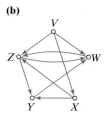 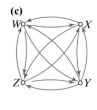

9. (a) 2   **(b)** 1   **(c)** 1   **(d)** 0

10. (a) 2   **(b)** 0   **(c)** 2   **(d)** 3

11. (a) A, B, D, E, F **(b)** A, B, D, E, C, F **(c)** B, D, E, B
   (d) $\text{outdeg}(F) = 0$ **(e)** $\text{indeg}(A) = 0$
   (f) B, D, E, B is the only cycle.

12. (a) A, E, C, D

   (b) $\text{indeg}(D) = 1$; hence, there is only one way to get to D and that is using the arc CD. Also, $\text{indeg}(C) = 1$, so there is only one way to get to vertex C and that is using the arc EC. Finally, $\text{outdeg}(B) = 0$, so the only way to get to D from A is the given path.

   (c) E, C, D, E **(d)** $\text{indeg}(A) = 0$ **(e)** $\text{outdeg}(B) = 0$
   (f) E, C, D, E is the only cycle.

13.

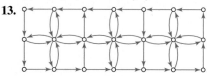

14.

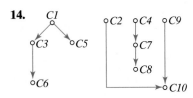

15. **(a)** B; that is the only person that everyone respects.

(b) A; that is the only person that no one respects.

(c) The individual corresponding to the vertex having the largest indegree.

16. **(a)** E. outdeg$(E) = 4$, meaning E defeated every other team.

(b) A. outdeg$(A) = 0$, meaning A didn't win any games.

(c) The indegree of a vertex represents the number of games that team lost. The outdegree represents the number of games that team won.

(d) Adding the number of games a team lost and the number of games won gives $N - 1$, the number of games played.

B   Project Digraphs

17. **(a)** $\mathcal{V} = \{START, A, B, C, D, E, F, G, H, END\}$;
$\mathcal{A} = \{START\text{-}A, START\text{-}H, AF, AC, B\text{-}END,$
$CB, CE, D\text{-}END, E\text{-}END, FB, GB, GD, HG\}$

(b)

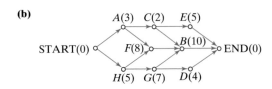

18. **(a)** $\mathcal{V} = \{START, A, B, C, D, E, F, G, H, END\}$;
$\mathcal{A} = \{START\text{-}C, START\text{-}G, AE, BE, CA, CB,$
$DB, DF, E\text{-}END, F\text{-}END, GD, GH, H\text{-}END\}$

(b)

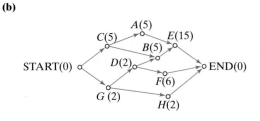

19.

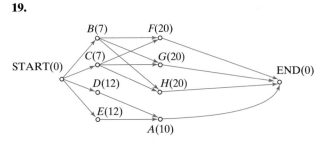

20.

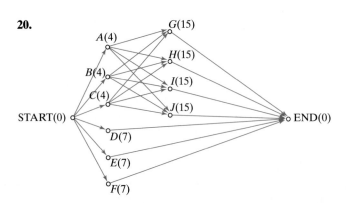

21.

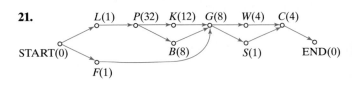

22.

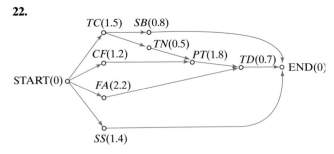

C   Schedules and Priority Lists

23. **(a)** 18 hours

(b) There is a total of 75 hours of work to be done. Three processors working without any idle time would take $\frac{75}{3} = 25$ hours to complete the project.

24. (a) 20 hours

(b) There is a total of 75 hours of work to be done. Five processors working without any idle time would take $\frac{75}{5} = 15$ hours to complete the project.

25. There is a total of 75 hours of work to be done. Dividing the work equally between the six processors would require each processor to do $\frac{75}{6} = 12.5$ hours of work. Since there are no $\frac{1}{2}$-hour jobs, the completion time could not be less than 13 hours.

26. (a) The completion time can never be less than 11 hours, since task C by itself takes 11 hours.

(b) There are 11 tasks, so clearly, having more than 11 processors makes no sense. To argue that 10 processors are enough, consider two cases: (i) if all tasks are independent, then we can give two "short" tasks [for example, $F(5)$ and $G(3)$] to one processor and each of the remaining nine tasks to a different processor and still meet the optimal completion time of 11 hours; (ii) if there is at least one precedence relation (call it $X \rightarrow Y$), then we can give tasks X and Y to one processor and each of the remaining nine tasks to a different processor. In this case, the finishing time will be the critical time for the project, and thus the schedule will be an optimal schedule.

27. ~~AD(8)~~, ~~AW(6)~~, ~~AF(8)~~, ~~IF(5)~~, ~~AP(7)~~, ~~IW(7)~~, ID(5), IP(4), ~~PL(4)~~, PU(3), HU(4), IC(1), PD(3), EU(2), FW(6)

28. ~~AD(8)~~, ~~AW(6)~~, ~~AF(5)~~, ~~IF(5)~~, ~~AP(7)~~, ~~IW(7)~~, ~~ID(5)~~, ~~IP(4)~~, ~~PL(4)~~, ~~PU(3)~~, ~~HU(4)~~, IC(1), PD(3) EU(2) FW(6)

29. Time: 0 1 2 3 4 5 6 7 8 9 10 11 12 13 14 15 16 17 18 19 20 21 22 23 24 25 26

P_1	$C(9)$ $E(6)$ $G(2)$ Idle F
P_2	$A(8)$ $B(5)$ $D(12)$ Idle

Fin = 26

30. Time: 0 1 2 3 4 5 6 7 8 9 10 11 12 13 14 15 16 17 18 19 20 21 22 23 24 25 26

P_1	$C(9)$ $E(6)$ $G(2)$ Idle F
P_2	$B(5)$ $A(8)$ $D(12)$ Idle

Fin = 26

31. Time: 0 1 2 3 4 5 6 7 8 9 10 11 12 13 14 15 16 17 18 19 20 21 22 23 24 25 26

P_1	$C(9)$ $E(6)$ $G(2)$ Idle F
P_2	$A(8)$ $D(12)$ Idle
P_3	$B(5)$ Idle

Fin = 21

32. Time: 0 1 2 3 4 5 6 7 8 9 10 11 12 13 14 15 16 17 18 19 20 21 22 23 24 25 26

P_1	$C(9)$ $E(6)$ $G(2)$ Idle F
P_2	$B(5)$ Idle $D(12)$ Idle
P_3	$A(8)$ Idle

Fin = 21

33. Both priority lists have all tasks in the top two paths of the digraph listed before any task in the bottom path.

34. Both priority lists have all tasks in the bottom path of the digraph listed before any task in the top two paths.

35. Time:

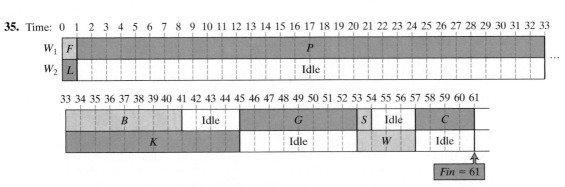

36. Time:

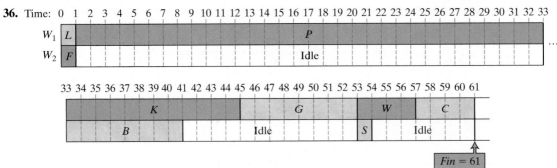

The Decreasing-Time Algorithm

37. Time:

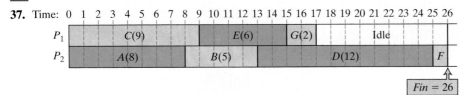

38. Time:

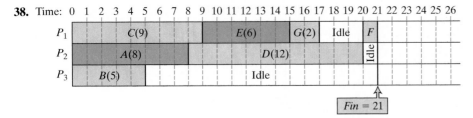

39. Time:

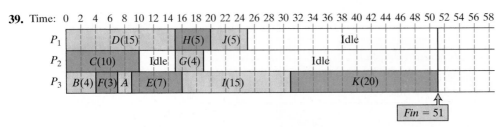

40. Time:

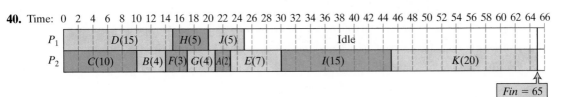

41. (a) Time: 0 1 2 3 4 5 6 7 8 9 10

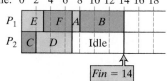

P_1 E(2) F(2) A B C D

P_2 G(2) H(2) Idle

Fin = 8

(b) Time: 0 1 2 3 4 5 6 7 8 9 10

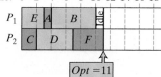

P_1 A B C D F(2)

P_2 E(2) G(2) H(2)

Opt = 6

(c) $33\frac{1}{3}\%$

42. (a) Time: 0 2 4 6 8 10 12 14 16 18

P_1 E F A B

P_2 C D Idle

Fin = 14

(b) Time: 0 2 4 6 8 10 12 14 16 18

P_1 E A B Idle

P_2 C D F

Opt = 11

(c) $27.\overline{27}\%$

43. (a) Time: 0 1 2 3 4 5 6 7 8 9 10 11 12 13 14 15 16 17 18 19 20 21 22 23 24 25 26

P_1 E(5) F(5) A(4) B(4) C(4) D(4)

P_2 G(5) H(5) Idle

P_3 I(5) J(5) Idle

Fin = 26

(b) Time: 0 1 2 3 4 5 6 7 8 9 10 11 12 13 14 15 16 17

P_1 A(4) B(4) C(4) D(4)

P_2 E(5) F(5) H(5) Idle

P_3 G(5) I(5) J(5) Idle

Opt = 16

(c) 62.5%

44. (a) Time: 0 1 2 3 4 5 6 7 8 9 10 11 12 13 14 15 16 17 18 19 20 21 22 23 24 25

P_1 I(5) J(6) A B(10)

P_2 G(4) H(7) Idle

P_3 E(3) F(8) Idle

P_4 C(2) D(9) Idle

Fin = 22

(b) Time: 0 1 2 3 4 5 6 7 8 9 10 11 12 13 14 15 16 17 18 19 20

P_1 C(2) A F(8) Idle

P_2 E(3) D(9) Idle

P_3 G(4) H(7) J(6)

P_4 I(5) B(10) Idle

Opt = 17

(c) $\frac{5}{17} \approx 29.4\%$

E   Critical Paths and the Critical-Path Algorithm

45. (a)

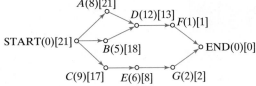

(b) START, A, D, F, END

(c) Time: 0 1 2 3 4 5 6 7 8 9 10 11 12 13 14 15 16 17 18 19 20 21 22 23 24 25 26

P_1	$A(8)$ $D(12)$ $G(2)$
P_2	$B(5)$ $C(9)$ $E(6)$ F Idle

$Fin = 22$

(d) There are a total of 43 work units, so the shortest time the project can be completed by two workers is $\frac{43}{2} = 21.5$ time units. Since there are no tasks taking less than one time unit, the shortest time in which the project can actually be completed is 22 hours.

46. (a) Time: 0 1 2 3 4 5 6 7 8 9 10 11 12 13 14 15 16 17 18 19 20 21 22 23

P_1	$A(8)$ $D(12)$ F
P_2	$B(5)$ Idle $E(6)$ $G(2)$ Idle
P_3	$C(9)$ Idle

$Fin = 21$

(b) The critical path of the project has length 21 hours, so the project cannot be completed in less than 21 hours.

47. (a) START, B, E, I, K, END

(b) Time: 0 2 4 6 8 10 12 14 16 18 20 22 24 26 28 30 32 34 36 38 40 42 44 46 48 50 52

P_1	B E F I K
P_2	A C Idle G Idle
P_3	D H J Idle

$Fin = 49$

48. (a)

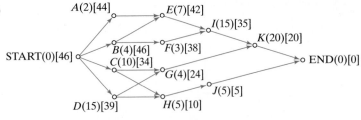

(b) Time: 0 2 4 6 8 10 12 14 16 18 20 22 24 26 28 30 32 34 36 38 40 42 44 46 48 50 52

P_1	B E F I H J Idle
P_2	A D C G K

$Fin = 51$

49. Finishing time is 15 hours and 15 minutes (61 units of 15 minutes).

Time: 0 3 6 9 12 15 18 21 24 27 30 33 36 39 42 45 48 51 54 57 60 63 66 69 72 75

P_1	L P(32) K(12) G(8) W C
P_2	F Idle B(8) Idle S Idle

Fin = 61

50.

Time: 0   0.4   0.8   1.2   1.6   2.0   2.4   2.8   3.2   3.6   4.0   4.4   4.8   5.2

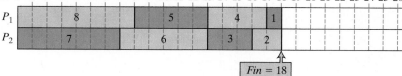

P_1	TC(1.5) TN(0.5) PT(1.8) TD(0.7)
P_2	CF(1.2) SS(1.4) Idle
P_3	FA(2.2) SB(0.8) Idle

Fin = 4.5

F **Scheduling with Independent Tasks**

51. (a) Time: 0 1 2 3 4 5 6 7 8 9 10 11 12 13 14 15 16 17 18 19 20 21 22 23 24 25 26

P_1	8 5 4 1
P_2	7 6 3 2

Fin = 18

(b) *Opt* = 18   **(c)** 0%

52. (a) Time: 0 3 6 9 12 15 18 21 24 27 30 33 36 39 42 45 48 51 54 57 60 63 66 69 72 75 78

P_1	16 13 12 9 8 5
P_2	15 14 11 10 7 6

Fin = 63

(b) *Opt* = 63   **(c)** 0%

53. (a) Time: 0 1 2 3 4 5 6 7 8 9 10 11 12 13 14 15 16 17 18 19 20 21 22 23

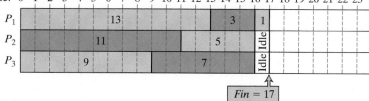

P_1	13 3 1
P_2	11 5 Idle
P_3	9 7 Idle

Fin = 17

(b) *Opt* = 17   **(c)** 0%

54. (a) Time: 0 1 2 3 4 5 6 7 8 9 10 11 12 13 14 15 16 17 18 19 20 21 22 23 24 25 26 27 28 29 30 31 32 33

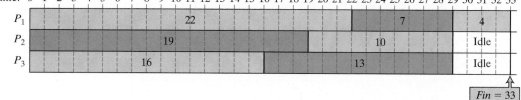

P_1	22 7 4
P_2	19 10 Idle
P_3	16 13 Idle

Fin = 33

(b) Time: 0 1 2 3 4 5 6 7 8 9 10 11 12 13 14 15 16 17 18 19 20 21 22 23 24 25 26 27 28 29 30 31 32

P_1	22	10	
P_2	19	7	4	Idle
P_3	16	13	Idle

Opt = 32

(c) 3.125%

55. (a) Time: 0 1 2 3 4 5 6 7 8 9 10 11 12 13 14 15 16 17 18 19 20

P_1	E(7)	A(4)	F(4)
P_2	I(7)	B(4)	Idle
P_3	D(6)	C(5)	Idle
P_4	H(6)	G(5)	Idle

Fin = 15

(b) Time: 0 1 2 3 4 5 6 7 8 9 10 11 12 13 14 15 16 17 18 19 20

P_1	E(7)	C(5)
P_2	I(7)	G(5)
P_3	D(6)	H(6)
P_4	A(4)	B(4)	F(4)

Opt = 12

(c) 25%

56. (a) Time: 0 1 2 3 4 5 6 7 8 9 10 11 12 13 14 15

P_1	D(8)	F(3)
P_2	E(5)	A(4)	Idle
P_3	G(5)	B(3)	C(2)	Idle

Fin = 11

(b) Time: 0 1 2 3 4 5 6 7 8 9 10 11 12 13 14 15

P_1	D(8)	C(2)
P_2	E(5)	G(5)
P_3	A(4)	B(3)	F(3)

Opt = 10

(c) 10%

57. (a) Time: 0 3 6 9 12 15 18 21 24 27 30 33 36 39 42 45 48 51 54 57 60 63 66 69 72 75 78

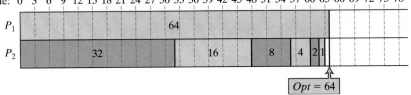

| P_1 | 64 |
| P_2 | 32 | 16 | 8 | 4 | 2 | 1 |

Opt = 64

(b) Time: 0   1/8   1/4   3/8   1/2   5/8   3/4   7/8   1

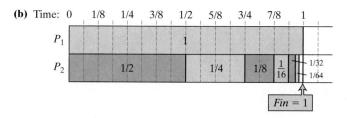

58. (a) Time: 0 3 6 9 12 15 18 21 24 27 30 33 36 39 42 45 48 51 54 57 60

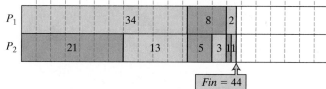

(b) Time: 0 3 6 9 12 15 18 21 24 27 30 33 36 39 42 45 48 51 54 57 60 63 66 69 72 75 78

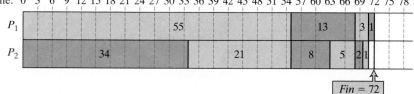

G   Miscellaneous

59. Each arc of the graph contributes 1 to the indegree sum and 1 to the outdegree sum.

60. (a) asymmetric   **(b)** symmetric   **(c)** neither   **(d)** symmetric   **(e)** asymmetric

61. $A, B, C, E, G, H, D, F, I$

62. $A, B, G, I, C, D, H, E, F$

63. Time: 0 2 4 6 8 10 12 14 16 18 20 22 24 26 28 30 32 34 36 38 40 42 44

64. Time: 0 2 4 6 8 10 12 14 16 18 20 22 24 26 28 30 32 34 36 38 40 42

65. Time: 0 2 4 6 8 10 12 14 16 18 20 22 24 26 28 30 32 34 36 38 40 42

66. Time: 0 2 4 6 8 10 12 14 16 18 20 22 24 26 28 30 32 34 36 38 40 42

P_1	AP IF IW IP HU $\frac{I}{C}$ FW	
P_2	AF AD Idle ID PU PD Idle	
P_3	AW Idle PL Idle EU Idle	

Fin = 34

JOGGING

67. (a)

N	ϵ
7	$\frac{6}{21} \approx 28.57\%$
8	$\frac{7}{24} \approx 29.17\%$
9	$\frac{8}{27} \approx 29.63\%$
10	$\frac{9}{30} = 30\%$

(b) Since $M - 1 < M$, we have for every M that

$$\frac{M - 1}{3M} \leq \frac{M}{3M} = \frac{1}{3}.$$

(c) Since $\frac{N - 1}{3N} = \frac{1 - (1/N)}{3}$, the Graham bound gets closer to $\frac{1}{3}$ as N gets larger.

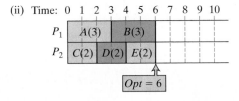

68. (a) (i) Time: 0 1 2 3 4 5 6 7 8 9 10

P_1	A(3) C(2) E(2)
P_2	B(3) D(2)

Fin = 7

(ii) Time: 0 1 2 3 4 5 6 7 8 9 10

P_1	A(3) B(3)
P_2	C(2) D(2) E(2)

Opt = 6

(iii) Graham bound: $\frac{N - 1}{3N} = \frac{2 - 1}{3 \times 2} = \frac{1}{6}$

Relative error of critical-path schedule: $\frac{7 - 6}{6} = \frac{1}{6}$

(b) (i) Time: 0 1 2 3 4 5 6 7 8 9 10 11 12 13 14 15

P_1	A(5) E(3) G(3)
P_2	B(5) F(3) Idle
P_3	C(4) D(4) Idle

Fin = 11

(ii) Time: 0 1 2 3 4 5 6 7 8 9 10 11 12 13 14 15

P_1	A(5) C(4)
P_2	B(5) D(4)
P_3	E(3) F(3) G(3)

Opt = 9

(iii) Graham bound: $\frac{N - 1}{3N} = \frac{3 - 1}{3 \times 3} = \frac{2}{9}$

Relative error of critical-path schedule: $\frac{11 - 9}{9} = \frac{2}{9}$

68. (c) (i) Time: 0 1 2 3 4 5 6 7 8 9 10 11 12 13 14 15 16 17 18 19 20

P_1: A(7), G(4), I(4)
P_2: B(7), H(4), Idle
P_3: C(6), E(5), Idle
P_4: D(6), F(5), Idle

Fin = 15

(ii) Time: 0 1 2 3 4 5 6 7 8 9 10 11 12 13 14 15 16 17 18 19 20

P_1: A(7), E(5)
P_2: B(7), F(5)
P_3: C(6), D(6)
P_4: G(4), H(4), I(4)

Opt = 12

(iii) Graham bound: $\dfrac{N-1}{3N} = \dfrac{4-1}{3 \times 4} = \dfrac{3}{12} = \dfrac{1}{4}$

Relative error of critical-path schedule: $\dfrac{15-12}{12} = \dfrac{3}{12} = \dfrac{1}{4}$

69. (a) Time: 0 1 2 3 4 5 6 7 8 9 10 11 12 13 14 15 16 17 18 19 20 21 22 23 24 25

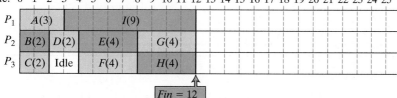

P_1: A(3), I(9)
P_2: B(2), D(2), E(4), G(4)
P_3: C(2), Idle, F(4), H(4)

Fin = 12

(b) Time: 0 1 2 3 4 5 6 7 8 9 10 11 12 13 14 15 16 17 18 19 20 21 22 23 24 25

P_1: A(3), H(4)
P_2: B(2), E(4), I(9)
P_3: C(2), F(4)
P_4: D(2), G(4)

Fin = 15

(c) An extra processor was used and yet the finishing time of the project increased.

70. (a) same as 69(a)

(b) Time: 0 1 2 3 4 5 6 7 8 9 10 11 12 13 14 15 16 17 18 19 20 21 22 23 24 25

P_1: A(2), E(3), H(3), Idle
P_2: B, D, F(3), I(8)
P_3: C, Idle, G(3), Idle

Fin = 13

(c) Each task was shorter and yet the finishing time of the project increased.

71. (a) same as 69(a)

(b) Time:

	0	1	2	3	4	5	6	7	8	9	10	11	12	13	14	15	16	17	18	19	20	21	22	23	24	25
P_1	$A(3)$			$F(4)$					$I(9)$																	
P_2	$B(2)$		$D(2)$		$G(4)$					Idle																
P_3	$C(2)$		$E(4)$				$H(4)$				Idle															

Fin = 16

(c) The project had fewer restrictions on the order of the assignments to the processors and yet the finishing time of the project increased.

72. Time:

	0	1	2	3	4	5	6	7	8	9	10	11	12	13	14	15	16	17	18	19	20	21	22	23	24	25
P_1	A		D		G		J		M		O			R												
P_2	B		E		H		K		N			Q		Idle												
P_3	C		F		I		L		P			Idle														

Fin = 21

73. (a) The finishing time of a project is always greater than or equal to the number of hours of work to be done divided by the number of processors doing the work.

(b) The schedule is optimal with no idle time.

(c) The total idle time in the schedule.

Mini-Excursion 2

A Graph Colorings and Chromatic Numbers

1. (a) Many answers are possible. **(b)**

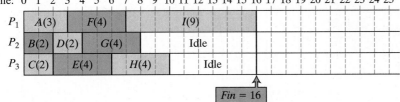

(c) $\chi(G) = 2$. At least two colors are needed, and (b) shows that G can be colored with two colors.

2. (a) Many answers are possible. **(b)**

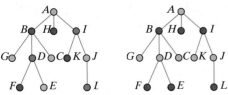

(c) $\chi(G) = 3$. At least three colors are needed because the graph has triangles, and (b) shows that G can be colored with three colors.

3. (a) Many answers are possible. **(b)**

 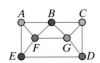

(c) $\chi(G) = 4$. The graph cannot be colored with three colors. If we try to color G with three colors and start with a triangle, say AEF, and color it blue, red, and green, then B is forced to be red, G is forced to be blue, C is forced to be green, and D will require a fourth color.

4. (a)

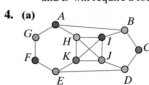

(b) $\chi(G) = 4$. Four colors are needed to color the vertices H, I, J, and K because they are adjacent to one another.

5. (a)

(b) $\chi(G) = 3$. Three colors are needed because A, B, C, D, and E form a circuit of length 5.

6. (a)

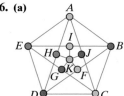

(b) $\chi(G) = 4$. First color $A, B, C,$ $D,$ and $E.$ Three colors are needed. Say A is blue, B red, C blue, D red, and E green. This forces G to be red, H to be green, and I to be blue, forcing K (the center of the spokes) to be a fourth color.

7. (a) Every vertex of K_n is adjacent to every other vertex, so every vertex has to be colored with a different color.

(b) $\chi(G) = n - 1$. If we remove one edge from K_n, there are two vertices that are not adjacent. They can be colored with the same color (say blue). The remaining vertices have to be colored with colors other than blue.

8. If a graph has an edge, two colors are needed just to color the two vertices joined by the edge. It follows that if $\chi(G) = 1$, then G has no edges.

9. (a) Adjacent vertices around the circuit can alternate colors (blue, red, blue, red, . . .).

(b) To color an odd circuit we start by alternating two colors (blue, red, blue, red, . . .), but when we get to the last vertex, it is adjacent to both a blue and a red vertex, so a third color is needed.

10. If n is even ($n \geq 4$), then $\chi(W_n) = 4$; if n is odd ($n \geq 5$), then $\chi(W_n) = 3$.

If n is even ($n \geq 4$), the "outer circuit" $v_1, v_2, v_3, \ldots, v_{n-1}$ is an odd circuit and it will require a minimum of three colors. The "center" of the wheel (v_0) is adjacent to all the other vertices, so it will require a fourth color. If n is odd ($n \geq 5$), then $v_1, v_2, v_3, \ldots, v_{n-1}$ is an even circuit and can be colored with just two colors. The center v_0 will require a third color.

11. $\chi(G) = 2$. Since a tree has no circuits, we can start with any vertex v, color it blue, and alternate blue, red, blue, red, blue, . . . along any path of the tree. Every vertex of the tree is in a unique path joining it to v and can be colored either red or blue.

12. Start with triangle AGF (a similar explanation would work if you start with any other triangle). Use any three colors to color $A, G,$ and $F.$ Let's say we choose red (A), green (G), and blue (F). This coloring forces the color of all the other vertices of the graph: First E has to be red, then C has to be blue, then D has to be green, then B has to be red, and finally H has to be green. This means that any other coloring will have to be equivalent (same pattern, different colors) to the one shown in the figure.

B **Map Coloring**

13. Answers may vary. One possible answer is shown below.

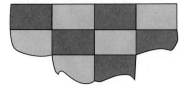

14. Many answers are possible. One possible answer is shown below.

15. (a) and (b)

Priority list of colors: Blue, Red, Green, Yellow.

(c) The chromatic number is 4 since Brazil, Bolivia, Argentina, and Paraguay are all adjacent to one another, and thus, require 4 colors.

16.

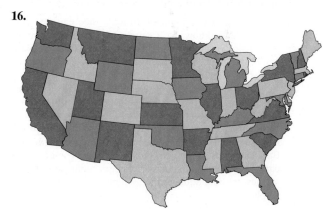

To show that it cannot be done with three colors, consider the subgraph of the dual graph formed by California, Oregon, Idaho, Utah, Arizona, and Nevada. This graph is W_6, the "wheel" with six vertices (see Exercise 10), and requires four colors.

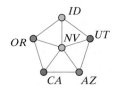

C Miscellaneous

17. (a) All the vertices in A can be colored with the same color, and all the vertices in B can be colored with a second color.

(b) Suppose G has $\chi(G) = 2$. Say the vertices are colored blue and red. Then let A denote the set of blue vertices and B the set of red vertices.

(c) True. If G had a circuit with an odd number of vertices, then that circuit alone would require three colors (see Exercise 9). But from (a), we know that $\chi(G) = 2$.

18. (a) The sum of the degrees of all the vertices in a graph must be even (see Euler's sum of degrees theorem in Chapter 5), so if every vertex has degree 3, n must be even. If $n = 2$, it is impossible for a vertex to have degree 3 unless there are multiple edges.

(b) From Brook's theorem, $\chi(G) \leq 4$.

(c) From the strong version of Brook's theorem, $\chi(G) = 4$ if G is complete (in which case $G = K_4$) or if G is disconnected (in which case it consists of several copies of K_4). The figure shows an example of a 3-regular graph G having $\chi(G) = 4$.

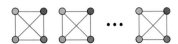

19. (a) Since the sum of the degrees of all the vertices must be even (see Euler's sum of degrees theorem in Chapter 5), it follows that the number of vertices of degree 3 must be even and thus n must be odd. If $n = 3$, there can be no vertices of degree 3 unless there are multiple edges.

(b) From the strong version of Brook's theorem, we have $\chi(G) \leq 3$. We know $\chi(G)$ cannot be 1 (see Exercise 8). Moreover, G cannot be a bipartite graph (see next paragraph), so $\chi(G)$ cannot be 2 (see Exercise 17). It follows that $\chi(G) = 3$.

A graph with $n - 1$ vertices of degree 3 and one vertex of degree 2 cannot be bipartite because the vertex of degree 2 must be in one of the two parts, say A. Then the number of edges coming out of A is 2 plus a multiple of 3. On the other hand, the number of edges coming out of B is a multiple of 3. But in a bipartite graph, the number of edges coming out of part A must equal the number of edges coming out of part B (they are both equal to the total number of edges in the graph).

20. The vertices listed in decreasing order of degrees are $H(4)$, $G(4)$, $C(3)$, $I(3)$, $D(2)$, $E(2)$, $A(2)$, $F(2)$, $B(2)$, $J(2)$. (There are many other ways to order the vertices by decreasing order of degrees, but a similar explanation will apply in all cases.) Using the greedy algorithm, H is colored with color 1 and then G must be colored with color 2. After that, C gets color 1, I gets color 2, D gets color 1, E gets color 2, A gets color 2, F gets color 1, B gets color 1, and J gets color 2.

21. Suppose the graph G is colored with $\chi(G)$ colors: color 1, color 2, ..., color K. For simplicity, we will use K for $\chi(G)$. Now make a list $v_1, v_2, \ldots, v_n$ of the vertices of the graph as follows: All the vertices of color 1 (in any order) are listed first (call these vertices group 1), the vertices of color 2 are listed next (call these vertices group 2), and so on, with the vertices of color K listed last (group K). Now when we apply the greedy algorithm to this particular list, the vertices in group 1 get color 1, the vertices in group 2 get either color 1 or color 2, the vertices in group 3 get either color 1, color 2, or at worst, color 3, and so on. The vertices in group K get color $1, 2, \ldots,$ or at worst, color K. It follows that the greedy algorithm gives us an optimal coloring of the graph.

22. (a) Each cell in a Sudoku grid shares a row with eight other cells, a column with an additional eight cells, and a box with eight cells, but two of these cells are in the same row (so we already counted them), and two more are in the same column. Thus, there are only four additional cells in the box that do not share a row or a column with our cell. Thus, the degree of every vertex of the Sudoku graph is $8 + 8 + 4 = 20$.

(b) The graph has 81 vertices of degree 20, giving a total of
$$\frac{81 \times 20}{2} = 810 \text{ edges.}$$

23.

6	2	4	8	7	1	9	5	3
1	9	3	4	6	5	8	7	2
7	5	8	3	9	2	6	1	4
2	1	9	6	4	3	5	8	7
5	8	6	7	2	9	3	4	1
4	3	7	1	5	8	2	6	9
3	4	5	2	1	6	7	9	8
8	6	1	9	3	7	4	2	5
9	7	2	5	8	4	1	3	6

Chapter 9

WALKING

A Fibonacci Numbers

1. (a) 55   **(b)** 57   **(c)** 144   **(d)** 27.5   **(e)** 5

2. (a) 144   **(b)** 143   **(c)** 89   **(d)** 36   **(e)** 2

3. (a) 12   **(b)** 610   **(c)** 6   **(d)** 144   **(e)** 2

4. (a) 21   **(b)** 987   **(c)** 11   **(d)** 1   **(e)** 233

5. (a) one more than three times the Nth Fibonacci number

(b) three times the $(N + 1)$st Fibonacci number

(c) one more than the $(3N)$th Fibonacci number

(d) the $(3N + 1)$st Fibonacci number

6. (a) three less than the $(2N)$th Fibonacci number

(b) the $(2N - 3)$rd Fibonacci number

(c) three less than twice the Nth Fibonacci number

(d) two times the $(N - 3)$rd Fibonacci number

7. (a) 39,088,169   **(b)** 9,227,465

8. (a) 2,178,309   **(b)** 5,702,887

9. I. This is an equivalent way to express that each term of the Fibonacci sequence is equal to the sum of the two preceding terms.

10. II. This is equivalent to $F_{N+1} = F_N + F_{N-1}$, which is another way to express that each term of the Fibonacci sequence is equal to the sum of the two preceding terms.

11. (a) $47 = 34 + 13$ **(b)** $48 = 34 + 13 + 1$

(c) $207 = 144 + 55 + 8$ **(d)** $210 = 144 + 55 + 8 + 3$

12. (a) $52 = 34 + 13 + 5$ **(b)** $53 = 34 + 13 + 5 + 1$

(c) $107 = 89 + 13 + 5$ **(d)** $112 = 89 + 21 + 2$

13. (a) $1 + 2 + 5 + 13 + 34 + 89 = 144$ **(b)** 22

(c) $N + 1$

14. (a) $2(13) - 21 = 5$ **(b)** 14   **(c)** N

15. (a) Answers will vary. One example is
$F_7 + F_4 = 13 + 3 = 16 = 2 \times 8 = 2F_6$.

(b) $F_{N+3} + F_N = 2F_{N+2}$

16. (a) $\dfrac{1 + 1 + 2 + \cdots + 55}{11} = 13$

(b) $\dfrac{(F_N + F_{N+1} + F_{N+2} + \cdots + F_{N+9})}{11} = F_{N+6}$

17. (a) $2F_6 - F_7 = 2 \times 8 - 13 = 3 = F_4$

(b) $2F_{N+2} - F_{N+3} = F_N$

18. (a) $F_8 \cdot F_{11} = 21 \cdot 89 = 55^2 - 34^2 = F_{10}^2 - F_9^2$

(b) $F_N F_{N+3} = F_{N+2}^2 - F_{N+1}^2$

B   The Golden Ratio

19. (a) 46.97871   **(b)** 46.97871   **(c)** 21

20. (a) 122.99187   **(b)** 122.99187   **(c)** 55

21. (a) 21   **(b)** 34   **(c)** 13

22. (a) 55   **(b)** 89   **(c)** 144

23. (a) $\phi^6 = \phi^5 \cdot \phi$
$$= (5\phi + 3)\phi$$
$$= 5\phi^2 + 3\phi$$
$$= 5(\phi + 1) + 3\phi$$
$$= 5\phi + 5 + 3\phi$$
$$= 8\phi + 5$$

(b) $a = 4, b = 9$

24. (a) $\phi^9 = \phi^8 \cdot \phi$
$$= (21\phi + 13)\phi$$
$$= 21\phi^2 + 13\phi$$
$$= 21(\phi + 1) + 13\phi$$
$$= 21\phi + 21 + 13\phi$$
$$= 34\phi + 21$$

(b) $a = 17, b = 38$

25. 1.394×10^{104}

26. 4.347×10^{208}

27. (a) 169   **(b)** 2.41429   **(c)** 2.41421   **(d)** 2.41421

28. (a) 360   **(b)** 3.30275   **(c)** 3.30278

C   Fibonacci Numbers and Quadratic Equations

29. (a) $x = 1 + \sqrt{2} \approx 2.41421, x = 1 - \sqrt{2} \approx -0.41421$

(b) $x = \dfrac{1 + \sqrt{6}}{5} \approx 0.68990, x = \dfrac{1 - \sqrt{6}}{5} \approx -0.28990$

(c) $x = \dfrac{4 + \sqrt{31}}{3} \approx 3.18925, x = \dfrac{4 - \sqrt{31}}{3} \approx -0.52259$

30. (a) $x = \dfrac{3 + \sqrt{13}}{2} \approx 3.30278, x = \dfrac{3 - \sqrt{13}}{2} \approx -0.30278$

(b) $x = \dfrac{3 + \sqrt{17}}{4} \approx 1.78078, x = \dfrac{3 - \sqrt{17}}{4} \approx -0.28078$

(c) $x = \dfrac{5 + \sqrt{89}}{16} \approx 0.90212, x = \dfrac{5 - \sqrt{89}}{16} \approx -0.27712$

31. (a) $55 = 34 + 21$ (55 is the sum of the two previous Fibonacci numbers)

(b) $x = -\dfrac{21}{55}$

32. (a) $21 = 55 - 34$ (21 is the difference of the two subsequent Fibonacci numbers)

(b) $x = \dfrac{55}{21}$

33. (a) Putting $x = 1$ in the equation gives $F_N = F_{N-1} + F_{N-2}$, which is a defining equation for the Fibonacci numbers.

(b) The sum of the roots of the equation is $-\dfrac{-F_{N-1}}{F_N} = \dfrac{F_{N-1}}{F_N}$, so the other root is $\dfrac{F_{N-1}}{F_N} - 1$.

34. (a) Putting $x = -1$ in the equation gives $F_N = -F_{N+1} + F_{N+2}$, or $F_{N+2} = F_{N+1} + F_N$, which is a defining equation for the Fibonacci numbers.

(b) The sum of the roots of the equation is $-\dfrac{-F_{N+1}}{F_N} = \dfrac{F_{N+1}}{F_N}$, so the other root is $\dfrac{F_{N+1}}{F_N} + 1$.

D Similarity

35. (a) 124.5 in.   **(b)** 945 in.^{2}

36. (a) 42π ft   **(b)** 13.5 gallons

37. (a) 156 m   **(b)** 2880 m^2

38. (a) 25   **(b)** 187.5

39. 3, 5

40. $3, \dfrac{1}{2}$

E Gnomons

41. $c = 24$

42. $x = 1.2$

43. $x = 4$

44. $x = 1.5$

45. 20 by 30

46. $x = 8$

47. (a) $m(\angle ACD) = 72°;\ \ m(\angle CAD) = 72°;\ \ m(\angle CDA) = 36°$

(b) $AC = 1; CD = \phi; AD = \phi$

48. $x = 12, y = 12$

49. $x = 12, y = 10$

50. $x = 20, y = 16$

JOGGING

51. $A_N = 5F_N$

52. (a) The expression $L_N = 2F_{N+1} - F_N$ satisfies the recursive rule:

$$L_N = 2F_{N+1} - F_N$$
$$= 2(F_N + F_{N-1}) - (F_{N-1} + F_{N-2})$$
$$= 2F_N + F_{N-1} - F_{N-2}$$
$$= 2F_N - F_{N-1} + (2F_{N-1} - F_{N-2})$$
$$= L_{N-1} + L_{N-2}$$

Also, this expression satisfies $L_1 = 2F_2 - F_1 = 2 \cdot 1 - 1 = 1$ and $L_2 = 2F_3 - F_2 = 2 \cdot 2 - 1 = 3$.

(b) $L_N = 2F_{N+1} - F_N$

$$= 2\left[\frac{\left(\dfrac{1 + \sqrt{5}}{2}\right)^{N+1} - \left(\dfrac{1 - \sqrt{5}}{2}\right)^{N+1}}{\sqrt{5}} \right]$$

$$- \left[\frac{\left(\dfrac{1 + \sqrt{5}}{2}\right)^{N} - \left(\dfrac{1 - \sqrt{5}}{2}\right)^{N}}{\sqrt{5}} \right]$$

$$= \frac{2}{\sqrt{5}}\left(\frac{1 + \sqrt{5}}{2}\right)\left(\frac{1 + \sqrt{5}}{2}\right)^{N} - \frac{1}{\sqrt{5}}\left(\frac{1 + \sqrt{5}}{2}\right)^{N}$$

$$- \frac{2}{\sqrt{5}}\left(\frac{1 - \sqrt{5}}{2}\right)\left(\frac{1 - \sqrt{5}}{2}\right)^{N} + \frac{1}{\sqrt{5}}\left(\frac{1 - \sqrt{5}}{2}\right)^{N}$$

$$= \left[\frac{2}{\sqrt{5}}\left(\frac{1 + \sqrt{5}}{2}\right) - \frac{1}{\sqrt{5}} \right]\left(\frac{1 + \sqrt{5}}{2}\right)^{N}$$

$$+ \left[-\frac{2}{\sqrt{5}}\left(\frac{1 - \sqrt{5}}{2}\right) + \frac{1}{\sqrt{5}} \right]\left(\frac{1 - \sqrt{5}}{2}\right)^{N}$$

$$= \left(\frac{1 + \sqrt{5}}{2}\right)^{N} + \left(\frac{1 - \sqrt{5}}{2}\right)^{N}$$

53. (a) $T_1 = 11, T_2 = 18, T_3 = 29, T_4 = 47, T_5 = 76, T_6 = 123,$
$T_7 = 199, T_8 = 322$

(b) $T_N = 7F_{N+1} + 4F_N$
$$= 7(F_N + F_{N-1}) + 4(F_{N-1} + F_{N-2})$$
$$= (7F_N + 4F_{N-1}) + (7F_{N-1} + 4F_{N-2})$$
$$= T_{N-1} + T_{N-2}$$

(c) $T_1 = 11, T_2 = 18, T_N = T_{N-1} + T_{N-2}$

54. (a) $T_1 = a + b,\ \ T_2 = 2a + b,\ \ T_3 = 3a + 2b,\ \ T_4 = 5a + 3b,$
$T_5 = 8a + 5b$

(b) $T_N = aF_{N+1} + bF_N$
$$= a(F_N + F_{N-1}) + b(F_{N-1} + F_{N-2})$$
$$= (aF_N + bF_{N-1}) + (aF_{N-1} + bF_{N-2})$$
$$= T_{N-1} + T_{N-2}$$

(c) $a = 6, b = -1$

55. $\dfrac{1 + \sqrt{5}}{2} + \dfrac{1 - \sqrt{5}}{2} = 1$

Since the first term is positive, the second term is negative, and the sum is a whole number, the decimal parts are the same.

56. (a) They approach 0.

(b) From Binet's formula,

$$F_N = \frac{\phi^N - [(1 - \sqrt{5})/2]^N}{\sqrt{5}}.$$

It follows from (a) that $F_N \approx \phi^N / \sqrt{5}$.

(c) $\dfrac{F_N}{F_{N-1}} \approx \dfrac{\phi^N / \sqrt{5}}{\phi^{N-1} / \sqrt{5}} = \phi$

57. $\dfrac{l}{s} = \dfrac{144\phi + 89}{89\phi + 55} = \dfrac{\phi^{12}}{\phi^{11}} = \phi$

58. $\dfrac{l}{s} = \dfrac{l + s}{l} \Leftrightarrow l^2 = sl + s^2$

$$\Leftrightarrow \frac{l^2}{s^2} = \frac{sl + s^2}{s^2}$$

$$\Leftrightarrow \left(\frac{l}{s}\right)^2 = \frac{l}{s} + 1$$

59. $\sqrt{2}$

60. Attaching a gnomon to the L-shaped figure must result in a new, larger L-shaped figure—it must be higher, wider, and thicker (all by the same proportions). Of all the (five) possible ways of attaching a square to the L-shaped figure, none produces a similar figure.

61. $x = 6, y = 12, z = 10$

62. $x = 65, y = 25$

63. $x = 3, y = 5$

64. From elementary geometry, $\angle AEF \cong \angle DBA$. So, ΔAEF is similar to ΔDBA. Thus, $AF\!:\!FE = DA : AB$, which shows that rectangle $ADEF$ is similar to rectangle $ABCD$.

65. (a) Since we are given that $AB = BC = 1$, we know that $m(\angle BAC) = 72°$ and so $m(\angle BAD) = 108°$. This makes $m(\angle ABD) = 36°$, so triangle ABD is isosceles having $AD = AB = 1$. Therefore, $AC = x - 1$. Using these facts and the similarity of triangle ABC and triangle BCD,

$\dfrac{x}{1} = \dfrac{1}{x - 1}$. That is, $x^2 = x + 1$, so $x = \phi$.

(b) $36°, 36°, 108°$

(c) $\dfrac{\text{longer side}}{\text{shorter side}} = \dfrac{x}{1} = \phi$

66. The result follows from Exercise 65(a) and a dissection of the regular pentagon into three golden triangles using any two non-intersecting diagonals.

67. (a) $\dfrac{10}{\phi}$ **(b)** $\dfrac{10r}{\phi}$

68. (a) ϕ **(b)** ϕ

69. The sum of two odd numbers is always even, and the sum of an odd number and an even number is always odd. Since the seeds in the Fibonacci sequence are both odd, every third number is even and the others are all odd.

Chapter 10

WALKING

A Percentages

1. (a) 0.0225 **(b)** 0.0075

2. (a) 0.0875 **(b)** 0.008

3. (a) $\dfrac{9}{400}$ **(b)** $\dfrac{3}{400}$

4. (a) $\dfrac{7}{80}$ **(b)** $\dfrac{1}{125}$

5. 80%

6. 77.5%

7. 215

8. 585

9. 9.5%

10. 5%

11. $12.5 trillion (i.e., $12,500,000,000,000)

12. $\dfrac{5}{9}$

13. 39.15%

14. 49.04%

15. 10.88%

16. 25.2%

17. an increase of 4.4%

18. a decrease of 6.5%

19. (a) $46,000,000 **(b)** approximately 0.0127%

20. (a) 3% **(b)** 2%

B Simple Interest

21. (a) $1024.80 **(b)** $1062.25 **(c)** $1249.50

22. (a) $1441.25 **(b)** $1632.50 **(c)** $2015.00

23. $12,125 ($2125 more than you paid for the bond)

24. $3442.80 ($442.80 more than you paid for the bond)

25. $4878.05

26. $8500

27. 6.75%

28. 4.5%

29. 8.33%

30. 10%

C Compound Interest

31. (a) $4587.64 **(b)** $5000.53

32. (a) $1569.74 **(b)** $1699.25

33. $5184.71

34. $3542.85

35. $1874.53

36. $2696.15

37. (a) $9083.48   **(b)** 12.6825%

38. (a) $1021.49   **(b)** 8.0573%

39. (a) $1451.67   **(b)** 6.9830%

40. (a) $1451.67   **(b)** 6.9830%

41. 6% APR compounded yearly. The APYs are 6%, 5.9040%, and 5.6541%, respectively.

42. 8.75% APR compounded monthly. The APYs are 9%, 9.1096%, and 8.9806%, respectively.

43.

Compounding	APY
Yearly	12%
Semiannually	12.36%
Quarterly	12.5509%
Monthly	12.6825%
Daily	12.7475%
Hourly	12.7496%
Continuously	12.7497%

44.

Compounding	APY
Yearly	8%
Semiannually	8.16%
Quarterly	8.24322%
Monthly	8.29995%
Daily	8.32776%
Hourly	8.32867%
Continuously	8.32871%

45. 8%

46. 6.875%

47. (a) 12 years   **(b)** 12 years

48. (a) 11 years   **(b)** 11 years

49. $3532.50

50. $550

D   **Geometric Sequences and the Geometric Sum Formula**

51. (a) $G_1 = 13.75$ **(b)** $G_6 = 11 \times (1.25)^6 \approx 41.962$
(c) $G_N = 11 \cdot (1.25)^N$

52. (a) $G_1 = 800$ **(b)** $G_5 = 327.68$ **(c)** $G_N = 1000 \cdot (0.8)^N$

53. (a) $c = \dfrac{3}{4}$ **(b)** $G_4 = 972$ **(c)** $G_N = 3072 \cdot \left(\dfrac{3}{4}\right)^N$

54. (a) $c = 1.5$ **(b)** $G_5 = 60.75$ **(c)** $G_N = 8 \cdot (1.5)^N$

55. (a) 1944   **(b)** $G_N = 256 \cdot (1.5)^N$
(c) approximately 14,762

56. (a) 32,768   **(b)** $G_N = 100,000 \cdot (0.8)^N$
(c) approximately 3518

57. (a) 3×2^{100} **(b)** $P_N = 3 \times 2^N$ **(c)** $3 \times (2^{101} - 1)$
(d) $3 \times 2^{101} - 3 \times 2^{50} = 3 \times 2^{50} \times (2^{51} - 1)$

58. (a) 2.5×4^{20} **(b)** $G_N = 2.5 \times 4^N$
(c) $2.5 \times \dfrac{(4^{21} - 1)}{3}$ **(d)** $2.5 \times \dfrac{(4^{21} - 4^8)}{3}$

59. $1006.28

60. $542,152.89

61. $165.47

62. $6,579.69

E   **Deferred Annuities**

63. $488,601.52

64. $15,002.95

65. $454,513.04

66. $14,903.59

67. $1133.56

68. $12,241.75

69. $76,932.20

70. $701.90

71. $179.11

72. $425

F   **Installment Loans**

73. (a) $38,350.07   **(b)** $34,960.75

74. (a) $827.61   **(b)** $868.79

75. $47,788.70

76. $692,198,046 (each year!)

77. (a) $933.72   **(b)** $176,139.20

78. (a) $340 each month   **(b)** $42,520

79. $163,104

80. $185,281

JOGGING

81. 100%

82. 4%

83. (a) $\left(1 - \dfrac{x}{100}\right)\left[\left(1 - \dfrac{y}{100}\right)P\right]$

(b) $\left(1 - \dfrac{y}{100}\right)\left[\left(1 - \dfrac{x}{100}\right)P\right]$

(c) Multiplication is commutative.

84. $7859.91

85. $601,477,274

86. 6%

87. As the periodic interest rate p goes up, more of the periodic payment must be used to pay interest. So, the value of the immediate annuity goes down.

88. $(P + cP + c^2P + \cdots + c^{N-1}P)(c - 1)$
$$= cP + c^2P + c^3P + \cdots + c^{N-1}P + c^NP$$
$$- P - cP - c^2P - c^3P - \cdots - c^{N-1}P$$
$$= -P + c^NP$$
$$= P(c^N - 1)$$

So, after dividing both sides by $(c - 1)$, we arrive at $P + cP +$

$c^2P + \cdots + c^{N-1}P = P\left(\dfrac{c^N - 1}{c - 1}\right)$.

89. **(a)** This is the geometric sum formula using $P(1 + p)$ in place of P, $(1 + p)$ in place of c, and T in place of N.

(b) $P(1 + p)\left[\dfrac{(1 + p)^T - 1}{p}\right] = P\left(\dfrac{1 + p}{p}\right)[(1 + p)^T - 1]$

$= P\left(\dfrac{1}{p} + 1\right)[(1 + p)^T - 1]$

90. **(a)** This is the geometric sum formula using $\dfrac{F}{1 + p}$ in place of P, $\dfrac{1}{1 + p}$ in place of c, and T in place of N.

(b) Looking at the sum from right to left, this is the geometric sum formula using $\dfrac{F}{(1 + p)^T}$ in place of P, $(1 + p)$ in place of c, and T in place of N.

(c) $\dfrac{F}{(1 + p)^T}\left[\dfrac{(1 + p)^T - 1}{p}\right] = F\left[\dfrac{\dfrac{(1 + p)^T}{(1 + p)^T} - \dfrac{1}{(1 + p)^T}}{p}\right]$

$= F\left[\dfrac{1 - \left(\dfrac{1}{1 + p}\right)^T}{p}\right]$

Chapter 11

WALKING

A   Reflections

1. **(a)** C **(b)** F **(c)** E **(d)** B

2. **(a)** C **(b)** A **(c)** D **(d)** E

3. **(a)**
(b)
(c)

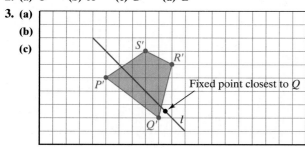

4. **(a)**
(b)
(c)

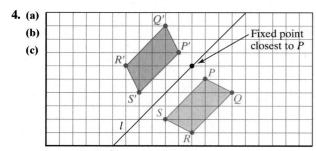

5. **(a)**
(b)
(c)
(d)

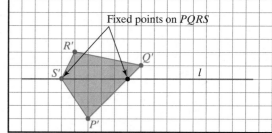

6. **(a)**
(b)
(c)
(d)

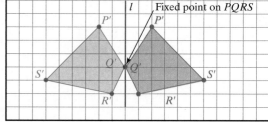

7. (a)
(b)

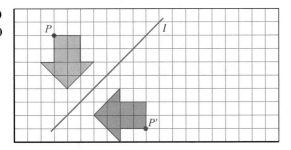

8. (a)
(b)

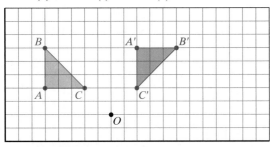

9.

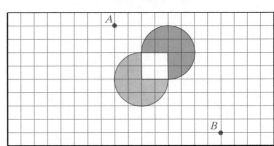

10.

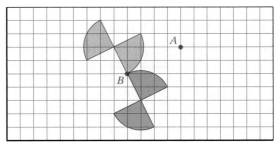

B Rotations

11. (a) *I*   **(b)** *G*   **(c)** *A*   **(d)** *F*   **(e)** *E*

12. (a) *G*   **(b)** *B*   **(c)** *C*   **(d)** *B*   **(e)** *C*

13. (a) $10°$ **(b)** $350°$ **(c)** $100°$ **(d)** $80°$

14. (a) $140°$ **(b)** $220°$ **(c)** $320°$ **(d)** $40°$

15. (a)
(b)

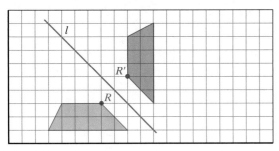

16. (a)
(b)

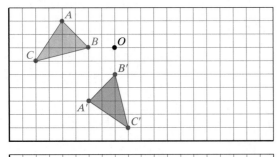

17. (a)
(b)
(c)

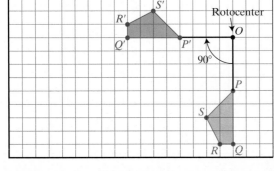

18. (a)
(b)
(c)

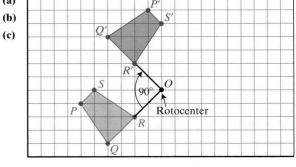

19.

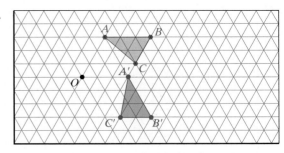

20.

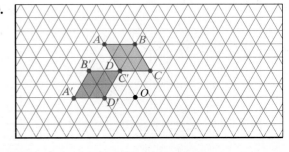

C   Translations

21. (a) *C*   (b) *C*   (c) *A*   (d) *D*

22. (a) *D*   (b) *A*   (c) *B*   (d) *C*

23. (a)
(b)
(c)

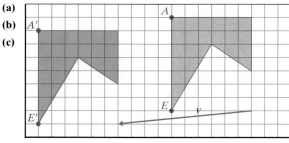

25.

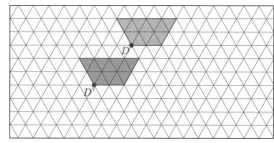

24. (a)
(b)
(c)

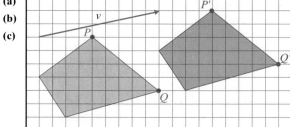

26.

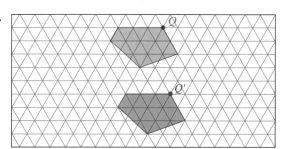

D   Glide Reflections

27.

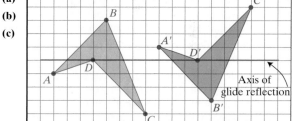

30. (a)
(b)
(c)

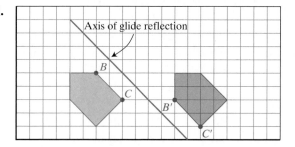

28.

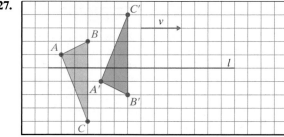

31.

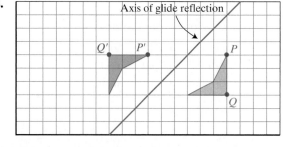

29. (a)
(b)
(c)

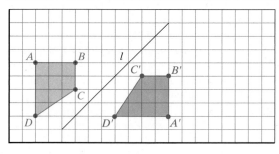

32.

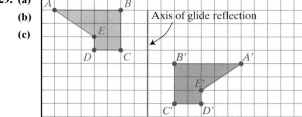

33.

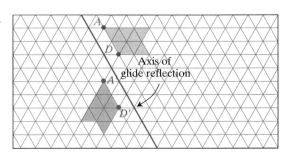

34.

E   Symmetries of Finite Shapes

35. (a) reflection with axis going through the midpoints of AB and DC; reflection with axis going through the midpoints of AD and BC; rotations of $180°$ and $360°$ with rotocenter the center of the rectangle

  (b) no reflections; rotations of $180°$ and $360°$ with rotocenter the center of the parallelogram

  (c) reflection with axis going through the midpoints of AB and DC; rotation of $360°$

36. (a) reflection with axis going through C and the midpoint of AB; rotation of $360°$

  (b) reflections (three of them) with axis going through a vertex and the midpoint of the opposite side; rotations of $120°$, $240°$, and $360°$ with rotocenter the center of the triangle

  (c) rotation of $360°$

37. (a) reflections (three of them) with axis going through pairs of opposite vertices; reflections (three of them) with axis going through the midpoints of opposite sides of the hexagon; rotations of $60°$, $120°$, $180°$, $240°$, $300°$, $360°$ with rotocenter the center of the hexagon

  (b) no reflections; rotations of $72°$, $144°$, $216°$, $288°$, and $360°$ with rotocenter the center of the star

38. (a) reflections (five of them) with axis going through pairs of opposite vertices; rotations of $72°$, $144°$, $216°$, $288°$, and $360°$ with rotocenter the center of the star

  (b) no reflections; rotations of $72°$, $144°$, $216°$, $288°$, and $360°$ with rotocenter the center of the shape

39. (a) D_2 **(b)** Z_2 **(c)** D_1

40. (a) D_1 **(b)** D_3 **(c)** Z_1

41. (a) D_6 **(b)** Z_5

42. (a) D_5 **(b)** Z_5

43. (a) D_1 **(b)** D_1 **(c)** Z_1 **(d)** Z_2 **(e)** D_2
  (f) Z_2

44. (a) Z_2 **(b)** Z_1 **(c)** Z_2 **(d)** D_4 **(e)** Z_1

45. (a) J   **(b)** T   **(c)** Z   **(d)** 0

F   Symmetries of Border Patterns

49. (a) $m1$ **(b)** $1m$ **(c)** 12 **(d)** 11

50. (a) 11 **(b)** mm **(c)** $m1$ **(d)** 12

51. (a) $m1$ **(b)** 12 **(c)** $1g$ **(d)** mg

46. (a) 5   **(b)** 3   **(c)** 96   **(d)** 8

47. Answers will vary.

  (a) Symmetry type D_5 is common among many types of flowers (daisies, geraniums, etc.). The only requirements are that the flower have five equal, evenly spaced petals and that the petals have a reflection symmetry along their long axis. In the animal world, symmetry type D_5 is less common, but it can be found among certain types of starfish, sand dollars, and in some single-celled organisms called diatoms.

  (b) The Chrysler Corporation logo is a classic example of a shape with symmetry D_5. Symmetry type D_5 is also common in automobile wheels and hubcaps.

  (c) Objects with symmetry type Z_1 are those whose only symmetry is the identity. Thus, any "irregular" shape fits the bill. Tree leaves, seashells, plants, and rocks more often than not have symmetry type Z_1.

  (d) Examples of human-made objects with symmetry of type Z_1 are plentiful: an umbrella, the letter J, a knife, and so on.

48. Answers will vary.

  (a) Snowflakes, some types of jellyfish, and a beehive cell are examples of natural objects with symmetry type D_6, often called hexagonal symmetry.

  (b) A hex nut, some hubcaps, and hexagonal tiles are examples of human-made objects with hexagonal symmetry.

  (c) A flower with three petals that have a slant to the side, the leafs of some plants.

  (d) Many (but not all) cards in a deck of cards. The six of diamonds does, but the six of clubs does not.

52. (a) 12 **(b)** $1g$ **(c)** $m1$ **(d)** mg

53. 12

54. 11

G **Miscellaneous**

55. (a) rotation   **(b)** identity motion   **(c)** reflection
(d) glide reflection

56. (a) rotation   **(b)** reflection   **(c)** glide reflection
(d) translation

57. (a) *D*   **(b)** *D*   **(c)** *B*   **(d)** *F*   **(e)** *G*

58. (a) *C*   **(b)** *P*   **(c)** *D*   **(d)** *D*

59. (a) improper   **(b)** proper   **(c)** improper   **(d)** proper

60. (a) (ii) exactly one fixed point   **(b)** (i) no fixed points

61. The combination of two reflections is a proper rigid motion [see Exercise 59(d)]. Since *C* is a fixed point, the rigid motion must be a rotation with rotocenter *C*.

62. *M* is a proper rigid motion [see Exercise 59(d)]. Moreover, *M* has no fixed point (see Exercise 60). It follows that *M* must be a translation.

JOGGING

63. (a) a clockwise rotation with center *C* and angle of rotation 2α

(b) a counterclockwise rotation with center *C* and angle of rotation 2α

64. (a) a translation by a vector perpendicular to l_1 and l_3 of length 2*d* going from left to right

(b) a translation by a vector perpendicular to l_1 and l_3 of length 2*d* going from right to left

65. (a)

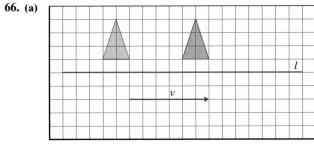

(b)

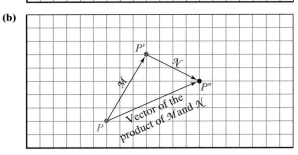

66. (a)

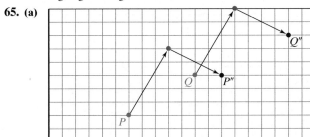

(b) The result of applying the same glide reflection twice is equivalent to a translation in the direction of the glide of twice the amount of the original glide. [See figure in part (a).]

67. (a)

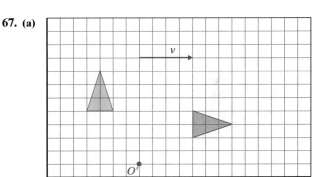

(b) Rotations and translations are proper rigid motions and hence preserve clockwise-counterclockwise orientations. Thus, the product of *M* and *N* is a proper rigid motion. However, the given motion cannot be a translation since the vertices of the triangle are each moved different distances. The product is a 90° clockwise rotation about rotocenter *O'*. [See figure in part (a)].

68. (a)

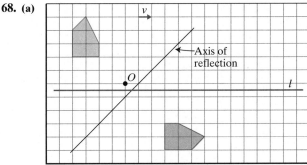

(b) Rotations and translations are proper rigid motions and hence preserve clockwise-counterclockwise orientations. Thus, the product of *M* and *N* is an improper rigid motion (it reverses the clockwise-counterclockwise orientation). Since performing the same rotation and glide reflection again results in the original figure, this is a reflection (rather than a glide reflection) whose axis is shown in (a).

69. (a) By definition, a border has translation symmetries in exactly one direction (let's assume the horizontal direction). If the pattern had reflection symmetry along an axis forming 45° with the horizontal direction, there would have to be a second direction of translation symmetry (vertical).

(b) If a pattern had reflection symmetry along an axis forming an angle of $\alpha°$ with the horizontal direction, it would have to have translation symmetry in a direction that forms an angle of $2\alpha°$ with the horizontal. This could only happen for $\alpha = 90°$ or $\alpha = 180°$ (since the only allowable direction for translation symmetries is the horizontal).

70. *mm:*

mg:

m1:

11:

12:

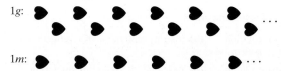

1*g:*

1*m:*

71. Rotations and translations are proper rigid motions. The given motion is an improper rigid motion (it reverses the clockwise-counterclockwise orientation). If the rigid motion was a reflection, then PP', RR', and QQ' would all be perpendicular to the axis of reflection and hence would all be parallel. It must be a glide reflection (the only rigid motion left).

72. (a) If a word has vertical reflection symmetry then the word remains unchanged when the word is reflected about a vertical line through the center of the word. Consequently, the word must be a palindrome.

(b) DAD

(c) Each letter must have vertical reflection symmetry.

(d) SOS

Chapter 12

WALKING

A **The Koch Snowflake and Variations**

1. (a)

	M	l	P
Start	3	1 cm	3 cm
Step 1	12	1/3 cm	4 cm
Step 2	48	1/9 cm	48/9 cm
Step 3	192	1/27 cm	192/27 cm
Step 4	768	1/81 cm	768/81 cm
Step 50	3×4^{50}	$\dfrac{1}{3^{50}}$ cm	53 km

(b) $M = 3 \cdot 4^N; l = \dfrac{1}{3^N}$ cm; $P = 3 \cdot \left(\dfrac{4}{3}\right)^N$ cm

2. (a)

	M	l	P
Start	3	6 cm	18 cm
Step 1	12	2 cm	24 cm
Step 2	48	2/3 cm	32 cm
Step 3	192	2/9 cm	128/3 cm
Step 4	768	2/27 cm	512/9 cm
Step 40	3×4^{40}	$6 \times \dfrac{1}{3^{40}}$ cm	$\approx$ 18 km

(b) $M = 3 \cdot 4^N; l = 6 \cdot \dfrac{1}{3^N}$ cm; $P = 18 \cdot \left(\dfrac{4}{3}\right)^N$ cm

3. (a)

	R	S	T	Q
Start	0	0	0	81 in.²
Step 1	3	9 in.²	27 in.²	108 in.²
Step 2	12	1 in.²	12 in.²	120 in.²
Step 3	48	$\dfrac{1}{9}$ in.²	$\dfrac{16}{3}$ in.²	$\dfrac{376}{3}$ in.²
Step 4	192	$\dfrac{1}{81}$ in.²	$\dfrac{64}{27}$ in.²	$\dfrac{3448}{27}$ in.²
Step 5	768	$\dfrac{1}{729}$ in.²	$\dfrac{256}{243}$ in.²	$\dfrac{31228}{243}$ in.²

(b) 129.6 in.²

4. (a)

	R	S	T	Q
Start	0	0	0	729 in.²
Step 1	3	81 in.²	243 in.²	972 in.²
Step 2	12	9 in.²	108 in.²	1080 in.²
Step 3	48	1 in.²	48 in.²	1128 in.²
Step 4	192	$\dfrac{1}{9}$ in.²	$\dfrac{64}{3}$ in.²	$\dfrac{3448}{3}$ in.²
Step 5	768	$\dfrac{1}{81}$ in.²	$\dfrac{256}{27}$ in.²	$\dfrac{31288}{27}$ in.²

(b) 1166.4 in²

5. (a)

	M	l	P
Start	4	9 cm	36 cm
Step 1	20	3 cm	60 cm
Step 2	100	1 cm	100 cm
Step 3	500	1/3 cm	$\dfrac{500}{3}$ cm
Step 30	4×5^{30}	$\dfrac{1}{3^{28}}$ cm	1628 km

(b) $M = 4 \cdot 5^N; l = \dfrac{1}{3^{N-2}}$ cm; $P = 100 \cdot \left(\dfrac{5}{3}\right)^{N-2}$ cm

6. (a)

	M	l	P
Start	4	a	$4a$
Step 1	20	$a/3$	$(20/3)a$
Step 2	100	$a/9$	$(100/9)a$
Step 3	500	$a/27$	$(500/27)a$
Step 4	2500	$a/81$	$(2500/81)a$

(b) $M = 4 \cdot 5^N; l = \dfrac{a}{3^N}; P = 4a\left(\dfrac{5}{3}\right)^N$

7.

	R	S	T	Q
Start	0	0	0	81
Step 1	4	9	36	117
Step 2	20	1	20	137
Step 3	100	$\frac{1}{9}$	$\frac{100}{9}$	$\frac{1333}{9}$
Step 4	500	$\frac{1}{81}$	$\frac{500}{81}$	$\frac{12{,}497}{81}$

8.

	R	S	T	Q
Start	0	0	0	243
Step 1	4	27	108	351
Step 2	20	3	60	411
Step 3	100	$\frac{1}{3}$	$\frac{100}{3}$	$\frac{1333}{3}$
Step 4	500	$\frac{1}{27}$	$\frac{500}{27}$	$\frac{12{,}497}{27}$

9. (a)

	M	l	P
Start	3	6 cm	18 cm
Step 1	12	2 cm	24 cm
Step 2	48	2/3 cm	32 cm
Step 3	192	2/9 cm	$\frac{128}{3}$ cm
Step 4	768	2/27 cm	$\frac{512}{9}$ cm
Step 30	3×4^{30}	$\frac{6}{3^{30}}$ cm	1 km

(b) $M = 3 \cdot 4^{N}$; $l = \frac{6}{3^{N}}$ cm; $P = 18 \cdot \left(\frac{4}{3}\right)^{N}$ cm

10. (a)

	M	l	P
Start	3	1 cm	3 cm
Step 1	12	1/3 cm	4 cm
Step 2	48	1/9 cm	48/9 cm
Step 3	192	1/27 cm	$\frac{192}{27}$ cm
Step 4	768	1/81 cm	$\frac{768}{81}$ cm
Step 50	3×4^{50}	$\frac{1}{3^{50}}$ cm	50 km

(b) $M = 3 \cdot 4^{N}$; $l = \frac{1}{3^{N}}$ cm; $P = 3 \cdot \left(\frac{4}{3}\right)^{N}$ cm

11. (a)

	R	S	T	Q
Start	0	0	0	81 in.^{2}
Step 1	3	9 in.^{2}	27 in.^{2}	54 in.^{2}
Step 2	12	1 in.^{2}	12 in.^{2}	42 in.^{2}
Step 3	48	$\frac{1}{9}$ in.^{2}	$\frac{48}{9}$ in.^{2}	$\frac{110}{3}$ in.^{2}
Step 4	192	$\frac{1}{81}$ in.^{2}	$\frac{192}{81}$ in.^{2}	$\frac{926}{27}$ in.^{2}
Step 5	768	$\frac{1}{729}$ in.^{2}	$\frac{768}{729}$ in.^{2}	$\frac{8078}{243}$ in.^{2}

(b) 32.4 in.^{2}

12. (a)

	R	S	T	Q
Start	0	0	0	729 in.^{2}
Step 1	3	81 in.^{2}	243 in.^{2}	486 in.^{2}
Step 2	12	9 in.^{2}	108 in.^{2}	378 in.^{2}
Step 3	48	1 in.^{2}	48 in.^{2}	330 in.^{2}
Step 4	192	$\frac{1}{9}$ in.^{2}	$\frac{192}{9}$ in.^{2}	$\frac{926}{3}$ in.^{2}
Step 5	768	$\frac{1}{81}$ in.^{2}	$\frac{768}{81}$ in.^{2}	$\frac{8078}{27}$ in.^{2}

(b) 291.6 in.^{2}

13. (a)

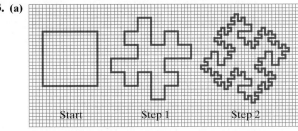

(b) 128   **(c)** 256   **(d)** 256   **(e)** 256

14. (a) same as exercise 13(a)   **(b)** $8a$ **(c)** a^2

(d) $16a$ **(e)** a^2

15. At each step, the area added is the same as the area subtracted.

16. Since the perimeter doubles at each step, the perimeter of the quadratic Koch island becomes infinitely large.

B   The Sierpinski Gasket and Variations

17.

	R	S	T	Q
Start	0	0	0	1
Step 1	1	1/4	1/4	3/4
Step 2	3	1/16	3/16	9/16
Step 3	9	1/64	9/64	27/64
Step 4	27	1/256	27/256	81/256
Step 5	81	1/1024	81/1024	243/1024

18.

	R	S	T	Q
Start	0	0	0	A
Step 1	1	$A/4$	$A/4$	$3A/4$
Step 2	3	$A/16$	$3A/16$	$9A/16$
Step 3	9	$A/64$	$9A/64$	$27A/64$
Step 4	27	$A/256$	$27A/256$	$81A/256$
Step 5	81	$A/1024$	$81A/1024$	$243A/1024$

19.

	U	V	W
Start	1	8 cm	8 cm
Step 1	3	4 cm	12 cm
Step 2	9	2 cm	18 cm
Step 3	27	1 cm	27 cm
Step 4	81	1/2 cm	81/2 cm
Step 5	243	1/4 cm	243/4 cm

20.

	U	V	W
Start	1	P	P
Step 1	3	$P/2$	$3P/2$
Step 2	9	$P/4$	$9P/4$
Step 3	27	$P/8$	$27P/8$
Step 4	81	$P/16$	$81P/16$
Step 5	243	$P/32$	$243P/32$

21. (a) $\left(\dfrac{3}{4}\right)^{N} A$

(b) The area of the Sierpinski gasket is smaller than the area of the gasket formed during any step of construction. That is, if the area of the original triangle is 1, the area of the Sierpinski gasket is less than $\left(\dfrac{3}{4}\right)^{N}$ for every positive value of N. Since $0 < \dfrac{3}{4} < 1$, the value of $\left(\dfrac{3}{4}\right)^{N}$ can be made smaller than any positive quantity for a large enough choice of N. It follows that the area of the Sierpinski gasket can also be made smaller than any positive quantity.

22. (a) $\left(\dfrac{3}{2}\right)^{N} P$

(b) Since the perimeter grows by a factor of 1.5 at each step, the perimeter of the Sierpinski gasket becomes infinitely large.

23. (a)

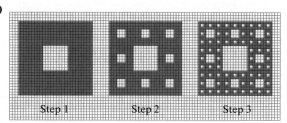

| Step 1 | Step 2 | Step 3 |

(b) 8/9 **(c)** $\left(\dfrac{8}{9}\right)^{2}$ **(d)** $\left(\dfrac{8}{9}\right)^{3}$ **(e)** $\left(\dfrac{8}{9}\right)^{N}$

24. (a) same as Exercise 23(a). **(b)** $\dfrac{8}{9} A$ **(c)** $\left(\dfrac{8}{9}\right)^{2} A$

(d) $\left(\dfrac{8}{9}\right)^{3} A$ **(e)** $\left(\dfrac{8}{9}\right)^{N} A$

25. (a) 16/3 **(b)** 80/9 **(c)** 496/27

26. (a) $\dfrac{16}{3} l$ **(b)** $\dfrac{80}{9} l$ **(c)** $\dfrac{496}{27} l$

27.

	R	S	T	Q
Start	0	0	0	1
Step 1	3	1/9	1/3	2/3
Step 2	18	1/81	2/9	4/9
Step 3	108	1/729	4/27	8/27
Step 4	648	1/6561	8/81	16/81
Step N	$3 \cdot 6^{N-1}$	$\dfrac{1}{9^{N}}$	$\dfrac{3 \cdot 6^{N-1}}{9^{N}}$	$\left(\dfrac{2}{3}\right)^{N}$

28.

	R	S	T	Q
Start	0	0	0	A
Step 1	3	A/9	A/3	2A/3
Step 2	18	A/81	2A/9	4A/9
Step 3	108	A/729	4A/27	8A/27
Step 4	648	A/6561	8A/81	16A/81
Step N	$3 \cdot 6^{N-1}$	$\dfrac{A}{9^{N}}$	$\dfrac{3 \cdot 6^{N-1}}{9^{N}} A$	$\left(\dfrac{2}{3}\right)^{N} A$

29.

	U	V	W
Start	1	9 cm	9 cm
Step 1	6	3 cm	18 cm
Step 2	36	1 cm	36 cm
Step 3	216	1/3 cm	72 cm
Step 4	1296	1/9 cm	144 cm
Step N	6^{N}	$\dfrac{1}{3^{N-2}}$ cm	$9 \cdot 2^{N}$ cm

30.

	U	V	W
Start	1	P	P
Step 1	6	P/3	2P
Step 2	36	P/9	4P
Step 3	216	P/27	8P
Step 4	1296	P/81	16P
Step N	6^{N}	$\dfrac{P}{3^{N}}$	$2^{N} P$

31.

	Q
Start	A
Step 1	3A/4
Step 2	9A/16
Step 3	27A/64
Step 4	81A/256
Step N	$\left(\dfrac{3}{4}\right)^{N} A$

32.

	W
Start	P
Step 1	P
Step 2	5P/4
Step 3	7P/4

C The Chaos Game and Variations

33.

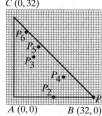

$C\,(0, 32)$
$A\,(0,0)$ $B\,(32, 0)$

34. $C\,(0, 32)$
$A\,(0,0)$ $B\,(32, 0)$

35.

Number rolled	Point	Coordinates
2	P_1	(0, 0)
6	P_2	(0, 16)
5	P_3	(0, 24)
1	P_4	(0, 12)
3	P_5	(16, 6)
6	P_6	(8, 19)

36.

Number rolled	Point	Coordinates
3	P_1	$(32, 0)$
1	P_2	$(16, 0)$
2	P_3	$(8, 0)$
3	P_4	$(20, 0)$
5	P_5	$(10, 16)$
5	P_6	$(5, 24)$

37. (a) 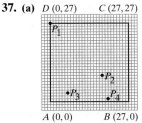   **(b)**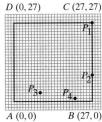

(c)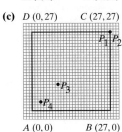

38. (a) 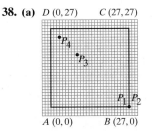   **(b)**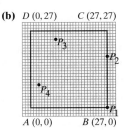

(c)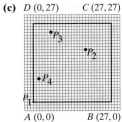

39. (a) $4, 2, 1, 2$ **(b)** $3, 1, 1, 3$ **(c)** $1, 3, 4, 2$
40. (a) $2, 3, 4, 1$ **(b)** $4, 2, 2, 4$ **(c)** $3, 1, 2, 4$

D   Operations with Complex Numbers

41. (a) $-1 - i$ **(b)** i **(c)** $-1 - i$
42. (a) $1 + 3i$ **(b)** $-7 + 7i$ **(c)** $1 - 97i$
43. (a) $-0.25 + 0.125i$ **(b)** $-0.25 - 0.125i$
44. (a) $-0.203 - 0.0625i$ **(b)** $-0.8 + 0.48i$
45. (a)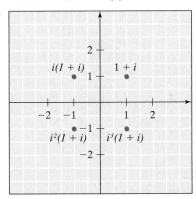

(b)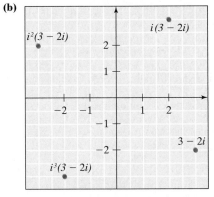

(c) It is a 90-degree counterclockwise rotation.
46. (a)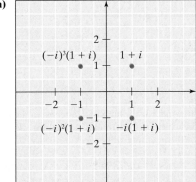

(b)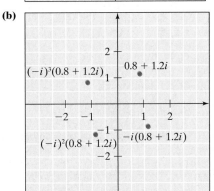

(c) It is a 90-degree clockwise rotation.

E **Mandelbrot Sequences**

47. (a) $2, 2, 2, 2$ **(b)** 2

(c) Attracted to 2. Each number in the sequence is 2.

48. (a) $s_1 = 6; s_2 = 38; s_3 = 1446; s_4 = 2,090,918$ **(b)** escaping

49. (a) $s_1 = -0.25; s_2 = -0.4375; s_3 \approx -0.3086; s_4 \approx -0.4048;$
$s_5 \approx -0.3362$

(b) -0.3660

(c) Attracted to -0.3660. From (a) and (b), the sequence gets close to -0.3660 and then $s_N = s_{N+1}$ (approximately).

50. (a) $s_1 = -0.1875, s_2 = -0.214844, s_3 = -0.203842,$
$s_4 = -0.208448, s_5 = -0.206549, s_6 = -0.207337,$

$s_7 = -0.207011, s_8 = -0.207146, s_9 = -0.207090,$
$s_{10} = -0.207114$

(b) -0.207107

(c) attracted to the number -0.207107 (rounded to six decimal places)

51. (a) $s_1 = -1 - i; s_2 = i; s_3 = -1 - i; s_4 = i; s_5 = -1 - i$

(b) Periodic; the odd terms are $-1 - i$ and the even terms are i.

52. $s_1 = 1 + 3i, s_2 = -7 + 7i, s_3 = 1 - 97i$

53. (a) 2 **(b)** 1

54. (a) $-3/2$ **(b)** 2

JOGGING

55. (a) Let $a^2 \dfrac{\sqrt{3}}{4} = A$.

	R	S	T	Q
Start	0	0	0	(A)
Step 1	3	$\dfrac{1}{9}(A)$	$\dfrac{3}{9}(A)$	$\dfrac{4}{3}(A)$
Step 2	12	$\dfrac{1}{9^2}(A)$	$\dfrac{12}{9^2}(A)$	$\dfrac{40}{27}(A)$
Step 3	48	$\dfrac{1}{9^3}(A)$	$\dfrac{48}{9^3}(A)$	$\dfrac{376}{243}(A)$
Step 4	192	$\dfrac{1}{9^4}(A)$	$\dfrac{192}{9^4}(A)$	$\dfrac{3448}{2187}(A)$
Step N	$3 \cdot 4^{N-1}$	$\dfrac{1}{9^N}(A)$	$\dfrac{3 \cdot 4^{N-1}}{9^N}(A)$	$(A) + \dfrac{3}{5} \cdot (A)\left[1 - \left(\dfrac{4}{9}\right)^N\right]$

56. (a)

	H
Start	0
Step 1	1
Step 2	$1 + 3$
Step 3	$1 + 3 + 3^2$
Step 4	$1 + 3 + 3^2 + 3^3$
Step 5	$1 + 3 + 3^2 + 3^3 + 3^4$

(b) Using the formula for the sum of the terms in a geometric sequence in which $r = 3$ (found in Chapter 10), we see that at the Nth step there will be $1 + 3 + 3^2 + 3^3 + \cdots +$
$3^{N-1} = \dfrac{1 - 3^N}{1 - 3} = \dfrac{3^N - 1}{2}$ holes.

57. (a)

	C	U	V
Start	0	0	1
Step 1	7	$1/27$	$20/27$
Step 2	20×7	$\left(\dfrac{1}{27}\right)^2$	$\left(\dfrac{20}{27}\right)^2$
Step 3	$20^2 \times 7$	$\left(\dfrac{1}{27}\right)^3$	$\left(\dfrac{20}{27}\right)^3$
Step 4	$20^3 \times 7$	$\left(\dfrac{1}{27}\right)^4$	$\left(\dfrac{20}{27}\right)^4$

(b) Since $20/27 < 1$, the Menger sponge has infinitesimally small volume.

58. (a)

	H
Start	0
Step 1	7
Step 2	$7 + 20 \times 7$
Step 3	$7 + 20 \times 7 + 20^2 \times 7$
Step 4	$7 + 20 \times 7 + 20^2 \times 7 + 20^3 \times 7$
Step 5	$7 + 20 \times 7 + 20^2 \times 7 + 20^3 \times 7 + 20^4 \times 7$

(b) Using the formula for the sum of the terms in a geometric sequence in which $r = 20$, we see that at the Nth step there will be $7[1 + 20 + 20^2 + 20^3 + \cdots + 20^{N-1}] = \dfrac{7}{19}[20^N - 1]$ holes.

59. (a) reflection with axis a vertical line passing through the center of the snowflake; reflections with axes lines making 30°, 60°, 90°, 120°, 150° angles with the vertical axis of the snowflake and passing through the center of the snowflake

(b) rotations of 60°, 120°, 180°, 240°, 300°, 360° with rotocenter the center of the snowflake

(c) D_6

60. (a) reflections with vertical and horizontal lines passing through the center of the carpet; reflections with 45° (clockwise and counterclockwise from horizontal) diagonal axes passing through the center of the carpet

(b) rotations of 90°, 180°, 270°, 360° with rotocenter the center of the carpet

(c) D_4

61. (a) Using the formula for adding the terms in a geometric sequence (with common ratio $r = 4/9$),

$$1 + \left(\frac{4}{9}\right) + \left(\frac{4}{9}\right)^2 + \cdots + \left(\frac{4}{9}\right)^{N-1} =$$

$$\frac{1 - \left(\frac{4}{9}\right)^N}{\left(\frac{5}{9}\right)} = \frac{9}{5}\left[1 - \left(\frac{4}{9}\right)^N\right].$$

(b) Using the result in (a),

$$\left(\frac{A}{3}\right) + \left(\frac{A}{3}\right)\left(\frac{4}{9}\right) + \left(\frac{A}{3}\right)\left(\frac{4}{9}\right)^2 + \cdots + \left(\frac{A}{3}\right)\left(\frac{4}{9}\right)^{N-1} =$$

$$\left(\frac{A}{3}\right)\left[1 + \left(\frac{4}{9}\right) + \left(\frac{4}{9}\right)^2 + \cdots + \left(\frac{4}{9}\right)^{N-1}\right] =$$

$$\left(\frac{A}{3}\right)\left(\frac{9}{5}\right)\left[1 - \left(\frac{4}{9}\right)^N\right] = \left(\frac{3}{5}\right)A\left[1 - \left(\frac{4}{9}\right)^N\right].$$

62. -0.5 is a solution to the quadratic equation $x^2 + 0.75 = x$. This means that if $x_n = -0.5$, then $x_{n+1} = x_n^2 + 0.75 = x_n = -0.5$. It follows that the Mandelbrot sequence is attracted to -0.5.

63. attracted to $1/2$

64. periodic

65. escaping

Mini-Excursion 3

A   Linear Growth and Arithmetic Sequences

1. (a) 225   **(b)** 185   **(c)** 186

2. (a) 40   **(b)** 26   **(c)** 26

3. (a) 3   **(b)** 158   **(c)** $P_N = 8 + 3N$

4. (a) 5   **(b)** 12   **(c)** $P_N = 12 + 5N$

5. (a) 497   **(b)** 24,950

6. (a) 413   **(b)** 12,369

7. (a) the 100th term   **(b)** 16,050

8. (a) the 301st term   **(b)** 406,651

9. (a) 213   **(b)** $137 + 2N$ **(c)** $7124   **(d)** $2652

10. (a) 1127   **(b)** $387 + 37N$ **(c)** $4024.80

   (d) $19,817.20

B   Exponential Growth and Geometric Sequences

11. (a) 13.75   **(b)** $11 \times (1.25)^9 \approx 81.956$

   (c) $P_N = 11 \times (1.25)^N$

12. (a) 1.5   **(b)** $8 \times (1.5)^9 \approx 307.547$

   (c) $P_N = 8 \times (1.5)^N$

13. (a) $P_N = 1.50P_{N-1}$, where $P_0 = 200$ **(b)** $P_N = 200 \times 1.5^N$

   (c) approximately 11,533

14. (a) $P_N = 0.8P_{N-1}$ where $P_0 = 100,000$

   (b) $P_N = 100,000 \times 0.8^N$ **(c)** approximately 3518

C   Logistic Growth Model

15. (a) $p_1 = 0.1680$ **(b)** $p_2 \approx 0.1118$ **(c)** 7.945%

16. (a) $p_1 = 0.1260$ **(b)** $p_2 \approx 0.06607$ **(c)** 3.702%

17. (a) $p_1 = 0.1680$, $p_2 \approx 0.1118$, $p_3 \approx 0.0795$,

   $p_4 \approx 0.0585$, $p_5 \approx 0.0441$, $p_6 \approx 0.0337$,

   $p_7 \approx 0.0261$, $p_8 \approx 0.0203$, $p_9 \approx 0.0159$,

   $p_{10} \approx 0.0125$

   (b) extinction

18. (a) $p_1 = 0.1260$, $p_2 \approx 0.0661$, $p_3 \approx 0.0370$,

   $p_4 \approx 0.0214$, $p_5 \approx 0.0126$, $p_6 \approx 0.0074$,

   $p_7 \approx 0.0044$, $p_8 \approx 0.0026$, $p_9 \approx 0.0016$,

   $p_{10} \approx 0.0009$

   (b) extinction

19. (a) $p_1 = 0.4320$, $p_2 \approx 0.4417$, $p_3 \approx 0.4439$,

   $p_4 \approx 0.4443$, $p_5 \approx 0.4444$, $p_6 \approx 0.4444$,

   $p_7 \approx 0.4444$, $p_8 \approx 0.4444$, $p_9 \approx 0.4444$,

   $p_{10} \approx 0.4444$

   (b) It stabilizes at 44.44% of the habitat's carrying capacity.

20. (a) $p_1 = 0.2400$, $p_2 \approx 0.2736$, $p_3 \approx 0.2981$,

   $p_4 \approx 0.3139$, $p_5 \approx 0.3230$, $p_6 \approx 0.3280$,

   $p_7 \approx 0.3306$, $p_8 \approx 0.3320$, $p_9 \approx 0.3327$,

   $p_{10} \approx 0.3330$

   (b) It stabilizes at 33.33% of the habitat's carrying capacity.

21. (a) $p_1 = 0.3570$, $p_2 \approx 0.6427$, $p_3 \approx 0.6429$,

   $p_4 \approx 0.6428$, $p_5 \approx 0.6429$, $p_6 \approx 0.6428$,

   $p_7 \approx 0.6429$, $p_8 \approx 0.6428$, $p_9 \approx 0.6429$,

   $p_{10} \approx 0.6428$

   (b) It stabilizes at $\frac{9}{14} \approx 64.29\%$ of the habitat's carrying capacity.

22. (a) $p_1 = 0.4$, $p_2 \approx 0.6$, $p_3 \approx 0.6$,

   $p_4 \approx 0.6$, $p_5 \approx 0.6$, $p_6 \approx 0.6$,

   $p_7 \approx 0.6$, $p_8 \approx 0.6$, $p_9 \approx 0.6$,

   $p_{10} \approx 0.6$

   (b) It stabilizes at 60% of the habitat's carrying capacity.

23. (a) $p_1 = 0.5200$, $p_2 \approx 0.8112$, $p_3 \approx 0.4978$,

   $p_4 \approx 0.8125$, $p_5 \approx 0.4952$, $p_6 \approx 0.8124$,

   $p_7 \approx 0.4953$, $p_8 \approx 0.8124$, $p_9 \approx 0.4953$,

   $p_{10} \approx 0.8124$

   (b) The population settles into a two-period cycle alternating between a high-population period at 81.24% and a low-population period at 49.53% of the habitat's carrying capacity.

24. (a) $p_1 = 0.8424$, $p_2 \approx 0.4660$, $p_3 \approx 0.8734$,

   $p_4 \approx 0.3880$, $p_5 \approx 0.8335$, $p_6 \approx 0.4872$,

   $p_7 \approx 0.8769$, $p_8 \approx 0.3788$, $p_9 \approx 0.8260$,

   $p_{10} \approx 0.5045$

   (b) The population settles into a four-period cycle with the following approximate percentages of the habitat's carrying capacity: 38%, 83%, 51%, 88%.

D **Miscellaneous**

25. **(a)** exponential   **(b)** linear   **(c)** logistic

(d) exponential   **(e)** logistic   **(f)** linear

(g) linear, exponential, and logistic

26. **(a)** linear   **(b)** logistic   **(c)** exponential

(d) logistic   **(e)** linear   **(f)** exponential

27. The first N terms of the arithmetic sequence are

$c, c + d, c + 2d, \ldots c + (N - 1)d$. Their sum is

$$\frac{(c + [c + (N - 1)d]) \times N}{2} = \frac{N}{2}[2c + (N - 1)d].$$

28. $10/3$

29. No. A constant population would require

$p_0 = p_1 = 0.8(1 - p_0)p_0$, so

$1 = 0.8(1 - p_0)$ or $p_0 = -0.25$.

30. $p_0 = \dfrac{r - 1}{r}$

31. **(a)** If $r > 4$ and $p_N = 0.5$, $p_{N+1} = r(1 - 0.5)(0.5) = 0.25r > 1$. This is impossible because the population cannot be more than 100% of the carrying capacity.

(b) If $r > 4$ and $p_N = 0.5$, $p_{N+1} = 4(1 - 0.5)(0.5) = 1$ and hence $p_K = 0$ for all $K > N + 1$ (i.e., the population becomes extinct.)

(c) $\frac{1}{4}$

(d) From part **(c)**, if $0 < p < 1, 0 < (1 - p)p < \frac{1}{4}$. So, if we also have $0 < r < 4$, then $0 < r(1 - p)p < 1$. Consequently, if we start with $0 < p_0 < 1$ and $0 < r < 4$, then $0 < p_N < 1$ for every positive integer N.

32. $p_0 = \dfrac{(r + 1) \pm \sqrt{(r + 1)(r - 3)}}{2r}$

33. If $P_0, P_1, P_2, \ldots$ is arithmetic, then $P_N = P_0 + N \cdot d$. So, the sequence $2^{P_0}, 2^{P_1}, 2^{P_2}, \ldots$ is given by $2^{P_0}, 2^{P_0+d}, 2^{P_0+2d}, 2^{P_0+3d}, \ldots$, which is the same as $2^{P_0} \cdot 1, 2^{P_0} \cdot (2^d), 2^{P_0} \cdot (2^d)^2, 2^{P_0} \cdot (2^d)^3, \ldots$. This is a geometric sequence with common ratio 2^d.

Chapter 13

WALKING

A **Surveys and Public Opinion Polls**

1. **(a)** the gumballs in the jar

(b) the gumballs drawn out of the jar

(c) 32%   **(d)** simple random sampling

2. **(a)** 12.5%   **(b)** 64

3. **(a)** 25%   **(b)** 7%

(c) Sampling variability. Simple random sampling, which does not suffer from selection bias, was used.

4. **(a)** The target population and the sampling frame are identical for this survey. Both consist of the jar full of gumballs.

(b) a census

5. **(a)** $\dfrac{680}{8325}$ **(b)** 45%

6. **(a)** the registered voters in Cleansburg

(b) the 680 registered voters that are polled by telephone

(c) simple random sampling

7. Smith: 3%; Jones: 3%; Brown: 0%

8. The sampling error should be attributed primarily to chance, since the sample was chosen randomly (this eliminates selection bias) and there was a 100% response rate (this eliminates nonresponse bias).

9. **(a)** the 350 students attending the Eureka High School football game the week before the election

(b) $\dfrac{350}{1250} = 28\%$

10. convenience sampling

11. **(a)** The population consists of all 1250 students at Eureka High School, whereas the sampling frame consists only of those

350 students who attended the football game the week prior to the election.

(b) Mainly sampling bias. The sampling frame is not representative of the population.

12. **(a)** Tomlinson: 58%, Garcia: 12%, Marsalis: 30%

(b) Tomlinson: 39%, Garcia: 39%, Marsalis: 0%

13. **(a)** all married people who read Dear Abby's column

(b) Dear Abby's target population appears to be all married people, but she is sampling from a (nonrepresentative) subset of the population—a sampling frame that consists of those married couples that read her column.

(c) self-selection

(d) A statistic. It is based on data taken from a sample.

14. **(a)** If we assume that the target population consisted of all married people then Dear Abby's readers are far from a representative sample. Second (and this applies even if we assume the target population consisted only of Dear Abby's married readers), the sample was self-selected—only those who took the trouble to write back were in the sample.

(b) A very small percentage (probably less than 10%) of Dear Abby's readers (in the millions) took the trouble to respond.

15. **(a)** 74%   **(b)** 81.8%

(c) Not very accurate. The sample was not representative of the target population.

16. Problem 1: Defining the target population. The target population of this survey consists of (presumably) all married Americans. A reasonably accurate list of this population could be compiled from state records (almost all marriages within a state are certified and recorded by the state).

Problem 2: Choosing a representative sample. We know that the best way to do this is by simple random sampling. Usually, a sample of about 1500–2000 individuals provides sufficient accuracy for a survey like this.

Problem 3: Getting truthful responses. It is obvious that how the question is asked is a critical issue in this survey. Many respondents will be reluctant to answer or will not answer truthfully unless complete confidentiality can be guaranteed. Mail questionnaires are likely to produce a high nonresponse rate, and telephone interviews are almost guaranteed to produce a lot of untruthful responses. So this pretty much leaves personal interviews as the only reasonable alternative. A few more things can be done to improve the chances of getting people to respond truthfully. For example, the interview should be held somewhere other than the home (office, a restaurant, etc.) and males should be interviewed by males, females by females.

17. **(a)** the citizens of Cleansburg

(b) The sampling frame is limited to that part of the target population that passes by a city street corner between 4:00 P.M. and 6:00 P.M.

18. **(a)** 475

(b) Yes, this survey is subject to nonresponse bias. The response rate was $\frac{475}{1313 + 475} \approx 0.266$.

19. **(a)** The choice of street corner could make a great deal of difference in the responses collected.

(b) Interviewer D. We are assuming that people who live or work downtown are much more likely to answer yes than people in other parts of town.

(c) Yes, for two main reasons: (1) People out on the street between 4 P.M. and 6 P.M. are not representative of the population at large. For example, office and white-collar workers are much more likely to be in the sample than homemakers and schoolteachers. (2) The five street corners were chosen by the interviewers, and the passersby are unlikely to represent a cross section of the city.

(d) No. No attempt was made to use quotas to get a representative cross section of the population.

20. Sending people out to interview passersby in a fixed location is not a good idea. Fixing a particular time of the day made it even worse. A telephone poll, based on a random sample of 500 registered voters in the city of Cleansburg would probably produce much more accurate data. Before asking the actual question ("Are you in favor, blah, blah, blah, . . .") the interviewer should

introduce himself as representing the City Planning Department and indicate that the survey consists of only one question and the survey will be used by the city to make important planning decisions. In this case, a better response rate can be expected.

21. **(a)** The target population and the sampling frame both consist of all TSU undergraduates.

(b) 15,000

22. **(a)** 1%   **(b)** $n = 135$

23. **(a)** In simple random sampling, any two members of the population have as much chance of both being in the sample as any other two, but in this sample, two people with the same last name—say Len Euler and Linda Euler—have no chance of being in the sample together.

(b) Sampling variability. The students sampled appear to be a representative cross section of all TSU undergraduates.

24. **(a)** 80%

(b) The most significant flaw with this survey is the size of the sample (too small). Other than that, a reasonable methodology is being used.

25. **(a)** stratified sampling   **(b)** quota sampling

26. **(a)** 80 oranges per tree

(b) A statistic. This is a numerical characteristic of a sample.

27. **(a)** convenience sampling   **(b)** stratified sampling

(c) simple random sampling   **(d)** quota sampling

28. **(a)** census   **(b)** stratified sampling

(c) quota sampling   **(d)** convenience sampling

29. **(a)** Label the crates with numbers 1 through 250. Select 20 of these numbers at random (put the 250 numbers in a hat and draw out 20). Sample those 20 crates.

(b) Select the top 20 crates in the shipment (those easiest to access).

(c) Randomly sample 6 crates from supplier A, 6 crates from supplier B, and 8 crates from supplier C.

(d) Select any 6 crates from supplier A, any 6 crates from supplier B, and any 8 crates from supplier C.

30. **(a)** Yes. One spoonful is representative of the entire bowl.

(b) No. This is a convenience sample.

(c) Yes. The blood selected is a representative sample.

(d) No. The sample may not be representative of that person's group of friends.

B   **The Capture-Recapture Method**

31. 2000

32. 20,500

33. 252

34. 261

35. $113.14

36. The capture-recapture method could be a reasonable way to estimate the number of coins in the jar. The coins are easy to "capture," they are equally accessible, and they are not disturbed by tagging. However, it may be hard to get a good random sample for both the capture and the recapture—the jar needs to be shaken well and the coins thoroughly mixed before both drawings.

37. approximately 660

38. approximately 225

39. Capture-recapture would underestimate the true population. If the fraction of those tagged in the recapture appears higher than it is in reality, then the fraction of those tagged in the initial capture will also be computed as higher than the truth. This makes the total population appear smaller than it really is.

40. Capture-recapture would overestimate the true population. If the fraction of those tagged in the recapture appears lower than it is in reality, then the fraction of those tagged in the initial capture will also be computed as lower than the truth. This makes the total population appear larger than it really is.

C   Clinical Studies

41. (a) anyone who could have a cold and would consider buying vitamin X (i.e., pretty much all adults)

(b) The sampling frame is a subset of the target population consisting of 50,000 college students in the San Diego area. The sample is 500 of these students, all of whom are suffering from a cold at the start of the study.

(c) Yes. This sample would likely underrepresent older adults and those living in colder climates.

42. (a) No, there was no control group.

(b) 1. San Diego college students are not typical of the population at large in several critical aspects (age, health, exposure to inclement weather, etc.).

2. The volunteers were paid to participate.

3. The subjects themselves determined the length of their cold.

4. There was no control group.

43. 1. Using college students—they are not a representative cross section of the population.

2. Using subjects only from the San Diego area.

3. Offering money as an incentive to participate.

4. Allowing self-reporting (the subjects themselves determine when their colds are over).

44. 1. Choose the subjects randomly from the population at large.

2. Divide the subjects randomly into a treatment group (gets vitamin X) and a control group (gets a fake pill).

3. Have trained professionals (nurses) measure the length of each subject's cold.

4. Neither the subjects nor the nurses should know who is getting the real vitamin X and who is getting a placebo.

45. all potential knee surgery patients

46. (a) 180 potential knee surgery patients at the Houston VA Medical Center who volunteered to be in the study

(b) 180   **(c)** No, the subjects volunteered.

47. (a) Yes. There was a control group receiving the sham surgery (placebo) and two treatment groups.

(b) The first treatment group consisted of those patients receiving arthroscopic debridement. The second treatment group consisted of those patients receiving arthroscopic lavage.

(c) Yes. The 180 patients in the study were assigned to a treatment group at random.

(d) blind

48. Answers will vary. One ethical issue to consider is that some patients that really do need surgery would not receive treatment if the study were continued.

49. all people having a history of colorectal adenomas

50. (a) 2586 participants all of whom had a history of colorectal adenomas

(b) $n = 2586$

51. (a) The treatment group consisted of the 1287 patients who were given 25 milligrams of Vioxx daily. The control group consisted of the 1299 patients who were given a placebo.

(b) This is an experiment because members of the population received a treatment. It is a controlled placebo experiment because there was a control group that did not receive the treatment, but instead received a placebo. It is a randomized controlled experiment because the 2586 participants were randomly divided into the treatment and control groups. The study was double blind because neither the participants nor the doctors involved in the clinical trial knew who was in each group.

52. The clinical study described is a double-blind randomized controlled placebo experiment. As such, a person suffering from arthritis and having a history of colorectal adenomas should avoid taking Vioxx as this study suggests the potential for an increase in the occurrence of cardiovascular events.

53. (a) women (particularly young women)

(b) The sampling frame consists of those women between 16 and 23 years of age who are not at high risk for HPV infection (i.e., those women having no prior abnormal Pap smears and at most five previous male sexual partners). Pregnant women are also excluded from the sampling frame due to the risks involved. It appears to be a rather large subset of the target population.

54. (a) 2392 women from 16 different centers across the United States

(b) No. Young women were encouraged to volunteer to participate through advertising on college campuses and in surrounding communities.

55. (a) the half of the 2392 women in the sample who received the HPV vaccine

(b) This is a controlled placebo experiment because there was a control group that did not receive the treatment, but instead received a placebo injection. It is a randomized controlled experiment because the 2392 participants were randomly divided into the treatment and control groups. The study was likely double blind because neither the participants nor the medical personnel giving the injections knew who was in each group.

56. This vaccine appears to be very effective at preventing HPV in young women. Therefore, widespread use of this vaccine could lead to fewer incidents of cervical cancer in women. What is not clear, however, are what side effects the HPV vaccine may have.

57. (a) all people suffering from diphtheria

(b) The sampling frame consists of those people having diphtheria symptoms serious enough to be admitted to one particular Copenhagen hospital between May 1896 and May 1897. It appears to be a rather large subset of the target population.

58. (a) 484 new diphtheria patients admitted to one particular Copenhagen hospital over a one-year period

(b) It is unlikely since humans were not involved in the selection of patients.

59. (a) The treatment group consisted of those patients admitted on the "odd" days who received just the standard treatment for diphtheria at the time. The control group consisted of those patients admitted on the "even" days who received both the new serum and the standard treatment.

(b) When received together with standard treatment, this new serum appears to be somewhat effective at treating diphtheria. Significantly more patients who did not receive the treatment in this study died (presumably many of them from diphtheria or related issues).

D Miscellaneous

61. (a) A clinical trial. A treatment was imposed (eating three meals at McDonald's every day for 30 days) on a sample of the population.

(b) the set of "average Americans"

(c) one person (Morgan Spurlock)

(d) a sample that is not representative of the population, small sample size, and the lack of a control group

62. (a) Merab Morgan herself

(b) Merab may have exercised more than usual during the study. Meal plans of fewer than 1400 calories a day may cause an average person to lose weight no matter what they eat. The choice of foods that Merab ate at McDonald's may also not be those of the average American.

(c) Since the experiment is far from legitimate, not many legitimate conclusions can be drawn. However, one might be able to say that combined with wise dietary behavior, it is possible to lose weight while eating only at McDonald's.

63. (a) This study was a clinical trial because a treatment was imposed (the heavy reliance on supplemental materials, online

60. The location of the hospital could be a confounding variable (patients at each hospital could come from different socio-economic and hereditary backgrounds). Treatment procedures at each hospital could be quite different. Patients admitted to each hospital may show different degrees of illness.

practice exercises, and interactive tutorials) on a sample of the population.

(b) (1) The instructors used in the treatment group may be more excited about this new curricular approach or they may be better teachers. (2) If students in this particular intermediate algebra class were self-selected, they may not be representative of the target population. (3) Students in the treatment group may have benefited simply by being forced to put more time into the course.

64. (a) Yes. Each student has a 20% chance of being selected.

(b) No. A group of 10 males has no chance of being selected. However, a particular group of six females and four males does have a chance of being selected.

65. (a) population parameter **(b)** sample statistic

(c) sample statistic **(d)** sample statistic

66. (a) association is not causation **(b)** placebo effect

(c) sampling variability **(d)** selection bias

JOGGING

67. (a) (i) the entire sky; (ii) all the coffee in the cup; (iii) the entire Math 101 class

(b) In none of the three examples is the sample random.

(c) (i) In some situations one can have a good idea as to whether it will rain or not by seeing only a small section of the sky, but in many other situations rain clouds can be patchy and one might draw the wrong conclusions by just peeking out the window. (ii) If the coffee is burning hot on top, it is likely to be pretty hot throughout, so Betty's conclusion is likely to be valid. (iii) Because Carla used convenience sampling and those students sitting next to her are not likely to be a representative sample, her conclusion is likely to be invalid.

68. Answers will vary.

69. (a) The question was worded in a way that made it almost impossible to answer yes.

(b) "Will you support some form of tax increase if it can be proven to you that such a tax increase is justified?" is better, but still not neutral. "Do you support or oppose some form of tax increase?" is bland but probably as neutral as one can get.

(c) Many such examples exist. A very real example is a question such as "Would you describe yourself as pro-life or not?" or "Would you describe yourself as pro-choice or not?"

70. Consideration of one exception made them less likely to consider a second exception.

71. (a) Under method 1, people whose phone numbers are unlisted are automatically ruled out from the sample. At the same time, method 1 is cheaper and easier to implement than method 2. Method 2 will typically produce more reliable data because the sample selected would better represent the population.

(b) Method 2. The two main reasons are (1) people with unlisted phone numbers are very likely to be the same kind of people who would seriously consider buying a burglar alarm, and (2) the listing bias is more likely to be significant in a place like New York City.

72. (a) An Area Code 900 telephone poll represents an extreme case of selection bias. People who respond to these polls usually represent the extreme view points (strongly for or strongly against), leaving out much of the middle of the road point of view. Economics also plays some role in the selection bias. (While 50 cents is not a lot of money anymore, poor people are much more likely to think twice before spending the money to express their opinion.)

(b) This survey was based on fairly standard modern-day polling techniques (random sample telephone interviews, etc.) but it had one subtle flaw. How reliable can a survey about the conduct of the newsmedia be when the survey itself is conducted by a newsmedia organization? ("The fox guarding the chicken-coop" syndrome.)

(c) Both surveys seem to have produced unreliable data—survey 1 overestimating the public's dissapproval of the role played by the newsmedia and survey 2 overestimating the public's support for the press coverage of the war.

(d) Any reasoned out answer should be acceptable. (Since Area Code 900 telephone polls are particularly unreliable, survey 2 gets our vote.)

73. (a) Fridays, Saturdays, and Sundays make up 3/7 or about 43% of the week. It follows that proportionately there are fewer fatalities due to automobile accidents on Friday, Saturday, and Sunday (42%) than there are on Monday through Thursday.

(b) On Saturday and Sunday there are fewer people commuting to work. The number of cars on the road and miles driven are significantly less on weekends. The number of fatalities due to accidents should be proportionally much less. The 42% indicates that there are other factors involved, and increased drinking is one possible explanation.

74. (a) Both samples should be a representative cross section of the same population. In particular, it is essential that the first sample, after being released, be allowed to disperse evenly throughout the population, and that the population should not change between the time of the capture and the time of the recapture.

(b) It is possible (especially when dealing with elusive types of animals) that the very fact that the animals in the first sample allowed themselves to be captured makes such a sample biased (they could represent a slower, less cunning group).

This type of bias is compounded with the animals that get captured the second time around. A second problem is the effect that the first capture can have on the captured animals. Sometimes the animal may be hurt (physically or emotionally) making it more (or less) likely to be captured the second time around. A third source of bias is the possibility that some of the tags will come off.

75. (a) People are very unlikely to tell IRS representatives that they cheated on their taxes even when they are promised confidentiality.

(b) The critical issue in this survey is to get the respondents to give an honest answer to the question. In this regard two points are critical: (1) the survey is much more likely to get honest responses when sponsored by a "neutral" organization (e.g., a newspaper poll); (2) a mailed questionnaire is the safest way to guarantee anonymity for the respondent, and in this case it is much better than a telephone or (heaven forbid) personal interview. This is a situation in which a trade-off must be made—some nonresponse bias is still preferable to dishonest answers. The nonresponse bias can be reduced by the use of an appropriate inducement or reward to those who can show proof of having mailed back the questionnaire (showing a post office receipt, etc.).

Chapter 14

WALKING

A **Tables, Bar Graphs, Pie Charts, and Histograms**

1.

Score	10	50	60	70	80	100
Frequency	1	3	7	6	5	2

2.

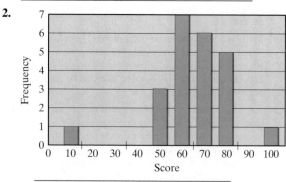

3. (a)

Grade	A	B	C	D	F
Frequency	7	6	7	3	1

(b)

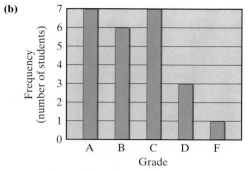

4. (a) 12.5%   **(b)** 45°

5.

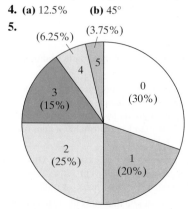

$N = 80$

6.

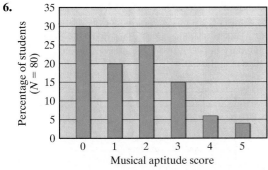

7.

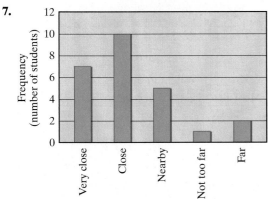

8.

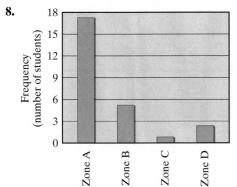

9.

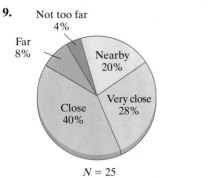

10.

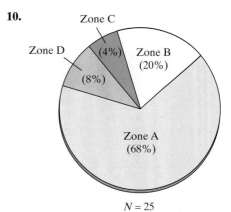

11. (a) 40   **(b)** 0%   **(c)** 57.5%

12.

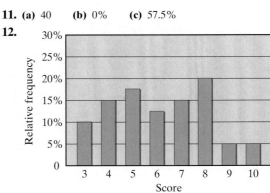

13. (a) qualitative   **(b)** 9231   **(c)** 58°

14. (a) 913

   (b) Accidents: 170°; homicide: 61°; suicide: 44°; cancer: 17°; heart disease: 10°; other: 58°

15. (a) $661,200,000,000 ($661.2 billion)

   (b) Defense: 82°; interest on the public debt: 30°; Medicare and Medicaid: 69°; Social Security: 77°; other: 102°

16. (a) $553,900,000,000 ($553.9 billion)   **(b)** 8 cm

17.

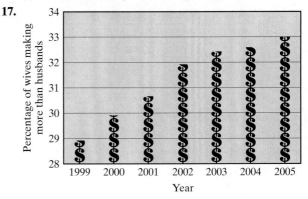

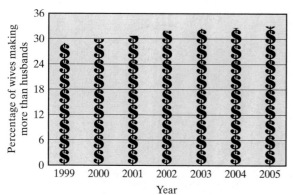

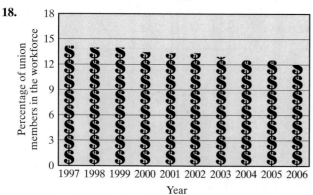

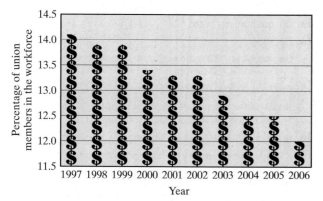

19. (a) 12 ounces

(b) The third class interval: "more than 72 ounces and less than or equal to 84 ounces." A value that falls exactly on the boundary between two class intervals belongs to the class interval to the left.

(c)

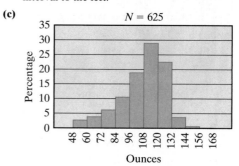

20. (a)

Weight in ounces		
More than	Less than or equal to	Frequencies
48	72	39
72	96	108
96	120	303
120	144	168
144	168	7

(b)

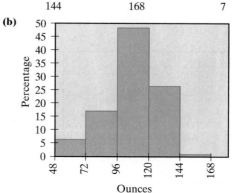

21. (a) 9 **(b)** 3
22. (a) 20 **(b)** 3

B **Means and Medians**

23. (a) 2 **(b)** 3 **(c)** average = 2, median = 2.5
24. (a) 0.1625 **(b)** −3.8
(c) average = −1.7, median = −3.8
25. (a) average = 4.5, median = 4.5
(b) average = 5, median = 5
(c) average = 5.5, median = 5.5
(d) average = 5.5a, median = 5.5a
26. (a) average = 1.5, median = 1.5
(b) average = 2.5, median = 2.5

(c) average = 3, median = 3
(d) average = $\dfrac{a + b + c + d}{4}$, median = $\dfrac{b + c}{2}$

27. (a) 50 **(b)** 50
28. (a) 500.5 **(b)** 500.5
29. (a) 1.5875 **(b)** 1.5
30. (a) 31.05 **(b)** 31
31. (a) 6.77 **(b)** 7
32. (a) 6.18 **(b)** 6

C   Percentiles and Quartiles

33. **(a)** $Q_1 = -4$ **(b)** $Q_3 = 7.5$ **(c)** $Q_1 = -3, Q_3 = 7$

34. **(a)** $Q_1 = -5.9$ **(b)** $Q_3 = 8.3$ **(c)** $Q_1 = -7.3, Q_3 = 8.3$

35. **(a)** 75th percentile = 75.5, 90th percentile = 90.5

   (b) 75th percentile = 75, 90th percentile = 90

   (c) 75th percentile = 75, 90th percentile = 90

   (d) 75th percentile = 74, 90th percentile = 89

36. **(a)** 10th percentile = 5.5, 25th percentile = 13

   (b) 10th percentile = 5, 25th percentile = 13

   (c) 10th percentile = 5, 25th percentile = 13

37. **(a)** $Q_1 = 29$ **(b)** $Q_3 = 32$ **(c)** 37

38. **(a)** $Q_1 = 5$ **(b)** $Q_3 = 8$ **(c)** 7

39. **(a)** the 747,266th score, $d_{747,266}$

   (b) the 373,633rd score, $d_{373,633}$

   (c) the 1,120,899th score, $d_{1,120,899}$

40. **(a)** halfway between the 732,872nd and 732,873rd scores

   (b) halfway between the 366,436th and 366,437th scores

   (c) halfway between the 1,099,308th and 1,099,309th scores

D   Box Plots and Five-Number Summaries

41. **(a)** $Min = -6, Q_1 = -4, M = 3, Q_3 = 7.5, Max = 8$

   (b)

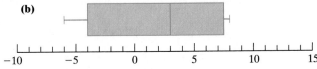

42. **(a)** $Min = -9.1, Q_1 = -5.9, M = -3.8, Q_3 = 8.3,$
      $Max = 13.2$

   (b)

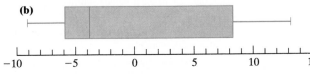

43. **(a)** $Min = 25, Q_1 = 29, M = 31, Q_3 = 32, Max = 39$

   (b)

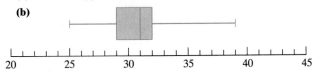

44. **(a)** $Min = 3, Q_1 = 5, M = 6, Q_3 = 8, Max = 9$

   (b)

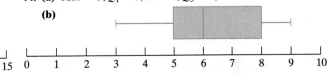

45. **(a)** between \$33,000 and \$34,000   **(b)** \$40,000

   (c) The vertical line indicating the median salary in the engineering box plot is to the right of the box in the agriculture box plot.

46. **(a)** 459   **(b)** 960

E   Ranges and Interquartile Ranges

47. **(a)** 14   **(b)** 11.5

48. **(a)** 22.3   **(b)** 14.2

49. **(a)** \$41,000   **(b)** at least 171 homes

50. **(a)** approximately \$31,000   **(b)** approximately \$8,500

   (c) at least 306 students

51. **(a)** 16.5   **(b)** 4.5   **(c)** 1 and 24

52. There are 10 outliers: 6 ages of 37 and 4 ages of 39.

53. heights less than 61 in. or greater than 77 in.

54. heights less than 57.25 in. or greater than 71.25 in.

F   Standard Deviations

55. **(a)** 0   **(b)** $\dfrac{5\sqrt{2}}{2} \approx 3.5$ **(c)** $\sqrt{50} \approx 7.1$

56. **(a)** 0   **(b)** 3   **(c)** 9

57. **(a)** $\sqrt{\dfrac{82.5}{10}} \approx 2.87$ **(b)** $\sqrt{\dfrac{82.5}{10}} \approx 2.87$ **(c)** $\sqrt{\dfrac{82.5}{10}} \approx 2.87$

58. **(a)** $\sqrt{\dfrac{28}{7}} \approx 2$ **(b)** $\sqrt{\dfrac{28}{7}} \approx 2$ **(c)** $\sqrt{\dfrac{21}{4}} \approx 2.29$

G   Miscellaneous

59. 0

60. 30

61. 4, 5, and 8

62. C

63. Caucasian

64. There is no mode.

JOGGING

65. 100

66. 58 points

67. 1,465,744

68. 1,419,007

69. (a) {1, 1, 1, 1, 6, 6, 6, 6, 6, 6} (average = 4, median = 6)

　　(b) {1, 1, 1, 1, 1, 1, 1, 6, 6, 6} (average = 3, median = 1)

　　(c) {1, 1, 6, 6, 6, 6, 6, 6, 6, 6} (average = 5, Q_1 = 6)

　　(d) {1, 1, 1, 1, 1, 1, 1, 1, 6} (average = 2, Q_3 = 1)

70. (a) 8　　**(b)** 48

71. From histogram (a) one can deduce that the median team salary is between \$70 million and \$100 million. From histogram (b) one can deduce that the median team salary is between \$50 million and \$80 million. It follows that the median team salary must be between \$70 million and \$80 million.

72. If the standard deviation is 0, then the average of the squared deviations is 0. The squared deviations cannot be negative, so if their average is 0, then they must all be 0. It follows that all deviations from the average must be 0. This means that each data value is equal to the average, which in turn implies that the data set is constant.

73. (a) 4　　**(b)** 0.4　　**(c)** 0.4

74.

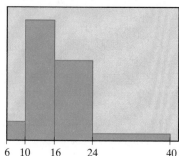

75. (a) Male: 10%; female: 20%　　**(b)** Male: 80%; female: 90%

　　(c) The figures for both schools were combined. Of a total of 1200 males who applied, 820 were admitted, an admission rate for males of approximately 68.3%. Similarly, of a total of 900 females who applied, 460 were admitted, an admission rate for females of approximately 51.1%.

(d) In this example, females have a higher percentage $\left(\frac{100}{500} = 20\% \right)$ than males $\left(\frac{20}{200} = 10\% \right)$ for admissions to the School of Architecture and also a higher percentage $\left(\frac{360}{400} = 90\% \right)$ than males $\left(\frac{800}{1000} = 80\% \right)$ for the School of Engineering. When the numbers are combined, however, females have a lower percentage $\left(\frac{100 + 360}{500 + 400} \approx 51.1\% \right)$ than males $\left(\frac{20 + 800}{200 + 1000} \approx 68.3\% \right)$ in total admissions. The reason this apparent paradox can occur is purely a matter of arithmetic: Just because $\frac{a_1}{a_2} > \frac{b_1}{b_2}$ and $\frac{c_1}{c_2} > \frac{d_1}{d_2}$, it does not necessarily follow that $\frac{a_1 + c_1}{a_2 + c_2} > \frac{b_1 + d_1}{b_2 + d_2}$.

76. (a) Two points will be added to each number in the five-number summary.

　　(b) Ten percent will be added to each number in the five-number summary.

77. (a) $A + c$

　　(b) Adding c to each value of a data set adds c to the average as well. Hence, subtracting the average A from each value of a data set (adding $-A$ to each value) subtracts A from the average, creating a new average of 0.

78. $M + c$; the relative sizes of the numbers are not changed by adding a constant c to every number.

79. (a) Let M be the maximum value and m the minimum value of $\{x_1, x_2, x_3, \ldots, x_N\}$. The maximum and minimum values of $\{x_1 + c, x_2 + c, x_3 + c, \ldots, x_N + c\}$ are $M + c$ and $m + c$, respectively. So, the range of $\{x_1 + c, x_2 + c, x_3 + c, \ldots, x_N + c\}$ is $M + c - (m + c) = M - m$ (the range of $\{x_1, x_2, x_3, \ldots, x_N\}$).

　　(b) Adding c to each value of a data set does not change how spread out the data set is. Since the standard deviation is a measure of spread, the standard deviation does not change. Specifically, the average of the data set will increase by c, but the deviations from the mean (and hence the standard deviation) will remain the same.

Chapter 15

WALKING

A **Random Experiments and Sample Spaces**

1. (a) {*HHH, HHT, HTH, THH, TTH, THT, HTT, TTT*}

　　(b) {*SSS, SSF, SFS, FSS, FFS, FSF, SFF, FFF*}

　　(c) {0, 1, 2, 3}

　　(d) {0, 1, 2, 3}

2. (a) {*HHHH, HHHT, HHTH, HHTT, HTHH, HTHT, HTTH, HTTT, THHH, THHT, THTH, THTT, TTHH, TTHT, TTTH, TTTT*}

(b) {*FFFF, FFFT, FFTF, FFTT, FTFF, FTFT, FTTF, FTTT, TFFF, TFFT, TFTF, TFTT, TTFF, TTFT, TTTF, TTTT*}

(c) {0%, 25%, 50%, 75%, 100%}

(d) {0%, 25%, 50%, 75%, 100%}

3. {*ABCD, ABDC, ACBD, ACDB, ADBC, ADCB, BACD, BADC, BCAD, BCDA, BDAC, BDCA, CABD, CADB, CBAD, CBDA, CDAB, CDBA, DABC, DACB, DBAC, DBCA, DCAB, DCBA*}

4. {$AA, AB, AC, AD, BB, BC, BD, CC, CD, DD$}

5. Answers will vary. A typical outcome is a string of 10 letters each of which can be either an H or a T. An answer like {$HHHHHHHHHH, \ldots, TTTTTTTTTT$} is not sufficiently descriptive. An answer like {$\ldots HTTHHHHTHTH, \ldots, TTHTHHTTHT, \ldots, HHHTHTTHHT, \ldots$} is better. An answer like {$(X_1X_2X_3X_4X_5X_6X_7X_8X_9X_{10})$: each X_i is either H or T} is best.

6. A typical outcome is a string of 10 letters each of which can be either a T or an F. An answer like {$FFFFFFFFFF, \ldots, TTTTTTTTTT$} is not sufficiently descriptive. An answer like {$\ldots FTTFFFTFTF, \ldots, TTFTFFTTFT, \ldots, FFFTFTTFFT, \ldots$} is better. An answer like {$(X_1X_2X_3X_4X_5X_6X_7X_8X_9X_{10})$: X_i is either T or F for each integer i between 1 and 10} is best.

7. Answers will vary. An outcome is an ordered sequence of four numbers, each of which is an integer between 1 and 6 inclusive. The best answer would be something like {(n_1, n_2, n_3, n_4): each n_i is 1, 2, 3, 4, 5, or 6}. An answer such as {$(1, 1, 1, 1), \ldots, (1, 1, 1, 6), \ldots, (1, 2, 3, 4), \ldots, (3, 2, 6, 2), \ldots, (6, 6, 6, 6)$} showing a few typical outcomes is possible, but not as good. An answer like {$(1, 1, 1, 1), \ldots, (2, 2, 2, 2), \ldots, (6, 6, 6, 6)$} is not descriptive enough.

8. An outcome consists of ten letters. The letters can be $A, B, C, D,$ or E only. Because the order in which the letters are listed is relevant, the best answer would then be of the form {$X_1X_2X_3X_4X_5X_6X_7X_8X_9X_{10}$: each X_i is one of the letters $A, B, C, D,$ or E}. Other acceptable answers involve listing a few typical outcomes separated by $\ldots$'s.

B **The Multiplication Rule**

9. **(a)** 158,184,000 **(b)** 1,757,600 **(c)** 70,761,600

10. **(a)** 4,569,760 **(b)** 456,976 **(c)** 4,394,000 **(d)** 3,906,250

11. **(a)** 62 ($2 \times 4 = 8$ heels/dress combinations plus $2 \times 3 \times 4 = 24$ heels/jeans/blouse combinations plus $2 \times 3 \times 5 = 30$ tennis shoes/jeans/T-shirt combinations)

(b) 92 [the 62 outfits in (a) plus $2 \times 3 \times 5 \times 1 = 30$ tennis shoes/jeans/T-shirt/sweater combinations]

12. **(a)** $3 \times 4 \times 9 = 108$ **(b)** $3 \times (2 + 4) \times 9 \times 5 = 810$ **(c)** $(2 + 4 + 8) \times 9 = 126$

13. **(a)** 40,320 **(b)** 40,319

14. **(a)** 40,320 **(b)** 20,160 **(c)** 576

15. **(a)** 39,270 **(b)** 2730 **(c)** 9570 **(d)** 29,700

16. **(a)** 1,256,640 **(b)** 221,760 **(c)** 79,800

17. **(a)** 4,500,000 **(b)** 1,800,000 **(c)** 360,000

18. **(a)** 3,265,920 **(b)** 90,000

C **Permutations and Combinations**

19. **(a)** 90 **(b)** 45 **(c)** 720 **(d)** 120

20. **(a)** 110 **(b)** 55 **(c)** 190 **(d)** 6840

21. **(a)** 10 **(b)** 45 **(c)** 200 **(d)** 19,900

22. **(a)** 21 **(b)** 210 **(c)** 201 **(d)** 20,100

23. **(a)** 330 **(b)** 330 **(c)** 19,600 **(d)** 19,600

24. **(a)** 715 **(b)** 715 **(c)** 2002 **(d)** 2002

25. **(a)** 495 **(b)** 495 **(c)** 1287 **(d)** 1287

26. **(a)** 330 **(b)** 330 **(c)** 2002 **(d)** 2002

27. 6.12×10^{106}

28. 1.2×10^{1550}

29. 3.95×10^{40}

30. 1.96×10^{413}

31. **(a)** 60,949,324,800 **(b)** 167,960 **(c)** 167,960

32. **(a)** 3,016,991,577,600 **(b)** 75,582 **(c)** 75,582

33. **(a)** $_{15}P_4$ **(b)** $_{15}C_4$

34. **(a)** $_{10}P_3$ **(b)** $_{10}C_7$

35. **(a)** $_{30}P_{18}$ **(b)** $_{30}P_{18}$

36. **(a)** $_{119}P_{25}$ **(b)** $_{119}P_8$ **(c)** $_{119}C_8$

D **General Probability Spaces**

37. **(a)** $\Pr(o_1) = 0.22, \Pr(o_2) = 0.24, \Pr(o_3) = 0.18, \Pr(o_4) = 0.18, \Pr(o_5) = 0.18$

(b) $\Pr(o_1) = 0.22, \Pr(o_2) = 0.24, \Pr(o_3) = 0.27, \Pr(o_4) = 0.17, \Pr(o_5) = 0.1$

38. **(a)** $\Pr(o_1) = 0.15, \Pr(o_2) = 0.35, \Pr(o_3) = 0.35, \Pr(o_4) = 0.15$

(b) $\Pr(o_1) = 0.15, \Pr(o_2) = 0.35, \Pr(o_3) = 0.22, \Pr(o_4) = 0.28$

39. $\Pr(A) = 1/5, \Pr(B) = 2/15, \Pr(C) = 4/15, \Pr(D) = 6/15$

40. $\Pr(A) = 1/4, \Pr(B) = 3/10, \Pr(C) = 1/20, \Pr(D) = 1/10, \Pr(E) = 1/10, \Pr(F) = 1/10, \Pr(G) = 1/10$

41. $\Pr(\text{red}) = 0.3, \Pr(\text{blue}) = \Pr(\text{white}) = 0.2, \Pr(\text{green}) = \Pr(\text{yellow}) = 0.25$

42. $\Pr(\text{blue}) = 0.4, \Pr(\text{red}) = 0.35, \Pr(\text{yellow}) = 0.15, \Pr(\text{white}) = 0.1$

E **Events**

43. **(a)** $E_1 = \{HHT, HTH, THH\}$ **(b)** $E_2 = \{HHH, TTT\}$

(c) $E_3 = \{\ \}$ **(d)** $E_4 = \{TTH, TTT\}$

44. **(a)** $E_1 = \{TTFF, TFTF, TFFT, FTTF, FTFT, FFTT\}$

(b) $E_2 = \{TTFF, TFTF, TFFT, FTTF, FTFT, FFTT, TTTF, TTFT, TFTT, FTTT, TTTT\}$

(c) $E_3 = \{FFTT, FTFT, FTTF, TFFT, TFTF, TTFF, FFFT, FFTF, FTFF, TFFF, FFFF\}$

(d) $E_4 = \{TTFF, TTFT, TTTF, TTTT\}$

45. **(a)** $E_1 = \{(1, 1), (2, 2), (3, 3), (4, 4), (5, 5), (6, 6)\}$

(b) $E_2 = \{(1, 1), (1, 2), (2, 1), (6, 6)\}$

(c) $E_3 = \{(1, 6), (2, 5), (3, 4), (4, 3), (5, 2), (6, 1), (5, 6), (6, 5)\}$

46. **(a)** $E_1 = \{QH, QD, QS, QC\}$

(b) $E_2 = \{AH, 2H, 3H, 4H, 5H, 6H, 7H, 8H, 9H, 10H, JH, QH, KH\}$

(c) $E_3 = \{JH, JD, JS, JC, QH, QD, QS, QC, KH, KD, KS, KC\}$

47. (a) $E_1 = \{HHHHHHHHH\}$

(b) $E_2 = \{HHHHHHHHHT, HHHHHHHHTH,$
$HHHHHHHTHH, HHHHHHTHHH,$
$HHHHHTHHHH, HHHHTHHHHH,$
$HHHTHHHHHH, HHTHHHHHHH,$
$HTHHHHHHHH, THHHHHHHHH\}$

(c) $E_3 = \{TTTTTTTTTH, TTTTTTTTHT, TTTTTTTHTT,$
$TTTTTTHTTT, TTTTTHTTTT, TTTTHTTTTT,$
$TTTHTTTTTT, TTHTTTTTTT, THTTTTTTTT,$
$HTTTTTTTTT, TTTTTTTTTT\}$

48. (a) $E_1 = \{AB, AC, AD, AE, BC, BD, BE, CD, CE, DE\}$

(b) $E_2 = \{ABC, ABD, ABE, ACD, ACE, ADE, BCD, BCE,$
$BDE, CDE\}$

(c) $E_3 = \{BCD, BCE, BDE, CDE\}$

F   Equiprobable Spaces

49. (a) 3/8   **(b)** 1/4   **(c)** 0   **(d)** 1/4

50. (a) 3/8   **(b)** 11/16   **(c)** 11/16   **(d)** 1/4

51. (a) 1/6   **(b)** 1/9   **(c)** 2/9

52. (a) 1/13   **(b)** 1/4   **(c)** 3/13

53. (a) 1/1024   **(b)** 5/512   **(c)** 11/1024

54. (a) 5/16   **(b)** 5/16   **(c)** 1/8

55. (a) 1/1024   **(b)** 1/1024   **(c)** 5/512   **(d)** 11/1024
(e) 0   **(f)** 7/128

56. (a) 1/16   **(b)** 3/8   **(c)** 1/2   **(d)** 1/2

57. (a) 4/15   **(b)** 11/15   **(c)** 1/1365

58. (a) 1/10   **(b)** 2/5   **(c)** 3/5

G   Odds

59. (a) 4 to 3   **(b)** 3 to 2

60. (a) 3 to 8   **(b)** 3 to 5

61. (a) 3/8   **(b)** 15/23

62. (a) 4/7   **(b)** 5/17   **(c)** 1/2

JOGGING

63. 6,492,640

64. (a) $\{AARRR, ARARR, ARRAR, RAARR, RARAR, RRAAR\}$

(b) $\{RRR, ARRR, RARR, RRAR, AARRR, ARARR, ARRAR,$
$RAARR, RARAR, RRAAR\}$

(c) $\{AARRR, ARARR, ARRAR, RAARR, RARAR, RRAAR,$
$RRAAA, RARAA, RAARA, ARRAA, ARARA, AARRA\}$

65. (a) $\{YXXXX, XYXXX, XXYXX, XXXYX\}$

(b) $\{YXXXX, XYXXX, XXYXX, XXXYX, XYYYY, YXYYY,$
$YYXYY, YYYXY\}$

(c) $\{YYXXXX, YXYXXX, YXXYXX, YXXXYX, XYYXXX,$
$XYXYXX, XYXXYX, XXYYXX, XXYXYX, XXXYYX,$
$XXXYYY, XYXYYY, XYYXYY, XYYYXY, YXXXYY,$
$YXYXYY, YXYYXY, YYXXYY, YYXYXY, YYYXXY\}$

(d) 72

66. 64

67. (a) 3,628,800   **(b)** 362,880   **(c)** 28,800   **(d)** 2880

68. (a) 28   **(b)** 56

69. 126,126

70. (a) 7/40   **(b)** 63/1000   **(c)** 7/120

71. (a) $\dfrac{46,189}{262,144} \approx 0.176$ **(b)** $\dfrac{285}{262,144} \approx 0.001$

(c) $\dfrac{1,048,365}{1,048,576} \approx 0.9998$

72. $\dfrac{1}{4165} \approx 0.0002$

73. $\dfrac{253}{4998} \approx 0.05$

74. $\dfrac{33}{16,660} \approx 0.00198$

75. $\dfrac{1}{2548} \approx 0.00039$

76. $\dfrac{6}{4165} \approx 0.00144$

77. Bet 1 is better.

78. (a) $(0.98)^{12} \approx 0.78$; we are assuming that the events are independent.

(b) $(0.98)^{12} + 12(0.98)^{11}(0.02) \approx 0.98$

79. $\dfrac{{}_5C_1 \cdot {}_{495}C_2}{{}_{500}C_5} + \dfrac{{}_5C_2 \cdot {}_{495}C_1}{{}_{500}C_5} + \dfrac{{}_5C_3 \cdot {}_{495}C_0}{{}_{500}C_5}$

Chapter 16

WALKING

A   Normal Curves

1. (a) 83 lb   (b) 83 lb   (c) 7 lb
2. (a) 253 m.   (b) 253 m.   (c) 61 m.
3. (a) 75 in.   (b) 3 in.   (c) 77 in.   (d) 73 in.
4. (a) 530 points   (b) 120 points   (c) 611 points
   (d) 449 points
5. (a) 89.6 lb   (b) 72.8 lb
6. (a) 2670 points   (b) 2038 points
7. (a) 500 points   (b) 100 points
8. (a) 310   (b) 120
9. 20 in.
10. $4,520
11. $\mu \neq M$
12. In a normal distribution, the first and third quartiles are the same distance from the mean. In this distribution, $\mu - Q_1 = 110$ and $Q_3 - \mu = 100$.
13. In a normal distribution, the mean μ must be at the median of the quartiles.
14. Rather than having a value of 35, the first quartile in this distribution should be approximately $47 - 0.675 \times 10 = 40.25$.

B   Standardizing Data

15. (a) 1   (b) −2   (c) 1.6   (d) −1.8
16. (a) −1   (b) 0   (c) 1.5   (d) −3.25
17. (a) 1.5   (b) −0.5   (c) −0.75   (d) 1.1
18. (a) −1   (b) 1.5   (c) −1.8   (d) 2.25
19. approximately −0.675
20. approximately 0.675
21. (a) 152.3 ft   (b) 199.1 ft   (c) 111.74 ft   (d) 183.5 ft
22. (a) 92.4   (b) 76.3   (c) 81.222   (d) 83.2
23. 17 lb.
24. 12
25. 5
26. 50
27. $\mu = 52; \sigma = 16$
28. $\mu = -10; \sigma = 30$

C   The 68-95-99.7 Rule

29. $\mu = 55; \sigma = 5$
30. $\mu = 51, \sigma = 16$
31. (a) $\mu = 92, \sigma = 3.4$ (b) $Q_1 = 89.705, Q_3 = 94.295$
32. (a) $\mu = 112, \sigma = 9.5$ (b) $Q_1 = 105.5875, Q_3 = 118.4125$
33. $\mu = 80, \sigma = 10$
34. $\mu = 150, \sigma = 20$
35. 56.3 cm
36. 6.1 cm
37. (a) 97.5%   (b) 13.5%
38. (a) 99.85%   (b) 2.35%
39. 59%
40. 83.85%

D   Approximately Normal Data Sets

41. (a) 52 points   (b) 50%   (c) 68%   (d) 16%
42. (a) 95%   (b) 2.5%   (c) 0.15%
43. (a) 44.6 points   (b) 59.4 points
44. (a) 50th percentile   (b) 84th percentile
    (c) 75th percentile   (d) 99.85th percentile
45. (a) 1900   (b) 1630
46. (a) 1950   (b) 1630
47. (a) 16th percentile   (b) 84th percentile
    (c) 99.85th percentile
48. (a) $Min \approx 86$ mm, $Max \approx 164$ mm.
    (b) $Min \approx 86$ mm, $Q1 \approx 116$ mm, $\mu = 125$ mm

$Q3 \approx 134$ mm, $Max \approx 164$ mm.

49. (a) 95%   (b) 47.5%   (c) 97.5%
50. (a) 68%   (b) 34%   (c) 16%
51. (a) 13   (b) 80   (c) 250   (d) 420
    (e) 488   (f) 499
52. (a) 202 or 203   (b) 510   (c) 510
    (d) 202 or 203   (e) 35
53. (a) 16th percentile   (b) 97.5th percentile   (c) 18.6 lb.
54. (a) 2.5th percentile   (b) 16th percentile   (c) 22.5 lb.
55. (a) 97.5th percentile   (b) 99.85th percentile   (c) 8 lb.
56. (a) 75th percentile   (b) 25th percentile   (c) 24.7 lb.

E The Honest- and Dishonest-Coin Principles

57. (a) $\mu = 1800, \sigma = 30$ **(b)** approximately 68%
   (c) approximately 34%   **(d)** approximately 13.5%
58. (a) $\mu = 3200, \sigma = 40$ **(b)** approximately 95%
   (c) approximately 50%   **(d)** approximately 13.5%
59. (a) approximately 68%   **(b)** approximately 16%
   (c) approximately 84%
60. (a) approximately 0.5   **(b)** approximately 0.475
   (c) approximately 0.025
61. (a) $\mu = 240, \sigma = 12$ **(b)** $Q_1 \approx 232, Q_3 \approx 248$
   (c) approximately 0.95

62. (a) $\mu = 900, \sigma = 15$ **(b)** $Q_1 \approx 890, Q_3 \approx 910$
   (c) approximately 0.5
63. (a) $\mu = 30, \sigma = 5$ **(b)** approximately 0.025
   (c) approximately 0.34
64. (a) approximately 0.5   **(b)** approximately 0.025
65. 0.08
66. 97.5%

JOGGING

67. (a) 20.55 lb   **(b)** 16.75 lb
68. (a) 83 points   **(b)** 49 points
69. (a) 60th percentile   **(b)** 30th percentile
70. (a) 99th percentile   **(b)** 20th percentile   **(c)** 5000
71. (a) 580 points   **(b)** 560 points   **(c)** 60th percentile
72. (a) 1.28   **(b)** -1.28 **(c)** 0.84
   (d) -0.84 **(e)** -1.04

73. (a) 78 points   **(b)** 70.4 points   **(c)** 60 points
   (d) 48.7 points
74. $n = 225$
75. $n = 400$
76. For $p = \frac{1}{2}, \mu = n \times p = \frac{n}{2}$, and

$$\sigma = \sqrt{n \times p \times (1 - p)} = \sqrt{n \times \frac{1}{2} \times \frac{1}{2}} = \sqrt{\frac{n}{4}} = \frac{\sqrt{n}}{2}.$$

Mini-Excursion 4

A Weighted Averages

1. 82.9%
2. 3.75

3. 16.4
4. 2,982

B Expected Values

5. 11
6. 24
7. (a)

Outcome	$1	$5	$10	$20	$100
Probability	1/2	1/4	1/8	1/10	1/40

   (b) $7.50   **(c)** $7.50
8. (a)

Outcome	0 points	1 point	2 points
Probability	0.0625	0.375	0.5625

   (b) 1.5 points
9. (a)

Outcome	0	1	2	3
Probability	1/8	3/8	3/8	1/8

   (b) 1.5 heads

10. (a) (-1.00)
   (b) Pay (-1.00). That is, in order to play the game, you must be offered $1.00.
11. (a) $-\frac{1}{19} \approx$ -0.05
   (b) $-\frac{1}{19}N \approx$ $-0.05N$. For every $100 bet on red, you should expect to lose about $5 ($5.26 to be more precise).
12. (a) $-\frac{1}{38} \approx$ -0.0263
   (b) $-\frac{1}{38}N \approx$ $-0.0263N$. For every $100 bet on 10, you should expect to lose $2.63.
13. (a) -0.50
   (b) Pay $0.50 to play a game in which you roll a single die. If an odd number comes up, you have to pay the amount of your roll ($1, $3, or $5). If an even number (2, 4, or 6) comes up, you win the amount of your roll.
14. (a) We solve $0.002 \times (P - 100{,}000) + 0.998 \times P = 0 for P, which gives $P - 200 = 0$ and $P = 200$.
   (b) $300

C Miscellaneous

15. It makes sense for Joe to purchse the extended warranty: His expected payoff is $\$320 \times 0.24 + \$(-80) \times 0.76 = \$16$.

16. Jackie should not buy this extended warranty since her expected payoff is negative: $\$31 \times 0.05 + \$(-19) \times 0.95 = -\$16.50$.

17. \$525 per policy

18. \$800 per policy

19. (a) There are three ways to select which of the three dice will not come up a 4. There are five possible numbers for this die (1, 2, 3, 5, or 6). By the multiplication rule there are $3 \times 5 = 15$ ways to select an outcome in which exactly two 4's are rolled. Since there are $6^3 = 216$ (again by the multiplication rule) possible outcomes in this random experiment, the probability of such an outcome is 15/216.

(b) There are three ways to select which of the three dice will come up a 4. There are five possible numbers for each of the other two die (1, 2, 3, 5, or 6). By the multiplication rule there are $3 \times 5 \times 5 = 75$ ways to select an outcome in which exactly one 4 is rolled. Hence, the probability of such an outcome is 75/216.

(c) There are five possible numbers for each die (1, 2, 3, 5, or 6). So, there are $5 \times 5 \times 5 = 125$ ways to select an outcome in which no 4 is rolled. So, the probability of not rolling any 4's is 125/216.

20. We have to solve $P_4 = P_0(1 + r)^4$ for r. Here $P_0 = \$10$ and $P_4 = \$15.25$. This gives $\dfrac{15.25}{10} = (1 + r)^4$ and thus, $r = \sqrt[4]{1.525} - 1 \approx 0.11$. That is, the expected annual yield is about 11%.

21. \$$-0.28$

22. \$$-0.64$

INDEX

PHOTO CREDITS